COLLABORATIVE INNOVATION IN DRUG DISCOVERY

Wiley Series on Technologies for the Pharmaceutical Industry

Sean Ekins, Series Editor

A complete list of the titles in this series appears at the end of this book.

COLLABORATIVE INNOVATION IN DRUG DISCOVERY

Strategies for Public and Private Partnerships

Edited by

Rathnam Chaguturu

iDDPartners
Princeton Junction, NJ, USA

WILEY

Published by John Wiley & Sons, Inc., Hoboken, New Jersey.
Published simultaneously in Canada.

Library of Congress Cataloging-in-Publication Data:

Collaborative innovation in drug discovery : strategies for public and private partnerships / edited by Dr. Rathnam Chaguturu.

p. ; cm.

Includes bibliographical references and index.

ISBN 978-0-470-91737-4 (cloth)

I. Chaguturu, Rathnam, editor of compilation.

[DNLM: 1. Drug Discovery. 2. Public-Private Sector Partnerships. QV 745]

RM301.25

615.1'9–dc23

2013035525

Printed in the United States of America.

10 9 8 7 6 5 4 3 2 1

To Sujatha, my wonderful wife and best friend.
For a myriad of reasons that Vamsee, Sreekanth, and Megan know all too well.

Collaborative Drug Discovery Ethos

I want to discover a drug
With collaborators so trusted I hug
        A once in a lifetime "Ah-ha"
        Thousand data points, insightfully claw
To separate wheat from chaff
Gold needle, overwhelming haystack
        Instead of just me, think of we
        Could collaborative drug discovery be the key?
Patents, funding, an approved IND
Clinical fast-track, collaboration catalyzed x 3
        My collaborators and me!

Barry Bunin

CONTENTS

FOREWORD

A book discussing the collaboration between industry, government, disease foundations, and academia (or the lack thereof) is important, timely, and welcome.

European pharmaceutical companies have generally been more active and public in their collaboration with academic laboratories. The effectiveness and transparency of their collaborations are evidenced by the frequent coauthorship of the resulting publications and the support provided to the investigator's laboratories.

The collaboration of American pharmaceutical companies with university laboratories has been much more variable and probably less productive for drug discovery and development than it could be. The relationships are often much less transparent. Projects and collaborations function more like contracts, with industry having sole ownership of the data and intellectual property. Data are often not mutually shared and there can even be elements of secrecy, with only some company data being shared with the academic scientists. For example, the structures of related compounds from a focus chemical library may not be shared, in order to decrease the risk that company information is leaked to competitors. I have learned years ago that good research ideas do not last long before your competitors are working on related projects.

Industry usually wants the option to review manuscripts before publication. Large multinational pharmaceutical companies are generally opposed to early publication of the work they are supporting. They tend to publish their work at two periods in the life of a project: when it has been terminated, in order to provide some recognition and reward to the project team; and most companies will also encourage publication with opinion leaders in the field before launching a product on the market, in order to influence physicians and promote sales.

If we review the current processes in drug discovery, it is usually the academic laboratory that identifies the interesting molecular targets that are important enzymes and proteins in various biochemical and physiological processes. Occasionally, the academic scientists may also identify chemical leads (drug leads) that function as agonists or antagonists with these molecular targets. More often, this process occurs in industry with high-throughput screening assays of large chemical libraries. The chemical leads then result in chemical focus libraries designed by medicinal chemists to identify the pharmacophores and more potent analogues of the leads; to identify design molecules through the use of computer modeling; and, using physicochemical studies, to identify compounds with more favorable pharmacokinetic properties, formulation requirements, easy low-cost synthesis and process chemistry, and, hopefully, low toxicity. Industry certainly has the appropriate staff to accomplish these difficult tasks. Most of the talented individuals necessary to accomplish this work cannot be found in academic laboratories. Medicinal chemists, computer modelers, process chemists, and toxicologists are usually not found in most medical centers, although some are found in pharmacy schools.

Today the biomarkers of disease identified in genomics, proteomics, and biochemical studies are generally coming from academic laboratories. They have been extremely important in identifying critical molecular targets and have provided important clues for drug discovery and development to both biotechnology companies and large multinational pharmaceutical companies.

Interestingly, about 20–25 years ago, the large multinational pharmaceutical companies in the United States were on average spending about 17% of their total revenue on R&D and about 30–35% on marketing and sales. They are now spending about 11% on research and development (R&D), while marketing and sales budgets have increased. How can technology-based companies decrease investments in technology? Furthermore, most of the R&D budgets is spent on clinical development and phase four marketing studies, which have less risk and are thought to have a greater and more rapid return on investment. As a result, there have been fewer novel molecular targets and novel drugs. So why are the large companies surprised that the pipelines have dried up and that they are having to pay premiums to acquire technology from other pharmaceutical and biotechnology companies?

About 20 years ago, some of us predicted that the big pharmaceutical companies would begin to have problems. Today, many pharmaceutical companies are merging with one another to fill their pipelines, decrease their costs, and presumably increase their profits. They are also acquiring biotechnology companies as a strategy to offset expiring patents on many of their multi-billion-dollar products. What is often not appreciated is that most of the technology in biotechnology companies has come from academic laboratories. Investments in collaborative research projects between pharmaceutical companies and academic research programs would have been needed at a much earlier stage, with greater risk and perceived cost to develop. Since time is money, and senior management in companies usually changes every few years, the acquisition of later-stage technology is generally preferred, as that permits senior management to enhance their equity positions and compensation.

Very often, academic scientists discover and patent a technology and license it to biotechnology companies. While this may also happen with multinational drug companies, it is probably less common. Many large companies often have the philosophy that if the technology is not invented in their company, it has less value or is suspect. Large companies also tend to prefer later stage technology that has less risk and less development time to reach the market.

Not only have many of the novel drug leads come from academic laboratories, but also much of the fundamental technology and discoveries has come from academic programs. Examples include physiochemical studies of molecular targets, genomics, proteomics, computer modeling, identification of new messenger molecules and their receptors, and identification of new biomarkers for various diseases, including cancers.

We are learning that many diseases, such as diabetes, hypertension, atherosclerosis, cancer, and other syndromes with multifactor causes, are heterogeneous diseases. In the future this will lead to personalized therapy for each patient, leaving little room for blockbuster drugs with multi-billion-dollar markets. Designing the most effective personalized treatment requires genomic and proteomic information for each patient. As only physicians and academic programs can provide the patient populations for study, a collaboration between industry and academia would be required to take this step forward.

Markets will be much smaller for these new generation therapies; however, clinical trials could also be smaller and target the most appropriate therapy, saving considerable amounts of clinical development time and cost. Currently, with heterogeneous populations of patients the benefit from an effective therapy is diluted among the larger pool of nonresponders. This requires extremely large numbers in clinical trials in order to see significant drug effects.

I realize that collaborations are often very difficult. Industry wants maximum ownership, control, and profit from the research. Academics, on the other hand, must demonstrate independence for their promotion and grant support. Thus, our current system tends to discourage collaborations between industry and academic programs.

In European countries, professors' research is often funded by university departmental budgets. Perhaps the collaborations in European countries between universities and industry are more effective because professors and their laboratories are less dependent on funding from research grants. Awards from companies are usually discretionary funds that are often used to supplement salaries. There is also a sense of co-ownership of the research. The competitiveness in the United States for funding and research recognition may be hindering effective collaborations between academics and industry. In the United States, the collaboration between academics and biotechnology companies is much more apparent, perhaps because the technology originated from the academic laboratories. Biotechnology companies have little or no revenue and smaller research budgets and staff, and therefore require collaborations for their research and product development. They often use stock options to reward their advisors and collaborators; this is rarely the case with large companies. Furthermore, outside of clinical studies, which require access to patients, large pharmaceutical companies often believe they are much less dependent on collaborations.

I believe the difficulties in collaborations between industry and academics can be summarized as the parties' concerns over ownership and profit, fear of the competitors, and lack of trust. These are common problems in many business relationships. Over the past 40 years I have been asked on numerous occasions to collaborate with others. As some of the collaborations could have been very important, I have often agreed. However, on only several occasions did a collaboration occur and did the collaboration lead to some exciting and important discoveries that affected my subsequent research. Most of the collaborations never materialized, usually because the other party committed very little effort to the project. The take-home message that comes across quite vividly and glaringly as you read through the chapters in this fascinating book is that the collaborating parties must plan carefully, take the project seriously, define who does what, and honor their commitments in a timely fashion. Without these elements a true and effective collaboration will not occur!

Ferid Murad
1998 Nobel Prize in Physiology and Medicine
Department of Biochemistry and Molecular Medicine
The George Washington University
Washington, DC

PREFACE

The pharmaceutical industry has primarily been the driver of drug discovery and development by managing knowledge for profit. Academia, for its turn, has been involved in advancing the frontiers of the fundamental and mechanistic aspects of disease biology and identifying relevant biochemical targets for therapeutic intervention. The resultant "golden nuggets" are then chosen by pharma, with or without a handshake, to drive its drug discovery engine. Academia has evolved in recent years from its traditional role of target identification and validation from a systems biology perspective to probing for tool molecules against these disease targets to explore their therapeutic relevance.

The pharmaceutical industry spends well over $135 billion on research and development (R&D) annually, but is nonetheless experiencing an innovation crisis, with no end in sight. Resurgence of the pharmaceutical industry to its preeminence can be realized only through strategic collaborative innovation in biomedicine. In the Foreword to this book, Ferid Murad, winner of 1998 Nobel Prize in Physiology and Medicine, states that most collaborations, open or otherwise, never materialize because of lack of commitment by the *partners* involved. He reminds us that the collaborating parties must plan carefully, take the project seriously, define who does what, and honor their commitments in a timely fashion. Without these elements a true and effective collaboration will *not* occur. As Winston Churchill so grandly stated, "If we are together nothing is impossible. If we are divided all will fail." While perhaps not as epic as Churchill's notion, it is undeniable that a major factor influencing the pharmaceutical industry's ability to turn today's innovation into tomorrow's life-saving medicine is establishing and maintaining strong and diverse partnerships in both the public and private sectors. With so many challenges faced in successfully bringing a medicine to market, we need all the members of the biomedical ecosystem to collaborate. This ecosystem includes academia, nonprofit and for-profit research institutions, government agencies, the pharmaceutical industry, and disease foundations.

The business model for pharmaceutical R&D is *evolving*. In the past, pharma relied mostly on internal efforts; in recent years it has moved toward establishing a greater number of partnerships with academic institutions spanning the entire spectrum of drug discovery and development activities. Given the complexities and the expense of R&D, we can accomplish much more, in a shorter time, and with less cost and duplication of effort when the private and public sectors bring their special strengths to bear. Publicly funded research, occurring at universities and the National Institutes of Health, over the years has produced a great majority of the key enabling discoveries underlying nearly 80% of the important drugs. Pharma then turned these enabling discoveries into viable medicines. The average time between the "key enabling discovery" and the introduction of a drug is 12–15 years, which reflects the significant work required to turn the discovery into a medically useful, market-oriented product. Thus, academia's enabling discoveries, while incredibly valuable, are not enough to meet the need without pharmaceutical industry's drug development capabilities.

Partnerships, no matter how well conceived and structured, are often fraught with tension. The public and private sectors must engage in a partnership wherein the core missions of each institution are recognized, respected, and accommodated, in order to choose the most disease-relevant therapeutic targets and thereby bridge the translational gap. The primary obligations for academia are in the areas of education of students, creation of knowledge, and dissemination of knowledge. For industry, the guiding principles are creation of value for investors/shareholders, provision of useful services and goods to customers, and competitive expansion of the state of the art. Intellectual property rights are of paramount importance to the pharmaceutical sector. The newly implemented "first-inventor-to-file" provisions of the Leahy-Smith America Invents Act brings the United States from its long practiced "first to invent" system to the "first to file" system practiced by the rest of the world. Academicians, who live in a world of "publish or perish," have now been made aware of the catch-all term "prior art" that includes anything that has been publicly disclosed, and the realization that protection of academic intellectual property rights is therefore a formidable task.

A successful pharma–academia collaboration should support the mission of each and incorporate the needs, opportunities, skills, expertise, and constraints that each brings to the table. *Any effort in conflict with the mission of either partner will ultimately fail.* Pharma and academia should focus on the benefits to each party that will result from collaborations by streamlining negotiations to ensure timely conduct of the research and the development of the research findings. When the guiding principles of each of the institutions are understood and respected, there is a tremendous opportunity for these institutions. Both the pharmaceutical industry and academia have now come to realize their respective strengths as well as their weaknesses. What we now see evolving is the much needed collaborative spirit between these two diverse institutions to close this risk–reward gap, as exemplified by the number of industry–academia collaborative agreements that are being put in place.

It is with much enthusiasm that we welcome the creation of "Partnerships to Accelerate Therapeutics" in September 2012 by the U.S. President's Council of Advisors on Science and Technology to identify and resolve bottlenecks and speed the development of life-saving medicines through synergistic alliances involving industry, academia, government, and disease foundations. Ten of the pharmaceutical giants have similarly responded with the creation of TransCelerate BioPharma, based on a nonprofit precompetitive model, to speed drug development by broad participation and collaboration across the global R&D community. Even more groundbreaking has been the 2014 launching of a collaborative enterprise under the auspice of the Foundation for the NIH (FNIH), the Accelerating Medicines Partnership (AMP). The $230 million over 5 years partnership between NIH, 10 pharmaceutical giants, and several nonprofit disease foundations calls for the drug-makers to collaborate, not compete, on drug discovery projects in four major diseases: Alzheimer's disease, type 2 diabetes, and the autoimmune disorders rheumatoid arthritis and lupus. The partners are mandated to share their research findings with each other to define a unified approach, and no commercial effort could be undertaken by any of the partners until the data are made public to the biomedical community. This is collaborative innovation at its best.

The new drug discovery paradigm is based on complementation, not competition, between the pharmaceutical industry and academia. The genesis for this new paradigm takes its roots in the United States from the Bayh-Dole Act of 1980. The passage of this act and its subsequent amendments in 1984 and augmentation in 1986 was primarily intended to induce "economic development by promoting investment by the private sector in the commercialization of federally funded research discoveries for the public good." This has resulted in the creation of offices of technology transfer within academic institutions to facilitate the execution of confidentiality agreements and material transfer agreements, to examine intellectual property rights, to license out technology, and to engage in collaborative agreements with industry. This has transformed academic faculty into entrepreneurs, managing their research projects and receiving support from federal or private sources. However, there is also a concern that this entrepreneurship will shift the focus from basic to more applied research and that there will be an inherent loss of scientific objectivity due to the conflicts of interest arising when institutions focus on equity management. While these are genuine concerns and have merit, the need for a strong, multidimensional public–private partnership to advance drug discovery is clearly gaining momentum. *The goal of the University-Industry Demonstration Partnership, a laudable initiative from the National Science Foundation, is to enhance the value of collaborative partnerships between university and industry.* The National Institutes of Health resources, especially the Roadmap Initiatives, have been integral to engaging academia in drug discovery research and have been effectively leveraged to help bridge the chasm between basic research activities and the commercialization of a drug.

The goal for this book has been to assess the current status of drug discovery in pharma and academia, to identify the areas where each of the institutions does its best and the areas where gaps exist, and to explore ways we can establish a mutually beneficial partnership to achieve our common goal of bringing improved and safer drugs to treat the diseases afflicting mankind in developed as well as developing countries. The book was originally slated to address the above topics in a dozen chapters, but the word about the thematic aspect of this book caught the attention and imagination of many who truly believe in the need for collaborative innovation in biomedicine. The expansive nature of the book in your hands is thus a true testament to its relevance in addressing the conundrum facing both the public and private sectors in the arena of drug discovery and development.

It is hoped that the book will be consulted not only by practicing scientists but also by the decision makers in both the public and private sectors. To make it interesting to read for the CEO, VP, the Director of Drug Discovery, and the Program Manager, the treatment of the discussion topics is both subjective as well as objective. The authors are leaders in the chosen topics, thus giving authenticity to the subject matter being discussed. This book is composed of several thematic concepts. Competition, the defining characteristic of life's existence (and at least capitalism), and its antithesis, symbiotic collaboration, are diametrically opposite but occasionally join forces in unexpected ways. An amalgamation of these divergent factors has led to the evolution of collaborative-innovation models as illustrated in the opening chapters by Rory Hume and William Mattes, and in the social scientists' perspective by Jim Shaeffer and Sarah MacDonald. Bernard Munos captures the painful "Kodak moment" faced by the pharmaceutical industry and provides a critical analysis in "Accelerating Innovation in the Bioscience Revolution." The chapters by Jason Cristofaro, Kate Marusina et al., and Anthony Boccanfuso deal with a governmental perspective on fostering university–industry collaborations. Much of the industrialization of academic drug discovery research in the United States can be directly linked to NIH's Roadmap Initiatives, as illustrated by Mark Scheideler. Anuradha Roy, Nathan Blow, David Camp, Horst Flotow and Alex Matter, T.C. Chung et al., and Philip Gribbon capture the current status of various academic centers engaged in high-throughput screening to drive early-stage drug discovery. Peter Covitz et al., Paul Eynott and Carole Fages, Monika Lessl and Khusru Asadullah, Litao Zhang and Carl Decicco, and Stefan Herzig et al. articulate the pharmaceutical industry's efforts to embrace the much needed and desired open and collaborative innovation in biomedicine to mitigate a future patent cliff. The pharmaceutical industry's business model of "low risk and high reward" generally leads to a void in combating the rare and neglected diseases that afflict mankind, and the nonprofit-based disease foundations and institutions are stepping in to fill this gap—as evidenced by the chapters by Barry Bunin and Sean Ekins, Gali Hagel, Allen Reitz and Kathleen Czupich, Sanchayita Kar and Tonny Johnson, Spencer Shorte, and Allan Tobin. The chapters by John Watson and Bhushan Patwardhan illustrate the recent trend in the globalization of drug discovery efforts, especially in Brazil, Russia, India, and China. Likewise, the el Baghdady brothers focus on unlocking the market potential of academic discoveries from a bench-to-bedside translational research perspective. Valentina Sartori et al. discuss how the pharmaceutical industry could unlock the value of its pipeline, and Les Mitscher drills deep into the value of consulting by the academic expert. Chris Lipinski, the author of the "Rule of Five," minces no words in his splendidly candid personal assessment of the challenges faced by the pharmaceutical industry in pursuing collaborative partnerships. According to Chris Paschall and Donna Marie De Carolis, issues pertaining to technology transfer and intellectual property rights take center stage during prospective partnership discussions and could make or break the partnership if they are not well thought through.

To conquer the "final frontier," Ibis Sánchez-Serrano depicts a novel, inter-institutional organizational paradigm for bringing academic discoveries to market through effective execution of translational science. Elizabeth Iorns extolls the virtues of market-driven collaboration to accelerate innovation, as a single scientist can no longer master all the techniques required to make major discoveries and solve complex problems. Deborah Collyar offers a candid glimpse into the (im)patient side of fundamental problems that thwart medical research efforts, confuse markets, discourage patients, and ultimately rescind profitability. She argues for placing the patients' needs first. Both Iorns and Collyar boldly highlight the cracks in the ivory towers that have publicly exposed the underbelly of drug development with issues like irreproducibility and scientific misconduct in preclinical research, by-products of a broken research enterprise that threaten to derail collaborative innovation and the partnerships involved. Finally, Hakim Djaballah offers a poignant personal perspective that it takes a village to develop a drug.

My own professional career has come full circle, from industry to academia to my current positions at iDDPartners, a nonprofit think-tank facilitating pharmaceutical innovation, and SRI International, a nonprofit organization whose mission is to apply science and technology for knowledge, commerce, prosperity, and peace. Forging collaborative partnerships has been my passion. The vast array of experiences that I gained and endured has provided me with a rare opportunity to see and assess, firsthand, the strengths and weaknesses of these divergent sectors. In the early 1980s, I crisscrossed the country scouting for academic investigators who may have complementary technologies to augment our own research at Dow Chemical. In the ensuing years I spent a great deal of effort in developing and managing a number of partnerships with academia and biotech companies in solidifying our biology-relevant, new lead discovery efforts. For the past 10 years, I have led panel discussions and organized tracks at drug discovery–focused national and international conferences discussing the need for collaborative innovation in drug discovery. These sessions resonated with the participants quite overwhelmingly and echoed the sentiment "The need is great, and the time is now."

My perspective on drug discovery in academia, after leading drug discovery efforts at the University of Kansas, is a mix of apprehension and appreciation. Academia's endeavor in the arena of drug discovery is a Mount Everest in the making, and requires careful nurturing to make it a successful reality. The pharmaceutical industry's best practice of a matrix-based multidisciplinary team approach is relatively nonexistent in academia, but should be instituted with intra-institutional experts of related disciplines, and active project management at its core. Academia's drug discovery efforts through industrialized high throughput screening could easily become an exercise in futility if no strategic help is sought from a competent medicinal chemistry faculty to separate the "wheat from the chaff."

I am greatly indebted to my good friend John Watson, Director of Cellular Health at Promega, for believing in my vision for the strategic and crucial role that academic drug discovery could play in the open source–driven biomedical ecosystem that has taken center stage in recent years. I am thankful to Jada Blinn, Ashleigh Price, Anupama Panikkar, Kimberly Hunter, Amanda Amanullah, Chandan Singh, and the Honorable Atta-ur-Rahman for their invaluable assistance. I am deeply thankful to Stephanie Sakson, Project Manager at Toppan Best-set Premedia, and Melissa Yanuzzi, Senior Production Editor at Wiley, for all their well-coordinated help when the publishing details became daunting, and at times Sisyphean. Jonathan Rose, Senior Editor at John Wiley and Sons, first asked and encouraged me to embark on this mission to develop this book, and I am grateful for his trust and confidence. The book would not be what it is without the time and energy of the contributing authors, and their vision of "the best is yet to come" through academia–industry partnerships. The authorship is truly global in representation and the viewpoints emphasize perspectives from academia, pharma, disease foundations, nonprofit and for-profit research institutions, and governmental agencies. I thank the authors for their valuable and timely contributions. I sincerely hope that the concepts expounded by the authors' personal experiences resonate with the key players and decision makers in the public and private sectors, and lead to truly collaborative innovation in drug discovery and biomedicine in general.

These are challenging times for drug discovery, but they also offer unprecedented opportunities for the public and private sectors to form complementary and synergistic partnerships and collaborative research enterprises with an overarching goal of advancing drug discovery breakthroughs and bridging the translational gap. Together, we can minimize our weaknesses and accentuate our strengths. A problem shared is a problem solved!

As Hakim Djaballah offers in his poignant personal perspective in the final chapter, it takes a village to develop a drug; "we never become our villages but merely guests of leisure, called upon in desperation and ignored during the blooming seasons." A stark reality that we must address with sincere conviction!

Rathnam Chaguturu
iDDPartners
Princeton Junction, NJ, USA

ABOUT THE BOOK

No one can whistle a symphony. It takes a whole orchestra to play it.

—*H.E. Luccock*

The public's perception of biomedical research to improve health care is a seamless process of harmony and data sharing, from discovery and development through delivery of care. If only it worked this way. There are about 7000 diseases that afflict mankind, yet treatments are available only for ~200 diseases. The genetic basis for most of these diseases is known, a large number of potential therapeutic targets have been uncovered, but the number of novel target drugs that enter the clinic remains about four per year over the last quarter century. The pharmaceutical industry is facing serious innovation crisis even after spending well over $135 billion on R&D annually. Resurgence of the pharmaceutical industry to its preeminence can only be realized through *collaborative innovation in drug discovery*, and placing patient needs first. Given the complexities and the expense of R&D, we can accomplish much more, in a shorter time and with less cost and duplication of effort, when the private and public sectors bring their special strengths to bear. These are challenging times, but they also offer unprecedented opportunities to form complementary and synergistic partnerships and collaborative research enterprises with an overarching goal of advancing drug discovery breakthroughs and bridge the translational gap. The partners must engage in a partnership, where the core missions of each institution are recognized, respected and accommodated, in order to conquer the so-called "valley of death." *Any effort in conflict with the mission of either partner will ultimately fail.*

The focus of the book is early stage drug discovery. It is the single most critical barrier in achieving a free flow of drug candidates in to the pipeline. The goal for this book is to assess the current status of drug discovery in pharma and academia, identify the areas where each of the institutions does its best and areas where gaps exist, and explore ways we can establish a mutually beneficial partnership to achieve our common goal of bringing improved and safer drugs to treat the diseases afflicting mankind in developed as well as developing countries. The authorship is truly global in reach, and represents every sector of the biomedical ecosystem—academia, pharma, disease foundations, for-profit and nonprofit research institutions, and governmental agencies. The jury is still out, but the thought leaders offer their personal insights in maximizing precompetitive knowledge management through sustainable public–private partnerships.

ABOUT THE EDITOR

Rathnam Chaguturu is the Founder and CEO of iDDPartners, a nonprofit think-tank focused on pharmaceutical innovation. He has more than 35 years of experience in academia and industry, managing new lead discovery projects and forging collaborative partnerships with academia, disease foundations, nonprofit agencies, and government agencies. Rathnam is the Founding President of the International Chemical Biology Society, a founding member of the Society for Biomolecular Sciences, and Editor-in-Chief of the journal *Combinatorial Chemistry and High Throughput Screening*. He is the author or coauthor of more than 60 publications and holds 11 U.S. patents. He serves on several editorial and scientific advisory boards, is the recipient of several awards, and is a sought-after speaker at major national and international conferences, passionately advocating the virtues of collaborative partnerships in addressing the pharmaceutical innovation crisis. Rathnam received his Ph.D. with an award-winning thesis from Sri Venkateswara University, Tirupati, India, and did postdoctoral work at the University of Wisconsin.

Forging collaborative partnerships has been his passion. The vast array of experiences gained and endured during his 35-year career has provided him with a rare opportunity to see and assess, with a mix of apprehension and appreciation, the strengths and weaknesses of collaborative partnerships, and hence the encyclopedic effort leading to the development of this book.

Jason V. Cristofaro, JD, PhD, Intellectual Property Advisor, Division of Cancer Treatment and Diagnosis, National Cancer Institute/NIH/DHHS, Bethesda, MD, USA

Kathleen M. Czupich, BS, MBA, Director and Co-Founder, Pennsylvania Drug Discovery Institute, Pennsylvania Biotechnology Center, Doylestown, PA, USA

Donna Marie De Carolis, PhD, Dean, Charles D. Close School of Entrepreneurship, Drexel University, Philadelphia, PA, USA

Carl Decicco, PhD, Senior Vice President, Lead Discovery, Bristol-Myers Squibb Company, Princeton, NJ, USA

Hakim Djaballah, Director, HTS Core Facility, Memorial Sloan-Kettering Cancer Center, New York, NY, USA

Sean Ekins, PhD, Chief Scientific Officer, Collaborative Drug Discovery, Inc., Burlingame, CA, USA

Matthias Evers, PhD, Principal, McKinsey & Company, Hamburg, Germany

Paul R. Eynott, PhD, MBA, Tech Launch Arizona, Dean's Office, College of Science Administration, University of Arizona, Tucson, AZ, USA

Carole Fages, PhD, Global Head, Partnering and Innovation, Early to Candidate Unit, SANOFI Group, Montpellier, France

Ingo Flamme, MD, Bayer HealthCare AG, Wuppertal, Germany

Horst Flotow, PhD, Head, Singapore Screening Centre, Experimental Therapeutics Centre, Singapore

Rafael A. Gacel-Sinclair, Associate Director, Technology Transfer Services, The Office of Research, University of California, Davis, Davis, CA, USA

Philip Gribbon, PhD, Chief Scientific Officer, European ScreeningPort GMBH, Hamburg, Germany

Betty P. Guo, PhD, Office of Research, University of California Davis School of Medicine, Sacramento, CA, USA

Gali Hagel, JD, MRF supporter and freelance writer, Burlingame, CA, USA

Stefan Herzig, PhD, Professor and Director, Department of Pharmacology, University of Cologne, Cologne, Germany

Wyatt R. (Rory) Hume, DDS, PhD, Executive Director, Education, Training and Development, Research Division, Qatar Foundation, Doha, Qatar

Elizabeth Iorns, PhD, Co-Founder and CEO, Science Exchange Inc., Palo Alto, CA, USA

Michael R. Jackson, PhD, Vice President, Drug Discovery and Development, Conrad *Prebys* Center for Chemical Genomics (CPCCG), Sanford-Burnham Medical Research Institute, La Jolla, CA, USA

Petra Jantzer, PhD, Principal, McKinsey & Company, Zurich, Switzerland

Tonny Johnson, PhD, CEO, Sciclips, Madison, WI, USA

Sanchayita Kar, PhD, President, Sciclips, Madison, WI, USA

Sundeep Khosla, MD, Dr. Francis Chucker and Nathan Landow Research Professor, Division of Endocrinology and Kogod Center on Aging, Mayo Clinic College of Medicine, Rochester, MN, USA

Monika Lessl, PhD, Director, Alliance Management, Bayer HealthCare Pharmaceuticals, Bayer Pharma AG, Alliance Management, Global Innovation Sourcing, Berlin, Germany

Christopher A. Lipinski, PhD, Scientific Advisor, Melior Discovery, Waterford, CT, USA

Sarah MacDonald, Director for Outreach and Engagement, James Madison University, Harrisonburg, VA, USA

Kate Marusina, PhD, MBA, Associate Director, Clinical and Translational Science Center, University of California, Davis, Sacramento, CA, USA

Alex Matter, PhD, CEO, Experimental Therapeutics Centre, Singapore

William B. Mattes, PhD, DABT, Diplomate, American Board of Toxicology, PharmPoint Consulting, Poolesville, MD, USA

Lester A. Mitscher, PhD, Distinguished Professor, Emeritus, Department of Medicinal Chemistry, University of Kansas, Lawrence, KS, USA

Bernard H. Munos, MBA, Founder and President, InnoThink Center for Research in Biomedical Innovation, Indianapolis, IN, USA

Ferid Murad, MD, PhD, Nobel Laureate, Director, Institute of Molecular and Cellular Signaling, Department of Biochemistry and Molecular Medicine, The George Washington University, Washington, DC, USA

Pamela Nagasawa, PhD, Measurement Specialist, Department of Medical Education and Biomedical Informatics, University of Washington, Seattle, WA, USA

Christopher Paschall, PhD, CLP, Licensing Manager, University of Virginia Licensing and Ventures Group, Charlottesville, VA, USA

Bhushan Patwardhan, PhD, Professor and Director, Interdisciplinary School of Health Sciences, University of Pune, Pune, Maharashtra, India

Allen B. Reitz, PhD, Chair and Founding Member, Pennsylvania Drug Discovery Institute, Pennsylvania Biotechnology Center, Doylestown, PA, USA

Lynn Rose, PhD, Director, Scientific Administration, Benaroya Research Institute and Institute of Translational Health Sciences, University of Washington, Seattle, WA, USA

Anuradha Roy, PhD, Director, High-Throughput Screening Laboratory, University of Kansas, Lawrence, KS, USA

Marion Rozowski, PhD, Professor, Department of Pharmacology, University of Cologne, Cologne, Germany

Terrence D. Ruddy, MD, Director, Nuclear Cardiology, University of Ottawa Heart Institute, Ottawa, Ontario, Canada

Peter G. Ruminski, MS, Executive Director, Center for World Health and Medicine, Doisy Research Center, St. Louis University, St. Louis, MO, USA

Ibis Sánchez-Serrano, MBA, Founder and President, The Core Model Corporation, S.A. (CMC), Panama City, Panama

Valentina Sartori, PhD, Associate Principal, McKinsey & Company, Zurich, Switzerland

Mark A. Scheideler, PhD, Founder, HumanFirst Therapeutics LLC, Silver Spring, MD, USA

James M. Shaeffer Sr., PhD, Associate Vice Provost for Outreach and Engagement, James Madison University, Harrisonburg, VA, USA

Spencer L. Shorte, PhD, Director, Plateforme d'Imagerie Dynamique (PFID), Imagopole, Institut Pasteur, Paris, France

Michael Steinmann, Principal, McKinsey & Company, Zurich, Switzerland

Allan J. Tobin, PhD, Eleanor Leslie Chair in Neuroscience, UCLA, Emeritus; formerly Senior Scientific Advisor to the CHDI Foundation, CHDI Management, New York, NY, USA

Pakou Vang, Intellectual Property Analyst, Technology Transfer Services, The Office of Research, University of California-Davis, Davis, CA, USA

John Watson, PhD, Director, Cellular Analysis and Pharma/Biotech, Promega Corporation, Madison, WI, USA

Dean J. Welsch, PhD, Head, Pharmacology, BioMed Valley Discoveries, Kansas City, MO, USA

Litao Zhang, PhD, Vice President, Lead Discovery, Bristol-Myers Squibb Company, Princeton, NJ, USA

Part I

PERSPECTIVES ON COLLABORATIVE INNOVATION

If I have seen further it is by standing on the shoulders of giants. —*Isaac Newton*

Collaborative Innovation in Drug Discovery: Strategies for Public and Private Partnerships, First Edition.
Edited by Rathnam Chaguturu.

1

PRODUCTIVE RELATIONSHIPS IN RESEARCH AND DEVELOPMENT BETWEEN GOVERNMENT, INDUSTRY, AND UNIVERSITIES

Wyatt R. Hume

Research Division, Qatar Foundation, Doha, Qatar

I approach this interesting and worthwhile topic from the perspective of someone focused primarily on helping universities to serve society in the best ways possible. There are other ways of considering the matter (naturally enough, from either the perspective of better government or that of benefits to industry), but I believe that each is very likely to reach similar conclusions and to create similar statements of principle for success. I also write optimistically, in the firm belief that productive and mutually beneficial research and development* relationships between government, industry, and universities can be developed and maintained, and that when they are developed well, they create broad economic, social, educational, and health benefits.

The benefits of productive relationships between the parties do not immediately accrue to universities, although universities may receive some benefits over time. Instead, the benefits fall primarily to the society that sustains the overall process and to the entrepreneurial individuals or groups who succeed in creating new business enterprises. If the relationships are appropriately structured, those benefits can substantially outweigh the costs.

*By "research" I mean the creation of new knowledge. By "development" I mean the social and commercial exploitation of new knowledge for economic and other social benefit.

Collaborative Innovation in Drug Discovery: Strategies for Public and Private Partnerships, First Edition. Edited by Rathnam Chaguturu.

As I see it, the ground rules for success in creating such beneficial relationships are not complex but in some cases may at first appear counterintuitive. However, many societies have now clearly shown that establishing and maintaining a strong set of enduring relationships, and therefore ensuring continuing success, is definitely achievable. This chapter will begin with a brief historical view of the development of productive relationships in research and development in two countries, Germany and the United States, and will then describe some ground rules that are likely to ensure success in the present day and near future.

UNIVERSITIES, GOVERNMENT, AND INDUSTRY: HISTORICAL BEGINNINGS IN PRUSSIA

Before the nineteenth century, universities in Europe and elsewhere principally functioned as communities of scholars devoted to personal intellectual development. University research, when it occurred, focused largely on the testing and refinement of ideas. Universities had, up to that point, played very little direct role in the creation or evolution of agriculture, mining, or industry. They were not expected to do so. That changed markedly, beginning in the early 1800s.

Prussia's early defeats in the Napoleonic Wars led to a desire to improve government, education, and industrialization within a society that was both relatively affluent, because of agriculture, and relatively well organized. In 1809, Prussia's Minister of Education, Wilhelm von Humboldt, initiated comprehensive educational reforms and in 1810 established Berlin University with the support of the King of Prussia. Humboldt had strong and, as it turned out, highly influential personal philosophies regarding education. He considered that university education should be centered on the discovery of knowledge—by students as well as by professors—something that was at that time a revolutionary concept. He stated as his ideal, "The university teacher is thus no longer a teacher and the student is no longer a pupil. Instead the student conducts research on his own behalf and the professor supervises his research and supports him in it."* This strong focus on research was a profound change from the accepted model and led to the development of what are now called research-intensive universities.

In large part, because of the effects on industrial and social development to be described later, the new paradigm proved to be influential first in Prussia, then elsewhere in Germany, and then in other countries. At about the same time, Wilhelm's brother, Alexander von Humboldt, pioneered, in many senses, quantitative scientific methodology in several fields. This advancement gave valuable substance to his brother Wilhelm's educational and research philosophies.

Berlin University,† and then others, added the relatively new fields of physics, chemistry, and biology. Professors in those disciplines and others were expected to be

*Quoted in Clark, C. *Iron Kingdom: The Rise and Downfall of Prussia, 1600–1947*. Belknap Press; 2009, p. 333.

†In 1949, the name of Berlin University was changed to Humboldt University in honor of the two brothers.

active in research, were required to educate and train both undergraduate and graduate students in research methodology, and were empowered both with financial support for their research and with a high degree of freedom of inquiry. The universities created both research-capable individuals and new knowledge. These products (the creatively trained people and the new intellectual property) became available to a society that, in the case of Prussia, was ready to adopt them in the development of new business enterprises, and that had the will and ability to invest in risk enterprises.

The result, within only a few decades, was not only the creation of new industries within Prussia and then in other societies that followed Prussia's example, but also the creation of a vibrant diversity within such new industries. Foremost among these were the development and industrial manufacture of chemicals, dyes, and pharmaceuticals. Similar advances occurred in precision manufacturing, mining, and metallurgy. A high level of capability and international competitiveness in aviation, weapons production, and automotive manufacturing followed. The strength of these areas of the German economy has continued since that time.

REFINEMENT AND BROAD ADOPTION: THE UNITED STATES

Other European countries began to follow the German example, but it was the vibrant, creative, and rapidly developing United States of America that most rapidly adopted the new German model and that has maintained and refined it ever since. Although many other countries now work to emulate what the United States has achieved, it is instructive to consider the key elements that have lead to this nation's success.

Rather than tracing in full historical detail the interesting development of the government–university–industry research ecosystem in the United States over the last 140 years, I will describe what I see as three key government actions that underpinned and continue to underpin its success. I will not describe or analyze the attempts, many of them successful, by other (usually smaller) nations and local economies to emulate the Prussian and U.S. models.

Land Grants

First, a system of "land grant" universities was established through federal legislation known as the Morrill Act in 1868 and in follow-up legislation in 1890. The purpose of the legislation was to establish universities that would "teach such branches of learning as are related to agriculture and the mechanic arts, in such manner as the legislatures of the States may respectively prescribe, in order to promote the liberal and practical education of the industrial classes in the several pursuits and professions in life."* This would be accomplished through grants of federal land to each of the states, revenues from which would support the universities. The effects on industrial and social

*First Morrill Act, 1862, Section 4, as amended April 13, 1926. Available at http://www.ans.iastate.edu/history/link/morrill1862.html.

development in the states that took advantage of the legislation (and almost all did) were profound and have continued up to the present time.

Investments in Research

The second factor contributing to success in the United States has been ongoing investments in research, on the order of 2–3% of the gross domestic product, through the National Science Foundation, the National Institutes of Health, and other agencies of the federal government, combined with additional research investments by individual states.

It became evident that those investments were most productive when the funds were awarded through competitive processes, where funding decisions are based on the outcome of peer review of the quality of proposals submitted in response to requests for such proposals by the respective agencies, with those requests in turn being responsive to priorities identified in consultation with government. The alternatives (i.e., allocation through "earmarking" based on personal or political considerations, or allocation based on institutional entitlement) resulted in lower quality research outcomes and therefore less social and commercial benefit.

It is clear that full and open competition among institutions and researchers is the most productive approach. As an example, within the constraints of national security considerations, the United States even follows this model in the design and continuing refinement of the reliability and safety of its nuclear weapons.

Intellectual Property

The third enabling factor has been legislation related to the ownership and exploitation of intellectual property derived from research supported by the federal government. It is, in my view, initially counterintuitive but both brilliant and highly effective.

In 1980, two U.S. Senators, Birch Bayh (a Democrat) and Robert Dole (a Republican), authored and carried an additional piece of federal legislation. The Bayh–Dole Act reversed the presumption of ownership (title) for intellectual property from the funding body (the federal government) to the entity supported (the university, nonprofit entity, or for-profit company or individual). Bayh–Dole allows such bodies to pursue and exploit ownership of an invention resulting from federally funded research rather than vesting such ownership in the federal government. This has proven to be a masterstroke, something that has been and continues to be highly beneficial to the U.S. economy. This assignment of title has conditions, but the conditions have proven to be productive rather than restrictive. The conditions are as follows:

- The funded entity must file for patent protection of intellectual property resulting from the supported research.
- The entity must report each disclosed invention to the funding agency.
- The entity must elect to retain title to such inventions in writing within a defined timeframe.

- The entity must grant the federal government a "non-exclusive, non-transferable, irrevocable, paid-up license to practice or have practiced on its behalf throughout the world."*
- The entity must undertake actively to promote and attempt to commercialize the invention but must not assign the rights to the technology, with some exceptions.
- The entity must give preference to U.S. industry and small business in commercial exploitation.
- The entity must share royalties with the inventor, and in the case of universities, use any remaining income for education and research.

Impact on U.S. Economy

As was the case in Prussia in the 1800s, the overall nature of the U.S. economy has provided individuals and groups with resources that might be used in the exploitation of new ideas, and with incentives (in the form of potential financial and social rewards) to bear some risk in attempting such exploitation.

In the United States over the last 150 years, these beneficial factors have led to substantial advances and commercial and social successes in such areas as agriculture (including most notably that of the Midwest and of California's Central Valley), mining, precision manufacturing, automotive manufacturing, electrical goods, and most recently, in information technology, telecommunications, and biotechnology. The pattern is likely to continue in new areas of materials science and nanoscience.

Research-intensive universities have proven to be very productive centers for the creation of high-level human capability as well as new intellectual property, and there are several geographic areas in the United States where substantial development of new industry has occurred close to successful universities in consequence.[†] Examples are California's Central Valley (in agriculture); the Boston Corridor (information technology and biotechnology); Silicon Valley (information technology and telecommunications); the North Carolina Research Triangle; Austin, Texas (information technology, pharmaceuticals); and San Diego (wireless telecommunications).

Universities can serve as training grounds for other parts of the beneficial ecosystem, such as intellectual property law and deal making, and can bring the different parties together. A great recent example is University of California, San Diego CONNECT®,[‡] which has helped to create communications and broker relationships

*Bayh–Dole Act, 37 CFR 401.14.

[†]It should be noted that universities are not "the only game in town" when it comes to research, and that they are generally not the best place for commercial development of new ideas or technologies to occur. In some cases, industry also invests directly in research, such as research being carried out either internally to companies or through contracts to universities or other research-capable entities. In that case, the company owns the resulting intellectual property, as it should.

[‡]See http://connect.org.

between scientists, entrepreneurs, and intellectual property lawyers, contributing significantly to the success of new industrial development in Southern California.

In summary, two pieces of U.S. federal legislation, and continuing federal government investment through the creation and support of federal research funding agencies, have promoted the development of new industry in the United States. Combined with the size and strength of the U.S. economy—and it is not unreasonable to say that the size and strength of the U.S. economy are in some part a consequence of the success of the relationships—productive relationships between government, universities, and industry in research and development have evolved, have succeeded, and continue to succeed.

GENERAL PRINCIPLES FOR SUCCESS

The following are the key elements for success:

1. Universities, whether privately or publicly funded, other not-for-profit entities, and private individuals or groups (often in the form of corporations) should be able to compete for government funding for the support of research. This will create new intellectual property.
2. The intellectual property should be available for transfer through open, marketplace competition to entrepreneurial individuals or groups prepared to put their capital and labor at risk, for use in the development of new industrial activity. This will create new goods and services.
3. The revenues resulting from the sale of the goods and services should accrue primarily to the benefit of the entrepreneurial individuals or groups.
4. Part of the revenues generated by the success of such exploitation should be returned to the government through taxes. This will allow further investments in research.
5. Part of the revenues generated should be returned to the inventor or inventors.

Governments

Governments should invest public funds on the order of 2–3% of gross domestic product in competitive research in areas of national or local priority. The full costs of research that the universities perform should be paid. Follow the example of the Bayh–Dole Act and aim to get revenues back from income tax, corporate tax, or tax on successful new enterprises, but not from the up-front sale or licensing of intellectual property. Be bold. It works.

Universities

Universities should be quick and be generous in licensing intellectual property to local or national entrepreneurial entities. It pays off; entrepreneurial entities can exploit new ideas better than universities can. Again, universities should not try to make money from licensing or selling intellectual property. Look after the interests of the inventor, but other than that, be as helpful as possible to outside parties who are likely to succeed. Successful businesses will be generous and attentive in return and will also provide continuing political support. Stick to doing what you do best.

Industry

Do what industry does best: take risk, be creative, and test and refine the value of new ideas in the real world of the marketplace. Pay taxes and employ people in the local economy rather than outsourcing to other countries. Remember that it was the local taxpayer who paid for the new intellectual property in the first place. And when you succeed, remember the benefits that the university gave to you, and give back in return. (Take a tax deduction for doing so, that's okay.)

2

DIVIDED WE FALL

William B. Mattes

PharmPoint Consulting, Poolesville, MD, USA

With an entire book devoted to exploring collaborations in drug discovery and development, there seems to be little question as to the value of such collaboration. There is a widespread call for interdisciplinary research as the key to future advances in research [1–3], and consortia between various public and private organizations have been advocated as key in addressing problems of technology and drug development [4, 5]. That call extends to data sharing that transcends the traditional intellectual property protection culture of companies and even universities [6]. Such precompetitive collaboration as advocated for biomedical research has proven precedence in Sematech, a research and development consortium formed by semiconductor firms and the U.S. government, with the purpose of improving semiconductor manufacturing technology [7]. Sematech served to pool semiconductor R&D for member companies and was shown to reduce the R&D spending of individual member companies, indicating an R&D efficiency achieved by the consortium. Thus, the intuition that pooling of expertise, experience, and resources from various otherwise competing groups can achieve efficiencies beneficial to all members can be objectively confirmed. Furthermore, recent consortia exploring toxicogenomics [4], preclinical and clinical biomarkers [8–10], gene and protein

Collaborative Innovation in Drug Discovery: Strategies for Public and Private Partnerships, First Edition.
Edited by Rathnam Chaguturu.

annotation [11, 12], and pharmacogenetics [13] have also shown the power of collaboration in addressing complex problems.

CHALLENGES

There are real issues and difficulties in establishing and maintaining large-scale cross-function and cross-organizational research collaborations. First and foremost is the illogical nature of cooperation. Given that competition between individuals within a species is a basic premise of evolution, its antithesis, cooperation, has been the subject of question for several decades [14, 15]. While much is made of market competition between private R&D companies, nonprofits must also take a competitive view in the face of limited funding [16]. Competition is not just a fact, but a norm, in academic science and in the training of scientists [17, 18]. Welsh and coworkers note that cross-disciplinary and collaborative research are not fostered by university cultures focused on "individual merit" with a "career hierarchy which fosters competition between individuals and between universities" [19]. Funding, academic positions, and influence are all dependent upon a scientist's ability to excel beyond colleagues in the same field of research, with metrics usually being that of publication priority and number. With first-author status as a prime component of the publication metric, issues of authorship and author order become problematic in research projects that demand a group effort [20–23]. Graduate students and postdoctoral fellows are trained in this environment and are keenly aware that their future careers depend on their ability to compete in it; as such, they rarely acquire collaboration and team skills. So despite the current allure of promoting collaborative efforts, the reality is that human nature, scientific training, and reward structures predispose participants to competition and the pursuit of individual goals.

COLLABORATION MODELS

Despite this undercurrent of competition, however, interindividual and interorganizational collaborations have a long history of successes (as well as failures). In fact, a rather large body of research literature on collaboration exists in social science and organizational research. Shrum and coworkers have examined collaborations in the physical sciences [24], but unfortunately, there have been relatively few studies of collaboration in biomedical research settings. One such study examined four multidisciplinary groups and examined forms of collaboration and the influence of personal compatibility, work connections, incentives, and infrastructure [25]. However, social science and organizational work has developed testable models, and it is well worth calling attention to this work and Shrum's findings concerning scientific collaborations, especially so as even large scientific collaborations are rarely constructed with an eye toward the principles of effective collaboration that come out of this research. It should be noted that there is a parallel body of research concerning effective team construction and activity, which in some cases provides overlapping insight into collaboration models.

Beginning with the work of Wood and Gray [26], theories and models have been advanced to define and analyze collaborations, both successful and unsuccessful. Thomson et al. have accordingly developed a quantitative model based on the definition: "Collaboration is a process in which autonomous or semi-autonomous actors interact through formal and informal negotiation, jointly creating rules and structures governing their relationships and ways to act or decide on the issues that brought them together; it is a process involving shared norms and mutually beneficial interactions" [27]. Their model incorporates five key dimensions: governance, administration, mutuality, norms, and organizational autonomy, with several components of each dimension. Governance entails how participants in collaborations jointly develop rules and roles for interactions and decision making. Thomson et al.'s study affirms the concept that each member in a successful collaboration should have a voice in decisions and planning that is heard by all. The administration dimension includes familiar concepts of roles and responsibilities, communication channels, and implementation procedures. As might be expected, the study supports the importance of clarity, coordination, and agreement on goals. The dimension of organizational autonomy explicitly acknowledges that while there may be collective interests embodied in the collaboration's goals, member organizations (and individuals) have very real self-interests. A corollary of this inherent tension will be the conflict between any individual's commitments to the collaboration and those to his/her parent organization. Likewise, there will be tension between the decision-making authorities in the two arenas. Clearly, it is advantageous to the collaboration if member organizations provide participating individuals with a concrete agreement as to the effort they can commit and the authority they have in decisions that affect the member organization. Mutuality, on the other hand, describes the synergistic values of collaboration, the benefits that a member uniquely derives from the whole, as well as the unique benefits that a member brings to the table. Both of these must be clear and real for the collaboration to be valuable. In addition, the mutuality dimension captures the concept of shared interests that members may have and goals that are valuable to, but somewhat beyond those of, each member organization. Finally, the norms dimension captures the concepts of trust, where each collaboration member believes that other partners will meet their obligations.

Both complementary and, in some ways, contradictory conclusions as to the elements of collaboration are reached by Shrum and coworkers in their study of 53 collaborations in physics and related sciences. These authors specifically examine "trust" as a factor and find that while the "trust" that collaborators will have something to contribute is foundational, "trust" on a personal level is not important. Conflict between organizations (or member teams) and between such member teams and an administrative management (e.g., project management) as discussed in this study call to mind the dimensions of governance and administration noted earlier, and echo the tension of those dimensions with organizational (or team) autonomy. Ironically, in these collaborations, those with the least conflict seemed to be those that offered the most autonomy to individual scientists, team and organizational members, and those that did not require participants to share data and analyses. Nonetheless, consensual decision making was viewed as a positive and "successful" element in collaborations, yet hierarchically organized projects were often viewed as successful.

While individual teams, even interdisciplinary ones, are not strictly comparable to interorganizational collaborations, the research on team dynamics can be instructive. Thus, the principles of team assembly that lead to a collaboration network, where there is a dynamic between newcomers and "incumbents" [28], can be considered to be supportive of interorganizational collaborations. Similarly, studies of team leadership can be considered, although as noted earlier, governance and administration of collaborations may function differently from that in a formally empowered team. "Transformational leadership" [29] has been cited as a critical factor in improving the function and outcomes of research and development teams with diverse membership [30]. The four elements of transformational leadership (individualized consideration, intellectual stimulation, inspirational motivation, and idealized influence) would seem to be applicable to collaboration leadership. Yet the governance and administration models for interorganizational collaborations all suggest the importance of member inclusion in developing rules and approaches to decision making. Of equal importance in collaborations is the need for some level of member autonomy. These concerns would not in and of themselves rule out a single point "transformational" collaboration leader, but would suggest that the choice of that leader would best be a collective choice and that the leader would need to be thoughtful of the characteristics of collaborations that are different from those of teams.

Not discussed in these studies is the critical issue of funding and resource procurement. More often than not, large collaborative projects require funding for, at the very least, common activities such as project management, and often for the projects themselves. Complicating such procurement may be poor or absent estimates of the true cost of the project prior to its initiation, suggesting the need for a two-phase approach to collaboration, with an initial exploratory phase to accurately determine costs and feasibility. Further complications may arise from the source of the funding. While it may be expected that industry or even foundation funding may be given with some expectation of a specific outcome, even government funding may have such expectations. When the expectations of the funding body are not fully aligned with the stated mission of the collaboration, the collaboration may find the directions of its operation constrained or may feel compelled to modify its mission, perhaps in an unspoken fashion. Such tensions can translate into disaffection of collaboration members and/or failures to expeditiously reach goals. Thus, clarity in an approach to funding and alignment in such funding with the mission of the collaboration is a key concept.

EXAMPLES OF SUCCESSFUL BIOMEDICAL COLLABORATIONS

In light of the analyses of collaborations discussed earlier, it is worthwhile to examine the salient characteristics of several well-publicized and successful biomedical collaborations.

Biomarkers Consortium

As a first example, the Biomarkers Consortium had as founding partners the U.S. National Institutes of Health (NIH), the Food and Drug Administration (FDA), and the Pharmaceutical Research and Manufacturers of America (PhRMA), under the auspices of the Foundation for the NIH, which itself is mandated by the U.S. Congress to support research of the NIH [9]. The NIH coordinates administration, funding, and staffing of the Biomarkers Consortium, which has three tiers of administration: executive committee, subject-matter steering committees, and project teams. The executive committee includes representatives of each of the founding partners, other industrial sectors (such as diagnostics and information technology), the Centers for Medicaid and Medicare Services, and the Biotechnology Industry Organization, as well as a public member. This committee is ultimately responsible for approving, funding, and implementing projects, which are suggested by the steering committees [31]. Thus, funding from a variety of sources is first put in place before project implementation. However, project plan development and management faced challenges in initial projects even though decision makers shared a great deal of trust in one another [32].

Predictive Safety Testing Consortium

A second collaboration example to consider is that of the Predictive Safety Testing Consortium (PSTC), launched by the nonprofit Critical Path Institute (C-Path) in 2006 [10]. The impetus for this effort came from a number of industry and FDA scientists who wished to enable the use of new biomarkers in FDA- and European Medicines Agency (EMA)-regulated safety studies. The PSTC charter (i.e., Consortium Agreement) was developed with C-Path as the sponsoring institution, and established governance with a Director designated by C-Path, a Co-Director elected from the membership, and an Advisory Committee of members. The Advisory Committee provides oversight of Consortium projects and funding. Membership is composed of industrial or nonprofit organizations willing to commit to contribute biomarkers and/or biomarker qualification work to the Consortium. As so envisioned, the PSTC is a true collaboration in that both resources and efforts are provided by members. The challenge of this approach is establishing equitable levels of contribution, when such resources and efforts can be diverse and not formally mandated. On the other hand, the approach empowers project teams with a great deal of autonomy and encourages open dialogue and trust between individual scientists in different organizations, including the FDA and EMA [33]. A critical goal of the PSTC was met with both establishing a process for FDA and EMA biomarker qualification [34] and regulatory acceptance of new renal injury biomarkers [35].

SNP and SAEC Consortia

A third set of examples are two consortia: the single nucleotide polymorphism (SNP) Consortium (TSC) [36] and the International Serious Adverse Event Consortium (SAEC) [8]. These consortia were consecutively developed by Arthur Holden, a biotechnology executive, to meet needs common to a number of pharmaceutical companies. Indeed, these efforts are analogous to many of those analyzed by Shrum and coworkers in that through large-scale research, these consortia developed resources widely acknowledged as important. These consortia were established as nonprofit entities, with a board of directors made up of representatives from member companies contributing funding. Individual committees answering to that board handle scientific management, legal issues, and communications, with each committee co-chaired by two representatives from member companies. By establishing overall missions that are supported at the highest level of company R&D management, funding, which of necessity needs to be substantial for such projects, is assured. Scientific subcommittees serve to bring in experts from industry, academia, and government [8]. Project efforts are accomplished through contracts with academic or private institutions, and project management is handled by a central administration. Nonetheless, member and associate scientist involvement is vital in the various committees, and it is at this level that the principles of collaboration clearly are relevant, as have been discussed earlier.

On the whole, biomedical collaborations and consortia have a good track record, and given the different models that have produced various results, it is certain that no one structure is appropriate for all such endeavors. However, given the research on collaborations and teams, and the experience of the many previous collaborations and consortia, it is well worth considering certain principles when considering the development of a new collaboration. Just as experiment design profits from hindsight and forethought, so should collaboration or consortia design, with the goal of providing the best possible structure to achieve the collective mission.

PRINCIPLES OF COLLABORATION

Principles that may be considered include mission, scope, scale, and support. Inasmuch as many of the principles of collaboration are interdependent, these four concepts are foundational. While no collaboration will proceed without some sort of support, how this support is structured depends upon the scale and scope of the collaboration. Furthermore, the overall collaboration may be negatively influenced if its mission is not fully, and exclusively, accepted by the funding source. Given that the mission can determine the scope and scale of the collaboration, a clearly defined mission becomes the critical underpinning of the collaboration. Ideally, such a definition of mission is jointly developed by the key participants [37], yet the collaboration may need to be supported by and grow with stakeholders or participants who accept the founding mission [27]. The scale and scope, as defined by the mission, must then be critically evaluated for achievability and level of support needed. Consider that the mission and goals of the SNP Consortium demanded a level of support necessarily provided at the top management levels of R&D organizations [36]. By contrast, the PSTC was envisioned as accomplishing its mission with purely peer-to-peer "sweat equity" [10]. In both cases, the missions were aligned with the expectations of the sources of funding, although it should be noted that assurance of "sweat equity" becomes an issue of governance, mutuality, norms, and trust as touched on in earlier discussions. As might be expected, substantial but equitable up-front financial support from top management, once established, provides for fewer complications in project management but requires more preliminary negotiation.

GOVERNANCE, ADMINISTRATION, AND AUTONOMY

The studies from Shrum and Thomson call attention to the real tension in collaborations between the needs for governance and administration, and the needs for member autonomy. The models used are tied to the scope and scale of the collaboration; large-scale "industrial" consortia can be successfully managed as entities independent from the member participants. In such a model, project management is mostly focused on contractors. By contrast, with more peer-to-peer collaborations, central project management may be dealing with teams made up of participants, and may face the conflicts noted by Shrum and coworkers. Regardless, the structure, style, and scope of governance and participation must be aligned with the mission and fully accepted by all members. A clear definition of the role and authority of project management should also be accepted, both by governing committees and by project teams. Given the dynamic of collaborative work, and the fluidity of organizations, the governance and administrative structure should be periodically reviewed to affirm its relevance to the mission and affirm the roles and accountabilities as defined.

HUMAN FACTORS

It goes without saying that collaborations, and even industrial scale consortia, are human endeavors and meet or miss their goals according to the dynamics of human interactions. While the analyses cited earlier describe collaborations in structural terms, the interview questions used and their responses address qualities of interpersonal relationships. Ultimately, collaborators can feel included or can feel that other participants are seeking self-focused agendas. Governance can mitigate, but cannot eliminate, the impact of negative human factors. The balance of commitment to mutuality and mission with focus on self-interest needs to be openly and intentionally acknowledged in collaboration. Humility must outweigh hubris. While the mission provides an organizing principle, and governance and administration a structure, a frank and open discussion of human factors such as respect, transparency, and civility can ground collaboration and set the stage for resolving conflicts that may arise. Collaboration that accepts and respects different personalities, and places value on positive interactions, cannot help but be more successful in the face of almost certain technical and scientific challenges. Indeed, collaboration is centered on human factors and ignores them at peril.

CONCLUSION

In summary, the biomedical scientific challenges of the twenty-first century echo those of the mid-twentieth-century challenges faced by the physical sciences. Multiorganizational, multidisciplinary collaborations served to address the needs of scale and diverse talent that the problems demanded. Just as science learns from the findings of earlier experiments, it stands to reason that as biomedical collaborations are formed now, they should learn from the experiences of prior efforts and of the very research into the social aspects of collaboration. Such hindsight and study into earlier systems was the very impetus for the writers of the United States Constitution to start the Preamble with "In Order to form a more perfect Union." Future biomedical collaborations could be similarly improved with such attention to the known dynamics of their human interactions.

REFERENCES

1. Goodman, S.R., Blake, C.A. The future of interdisciplinary research and training: how to conquer the silo guardians. *Exp Biol Med (Maywood)*. 2006;231:1189–1191.
2. Disis, M.L., Slattery, J.T. The road we must take: multidisciplinary team science. *Sci Transl Med.* 2010;2:22cm29.
3. Portilla, L.M., Alving, B. Reaping the benefits of biomedical research: partnerships required. *Sci Transl Med.* 2010;2:35cm17.
4. Mattes, W.B. Public consortium efforts in toxicogenomics. *Methods Mol Biol.* 2008; 460:221–238.
5. Woodcock, J., Woosley, R. The FDA critical path initiative and its influence on new drug development. *Annu Rev Med.* 2008;59:1–12.
6. Munos, B.H., Chin, W.W. A call for sharing: adapting pharmaceutical research to new realities. *Sci Transl Med.* 2009;1:9cm8.
7. Irwin, D.A., Klenow, P.J. Sematech: purpose and performance. *Proc Natl Acad Sci U S A.* 1996;93:12739–12742.
8. Holden, A.L. The innovative use of a large-scale industry biomedical consortium to research the genetic basis of drug induced serious adverse events. *Drug Discov Today Technol.* 2007;4:75–87.
9. Mayrand-Chung, S. The Biomarkers Consortium: a novel public–private partnership leading the cutting-edge of biomarker research. *Biomarkers Med.* 2008;2:617–620.
10. Goodsaid, F., Frueh, F., Mattes, W. The Predictive Safety Testing Consortium: a synthesis of the goals, challenges and accomplishments of the Critical Path. *Drug Discov Today Technol.* 2007;4:47–50.
11. King, O.D., Foulger, R.E., Dwight, S.S., White, J.V., Roth, F.P. Predicting gene function from patterns of annotation. *Genome Res.* 2003;13:896–904.
12. UniProt Consortium. The Universal Protein Resource [UniProt] 2009. *Nucleic Acids Res.* 2009;37:D169–D174.
13. Relling, M.V., Klein, T.E. CPIC: clinical pharmacogenetics implementation consortium of the pharmacogenomics research network. *Clin Pharmacol Ther.* 2011;89:464–467.

14. Nowak, M.A. Five rules for the evolution of cooperation. *Science*. 2006;314:1560–1563.
15. Axelrod, R., Hamilton, W.D. The evolution of cooperation. *Science*. 1981;211:1390–1396.
16. Dees, J.G. Enterprising nonprofits. In: Harvard Business Review, editor. *On Nonprofits*. Boston: Harvard Business School Press; 1999, p. 209.
17. Atkinson, P., Batchelor, C., Parsons, E. Trajectories of collaboration and competition in a medical discovery. *Sci Technol Hum Values*. 1998;23:259–284.
18. Anderson, M.S., Ronning, E.A., De Vries, R., Martinson, B.C. The perverse effects of competition on scientists' work and relationships. *Sci Eng Ethics*. 2007;13:437–461.
19. Welsh, E., Jirotka, M., Gavaghan, D. Post-genomic science: cross-disciplinary and large-scale collaborative research and its organizational and technological challenges for the scientific research process. *Philos Transact A Math Phys Eng Sci*. 2006;364:1533–1549.
20. Grieger, M.C. Authorship: an ethical dilemma of science. *Sao Paulo Med J*. 2005;123: 242–246.
21. Strange, K. Authorship: why not just toss a coin? *Am J Physiol Cell Physiol*. 2008;295: C567–C575.
22. Kennedy, D. Multiple authors, multiple problems. *Science*. 2003;301:733.
23. Floyd, S.W., Schroeder, D.M., Finn, D.M. Only if I'm first author: conflict over credit in management scholarship. *Acad Manag J*. 1994;37:734–747.
24. Shrum, W., Genuth, J., Chompalov, I. *Structures of Scientific Collaboration*. Boston: The MIT Press; 2007.
25. Hara, N., Solomon, P., Kim, S.-L., Sonnenwald, D.H. An emerging view of scientific collaboration: scientists' perspectives on collaboration and factors that impact collaboration. *J Am Soc Inf Sci Technol*. 2003;54:952–965.
26. Wood, D.J., Gray, B. Toward a comprehensive theory of collaboration. *J Appl Behav Sci*. 1991;27:139–162.
27. Thomson, A.M., Perry, J.L., Miller, T.K. Conceptualizing and measuring collaboration. *J Public Adm Res Theory*. 2009;19:23–56.
28. Guimera, R., Uzzi, B., Spiro, J., Amaral, L.A. Team assembly mechanisms determine collaboration network structure and team performance. *Science*. 2005;308:697–702.
29. Bass, B.M. From transactional to transformational leadership: learning to share the vision. *Organ Dyn*. 1990;18:19–31.
30. Kearney, E., Gebert, D. Managing diversity and enhancing team outcomes: the promise of transformational leadership. *J Appl Psychol*. 2009;94:77–89.
31. Altar, C.A. The Biomarkers Consortium: on the critical path of drug discovery. *Clin Pharmacol Ther*. 2008;83:361–364.
32. Wagner, J.A., Prince, M., Wright, E.C., Ennis, M.M., Kochan, J., Nunez, D.J., Schneider, B., Wang, M.D., Chen, Y., Ghosh, S., Musser, B.J., Vassileva, M.T. The Biomarkers Consortium: practice and pitfalls of open-source precompetitive collaboration. *Clin Pharmacol Ther*. 2010;87:539–542.
33. Woosley, R.L., Myers, R.T., Goodsaid, F. The Critical Path Institute's approach to precompetitive sharing and advancing regulatory science. *Clin Pharmacol Ther*. 2010;87: 530–533.
34. Mattes, W.B., Walker, E.G., Abadie, E., Sistare, F.D., Vonderscher, J., Woodcock, J., Woosley, R.L. Research at the interface of industry, academia and regulatory science. *Nat Biotechnol*. 2010;28:432–433.

35. Dieterle, F., Sistare, F., Goodsaid, F., Papaluca, M., Ozer, J.S., Webb, C.P., Baer, W., Senagore, A., Schipper, M.J., Vonderscher, J., Sultana, S., Gerhold, D.L., Phillips, J.A., Maurer, G., Carl, K., Laurie, D., Harpur, E., Sonee, M., Ennulat, D., Holder, D., Andrews-Cleavenger, D., Gu, Y.Z., Thompson, K.L., Goering, P.L., Vidal, J.M., Abadie, E., Maciulaitis, R., Jacobson-Kram, D., Defelice, A.F., Hausner, E.A., Blank, M., Thompson, A., Harlow, P., Throckmorton, D., Xiao, S., Xu, N., Taylor, W., Vamvakas, S., Flamion, B., Lima, B.S., Kasper, P., Pasanen, M., Prasad, K., Troth, S., Bounous, D., Robinson-Gravatt, D., Betton, G., Davis, M.A., Akunda, J., McDuffie, J.E., Suter, L., Obert, L., Guffroy, M., Pinches, M., Jayadev, S., Blomme, E.A., Beushausen, S.A., Barlow, V.G., Collins, N., Waring, J., Honor, D., Snook, S., Lee, J., Rossi, P., Walker, E., Mattes, W. Renal biomarker qualification submission: a dialog between the FDA-EMEA and Predictive Safety Testing Consortium. *Nat Biotechnol.* 2010;28:455–462.
36. Holden, A.L. The SNP Consortium: summary of a private consortium effort to develop an applied map of the human genome. *Biotechniques Suppl.* 2002;22–24:26.
37. Covey, S.R. *The Seven Habits of Highly Effective People.* New York: Simon & Schuster; 1989.

3

INNOVATION: OPEN SOURCE AND NONPROFIT MODELS IN DRUG DISCOVERY

James M. Shaeffer and Sarah MacDonald

Outreach and Engagement, James Madison University, Harrisonburg, VA, USA

Occasionally, it can be helpful to have someone who is not an expert in your field provide some perspective. While neither of us have scientific expertise in drug discovery, we do have experience with encouraging, inspiring, and leading innovation. As social scientists, we can provide perspective on leading and managing organizational change, as well as suggest models or approaches to change and innovation.

This chapter is the result of a presentation made at the 4th International Drug Discovery and Therapy Conference in Dubai, United Arab Emirates, called "Staying Competitive: Creating and Fostering Innovation in the Workplace." This chapter fleshes out that concept by incorporating suggestions of models that can be applied to encourage and sustain that innovation. The goal of this article is to provide a viewpoint on models and approaches to enhance innovation that have been successful in other contexts, and to apply those models to the drug discovery process. These models could be applied to a number of systems and processes, as they can be useful analogies or metaphors in a range of contexts. We are also approaching this from the point of view of a consumer of drug discovery, rather than as a producer of drug discovery. Others before us have argued that some of these models might actually be appropriate in the context of drug discovery [1–3], but the additional framework of innovation here contributes a further level of complexity.

Collaborative Innovation in Drug Discovery: Strategies for Public and Private Partnerships, First Edition.
Edited by Rathnam Chaguturu.

In their seminal book, *Innovation: The Five Disciplines for Creating What Customers Want*, Carlson and Wilmot define innovation as "the process of creating and delivering new customer value in the marketplace." [4] Innovation is not a magic process, or like a bolt of lightning, but is the result of a disciplined, continuous improvement process with an unrelenting focus on creating the highest customer value. The sustained and disciplined approach to finding solutions to problems through innovation requires time, dedication, and support, but being open to the recursive, trial and error nature of this process can yield measurably different results that solve real problems. Innovation is an intensely collaborative activity by nature and focuses on identified customer or consumer needs.

WHY DO PEOPLE INNOVATE?

Carlson and Wilmot's definition [4] implies that the primary motive for innovation is profit, since they describe it as delivering customer value in the marketplace. However, profit is not the only underlying motive for innovation. Recent research has suggested that innovation can still thrive even when financial profit is not the primary goal, because people are driven by mission-related motives and the goal to make the world a better place. Von Hippel suggests that a new model of innovation is focused not on the companies that introduce a product but by the "users" that modify the product for their private return and not for profit [5]. For example, when Microsoft released their interactive Kinect product, users began modifying and improving it. Microsoft initially described this end-user activity as "hacking," but soon came to recognize the potential for profit, mutual gains, and innovation. While the idea of end users or consumers modifying pharmaceutical drugs may seem far-fetched, this notion of giving freely as part of a shared innovation may provide ground for discussion of alternative models of innovation. In particular, scientists "thrive on solving great problems and live to tackle big challenges" [6]. Science, like software, relies on teams of volunteers, notably graduate students and young professionals, who have an incentive to get involved because it will enhance people's lives and make the world a better place, as well as enhance their own professional careers. Human beings are inherently motivated to solve problems and address challenges. One specific model that highlights a motive for innovation not related to profit is the Gandhian innovation model proposed by Prahalad and Mashelkar [7]. They demonstrate innovation strategies used in India as a specific example of how pioneers have figured out how to accomplish more with fewer resources for a greater number of people. For example, Tata Motors in India has developed a $2000 Nano vehicle by creating a number of international partnerships to obtain technical expertise. The resulting vehicle fits the needs of the rapidly growing population in India while reducing the size of the vehicle to meet emerging pollution standards. The five tenets of the Gandhian innovation model developed by Pralahad and Mashelkar [7] are a deep commitment to serving the underserved, a clearly articulated vision, audacious goals, recognition of constraints, and a primary focus on people rather than profits. The example of Tata Motor's Nano car is a clear demonstration of those principles.

However, it is not just employees (in this case, researchers) who provide and sustain innovation in an organization; successful and long-term innovation requires leadership. Without support from leaders, innovative efforts are destined to fail, either in the short term or the long term. That support can include a range of strategies and tactics, such as decentralizing decision-making authority, encouraging risk taking by nurturing innovation-supportive culture, accepting failure as a natural consequence, and rewarding both failures and successes [8].

In addition to encouraging innovation, leaders need to be aware of and overcome possible impediments to innovation. Employees are often resistant to significant change, and are therefore often resistant to innovation in ways that are not immediately apparent. But successful innovation requires collaboration and a significant amount of trust, and leaders can help their employees reach that goal by reducing fear of the unknown. Employees have indicated that they choose to uphold the status quo because they fear unfavorable social impressions with peers, or they have a lack of support for innovation by leadership [9]. Carlson and Wilmot describe what they call the "FUD" factor, which stands for fear, uncertainty, and doubt, which are people's natural reactions to significant change [4].

HOW DO WE ENCOURAGE INNOVATION?

There are multiple strategies that leaders can employ to encourage innovation and navigate around some of the possible impediments. Wiseman and McKeown describe two types of leaders: diminishers and multipliers [10] (see Table 3.1).

Successful leaders use these concepts of diminishing versus multiplying to empower employees to innovate and collaborate. Another strategy, as suggested by Carlson and Wilmot [4] is to reframe statements related to the "FUD" factor as clues for how to move toward a solution. For example, if an employee says "I'm too busy to do that," a strong leader can interpret that as a clue that the employee is really saying "I'm out of energy, and need some help." Finally, strong leaders who want to encourage

TABLE 3.1. Characteristics of Leaders, Diminishers, and Multipliers

Diminisher	Multiplier
Empire builder	Talent magnet
Hoarding resources	Giving ownership
Tyrant that suppresses thinking	Liberator
Know it all	Challenger
Make all the decisions	Create debate
Micromanager	Invest in others

innovation need to ensure that any misperceptions about the vision or support to reach that vision have been addressed [4].

WHY IS INNOVATION IMPORTANT?

If you are stationary or moving forward slowly, and everyone else is accelerating rapidly, you will quickly be left behind. [3]

Innovation is important in all sectors, but Bennani reminds us that it is critically important in the health-care industry, because of the high stakes and human costs involved [6]. There is no shortage of human health problems to solve, and each of those problems bears enormous cost. Innovative strategies for addressing those challenges will enhance human well-being globally, increasing productivity in other sectors.

We need innovation to address the problem that "pharmaceutical drug development costs are increasing, yet the pharma success rate is decreasing." [3] Innovation is needed to take advantage of new technologies that may assist with the high cost of drug discovery. Innovation is needed to provide drug therapy for those who are most in need, including millions of children who die each year of diseases that are preventable, according to the World Health Organization [11]. This includes the "massive global concern of tuberculosis, since one in three are infected around the world—more than two billion people!" [3] Innovation is needed in drug discovery to continue research on nonpatentable compound and drugs whose patents have expired. Without encouraging innovation, we may very well miss out on the discovery of secondary uses for existing drugs. Finally, we need innovation to move beyond the "patent thickets" that have led to the "tragedy of the anti-commons." [11] In other words, the proliferation of patents is blocking fundamental tools in biotechnology research which may have led to the underutilization of existing knowledge, due to high costs and lack of cooperation by patent holders [11].

The need for innovation in the pharmaceutical industry, according to Bennani, is that we are in a perfect storm with "falling success rates in the development of innovative therapeutics, pending patent expirations for major drug classes" as well as stricter regulatory environments and the general downturn in the economy [6].

WARNINGS REGARDING INNOVATION IN DRUG DISCOVERY

One of the impediments to innovation, according Bennani is that multiple companies are working on similar "blockbuster" drugs, which means there is duplication of efforts and nonproductive investments [6]. The economic model of companies competing in a small market space for the blockbuster drugs has actually backfired and led to the loss of thousands of jobs. However, Bennani also provides a useful warning when examining innovation in the pharmaceutical sector; it is different from other sectors because "when researchers embark on a truly new project, they generally have no idea what the final product will look like." [6]

One more cautionary note regarding encouraging innovation in the pharmaceutical industry: we need to foster innovation and creativity while at the same time not encourage unethical or overtly risky behavior. Asking employees to engage in innovative practices in any industry can lead them to ethical dilemmas, including breaking rules and standard operating procedures; challenging authority and avoiding tradition; creating conflict competition and stress; and taking unnecessary and dangerous risks. Innovation requires that rules be broken, but especially in health care, there are particular rules and procedures that cannot and should not be broken. The balance of encouraging risks, but not unwarranted risks, is to help employees "understand the types of risks they can take, the scope of risk undertaken, and the potential negative consequences of the organization's stakeholders." [12]

OPEN SOURCE MODELS

One approach is to incorporate open source models into drug discovery. Open source concepts initiated with the software industry and have seen enormous success, including projects such as the Linux operating system or the Mozilla browser. Other terms related to, but slightly different from, open source concepts are open access, open knowledge, and open innovation. The Open Source Initiative defines open source with an elaborate, 10-point definition, but the main components of that definition involve access to the source code, free redistribution, and intellectual property. The question then becomes whether the open source model can be applied to other industries with similar success. As Rai et al. asked, "Can goodwill, aggregated over the internet, produce good medicine?" [1]

Open source efforts have already taken place in scientific fields; for example, the Human Genome project completed in 2003 [13] shared a number of characteristics with open source software development. Both fields attract the same sort of people, with the motivation (as discussed earlier) to solve challenging problems and make a difference in the world. Applying the concept of open source development to drug discovery could decrease transaction costs, leverage costly investments in technology and resources, and allow researchers the ability to access the intelligence, creativity, and goodwill of their colleagues.

However, the dissimilarities between open source software development and pharmaceutical research are profound and not easily overcome. Software can be developed very quickly, sometimes in a matter of weeks or months, and the technology required is often available to everyday consumers, whereas pharmaceutical research is more often measured in years, and requires advanced scientific degrees and expensive laboratory equipment. The most cost-prohibitive part of drug discovery is not the initial scientific work, but instead the enormous expense of clinical trials and navigating the regulatory approval process. "The application of the open-source approach to drug development may prove to be more useful as an analogy than an application," notes Janet Hope, a lawyer completing a doctorate on open source biotechnology at the Australian National University, in Canberra. "One reason is that different intellectual property rights apply, and are protected differently. Software usually falls under copyright, which arises automatically and without cost to the author. Biomedical discoveries are generally protected by patents, which are costly to obtain." [1]

In general, patents are directly linked to innovation [14]. Producing innovative ideas and solutions requires protection of intellectual property, and patents and copyright are the two main tools to accomplish that. However, confounding issues surrounding intellectual property, innovation, and international development have been raised. For patents, these issues include whether patents are being granted for truly novel inventions or for slightly different iterations of existing technologies; when patent protection should be overridden for humanitarian reasons; and the enormous cost of both time and money in securing patents. As an example, discoveries related to the human genome are vital to future biomedical innovation, but it is estimated that 20% of the human genome is claimed by patents [11]. Two-thirds of these patents are owned by private companies, and a similar fraction may be legally questionable on the grounds that they are too broad, not disclosed properly, or overlap other patent claims [15]. One possible solution to this conundrum is the idea of patent pools, which are "consortiums that agree to cross-license patents relating to a particular technology. They are beginning to be used to stimulate research in neglected diseases, allowing both access to select technologies and competitive business practices." [16] Certainly, patent pools are far more difficult to put into practice than they are to describe, but the idea holds merit and potential.

Masum et al. present an overview of two outstanding examples of open source biotechnology platforms: Cambia and India's Open Source Drug Discovery (OSDD) [11]. Cambia is a private, nonprofit institute in Australia, and will be discussed later in this chapter. OSDD is a consortium launched in 2007 by India's Council of Scientific and Industrial Research that "aims to achieve affordable health care through a platform where talented minds can collectively discover novel therapies, as well as bring openness and collaboration to the drug discovery process, and keep drug costs low." [11] OSDD incorporates a web-based portal that allows participants access to bioinformatics tools, biological information, data on pathogens, and discussion forums, with the aim of breaking down drug discovery into smaller activities with clear deliverables. Participants can contribute anything that helps solve these problems and achieve the goals of these smaller activities, including ideas, software, articles, and intellectual property.

Users must grant worldwide nonexclusive rights to OSDD for the use of any IP rights acquired, which helps contribute to the overall collaboration with all users while still protecting the intellectual property rights of all involved. OSDD's high-profile backers may also increase the incentive for researchers to participate in the network of collaboration and open source sharing of ideas and resources [11].

As a result of the efforts of OSDD and their partners, there has been a renewed focus on solutions for diseases like tuberculosis and malaria, diseases that disproportionately affect third world and developing countries. More specific information about those accomplishments can be found at OSDD's website, http://osdd.net. The success of OSDD's initiatives suggests possible applications of open source development concepts that sidestep some or all of the challenges described earlier. Masum et al. describe such application as including three aspects of open source concepts: open access to underlying information, open licensing practices, and open collaborative methods and platforms [11]. Beyond just sharing software code for collaborative development, open source implies a set of cultural practices, licenses, and innovative collaboration methods that suggests future potential applications for drug discovery. While not a panacea, open source models could encourage collaboration and innovation, and are worth considering for the pharmaceutical industry.

NONPROFIT MODELS

A second possible approach to the current state of pharmaceutical research is the idea of nonprofit organizations. Nonprofit organizations are chartered to address a need and accomplish a mission, rather than to create profits; they are eligible for tax-exempt status, and are often called the "third sector" (the other two being public and private organizations). Nonprofits can play a pivotal role in addressing some of the challenges of drug development because they tend to be focused, nimble, and entrepreneurial. Being nonprofit-focused by law, such organizations conduct cutting-edge research often funded by competitive grants from governments and foundations that are traditionally not accessible to for-profit companies. Nonprofits freely embrace research areas that are shunned by the mainstream industry, as long as those areas serve the mission of the organization, even when the risks are greater and future profits are lower [3].

Stanford Research Institute (SRI) provides a prime example of nonprofit contributions to drug discovery. Founded as the Stanford Research Institute in 1946, it has grown to more than 2000 employees at 20 locations around the world, including a new biosciences laboratory in the Shenandoah Valley of Virginia, where James Madison University is located. By reinvesting all of its profits in internal research and development, SRI has collaborated with the National Institutes of Health, other government agencies, universities, and other organizations to advance new therapies and vaccines to human clinical trials. SRI describes the collaboration process as both an art and a science, and that the best collaborations are "mutually flexible, seek win/win solutions, leave their egos at home, listen first to understand, and play to each other's strengths." [3] They also describe numerous ways to collaborate, including fellowships, innovation workshops, in- or out-licensing, philanthropy, outsourcing of research and development, and partnering on funding proposals.

Another example of nonprofit organizations partnering in the drug discovery process is Cambia, a private nonprofit based in Australia. Cambia's primary achievement has been the launching of Biological Innovation for Open Society (BiOS), an initiative that aims to create a "protected commons" that allows participants to "access, improve, and modify existing technologies without infringing on proprietary rights." [11] This is done through BiOS licenses, which are similar to Creative Commons licenses used in cultural centers online for creations like photographs, blogs, and other writing, or General Public Licenses (GPLs), used in software development. BiOS licenses also share a great deal in common with the work done by OSDD in India. According to Masum et al.,

> *BiOS licensees must sign a detailed legal contract to preserve the right of others to use the technology—e.g. by agreeing not to assert IP rights against others who have also signed the contract. In exchange, they gain access to the technology (BiOS, 2009). Unlike some other open source licenses, BiOS licenses do not prohibit licensed technology from being used to develop downstream proprietary products [11].*

It seems that the model used by Cambia offers a useful platform for connecting researchers and scientists, while at the same time avoiding some of the patent crisis that has plagued other efforts. There are a number of advantages and benefits to the model under which Cambia's BiOS licenses were developed, but above all is the ability to access the intelligence, goodwill, and laboratory and testing facilities of a broader community of researchers. Cambia's model may be the answer to our earlier question, "Can goodwill, aggregated over the internet, produce good medicine?"

A final example of a nonprofit organization offers an exploration of what happens when a nonprofit organization partners with a government organization and a for-profit company, all the while trying to work across international borders. In a recent issue of the *Chronicle of Higher Education*, Steven Ambrus describes a cross-border partnership between the University of Arizona College of Medicine, the National Autonomous University of Mexico, and the Bioclon Institute (a Mexican pharmaceutical company). This partnership was designed to bring a newly developed antivenom developed in Mexico through the U.S. approval process from the Food and Drug Administration (FDA). Because of the international nature of this collaboration, along with the increasingly stringent regulations on customs and international travel, the process of seeking FDA approval for this antivenom and of undergoing a rigorous clinical trial protocol were even more difficult than for typical drugs developed in the United States. But the end result is a vast improvement over previous antivenom treatments available in the United States, as well as additional recognition of the value of reversing the typical process of developing drugs in the first world and testing them in the third world [17].

RECOMMENDATIONS

The goal of this chapter is to examine the drug discovery process, both from the perspective of social scientists as well as end users, or customers, in the drug discovery process. It would seem that one of the important links in the drug discovery process, the customer, is often seen only as the receiver of the drug therapy. In reality, consumers can become advocates for change, if and when they can be fully educated about the challenges of the drug discovery process, the cost versus the revenue, the time to completion due to the mountain of regulations, and the anti-commons created in the pursuit of patents. We recommend that a greater effort be made by the pharmaceutical industry to inform the general public of the challenges and cultivate an ally.

Another strategy that may provide opportunities for testing as well as a feedback loop is engaging the volunteer health associations, such as the American Heart Association or the international Red Cross. These associations have created a network of health professionals who are passionately working with their colleagues to find therapies to address the needs of their patients. The question we raise is whether the drug discovery process can better utilize the resources of volunteer health associations in the drug discovery process, thereby lowering the cost of drug discovery and possibly speeding up the process of discovery and having an effective drug on the market.

We have also made the argument in this chapter that exploring and discussing alternative approaches to drug discovery is just as important as the science of drug discovery. Utilizing open source discovery and nonprofit organizations show promise in assisting with the cost, the speed, and sharing of information in drug discovery. Open source and nonprofit organizations can assist in addressing drug discovery for diseases impacting a small population, the need for drug therapy for impoverished countries, and in furthering the work on drugs that are long on patent. Open source and nonprofit organizations have the potential to fulfill the Gandhian approach of making drugs that are affordable and sustainable.

Finally, as consumers and social scientists, we've learned through our research for this book chapter and through interactions with scientists that there are many dedicated professionals who are passionate about drug discovery, and that as consumers we can best assist and advocate for changes when we more fully comprehend the challenges faced in drug discovery.

I would prize every invention of science made for the benefit of all.

—Mahatma Gandhi

REFERENCES

1. Rai, A., Sali, A., Maurer, S. An open source shot in the arm? *Economist* 2004; 371(8379):17–19.
2. Ardal, C., Alstadsaeter, A., Rottingen, J. Common characteristics of open source software development and applicability for drug discovery: a systematic review. *Health Res Policy Syst.* 2011;9(36):1–16.
3. Moos, W.H., Kodukula, K. Nonprofit pharma: solutions to what ails the industry. *Curr Med Chem.* 2011;18:3437–3440.
4. Carlson, C.R., Wilmot, W.W. *Innovation: The Five Disciplines for Creating What Customers Want.* New York: Random House; 2006.
5. Von Hippel, E. People don't need a profit motive to innovate. *Harv Bus Rev.* 2011; November:36–37.
6. Bennani, Y.L. Drug discovery in the next decade: innovation needed ASAP. *Drug Discov Today* 2011;16/17:779–792.
7. Pralahad, C.K., Mashelkar, R.A. Innovation's holy grail. *Harv Bus Rev.* 2010;July–August: 132–141.
8. Wan, D., Ong, C.H., Lee, F. Determinants of firm innovation in Singapore. *Technovation* 2005;25(3):261–268.
9. Yuan, F., Woodman, R.W. Innovative behavior in the workplace: the role of performance and image outcomes expectations. *Acad Manag J.* 2010;53(2):323–342.

10. Wiseman, L., McKeown, G. Bringing out the best in your people. *Harv Bus Rev*. 2010; May:117–121.
11. Masum, H., Schroeder, K., Khan, M., Daar, A. Open source biotechnology platforms for global health and development: two case studies. *Inf Technol Int Dev*. 2011;7(1):61–69.
12. Baucus, M.S., Norton, W.I., Jr., Baucus, D.A., Human, S.E. Fostering creativity and innovation without encouraging unethical behavior. *J Bus Ethics* 2008;81:97–115.
13. Human Genome Project Information. 2013. Available at http://www.ornl.gov/sci/techresources/Human_Genome/home.shtml. Accessed December 19, 2013.
14. Netanel, N.W., editor. *The Development Agenda: Global Intellectual Property and Developing Countries*. Oxford and New York: Oxford University Press; 2009.
15. Cukier, K.N. Navigating the future(s) of biotech intellectual property. *Nat Biotechnol*. 2006;24:249–251.
16. Van Overwalle, G., editor. *Gene Patents and Collaborative Licensing Models: Patent Pools, Clearinghouses, Open Source Models, and Liability Regimes*. Cambridge, UK: Cambridge University Press; 2009.
17. Ambrus, S. Border crossing research lessens the sting. *The Chronicle of Higher Education* 2012. Available at http://chronicle.com/article/Drug-Developed-in-Latin/130932/. Accessed December 19, 2013.

4

THE CHANGING FACE OF INNOVATION IN DRUG DISCOVERY

Litao Zhang and Carl Decicco

Bristol-Myers Squibb Company, Princeton, NJ, USA

INTRODUCTION

Over the past 30 years, pharmaceutical companies have contributed many innovative medicines for the treatment of human diseases. Scientific and technological advances have enabled the discovery of important drugs in critical areas such as infectious diseases, cancer, cardiovascular disease, and metabolic disorders such as diabetes. Since 1982, the U.S. Food and Drug Administration (FDA) has approved about 500 new chemical entities (NCEs) [1]. Among these, more than 100 drugs [2–5] work through new mechanisms of action (Table 4.1). Our knowledge, facilitated in large part by advances in genomics and a plethora of new approaches aimed at uncovering the essential facts of cellular and physiological pathways, may finally be delivering on its promise, as evidenced by the significant increase in first-in-class drugs between 2003 and 2012 (Table 4.1). Cancer drug discovery, for example, has evolved rapidly, thereby shifting the focus toward targeted medicines with improved efficacy and safety. The field of immuno-oncology [6, 7], which targets the immune system in certain types of cancer, holds great promise, as illustrated by the approval in 2012 of the first true immune-oncologic, Yervoy [8], which has been bringing great hope to many melanoma

Collaborative Innovation in Drug Discovery: Strategies for Public and Private Partnerships, First Edition.
Edited by Rathnam Chaguturu.

TABLE 4.1. New Chemical Entity (NCE) Approvals from 1982 to 2012

	1982 to 1992	1993 to 2002	2003 to 2012
# NCEs approved by FDA	120	131	248
First-in-class [2–5]	19	24	60
Percentage of the first launched in the United States versus in the world [1]	10%	30%	50%

patients. Biopharmaceutical research has also witnessed increased efforts across the industry in the discovery of drugs intended for small populations of patients with rare (or "orphan") diseases. These efforts were highlighted in 2011 [3], with 11 of the 30 new molecular entities (NMEs) approved by the FDA going to first-in-class drugs having a rare disease as the primary indication. Clearly, pharmaceutical companies are leading the way in the discovery of novel medicines, with the United States accounting for 50% of first-in-class drugs discovered between 2004 and 2012 [1]. In 2012 alone, the FDA approved a total of 37 NMEs, an output unseen since the late 1990s. The most exciting result for the pharmaceutical industry is that about 28 first-in-class NMEs were approved by the FDA in 2012. These data clearly show that innovation is resurging across the industry.

The number of NME approvals is one measure of success for pharmaceutical companies. Yet a commercial reality that constantly challenges continued investment in drug research is that after thorough clinical testing and the meeting of the appropriate regulatory requirements for approval, only about 8–12 years of patent exclusivity typically remain following the launch of a new drug. Patent expiration in most cases leads to a rapid decline in profit, as generic companies essentially commoditize the innovations of pharma companies. Moreover, modern drug discovery operates in a relatively new reality: the essential requirement that new drugs need to be differentiated from available therapies if they are to have any promise of delivering a return on investment. Because differentiation can only be demonstrated relatively late in the drug discovery process via carefully controlled human clinical trials, an even higher bar is now set for discovery organizations to provide a data package that supports investment.

Modern drug discovery is proceeding in an environment that has been radically altered to accommodate the explosion of new biological information. Engineering advances are steadily opening up exciting new possibilities for innovations, which in turn are leading to new and more efficient ways to identify and qualify new drug candidates. Technological advances have become a great facilitator for testing new ideas and thus now play an essential role in driving drug innovation. New technologies have led to new challenges and to questions like this one: How do we test, assimilate, and apply new technologies in a manner that efficiently enables future drug discovery? The industry, in its rapid evolution, has encountered several pitfalls related to overpromising results from technology. A hard lesson learned is that no single approach, such as high-throughput screening (HTS) or combinatorial chemistry, will deliver simply on its own. On the contrary, an integrated approach to leveraging technology in a cross-functional fashion has proven to be one important way of generating data that, used appropriately, can positively impact both the capacity and the cycle time needed for discovering new drug candidates.

The drug discovery process entails a broad range of disciplines—from genomics and target discovery to the identification and optimization of lead compounds for preclinical testing. Innovative technologies play an important role in enabling the integration of those disciplines. Over the past two decades, most companies have invested a great deal of human as well as economic capital to deploy advanced technology platforms, and with varying degrees of success. For example, technologies related to automation, informatics, and high-throughput assays have all played important roles in creating better data faster—data that guide the iterative nature of drug discovery.

In this chapter, we will focus on three themes:

1. Highlights of the key technological innovations that have revolutionized the drug discovery process in recent years
2. Important lessons learned from the process of creating and applying new technologies, and
3. New trends that will drive future innovations in drug discovery research.

CORE OBJECTIVES OF THE DRUG DISCOVERY PROCESS

The pharmaceutical industry started on its long evolutionary journey in the nineteenth century with essentially "accidental" discoveries of chemicals derived from tars or natural derivatives, chemicals that could be used to treat human diseases [9]. Drug discovery as a formal process did not become integral to the pharmaceutical industry until the early years of the twentieth century. In the ensuing years—especially in the past half-century—drug discovery has joined clinical development, manufacturing, and regulatory sciences to become one of the most fundamental and critical business units in the industry. This has happened as the principles of chemistry have been increasingly applied to problems beyond chemistry itself, as pharmacology—the medical and biological study of drug activity—has become a well-defined scientific discipline in its own right, and as clinical research has yielded in-depth knowledge of diseases and their treatment.

Today, although some aspects of drug discovery vary from one company to another, the basic processes are the same everywhere. These processes can be divided into four interconnected processes:

- *Target identification and target validation* focus on discovering and confirming molecular targets that play a fundamental role in both normal and disease states.
- *Hit identification* entails using the information derived from target identification and validation to discover multiple molecules active against potential therapeutic targets.
- *Lead optimization* is a process designed to help optimize potency (a function of how tightly the molecules interact with the target), improve selectivity (the

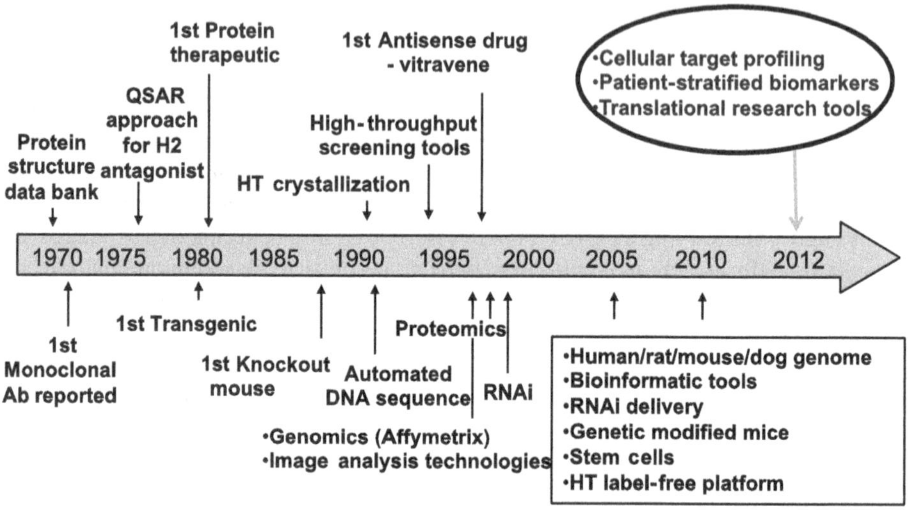

Figure 4.1. Innovative technology platforms from target discovery to preclinical assessment launched over the past three decades have fundamentally changed the drug discovery process.

degree to which a dose of a drug produces the desired effect against a specific target), and evaluate the safety of leading molecules by applying iterative cycles of "medicinal chemistry" efforts.

- *Preclinical assessment* is performed to select the best candidate molecules for clinical development based on diverse *in vitro* and *in vivo* data packages.

For several decades, as pharmaceutical companies have used target identification and target validation, hit identification, lead optimization, and preclinical assessment, they have continuously monitored and evaluated these processes to improve them and thereby close any gaps in drug discovery. Improvement has come in the form of innovative solutions. Figure 4.1 highlights some key technological breakthroughs that have significantly improved the quality and capacity of these core drug discovery processes. These innovative solutions have indeed improved the quality and capacity of the work performed by drug discovery organizations across the industry.

One of the key lessons we have learned is that science must set the direction for technological innovations in discovery. Thus, the next phases of innovation need to be based on the challenges we are facing today. This is because the drug discovery process is an extremely challenging endeavor due to the fact that numerous hurdles beyond activity and selectivity have to be overcome. Recent studies [10, 11] indicate that failure appears to be occurring most often in Phase 2/3 and primarily for reasons of toxicity and insufficient efficacy. During preclinical assessment, therefore, thoughtful integration of toxicology work plans is critical to understand drug development risks, develop appropriate levels of testing schemes, and identify high-quality drug candidates.

Effective solutions to improve drug discovery success rates depend on many factors. First and foremost, effective tools for predicting efficacy and toxicity are urgently needed. Indeed, the judicious selection of therapeutic modalities with high success probabilities has become the most critical task for the discovery organization. In summary, technological innovations for drug discovery processes must transition from simply targeting capacity gaps to searching for solutions that can produce high-quality data, provide enriched information, and establish better connectivity between *in vitro* and *in vivo* models.

MAJOR INNOVATIONS IN TECHNOLOGY HAVE ADVANCED DRUG DISCOVERY

Innovations in technological advancement for discovery have been driven by the need to close substantial gaps in discovery processes. This section shows how innovations targeted at these gaps have produced the breakthroughs that greatly enhance drug discovery processes.

Target Identification and Target Validation

The sequencing of the human genome represents the largest undertaking in the history of biological science, and the completion of the Human Genome Project (HGP) created a new landmark in the life sciences [12]. While the predicted strong impact from HGP may not yet be felt in daily clinical practice, that day is definitely accelerating toward us. For example, researchers have already identified single genes associated with a number of diseases, such as cystic fibrosis, Duchenne muscular dystrophy, neurofibromatosis, and retinoblastoma [13–16]. As research progresses, investigators will decode the disease-associated mechanisms that are caused by a single gene or multiple genes interacting with environmental factors.

More important and relevant, the HGP has transformed the way a great deal of research is done in the pharmaceutical industry. Understanding unknown protein function in the human genome is continuously enriching the druggable target landscape [17–20]. The best-known technological advancements, such as genomics [21, 22] and proteomics [23], have significantly enhanced the capability for target identification and validation in the industry. Moreover, the measurement of the differential expression of messenger RNAs and proteins in both diseased and normal tissues has become an important technique widely used in the drug-target phases of discovery research.

Among the breakthrough technology platforms used in genomics-based research are gene microarray chips [24] containing a collection of oligonucleotides. Today, Affymetrix chips [25] can be used to measure the expression of more than 10,000 human genes in a single experiment. Furthermore, microarray chips for the determination of gene expression in other species, such as rats, mice, *Escherichia coli*, *Drosophila melanogaster*, *Caenorhabditis elegans*, and *Saccharomyces cerevisiae*, are also commercially available. These tools significantly accelerate the identification of novel therapeutic target genes that are expressed differently in disease tissues and models.

Yet the major challenge associated with this approach is that the number of "hits" is dependent on the magnitude of significant biological manipulations. As a result, the "hits" are typically limited to hundreds of genes. Furthermore, those changes at the mRNA level cannot always be automatically translated into corresponding alterations in protein expression activity in the disease states. Nonetheless, the next gene sequence technologies [26, 27] have enabled us to identify mutations in the abnormal genes in disease states accurately and rapidly. The applications of next-generation sequencing (NGS) technologies and gene microarray chips have opened a possible channel for personalized medicine by determining whole-genome and gene function information just-in-time.

To facilitate target identification and target validation, discovery research has been increasingly turning in the past 5 years or so to the task of converting DNA sequence information into disease-associated protein functions. Analysis of a proteome—the entire set of proteins expressed in an organism, cell, or organelle—is very challenging because protein functions are in constant flux. Proteomics has to deal with the separation of highly complex protein mixtures. In addition, the difference in concentration between low-abundance and high-abundance proteins can exceed more than five orders of magnitude; hence, proteins in high abundance can mask those in low abundance. This situation leads to big challenges in protein detection and protein separation. To help meet this challenge, innovative advances in liquid chromatography–mass spectrometry (LC-MS) technology [28, 29] have transitioned proteomics from identifying functional proteins to illuminating quantitative changes in key protein components within cellular systems. Quantitative analysis of global protein levels, termed "quantitative proteomics," [23] is important to the system-based understanding of both the molecular functions and molecular mechanisms of each protein component in the various biological processes and systems.

Although genomic and proteomic approaches have provided more target options, so far the pharmaceutical industry has only investigated a subset of human genome encoding ~800 proteins [30]. It is estimated that at least 3000 druggable targets are in the human genome [31, 32]. The overlap space between the disease-relevant human genome pool and drug-like molecules [17–20] is referred to as druggable space [19] (Figure 4.2). Remarkably, medicinal chemists working over the decades and across industry have made millions of diverse chemical structures. However, only about 1000 of these molecules have been approved as drugs. Currently, the great challenge related to druggable space is that the "low-hanging fruit" among targets has largely been picked, and the druggable space has begun to shrink. Thus, to expand the druggable genome and accelerate drug-target discovery processes, new and innovative target discovery approaches have to be explored and applied.

In recent years, chemogenomics [33–36] has become an increasingly important approach to target identification. With this approach, large collections of compounds are screened for parallel identification of biological targets and biologically active compounds. The success of the chemogenomics approach depends on three critical components: (1) a compound library, (2) a representative biological system (a target library, a single cell, and a whole organism), and (3) a reliable readout (e.g., gene/protein expression, high-throughput binding, or a functional assay).

(a)

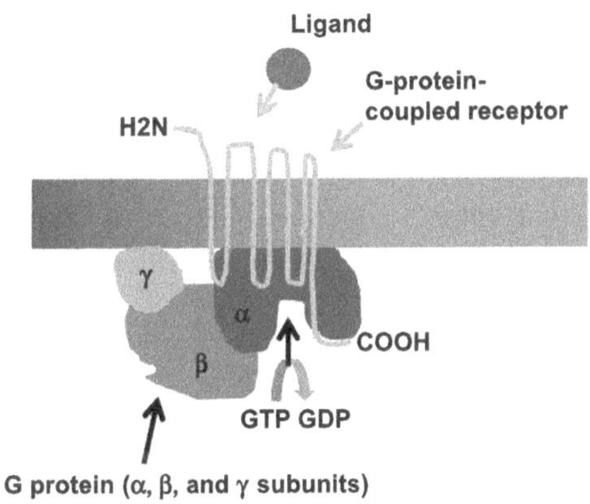

(b)

Receptor	Binding affinity
G protein	GTPyS
G-protein dependent	cAMP
	Calcium mobilization
	IP3
	ERK1/2
	Internalization
G-protein independent	β-arrestin
	ERK1/2
	Akt
Cellular response	Phenotypic changes

Figure 4.2. Advanced detection systems for G-protein-coupled receptor signaling provide novel drug discovery tools along the GPCR signaling pathway; these tools have significantly accelerated the progress in understanding the mechanisms modulated by GPCRs. (a) GPCRs are seven-transmembrane receptors containing the intervening portions loop both inside and outside the cell membranes. When an external signaling molecule or a ligand binds to a GPCR, it causes a conformational change in the GPCR. This change triggers the interaction between the GPCR and a nearby G protein, a type of specialized protein with the ability to bind the nucleotides guanosine triphosphate (GTP) and guanosine diphosphate (GDP). The G proteins have three different subunits: alpha, gamma, and beta. The alpha and gamma subunits are attached to the plasma membrane by lipid anchors. Exchange of GDP with GTP on the alpha subunit activates G proteins and leads to formation of the GTP-bound alpha subunit and a beta-gamma dimer. Activated G proteins then interact with other molecules and serve as a switch to turn on or off the GPCR signaling cascades. (b) A comprehensive GPCR pharmacology tool kit includes GPCR activities from receptors, effectors that are G-protein-dependent and independent to cellular phenotopic responses.

An example of the application of chemogenomics is to identify novel and disease-associated G-protein-coupled receptors (GPCRs) [33, 37], which constitute a large and diverse family of proteins whose primary function is to transduce extracellular stimuli into intracellular signals. GPCRs represent one of the most important drug-target families because more than 50% of the current therapeutic agents on the market target these receptors [38, 39]. In the human genome, over 800 GPCRs [40] have been identified. Yet more than half of these have sensory functions and are generally not relevant for drug discovery. Moreover, only 300 GPCRs are considered druggable [19, 40]. Approved drugs and ongoing clinical trials account for about 100 out of 300 GPCRs. Today, there are still over 100 nonsensory orphan druggable GPCRs [41, 42]. So it is not surprising that across the drug discovery industry, there is a great deal of interest in exploring novel GPCRs.

Chemogenomic approaches can be used for novel GPCR target discovery for three main reasons. First, due to the important role that GPCRs play in drug discovery, diverse GPCR compound libraries have been designed and generated at most of the leading pharmaceutical companies [43]. The major sources of GPCR compound collections are based on (1) the homology models derived from the rhodopsin crystal structure [44], (2) the use of site-directed mutagenesis in relation to ligand structure–activity relationships (SARs), and (3) the accumulated knowledge from past and ongoing clinical compounds. The GPCR compound collections at pharmaceutical companies have become one of the critical drivers in the contemporary chemogenomic environment.

Second, our knowledge of GPCR functions in biological systems is no longer limited to the isolated receptor level. In fact, the dynamic functions of GPCRs in a cellular context have been extensively investigated. It is well known that more than half of all drugs target GPCRs by either activating or inhibiting GPCR activity. The binding of specific ligands—such as hormones, neurotransmitters, chemokines, lipids, and glycoproteins—activates GPCRs by inducing or stabilizing a new conformation in the receptor [45–47] (Figure 4.2a). Activated receptors (R*) can then activate heterotrimeric G proteins (composed of α.GDP, β, and γ subunits) on the inner surface of the cell membrane. The GPCR signaling process is subsequently triggered by a specific ligand β that binds and activates the receptor, thereby inducing a conformational change in the receptor. The receptor, thus activated, stimulates a specific G protein by catalyzing the exchange of guanine nucleotides in the α-subunit. The GTP-bound α-subunit and the $\beta\alpha$ subunits disassociate from the receptor and activate their respective downstream GPCR effector proteins. Unfolding GPCR signaling pathways provide researchers with molecular insights into the key intermediates lying between ligand-mediated cell surface receptors and biological responses (Figure 4.2b). As a result, diverse biological systems can be used today to identify novel GPCR targets. Physiologically relevant systems, such as primary cells and tissues, have become the leading option for identifying novel GPCRs and building connectivity between targets and disease states via chemogenomic approaches.

Finally, high-throughput technology platforms [48–51] have laid a solid foundation for measuring GPCR compound activities in the high-throughput mode at the receptor level, G-protein level, and G-protein-dependent second messenger level (Figure 4.2b). Phenotypic technology platforms, such as high-content image systems and label-free platforms, have provided powerful detection systems for chemogenomic approaches to support GPCR target research [52–54].

Phenotypic screens play a critical role in target identification and validation. The concept of a phenotypic screen is not new; in fact, phenotypic screening approaches have been used in the drug discovery business for decades. A recent review paper [4] provides an in-depth analysis of NMEs approved between 1999 and 2008. The data demonstrate that during this period, 45 approved drugs were first-in-class, and out of those drugs, 27 (60%) were discovered using phenotypic assays, and 18 (40%) were discovered using target-based approaches. These data clearly illustrate the important role of phenotypic screening.

Phenotypic approaches do not require prior understanding of the molecular mechanism of action, but they can reflect compound activity in physiologically relevant systems. Therefore, they can be translated into a given disease state more effectively than the nonphenotypic approaches, target-based assays, which are often more artificial. The main disadvantage of phenotypic screening approaches lies in the challenge of optimizing the molecular properties of candidate drugs in the absence of target-mechanism knowledge. The sensitivity and reproducibility of phenotypic screens can also form a bottleneck in the process of assessing target–activity relationships (TARs) precisely. In any case, the integration of phenotypic screening approaches and target-based assay technologies will bring about a future paradigm shift in target identification and target validation [55–57].

Hit Identification: The First Stage in Drug Hunting

For every discovery program, the entry point is the identification of specifically active molecules—"hits"—that have adequate activity in a suitable target assay. Such initial hits can be generated in a number of ways. One of the common approaches is HTS [58–61]. Today, HTS is defined as the process of testing batches of compounds for binding activity against relevant disease targets. Since the 1980s, innovations and enhancements in screening technologies have yielded remarkable results. About 30 years ago, the *annual* number of throughputs was a about 10,000 compounds. Today, the *daily* number using ultrahigh-throughput screening approaches is more than 100,000 compounds—a staggering 1000-fold increase. The increase can be ascribed to a number of important technological innovations.

At present, HTS assays are typically performed on "automation-friendly" microtiter plates with a 96-, 384-, or 1536-well format [58]. The remarkable transition from a Petri dish to a 1536-microtiter plate reflects over a century of evolution in drug discovery screening approaches. Figure 4.3 shows the path that innovations in basic lab equipment have taken over in the past 100 years. Such innovative solutions have made today's low-cost and rapid HTS assays possible.

The rapid expansion and maturation of HTS in the relatively recent past is a good example of how scientific needs drive technology innovations in the pharmaceutical world. As suggested earlier, there was, in the early 1990s, a rapidly growing need to perform many biochemical or cellular assays on many compounds in an efficient and quantitatively rigorous manner. At that time, however, screening thousands of compounds in a single day presented an impossible task. Researchers at pharmaceutical companies soon realized that the gaps in drug screening required that the industry work closely with equipment and software vendors to develop a wide range of enhanced technologies. The industry's willingness to champion cutting-edge technology platforms gave vendors the incentive to take risks in researching, developing, and licensing powerful new products. The exciting cooperation between pharmaceutical companies and vendors resulted in a wide spectrum of hardware and software innovations that allowed the lofty HTS goals to be met and, in time, surpassed.

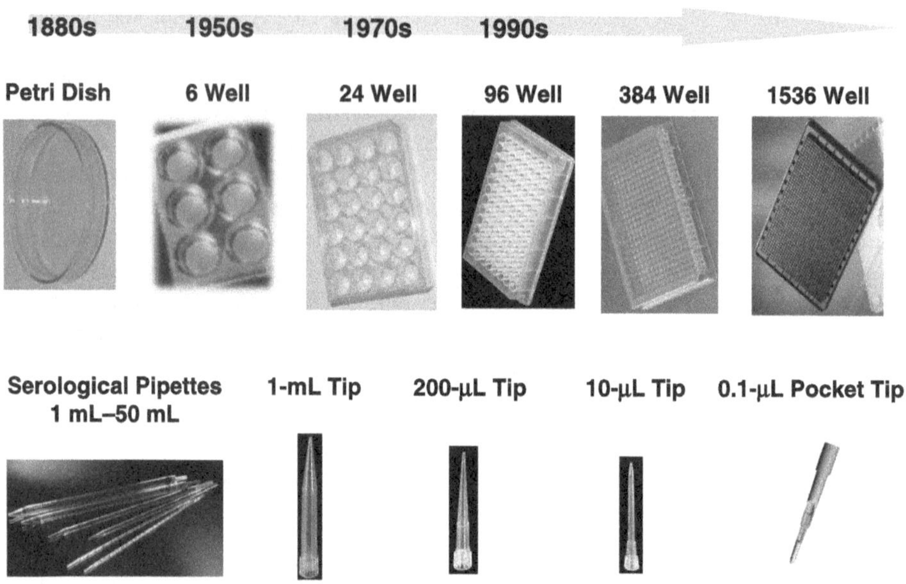

Figure 4.3. The evolution in industry of plate types and tip types reflects the needs that are driving the technology solutions. Over the past decades, these tools have enabled assay miniaturization and high-throughput screening. Moreover, both assay cost and capacity have been dramatically improved by these innovative solutions.

Figure 4.4 illustrates how, during this period, a number of optical concepts that had been invented, evaluated, and demonstrated at academic laboratories—concepts such as absorbance, fluorescence, radiometry, impedance, and luminescence—were used to create "microplate readers," which pharmaceutical scientists soon adopted as an important technological advance. When coupled with advanced liquid-handling systems and sophisticated software for data collection and analysis, microplate readers generated much of the rapid growth of the HTS infrastructure in pharmaceutical companies. As more and more "hits" were found, this sophisticated equipment was continuously improved to provide greater sensitivity and speed. So, throughout the early 2000s, technology providers and pharmaceutical companies alike invested a large amount of time and money in the development and implementation of microplate readers and other innovative technology platforms.

Technology enhancements in HTS have resulted in a paradigm shift from just one HTS screen per target to multiscreens [34]. Figure 4.5 illustrates the evolution of HTS strategy in most pharmaceutical companies over the past two decades. Diverse options for technology platforms are now enabling researchers to have more chances to achieve important goals. As a result, the quality of hits is being increased significantly, and at the same time, false positives are being eliminated early.

Due to several decades of rapid growth in the compound collection—from approximately 10,000 to 1,000,000—an urgent need to develop a low-liquid dispensing tool

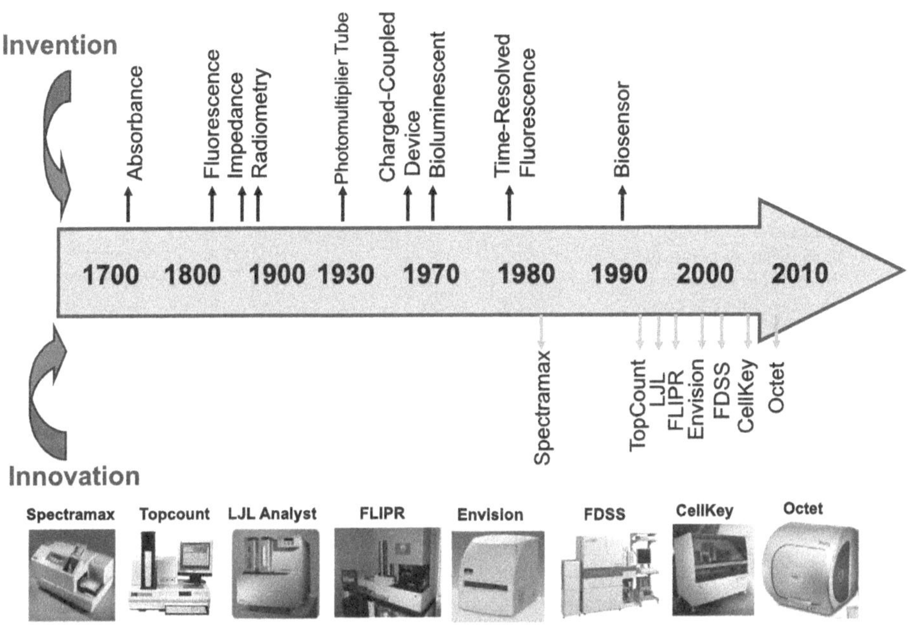

Figure 4.4. The biologic activity detection system revolution from concept to application is a prime example of how to transition inventions created in academic laboratories to innovation technology platforms that can be used to enhance drug discovery capabilities in industry. Advanced technology platforms such as Spectramax, Topcount, LJL Analyst, FLIPR, FDSS, CellKey, and Octet have played a critical role in today's drug discovery processes.

arose about 10 years ago. Such a tool represented the only way to keep assay costs low and complete HTS activities within a matter of days. Hence, the field of nanodispensing was developed with the goal of dispensing picoliter-to-nanoliter volumes. The establishment of acoustic dispensing systems (i.e., Echo Systems) [62] represents an important breakthrough. Echo Systems has made nanodispensing much more robust and practical for HTS—from compound dispensing to assay assembly on automated systems. This technology has surely produced a quantum leap in HTS capacity.

Across the industry during the past 5 years, however, the investment in HTS infrastructure has noticeably declined. This is because global economic problems and especially difficult market conditions have required pharmaceutical companies to monitor investments in technology closely before continuing to fund additional or enhanced platforms. Decreased spending on new equipment, combined with reduced investments in the human resources needed to manage technology assessment and implementation, has resulted in the much more stringent cost–benefit assessment of any technology being introduced to the market. Therefore, the next winning technological innovations will definitely have to accommodate a broad range of requirements: system flexibility, information content, and the ability to test physiologically relevant cellular systems, along with cost, speed, sensitivity, and dynamic range.

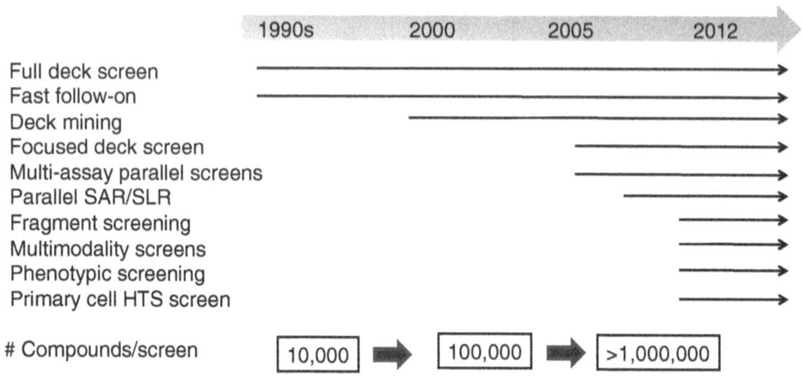

Figure 4.5. Dynamic high-throughput screening (HTS) strategies have resulted from both the growth of scientific knowledge and advances in laboratory technologies. The ability to pursue different HTS strategies in the hit identification process has become the key factor in accelerating hit identification, expanding chemical space, and improving the quality of hit molecules. In addition, assay miniaturization has made the economics of full deck screens feasible. Currently, fast follow-on, deck mining, and focused deck screening are the primary options for knowledge-driven HTS campaigns. Multi-assay parallel screens and parallel structure–activity relationship (SAR)/structure–liability relationship (SLR) approaches have enhanced our ability to gain selectivity and identify off-target liability early on. Fragment screening is an alternative screening strategy used to search for small chemical fragments (MW <300) that exhibit much weaker binding interactions against specific targets when compared with the traditional hits identified by high-throughput screening but that are more likely to fit into a binding pocket. The goal of fragment screening is to leave room for chemical expansion and allow the hits to evolve into promising leads with the desired pharmacological properties. Multi-modality screens represent the advance of HTS technology beyond small-molecule hit searching. The HTS technology infrastructure has ability researchers to handle DNA, RNA, and antibodies as well. Phenotypic screening and primary cell HTS screens reflect the possibility of linking hit activity directly to physiologically relevant models.

Lead Optimization: A Field to Which Companies Must Commit Major Amounts of Human and Financial Resources

The lead optimization process is the one of the most important landmarks in the entire discovery processes because the discovery organization must be ready to commit its major resources to support the work of lead optimization [63]. Throughout lead optimization, discovery scientists must face many scientific, medical, and business challenges—challenges that are often difficult or impossible to predict. Therefore, the transition phase from hit identification to lead optimization—called "hit assessment"—is critically important. Hits derived from HTS campaigns provide the essential starting points for lead optimization.

Yet due to the large size of compound collections and the absence of in-depth knowledge about the targets of interest, HTS campaigns often have to balance the cost of the work to be done against the likely viability of the assays. Consequently, hit selection criteria can be biased and, in some cases, researchers will put too much emphasis on potency. The tendency toward overweighting potency can hamper and delay the progress of discovery projects in the exploratory phases. Therefore, both the quality and the quantity of hits available to medicinal chemists before they initiate the lead optimization step represent one of the key contributors to the discovery of best-in-class medicines.

Figure 4.6a depicts target identification, hit identification, hit assessment, and lead optimization. The figure shows how the magnitude of the compounds varies greatly from target identification to lead optimization. Because different hit identification strategies are broadly applied (Figure 4.5), thousands of hits with diverse profiles are identified at the conclusion of the hit identification process.

To direct chemistry resources to the most promising chemical series (chemotypes) for lead optimization, the concept of hit assessment was introduced and soon became

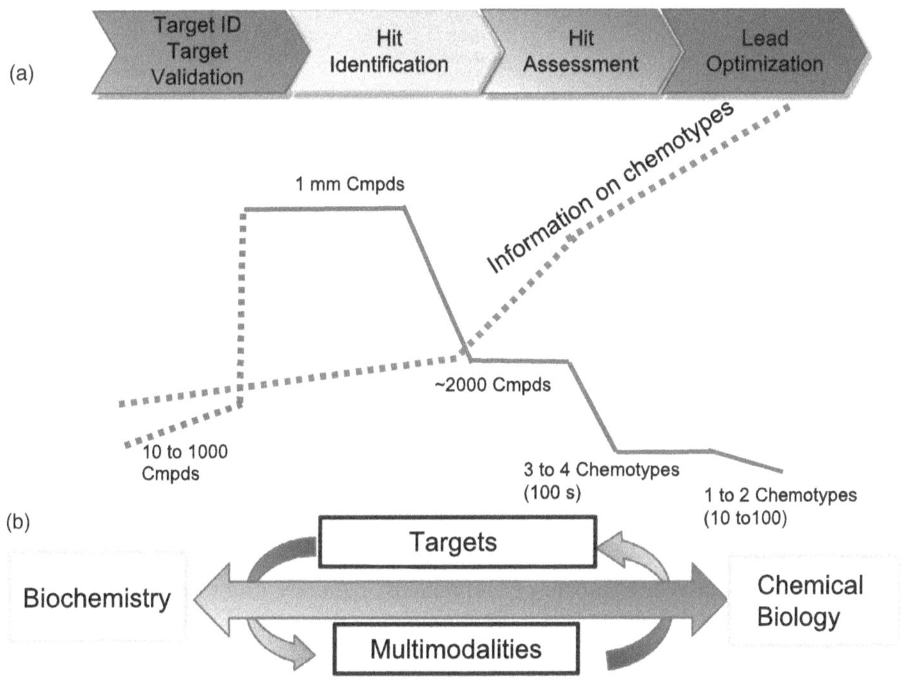

Figure 4.6. The transition from leveraging improved technology to leveraging expanded information is part of the critical path for future drug discovery endeavors. The investment in technology should play a critical role, going forward, in capacity enhancement as well as quality enhancement. (a) Compounds involved in each drug discovery process are dramatically different. Information generated from each step can be used to provide valuable guidance for drug discovery programs. (b) Chemical biology and biochemistry concepts provide an iterative feedback loop between drug candidate selection and target discovery.

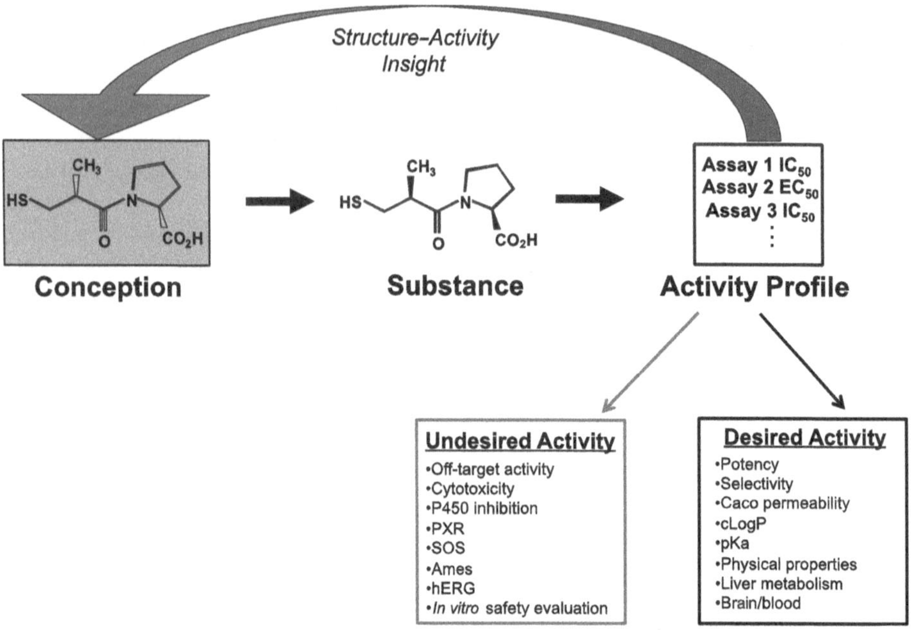

Figure 4.7. Lead optimization is an iterative process and represents a key step in gathering data relevant to both the desirable and undesirable aspects of drug candidate profiles. Such information is critical to guiding chemistry-based activities during lead optimization.

a key step in gaining more information about the hits identified in the HTS campaign. To increase understanding of the properties of each molecule, it is essential to establish selection criteria, based on both desired and undesired activities. To identify the most viable leading series (Figure 4.7), a paradigm shift in terms of assay options is necessary to collect as much information as possible and as early as possible [63].

Furthermore, the screening challenges related to lead optimization are significantly different from those related to hit identification [63]. In hit identification, one bioassay tests many hundreds of thousands of compounds in a short period of time. Today, hit identification needs only weeks, whereas lead optimizations can take years. In lead optimization, far smaller numbers of compounds, typically two to three chemotypes containing 10 to 100 compounds, are put through an array of assays weekly. Discovery scientists commonly modify *in vitro* screening paradigms to respond to and solve issues that occur during the lead optimization.

Although a series of assay technologies used for hit identification can be modified for use in lead optimization, the assays used for lead optimization must not only provide the chemistry teams with just-in-time feedback, but they must also build connectivity between *in vitro* and *in vivo* studies. Therefore, the factors that dictate the selection of the assays for hit identification and lead optimization differ. Results from lead optimization assays must be used to guide the decisions on which compounds to synthesize and which compounds to advance to *in vivo* efficacy tests. Providing a full array of safety and selectivity profiles on each and every project compound as quickly as possible after synthesis allows scientists to compare and rank compounds according to a multiparameter matrix (i.e., selectivity and efficacy), not just according to potency. The appropriate integration of HTS technologies and the selection of suitable, biologically relevant *in vitro* assays for lead optimization are critical to enabling project teams to track a full package of essential information throughout the entire course of drug discovery programs.

As indicated earlier, the increased number of novel targets demands that scientists search for alternative approaches. As a result, the drug discovery process should not be viewed as a sequential process from target ID to lead optimization. Instead, it should be seen as an iterative process designed to yield information that can help guide the target identification and target validation processes.

Recently, chemical biology has emerged as a new scientific discipline that supports drug discovery. Chemical biology, which uses chemical tools to study and manipulate biological systems, is very different from biochemistry, which is classically defined as the study of the chemistry of biomolecules. Both chemical biology and biochemistry are important approaches for lead optimization. For example, a biochemist seeks to understand the three-dimensional structure of a protein and the protein's activity relationship with the molecules. In contrast, a chemical biologist aims to modulate biological- or disease-associated systems by applying chemical principles (Figure 4.6b). The combination of small molecules and large molecules provides broad options for investigating drug targets. Moreover, the ongoing expansion of the chemical space (Figure 4.8) will extend therapeutic opportunities and strongly impact the delivery of first-in-class or best-in-class drug candidates. In conclusion, compound collections during lead optimization can be used as annotated chemical tools. These tools have proved exceptionally capable of influencing and refining decisions on both target identification and target validation [64].

One good example of how advanced technological innovation and enriched data sets can facilitate both hit assessment and lead optimization is kinase drug discovery. Protein kinases regulate many essential aspects of cell life, including metabolism, cell cycle progression, cytoskeletal rearrangement and cell movement, apoptosis, and differentiation processes through the reversible phosphorylation of protein kinases

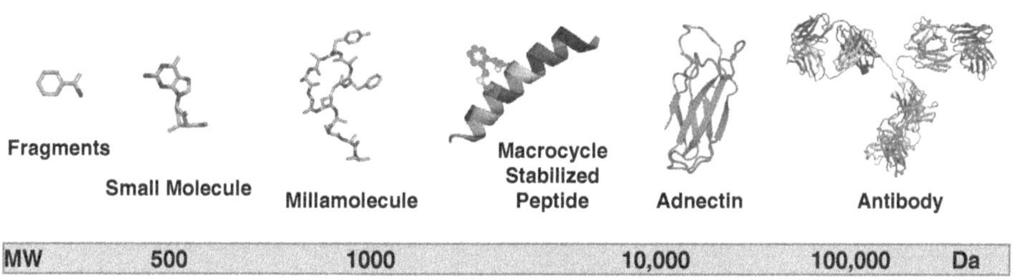

Figure 4.8. The evolution of the chemical space is essential to expand the druggable genome and identify potential therapeutic solutions. See color insert.

[65–68]. Today, more than 518 protein kinases have been shown to be present in the human genome [53]. Almost all protein kinases contain a catalytic domain (250 to 300 amino acids) for adenosine triphosphate (ATP), a chemical compound that cells use to store or release energy. All catalytic domains of various kinases share homologous structural features for binding ATP. This common catalytic mechanism across the many families of kinases leads to the question as to whether any selectivity and specificity can ever be achieved by protein kinase inhibitors competitively targeting the ATP binding site. Considering that there are over 500 protein kinases, it is not surprising that selectivity is a major obstacle for kinase drug discovery [69].

As recently as a decade ago, protein kinases were not considered the most druggable target class, but this notion is changing rapidly. In fact, up to the present, 20 kinase inhibitors have been approved and launched for the treatment of cancer [70]. Although the primary indications so far are various types of cancer, at present, hundreds of kinase inhibitors are in clinical trials for diseases other than cancer [70]. Thus, protein kinases as drug targets have traveled a long way evolutionally (Figure 4.9). In 1978, the first oncogene [71], the transforming factor of the Rous sarcoma virus, was shown to be a protein kinase. In 1986, staurosporine, an antifungal agent, was discovered as one of the nanomolar protein kinase C inhibitors, and this discovery led to great excitement about using staurosporine as an antitumor drug. Unfortunately, the excitement

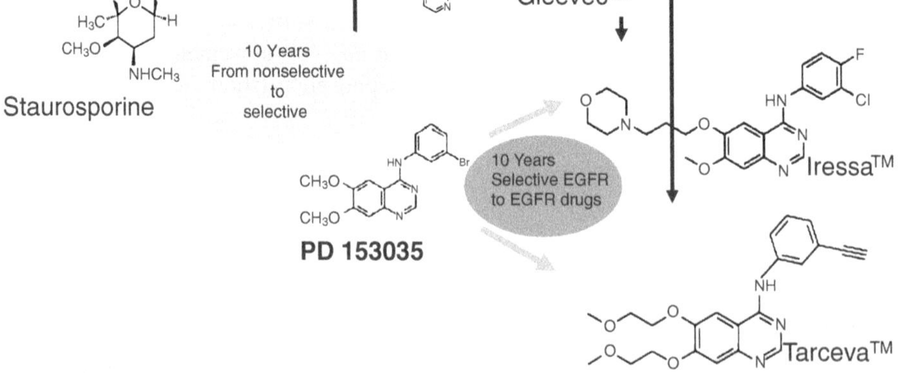

Figure 4.9. A historic overview of kinase inhibitor discovery reveals the challenges that kinases present as a drug target class. On the other hand, the discovery of the breakthrough cancer drug Gleevec™ reveals the possibility of identifying selective kinase inhibitors for use in treating diseases. Reprinted with permission from Fry, D. *Science* 1994;265:1093.

was soon replaced by massive disappointment because staurosporine is a nonselective inhibitor. In 1994, after about 10 years of intense effort, the first selective epidermal growth factor receptor inhibitor was reported in *Science* [72]. The success of the first-launched kinase drug, Gleevec™ [73, 74], for the treatment of chronic myelogenous leukemia, has validated the concept that a protein kinase can be targeted specifically for clinical use.

Preclinical Assessment: Select the First-in-Class and/or Best-in-Class Drug Candidate for Clinical Development

The fundamental goal of preclinical assessment is to reduce the attrition rate among molecules as they advance through R&D by identifying the drug candidates with the highest probability of being safe and effective. The reason for this critical mission is that toxicity and safety issues are by far the most common causes of compound attrition. For example, based on a recent analysis [11, 12, 75], about 30% of NCE attrition is attributed to unexpected toxic effects. Moreover, in the past 10 years, nearly 63% of drug withdrawals from the market have been due to drug-induced hepatotoxicty or cardiotoxicity [76, 77]. Battling a high attrition rate is very difficult during the discovery phases because, very often, many assumptions have to be made in the absence of clinical validation. To address this problem, scientists have formulated some very useful guiding principles—as illustrated by the following three examples.

First, applying lessons learned from clinical research, teams performing preclinical assessment should focus on the toxicity and safety issues that most often lead to compound failure. These issues include liver metabolism problems, cardiovascular liability, teratology, compound-induced toxicity, and target-based toxicity.

Second, preclinical assessment needs to be supported by a screening tree (a series of biological tests) with the goal of bridging *in vitro* and *in vivo* connectivity. Therefore, functional readouts, *ex vivo* assays, mechanistic assays, compound half-life assessment, mutagenicity, PK/PD activity, drug metabolism, and pharmaceutical properties must be considered to guide the go/no go decisions for drug-candidate selection.

Third, research teams must achieve the right balance between rigorous scientific evaluation and reasonable levels of risk taking. Putting too much weight on either evaluation or risk taking will not lead to an excellent outcome. For example, allowing room for hypothesis testing can sometimes be more valuable in terms of generating knowledge than stopping work on a molecule to avoid a difficult issue that might arise later. Identifying and assessing the potential risks involved in moving first-in-class compounds along the R&D pathway can be the most challenging test for a knowledge-centric drug discovery organization. Under these guidelines, the pharmaceutical industry has made transformative progress in delivering breakthrough drugs for highly unmet medical needs during the past decades.

The following Gleevec success story illustrates such important roles as target ID, target validation, hit ID, lead optimization, and preclinical assessment.

In 2001, Novartis's breakthrough drug, Gleevec, revolutionized the treatment of chronic myeloid leukemia (CML), a cancer of the white blood cells. The discovery of Gleevec demonstrated that it is possible to disrupt one specific protein kinase that causes the cancer. It further validated that molecular targeting is a highly promising treatment approach in cancer. What makes this drug especially fascinating is the story behind its discovery [78–80].

- In 1960, scientists at the University of Pennsylvania first observed chromosomal abnormalities in the blood of CML patients.
- By the 1980s, scientists from the University of Chicago were able to use genetic mapping to show that the two ends of the broken chromosomes produced a cancer-causing protein, known as Bcr-Abl.
- Novartis researchers synthesized and tested 400 molecules to find one that would target this enzyme without disrupting any of the hundreds of similar kinases in a healthy cell. After 2 years of testing, they developed the molecule that would become Gleevec.
- Following another 8 years of preclinical assessment and development, scientists were ready to try the drug in patients. When they did, they saw dramatic results. Nearly every CML patient who took the drug responded. One of the key pioneers responsible for the discovery of Gleevec is Dr. Brian Druker of Oregon Health & Science University. He was the first scientist to champion the concept of using Gleevec as a target-based chemotherapy for CML. His hypothesis was based on three factors [81–83]:
    1. Physicians understood the causes of CML better than those of any other cancer types.
    2. CML can be diagnosed in an early malignant or nearly premalignant phase.
    3. CML is driven by one main gene alteration, Bcr-Abl.

Dr. Druker's persistence and engagement convinced Novartis to initiate the Phase 1 trials. Eventually, the human trials led to the successful launch of Gleevec. Due to his outstanding work, Dr. Druker won the Lasker-DeBakey Clinical Medical Research Award, often called the "American Nobel Prize." The Gleevec story is also a compelling example of how academia and industry can effectively work together to discover novel medicines.

AN IMPORTANT TREND: ACADEMIC AND PHARMACEUTICAL ORGANIZATIONS ARE COLLABORATING IN SUPPORT OF DRUG DISCOVERY

In recent years, researchers from academic institutions and pharmaceutical companies have been looking for ways to increase synergy as they create the next waves of research collaboration [84–86]. Universities and companies alike are both very eager to increase the efficiency and impact of the discovery process and thereby help produce new treatments for unmet medical needs. Two main factors are driving this substantial new trend toward partnering between academia and industry [87–89].

First, weaknesses in the global and national economies could compel government agencies to reduce the size and number of the academic and scientific grants they make. Researchers in academic settings are feeling the pressure to explore alternatives to government funding to continue the theoretical and pioneering research that will enable them and their peers in industry to transform their findings into medicines.

Second, pharmaceutical companies need to gain knowledge about complex chemical and biological processes to understand the hitherto unknown mechanisms of action of novel drug targets. Only with such knowledge can they refresh their drug discovery pipelines. The end result is that top research scientists in both realms have become catalysts for knowledge sharing between academia and industry.

Certainly, researchers in academic centers and pharmaceutical companies alike have partnered for many decades to study and solve scientific problems. But the recent partnerships are, in important ways, quite different from the historic partnerships. Here are three examples of how the partnerships have changed:

- Before establishing pathways for collaboration, industry is no longer waiting to identify precisely what information and knowledge its researchers need. Instead, companies are receptive to meeting with academic research institutions early to look for mutually useful opportunities for collaboration.
- Previously, industrial-academic collaborations focused on specific targets, and the scope of the partnership agreements was typically small. In contrast, the scope of recent partnership agreements has expanded substantially. Moreover, annual corporate funding for research alliances has increased significantly across industry.
- Especially over the past 3 years or so, these partnerships have become much more common. Of course, drug development remains a business characterized by exceptionally difficult challenges spread along lengthy timelines. As a result, the long-term impact of early-stage discovery collaborations remains difficult to predict with a high degree of confidence.

Figure 4.10 highlights some pharmaceutical and academic partnerships in 2012. The data demonstrate the strong commitment from both pharmaceutical companies and academic institutions. Moreover, the effectiveness and durability of these alliances have largely derived from good strategic planning and highly developed management. This trend indicates a fundamental change in mindset: innovation in discovering novel medicines can proceed most effectively as a series of collaborative efforts across industry and academic institutions.

Successful partnership naturally relies on active engagement and management from both parties. For example, over the past decade, Bristol-Myers Squibb (BMS) has formed partnerships with numerous university-based researchers. Important lessons that BMS has learned from these partnerships include the following:

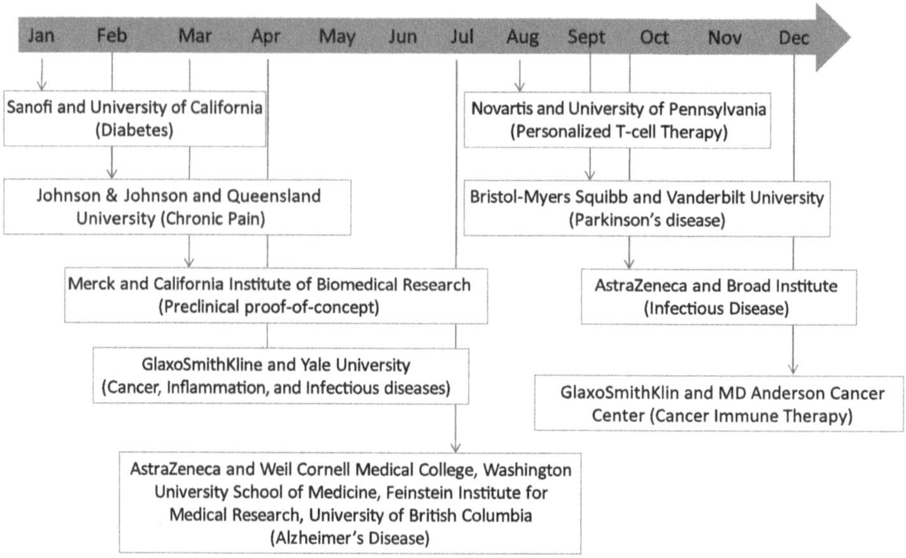

Figure 4.10. Highlights of pharmaceutical and academic partnerships in 2012.

- As soon as possible, establish strategic partnerships that will improve access to extensive, expert knowledge of diseases and patient populations.
- Discover and develop needed medicines more successfully by pioneering advanced technology platforms and by being willing to implement open-innovation models, such as unrestricted business models, to accelerate innovation.
- Manage joint decisions transparently and equitably.
- Foster and, as appropriate, share expertise from both sides of the partnership.
- Be willing to change a project's direction when data and scientific thinking call for it.
- Address any disagreements between the partners proactively, candidly, and reasonably.

Although the impacts of the new pharmaceutical/academic partnerships are still too early to assess, it seems safe to say that they are revolutionizing thinking about where drug discovery should reside and how it should be conducted. This trend is surely leading to a new wave of activity in which drug discovery is no longer the nearly exclusive mission of pharmaceutical companies.

FUTURE PERSPECTIVES: WHERE IS INNOVATION LEADING?

The next innovations in pharmaceutical companies will be significantly influenced by two trends. One trend is the growth in pharmaceutical and academic partnerships described earlier. Such partnerships will help expand the druggable genome, generate novel targets, and provide mechanistic understandings of disease-associated targets. The second trend is related to translational research [90–93]. By making it possible to observe—at the molecular level—the nature and progression of diseases, translational research holds much promise. It represents a new approach to connecting basic research discoveries with their clinical applications as well as to stimulating new forms of research.

Translational research entails two important dynamic processes—each the mirror image of the other. First, it involves taking the knowledge gained from the basic science and integrating it into the clinical setting. Second, it also involves taking the knowledge gained from the clinical science and using it to make the basic and investigative science more precise and accurate.

The concept of tailoring treatment to a patient's individual needs by using diagnostic or predictive biomarkers is, of course, not new. Physicians have always sought to understand the causes and course of an individual patient's disease and treat it accordingly. What *has* changed in recent years is the rapid advancement of science, technology, and of the understanding of diseases at the molecular level.

Transforming research from "the bench to the bedside" introduces several challenges. The genesis of these challenges can take many forms. However, one of the major hurdles entails the differences in "scale" between the models used for discovery research as opposed to those used in clinical trials. A few examples are as follows:

- Humans have 35,000+ genes, 100,000 proteins, 300+ cell types, 4 tissue types, and 12 organ systems.
- The research model is ~1 nm proteins whereas the clinical setting is greater than ~1 m in the entire human body.
- A laboratory result is derived from 10^{-6} second molecular events, but drug safety and efficacy need to expand into the 10^{9} seconds of human lifetime.

As discussed earlier, the sequencing of the human genome—along with the development and application of multiple technology platforms across the discovery process—has opened up a new era in translational research. A vast amount of information is now available at the level of genes, proteins, cells, tissues, organs, and humanized animal models. A "humanized mouse" is a mouse carrying functioning human genes, cells, and/or tissues. These data can be used to understand the mechanisms involved in disease biology and validate scientific hypotheses across patient populations.

The new genetic tools are enabling physicians and scientists to gain a far greater understanding of the underlying nature of disease states. They are also increasingly illuminating how certain patient groups might respond to various treatment options. In turn, understanding the patterns of clinical response better is enabling scientists and physicians to refine treatment regimes. As a result of all these contributions to basic as well as to clinical science, the translational approach is taking center stage in the creation of medicines to treat diseases. Moreover, future innovations will greatly help scientists close the remaining gaps in translational research and thereby help them in their mission of improving human health.

REFERENCES

1. Innovative Drug Approvals, FY 2012. Available at http://www.fda.gov/AboutFDA/ReportsManualsForms/Reports/ucm276385.htm. Accessed December 20, 2013.
2. Grabowski, H.G., Wang, Y.R. The quantity and quality of worldwide new drug introductions, 1982–2003. *Health Aff (Millwood)* 2006;25(2):452–460.
3. New Drug Review: an update and a look ahead. 2012. Available at http://www.fda.gov/downloads/AboutFDA/.../CDER/UCM309983.pdf. Accessed December 20, 2013.
4. Swinney, D.C., Anthony, J. How were new medicines discovered? *Nat Rev Drug Discov.* 2011;10:507–519.
5. Jenkins, J.K. CDER new drug review. 2012. Available at http://www.fda.gov/downloads/AboutFDA/.../CDER/UCM331454.pdf. Accessed December 20, 2013.
6. Hoos, A., Britten, C. The immuno-oncology framework: enabling a new era of cancer therapy. *Oncoimmunology* 2012;1(3):334–339.
7. Mellman, I., Coukos, G., Dranoff, G. Cancer immunotherapy comes of age. *Nature* 2011;480:480–489.
8. Graziani, G., Tentori, L., Navarra, P. Ipilimumab: a novel immunostimulatory monoclonal antibody for the treatment of cancer. *Pharmacol Res.* 2012;65(1):9–22.
9. Pina, A.S., Hussain, A., Roque, A.C. An historical overview of drug discovery. *Methods Mol Biol.* 2009;572:3–12.
10. Arrowsmith, J. Trial Watch: phase II failures: 2008–2010. *Nat Rev Drug Discov.* 2011;10(5):328–329.
11. Arrowsmith, J. Trial Watch: phase III and submission failures: 2007–2010. *Nat Rev Drug Discov.* 2011;10(2):87–90.
12. Brunschweiger, A., Hall, J. A decade of the human genome sequence—how does the medicinal chemist benefit? *ChemMedChem.* 2012;7(2):194–203.
13. Hoffman, L.R., Ramsey, B.W. Cystic fibrosis therapeutics: the road ahead. *Chest* 2013;143(1):207–213.
14. Yokota, T., Duddy, W., Echigoya, Y., Kolski, H. Exon skipping for nonsense mutations in Duchenne muscular dystrophy: too many mutations, too few patients? *Expert Opin Biol Ther.* 2012;12(9):1141–1152.
15. Pasmant, E., Vidaud, M., Vidaud, D., Wolkenstein, P. Neurofibromatosis type 1: from genotype to phenotype. *J Med Genet.* 2012;49(8):483–489.
16. Reis, A.H., Vargas, F.R., Lemos, B. More epigenetic hits than meets the eye: microRNAs and genes associated with the tumorigenesis of retinoblastoma. *Front Genet.* 2012;3:284.
17. Landry, Y., Gies, J.P. Drugs and their molecular targets: an updated overview. *Fundam Clin Pharmacol.* 2008;22(1):1–18.
18. Sakharkar, M.K., Sakharkar, K.R. Targetability of human disease genes. *Curr Drug Discov Technol.* 2007;4(1):48–58.
19. Hopkins, A.L., Groom, C.R. The druggable genome. *Nat Rev Drug Discov.* 2002;1(9):727–730.
20. Hopkins, A.L., Bickerton, G.R. Drug discovery: know your chemical space. *Nat Chem Biol.* 2010;6(7):482–483.
21. Fields, S., Kohara, Y., Lockhart, D.J. Functional genomics. *Proc Natl Acad Sci U S A.* 1999;96(16):8825–8826.

22. Subramanian, G., Adams, M.D., Venter, J.C., Broder, S. Implications of the human genome for understanding human biology and medicine. *JAMA*. 2001;286(18):2296–2307.
23. Marcilla, M., Albar, J.P. Quantitative proteomics: a strategic ally to map protein interaction networks. *IUBMB Life* 2013;65(1):9–16.
24. Hardiman, G. Microarray platforms—comparisons and contrasts. *Pharmacogenomics* 2004;5(5):487–502.
25. Reimers, M. Statistical analysis of microarray data. *Addict Biol.* 2005;10(1):23–35.
26. Bai, Y., Sartor, M., Cavalcoli, J. Current status and future perspectives for sequencing livestock genomes. *J Anim Sci Biotechnol.* 2012;3(1):8–13.
27. Lin, X., Tang, W., Ahmad, S., Lu, J., Colby, C.C., Zhu, J., Yu, Q. Applications of targeted gene capture and next-generation sequencing technologies in studies of human deafness and other genetic disabilities. *Hear Res.* 2012;288(1–2):67–76.
28. Gillespie, T.A., Winger, B.E. Mass spectrometry for small molecule pharmaceutical product development: a review. *Mass Spectrom Rev.* 2011;30(3):479–490.
29. Liang, Y., Wang, G., Xie, L., Sheng, L. Recent development in liquid chromatography/mass spectrometry and emerging technologies for metabolite identification. *Curr Drug Metab.* 2011;12(4):329–344.
30. Paolini, G.V., Shapland, R.H., van Hoorn, W.P., Mason, J.S., Hopkins, A.L. Global mapping of pharmacological space. *Nat Biotechnol.* 2006;24:805–815.
31. Russ, A.P., Lampel, S. The druggable genome: an update. *Drug Discov Today* 2005;10(23–24):1607–1610.
32. Overington, J.P., Al-Lazikani, B., Hopkins, A.L. How many drug targets are there? *Nat Rev Drug Discov.* 2006;5(12):993–996.
33. Van der Horst, E., Peironcely, J.E., van Westen, G.J., et al. Chemogenomics approaches for receptor deorphanization and extensions of the chemogenomics concept to phenotypic space. *Curr Top Med Chem.* 2011;11(15):1964–1977.
34. Medina-Franco, J.L., Giulianotti, M.A., Welmaker, G.S., Houghten, R.A. Shifting from the single to the multitarget paradigm in drug discovery. *Drug Discov Today* 2013; 19:S1359–S6446.
35. Jacoby, E., Mozzarelli, A. Chemogenomic strategies to expand the bioactive chemical space. *Curr Med Chem.* 2009;16(33):4374–4381.
36. Pirard, B. Structure-based chemogenomics: analysis of protein family landscapes. *Methods Mol Biol.* 2009;575:281–296.
37. Garland, S.L., Gloriam, D.E. Methods for the successful application of chemogenomics to GPCR drug design. *Curr Top Med Chem.* 2011;11(15):1870–1871.
38. Latek, D., Modzelewska, A., Trzaskowski, B., Palczewski, K., Filipek, S. G protein-coupled receptors—recent advances. *Acta Biochim Pol.* 2012;59(4):515–529.
39. Trzaskowski, B., Latek, D., Yuan, S., Ghoshdastider, U., Debinski, A., Filipek, S. Action of molecular switches in GPCRs: theoretical and experimental studies. *Curr Med Chem.* 2012;19(8):1090–1109.
40. Fredriksson, R., Lagerström, M.C., Lundin, L.G., Schiöth, H.B. The G-protein-coupled receptors in the human genome form five main families: phylogenetic analysis, paralogon groups, and fingerprints. *Mol Pharmacol.* 2003;63(6):1256–1272.
41. Singh, N., Guha, R., Giulianotti, M.A., Pinilla, C., Houghten, R.A., Medina-Franco, J.L. Chemoinformatic analysis of combinatorial libraries, drugs, natural products, and molecular libraries small molecule repository. *J Chem Inf Model.* 2009;49(4):1010–1024.

42. Morris, I., Williams, R. Research strategies for orphan G-protein-coupled receptors. *Drug News Perspect.* 2002;15(4):249–252.
43. Jimonet, P., Jäger, R. Strategies for designing GPCR-focused libraries and screening sets. *Curr Opin Drug Discov Devel.* 2004;7(3):325–333.
44. Mason, J.S., Bortolato, A., Congreve, M., Marshall, F.H. New insights from structural biology into the druggability of G protein-coupled receptors. *Trends Pharmacol Sci.* 2012; 33(5):249–260.
45. Jacobson, K.A., Costanzi, S. New insights for drug design from the X-ray crystallographic structures of G-protein-coupled receptors. *Mol Pharmacol.* 2012;82(3):361–371.
46. Leff, P., Scaramellini, C., Law, C., McKechnie, K. A three-state receptor model of agonist action. *Trends Pharmacol Sci.* 1997;18:355–362.
47. Audet, M., Bouvier, M. Restructuring G-protein-coupled receptor activation. *Cell* 2012; 151(1):14–23.
48. Cvijic, M.E., Chen, J., Kubala, S., Nolfo, J., Kolodin, G., Zhang, L. GPCR assay automation for leveraging lead optimization. *J Lab Autom.* 2012;17(4):266–274.
49. Waller, A., Simons, P., Prossnitz, E.R., Edwards, B.S., Sklar, L.A. High throughput screening of G-protein coupled receptors via flow cytometry. *Comb Chem High Throughput Screen* 2003;6(4):389–397.
50. Degorce, F., Card, A., Soh, S., Trinquet, E., Knapik, G.P., Xie, B. HTRF: a technology tailored for drug discovery: a review of theoretical aspects and recent applications. *Curr Chem Genomics* 2009;3:22–32.
51. McGuinness, R.P., Proctor, J.M., Gallant, D.L., van Staden, C.J., Ly, J.T., Tang, F.L., Lee, P.H. Enhanced selectivity screening of GPCR ligands using a label-free cell-based assay technology. *Comb Chem High Throughput Screen* 2009;12(8):812–823.
52. Zhang, R., Xie, X. Tools for GPCR drug discovery. *Acta Pharmacol Sin.* 2012;33(3): 372–384.
53. Bamborough, P. System-based drug discovery within the human kinome. *Expert Opin Drug Discov.* 2012;7(11):1053–1070.
54. Bickle, M. High-content screening: a new primary screening tool? *IDrugs* 2008;11(11): 822–826.
55. Haney, S.A., LaPan, P., Pan, J., Zhang, J. High-content screening moves to the front of the line. *Drug Discov Today* 2006;11(19–20):889–894.
56. Blake, R.A. Target validation in drug discovery. *Methods Mol Biol.* 2007;356:367–377.
57. Fox, S., Farr-Jones, S., Sopchak, L., Boggs, A., Nicely, H.W., Khoury, R., Biros, M. High-throughput screening: update on practices and success. *J Biomol Screen* 2006;11(7): 864–869.
58. Mayr, L.M., Bojanic, D. Novel trends in high-throughput screening. *Curr Opin Pharmacol.* 2009;9(5):580–588.
59. Liu, B., Li, S., Hu, J. Technological advances in high-throughput screening. *Am J Pharmacogenomics* 2004;4(4):263–276.
60. Entzeroth, M. Emerging trends in high-throughput screening. *Curr Opin Pharmacol.* 2003;3(5):522–529.
61. Sharff, A., Jhoti, H. High-throughput crystallography to enhance drug discovery. *Curr Opin Chem Biol.* 2003;7(3):340–345.

62. Griffith, D., Northwood, R., Owen, P., Simkiss, E., Brierley, A., Cross, K., Slaney, A., Davis, M., Bath, C. Implementation and development of an automated, ultra-high-capacity, acoustic, flexible dispensing platform for assay-ready plate delivery. *J Lab Autom.* 2012;17(5): 348–358.

63. Zhang, L., Cvijic, M.E., Lippy, J., Myslik, J., Brenner, S.L., Binnie, A., Houston, J.G. Case study: technology initiative led to advanced lead optimization screening processes at Bristol-Myers Squibb, 2004–2009. *Drug Discov Today* 2012;17(13–14):733–740.

64. Petrone, P.M., Simms, B., Nigsch, F., Lounkine, E., Kutchukian, P., Cornett, A., Deng, Z., Davies, J.W., Jenkins, J.L., Glick, M. Rethinking molecular similarity: comparing compounds on the basis of biological activity. *ACS Chem Biol.* 2012;7(8):1399–1409.

65. Noble, M.E.M., Endicott, J., Johnson, L.N. Protein kinase inhibitors: insight into drug design from structure. *Science* 2004;19(3):1800–1804.

66. Cohen, P. Protein kinases: the major drug target of the twenty-first century? *Nat Rev Drug Discov.* 2002;1:309–315.

67. Manning, G., Whyte, D.B., Martinez, R., Hunter, T. Sudarsanam, S. The protein kinase complement of the human genome. *Science* 2002;298:1912–1934.

68. Hanks, S.K., Hunter, T. The eukaryotic protein kinase superfamily: kinases (catalytic) domain structure and classification. *FASEB J.* 1995;9:576–596.

69. Knight, J.D., Pawson, T., Gingras, A.C. Profiling the kinome: current capabilities and future challenges. *J Proteomics.* 2012;12:S1874–S3919.

70. Cohen, P., Alessi, D.R. Kinase drug discovery: what's next in the field? *ACS Chem Biol.* 2013;8(1):96–104.

71. Marcos, M., Mariano, B. RAS oncogenes: the first 30 years. *Nat Rev Cancer* 2003;3:7–13.

72. Fry, D.W., Kraker, A.J., McMichael, A., Ambroso, L.A., Nelson, J.M., Leopold, W.R., Connors, R.W., Bridges, A.J. A specific inhibitor of the epidermal growth factor receptor tyrosine kinase. *Science* 1994;265:1093–1095.

73. Panjarian, S., Iacob, R.E., Chen, S., Engen, J.R., Smithgall, T.E. Structure and dynamic regulation of Abl kinases. *J Biol Chem.* 2013;288:5443–5450.

74. Gora-Tybor, J., Robak, T. Targeted drugs in chronic myeloid leukemia. *Curr Med Chem.* 2008;15(29):3036–3051.

75. Kola, I., Landis, J. Can the pharmaceutical industry reduce attrition rates? *Nat Rev Drug Discov.* 2004;3(8):711–715.

76. DiMasi, J.A., Fedlman, L., Seckler, A., Wilson, A. Trends in risks associated with new drug development: success rates for investigational drugs. *Clin Pharmacol Ther.* 2010;87(3): 272–277.

77. Shah, R.R. Can pharmacogenetics help rescue drugs withdrawn from the market? *Pharmacogenomics* 2006;7(6):889–908.

78. Moen, M.D., McKeage, K., Plosker, G.L., Siddiqui, M.A. Imatinib: a review of its use in chronic myeloid leukaemia. *Drugs.* 2007;67(2):299–320.

79. Baran, Y., Saydam, G. Cumulative clinical experience from a decade of use: imatinib as first-line treatment of chronic myeloid leukemia. *J Blood Med.* 2012;3:139–150.

80. Gray, J., Druker, B. Genomics: the breast cancer landscape. *Nature* 2012;486:328–329.

81. Druker, B.J. Perspectives on the development of imatinib and the future of cancer research. *Nat Med.* 2009;15(10):1149–1152.

82. Druker, B.J. Chronic myeloid leukemia in the imatinib era. *Semin Hematol.* 2003;40:1–3.

83. Buchdunger, E., Matter, A., Druker, B.J. Bcr-Abl inhibition as a modality of CML therapeutics. *Biochim Biophys Acta* 2001;1551(1):M11–M18.

84. Hughes, B. Pharma pursues novel models for academic collaboration. *Nat Rev Drug Discov.* 2008;7(8):631–632.

85. Vallance, P., Williams, P., Dollery, C. The future is much closer collaboration between the pharmaceutical industry and academic medical centers. *Clin Pharmacol Ther.* 2010;87(5):525–527.

86. Hoffmann, T., Bishop, C. The future of discovery chemistry: quo vadis? Academic to industrial—the maturation of medicinal chemistry to chemical biology. *Drug Discov Today* 2010;15(7–8):260–264.

87. Bernstein, R. Science for sale: academic meets industry. *Cell* 2012;150(2):239–240.

88. Chin-Dusting, J., Mizrahi, J., Jennings, G., Fitzgerald, D. Outlook: finding improved medicines: the role of academic-industrial collaboration. *Nat Rev Drug Discov.* 2005; 4(11):891–897.

89. Rothman, D.J., Chimonas, S. New developments in managing physician-industry relationships. *JAMA.* 2008;300(9):1067–1069.

90. Wist, A.D., Berger, S.I., Iyengar, R. Systems pharmacology and genome medicine: a future perspective. *Genome Med.* 2009;1(1):11–19.

91. Khoury, M.J., Clauser, S.B., Freedman, A.N., Gillanders, E.M., Glasgow, R.E., Klein, W.M., Schully, S.D. Population sciences, translational research, and the opportunities and challenges for genomics to reduce the burden of cancer in the 21st century. *Cancer Epidemiol Biomarkers Prev.* 2011;20(10):2105–2114.

92. Prabhakar, S. Translational research challenges: finding the right animal models. *J Investig Med.* 2012;60(8):1141–1146.

93. Zhang, L., Schweizer, L. Building the bridge from drug discovery to clinical research using pathway approaches. *Int Drug Discov.* 2011;1:60–67.

5

CURRENT TRENDS IN COLLABORATIVE DRUG DISCOVERY AND STRATEGIES TO DE-RISK PRECOMPETITIVE INITIATIVES

Anuradha Roy[1] and Rathnam Chaguturu[2]

[1]*High-Throughput Screening Laboratory, University of Kansas, Lawrence, KS, USA*
[2]*iDDPartners, Princeton Junction, NJ, USA*

Academic research has contributed immensely in advancing the frontiers of disease biology, especially from a systems biology context, and in identifying potential therapeutic targets. The biotechnology and the pharmaceutical industries, standing on the shoulders of academic research and savoring the fruits of academic innovations, contributed to the discovery and development of new drugs [1]. Historically, academic research into basic biology unraveled new targets and pathways of potential therapeutic value. The pharmaceutical companies followed, for the most part, closed business models in pursuing therapeutic targets. In recent years, a number of factors have impacted the success of closed business models; these include, among others, expiring patents on blockbuster drugs, the flooding of the market with low-cost generics, escalating research and development (R&D) costs, and the introduction of a large number of "me-too" drugs. At the same time, dwindling funding to academic basic research projects has resulted in changes in the academic mind-set, making academia now more open to focused industry-funded projects and mutually beneficial technology transfer arrangements. Pharma has now adopted a more open business model and is making an increased capital and personnel investment in setting up collaborations with academia

Collaborative Innovation in Drug Discovery: Strategies for Public and Private Partnerships, First Edition. Edited by Rathnam Chaguturu.

in the areas of target identification and validation, probe development, and early drug discovery. In addition to collaborations, pharma has made available to public consortia various complex data sets for public access and mining. This open innovation paradigm is designed to generate new ideas and seek global expertise in bringing more effective and safe drugs to the market. This new developing relationship between industry and academic partners with divergent ideology and overall goals is still in the primordial stage; its success requires a constant evaluation of work flows, timelines, and communication infrastructure at the pharma–academia interface, to further strengthen the open innovation and collaborative research.

Against the backdrop of economic recession that we have just experienced, the venture capital investment in start-ups with new innovation technologies has sharply declined, and the funding to existing biotech companies was significantly curtailed. This purported lack of promising biotech ventures has affected the earlier biotech acquisition strategies of big pharmaceutical companies to feed their drug pipelines. In addition, one-third of the patents on blockbuster drugs have expired in the last couple of years. These and other factors have pressured big pharmaceutical companies to undergo major restructuring and to redefine their business models. A large number of pharmaceutical companies have also avoided the high costs associated with in-house R&D projects [2]. Pharma, over the last few years, has either outsourced R&D to other global markets, especially China and India, or has extended collaborations with domestic academic institutions. With cuts in federal funding for basic research, academic institutions are seeking investments from all sources, including pharmaceutical companies, nonprofit organizations, and philanthropies as well as the National Institutes of Health (NIH). An increasing number of academicians are willing to invest intellect and time to procure funding for early and translational drug discovery research. In the environment of financial and innovation challenges, this decade is witnessing a new phase of intense pharma–academia collaborations for the advancement of drug discovery. The collaborations, which range from small contracts to large investments, are still a work in progress. This learning process has highlighted the need for a better understanding of the core missions and driving ideology of collaborating partners.

PHARMACEUTICAL INDUSTRY'S PERSPECTIVE

Research in the early discovery phase has focused primarily on target validation and translation of basic research into applied drug discovery. In this aspect, drug discovery research in the pharmaceutical sector has relied heavily on the information published based on academia's long-term, basic research studies on biomolecules and their roles in normal physiology and in disease. Pharmaceutical companies have revolutionized modern drug discovery and are responsible for evolving compound screening to a highly integrated and automated process, capable of delivering large quantities of high-quality data output. Pharma has developed reliable platforms for hit-to-lead development and associated large-scale data processing and management over the last 25 years. Pharmaceutical sector research has contributed largely to the applied science of drug discovery, through improvements in drug candidates' clinical performance, and to a limited number of significant scientific discoveries in basic science. The arduous and costly process of drug discovery and development includes long periods of drug candidate optimization and preclinical development. The time from drug discovery to approval can take up to 15 years, and, based on drug approval data from 2010, the average cost of bringing a pharmaceutical product to market is between $0.8 and $1.3 billion [3]. This cost includes the cost of failed candidates since only 1 out of 10,000 to 50,000 potential candidates succeeds in late-stage drug approval. Today's expenditures on drug discovery are at least 10 times more than that of the 1990s. Project failures and high attrition rates in pharma may be traced to a lack of safety and efficacy and selection of therapeutic targets with unattractive toxicobiology properties.

A number of big pharmaceutical companies have succeeded in marketing a very limited set of blockbuster drugs in the last two decades [1]. Although an average of 30% of marketed drugs recoup the return from R&D investment, the majority of drugs lose out on market competitiveness [1]. The return on investment equation is also influenced by the redundancy of "me-too" drugs as well as by the patent's life span. The U.S. Food and Drug Administration classified three-fourths of the 119 drugs it approved in 2004 as similar to existing ones in chemical makeup or therapeutic value [4]. Also, the patents for most of the marketed drugs, which would previously have allowed a competitive advantage and therefore a return on investment, are due to expire in the next few years and are projected to challenge the bottom line of most of the pharmaceutical companies [5]. The increased regulatory demands for greater safety and efficacy, and the need for increased postmarketing surveillance, have also added to overall lower productivity [6]. In addition to the return-on-investment problems, pharma faces innovation challenges, since the low-hanging fruit (i.e., drugs for all easily druggable targets) have been discovered and developed [7]. The less tractable targets have remained unexplored due to the associated issues of R&D time and cost. The big pharmaceutical companies are trying to circumvent drying pipeline and productivity issues by following new business models, including acquisitions of micropharma [8] and in-licensing biological drugs/biologics that have proven marketability,* as well as by collaborating in preclinical research with academicians.

ACADEMIA'S PERSPECTIVE

The mission of academic universities is based on conducting research to acquire and expand new knowledge through study, discovery, and inquiry. Academia has always been involved in the education and training of students and in advancement of science through freedom of learning and not through return on investments or to solve real-world problems. The major acceptable outcome from federally funded research is scientific reports and publications that are made available through public databases. Despite the freedom of research in any area of interest, not strictly those areas dictated by commercial interests, academia has made significant direct and indirect contributions to the field. These contributions have been oriented primarily toward understanding of

*See http://www.icis.com/Articles/2008/09/29/9158479.

diseases and secondarily toward drug discovery. A detailed study recently reported that in the last four decades, 153 new Food and Drug Administration (FDA)-approved drugs, vaccines, or new indications for existing drugs were discovered through research carried out by publicly funded research investigators [9]. These drugs included 93 small-molecule drugs, 36 biologic agents, 15 vaccines, 8 *in vivo* diagnostic materials, and 1 over-the-counter drug, most of which were used in the treatment of cancer or infectious diseases treatment. There was always the recognition that academia's basic research provided pharma with a continuous flow of essential basic information for translation into applied and clinical drug development, but the evidence shows that academic research has not only provided basic information on biomolecules and pathways, but has also resulted in competitive patents over the biologics used in the treatment of certain diseases. The traditional academic disdain for partnerships with industry and corporate-dictated research has also undergone a major change of perspective in the current fiscal environment. Academic universities and institutes recognize the need for collaborating with industry for the much-needed dollars and resources for their own survival and are reaching out to the pharma to establish new alliances [10].

The formation of academic partnerships with industry was also facilitated by changes in intellectual property rights to accommodate the two ideologically distinct collaborating partners. Earlier, the government solely owned the intellectual property rights for all discoveries arising from federally funded research. For over 20 years, only a small fraction of the patents owned by federal government were commercially licensed. With the promulgation of the Bayh–Dole Act, universities, hospitals, and nonprofit and federal research institutes obtained complete ownership and control over their inventions and intellectual property resulting from federally funded research [11]. Under this new approach, inventions from nonprofits and universities were converted into intellectual property and transferred through license agreements to the private sector for commercialization and public use. Inventions could also be published in scientific literature after the patent was filed. The ability to acquire patent rights encouraged the establishment of technology transfer offices in many university campuses. These offices mediate all legal negotiations between the investigators or departments and the pharmaceutical companies.

OPEN (COLLABORATIVE) INNOVATION: A NEW BUSINESS MODEL

An analysis of the drugs approved in Europe and the United States reveals that a relatively small percentage of new chemical entities add to therapeutic value, while a vast majority of approved drugs fall in the category of "me-too" drugs having redundant properties of previously approved drugs. Pharma's drug discovery pipelines were routinely fed through internal target validation as well as through mergers and acquisitions of biotech companies. With the reduced venture capital investment in the biotechnology sector and small start-ups and the reduced R&D investment in big pharma, a new mutually beneficial business model has emerged, which promotes collaborations between industry, the government, and academia for bringing better, more diverse, and cost-effective drugs to the market. In the current fiscal environment, it is not profitable for the industrial sector to perform closed-door comprehensive basic research on potential therapeutic targets. Since the core strength of academia lies in its pursuit of all known and unknown aspects of biology, the academic scientists have an edge in identifying, understanding, and validating molecular targets for various diseases, developing assays, and to some extent, in probe discovery [10, 12]. These diverse, yet complementary sets of mission interests warrant active and profitable collaboration between industry and academia. To bridge this innovation gap while still competing for products, pharma needs newer strategies and business models for working on efficacious and novel therapeutic targets, finding molecules to query new drug targets, and developing smart lead optimization strategies [13, 14]. Pharmaceutical companies are leaning increasingly toward an "open innovation" approach, a concept coined by Professor Henry Chesbrough, University of California at Berkeley [15]. Open innovation is a business model in which the companies seek diverse paths to market through collaborative and open exchange of information and ideas. To be more profitable and productive, and to increase innovation toward developing new therapeutics, the pharmaceutical industry has embraced the open innovation approach as part of its new business plan to share the drug discovery processes and data with academia [16]. The open innovation approach is designed to tap into global expertise and talent to generate new ideas, new interpretations, and a new way to approach bottlenecks and problems. For example, Transparency Life Sciences,* a New York-based biopharmaceutical company, has launched an open innovation approach to design more patient population-centered, effective, and productive clinical trials for unmet medical needs by incorporating inputs from doctors, patients, and their families. The goal of the company is to revitalize and design productive clinical trials for previously shelved compounds with demonstrated efficacy and safety profiles.

DRUG DISCOVERY THROUGH COLLABORATIONS

The focus of the seemingly diverse pharma–academia collaborations covers target identification and validation, bioinformatics analysis of complex data sets, treatment of diseases using stem cells, and development of biologics and natural products. Treatments are only available for less than 3% of the numerous diseases that afflict mankind. FDA-approved drugs target less than 0.5% of the entire human genome, with 289 drugs targeting just 133 genes. A large proportion of human genome remains unexplored for druggability [17–19]. The target selection in the pharmaceutical sector has largely been decided by risk tolerance, druggability, return on investment, and shareholder profitability. Diseases for which a large patient population exists bring in more revenue for the pharmaceutical industry and are preferred over rare disorders afflicting small population groups and projecting smaller revenues [20]. In general, the low-risk and high-reward approach has resulted in pharma being too conservative and restrictive in its therapeutic drug target selection, and has led to the development of several "me-too" drugs [20]. On the other hand, academic research is not defined by any boundaries, and

*See http://www.transparencyls.com.

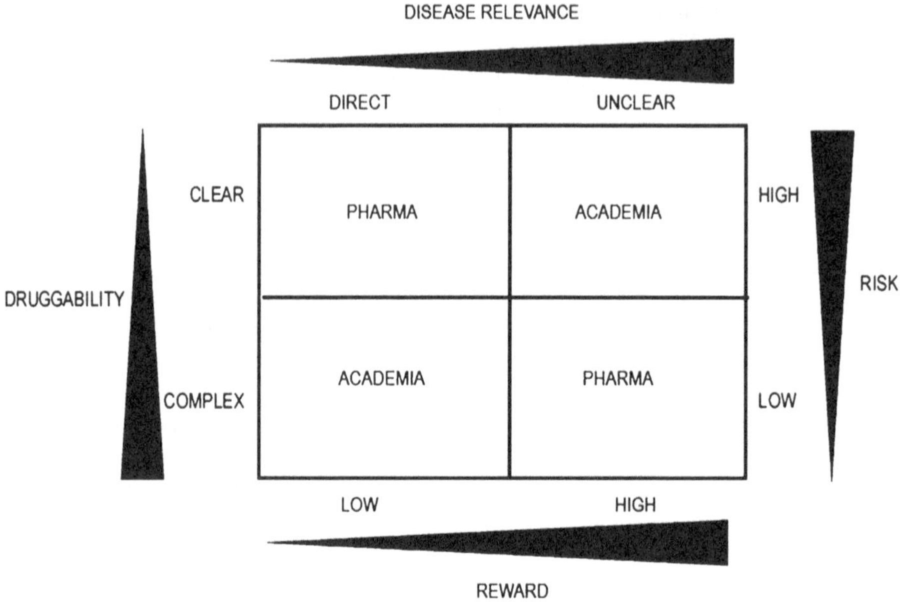

Figure 5.1. Key consideration in target selection. Academia and pharma operate in well-defined target areas characterized by risk and return on investment. Targets that are difficult to decipher and/or affect small population size are generally ignored by pharma as drug discovery targets.

research includes not only the widely pursued biological targets but also high-risk and low-reward biological targets from which pharma shies away (Figure 5.1). The continuously expanding knowledge pool from systems biology, genomics, molecular biology, and cell signaling has brought into focus the complex processing of biological information and biological responses via both hierarchical and parallel networks. Thus, any target biomolecule as well as chemical or biological probes and compounds have to be evaluated in the context of the complex biological networks. These studies will fill the innovation gap, promote design of drug candidates that are more efficacious, and lend confidence to drug discovery partnerships between academia and pharma. While academic input in data interpretation and basic mechanistic research is of value to pharma–academia collaborations, academia is also required to consider and value the experience of pharma in developing adaptable, high-quality assay technology formats to de-orphanize the highly refractory targets and make them amenable for standard drug discovery processes. In the new relationship, academia performs more focused, reproducible, and timeline-based research to lay a solid foundation for transition to late drug discovery workflow.

In the majority of collaborations financed by pharmaceutical companies, the academic scientists, either alone or in collaboration with scientific personnel from pharma, work toward target or novel pathway identification, characterization, and disease relevancy. Pharma then uses the information to translate basic science into drug screening and optimization campaigns. For example, Merck and Harvard Medical School are collaboratively studying novel signaling pathways in the bone-growth pathway. The AstraZeneca and Columbia University partnership includes collaborative development of research plans and protocols for identification of the biology and mechanism of targets in diabetes and obesity. In 2010, Pfizer, Inc. established the Centers for Therapeutic Innovation (CTI) to promote innovative translational research partnerships with academic medical centers on a global scale. The academic scientists work collaboratively with Pfizer's staff and have access to Pfizer's proprietary tools and technology. At least 20 such centers have been set up with academic institutions in Boston, New York City, Philadelphia, San Francisco, and San Diego.* In other instances, collaborations are centered on applying stem cell therapies or development of biologics for treatment of diseases or in rekindling research around previously shelved projects. For example, the University College London (UCL) established two distinct collaborations exploiting its expertise on stem cell replacement therapies. AstraZeneca and UCL are collaborating on stem cell therapies for the treatment of diabetic retinopathy. Collaboration between UCL and Pfizer was set up to utilize Pfizer's expertise in drug design and delivery to advance the work of UCL researchers in the field of stem cell-based therapies for age-related macular degeneration. Much of such collaboration is listed in Table 5.1. In another approach, Eli Lilly has established the PD2 Initiative,† a free service that offers to test the therapeutic potential of novel compounds synthesized in university/biotechnology laboratories in Lilly's disease-relevant phenotypic drug discovery platform. Eli Lilly analyzes the data, and compounds that show favorable effects on their phenotypic models are candidates for optimization. In this collaboration, Eli Lilly shares the phenotypic profiles and the secondary assay information with the academic investigator. In this mutually beneficial relationship, global researchers have the opportunity to test the therapeutic potential of their previously uncharacterized compounds in Lilly's drug discovery and development process, and Eli Lilly, in return, has access to potential hit-to-lead development candidates. Future collaborations are defined by the biological activity of these test compounds. While intellectual property (IP) rights remain with the submitting researcher, Lilly seeks the first rights to exclusively negotiate a licensing agreement. If the compound lacks promise, the researcher is granted no-strings-attached ownership of the data. Like Eli Lilly, Bayer Healthcare has launched a "Grants4Targets" initiative‡ to foster collaborations with academia or start-ups [21]. In this relationship, Bayer financially supports funding to validate and develop novel drug targets in oncology, gynecology, cardiology, and hematology. The overall expectation from partnerships between industry and pharma is that the collaborative intellectual and technological input would lead to new treatments, identification of disease-relevant targets, and evaluation of target safety and efficacy based on its network biology. In this new relationship, the research function would no longer be driven by a quest for new knowledge for knowledge's sake but by a more focused, dollar-driven need to solve current problems in the real world.

*See http://www.pfizer.com/research/rd_works/centers_for_therapeutic_innovation.

†See https://pd2.lilly.com.

‡See http://www.grants4targets.com.

TABLE 5.1. Collaborations between Pharma and Academia to Accelerate Drug Discovery

Private Sector	Public Sector	Drug Target Research Area
Agilent	UC, Berkeley	Synthetic biology institute
Astra Zeneca	Broad Institute	Infectious disease
Astra Zeneca	University of Manchester	Inflammatory disease
Astra Zeneca	University College London	Stem cells
Astra Zeneca	Weil Cornell Medical College, Washington School of Medicine, Feinstein Institute for Medical Research, University of British Columbia	Alzheimer's Disease
Bayer	UCSF	R&D agreement (10 years)
Bristol-Myers Squibb	Vanderbilt University	Parkinson's disease
Gilead	Yale	Oncology ($40 M)
GlaxoSmithKline	MD Anderson Cancer Center	Cancer immune therapy
GlaxoSmithKline	Harvard Stem Cell Institute	Heart disease and cancer ($25 M)
GlaxoSmithKline	University College, London	Amyloidosis: Transthyretin protein stabilization
GlaxoSmithKline	Vanderbilt University	Obesity
GlaxoSmithKline	Emory Institute for Drug Discovery	Drug discovery for rare diseases such as malaria
GlaxoSmithKline	Yale University	Cancer, inflammation, infectious disease
Genentech	UCSF	Neurodegenerative diseases
Johnson&Johnson	MIT	Oncology early academic research
Johnson&Johnson	Sanford-Burnham	Alzheimer early academic research
Johnson&Johnson	Queensland University	Chronic pain
Merck	California Institute for Biomedical Reserach	Preclinical proof of concept
Novartis	Harvard	Stem cells
Novartis	University of Pennsylvania	Center for Advanced Cellular Therapies ($20 M)
Pfizer	University of Pennsylvania	Research, clinical development, and policy ($15 M)
Pfizer, Entelos	University of California, University of Massachusetts, MIT	Energy metabolism, diabetes, and obesity ($14 M)
Pfizer	University College London	Stem cells based therapies
Pfizer	Washington University, St. Louis	Immunology, indications discovery
Sanofi Aventis	Columbia University	Diabetes
Sanofi Aventis	Salk institute	Gene therapy
Sanofi Aventis	Stanford	Stem cells
Sanofi Aventis	Harvard	Cancer, diabetes, inflammation
Sanofi Aventis	University of California	Diabetes
Takeda	Sanford-Burnham	Obesity
Veridex	Massachusetts General Hospital	Circulating tumor cells
Zambon	UCSF	Drug delivery

DRUG DISCOVERY THROUGH CONSORTIA

As a first step toward open innovation, several of the big pharmaceutical companies are actively participating in consortia previously shared by public sector academics and nonprofits [22] (Table 5.2). The high quantity of data generated by some pharma screening campaigns, ADME/tox studies, and genomic and expression arrays have been deposited to various public databases and are available for interpretation, for annotation, and for further drug or probe development or for diagnostics. A few representative

TABLE 5.2. Drug Discovery through Pharma–Academia Consortia

Private Sector	Public Sector	Drug Target Research Area
AstraZeneca, Eli Lilly, GlaxoSmithKline, Janssen Pharmaceutica, Novartis, Orion, Pfizer, Roche, Servier, and Wyeth.	Karolinska Institutet (Sweden), The University of Cambridge (United Kingdom), Central Institute of Mental Health (Germany), CSIC (Spain), University of Manchester (United Kingdom), and the Bar Ilan University (Israel).	Novel Methods leading to New Medications in Depression and Schizophrenia (NEWMEDS)
Pfizer CTI	NYC, Boston , San Francisco University Consortium	Multiple areas
GlaxoSmithKline, Pfizer Inc., and AstraZeneca	Critical Path Institute, a nonprofit partnership with the FDA	Coalition Against Major Diseases to share data on thousands of Alzheimer's patients in hopes that the extra information will spark new ideas for treatments
Eli Lilly	Open-access submission of compounds	Phenotypic Drug Discovery platform
GSK	Open-access database	Genomic and protein expression profiling data for over 300 cancer cell lines via the NCI's Cancer Bioinformatics Grid™ for academia to mine
Merck-SAGE	Open-access database	Sharing and disseminating complex data representing disease biology (genetic [SNP, copy number variations], RNA expression [mRNA, miRNA, other noncoding RNA])
Pharma Consortium	CTSA Pharmaceutical Assets Portal-NIH	To improve information exchange regarding drugs available for repurposing—proactively engage pharma in data sharing

consortia between Eli Lilly, GlaxoSmithKline (GSK), Merck, and the Clinical and Translational Science Awards portal are highlighted as follows.

1. *Merck-Sage Bionetworks.* This is an open-access, nonprofit, organization established by Merck,* which consists of data from human and mouse disease models from the Rosetta subsidiary of Merck. The public can access Merck data on systems biology networks and can also download the essential computational disease biology software tools for data access and analysis. The Sage program is an open-access approach designed to bring academia and pharma together to develop comprehensive human disease biology models to predict network biology in normal physiology and disease and for all sources to deposit integrated networks of biological data. This aims to facilitate the integration of diverse molecular mega-data sets to build predictive bionetworks and to establish a precompetitive position for human disease biology. The long-term goal of network biology is aimed at designing better and more targeted drugs based on systems network models.
2. *GSK-caBIG Collaboration.* GSK released genomic profiling data for over 300 cancer cell lines via the National Cancer Institute's Cancer Bioinformatics Grid™ (caBIG), a network of infrastructure and tools that enables the collection, analysis, and sharing of data. The site provides valuable information for genomic profiles for a wide variety of tumors, including breast, prostate, lung, and ovarian cancers. The public access allows any researcher to download the open-source GSK cancer data through caArray.
3. *ChEMBL-Neglected Tropical Disease Archive.* Large volumes of primary screening and medicinal chemistry data targeting neglected diseases are available through ChEMBL-Neglected Tropical Disease archive.† The open source repository contains the structures and screening data against the malarial parasite from GSK (13,500 compounds), Novartis (>5600 compounds), and St. Jude's Children's Research hospital (310,000 compounds). The data are available to all for annotation and analysis in the hope of expediting development of new therapeutics against neglected diseases.
4. *EU Innovative Medicines Initiative-Joint Undertaking (*IMI-JU*).* This initiative represents a public–private partnership between the European Communities and the pharmaceutical industry (the European Federation of Pharmaceutical Industries and Associations [EFPIA]). The goal of the SAFE-T consortium‡ is to identify sensitive translational safety biomarkers for drug-induced kidney, liver, and vascular injury. The goal is to predict drug toxicity in humans by minimizing failures related to drug toxicity data from species differences in preclinical safety tests and nonrepresentative patient populations in clinical trials. The

*See http://sagebase.org.

†See https://www.ebi.ac.uk/chemblntd.

‡See http://www.imi-safe-t.eu.

eTOX program of IMI aims to create a database of high-quality *in vitro* and *in vivo* data and predictive toxicological models. The data are available to pharma and academia and serves to highlight the ADME/tox issues associated with compounds and have the potential to greatly reduce the high attrition rates associated with many lead scaffolds.

5. *Innovative Medicine Initiative.* Europe's Innovative Medicine Initiative,* a public–private consortium, has committed US$271 million investment to support discovery of novel drug candidates. The consortium currently consists of 30 academic and corporate collaborators [23].
6. *Structural Genomics Consortium (SGC).* This is an open-access, not-for-profit, public–private partnership to contribute to basic biology relevant to drug discovery using large-scale 3-D structures of proteins of therapeutic importance from humans and their parasites. The SGC includes collaborations between the Universities of Toronto and Oxford and the Karolinska Institute in Stockholm with GSK, Eli Lilly, Pfizer, the Novartis Research Foundation, the Wellcome Trust, and Canadian agencies. The SGC has also collaborated with the chemistry and biochemistry departments of University of Oxford, the NIH Chemical Genomics center in Washington, and GSK to set up open-access chemical probes for epigenetic proteins.

The data sharing between major pharmaceutical companies and academic researchers is designed to attract global intellect, at no cost, in dissecting, annotating, and predicting complex data sets. The hope from collaborative data sharing and analysis is that it will expedite translation of complex data into drug discovery pipeline. The large, costly, and high-quality data sets shared by pharma in public consortia are of immense benefit to academicians, who normally cannot undertake high-throughput biomarker, molecular, and metabolism studies. The approach also helps academic institutions, small research facilities, and nonprofit organizations by allowing them to avoid cost- and time-prohibitive research and instead redirecting their efforts toward nonredundant areas for future therapeutic research.

DRUG DISCOVERY THROUGH REPURPOSING STRATEGIES

The majority of biological events do not occur in isolation, but involve orchestration between various pathways and networks, which collectively are perturbed in response to disease or a drug treatment. The drugs or biologics that target one protein or gene or pathway may have therapeutic benefits in other diseases, and can act at known and unknown sites, resulting in "on-" and "off-target" effects. The candidates for repurposing are drugs that were previously approved by the FDA, drugs in clinical development, drugs that have failed in clinical trials for reasons other than safety, and drugs launched

*See http://www.imi.europa.eu.

TABLE 5.3. Examples of FDA-Approved Repurposed Drugs

Drug	First Indication	New Indication
Thalidomide (Celgene)	Antinausea	Immunomodulatory amd antiangiogenic in multiple myeloma
Finasteride (Merck)	Enlarged prostate	Male baldness
Sildenafil (Pfizer)	Hypertension	Erectile dysfunction
Ropinirole (GSK)	Parkinson's disease	Restless leg syndrome
Pramipexole (Boehringer Ingelheim)	Parkinson's disease	Restless leg syndrome
Raloxifene (Eli Lilly)	Osteoporosis	Breast cancer clinical trials
Minoxidil	Hypertension (Pharmacia)	Hair loss (Pfizer, Rogaine)
Buproprion	Depression (GSK)	Smoking cessation (GSK)
Plerixafor (Genzyme)	HIV infection	Multimyeloma
Dalcipran (Cypress Bioscience)	Antidepressant	Fibromylagia
Cymbalta (Eli Lilly)	Antidepressant	Fibromylagia
Fluoxetene (Eli Lilly)	Prozac (antidepressant)	Sarafem (premenstrual dysphoria)
Topiramate	Epilepsy (J&J)	Obesity (J&J)
Polidocanol (Asclera)	Reticular veins (lower extremities)	Neovessels/increased blood flow at the point of pain (experimental, not FDA approved)
Bimatoprost (Lumigan, Allergan)	Open-angle glaucoma/ocular hypertension	Eyelash hypotrichosis (Latisse, Allergan)

in emerging markets but not launched in the United States. Pharma is actively pursuing this low-risk and low-cost strategy of repurposing drugs or drug-like leads for new therapeutic indications to fill product pipelines [24]. Drug repurposing for marketed drugs for which patents are expiring or which have generic substitutes is an effective approach to extend the life cycle and commercial productivity in an extremely short period of time. As an example, Eli Lilly's Cymbalta (duloxetine), a serotonin-norepinephrine reuptake inhibitor for fibromyalgia, was initially used as an antidepressant. Pfizer's sildenafil, which was originally intended for treatment of hypertension, was repurposed for erectile dysfunction. Other examples of repurposed drugs are shown in Table 5.3. Drug repurposing also highlights the fact that the compounds that are good fits for repurposing are not specific enough to begin with, and that may be the reason why most compounds fail during Phase 2 clinical trials. Pfizer (Indications Discovery Unit) and Novartis are known to have internal departments devoted to drug repurposing. While companies such as Melior Discovery and Ore Pharmaceuticals are discovering new indications for shelved compounds, Zalicus, formerly CombinatoRx, is investigating new uses for off-patent drugs.

Nonprofits and academia have also joined the race for finding new uses for compounds known to have activities in other diseases [25]. One of the goals of the NIH-funded Clinical and Translational Science Awards* (CTSA) is to promote drug repurposing collaborations between pharma and the CTSA/NIH consortium, comprising nonprofits and academics. The emphasis is on finding new targets for discontinued or failed compounds that have been previously characterized. The CTSA facilitates interactions between matched academic investigators and the pharmaceutical industry to develop research partnerships based on compounds shelved at the clinical stage. CTSA mediates one of the largest pharma–academia integration efforts involving 348 researchers nationwide, with diverse target-disease interests, and eight pharmaceutical companies (Pfizer, Merck, GSK, Novartis, Genentech, Abbott, Eli Lilly, and AstraZeneca). CTSA-sponsored researchers from the University of California-Davis have established a Pharmaceutical Assets portal,† which allows academic researchers to gain access to compounds from pharma that were abandoned during clinical development [26].

Pharma can benefit commercially from repurposing compounds that previously failed in clinical trials due to a lack of specificity but had clean safety profiles. The patentability of an FDA-approved drug for a new indication would depend on many factors such as operational space around patents or around improvements in existing drugs, and their new indication value.

PROBE DISCOVERY IN PUBLIC DOMAIN

The new collaborative relationship between academia and pharma in the area of drug discovery was also bolstered by the Molecular Libraries Probe Production Centers Network (MLPCN) set up by the NIH via its investment in the Roadmap initiative [27, 28]. The funding enabled by the NIH opened doors to the science of high-throughput screening of diverse chemical libraries in academia, and greatly enhanced bioassay development and small molecule discovery. A number of high-throughput screening centers arose on university campuses and made generic or specialized screening services available to institutional investigators [29]. This change in academic philosophy offered great impetus to drug discovery in academia as well as provided infrastructure for pharma–academia collaborations. The Roadmap initiative also endorsed the term "probe," a chemical compound, an antibody or a biomarker that can be used as a tool to study a biological event or a pathway [30]. The academic screening centers, funded by the academic institution itself or by the NIH Roadmap initiative, promoted the identification of probes that are reasonably specific to trace a pathway or are more drug-like with proven chemical identity, defined structure, high potency, and selectivity. The outcome for a screen also provided scaffolds for hit-to-lead optimization studies [31]. The academic high-throughput centers built up their compound collections by purchasing their choice of chemical libraries of almost 11 million compounds available from over 20 commercial vendors [32]. In addition to diverse and focused commercial libraries of chemical compounds, siRNAs, and peptides, many academic centers also

*See https://www.ctsacentral.org/.

†See https://www.ctsacentral.org/.

possess institutional chemistries as part of the chemical library holdings. The institutional screening centers acquired sensitive and appropriate technology platforms, large-scale data storage, and the processing systems necessary for running high-throughput screening campaigns. In addition to the primary and secondary screening facilities, some academic centers also provide medicinal and computational chemistry support for hit-to-lead optimization and development. Some high-throughput screening centers also incorporated pharma's best practices, including active project management by employing industry-trained project managers to manage multidisciplinary teams, conduct regular project team meetings, facilitate go/no go decisions, pursue IP activities, and maintain the overall momentum of the project.

EU-OpenScreen,* the European counterpart of the U.S. NIH Roadmap initiative, has a membership of 26 academic and nonprofit institutions from 14 European countries. This European Infrastructure of Open Screening Platforms for Chemical Biology integrates high-throughput screening platforms, chemical libraries, chemical resources for hit discovery and optimization, bio- and cheminformatics support, and a database containing screening results, assay protocols, and chemical information. These platforms, offering the most advanced technologies, will be used by European researchers to identify compounds affecting new targets. Open access to an integrated infrastructure for chemical biology will thus satisfy the needs for new bioactive compounds in many fields of the life sciences (e.g., human and veterinary medicine, systems biology, biotechnology, agriculture, and nutrition). Unlike the U.S. MLPCN initiative, the EU-OpenScreen has a well-established framework for IP management.

Another significant open source consortium is the International Genomics Consortium† (IGC), a nonprofit medical research organization that includes the NIH's The Cancer Genome Atlas project (TCGA), which generates genomic and clinical data for specific cancer types for use by the cancer research community. The overall goal of IGC is to acquire data for diagnostics, treatment, and prevention of cancer and complex diseases by molecular profiling, and characterization of differential biomarker expression for personalized medicine.

EFFECTIVE MANAGEMENT OF ACADEMIA–PHARMA COLLABORATIONS

Successful academia–pharma collaboration is a form of sponsored, focused research with an overall objective of generating new and better drugs and viable intellectual property. The collaborations are recent and evolving, and there are no reports of successes to strongly corroborate expectations. Academic–pharma alliances may be financed via a private grant with freedom to operate in a limited research space, via a fee for service contract for focused research, or in a true collaboration with management by all partners. Successful academia–pharma partnership requires mutual understanding of institutional objectives, ideologies, and bottom lines. Problems may arise from

*See http://www.eu-openscreen.de/.

†See http://www.intgen.org.

divergent interests or work cultures of the public and private domains; missing professional skill sets; lack of delivery, execution, or robust timelines; poor project management; inadequate compensation for indirect costs; poor negotiation skills; and partial or incomplete understanding of legal concepts, IP, contract law, federal and state regulations, and policy positions. Academic interests in publication of research data in a timely manner as well as grant support for ongoing academic financial needs have to be addressed upfront while negotiating IP policies and ownership of results and exclusivity. Pharma-funded academic research requires higher accountability, greater data reproducibility, and timely delivery of expected data. Through pharma alliances, academic faculty and students have a unique opportunity to be trained in project management and timeline-based, focused research, and be directly involved in drug discovery and development activities. The training provided to university students through such alliances has the potential to feed employment searches for skilled hires. University technology transfer offices provide academia with a means to secure product development and circumvent traditional licensing outcomes. In most negotiations, pharma shares the IP rights with academic partners, with their prerogative for the first right of refusal. Academia should agree with pharma in providing opportunities for "product" development, and pharma should readily support academia's need for timely research publications and data for seeking additional grant support. A long-term value-based productive outcome favoring both academia and pharma is bound to face challenges because of inherent differences in the approach and mind-set of the interacting partners, but can be managed when the partners with competing interests understand and respect each other's policy positions.

CONCLUDING REMARKS

In the current environment of patent cliffs, high R&D costs, and paucity of innovative drug discovery, a number of new business models have emerged. Pharma, micropharma, disease foundations, the NIH, and academia have now come to realize the value of open collaborations in easing the drug discovery bottlenecks. The common underlying theme of new business approaches is to finance collaborations that bring together global public and private brainpower, their data sets, and experience, with the goal of unraveling biology for improving quality and quantity of drugs while still competing on products.

REFERENCES

1. Zycher, B., DiMasi, J.A., Milne, C.P. Private sector contributions to pharmaceutical science: thirty-five summary case histories. *Am J Ther*. 2010;17:101–120.
2. Vernon, J.A., Golec, J.H., Dimasi, J.A. Drug development costs when financial risk is measured using the Fama-French three-factor model. *Health Econ*. 2010;19:1002–1005.
3. DiMasi, J.A., Hansen, R.W., Grabowski, H.G. The price of innovation: new estimates of drug development costs. *J Health Econ*. 2003;22:151–185.

4. Gagne, J.J., Choudhry, N.K. How many "me-too" drugs is too many? *JAMA*. 2011;305: 711–712.
5. Harrison, C. The patent cliff steepens. *Nat Rev Drug Discov*. 2011;10:12–13.
6. Allarakhia, M., Walsh, S. Managing knowledge assets under conditions of radical change: the case of the pharmaceutical industry. *Technovation* 2011;31:105–117.
7. Williams, M. Productivity shortfalls in drug discovery: contributions from the preclinical sciences? *J Pharmacol Exp Ther*. 2011;336:3–8.
8. Barden, C.J., Weaver, D.F. The rise of micropharma. *Drug Discov Today* 2010;15:84–87.
9. Stevens, A.J., Jensen, J.J., Wyller, K., Kilgore, P.C., Chatterjee, S., et al. The role of public-sector research in the discovery of drugs and vaccines. *N Engl J Med*. 2011;364:535–541.
10. Frye, S., JanZen, B. Discovering Academia. Available at https://pharmacy.unc.edu/research/centers/center-for-integrative-chemical-biology-and-drugdiscovery/publications/dd_stephenfry_editorial.pdf.
11. Bremer, H.W. University technology transfer: evolution and revolution. In *Council on Governmental Relations: 50th Anniversary: Journal of Papers*. Washington, DC: Council on Governmental Relations; 1998:25.
12. Chin-Dusting, J., Mizrahi, J., Jennings, G., Fitzgerald, D. Outlook: finding improved medicines: the role of academic-industrial collaboration. *Nat Rev Drug Discov*. 2005;4: 891–897.
13. Tralau-Stewart, C.J., Wyatt, C.A., Kleyn, D.E., Ayad, A. Drug discovery: new models for industry-academic partnerships. *Drug Discov Today* 2009;14:95–101.
14. Paul, S.M., Mytelka, D.S., Dunwiddie, C.T., Persinger, C.C., Munos, B.H., et al. How to improve R&D productivity: the pharmaceutical industry's grand challenge. *Nat Rev Drug Discov*. 2010;9:203–214.
15. Chesbrough, H. *Open Innovation: The New Imperative for Creating and Profiting from Technology*. Cambridge, MA: Harvard Business School Press; 2003.
16. Wagner, J.A. Open-minded to open innovation and precompetitive collaboration. *Clin Pharmacol Ther*. 2010;87:511–515.
17. Lindsay, M.A. Target discovery. *Nat Rev Drug Discov*. 2003;2:831–838.
18. Overington, J.P., Al-Lazikani, B., Hopkins, A.L. How many drug targets are there? *Nat Rev Drug Discov*. 2006;5:993–996.
19. Hopkins, A.L., Groom, C.R. The druggable genome. *Nat Rev Drug Discov*. 2002;1: 727–730.
20. Wellman-Labadie, O., Zhou, Y. The US Orphan Drug Act: rare disease research stimulator or commercial opportunity? *Health Policy (New York)* 2009;24:24.
21. Lessl, M., Schoepe, S., Sommer, A., Schneider, M., Asadullah, K. Grants4Targets—an innovative approach to translate ideas from basic research into novel drugs. *Drug Discov Today* 2011;16:288–292.
22. Hunter, J., Stephens, S. Is open innovation the way forward for big pharma? *Nat Rev Drug Discov*. 2010;9:87–88.
23. Baker, M. Europe bets on drug discovery. *Nature* 2013;494:20.
24. Sleigh, S.H., Barton, C.L. Repurposing strategies for therapeutics. *Pharmaceut Med*. 2010;24:151–159.
25. Dolgin, E. Nonprofit disease groups earmark grants for drug repositioning. *Nat Med*. 2011;17:1027.

26. Schubert, C. Matchmaking service links up researchers to wallflower drugs. *Nat Med.* 2010;16:7–7.
27. Austin, C.P. The impact of the completed human genome sequence on the development of novel therapeutics for human disease. *Annu Rev Med.* 2004;55:1–13.
28. Austin, C.P., Brady, L.S., Insel, T.R., Collins, F.S. NIH molecular libraries initiative. *Science* 2004;306:1138–1139.
29. Gordon, E.J. Small-molecule screening: it takes a village. *ACS Chem Biol.* 2007;2:9–16.
30. Roy, A., McDonald, P.R., Sittampalam, S., Chaguturu, R. Open access high throughput drug discovery in the public domain: a Mount Everest in the making. *Curr Pharm Biotechnol.* 2010;11:764–778.
31. Oprea, T.I., Bologa, C.G., Boyer, S., Curpan, R.F., Glen, R.C., et al. A crowdsourcing evaluation of the NIH chemical probes. *Nat Chem Biol.* 2009;5:441–447.
32. Roy, A., Taylor, B., McDonald, P., Price, A., Chaguturu, R. Hit-to-probe-to-lead optimization strategies: a biology perspective to conquer the valley of death. In: Seethala, R., Zhang, L., editors. *Handbook of Drug Screening*. New York: Informa Healthcare; 2009, pp. 21–55.

6

A PERSPECTIVE ON THE EVOLUTION OF COLLABORATIVE DRUG DISCOVERY AND FUTURE CHALLENGES

Christopher A. Lipinski

Melior Discovery, Waterford, CT, USA

In this chapter I would like to bring in a personal perspective illustrating how, in just a decade, collaborative drug discovery (CDD) went from an idea with little traction to one warranting an entire book. I consider drug repurposing and the discovery of treatments for rare and orphan diseases as major opportunity factors for CDD. CDD *per se* will not solve any of the myriad problems evident in mainline big pharma drug discovery but with respect to academic drug discovery does pose new challenges that are more organizational and cultural rather than technical and scientific.

EVOLUTION OF COLLABORATIVE DRUG DISCOVERY

In the year 2000, the Pfizer laboratories in Sandwich, UK, produced a white paper on drug repurposing (the Pfizer terminology for drug repurposing was "indications discovery"). Widely viewed within the Pfizer organization as scientifically sound and eminently reasonable, the proposal to engage in drug repurposing went nowhere. To implement the proposal, some sort of centralized screening group or facility would have to be set up to run *in vitro* and *in vivo* assays across the tremendous breadth of Pfizer research across multiple therapeutic areas. A specific small example gives an idea of

Collaborative Innovation in Drug Discovery: Strategies for Public and Private Partnerships, First Edition. Edited by Rathnam Chaguturu.

the organizational hindrance to setting up this type of effort. In 2000, when a therapeutic area anywhere in the Pfizer world nominated a compound for clinical development, an information freeze went into effect on any information exchange that required testing of the compound. The nominated clinical candidate could not be obtained from the central Pfizer sample repository for testing anywhere within Pfizer without the explicit permission of a senior director or VP from the originating therapeutic area. In 2000, the Pfizer organizational structure and cultural mind-set in drug discovery was just not ready for the concept of discovering a new use for an existing drug.

In mid-2002 I retired from the Pfizer discovery organization, and over the next 4 years, the corporation kindly allowed me to continue my scientific activities in an office in the computational suite. While in my Pfizer office, I was contacted by Andrew Reaume, a then Pfizer employee working in the Pfizer technology zone who had very interesting ideas on using automated *in vivo* phenotypic rodent screening to find new uses for existing drugs. Andrew was struggling to set up drug repurposing within Pfizer using technology he had learned and developed within Pfizer. The upshot is that up to 2005, the internal attempt at drug repurposing within Pfizer failed. The positive aspect is that Pfizer legal saw no problem with Andrew leaving Pfizer with his technology since his automated *in vivo* rodent assays were all based on literature assays with high citation count that the biologists typically used to convince themselves that the activity was real. Andrew left Pfizer on good terms and cofounded Melior Discovery, an Exton PA drug repurposing start-up, with Mike Saporito in 2005. Before he left Pfizer, Andrew had recruited me for the Melior Discovery Scientific Advisory Board (SAB), and we had picked CP-26154 as our first drug to explore in drug repurposing. CP-26154 was a compound I had made at Pfizer in 1972 as an antiulcer drug and which had reached Phase 2 studies in a time era where compounds entered the clinic without knowing mechanism. As it turned out, CP-26154 (now renamed as MLR-1023) was found by Melior Discovery to have unexpected activity in four phenotypic models, suggesting utility in type II diabetes and later was found by Melior to be an allosteric Lyn kinase activator. Subsequent to the phenotypic discovery, the investigational new drug (IND) application for the diabetes indication was approved by the Food and Drug Administration (FDA). As of 2005, Pfizer, the largest pharmaceutical organization in the world, was not yet ready for internal drug discovery in drug repurposing.

In 2004, I was the cochair of the Society for Biomolecular Screening (SBS) national conference in Orlando, Florida. SBS was the big high-throughput screening (HTS) conference in the United States, and I remember that we had no more than 10% academic attendees. This was a big contrast to the SBS meeting in 2007 in Montreal with about 30% academic attendees. With Mel Reichman (LIMR Chemical Genomics Center, Wynnewood, PA) at this later conference, I participated in a Point-Counterpoint entitled "Is the NIH Roadmap a New Model for Drug Discovery: Pipelines or Pipedreams of Future Drugs." Mel was the pipeline spokesman and I was the pipedreams proponent, and we both felt strongly about our respective positions.

On the lecture circuit I had been speaking on how I felt that intellectual property (IP) considerations were hindering information exchange. In the PowerPoint slides in my talks I had used the terminology of "anti-intellectual effects of intellectual property." In 2005, I was approached by the editors of an edition of *Current Opinion in Compu-*

tational Biology (*COCB*) to write on the subject of "anti-intellectual effects of intellectual property." I agreed to write an opinion piece, and over the next 9 months, every time I visited a university I asked to speak to the university intellectual property people. The "Anti-Intellectual Effects of Intellectual Property" was published in *COCB* in August of 2006 [1]. From the academic researchers' viewpoint, IP was at best something they learned to live with and at worst was a royal pain in the neck. In their opinion, IP considerations, at that time period, did nothing to enhance their scientific productivity.

The safe exchange of information about compounds without actually revealing drug structure can be an important component to collaborative drug discovery. Tudor Oprea at the University of New Mexico and I jointly organized a session cosponsored by the CINF and COMP sections at the spring 2005 ACS national meeting in San Diego. A wide range of opinions on computational feasibility were presented, and although the session received reasonable publicity [2, 3], to my knowledge, there was no direct outcome from this session that enhanced collaborative drug discovery. In this time period, organizations facilitating information exchange began to appear, for example, in 2004 Collaborative Drug Discovery (CDD) (*vide supra*) was spun out of Lilly, and the formation of the Clinical and Translational Science Award Pharmaceutical Assets Portal (CTSA portal) was completed in 2008. The history of Collaborative Drug Discovery is described in this book in Chapter 19. The very interesting history of the evolution of the CTSA consortium launched by the NIH in 2006 into the CTSA portal has been recently described by Kate Marusina and colleagues in academia and industry in a special edition of *DDT Therapeutic Strategies* dedicated to drug repurposing [4], and in a chapter later in this book. As the editor of this special issue, I read the fascinating details of the very divergent views to data sharing and drug repurposing that were encountered among major pharmaceutical organizations. Of particular note is that Pfizer became a major sponsor of the CTSA portal. Taking Pfizer, as an example, there was a marked change between 2005 and 2008 in viewpoint toward collaborative drug discovery from caution in 2005 to advocacy in 2008.

In the summer of 2005, I spoke at a conference in central London organized by the Wellcome Trust, and over the 2005 Christmas break, I was one of the reviewers for a new innovative Wellcome Trust plan to sponsor academic drug research. This was the beginning of the successful grant program that was called "seeding drug discovery," key elements of which have been summarized in *Nature Reviews Drug Discovery* (*NRDD*) [5].

From 2000 through 2006, I had been on the SAB of the Global Alliance for TB Drug Development (GATB), and in my first 2 years before the GATB recruited a resesearch director, I had sat in on SAB meetings where effectively the research policy was set by academic committee. With respect to any of the usual industry metrics regarding project planning, stage gates, and so on, this was a dysfunctional group. The situation radically changed for the better when the GATB hired an experienced research director with industrial experience. As a scientist at Pfizer, I had been somewhat dismissive of the management project planning process, but the early GATB SAB experience converted me into a believer to a sensible form of project planning. So it was with keen interest that I noted that the Wellcome Trust "seeding drug discovery" grants provided for external project planning through the Wellcome Trust. In light of my early GATB experience, I interpreted this as "We (the Wellcome Trust) know that academics are inexperienced in project discipline and because we know this is so important we will provide this ourselves." In more recent times, it is really gratifying to me that this lesson has been learned by the better academic drug discovery groups. I would single out the Center for Drug Discovery at the KU Leuven in Belgium which has an outstanding IP licensing and development organization with initiatives relating to collaborations among the CD3 researchers, the academic biologists, and the little spin-out companies that I have not seen duplicated in the United States. As a member of the SAB of the KU Leuven CD3, I personally have seen the importance of "ring fencing" the medicinal chemistry so that the medicinal chemistry focus is directed entirely to drug discovery and not to the generation of peer-reviewed publications (more on this later).

The Wellcome Trust had a major role in opening up previously proprietary databases to the public. In 2008, the Trust awarded a 5-year, UK£4.7 million grant to transfer well-structured chemogenomics data from the publicly listed company Galapagos to the European Molecular Biology Laboratory's European Bioinformatics Institute (EMBL-EBI) [6]. The databases in question were those developed by the former Inpharmatica, which was then owned by Galapagos. The Trust, in a senior management key strategic decision, decided to make the Inpharmatica data available to everyone whether in academia or industry. The Inpharmatica data and servers were moved from central London to the EBI Genome campus in Hinxton, UK, a few miles outside of Cambridge, and have now been accessed by hundreds of research groups worldwide.

In 2006, I had exchanged e-mails and had conversations with Barry Bunin (then at Libraria) who was to become the President and Founder of Collaborative Drug Discovery, and in 2007, I gave a keynote lecture at the March Collaborative Drug Discovery community meeting at the University of California, San Francisco, Mission Bay campus that was entitled "Academic Drug Discovery: The Chemistry Challenges of Target Choice and Screening Library Selection." As the title suggests, I was beginning to come around to the notion that academic screening might have value, although I was still skeptical that the academics could get the medicinal chemistry part to work well. I still retain some of this skepticism and think that the biology–medicinal chemistry interface is among the most difficult parts of academic drug discovery and one in which fortunately there is notable recent progress. However, what is also very true is that there is a wide range in the skills sets in academic drug discovery. Some groups are very competent in true drug discovery and some are more at the other extreme.

The goals of chemical biology and drug discovery are quite different, with implications for the types of compounds one would like to screen. In chemical biology, one is seeking a tool or probe with known selectivity to learn about some biological process. We now know that few, if any, biologically active compounds are truly selective. So the ligand selectivity for a single target metric is likely seldom met, and what is important is to have a sensible appreciation of the ligand's known spectrum of activities and to keep an open mind as to the likely yet to be discovered activities.

In May 2002, Elias Zerhouni became director of the NIH with a goal to enhance the agency's capacity to foster state-of-the-art science. The initiative to accomplish this goal was the NIH Roadmap for Medical Research that was announced in October 2003 after input from more than 300 biomedical leaders in late 2002. Funding started in 2004 and so began the NIH's encouragement and support for academic HTS through the Molecular Libraries Screening Center Network (the biology screening side) and the acquisition by purchase of ultimately hundreds of thousands of chemical compounds that would be screened in a pilot phase in nine academic and one intramural NIH center. At the time of the start-up of the screening library collection, the NIH had little medicinal chemistry expertise, and so, as an organization responsive to the U.S. public and Congress, it sought expert assistance. I was a member of an early teleconference to discuss what types of compounds should be acquired by the NIH for HTS assays to discover chemical biology tools and probes. Our teleconference group was about evenly split between industry people and academics. I have to say that I felt that there was something like a Grand Canyon separating the views of the academics and industrial people. The academics talked about innovation, thinking out of the box, maximum chemical diversity, and not being limited by preconceived rules and filters. The industry people talked about pragmatism, the lessons learned (aka rules and filters), and about worthless compounds that could appear active in HTS screens. The NIH was faced with two irreconcilable viewpoints. They had to pick one, and after a period of study and deliberation, they chose the academic viewpoint. Subsequently, I was fortunate to be a coauthor with Tudor Oprea and many others on a crowdsourcing analysis of the value of the first 75 tools and probes from the NIH Roadmap screening effort [7]. Summarizing, charitably about 75% of the probes had value, 25% were worthless. The wording in the *Nature Chemical Biology* paper was not nearly as blunt as I am being here. The published politeness was forced on Tudor Oprea by the *Nature* editors, since those who know him well will recognize that excessive politeness is not Tudor's style. Hidden in the paper is the recitation of a series of backtracks made by the NIH in the compound purchasing process. Faced with really atrocious HTS screening data, the compound quality criteria in a series of three steps were moved back from the early academic viewpoint to one more aligned with the industrial viewpoint. Again, it is gratifying in recent times to see the evolution and progress toward higher quality chemical screening compounds.

CHALLENGES FOR COLLABORATIVE DRUG DISCOVERY

I have previously discussed how academics talk about innovation, thinking out of the box, maximum chemical diversity, and not being limited by preconceived rules and filters. Challenges for CDD are more organizational and cultural than technical and scientific. While industry people talk about pragmatism, the lessons learned (aka rules and filters), and about worthless compounds that could appear active in HTS screens, I firmly believe that well-meaning but nevertheless incorrect ideas about the nature of chemistry space can loom as large impediments to academic drug discovery and thus to any collaborative drug discovery that relies on academic resources.

The physicochemical properties consistent with useful drugs are of low dimensionality in that their properties can be described by a small number of independent variables [8]. This is why the quite simple rule of five works. By contrast, biological activity at a target is of high dimensionality. Something as simple as the rule of five would never work to describe a biological structure–activity relationship. The biology world is incredibly complex, and to be really certain about activity, one has to run an experimental assay. The chemistry world is far simpler, with rules and principles and decades of lore relating chemistry structure to chemical reactivity. Thus, it is possible to classify which types of chemical functionality are more likely to lead to promiscuity in chemical biology or drug discovery assays. The rhodanine moiety (CAS 141-84-4) is a perfect example. This moiety has been discussed by Jonathan Baell in the context of a promiscuous substructure that one should be wary of [9, 10]. A substructure search in CAS SciFinder for compounds with a rhodanine moiety resulted in 109,919 hits, of which 69,668 are commercially available. Clearly, the compound vendor community has recognized the value in making rhodanines available for sale as screening compounds. A recent discussion by an academic group on the screening of rhodanines and the propensity of the rhodanine C = S moiety to engage in polar interactions offers a different and perhaps more nuanced view [11]. Yes, rhodanines will likely show up more often in HTS screens, but do we really want to remove a moiety from the screening deck that could be useful in a chemical biology sense? Rhodanine screening, as an example, illustrates an important conflict between longer term societal and scientific interests on the one hand and the shorter term interests of academic chemical biologists on the other.

Chemistry space is not uniformly occupied by biologically active compounds. Rather, biologically active compounds occur in small tight clusters infrequently through chemistry space. The similarities in the biology and chemistry worlds can be quite surprising. In both biology and chemistry, the targets and their ligands occupy, respectively, an infinitesimally tiny portion of the possibly available target or chemistry space. Evolution prunes down the choices on the target side with a corresponding pruning down of ligand choices on the chemistry side. One implication of this is that screening a truly diverse library is the worst possible way to discover a drug. Both the universe of biology targets and the universe of their chemical ligands exist as small world networks. Both the targets and their chemistry ligands are tightly clustered rather than diverse, and interestingly, the networks of targets and ligands do not overlap particularly well. Target motifs tend to be reused across different types of biological activity, and chemistry motifs tend to reoccur in ligands across different types of biological activity. For example, there are distinct structural preferences for fragments that can hydrogen-bond to hydrogen bonding amino acid side in proteins [12]. The net effect is that similar ligand structural motifs tend to recur across targets that in a gross biological sense have no similarity. The medicinal chemistry implication is that it is very important to understand the history of any lead structure. What has been tried before in terms of structural modification might just work again although almost certainly not in any truly predictive sense in a new biology situation.

I have so far described what is a scientific scenario. Now, what happens when this scenario plays out in an academic setting? I will use rhodanine screening as an example. The academic biologist's professional survival literally depends in the United States on NIH R01 grants. This means peer-reviewed biology papers. With a few exceptions, there is no medicinal chemistry quality refereeing in the biology journal world. Medicinal chemistry quality considerations are unwelcome if they hinder getting the publication. The vendor community provides screening compounds that they hope will sell. A compound structural class containing functionality capable of strong interactions with protein target functionality is more likely to provide a hit in a screen. There is a long history of compounds like this. If I wanted a library capable of giving hits in most any HTS screen, I would test sulfonic acid dyes. There is a long history on "warhead groups," replace the scissile amide bond in a peptide substrate with an alpha halo ketone, or aldehyde or epoxide, or boronic acid. The common feature of these warheads is their ability to strongly interact with an enzyme nucleophile such as a cysteine sulfur or serine hydroxyl or, in the case of a boronic acid, with a cis diol-like functional array. Enough time has passed for even the most chemistry-phobic biologist to become sensitized to medicinal chemistry quality complaints on the extant warheads. In the current era, we replace well-known problematic functionality with newer and more subtle problem compounds. Long laundry lists of such problematic chemistry functionality are in the current medicinal chemistry literature. This does not matter to the publication-centered, R01-seeking academic biologist. Medicinal chemistry quality concerns pale when the choice is publish or (literally professionally) perish. This is why it is so important to "ring-fence" medicinal chemistry in an academic drug discovery setting. Without protection, medicinal chemistry is all too likely to be subverted to the goals of flawed chemistry, leading to an easier pathway to publication. It is really telling that in industry, many publication committees in an industrial drug discovery company will not approve publication until Phase 2 proof of concept. People will respond to organizational drivers. If progress toward drug discovery is rewarded, that is what you will get. If publications are rewarded, that is what you will get.

Hypothesis-driven research is a big concern in the future of collaborative drug discovery. When a researcher is driven to publish and the biology research is driven by a hypothesis, then guess what happens? There is a high probability that experiments will support the hypothesis whether the hypothesis is correct or not. Hypothesis-driven research is aided in the pathway to bias and error as I have discussed previously by promiscuous chemistry. The outcome is as reported by workers at Bayer Health Care [13], commented on in *NRDD* [14] and supported by a VC commentary [15]. Fifty percent of academic target identification is wrong, and another 25% is flawed in some way. The general tendency of research toward bias and error is well documented [16, 17] by Stanford's John Ioannidis and has even made it to the popular press [18]. In the clinical trials realm, methodology has been developed to as best as possible control for this unfortunate tendency. Even in the clinic, though, it is not that uncommon for preliminary positive Phase 2 trials to not hold up in larger Phase 3 trials [19]. How does industry compare with academia with respect to bias and error? We do know that in industrial drug discovery figures, of one in four or of one in five projects surviving through to end of preclinical trials are quoted. It is hard to get an exact fix on preclinical attrition because preclinical drug discovery can be run in rich or lean mode. Rich mode applies to both dollar cost and quantity and quality of data. Lean is the opposite. As you might expect, rich mode results in earlier and thus higher preclinical attrition. Lean is the opposite. Rich mode might cost 20 million USD to get a compound from earliest biology idea to IND stage. Lean mode might be closer to 4 million dollars. The pros and cons of the two approaches have been hotly debated in major pharma. For example, through most of my career, Pfizer ran in rich mode but heading into the last decade was increasingly exploring lean mode. There are far more handoffs and greater breadth of discipline expertise in traditional industrial versus traditional academic drug discovery. My guess is that exposure to a breadth of disciplines as found in traditional industrial drug discovery is more likely to correct for bias and error than a narrower discipline breadth, but this is only a guess. To my knowledge, there is no hard data on how to compare bias and error across industrial and academic drug discovery groups.

Both industry and academia have contributed to error in drug discovery. Sequencing the genome enhances drug discovery? I remember the collaboration between Bayer and Millenium in the early 2000 time period. Hundreds of genomically based drug discovery screens were run; nothing positive came out of this screening effort. It is hard to believe that just a little more than a decade ago there were robots spitting out DNA sequences for electronic transfer to the U.S. Patent and Trade Office, and people really believed that the drug discovery organizations would run out of intellectual property opportunities in drug targets. There were collaborations to mine genomic targets; massive HTS campaigns to discover ligands; and discussions on 500 different targets, a million data points, and "a wish to screen 100,000 compounds per day in a drug discovery factory and a wish to make a drug for each target" [20, 21]. "The DNA reductionist viewpoint of the molecular genetics community has set drug discovery back by 10–15 years" is a quote attributed to Craig Venter. "In 1992–1997 if you had stored combinatorial chemistry libraries in giant garbage dumpsters you would have much improved drug discovery productivity" is a Chris Lipinski quote that appeared in *Chemical & Engineering News*, among other places.

Innovation, and the thinking out of the box mind-set, is one area that is ideally suited for academia. The evolutionary warfare between bacteria, fungi, plants, and their predators has likely been playing out since the origins of life. For the simpler organisms, the playing field is chemistry. As predator and prey adjust to competition with each other, increasingly more complex targets pathways and signaling networks are probed. As discussed by Schreiber, there is a tendency for natural products to target increasingly more complex network nodes, which may not be so good unless one is after a cytotoxic effect. After all, in evolutionary warfare, the organism is not interested in selectivity and does not care about mechanism but rather cares about the outcome. Having said this, natural products are the largest test on what the possibilities are for biological activity in chemistry space. If something can be done on a certain target, then natural products provide the best test. However, there is a caveat. The buildup of natural products uses a set of component chemistries very unlike those used in normal synthetic medicinal chemistry.

It is difficult to quantitate the difference in the overall chemical space distribution between the natural product and synthetics classes. For example, it is much more difficult to classify natural products by chemical structural descriptors than is the case for synthetic compounds. The chemistry scaffold classification, which works so well for synthetics to explore scaffold-related polypharmacology [22], has not been described in a systematic manner for real natural products. Natural products may contain ring structures as a consequence of covalent chemistry but can also form ring structures or internal conformational constraints as a consequence of intramolecular hydrogen bonding or salt bridge formation. The possibility of ring formation is important because of the clear demonstration that ring structures are inherently richer in biological content than their acyclic counterparts [23]. Our current inability on a large scale to reliably predict energetically preferred conformations and hydrogen-bonded ring structures in complex natural product structures prevents the types of analyses that have been performed on synthetic compounds. Another issue relates to the confounding effects of something as simple as molecular weight (MWT). Compound collections rich in sp3 carbon centers (natural products) have different protein target profiles than synthetics [24]. However, complex natural products also have higher MWT. Chemistry possibilities expand exponentially as MWT increases, and this greatly confuses the analysis of whether natural products cluster or are diverse in chemistry space.

CONCLUSION

I have a wish in chemical biology and drug discovery for something that is most likely to arise from a collaborative effort and that is most likely to be academically oriented. We need a dictionary or atlas of three-dimensional natural product structures annotated as to biological activity. We are progressing along the path to something like this in the small molecule synthetic medicinal chemistry world. The herculean task in the natural products world is our opportunity to learn from more evolutionary experiments than we with our current technology will ever be able to entertain.

REFERENCES

1. Lipinski, C.A. The anti-intellectual effects of intellectual property. *Curr Opin Chem Biol.* 2006;10(4):380–383.
2. Wilson, E.K. Is safe exchange of data possible? *Chem Eng News* 2005;83(17):24–29.
3. Bradley, D. Share and share alike. *Nat Rev Drug Discov.* 2005;4:180.
4. Marusina, K., Welsch, D.J., Rose, L., Brock, D., Bahr, N. The CTSA pharmaceutical assets portal—a public-private partnership model for drug repositioning. *Drug Discov Today Ther Strateg.* 2011;8(3–4):77–83.
5. Hughes, B. A Wellcome experiment in seeding drug discovery. *Nat Rev Drug Discov.* 2010;9: 178–180.
6. Houlton, S. Wellcome boost for open-access chemistry. *Nat Rev Drug Discov.* 2008;7: 789–790.
7. Oprea, T.I., Bologa, C.G., Boyer, S., Curpan, R.F., Glen, R.C., Hopkins, A.L., Lipinski, C.A., Marshall, G.R., Martin, Y.C., Ostopovici-Halip, L., Rishton, G., Ursu, O., Vaz, R.J., Waller, C., Waldmann, H., Sklar, L.A. A crowdsourcing evaluation of the NIH chemical probes. *Nat Chem Biol.* 2009;5(7):441–447.
8. Bruestle, M., Beck, B., Schindler, T., King, W., Mitchell, T., Clark, T. Descriptors, physical properties, and drug-likeness. *J Med Chem.* 2002;45(16):3345–3355.

9. Baell, J.B., Holloway, G.A. New substructure filters for removal of pan assay interference compounds (PAINS) from screening libraries and for their exclusion in bioassays. *J Med Chem.* 2010;53(7):2719–2740.
10. Baell, J.B. Observations on screening-based research and some concerning trends in the literature. *Future Med Chem.* 2010;2(10):1529–1546.
11. Mendgen, T., Steuer, C., Klein, C.D.P. Privileged scaffolds or promiscuous binders—a comparative study on rhodanines and related heterocycles in medicinal chemistry. *J Med Chem.* 2012;55(2):743–753.
12. Chan, A.W.E., Laskowski, R.A., Selwood, D.L. Chemical fragments that hydrogen bond to Asp, Glu, Arg, and his side chains in protein binding sites. *J Med Chem.* 2010;53(8): 3086–3094.
13. Prinz, F., Schlange, T., Asadullah, K. Believe it or not: how much can we rely on published data on potential drug targets? *Nat Rev Drug Discov.* 2011;10(9):712–713.
14. Mullard, A. Reliability of "new drug target" claims called into question. *Nat Rev Drug Discov.* 2011;10:643–644.
15. Booth, B. Academic bias & biotech failures. 2011. Available at http://blogs.nature.com/tradesecrets/2011/04/22/academic-bias-and-biotech-failures. Accessed December 21, 2013.
16. Ioannidis, J.P. Molecular evidence-based medicine: evolution and integration of information in the genomic era. *Eur J Clin Invest.* 2007;37(5):340–349.
17. Ioannidis, J.P. An epidemic of false claims. Competition and conflicts of interest distort too many medical findings. *Sci Am.* 2011;304(6):16.
18. Freedman, D.H. Lies, damned lies, and medical science. *The Atlantic* November 2011. Available at http://www.theatlantic.com/magazine/archive/2010/11/lies-damned-lies-and-medical-science/8269/. Accessed December 21, 2013.
19. Ioannidis, J.P. Contradicted and initially stronger effects in highly cited clinical research. *JAMA.* 2005;294(2):218–228.
20. Sehgal, A. Drug discovery and development using chemical genomics. *Curr Opin Drug Discov Devel.* 2002;5(4):526–531.
21. Archer, R. The drug discovery factory: an inevitable evolutionary consequence of high throughput parallel processing. *Nat Biotechnol.* 1999;17(9):834.
22. Hu, Y., Bajorath, J. Polypharmacology directed compound data mining: identification of promiscuous chemotypes with different activity profiles and comparison to approved drugs. *J Chem Inf Model.* 2010;50(12):2112–2118.
23. Kim, Y.K., Arai, M.A., Arai, T., Lamenzo, J.O., Dean, E.F., III, Patterson, N., Clemons, P.A., Schreiber, S.L. Relationship of stereochemical and skeletal diversity of small molecules to cellular measurement space. *J Am Chem Soc.* 2004;126(45):14740–14745.
24. Clemons, P.A., Bodycombe, N.E., Carrinski, H.A., Wilson, J.A., Shamji, A.F., Wagner, B.K., Koehler, A.N., Schreiber, S.L. Small molecules of different origins have distinct distributions of structural complexity that correlate with protein-binding profiles. *Proc Natl Acad Sci U S A.* 2010;107(44):18787–18792.

Part II

GOVERNMENTAL INITIATIVES ACCELERATE PRECOMPETITIVE COLLABORATION

Governments will always play a huge part in solving big problems. . . . They also fund basic research, which is a crucial component of the innovation that improves life for everyone. —Bill Gates

Collaborative Innovation in Drug Discovery: Strategies for Public and Private Partnerships, First Edition.
Edited by Rathnam Chaguturu.

7

THE VALUE OF UNIVERSITY–INDUSTRY PARTNERSHIPS

Anthony M. Boccanfuso

The National Academies University–Industry Demonstration Partnership, Washington, DC, USA

Leaders in the public and private sectors have long recognized that strategic university–industry (U-I) research collaborations are critical to advancing a nation's commercialization portfolio and its overall economic competitiveness and well-being. According to a 2012 report from Battelle and *R&D Magazine*, U.S. corporate investments in research exceed all other sectors, with industrial funds for research and development (R&D) estimated to be \$262 billion in 2013, an increase of 2.3% over their 2012 estimate [1]. The majority of these funds is directed toward applied R&D and constitutes approximately 62% of all U.S. R&D funding. Industry's R&D investments for academic R&D declined steeply after the 1990s (above 7% in 1999 down to about 5% by 2004), but has seen a 5-year increase to approximately 6% in 2009 [2]. Recent trends, especially pronounced in the life sciences, indicate a changing approach to commercialization activities where large innovation firms are looking to make more efficient use of R&D investments by changing their R&D model; these changes include reducing overall R&D spending, outsourcing basic research (discovery) to universities and other research organizations such as nonprofit institutions and small businesses, and focusing by limiting the number of research collaborators to a smaller set of strategic partners [3].

Collaborative Innovation in Drug Discovery: Strategies for Public and Private Partnerships, First Edition. Edited by Rathnam Chaguturu.

While few challenge this hypothesis that the idea-to-innovation process is bolstered by healthy U-I relations, the type of strategic and high value U-I collaborations are difficult to implement and are negatively impacted by issues related to grant and contract conditions (both financial and nonfinancial), the treatment of intellectual property (IP), and compliance issues such as export control and conflict of interest [4].

GUIDING PRINCIPLES FOR UNIVERSITY–INDUSTRY COLLABORATIONS

A 2006 publication developed by leaders of the nascent University-Industry Demonstration Partnership (UIDP) movement outlined a series of Guiding Principles to support healthy and sustainable U-I partnerships [5].

Guiding Principle # 1: **A successful U-I collaboration should support the mission of each partner. Any effort in conflict with the mission of either partner will ultimately fail.**

This seems obvious, but this concept is sometimes forgotten when universities and companies seek to collaborate on a project. While a conflict in mission can exist for a single project, this situation is not viable for sustainable, long-term collaborations.

Guiding Principle # 2: **Institutional practices and national resources should focus on fostering appropriate long-term partnerships between universities and industry.**

To maximize the benefits to each party and reduce the amount of time devoted to non-value-added activities, more companies are identifying a smaller set of institutions to develop deeper and more rewarding collaborations. Alternatively, this means that more institutions will be seeking to become a company's "strategic partner" and gain access to more resources and benefits; many institutions will not be anointed as a core corporate partner.

Guiding Principle #3: **Universities and industry should focus on the benefits to each party that will result from collaborations by streamlining negotiations to ensure timely conduct of the research and the development of the research findings.**

The time and energy devoted to sponsored research agreement negotiations can lead to significant frustrations for researchers and administrators alike. Companies and universities have different perspectives on the desired outcomes from a research project as well as differing missions, and these factors can lead to protracted negotiations. For example, Dow Chemical has released data that show that the negotiation of specific contract terms was the most time-intensive aspect of coming to agreement. In their review of Dow agreements, it took over 5 months, on average, to reach a formal agreement (Figure 7.1).

THE JOHNSON CONTINUUM

While at Hewlett-Packard (HP), Wayne Johnson introduced the Johnson Continuum [6] that describes the myriad of ways in which HP engaged universities (Figure 7.2).

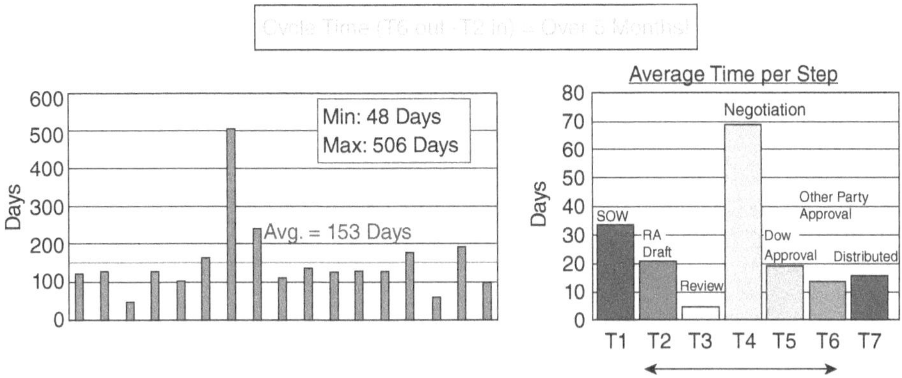

Figure 7.1. Dow Six Sigma study of cycle time for sponsored research agreements in the United States. Courtesy of Susan Butts, Dow Chemical Co. See color insert.

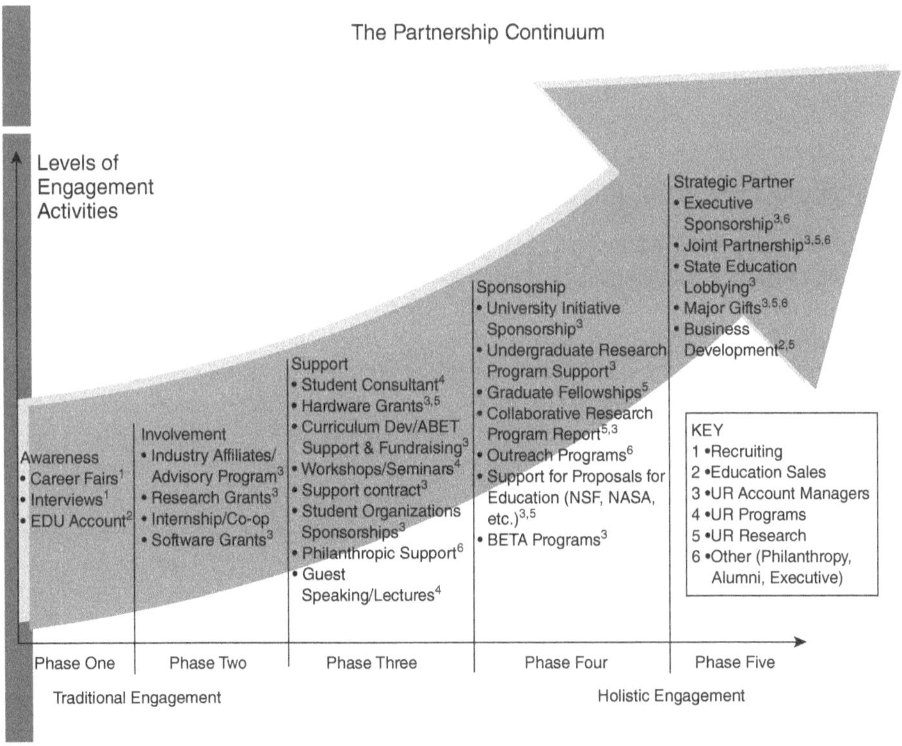

Figure 7.2. The Johnson Continuum.

This continuum, while developed for one company, has broad applicability to others and can be used by other companies to advance their respective missions. Frequently, universities seek to quickly form a "strategic collaboration" with a company in the hopes of earning significant external funding from the industry partner. Several well-publicized examples have resulted in multimillion dollar arrangements; these high investment relationships are appealing to universities but are typically the result of foundational work that has taken place over a number of years and at much more modest levels of support where the academic institution has demonstrated its capabilities to meet or exceed a company's expectations and provide tangible benefit to the corporate partner.

ROLE OF COCREATION: COMPANIES, GOVERNMENT, AND UNIVERSITIES

Recognizing the value and potential impact of strategic U-I research engagement, governments—both at the state and federal levels—have been increasing their support for these partnerships. Some states (such as California and Maryland) have had long-standing programs to promote these collaborations with the idea that this investment of resources can bolster job creation through company relocation, retention, or expansion.

Maryland Industrial Partnerships (MIPS) Program

Maryland Industrial Partnerships (MIPS) promotes the development and commercialization of products and processes through U-I research partnerships [7]. MIPS provides matching funds to help Maryland companies pay for the university research. Projects are initiated by the companies to meet their own research and development goals.

Through MIPS, Maryland firms have the opportunity to leverage their research and development funds and gain access to the creative talents and extensive research base of the University System of Maryland. MIPS matching funds are awarded on a competitive basis for projects based on proposals submitted jointly by Maryland companies and researchers from any of the 13 University System institutions.

The maximum MIPS award for any single project is $100,000 per year for large and small companies and $90,000 for start-up firms. According to the program's website,

> With more than 400 Maryland companies participating in project awards since 1987, worth over $180 million-MIPS projects create results. MIPS-supported products have generated more than $25.2 billion in sales, added jobs to the region, and infused state-of-the-art technology into the global marketplace.*

*http://www.mips.umd.edu/.

FUNDING FOR COLLABORATIONS WITH ACADEMIA

Unlike other countries, the United States doesn't have a formal industrial policy. Direct, financial support for collaborations between academia and large companies is typically limited to funding the university side of the collaboration. Among the existing programs supporting these collaborations are

- Department of Commerce i6 Challenge
- National Institutes of Health Clinical and Translational Science Award (CTSA) program
- National Science Foundation's Grant Opportunities for Academic Liaison with Industry, Innovation Corps (I-Corps), and Partnership for Innovation (PFI)
- Small business–academic interactions greatly benefit from the Small Business Innovation Research (SBIR) and the Small Business Technology Transfer (STTR) programs, which directly fund research projects from designated small firms and universities.

A 2013 study [8] has examined how companies benefit from these interactions, and found that small- and medium-sized firms receive direct R&D value from these partnerships and that larger firms are more likely to receive indirect benefits (rent spillovers).

NEW INITIATIVES TO TRANSFORM UNIVERSITY–INDUSTRY COLLABORATIONS

Intellectual property (IP) rights continue to be the major stumbling block, although only a small percentage of research projects produce any IP that generates revenue.

Kauffman Foundation Free Agent Model

In the Kauffman Foundation's "free agent" model, researchers are given the power to choose a third party (or themselves) to negotiate license arrangements for entrepreneurial activities, provided that they return some portion of their profits to the university [9]. The researcher's home institution technology transfer office can be one of the third parties offering services, but other parties can also compete on a range of services and experience offered.

Since being proposed, this proposal has met with stiff resistance in the academic community; however, other models that incorporate UIDP values and were created by UIDP academic members have been implemented.

Penn State Keystone Edge

Under Pennsylvania State University's Keystone Edge program [10], the university no longer mandates ownership of IP associated with industry-funded research. With the adoption of this plan, Penn State is adopting an alternative approach to industry-sponsored research, one that argues that the real value is not in IP ownership, but rather in the contact students and faculty have with real problems in the world.

Industry-Sponsored Research: Minnesota-Innovation Partnerships (MN-IP)

Like Penn State's Keystone Edge program, MN-IP utilizes a new and unique approach to the way the university handles IP arising from research projects funded by business and industry partners. This approach eliminates the need for protracted negotiations over rights to IP that may result from industry-funded research.

Companies sponsoring research at the university can pre-pay a fee and receive an exclusive worldwide license, with royalties taking effect only in cases of significant commercial success. Agreements are available immediately to industry research sponsors.

Basic Information

- Involves prepaid exclusive option fee amounting to 10% of sponsored research contract or $15,000, whichever is greater
- Includes option to exclusive license with preset terms: no annual minimums or other fees; no time limits or milestones; sponsor is free to sublicense/cross-license technology; if annual sales involving licensed IP exceed $20 million, licensee pays 1% royalty fee; no cap on royalties unless invention improves on sponsor's preexisting product or processes
- Sponsor pays patent costs and has the benefit of driving prosecution while collaborating with the university on patent claims.

ISSUES AFFECTING UNIVERSITY–INDUSTRY COLLABORATIONS

While U-I collaborations are generally supported, there are a number of issues that have a profound impact on these partnerships, and a few of these items are highlighted.

Conflict of Interest

There have been a number of well-publicized cases where academic investigators did not sufficiently disclose their relationships with industry or operated outside of their institution's conflict of interest policies and procedures. Most of these cases were in the life sciences arena, and many of these transgressions were deliberate actions (for financial or nonfinancial reasons) taken by researchers. To minimize the likelihood of future problems, a strategic set of federal agencies have adopted new regulations to "manage" the situation. Concern has been raised in numerous quarters that these recent efforts are damaging the relationship among investigators, academic institutions, companies, and federal agencies, and harming the nation, the research enterprise, and the nation's ability to compete within the global marketplace.

A recent Institute of Medicine report [11] addressed this topic and made a strategic set of recommendations for addressing the challenges posed in addressing this issue.

Export Control

Information and technology that is not in the public domain or created with the intent to put it in the public domain may be subject to restrictions on export to foreign persons, including those in the United States. In U-I research collaborations, projects can produce technology developments with findings that will not be immediately published but will be protected for a time so that they can be commercialized for public benefit. In some cases, either party may provide the other party with export-controlled background IP or proprietary information for use in the project. Such information is not in the public domain and is generally protected by a nondisclosure agreement. Finally, some projects are undertaken by U-I collaborations with sponsorship from the U.S. government in which there are limited restrictions on publications. Universities are inherently open in how they pursue scholarly activities; companies seek to ensure that they maintain their competitive advantage, and keep things from reaching the public domain until such time as it is warranted.

DIFFERENCES BETWEEN LARGE- AND SMALL-/MEDIUM-SIZED COMPANIES

While we commonly refer to U-I relations, it would be more accurate to divide the industry category into two parts: large companies and small-/medium-sized companies.

LARGE COMPANIES SHARE CERTAIN CHARACTERISTICS

These companies invest in innovation and may spend some resources on basic research/discovery. Large companies (such as Boeing, Hewlett-Packard, Novo Nordisk, and Pfizer) invest significant resources on the discovery to innovation process to ensure that they stay competitive. Most of these funds are spent internally and on product development.

External funding represents a small share of a company's R&D investment. A very modest percentage of a firm's R&D is devoted to external research at universities, national labs, or other research organizations. While there are no publicly available resources that document what percentage of a company's R&D funding is spent externally, a representative set of UIDP companies have stated that 1–3% of all research dollars are spent outside their firms.

Open Innovation. A term coined by Henry Chesbrough, open innovation has been adopted by many large companies as a way to extract more from their research investments [12]. Through its Connect and Develop program, a majority of Proctor and Gamble's product initiatives involve significant collaboration with outside innovators.

SMALL- AND MEDIUM-SIZED ENTERPRISES (SMES) OFFER A CONTRASTING PERSPECTIVE

Most SMEs Are Not Making Major Investments in R&D

Some SMEs are research focused but most are not. When they seek to partner with universities, it is typically due to a short-term need. They also have limited financial resources to provide universities. With the increasing emphasis on the role universities play in advancing economic development in their communities, some schools will "subsidize" work needed by SMEs; this is especially true when this work is done in conjunction with a university-affiliated incubator program.

SMEs May Have a Single Technology Platform That Is the Basis of the Company

When this is the case, companies are keenly interested in protecting their IP and ensuring that any university interactions do not inhibit their ability to commercialize the company's technology.

SMEs Have Limited Administrative Support/Infrastructure

SMEs are lean organizations that don't possess a deep stable of administrators or staff to negotiate agreements with universities. Many times, these firms have never worked with a university until they have a unique need that is best addressed by a university. Some government programs also encourage U-I collaborations; when SMEs pursue and receive federal or state funds, these firms may interact with a university for the first time and have a significant learning curve in negotiating agreements with universities.

THE NATIONAL ACADEMIES' UNIVERSITY–INDUSTRY DEMONSTRATION PARTNERSHIP

To address many of these issues affecting U-I collaborations, the UIDP was created in 2006. The UIDP is a membership organization that supports mutually beneficial U.S. U-I collaborations, encouraging U.S. competitiveness by developing and disseminating strategies for addressing common issues between the two sectors.

UIDP Operational Model: Projects and Demonstrations

Demonstration Project: Develops, Tests, or Demonstrates New Technology or Approaches in Fulfillment of the UIDP Mission. As an organization committed to developing, identifying, and testing new approaches to strategic U-I relationships, the UIDP performs projects related to the processes used to advance these collaborations. These projects are typically identified by members as addressing issues that can either cause problems for high-value relationships or be resolved by new approaches that can catalyze and bolster U-I interactions.

These projects can be short- or long-term in duration and are typically staffed and conducted by volunteers from the membership and the broader academic and corporate communities. Results or products from these projects are first shared with the UIDP Board, the general membership, and then the public. These projects are meant to provide guidance, insights, and strategies for making sure that U-I research relationships are able to operate in a high-performance environment not inhibited by extraneous matters or other non-value-added activities. UIDP project products range from print materials to online services.

Since its inception, the UIDP has been at the forefront of producing a number of high-value products, including

1. *Guiding Principles for U-I Endeavors.* These Guiding Principles emphasize the long-term returns that can accrue to U-I partnerships that are constructed fairly and conducted with competence and sophistication.
2. *Living Studies Compendium* [13]. This compendium describes in concrete terms how the Guiding Principles apply to day-to-day IP negotiations.
3. *Contract Accords for U-I Sponsored Research Agreements* [14]. These accords provide an in-depth understanding of the various clauses found in research agreements, the rationale for these terms, and ways to address the concerns of the academic and corporate communities.
4. *Researcher Guidebook: A Guide for Successful Institutional-Industrial Collaborations* [15]. This guidebook specifically targets active researchers (at all career stages) in institutions and industry, and is the product of the collective expertise of knowledgeable member representatives. It presents the perspective of a senior researcher, that is, one with extensive experience in this arena, whose insights may benefit a researcher who may have little experience in forging and maintaining institutional–industrial collaborative relationships.

STRATEGIES FOR BUILDING HIGH-VALUE, HIGH RETURN U-I PARTNERSHIPS

The Guiding Principles offer strategies on how to build these relationships. This section considers specific approaches:

- *Building These Relationships Requires Champions.* Academic institutions and companies are things and can't build or sustain relationships. U-I collaborations are only possible when individuals are committed to their success. More universities have established "one-stop shops" for industry engagement; the Massachusetts Institute of Technology Industrial Liaison Program is well known and long-standing; more recent efforts include Ohio State's Industry Liaison Office, launched in 2008. These organizations are successful when the individuals that support them have the needed expertise and skills to develop strong collaborations. Industry has also moved in a similar direction, although much of industry's efforts are human resource–based.

- *High Value Relationships Can Be Realized Only after the Needed Investment of Time and Energy Has Been Made.* The Partnership Continuum describes the various phases of U-I collaborations, and while many individuals want to leapfrog to the higher, more integrated forms of collaborations, very few if any, can do this without going through the more transactional phases.
- *Know Thyself.* Universities will argue that they are world class in many research arenas, while in reality this is not the case. Companies appreciate it when universities have a clear understanding of their skills and assets. Many times, companies do not need to engage the world's premier academic researcher for a specific project. Sometimes, a company seeks to meet a modest but urgent need that can be handled by trusted partners.
- *Regionality Does Matter.* Companies continually seek out talent globally, but for many activities, companies prefer to work with academic institutions that are close to their R&D facilities and manufacturing plants. Workforce needs are typically met by local schools, and companies benefit from building linkages with these institutions.

CONCLUSION

High-value, high-return U-I collaborations, in whatever form they are pursued, can provide benefits to both parties as well as the regions and states where they are located. Because of these benefits and the evolving nature of how R&D is conducted in the United States, more companies are creating and implementing strategic approaches to academic engagement and devoting more resources (financial and nonfinancial) to identifying partners that can help them achieve their business objectives.

NOTE ON THIS CHAPTER

Anthony Boccanfuso is director of the University Industry Demonstration Partnership (UIDP) of the National Academies' Government–University–Industry Research Roundtable (GUIRR). The opinions expressed here are those of the author and do not necessarily reflect the views of the UIDP, GUIRR, or the National Academies.

REFERENCES

1. Grueber, M., Studt, T. The uncertain state of U.S. R&D. *R&D Magazine* December 2012. Available at http://www.rdmag.com/articles/2012/12/uncertain-state-us-r-d. Accessed December 21, 2013.

2. National Science Board. 2012. *Science and Engineering Indicators 2012*. NSB 12-01. Arlington, VA: National Science Foundation.
3. Goldstein, J.. Merck exec on R&D: "Nothing is safe from cuts." 2009. Available at http://blogs.wsj.com/health/2009/04/15/merck-exec-on-rd-nothing-is-safe-from-cuts/. Accessed December 21, 2013.
4. Boccanfuso, A.M. Why university-industry partnerships matter. *Sci Transl Med*. 2010; 2(51):51cm25.
5. Guiding Principles for University-Industry Endeavors. Washington, DC, National Council of University Research Administrators; 2006.
6. Southerton, J., Umberger, G., Matijasevic, G., Steele, S., Johnson, W. *Partnership Continuum*. Atlanta, GA: Georgia Technology Research Corporation; 2012.
7. Maryland Industrial Partnerships. Additional information Available at http://www.mtech.umd.edu/mips/index.php. Accessed February 10, 2014.
8. Nishimura, J., Okamuro, H. Knowledge and rent spillovers through government sponsored R&D consortia. Hitotsubashi University Repository; 2013.
9. Litan, R., Mitchell, L., Reedy, E. The university as innovator: bumps in the road. *Issues Sci Technol*. 2007;23(4):57–66.
10. Foley, H.C. A new approach to intellectual property management and industrially funded research at Penn State. *Research Technology Management*, September 2012.
11. Lo, B., Field, M., eds. *Conflict of Interest in Medical Research, Education, and Practice*. Washington, DC: Institute of Medicine; 2009.
12. Chesbrough, H.W. *Open Innovation: The New Imperative for Creating and Profiting from Technology*. Boston: Harvard Business School Press; 2003.
13. Casey, J., Jr., Kramer, B. *Living Studies in University-Industry Negotiations*. Washington, DC: National Council of University Research Administrators; 2006.
14. *Contract Accords for University-Industry Sponsored Research Agreements*. Atlanta, GA: Georgia Technology Research Corporation; 2012.
15. *Researcher Guidebook: A Guide for Successful Institutional-Industrial Collaborations*. Atlanta, GA: Georgia Technology Research Corporation; 2012.

8

TRENDS IN THE PUBLIC SECTOR ADOPTION OF TRANSLATIONAL RESEARCH APPROACHES

Mark A. Scheideler*

HumanFirst Therapeutics LLC, Silver Spring, MD, USA

The past decade has witnessed strong growth in the adoption and funding of translational research strategies by the public sector [1]. This has been driven by an increasing awareness within the scientific community of the need to engage in the high-risk and consortium-driven steps required to move basic science discoveries toward the clinic. Early drug discovery activities, such as target discovery and pharmacological validation, are increasingly understood to be precompetitive, community-based efforts in which information sharing can accelerate the formation of collaborative alliances between the complementary capabilities of biology, chemistry, and informatics. Chemical biology approaches, in particular, have attained new prominence as a vital enabler of novel pharmacological investigation. These project-based team efforts allow, for example, preclinical testing and validation of a new therapeutic approach. This leverages investment in investigator-initiated research, and motivates directed preclinical development activities that aim to progress a therapy into the clinic. A model for the

*Mark Scheideler, PhD, was a Senior Scientific Officer at the National Institute of Neurological Disorders and Stroke of the National Institutes of Health USA, from 2005 to 2011. He had responsibility for forming and directing translational research initiatives discussed in this chapter.

Collaborative Innovation in Drug Discovery: Strategies for Public and Private Partnerships, First Edition. Edited by Rathnam Chaguturu.

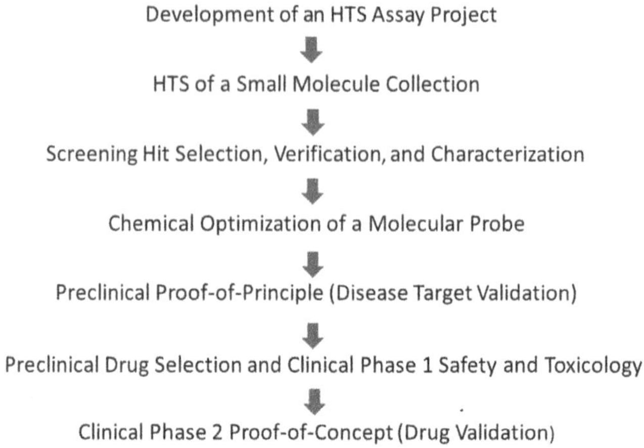

Figure 8.1. Path of activities leading from lab bench to clinical testing.

path of these activities, leading from lab bench to clinic, is shown in Figure 8.1. A more elaborate academic view of translational research process is described by Trochim et al. [2]. Both the National Institutes of Health (NIH) and nonprofit disease foundations are stakeholders in this process and have made efforts to support engagement of the scientific community in translational research.

Particular note will be made of best practices learned from translational research initiatives launched by the NIH. The NIH Molecular Libraries Initiative has been an early driver of work to ascribe function to new genomic information by building a compendium of molecular probes (pharmacological tools) that could be used by the scientific community to better understand and validate possible disease targets, pathways, and phenotypes. The molecular probes resulting from the chemical biology effort are beginning to drive preclinical proof-of-principle (POP) work aimed at validating strategies to therapeutically alter processes that lead to disease initiation or progression [3]. They can also be employed to identify similarities in the molecular pathology of diseases for which there was not previously a direct association. At the other end of the spectrum lies the NIH Clinical Translational Science Awards (CTSA) initiative,* a collaborative consortium of 60 medical research institutes with efforts centered on early investigative clinical trials, both in terms of clinical methods development and training, comparative effectiveness research, and human proof-of-concept (POC) testing of new therapeutic approaches [4–6].

The difficulty in connecting research success to clinical testing is often voiced as a prominent concern of the medical research community [7, 8]. Whereas a number of medical research institutes have robust and innovative research activity (supported primarily by NIH funding) and large clinical trials enterprises, it is rare for research discoveries to progress from bench to bedside within the walls of an institution. Key factors include funding, mentorship, and training, and the need to partner with external

*Available at https://www.ctsacentral.org/.

organizations (e.g., pharmaceutical companies, disease foundations), which have competencies that are generally missing from the academic environment. This chapter will examine several public efforts aimed at facilitating the migration of research innovation from bench-to-bedside (and POP-2-POC). This path typically requires a handoff of investigator-initiated activities at the lab bench to large-scale consortium efforts (e.g., high-throughput screening [HTS]), before returning to activities such as molecular probe optimization and preclinical experimentation that are executed by individual investigator teams or comparatively small collaborating teams.

The discussion will center on the first step in many translational efforts: the adaptation by investigators of lab-based assays for novel biological targets, pathways, and cellular phenotypes as a means of initiating a subsequent chemical biology effort to design molecular probes for use in investigative POP research. Initiatives to perform large-scale HTS of small molecule compound libraries using these assays, by the NIH Molecular Libraries Centers and by other Centers, are discussed elsewhere [9, 10]. A model NIH program to facilitate investigator-driven optimization of chemical leads coming from HTS will also be covered to illustrate unique issues encountered in assembling small multidisciplinary teams in an effective manner. Finally, mention will be made of a CTSA Public–Private Partnership efforts to identify, aggregate, and partner tools and know-how comprising a POP-2-POC toolkit for therapy development projects, via the open-access web resource CTSA•IP.*

STEPS TO SUCCESSFUL ASSAY DEVELOPMENT FOR HTS

The NIH Molecular Libraries Assay Development for HTS Initiative began with the purpose of funding independent investigators to miniaturize and characterize biological assays that could be used in small molecule screening campaigns by the pilot phase Molecular Libraries Screening Center Network (MLSCN). The data would then be posted publically in *PubChem*.[†] The Initiative quickly expanded to encompass a broader scope and mandate of assembling complete assay projects that could be used for HTS and probe development in the Molecular Libraries Probe Production Centers Network (MLPCN).[‡] In this section of the review, a Best Practices in Assay Development is discussed that has been developed and used to guide investigators seeking to prepare assay development proposals and forge working relationships with Molecular Libraries Centers. Many of these details are embedded in prior NIH Program Announcements for the Assay Development for HTS program,[§] and in the Fast Track program guidance for the entry of completed HTS projects into the MLPCN.[¶]

*CTSA•IP: building discoveries and medical practice. Available at http://www.ctsaip.org/.

[†]Accessing MLSCN Data in PubChem. Available at http://mli.nih.gov/mli/secondary-menu/mlscn/pubchem/accessing-mlscn-data-in-pubchem/.

[‡]MLPCN Centers. Available at http://mli.nih.gov/mli/mlpcn/mlpcn/.

[§]Assay Development for High Throughput Molecular Screening (R21). Available at http://grants.nih.gov/grants/guide/pa-files/par-10-182.html.

[¶]Notice of Opportunity for Fast Track Entry of Assay Projects for High Throughput Screening into the NIH Roadmap Molecular Libraries Probe Production Centers Network. Available at http://grants.nih.gov/grants/guide/notice-files/NOT-RM-11-018.html.

The Assay Development for HTS program worked to adapt assays from the research programs of investigators for use in HTS and molecular probe design. Resources were provided to configure and characterize assays with the goal of their entering screening in the MLSCN (and MLPCN) within 12–18 months. A particular emphasis was given to the development of novel assays and assay technologies. The effort successfully assembled a portfolio of diverse small molecule screening projects well suited to HTS.

Principal technical challenges in the preparation of a biological assay for screening against a small molecule compound library such as the NIH Molecular Libraries Small Molecule Repository (MLSMR)* include the requirements for miniaturization, the availability of consistent and stable reagent preparations (e.g., cells, vectors, biochemical substrates), a low number of handling steps to accommodate full automation, and an assay readout with robust signal and low background. These are default minimum requirements for an HTS assay but are not sufficient to drive a screening project to successful completion.

Academic assay projects intending to advance to screening can often be open-ended, be performed with the intent of adapting the approach to the results of screening, and may lack a desired outcome, specified at the outset to guide the project. Played out in the context of a multidisciplinary screening center environment, which employs a team approach encompassing a number of skill sets and coordinated activities, open-ended projects can experience long delays as different methodologies are adopted, or new assays inserted to accommodate a change in goals. In addition, at the handoff of a project from the Investigator to the Center, it is necessary to establish the criteria guiding the selection and progression of screening positives (hits). Equally important is agreement on when a project should be stopped, with the understanding that HTS resources are typically constrained, and that early termination of a project not meeting its goals would allow centers to expedite new project efforts.

To resolve project start-up and decision-making issues, the Molecular Libraries Initiative adopted guidelines for interrogating the HTS-readiness of a project prior to its acceptance into the network. These Fast Track program guidelines were originally designed to allow NIH staff to evaluate Assay Development for HTS projects that had neared completion, and were later broadly applied to assay projects originating in diverse NIH-funded activities. The assay project guidelines "start at the end" by first asking the Investigator to visualize a successful outcome, and then describe the "critical path" of assays and compound selection criteria to be used at each step in the path. A simple model of a project's critical path is illustrated in Figure 8.2.

Investigators are prompted to answer the following questions in a project proposal:

- What experiments (e.g., mechanism-of-action, POP) would be enabled by having a molecular probe?
- What features would this probe need to have in order for it to be useful (e.g., affinity, selectivity, cell penetrance, stability to metabolism)?

*Available at http://mlsmr.glpg.com/MLSMR_HomePage/.

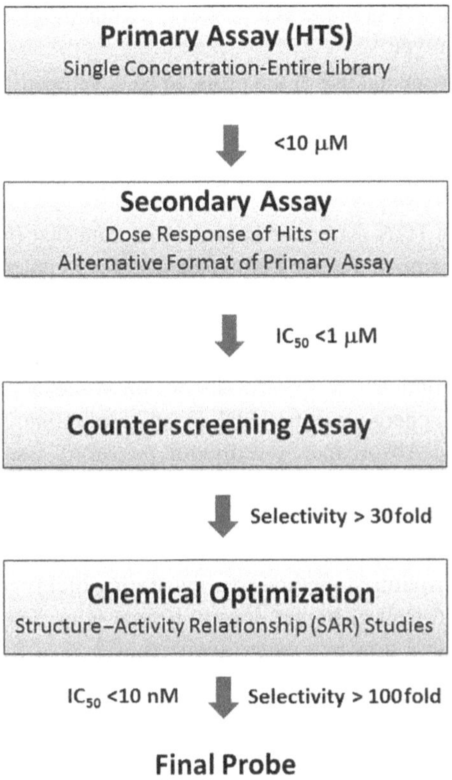

Figure 8.2. Simple model for *critical path* of screening assays.

- What assays would be used to measure these probe attributes?
- In what order would you use these assays? What would the assay cutoff values need to be to select compounds in the test?

An evaluation using technical screening criteria is then employed to judge the robustness of the proposal and allow determination of the NIH institute-specific interest (basic science or disease relevance) in a project. A project ranking is accomplished using the following scorecard as guide:

1. Feasibility of the proposed screening assays
2. Completeness of the screening plan
3. Availability of key reagents required for execution of the HTS project
4. Completion of a pilot screen.*

*Typically, a small compound diversity collection is used. In addition to the commercially available compound sets commonly used, NCI offers compound plates free of charge through their Developmental Therapeutics Program (available at http://dtp.nci.nih.gov/branches/dscb/repo_open.html). The NIH Clinical Collection is also available at modest cost (available at http://www.nihclinicalcollection.com/).

5. Current availability of molecular probes (pharmacological tools) to the proposed target or phenotype
6. Intended experimental use of the probe to be developed, to include a description of values for the attributes that are required in a useful probe (e.g., affinity value as measured in the primary assay, fold-selectivity versus an undesired target interaction).

A proposal for an HTS assay needs to demonstrate that the assay can be further miniaturized and implemented in an automated screening center environment. The following detailed information is needed to describe the primary HTS assay:

- A description of reagents and readouts that can be translated to an automated HTS environment. Common assay readouts are fluorescence, luminescence, fluorescence resonance energy transfer (FRET), and bioluminescence resonance energy transfer (BRET). Absorbance, scintillation proximity assay (SPA) or biophysical readouts, and whole cell imaging are also used. Assays requiring only addition of reagents (e.g., "mix and measure") are preferable, and steps such as centrifugation, filtration, and extraction are to be avoided in an automation environment.
- A detailed description of methods demonstrating highly robust and reproducible behavior in a 96-well or higher density format (e.g., 384- or 1536-well plates), or adaptability to a flow cytometry approach.
- Performance data for the primary assay, to include the following:
  - A signal of sufficient intensity that can be easily measured (using, e.g., a microtiter plate of low volume).
  - A signal-to-background ratio of at least 5 and a coefficient of variation (CV) below 10% (determined from measurements across an entire plate). These factors can be expressed individually and as the statistical parameter Z, which has an acceptable lower limit of 0.5 [11].
  - Reproducible, dose-dependent responses in the primary assay to a small collection of pharmacologically active standards, reaction substrates, or inhibitors. Alternatively, data from pilot screening of a structurally diverse collection of small molecules could be used. The intent of this characterization is to demonstrate that the assay can detect interaction between a small molecule and the target.
  - Control parameters, including time, temperature, and protein concentration dependencies. Additional information that should also be provided in a proposal includes tolerance to the effect of 0.1–1% dimethyl sulfoxide (DMSO), between-plate and day-to-day variations, and a determination of reagent stability to storage and assay conditions.

The critical path of activities constituting a complete screening project plan includes the following: a primary HTS assay to identify compounds having activity toward the target, phenotype, or pathway that is the subject of molecular probe design; a secondary assay used to confirm the activity of the chemical hits and additionally provide detail about their activity; and counterscreening assays to test for the activity of hits at other targets. These assays can also be used to characterize additional hit attributes such as cellular penetrance or toxicity.

Information provided to describe the critical path assays should include the following:

- A description of secondary screens that can be used to confirm that the activity of compounds in the primary assay is reproducible and rule out artifacts (e.g., compound impurity or fluorescence interference of the test compound). This test can be performed using the primary assay to establish dose-dependent activity, or an assay to the target that employs a different measurement of activity. The secondary assay can also be used to rank order the potency or affinity of hits, and assess agonism or antagonism.
- A description of counterscreening assays allowing the prioritization of hits for further testing (e.g., on the basis of selectivity or potential toxicity). It is likely that several hundred or more active compounds may be identified in an HTS effort that employs a library of $>10^5$ compounds.
- Recommendations of appropriate test concentrations, together with concentration cutoffs to be used in selecting active compounds.
- Information describing reagent availability, equipment required for performing the assays, and estimated cost per well of an assay.
- Description of the throughput rate (compounds per day in the format provided), the current plate format (384-well, 96-well, cuvette), and the average assay cycle time (hours or days to complete an assay).

Last, it is worth noting that phenotypic assay projects represent an exception to the standard formatting of a critical path, as followed for molecular target assay projects. The goal of developing a probe in this instance often supports target discovery rather than target characterization efforts. Thus, successful hit deconvolution presents a challenge. Phenotypic assays entering the MLPCN for screening have, in general, used one of two approaches to verifying and sorting hits from the primary (phenotypic) HTS assay:

1. The secondary and counterscreen assays used are alternative phenotypes. For instance, the goal may be to target cancer cells but not isogenic cells lacking oncogene [12].
2. The characterization assays represent cellular signaling pathways in a cell that can be used to narrow the potential activity of a compound to a specific pathway or member of that pathway [13, 14].

IMPACT OF THE NIH MOLECULAR LIBRARIES ASSAY DEVELOPMENT FOR HTS INITIATIVE

The NIH Molecular Libraries Assay Development for HTS initiative was started in 2004 in anticipation of the launch of the MLSCN, and the need to supply a pipeline of screening projects to the member screening centers of this roadmap consortium. A

TABLE 8.1. Metrics Describing Success Rates and Impact for the NIH Molecular Libraries *Assay Development for HTS* Initiative (2004–2009)

Key Indicators of Success Assay Development for HTS	
Total number of applications/awards	949/221
Projects successfully completing assay development	121 (55%)
Projects completing a pilot screen	109
Projects entering the MLSCN and MLPCN (of assay projects successfully developed)	56 (46%)
HTS projects initiated at other centers	31
High-throughput screens completed	66
Publications supported by the award	146
New grant applications resulting from the award	424
New grant applications awarded	157 (37%)
Institutions filing investigator-initiated applications to the program	265

series of metrics were collected in the first 5 years of the initiative, which allow progress and impact to be assessed.* These "key indicators of success," covering the period of 2004–2009,† are the subject of review and discussion in this section.

The information in Table 8.1 allows an assessment of the public interest in assay development as a discipline supporting high-throughput library screening and its impact on the academic scientific community. Interest in applying to the Assay Development for HTS program announcements was high, with close to 1000 applications in the first 5 years. Of the 221 projects that were funded, almost all were extramural projects, although several NIH intramural applications were also funded.‡ A high rate of successful project completion was observed (55%), and most of these projects had also performed a pilot small molecule screen. The typical project required more time to complete (~18–24 months) than the 1 year projected by the program announcement. Nonetheless, almost half of the projects that succeeded in the assay development phase had entered the MLSCN (or MLPCN) for HTS by the end of 2010. Most of these projects had initiated collaboration with a center outside of the NIH network first, and completed a preliminary HTS, before applying for entry to the Molecular Libraries consortium. Thus, a conclusion to be drawn is that investigators prefer to develop and screen locally prior to migrating projects into centers where they may have access to larger compound collections.§

*Sources of Data: PubMed; PubChem; NIH RePORTER; NIH QVR database; NIH Molecular Libraries CARS Database.

†These data were previously presented and discussed at Drug Discovery: Easing the Bottleneck-Opal Events' Second Annual Conference, Philadelphia, PA (August 23–24, 2010), and Next Generation Pharmaceutical Summit, Miami, FL (October 25–27, 2010).

‡NIH intramural investigators are eligible to apply for NIH Common Fund Initiatives.

§See the Society for Biomolecular Sciences website for a list of academic screening facilities. Available at http://www.slas.org/screeningFacilities/facilityList.cfm.

The process to fast-track a completed assay development for HTS project into the Molecular Libraries network, via expedited programmatic review (to evaluate technical readiness of the project, see footnote on p. 101 on MLPCN Centers), has been extensively used by investigators. This program, started in 2007, effectively shortens the time to progress from investigator to consortium effort, and was subsequently extended to all NIH-funded efforts that had undergone formal peer review of the goals of assay development and HTS. The Assay Development for HTS awards allowed for both Fast Track entry into the Molecular Libraries Centers, and ensured investigator engagement in the HTS effort that followed by also providing a grant supplement to facilitate collaboration with the center assigned the project. The availability of the supplement proved to be highly enabling to investigators wishing to pursue HTS work, and was followed by similar supplement programs initiated by individual NIH Institutes such as the National Institute for Neurological Disorders and Stroke (NINDS),* and institute consortia such as the NIH Blueprint for Neuroscience.† This ensured that the investment of time and resources for assay development by institute-supported investigators could be leveraged forward by facilitating the participation of their HTS-ready projects in a multidisciplinary probe development effort.

Investigators with Assay Development for HTS grant awards have continued to publish their study results in numbers reflective of the project success rate (Table 8.1), and cite the award in subsequent grant applications. A most striking observation from the data was that the success rate for the new grant applications exceeded 30%. The result strongly suggests that this advanced technical work enables investigator-led projects that seek to employ molecular probes in basic science and translational research studies.

Strong community engagement and interest in assay development was assessed in two different ways: First, a large number of different institutions (265) filed investigator-initiated applications to the Assay Development for HTS program announcements (Table 8.1), indicating that the U.S. academic community (and, to a small extent, the biotech community of small businesses) has responded to the effort and have disseminated information about the program to their investigators. Second, 15 of the 27 NIH institutes and centers were associated (via the funded investigator and/or research area of the proposal) with Assay Development for HTS awards (Figure 8.3). Notably, those institutes with funding efforts in translational research were most strongly represented.

Project applications needed to strike a balance between technology innovation, target novelty, and feasibility. A useful middle ground was to prioritize funding of those projects that were likely to advance to HTS with 18–24 months. Reflective of work in the scientific community, proposed assay formats strongly centered on cell-based assays

*Notice of Availability of Administrative Supplements for Roadmap Assay Development for HTS Awards (National Institute for Neurological Disorders & Stroke). Available at http://grants.nih.gov/grants/guide/notice-files/NOT-NS-10-012.html.

†Notice of Availability of Administrative Supplements for Roadmap Assay Development for HTS Awards (NIH Blueprint for Neuroscience Research). Available at http://grants.nih.gov/grants/guide/notice-files/NOT-NS-07-001.html.

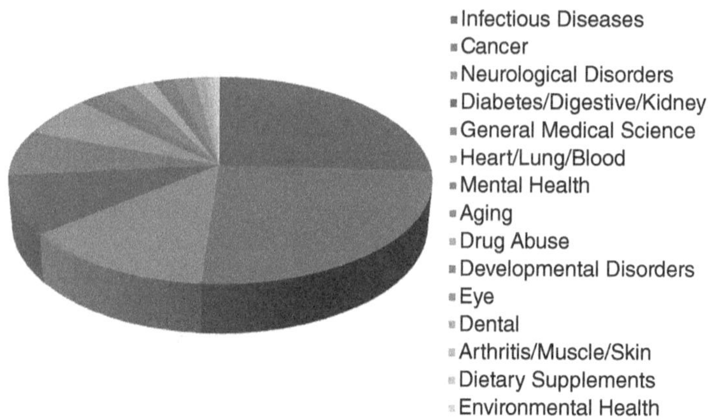

Figure 8.3. NIH Institute representation in the *Assay Development for HTS* Awards. See color insert.

TABLE 8.2. Detection Technologies Proposed by NIH Molecular Libraries *Assay Development for HTS* Project Awards (2004–2009)

Detection Technology	% of Projects
Fluorescence intensity	38
Luminescence	20
FRET-BRET	14
Fluorescence polarization	10
Cell growth-toxicity-viability	7
Absorbance-spectral shift	6
ELISA	3
Scintillation proximity assay	2
Other	1

(59% of awards) and assays employing pure enzyme or protein reagent preparations (33% of awards). Model organism assays were perceived as difficult to miniaturize and automate, and represented only 4% of the awards even though several MLSCN centers possessed screening capability.

As shown in Table 8.2, fluorescence and luminescence were preferred signal detection methods within the funded Assay Development for HTS projects, reflecting current laboratory approaches and available commercial kit formats. Both enzyme-linked immunosorbent assay (ELISA) and SPA readouts were poorly represented, due to perceived protocol complexity and number of handling steps (ELISA), and reliance upon isotopic detection (SPA).

The majority of assay and probe development effort was directed at molecular targets (Table 8.3), whereas phenotypic assay projects (Table 8.4) represented a minor proportion (<15%) of the 221 award total. Well-established molecular target assay

TABLE 8.3. Molecular Target Assays Proposed by NIH Molecular Libraries *Assay Development for HTS* Project Awards (2004–2009)

Molecular Target Assays	% of Projects
Protein–protein interaction	29
Enzyme	27
Receptor	15
Transcription and replication	10
Channels and transporters	7
Protein synthesis and modification	5
Translocation and secretion	3
Protein folding	2
Cell survival and regeneration	2

TABLE 8.4. Phenotypic Assays Proposed by NIH Molecular Libraries *Assay Development for HTS* Project Awards (2004–2009)

Phenotypic Assays	% of Projects
Cytotoxicity-apoptosis-degeneration	24
Cell growth-viability	20
Infectivity-replication	18
Differentiation-morphology	16
Antigen expression	11
Stress response-heart rate	11

categories (enzyme, receptor, and channels and transporters) represented >50% of the HTS assays. However, a high level of novelty was observed in the remaining half of the funded projects. Protein–protein interaction projects had the strongest representation in the molecular target assay category overall (29%), a reflection of strong community interest and current research funding. These projects were also perceived as being very challenging, with a high hurdle for success in chemical biology efforts to develop small molecule molecular probes. The most striking (and well received) of the protein–protein interaction projects were those for which there was evidence of a discrete domain that could be targeted by a small molecule using the proposed assay. By contrast, most phenotypic assays thus far have been relatively simple measures of cell viability or death (approaches to deconvoluting hits from these experiments were discussed earlier in this chapter). A key technology development issue for this assay category concerns the ability to miniaturize single cell imaging to a screening plate format, and effectively process the data-rich information.

RESOURCING THE CHEMICAL OPTIMIZATION OF SMALL MOLECULE PROBES

A project that enters a screening Center can require 1–2 years of effort for assay optimization, HTS, hit verification and characterization, chemical lead selection, and initial structure–activity relationship (SAR) studies. The time and dedicated medicinal chemistry effort required for advanced chemical optimization afterwards is often beyond the scope of a screening center's capability and resources. In particular, for CNS-active agents requiring brain penetration, 1–2 years of additional work is often needed to develop the optimized molecular probe originally outlined in the assay development proposal by the investigator at the inception of the project. Investigators that came away from an HTS effort with knowledge of chemical leads and optimized screening assays were frequently challenged to find public sources of funding for this work. Further, the medicinal chemistry expertise needed, while well represented in the pharmaceutical industry, is mostly not available in the academic community.

For these reasons, an experimental initiative was designed to specifically relieve the aforementioned constraints. A grant program titled Chemical Optimization of Small Molecule Probes for the Nervous System was first launched in 2009 and has had six NIH Institutes participating.* This section will briefly use the initiative as an example of how solutions might be proposed that lead to effective public funding of this research.

Key issues the initiative has worked to identify and resolve were as follows: (1) chemical synthesis requires both infrastructure and time; (2) the cost of the work can rapidly exceed available funding structures, particularly if the iterative design of analogues follows several series of lead structures; and (3) an experienced medicinal chemist is needed to regularly lead analogue design. To be fully incentivized, the chemist should also be able to participate in the filing of any university inventions derived from the project.

Eligible projects would ideally start with lead scaffold series that had been identified in an HTS effort and extended through initial SAR efforts that employed, in most cases, an analogue-by-catalogue approach. The biologist proposing the optimization would have a complete critical path of screening assays available in the laboratory that could be used at a moderate throughput level not requiring automation. The chemistry model used for the program was based upon biased array synthesis as an iterative design cycle, in which ~20–40 analogues would be designed, synthesized, and tested monthly to reach the goal of a final molecular probe within 2 years. A project would ideally have at the outset several scaffolds with which to work, such that premature termination of the project would not occur if the SAR on a single series failed to show promise.

The first issue mentioned was addressed in the initiative by encouraging investigators to take advantage of a cost-effective compound outsourcing approach to extend their synthesis capability where possible. The National Cancer Institute (NCI) had sup-

*Optimization of Small Molecule Probes for the Nervous System (R21). Available at http://grants.nih.gov/grants/guide/pa-files/PAR-09-251.html. The Initiative has also been issued as a STTR Program (available at http://grants.nih.gov/grants/guide/pa-files/PAR-09-259.html), and as an SBIR Program (available at http://grants.nih.gov/grants/guide/pa-files/PAR-09-260.html).

ported the development of ChemNavigator for this purpose, and this web-based service has been available for compound searching to NIH-funded investigators.* This facility further provides a compound acquistion service. ChemNavigator also provides a Semi-Custom Synthesis On-Line Request Service (SCSORS, also developed with NCI support), which allows investigators to solicit bids from chemistry companies for compounds not in the online catalogue. Substantial use of compound outsourcing approaches such as this can speed the process of compound testing and reduce the need for infrastructure investment.

The second issue of cost containment was addressed by separating the expensive and expert process of analogue design from the task of compound synthesis. An informal survey of centers practicing external small molecule outsourcing indicated that compound acquisition by these means can be highly cost-effective.

The remaining issue, which concerned the availability of qualified medicinal chemistry design expertise, was tackled by facilitating the partnering of university-based biologists and their projects with expert medicinal chemists, who typically possessed significant pharmaceutical industry drug discovery expertise. Chemisty consultants were not contracted for assignment to projects in this initiative; rather, expert medicinal chemists were identified and prequalified by NIH staff. A list of these interested experts was then provided to biologists looking for a chemistry partner for their project. Each of the medicinal chemists had agreed to participate in a grant proposal, could be included as an investigator on the application, and would be expected to participate in intellectual property (IP) filings that might result from their contribution to the project.

USING CTSA•IP TO IDENTIFY AND ASSEMBLE TRANSLATIONAL PROJECT TEAMS

Several priority areas of effort were identified at the inception of the Clinical-Translational Science Awards (CTSA) Public-Private Partnerships Key Function Committee (3P KFC).† One of the priority areas identified was the need to aggregate and market technologies from CTSA institutions as well as those of the NIH, with the goal of enhancing research and partnership activity across the CTSA consortium. This began an effort led by the University of Rochester to develop the web-based search engine CTSA•IP. A review of this initiative's purpose and outcomes to date has recently been published [15], and this chapter briefly summarizes key points that are particularly germane to this discussion.

This website, which aggregates publically available IP across participating institutions of the CTSA consortium, aims to address several strategic needs in the formation of translational research projects:

*ChemNavigator website for NIH researchers. The online catalogue currently contains >90 × 10^6 chemicals, sourced through >300 suppliers (available at http://www.chemnavigator.com/nih.asp).

†CTSA Public Private Partnerships. Available at https://www.ctsacentral.org/committee/public-private-partnerships-0.

- The toolkit required for a POP-2-POC effort, starting with POP studies in animal models employing molecular probes and ending with POC studies, in many cases cannot be assembled within one institution for a specific project. IP around screening technologies, possible therapeutic candidates, markers of drug response, and preclinical models needs to be identified and aggregated.
- For small molecule therapeutic development projects, a sponsored research effort is necessary to advance a project from the bench and closer to an Investigational New Drug (IND) application proposing clinical studies. Thus, investigators need to be identified that can collaborate to bring research, preclinical, and clinical trials expertise to a specific therapy project opportunity.
- Molecular probes may not be a sufficient source of new chemical entities for clinical development in many cases, but may have advanced the state of validation for a project to a point at which private sector partnering becomes possible. Partnering can provide the pharmaceutical skill sets and capacity needed for clinical drug candidate selection, and characterization steps such as chemistry process development, safety, and toxicology.

CTSA•IP currently provides the ability to identify and aggregate the IP needed for a cross-institutional translational research effort. The aim of the site is to aid the formation of a commercial proposition that includes domain expertise, proprietary technology, models, and (in some cases) drug leads, as a means of facilitating partnering with private sector drug development resources. The website effort is currently evolving by expanding membership to additional CTSA member institutions, collaboration with other tech transfer communities such as iBridge,* and the incorporation of researcher profiles via collaboration with focused efforts such as VIVO.†

CONCLUDING REMARKS

Recent years have witnessed a migration of high-risk, precompetitive drug discovery activity into the academic community. Most notably, research to identify and pharmacologically validate drug targets and potential drug response markers through a POP demonstration in animal models is being driven by the availability of new molecular probes. This work has the potential to inform both drug candidate and disease area selection, markedly lowering the risk of drug development by providing decision-making tools at an earlier stage than previously employed. Involvement of the CTSA consortium in public–private partnering efforts such as CTSA•IP has the potential to aid POP-2-POC project formation, and at an early stage engage clinicians and key opinion leaders (KOL) with disease area expertise with researchers working to move laboratory-based innovation forward to clinical testing.

*iBridge CTSA community. Available at http://www.ibridgenetwork.org/community/ctsa.

†VIVO: An interdisciplinary national network. Enabling collaboration and discovery among scientists across all disciplines. Available at http://vivoweb.org/.

REFERENCES

1. Frye, S., Crosby, M., Edwards, T., Juliano, R. US academic drug discovery. *Nat Rev Drug Discov*. 2010;10:409–410.
2. Trochim, W., Kane, C., Graham, M.J., Pincus, H.A. Evaluating translational research: a process marker model. *Clin Transl Sci*. 2011;4:153–162.
3. Oprea, T.L., Bologna, C.G., Boyer, S., Curpan, R.F., Glen, R.C., Hopkins, A.L., Lipinski, C.A., Marshall, G.R., Martin, Y.C., Ostopovici-Halip, L., Rishton, G., Ursu, O., Vaz, R.J., Waller, C., Waldman, H., Sklar, L.A. A crowdsourcing evaluation of the NIH chemical probes. *Nat Chem Biol*. 2009;5:441–447.
4. Shurin, S.B. Clinical Translational Science Awards: opportunities and challenges. *Clin Transl Sci*. 2008;1:4.
5. Califf, R.M., Berglund, L. Linking scientific discovery and better health for the nation: the first three years of the NIH's Clinical and Translational Science Awards. *Acad Med*. 2010;85:457–462.
6. Selker, H.P., Strom, B.L., Ford, D.E., Meltzer, D.O., Pauker, S.G., Pincus, H.A., Rich, E.C., Tompkins, C., Whitlock, E.P. White paper on CTSA consortium role in facilitating comparative effectiveness research: September 23, 2009 CTSA consortium strategic goal committee on comparative effectiveness research. *Clin Transl Sci*. 2010;3:29–37.
7. Finkbeiner, S. Bridging the valley of death of therapeutics for neurodegeneration. *Sci Transl Med*. 2009;16:1227–1232.
8. Coller, B.S., Califf, R.M. Traversing the valley of death: a guide to assessing prospects for translational success. *Sci Transl Med*. 2009;1(10):1–11.
9. Macarron, R., Banks, M.N., Bojanic, D., Burns, D.J., Cirovic, D.A., Garyantes, T., Green, D.V.S., Hertzberg, R.P., Janzen, W.P., Paslay, J.W., Schopfer, U., Sittampalam, S. Impact of high-throughput screening in biomedical research. *Nat Rev Drug Discov*. 2011;10: 188–195.
10. Zhang, J., Lushington, G.H., Huan, J. Characterizing the diversity and biological relevance of the MLPCN assay manifold and screening set. *J Chem Inf Model*. 2011;51: 1205–1215.
11. Zhang, J.H., Chung, T.D.Y., Oldenburg, K.R. A simple statistical parameter for use in evaluation and validation or high throughput screening assays. *J Biomol Screen*. 1999;4: 67–73.
12. Bittker, J.A., Weiwer, M., Shimada, K., Yang, W.S., MacPherson, L., Dandapani, S., Munoz, B., Palmer, M., Stockwell, B.R., Schreiber, S.L. Screen for RAS-Selective Lethal Compounds and VDAC Ligands. 2010: Probe Reports from the NIH Molecular Libraries Program. Bethesda (MD): National Center for Biotechnology Information (US). Available at http://www.ncbi.nlm.nih.gov/books/NBK55069/#ml162.s36. Accessed December 23, 2013.
13. Khan, P.M., Correa, R.G., Divlianska, D.B., Peddibhotla, S., Sessions, E.H., Magnuson, G., Brown, B., Suyama, E., Yuan, H., Mangravita-Novo, A., Vicchiarelli, M., Su, Y., Vasile, S., Smith, L.H., Diaz, P.W., Reed, J.C., Roth, G.P. Identification of inhibitors of NOD1-induced nuclear factor-κB activation. *ACS Med Chem Lett*. 2010;2:780–785.
14. Magnuson, G., Khan, P., Yuan, H., Brown, B., Divlianska, D.B., Stonich, D., Peddibhotla, M., Su, Y., Dad, S., Sergienko, E., Chung, T.D.Y., Roth, G.P., Wimer, C., Diaz, P., Correa, R.G., Reed, J.C. High throughput screening assays for NOD1 inhibitors. 2010: Probe

Reports from the NIH Molecular Libraries Program. Bethesda (MD): National Center for Biotechnology Information (US). Available at http://www.ncbi.nlm.nih.gov/pubmed/21433392. Accessed December 23, 2013.

15. Hazard, M., Steele, S., Wang, D., Pearson, T., Scheideler, M., Dewhurst, S. CTSA-IP: a solution to identifying and aggregating intellectual property across the NIH Clinical Translational Science Award (CTSA) consortium of biomedical research institutes. *Clin Transl Sci*. 2011;4:328–331.

9

PARTNERSHIPS FOR DRUG REPOSITIONING: LESSONS FROM THE CTSA PHARMACEUTICAL ASSETS PORTAL

Kate Marusina,[1] Dean J. Welsch,[2] Lynn Rose,[3] Doug Brock,[4] Nathan Bahr,[5] Aaron M. Cohen,[5] Rafael A. Gacel-Sinclair,[6] Pakou Vang,[6] Peter G. Ruminski,[7] Bruce E. Bloom,[8] Pamela Nagasawa,[4] and Betty P. Guo[9]

[1]*Clinical and Translational Science Center, University of California, Davis, Sacramento, CA, USA*

[2]*BioMed Valley Discoveries, Kansas City, MO, USA*

[3]*Benaroya Research Institute and Institute of Translational Health Sciences, University of Washington, Seattle, WA, USA*

[4]*Department of Medical Education and Biomedical Informatics, University of Washington, Seattle, WA, USA*

[5]*Department of Medical Informatics and Clinical Epidemiology, Oregon Health and Sciences University, Portland, OR, USA*

[6]*Technology Transfer Services, University of California, Davis, CA, USA*

[7]*Center for World Health and Medicine, Doisy Research Center, St. Louis University, St. Louis, MO, USA*

[8]*Cures within Reach, aka Partnership for Cures, Chicago, IL, USA*

[9]*Office of Research, University of California Davis School of Medicine, Sacramento, CA, USA*

Collaborative Innovation in Drug Discovery: Strategies for Public and Private Partnerships, First Edition.
Edited by Rathnam Chaguturu.

Despite remarkable medical advances, many conditions lack safe and effective therapies. More than 90% of drug candidates fail to make it to market [1, 2], and those that are successful gain Food and Drug Administration (FDA) approval after an average of 13 years in development and a cost of ~$1 billion for a single drug [2]. Novel approaches to drug development would greatly facilitate the discovery of therapies for unmet medical needs. One such novel approach is drug repositioning, or finding new uses for existing compounds [3, 4]. By starting with compounds that have already undergone significant testing, drug repositioning dramatically cuts the time and cost necessary to bring a drug to market. This strategy is particularly applicable to therapeutics discovery for rare or neglected diseases, whose limited commercial potential attracts scarce funding from industry.

The Clinical Translational Science Awards (CTSA) Pharmaceutical Assets Portal was created with the goal of facilitating collaborative partnerships between industry and academic investigators for discovery of new indications for discontinued compounds [5]. The project focused on providing a mechanism for academic researchers to access compounds that are no longer being actively developed by pharmaceutical companies. *De novo* drug development is a lengthy, expensive process during which compounds are progressively evaluated for their therapeutic potential. Compounds can be discontinued at any stage for a variety of reasons, including lack of efficacy, insufficient therapeutic index, lack of commercial potential, and strategy-driven changes. Although it is difficult to estimate the exact number of shelved, discontinued compounds, clinical trial attrition rates suggest that, collectively, thousands of such compounds from decades of clinical trials may exist in industry repositories worldwide. Discontinued compounds that have been tested in humans represent a valuable scientific resource; however, to date, many of these compounds remain unknown to the academic research community.

The NIH Chemical Genomics Center (NCGC) has embraced the concept of drug repurposing by creating a comprehensive list of small-molecule drugs as well as a physical collection of these entities [6]. This invaluable resource encompasses FDA-approved compounds known through publicly available sources. The CTSA Pharmaceutical Assets Portal complements the NCGC's approach by focusing on making available compounds that are unknown to the public because they were discontinued prior to FDA approval. Historically, identifying new indications for existing drugs has occurred through serendipity or chance observation. Notable examples include the blockbuster drugs sildenafil (Viagra) and minoxidil (Rogaine) [3]. Drug repositioning is significantly less costly than *de novo* drug development and has an estimated success rate of 30% [4]. However, fully exploiting this opportunity relies on collaboration between pharmaceutical companies and the researchers interested in testing these compounds.

Through public outreach, the Pharmaceutical Assets Portal brought multiple parties to the table, including academic institutions, pharmaceutical companies, government entities, and disease-specific nonprofit organizations. To date, the Pharma Portal has established an infrastructure capable of supporting future public–private partnerships in drug repositioning, including a national database of principal investigators with their respective biological models and delivery systems; a collaboration browser tool; a facility to house and maintain a compound repository; and a funding partner for repositioning studies. While the ultimate goal of creating a shared repository accessible to the academic community remains to be achieved, the CTSA Pharmaceutical Assets Portal has created a strategy and a mechanism to facilitate future opportunities for drug repositioning.

THE CTSA PHARMACEUTICAL ASSETS PORTAL

The CTSA Pharmaceutical Assets Portal (http://www.ctsapharmaportal.org) was initiated by the consortium of universities linked by the CTSA [5, 7]. The project's mission was to create a public–private partnership based on collaboration among academic, government, foundation, and industry scientists to facilitate repositioning efforts by effectively leveraging the knowledge/expertise important to drug discovery and development. These matches are envisioned to ultimately result in an increased number of approved drugs for new indications and considerable public benefit.

In collaboration with Pfizer's Indications Discovery Unit (IDU) and the National Center for Research Resources (NCRR), the Pharma Portal has made significant progress in establishing the infrastructure for such partnerships. Specifically, key elements comprising the Portal include (1) membership from 50 universities, (2) a tool for connecting partners and members (Foci-of-Expertise, [FoX]), (3) a funding arm (Partnership for Cures), (4) an Intellectual Property (IP) arm (the University-Industry Demonstration Partnership, UIDP), and (5) facilities/resources to house, maintain, and distribute the discontinued compounds (the Center for World Health and Medicine, CWHM) (Figure 9.1). Establishing an infrastructure with these elements was considered important to minimize barriers to collaboration. The resulting infrastructure provides a platform for promoting shelved compounds for further investigation and medical use. This chapter describes the evolution of the CTSA Pharmaceutical Assets Portal, the real and perceived barriers to creating public–private partnerships, and potential opportunities for moving forward.

THE PHARMA PORTAL MEMBERS

The success of the Pharmaceutical Assets Portal rests both with industry partners who provide access to compounds and the Pharma Portal members who seek access to these compounds. In an online, 28-question survey to recruit members, the "Portal" was described as a tool that would connect investigators to researchers with complementary knowledge and skills, as well as to pharmaceutical companies wishing to explore new indications for their active or discontinued drug programs. The survey was divided into categories related to demographics, research interests (by MeSH headings), and previous experience with transfer of compounds. It also tracked the sources from which investigators obtain information about investigational compounds. In addition, respondents were asked to provide their primary reason for seeking compounds, for example, for use *in vitro* or in animal models, to conduct clinical trials, or a combination of these

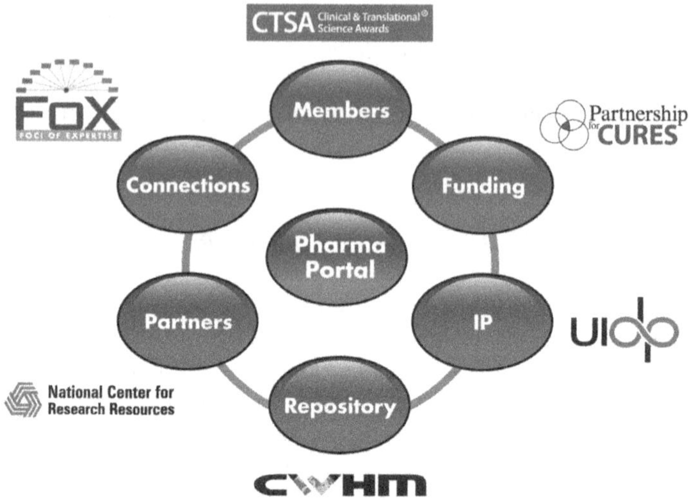

Figure 9.1. The Infrastructure of the CTSA Pharmaceutical Assets Portal. Shown are the components necessary for achieving the Portal's mission. Reprinted with permission from Marusina et al. 2011.

reasons, and probed the success rates of obtaining compounds from industry. Finally, the survey sought to identify the barriers to such collaborations.

Between 2008 and 2011, a total of 696 individuals from 47 (85%) CTSA institutions, industry, and the NIH responded to the survey. The majority of respondents were physicians, with most self-identifying as institutional leaders or senior faculty. The primary focus of respondents was on clinical applications, with 80% stating that they would conduct clinical trials if the compounds of interest were made available. This response underscores the overwhelming interest in gaining access to compounds for clinical testing rather than for basic research purposes. The four most commonly reported diseases of interest included cancer and noncancerous diseases of the nervous, cardiovascular, and immune systems (Table 9.1). Approximately half of those surveyed had access to unique experimental models supporting their clinical research hypotheses, including animal models of disease; high-throughput, cell-based screening assays; or assays for specific molecular targets in the nucleus and cytoplasm. Many investigators expressed an interest in new routes of administration, novel delivery systems, and compounds/delivery systems that can cross the blood–brain barrier (Table 9.2).

Most respondents (58%) had previously requested compounds directly from the pharmaceutical industry (Table 9.3). More than half of the respondents had been "successful" (57%), while another 37% reported being "sometimes successful." The relatively small fraction (4%) of those who had prior experience requesting compounds but were "unsuccessful" often reported that their success was limited primarily by contractual issues. Importantly, those who sought compounds desired a more efficient process by streamlining the negotiation process. The survey did not seek specific details of the contractual limitations, although our experience suggests that intellectual property and

TABLE 9.1. Disease Areas Most Relevant to Respondents' Research[a]

Category	All Respondents ($n = 969$) Number	Percent[b]
Nervous system diseases [C10]	236	34%
Neoplasms [C04]	181	26%
Cardiovascular diseases [C14]	171	25%
Immune system diseases [C20]	166	24%
Nutritional and metabolic diseases [C18]	127	18%
Bacterial infections and mycoses [C01]	112	16%
Endocrine system diseases [C19]	110	16%
Respiratory tract diseases [C08]	103	15%
Viral diseases [C02]	98	14%
Musculoskeletal diseases [C05]	98	14%
Digestive system diseases [C06]	94	14%
Hemic and lymphatic diseases [C15]	72	10%
Congenital, hereditary, and neonatal diseases and abnormalities [C16]	72	10%
Skin and connective tissue diseases [C17]	71	10%
Eye diseases [C11]	48	7%
Female urogenital diseases and pregnancy complications [C13]	52	7%
Male urogenital diseases [C12]	45	6%
Parasitic diseases [C03]	43	6%
Otorhinolaryngologic diseases [C09]	23	3%
Stomatognathic diseases [C07]	13	2%

[a]Survey respondents were asked to "Please check all general disease MeSH headings that are relevant to your research" to categorize their research interests using MeSH headings to facilitate future connections with the Foci-of-Expertise (FoX) Synergy Browser.
[b]The majority (62.8%) of investigators listed multiple interest areas. The total percentages will not add up to 100%. MeSH index numbers are shown in brackets. Reproduced with permission from Marusina et al. 2011.

TABLE 9.2. Highest Priority Drug Delivery Routes for Respondents' Research[a]

Delivery Mechanism	Number	Percent[b]
Oral	343	49%
Intravenous	263	38%
Transdermal	152	22%
Blood–brain barrier	110	16%
Inhalation	94	14%
Topical cream	73	11%

[a]Survey respondents were asked, "For your research studies do you have any specific drug delivery requirements?"
[b]The majority (51.3%) of investigators listed multiple interest areas. The total percentages will not add up to 100%. Reproduced with permission from Marusina et al. 2011.

TABLE 9.3. Prior Experience with Obtaining Investigational Compounds[a]

Experience		Number	Percent w/ Experience	Percent Successful
Yes[b]		397	58%	
	Yes, successful	227		57%
	Yes, sometimes successful	146		37%
	Yes, but not successful	17		4%
	Yes, not answered (2)	7		2%
No		293	42%	

[a]Survey respondents were asked, "Do you have prior experience in obtaining or seeking to obtain pharmaceutical compounds for your research?"
[b]Survey respondents with experience seeking compounds were asked, "Were your requests approved?"
Reproduced with permission from Marusina et al. 2011.

publication language are typically the most contentious. Lack of sufficient compound quantities or lack of a compelling business case were also frequently cited as reasons for refusal. Most investigators identified compounds of interest through public sources, such as the scientific literature, professional conferences, web searches, or by word of mouth. Fewer than 40% reported using commercial drug information databases—the subscription fees associated with these services may discourage their use. Our own informal research revealed that commercial databases capture only a portion of the potentially useful shelved compounds, perhaps 50%. This underscores the importance of projects such as the Pharma Portal, which aims to make available compounds not described in publicly accessible literature. The goals of the Portal project appear to be in alignment with the needs of this community, as evidenced by the fact that the majority of respondents (82%, 570/696) signed on to become members of the Pharma Portal.

USING THE FOCI-OF-EXPERTISE TO IDENTIFY POTENTIAL COLLABORATORS

One barrier to forming productive public–private partnerships is the difficulty in identifying companies and academic investigators with complementary interests. We specifically developed the Foci-of-Expertise (FoX) Synergy Browser, a graphical scientific search tool, to address this challenge. The browser extracts information from online biomedical resources and visually displays the relationships between researchers, diseases, and gene/protein targets. Moreover, search results specifically "highlight" members of the Pharmaceutical Assets Portal, allowing easy identification of researchers who wish to establish collaborative drug repositioning partnerships (Figure 9.2).

Currently existing browsers, such as BiomedExpert's Collexis (http://www.biomedexperts.com/) and Harvard's Catalyst (http://catalyst.harvard.edu/), mine the public literature, such as MEDLINE, and present author-centric profiles containing an individual's publications, coauthors, affiliations, and frequently associated keywords.

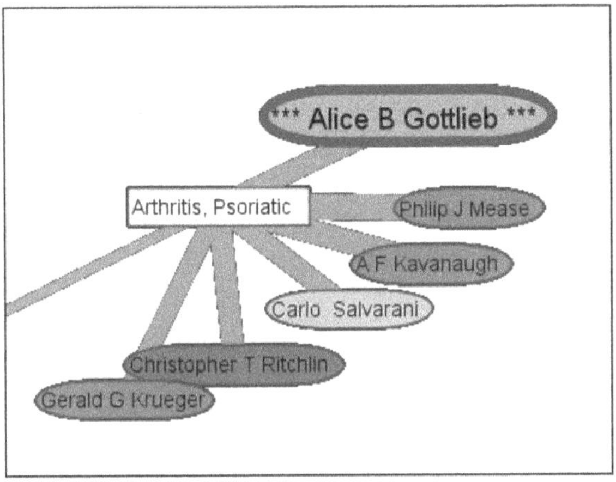

Figure 9.2. Pharma Portal members are highlighted by triple asterisks (***) to quickly identify them for collaboration.

These systems retrieve a collection of experts whose research most closely matches the search criteria. In other words, they are found through similarly associated keywords, which can imply that they have similar specialties. A productive team, however, requires not researchers with similar skills, but those with complementary, nonoverlapping knowledge. This led us to develop the FoX tool, which shows an alternative view of authors connected to one another through interrelated topics.

With assistance from researchers at Pfizer, FoX was developed in an iterative fashion [8] to browse three linked search categories: target genes/proteins, authors, and diseases. Links are established when search topics co-occur in the same document frequently. These documents come from MEDLINE (publications), NIH RePORTER (funded grant abstracts), and Reactome (gene–gene interactions). Linking these categories (target genes/proteins, authors, and diseases) together enables users to explore the collaboration space and discover relationships that may not be immediately obvious when a search is initiated. A search, for example, for the disease "idiopathic pulmonary fibrosis" displays multiple authors whose strength of association with the search term is indicated by the thickness of the connecting lines. Authors are color-coded by university affiliation, and each author can be individually expanded to reveal target genes or proteins associated with their research (Figure 9.3). The tool also has filtering capabilities to limit the number of results presented from a given query. The ease and visual appeal of FoX to quickly identify both direct and indirect connections between researchers, diseases and gene/protein targets distinguishes this tool from currently available browsers.

Another example uses FoX to expand on a disease search for "rheumatoid arthritis" (RA): we found 3512 authors (CTSA authors disambiguated by coauthor inclusion [9], 316 related diseases, and 764 targets. We expanded RA to show the top 13 targets, then expanded on PIP4K2C (phosphatidylinositol-5-phosphate 4-kinase, type II, gamma) and ATP6V1G1 (ATPase, H^+ transport, lysosomal 13 kDa, V1 subunit G1) to find

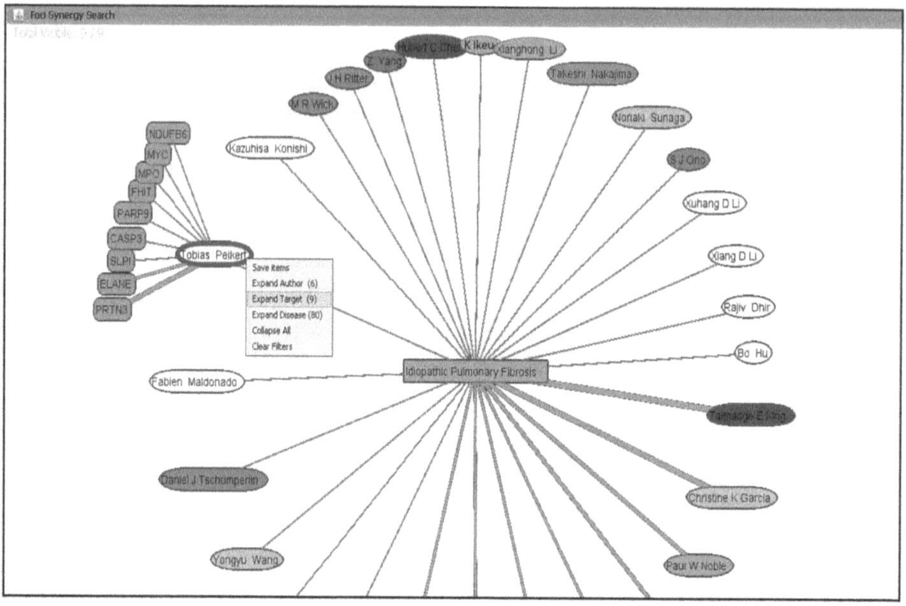

Figure 9.3. Dataflow diagram showing the disease node, idiopathic pulmonary fibrosis, in the center, surrounded by author nodes color-coded by their affiliated institutions. Connections are shown in different line thicknesses to help visually differentiate between weak and strong connections.

related diseases. Performing an "AND" expand on the related diseases revealed authors who have published on both the target and the disease in question (Figure 9.4). Four authors have published on PIP4K2C and RA [10, 11], and two have published on PIP4K2C and diabetes [12]. In certain circumstances, such expertise may be important to further a specific area of translational research. Similarly, ATP6V1G1 was studied by four investigators, but was studied by only one in the context of RA, while others investigated ATP6V1G1 in the context of finding gene orthologs [13] or haplotypes [14]. Although some of the author assignments were ambiguous (multiple investigators were assigned to a single "M Martin" in one paper [10], and Joel D Cooper had a variant name at another institution, J D Cooper [12]), investigators with important complementary experience could be clearly seen.

Notable features of the FoX system include the ability to (1) identify and highlight researchers working at CTSA institutions and participating in the online CTSA Pharmaceutical Assets Portal (http://www.ctsapharmaportal.org/); (2) support visualizing "AND/OR" relations with a chosen set of authors, diseases, and targets; (3) filtering capabilities to narrow query results; (4) visually show the strength of linkages through the thickness of connecting lines; (5) extract results into spreadsheet-compatible forms for further use and analysis or render results in another graphing package; and (6) export data or graphs as images or text.

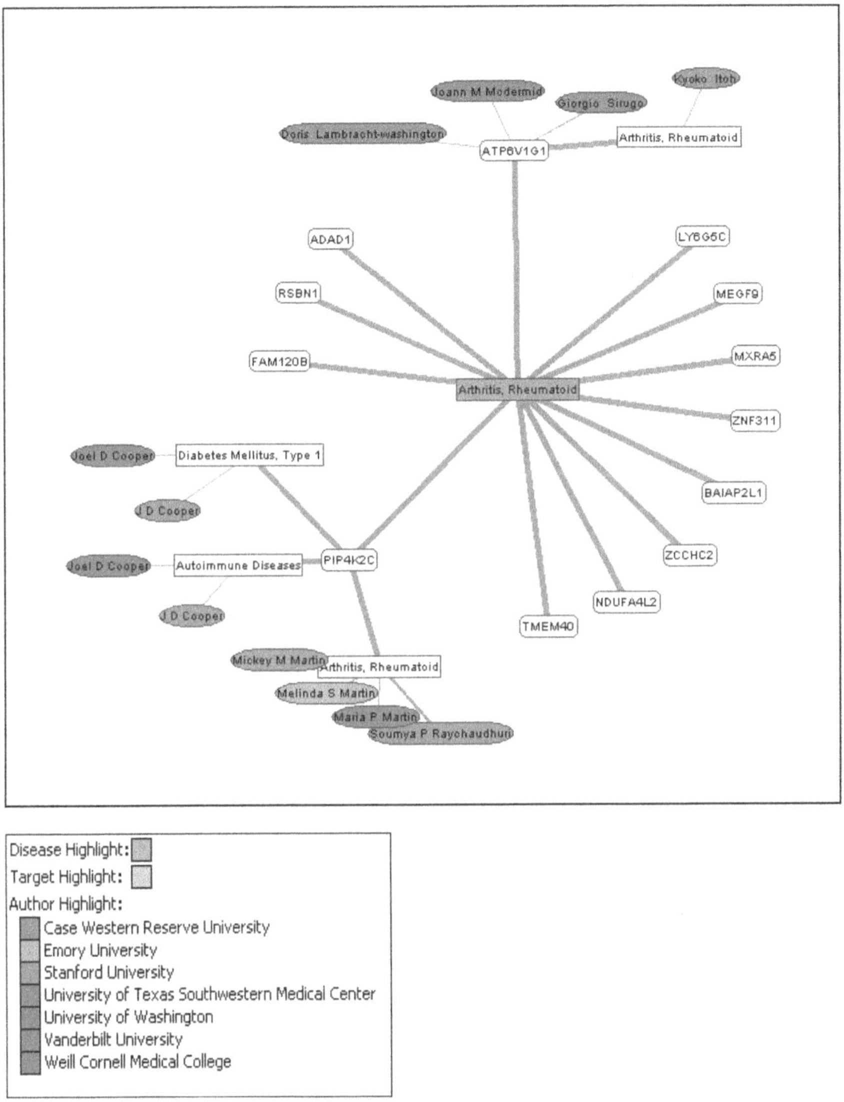

Figure 9.4. Shown are authors working on PIP4K2C and ATPV1G1 in the context of RA and other diseases. See color insert.

Examples of the many potential uses for the FoX system include facilitating organizing a symposium for collaborative research on a given disease at a specific institution; identifying potential members of a translational research team to investigate a new pharmacological agent; identifying key opinion leaders in certain disease areas; identifying researchers with expertise in a specific area of interest; identifying collaborators for a public–private partnership, and so on. It is not straightforward to use traditional search tools such as PubMed to make these connections. This can be an open-ended process, as the users' requirements and observations from the search results will continuously shape how they query and browse relationships. Traditional tools are useful for finding material if the subject matter is known beforehand, but it can be difficult to sift through large numbers of publications to discover connections that were not defined at the start of the search. For example, the search for RA resulted in >100,000 publications. The capabilities of FoX enabled progressive queries to narrow the search results and quickly identify both potentially complementary areas of exploration as well as researchers with very specific expertise in these areas. FoX is a knowledge discovery system, designed to help users tease out interesting connections in the literature instead of searching paper by paper.

While FoX can be used broadly to develop new collaborations, for the purposes of the Pharma Portal, the tool was used in a more focused manner—to find complementary overlap between drug characteristics (including the biological target and its clinical history) and existing academic expertise, including those of Portal members. For instance, if a drug's primary mechanism of action is the inhibition of a particular gene, then a collaboration can be built by engaging researchers studying that gene pathway, irrespective of the disease indication. Using FoX, we can identify different avenues of collaborative research that may not have been obvious initially, and may not have been recognized using existing browsers and simply seeking other researchers working in similar areas.

PARTNERSHIP FOR CURES

The lack of dedicated funding for repositioning studies was another important challenge that the Pharma Portal overcame by identifying a nimble funding mechanism from the nonprofit Partnership for Cures (http://www.4cures.org). Partnership for Cures began the Patient Impact Initiative (PII) in November of 2011 as a collaboration of donors, researchers, institutions, clinicians, and patients working together to find, fund, and finish Rediscovery Research™ to repurpose "existing science and medicine to create quick, safe and inexpensive treatments for patients." The PII provides 90% of a project's direct costs, ranging from $25–$250K, for 1- to 3-year projects.

Rediscovery Research projects can focus on any disease, any therapy, and impact any group of patients to create better treatment options, improve early diagnosis, and lengthen disease remission, mostly through repurposing FDA-approved drugs and devices to treat "off-label" diseases; combining older drugs with newer ones to increase the newer drug's effectiveness; modifying current treatment protocols to improve efficacy and to help more patients for longer periods of time; and testing anecdotal successes reported in clinical practice, including those from integrative medicine. The projects are primarily human pilot clinical trials. A successful end result would be a "new" treatment, or evidence that changing the current standard of care would benefit patients without increasing risks or costs. Some results will provide the foundation for larger clinical trials, and perhaps FDA approval of a repositioned drug. This path has enabled many investigators to obtain pilot funding for their drug repositioning projects.

Thanks in large part to the support of the CTSA Pharmaceutical Portal, the PII was able to quickly present to and answer questions from a large number of CTSA and other potential research institutional partners. By early 2011, over 20 institutions entered in Memoranda of Understanding (MOU) and became Charter Partners in the PII with Partnership for Cures. Under the MOU, a Charter Partner agrees to establish the process for selecting the appropriate studies for submission to the PII, and to provide 10% of direct costs.

Charter Partners are primarily responsible for motivating their faculty and staff scientists, clinicians, and lay members to develop and submit Rediscovery Research projects, and for cofunding the research and to assist with improving Rediscovery Research project submission, review, ranking, and selection processes. Charter Partners agree to encourage participation by other research institutions, researchers, voluntary health organizations, industry, and media and patient advocacy groups.

The PII announced its first Rediscovery Research Request for Applications to Charter Partners, including multiple CTSA universities, in February of 2011 and received 126 letters of intent, more than double the number expected. Most projects met the requirements for the PII and most investigators were encouraged to submit complete proposals. Eighty full proposals were received. The PII's Science Advisors and Lay Board members reviewed and ranked these proposals to populate the PII Model Portfolio, sample projects that show the many ways to create better treatments for patients through Rediscovery Research. At this writing, the PII Model Portfolio contains 22 Rediscovery Research projects (Table 9.4) and Partnership for Cures has already secured more than $200K in funding for two projects, plus partial funding for a third project.

A number of factors contributed to the rapid success of the PII. First, researchers and clinicians had a backlog of Rediscovery Research ideas that were not being pursued due to lack of funding. As these projects do not have the same economic and academic advancement incentives as "new discovery" research, industry cannot afford to support this type of work and most philanthropists are unaware of its benefits and potential impact. The PII's request for proposals opened the floodgates for such ideas. Second, the Partnership for Cures has a proven track record of success with Rediscovery Research.

There are dozens of examples of drugs, devices, and nutraceuticals that have serendipitously been repurposed to create new treatments [15], and the PII aims to organize this serendipity. For example, PII-sponsored clinical evaluations of the drug sirolimus and a nutraceutical established efficacy in alleviating symptoms of the deadly disease autoimmune lymphoproliferative syndrome [16] and familial dysautonomia [17, 18], respectively.

Third, while "new discovery" research is vitally important to progress in medicine, it is nevertheless expensive, time-consuming, and has a poor success-to-cost ratio. New drugs are also expensive to the consumer and require many years of use before all side effects become known. By comparison, Rediscovery Research, by virtue of starting

TABLE 9.4. Rediscovery Research Projects

Project Title	Institution
The Use of Drug X for Treatment of Seven Pediatric Autoimmune Diseases	Children's Hospital of Philadelphia
The Use of Drug X for Treatment of Children with Somatic Variant ALPS and RALD	Children's Hospital of Philadelphia
Immunoglobulin Therapy for Small Fiber Neuropathy in Sarcoidosis	Cleveland Clinic
Investigation of the Generic Drug X in Sjögren's Syndrome	Harvard Medical School
Repurposing a Nutraceutical for the Treatment of Intractable Epilepsy	IWK Health Centre in Nova Scotia
Fecal Transplantation for Recurrent and Relapsing *Clostridium difficile* Infection	Mayo Clinic
Predicting Flare-Ups of Systemic Lupus Erythematosus	Ohio State University
Nicotine Treatment for Pulmonary Sarcoidosis	Ohio State University
The Use of a Repurposed Immune Cell Drug X for the Treatment of *Burkholderia cepacia* Infection in Cystic Fibrosis Patients	Ohio State University
A Different Antibiotic for Early Treatment of Bronchiolitis Obliterans Syndrome	Ohio State University
Three Common Chemicals as Renal Protectors during Cisplatin Chemotherapy	Ohio State University
Treating the Developmental Consequences of Prematurity with a Nutraceutical	Ohio State University
Movement Therapy in Cerebral Palsy	Ohio State University
Early Treatment with an FDA-Approved Drug Attenuates Cardiomyopathy in Duchenne Muscular Dystrophy	Ohio State University
Drug X as a Therapeutic Intervention for Sickle Cell Vaso-Occlusive Pain Crisis	Saint Louis University School of Medicine
Repurposing a Foreign Approved Drug for Inflammatory Research Institute	Sanford-Burnham Medical Bowel Disease (IBD)
Improving Survival of Encapsulated Islets by Repurposing a Clinically Approved Imaging Agent as an Oxygen Carrier	Sanford-Burnham Medical Research Institute
Targeting Metastatic Breast Cancer with Non-Toxic Radiation Sensitizers	University of Chicago
Impact of Pet Ownership on Blood Sugar Control in Youths with Type 1 Diabetes Mellitus	University of Massachusetts Medical School
Targeting Leukemic Stem Cells in Chronic Myeloid Leukemia (CML)	University of Massachusetts Medical School
Commonly Used Drug Group Repurposed to Correct Cystic Fibrosis Immune Defect	University of Massachusetts Medical School
Nutraceutical Adjunct Therapy in Patients with *Clostridium difficile*	University of Texas Medical Branch

with a known entity, requires relatively little time and money, and has a much higher success-to-cost ratio. This form of drug discovery creates relatively inexpensive treatments using familiar drugs or devices whose side effects are known.

The biggest challenge for the PII is convincing patient support groups, voluntary healthcare organizations, foundations, and other philanthropic supporters to focus some of their funding on Rediscovery Research. Supporting the many Rediscovery Research ideas that are waiting to be tested could potentially help millions of patients waiting desperately for better treatments, and save billions of dollars in healthcare costs.

MATERIAL TRANSFER AGREEMENTS FOR COMPOUND TRANSFERS

One of the Portal's specific goals was to evaluate potential barriers to technology transfer. One such barrier is the negotiation of Material Transfer Agreements (MTAs). MTAs for pharmaceutical compounds are often exceptionally challenging and time-consuming for both parties to draft, negotiate, and reach consensus on compared with other types of MTAs. For example, from 2001 to 2010, the UC Davis technology transfer office successfully executed nearly 200 drug transfer MTAs with approximately two dozen pharmaceutical companies. Most of these documents were several pages long and required more than 4 months each to negotiate and execute, approximately three to four times longer than that required for other MTAs. Negotiations are commonly most heated surrounding issues of limiting the use of the compound to noncommercial research purposes; limiting company liability; delaying academic publications or presentations to protect confidential information and IP; defining confidential information; reporting research results to the company; and IP provisions.

IP terms are by far the most difficult to negotiate historically. Companies invest considerable resources in developing proprietary drugs, and understandably want to protect their freedom to use their own compounds. Universities, on the other hand, aim to maintain ownership and receive fair consideration of their discoveries and inventions, while making them available for public benefit. Technology transfer offices at some universities may compromise and agree to grant companies nonexclusive, royalty-free commercial licenses for certain discoveries and inventions. As part of the agreement for such a license, the university must keep a close nexus between the patentable invention and the drug; avoid involvement in past and future inventions; ensure that the university will be reimbursed for out-of-pocket patent prosecution costs; and be satisfied that provision of the drug is fair consideration for such a license. When these challenging and often time-consuming issues are carefully addressed, companies normally accept the terms in which the university grants the company a nonexclusive, royalty-free commercial license and a first right to negotiate an exclusive license for a patentable invention or discovery that uses or incorporates the proprietary compound.

Taking into consideration the potential IP restrictions, the Portal chose to focus initially on compounds that were either nearing patent expiration or had already lost their patent protection. We anticipated that such compounds would present with diminished IP barriers. The Portal's leading partner, the Center for World Health and

Medicine (CWHM) at St. Louis University, took the lead in developing the first comprehensive MTA template to negotiate transfers. Under the proposed structure, the CWHM would become a shared resource for housing, maintaining, and distributing the discontinued compound library. Consolidating all donated compounds into one shared resource relieves companies of the burden of negotiating MTAs with individual compound seekers. Companies would therefore only need to negotiate one MTA with the CWHM, which in turn would distribute donated compounds to other universities. The CWHM has achieved significant progress in defining the IP provisions of a template MTA. Based in part on this progress, the UIDP, a group of 35 selected individuals brought together by the Industrial Research Institute and the National Council of University Research Administrators (NCURA), chose the Pharma Portal as one of its projects in 2010. The UIDP plans to build on the Portal's initial achievements to ultimately define mutually agreeable conditions related to drug repositioning.

ESTABLISHING AN INTERNATIONAL CLINICAL COMPOUND LIBRARY AT CWHM

A freely available physical library of discontinued clinical compounds is an integral part of the Pharma Portal concept. Such a library would allow for empirical discovery of new drug uses and undoubtedly increase the chances of finding new therapies in a cost-effective manner. The not-for-profit CWHM (http://www.cwhm.org), which runs multiple drug discovery and repurposing projects for neglected diseases, is establishing such a library, called the International Clinical Compound Library (ICCL). This library is envisioned to be a high-quality pharmaceutical screening library made up of all the drug molecules that have been in human clinical trials (Phases 1 through 3), largely composed of those that were shelved (discontinued) and never advanced as marketed drugs.

The CWHM is staffed with highly skilled ex-industry scientists who offer comprehensive drug development expertise to the international academic community. These scientists collectively possess knowledge seldom found concurrently in academic and government settings. The Center utilizes translational pharmacology and medicinal chemistry techniques to create new molecular entities based on the discontinued off-patent compounds. This legacy pharmaceutical industry know-how uniquely positions the CWHM to become a valuable interface between academic research discoveries and the knowledge of what is required to transform these discoveries into viable clinical entities.

In the next three years, CWHM plans to accumulate approximately 1000 compounds in the ICCL using a dual, parallel approach: (1) by synthesizing all clinical compounds identified from CWHM databases in collaboration with international partners with synthetic chemistry capabilities, and (2) by seeking to form a Pharmaceutical Company Consortium, from which pharmaceutical companies will "donate" discontinued compounds (and compounds that are no longer of strategic interest to their company) to the ICCL. The compounds would be made available to all international clinical investigators, researching many different diseases.

An ICCL Consortium of international partner institutions has been established to undertake the ambitious task of synthesizing these identified discontinued compounds (numbering in the many thousands). The growing Consortium already consists of institutions in China, South Africa, and several universities in the United States, with interest from Brazil and other countries globally. The expectation is that an entire network of partners will contribute toward this effort, with the CWHM at Saint Louis University being the overall coordinator and central core of the ICCL. It is envisioned that partnering university chemistry departments can utilize the synthesis of compounds as a teaching tool in synthetic organic chemistry, where students, under the supervision of faculty advisors, will gain a sense of accomplishment that their efforts are contributing to a valuable biomedical resource. Partnering institutions in the developing world (especially in Africa and South America) will also benefit from interaction with the CWHM and its vast experience in all aspects of drug development.

Despite being dropped from further development, discontinued compounds have advanced drug characteristics that would make them excellent leads for diseases other than those for which they were originally intended. The ICCL will offer the opportunity to identify alternate indications for these molecules either via direct repurposing or by additional optimization of drug properties. Once established, the ICCL will become a valuable resource for translation of basic research discoveries into clinically useful therapies and provide a mechanism to accelerate drug repurposing efforts.

INDUSTRY PERSPECTIVE

The pharmaceutical industry has directed considerable effort over the last several years at implementing novel business models to address diminishing returns on R&D investments. Previously, companies focused almost exclusively on internal R&D to drive opportunities from conception to therapeutics. More recently, this model has rapidly evolved to recognize and intentionally seek to leverage expertise outside the company. Recognition that a single biological target may be important in multiple disease pathologies has also spurred interest in exploring the full medical potential of each clinical candidate.

The interest of Pfizer's Indications Discovery Unit (IDU) in leveraging scientific expertise beyond Pfizer closely matched the Pharma Portal's goals, leading to significant contributions of the IDU to the Portal strategy. To enable collaborative dialog, academic researchers (Pharma Portal members) were provided a mechanism to inquire whether the IDU had a compound for a specific target of interest. Inquiries were submitted as short proposals through the Pharma Portal website. Investigators unfortunately had no context and little guidance regarding what would constitute a successful submission, thus this open-ended strategy proved to be of limited value. Requests were denied because of lack of a suitable compound; because access was encumbered by business agreements; or because investing in the proposed development path could not be justified.

As pharmaceutical companies are key players in the success of public–private partnerships in drug repositioning, the Portal aimed to understand the factors that affect a company's willingness to participate in the Pharmaceutical Assets Portal. We discovered that different companies had vastly different perspectives on compound repositioning. Some disagreed with the idea that many clinical compounds were "sitting on the shelf." Furthermore, not all companies saw compounds "sitting on the shelf" as a significant opportunity for drug development. Some companies refused to repurpose compounds that had been discontinued for safety-related reasons, even though the risk in other patient populations might be acceptable. Others chose to out-partner the development of compounds shelved for strategic reasons (i.e., priority or funding); these compounds were therefore not considered for in-house efforts to identify new indications. Compounds that were shelved for lack of efficacy might be considered for repurposing, although many were also out-licensed or even sold at open auction. Some companies chose a narrow repurposing strategy only in specific disease areas. Still others did not feel ready to engage in discussions, as they had not yet developed a repositioning strategy. It is clear that striking the right balance between both protecting the company's bottom line and aggressively embracing new, collaborative models is a key consideration that will shape the future of the Pharma Portal project.

CONCLUSIONS

The advantages of a global public–private partnership for drug repositioning are well understood by all parties involved. Most pharmaceutical companies have publicly expressed a desire to balance commercial viability with social responsibility, an objective that can be facilitated through the Pharma Portal. Some companies have already initiated internal drug repositioning programs, with the scope of such programs guided by the company's overall strategy and priorities. Making their discontinued compounds available to the academic network, while retaining certain safeguards, is a form of R&D outsourcing. Pharmaceutical companies would gain access to a diverse community of global researchers with expertise in many disease areas, potentially leading to newfound utility for drug candidates that would otherwise remain dormant.

Although many challenges to taking full advantage of this opportunity remain, the most significant achievements of the Pharma Portal were identifying and systematically addressing the barriers to collaborative drug repositioning, such as (1) pharmaceutical companies' hesitation to release compounds to outside entities; (2) difficulties in identifying mutually beneficial matches between academic and industry investigators; and (3) prolonged MTA negotiations (Table 9.5).

Lack of information about discontinued compounds in particular, and inconsistencies in the nomenclature of molecular entities in general, is another significant barrier. This obstacle was clearly recognized by the NCGC, and led to creation of a new drug knowledge database now made available to the academic community [6]. The NCGC accepts applications for screening of the compounds in-house. If in the future NCGC is able to make some of these compounds available for distribution, we believe that the CWHM's internal expertise and its experience with the Pharma Portal would make it a strong candidate as a compound distribution center for the NCGC's pharmaceutical collection.

TABLE 9.5. Barriers to Collaboration: Current Status, Future Avenues, and the Contribution of the CTSA Pharmaceutical Assets Portal

Barriers to Collaborative Initiatives in Drug Repositioning	Current Status	Possible Future Avenues	Pharma Portal Contribution
Reluctance of pharmaceutical companies to release compounds for use as a shared resource	Companies at different stages of implementing internal repositioning strategies Few examples to date of compound transfers	Approach companies regarding molecular entities that are near or past patent expiration Utilize NCGC database as a source of knowledge about compounds	Initiated dialogue with several companies Brought to light possibility of large public–private partnership Elevated awareness of collaborative opportunity
Difficulty identifying potential collaborators	Considerable human resources are engaged in both academia and industry to understand scientific expertise and identify potential partners	Online tools offer opportunity to streamline search for complementary skills and knowledge Combining several databases may produce a powerful collaboration tool	OHSU together with Pfizer IDU developed the first prototype of the collaborative research visualization tool (FoX) UW created a readily deployed database of academic researchers as a source of potential collaborators
Prolonged drug MTA negotiations	Many MTA provisions have been standardized, including widely accepted IP license provisions	Consolidate MTA negotiations to a single shared resource vested with task of maintaining and distributing compound library	CWHM has agreed to be a compound repository site and negotiate MTAs Considerable progress in developing MTA provisions
Lack of funding for drug repositioning research	Many nonprofit organizations fund specific disease areas	A disease-independent funding source with a nimble funding mechanism	Over 20 CTSA universities signed with Partnership for Cures

UW, University of Washington; OSHU, Oregon Health and Sciences University.

In conclusion, the Pharmaceutical Assets Portal brought to light many perceived and real barriers to global public–private partnerships in drug repositioning. The Pharma Portal deployed a bottom-up strategy, focusing its limited resources on factors directly affecting investigators, such as funding, agreements, and industry outreach. This project popularized the notion that it is, indeed, possible to bring a broad consortium of academic researchers into drug repositioning efforts. We hope that together with the NCGC, the Pharma Portal and other national collaborative resources may serve the common purpose of facilitating discovery of safe and effective therapies for unmet medical needs.

ACKNOWLEDGMENTS

This publication was made possible by Grant Number UL1 RR024146 from the National Center for Research Resources (NCRR), a component of the National Institutes of Health (NIH), and NIH Roadmap for Medical Research. Its contents are solely the responsibility of the authors and do not necessarily represent the official view of NCRR or NIH.

REFERENCES

1. Adams, C.P., Brantner, V.V. Estimating the cost of new drug development: is it really 802 million dollars? *Health Aff (Millwood)* 2006;25(2):420–428.
2. DiMasi, J.A., Grabowski, H.G. The cost of biopharmaceutical R&D: is biotech different? *Manage Decis Econ*. 2007;28:469–479.
3. Ashburn, T.T., Thor, K.B. Drug repositioning: identifying and developing new uses for existing drugs. *Nat Rev Drug Discov*. 2004;3(8):673–683.
4. Kaiser, J. Biomedicine. NIH's secondhand shop for tried-and-tested drugs. *Science* 2011; 332(6037):1492.
5. Marusina, K., Welsch, D.J., Rose, L., Brock, D., Bahr, N. The CTSA Pharmaceutical Assets Portal—a public-private partnership model for drug repositioning. *Drug Discov Today Ther Strateg*. 2011;8(3–4):77–83.
6. Huang, R., Southall, N., Wang, Y., Yasgar, A., Shinn, P., Jadhav, A., Nguyen, D.T., Austin, C.P. The NCGC pharmaceutical collection: a comprehensive resource of clinically approved drugs enabling repurposing and chemical genomics. *Sci Transl Med*. 2011;3(80): 80ps16.
7. Reis, S.E., Berglund, L., Bernard, G.R., Califf, R.M., Fitzgerald, G.A., Johnson, P.C. National Clinical and Translational Science Awards Consortium. Reengineering the national clinical and translational research enterprise: the strategic plan of the National Clinical and Translational Science Awards Consortium. *Acad Med*. 2010;85(3):463–469.
8. Sommerville, I. *Software Engineering*. Boston: Addison-Wesley; 2006.
9. Wooding, S., Wilcox-Jay, K., et al. Co-author inclusion: a novel recursive algorithmic method for dealing with homonyms in bibliometric analysis. *Scientometrics* 2006;66(1): 11–21.

10. Auger, I., Balandraud, N., Rak, J., Lambert, N., Martin, M., Roudier, J. New autoantigens in rheumatoid arthritis (RA): screening 8268 protein arrays with sera from patients with RA. *Ann Rheum Dis.* 2009;68(4):591–594.

11. van der Linden, M.P., Feitsma, A.L., le Cessie, S., Kern, M., Olsson, L.M., Raychaudhuri, S., Begovich, A.B., Chang, M., Catanese, J.J., Kurreeman, F.A., van Nies, J., van der Heijde, D.M., Gregersen, P.K., Huizinga, T.W., Toes, R.E., van der Helm-Van Mil, A.H. Association of a single-nucleotide polymorphism in CD40 with the rate of joint destruction in rheumatoid arthritis. *Arthritis Rheum.* 2009;60(8):2242–2247.

12. Cooper, J.D., Walker, N.M., Healy, B.C., Smyth, D.J., Downes, K., Todd, J.A., Type I Diabetes Genetics Consortium. Analysis of 55 autoimmune disease and type II diabetes loci: further confirmation of chromosomes 4q27, 12q13.2 and 12q24.13 as type I diabetes loci, and support for a new locus, 12q13.3-q14.1. *Genes Immun.* 2009;10(1):S95–S120.

13. Lambracht-Washington, D., Fischer Lindahl, K. Does the rat have an H2-D orthologue next to Bat1? *Immunogenetics* 2002;53(12):1039–1046.

14. Price, P., Bolitho, P., Jaye, A., Glasson, M., Yindom, L.M., Sirugo, G., Chase, D., McDermid, J., Whittle, H. A Gambian TNF haplotype matches the European HLA-A1,B8,DR3 and Chinese HLA-A33,B58,DR3 haplotypes. *Tissue Antigens* 2003;62(1):72–75.

15. Hategan, G., Polozov, A.M., Zeller, W., Cao, H., Mishra, R.K., Kiselyov, A.S., Ramirez, J., Halldorsdottir, G., Andresson, T., Gurney, M.E., Singh, J. Heterocyclic 1,7-disubstituted indole sulfonamides are potent and selective human EP3 receptor antagonists. *Bioorg Med Chem Lett.* 2009;19(23):6797–6800.

16. Teachey, D.T., Greiner, R., Seif, A., Attiyeh, E., Bleesing, J., Choi, J., Manno, C., Rappaport, E., Schwabe, D., Sheen, C., Sullivan, K.E., Zhuang, H., Wechsler, D.S., Grupp, S.A. Treatment with sirolimus results in complete responses in patients with autoimmune lymphoproliferative syndrome. *Br J Haematol.* 2009;145(1):101–106.

17. Anderson, S.L., Qiu, J., Rubin, B.Y. EGCG corrects aberrant splicing of IKAP mRNA in cells from patients with familial dysautonomia. *Biochem Biophys Res Commun.* 2003; 310(2):627–633.

18. Anderson, S.L., Rubin, B.Y. Tocotrienols reverse IKAP and monoamine oxidase deficiencies in familial dysautonomia. *Biochem Biophys Res Commun.* 2005;336(1):150–156.

10

DEVELOPMENT PROGRAMS AT THE U.S. NATIONAL CANCER INSTITUTE: USE OF PUBLIC–PRIVATE PARTNERSHIPS AS A CATALYST TO ADVANCE CANCER THERAPY

Jason V. Cristofaro

Division of Cancer Treatment and Diagnosis, National Cancer Institute/NIH/DHHS, Bethesda, MD, USA

The history of drug development and screening at the National Cancer Institute (NCI) spans over 70 years; the NCI has had a screening program to test experimental anticancer drugs since 1937. Over the ensuing decades, the NCI has had a profound impact on the development of cancer therapeutics. The NCI reports that approximately one-half of the chemotherapeutic drugs currently used by oncologists were discovered and/or developed at the NCI [1]. Some of the most prominent examples where the NCI was involved in early-stage development include cisplatin, paclitaxel, and fludarabine phosphate for treating solid tumors and hematological malignancies. In these cases, important technology was created or licensed by the NCI that enabled these drugs to reach the market. The NCI has also played a role in later-stage development activities; tamoxifen, trastuzumab, imatinib, and cetuximab provide examples of NCI contributions to obtaining FDA approval for cancer therapeutics.

It has been recognized for over a quarter century that the most difficult stage of drug development is translating basic mechanistic research and early-stage targeted compounds into therapeutics suitable for clinical development—this late preclinical/

Collaborative Innovation in Drug Discovery: Strategies for Public and Private Partnerships, First Edition. Edited by Rathnam Chaguturu.

early clinical phase is commonly known as "the valley of death." Only 11% of agents entering this stage of development eventually show clinical success, with only an 8% rate in oncology [2]. This poor success rate is attributable to a wide variety of factors, but the most significant (causing over 40% of failures) is adverse pharmacokinetics and bioavailability [2]. In addition to early-stage screening, over the past 25 years, NCI's Division of Cancer Treatment and Diagnosis (DCTD) developed a variety of programs, including an expansive pharmacodynamics program, to move cancer therapeutics through late-stage preclinical and early-stage clinical development to address "the valley of death" in drug development.

DRUG SCREENING PROGRAMS AT THE NCI

In 1955, the NCI formalized its screening program by establishing the Cancer Chemotherapy National Service Center's (NSC) screening program [3]. The NSC program functioned until the mid-1980s, and most of the screening focused on the use of *in vivo* murine P1388 or L1210 leukemia cell lines [4]. These cell lines were chosen because hematologic murine tumors were inexpensive, stable, reproducible, and easily handled. Unfortunately, screening with rapidly growing leukemic cells biased the agent pool toward those compounds with activity against rapidly growing tumors and those with high growth fractions. This resulted in a relative lack of success in identifying agents with activity against common human solid tumors [5].

To address these limitations, in 1989 DCTD's Developmental Therapeutics Program (DTP) moved from using hematologic murine tumors to an array of 60 cell lines derived from a wide variety of human solid tumors [6]. The "60-cell-line screen," as it became known, remains the centerpiece of extramural DTP's screening program; over 80,000 compounds have been screened since 1990, using the current screening system [1]. Data generated from 60-cell screens are analyzed via an algorithm known as COMPARE, a program that categorizes different groups of agents based on their patterns of cytotoxic activity [7]. The program is able to classify anticancer agents based solely on the cytotoxic pattern exhibited in the 60-cell-line screen [8]. The open screening program functions to this day; applicants may submit compounds for screening through DTP's online screening application found on its website [9]. In recent years, the NCI has used the 60-cell-line screen as part of a multi-pronged approach to identify promising combinations of approved cancer therapeutics [10].

THE RAPID ACCESS TO INTERVENTION DEVELOPMENT PROGRAM

In 1998, the NCI pioneered the Rapid Access to Intervention Development (RAID) Program as a partner to the Drug Development Group (DDG, see later discussion). The goal of this program was to use NCI contract support to provide critical early research to assist academic investigators in initiating clinical trials based on their own research. The investigator held an investigational new drug application (IND), while the NCI supplied early-stage resources, including IND-directed toxicology studies, pharmacologic assay development, clinical drug formulation, and good manufacturing practice (GMP)-quality agent for use in the investigators' initial clinical trials. During its existence, the RAID Program approved 126 projects, through which 15 small molecules and 17 biologic agents later entered clinical trials [11].

In 2001, the NCI initiated a sister program to RAID—RAND (Rapid Access to NCI Discovery Resources)—to assist academic investigators with earlier stages of drug development. The RAND Program was intended to provide support for earlier drug discovery efforts than RAID, and offered resources to conduct high-throughput screening efforts, medicinal chemistry activities, and formulation studies. RAID and RAND were eventually phased out and their activities subsumed into the NCI Experimental Therapeutics (NExT) program. The impetus for these changes was recommendations by review committees that RAID and RAND would be more productive if the focus was shifted to the overall development of a therapeutic rather than as an assistance mechanism for individual investigators [12].

THE DECISION NETWORK/DRUG DEVELOPMENT GROUP

"Internal" extramural drug development prior to the NExT Program was approved and conducted by the DDG (and prior to that, its earlier incarnation, the Decision Network). DDG projects differed from RAID projects in that the NCI held the IND and was the main sponsor of the clinical development of DDG agents. DDG resources spanned a wide range of development, from early-stage screening through initiation of Phase 1 clinical trials. The DDG served as a bridge between preclinical and clinical resources provided by the NCI and had several successes, including the work conducted on halichondrin B (see case studies).

NATURAL PRODUCTS

The use of natural products as antitumor medications has a very long historical background [13]. In conjunction with the revision of the NCI's screening program in the 1980s, the NCI began a concerted effort to collect natural products to test extracts in NCI 60-cell-line screen. Over the ensuing 30 years, the DCTD Natural Products Branch has collected more than 170,000 extracts from samples of more than 70,000 plants and 10,000 marine organisms. This was facilitated by the use of letters of collection with over 25 host countries. Letters of collection are a unique mechanism by which DCTD collects natural products in partnership with host countries. As part of the agreement to allow collection, any investigator who derives an invention from the natural product is required to offer rights back to the host country, with the terms varying from country to country. In some cases, this takes the form of a royalty, and in others it may take the form of agent development in the provider country. As of 2011, the NCI holds the most diverse natural products collection in the world and is currently in the process of reformatting the distributable collection into 384-well plates to facilitate high-throughput screening activities.

CANCER THERAPY EVALUATION PROGRAM: SPONSORED CLINICAL TRIALS

In 1955, the NCI's Clinical Trials Cooperative Group Program was established by Congress at a cost of $5 million. Over the ensuing half century, the program grew to become the largest sponsor of cancer clinical trials in the world. The program functions by funding networks of researchers who develop and conduct large multicenter trials. Trials are supported via contract and grant mechanisms to the cooperative groups. Funding for the groups is not linked to a specific trial; rather, the NCI funds the infrastructure, and the cooperative groups themselves generate trials compatible with their particular areas of interest and expertise.

The NCI Cancer Therapy Evaluation Program (CTEP) is the NCI program responsible for funding and oversight of the Clinical Trials Cooperative Group Program. Since the mid- to late-1990s, CTEP has stood as the largest sponsor of cancer-related clinical trials in the world. As of 2011, CTEP sponsors over 100 INDs, has approximately 11,000 registered investigators in over 3300 institutions, has over 1000 active clinical protocols that accrue over 33,000 patients, and has entered into over 80 collaborative agreements with both large and small pharmaceutical concerns for the development of experimental cancer therapies.

NCI'S EXPERIMENTAL THERAPEUTICS PROGRAM

Over the 50 preceding years, the decision-making processes for the resources described earlier gradually became more and more complex. It was often difficult for the extramural community to gain access to or even acquire information about the many resources that the NCI had available for therapeutics development. As shown in Figure 10.1, resources were housed all over the institution and resided with multiple decision-making bodies, many of which did not coordinate their activities.

In 2009, the NCI launched the NExT Program. The primary impetus for the new program was to bring together the NCI's disparate extramural development resources into a robust, balanced, goal-driven therapeutics pipeline. This pipeline focuses on developing a broad portfolio of molecularly targeted agents to pursue new and/or challenging pathways that have not been fully exploited by available cancer therapeutics. Combined, DCTD's resources are capable of supporting a drug discovery and development continuum from initial discovery through clinical trial evaluation. Whereas in the past resources were awarded separately for different phases of development, now all the resources necessary for moving high-priority projects through the pipeline exist in a single structure, so access to the various components does not require additional research and negotiation by investigators.

It is important to emphasize that the NExT Program is not a grant-awarding mechanism, but rather an opportunity for investigators to partner with the NCI for drug discovery and development activities. Investigators gain access to resources rather than funding. In addition, by serving as scientific advisors to the project team, they enable the success of their individual projects and, collectively, the entire NExT Program.

pharmacology, toxicology, and GMP scale-up for small molecules and biologics, as well as early-stage regulatory support. In addition, concurrent molecular imaging and/or pharmacodynamic assay development is provided by the NCI's Cancer Imaging Program, National Clinical Target Validation Laboratory, Pharmacodynamic Assay Development and Implementation Section, and the Center for Cancer Research to allow early assessment of potential clinical biomarkers of drug effect. An important element of this support is providing validated assays as well as standardized specimen handling and testing procedures. Validated assays are particularly important for molecularly targeted drugs because they can demonstrate proof-of-concept that the drug is modulating its intended target in a patient's tumor, ideally using nontoxic doses. For example, the NCI validated an immunoassay for use in the first phase 0 clinical trial in oncology of a therapeutic agent under the FDA's Exploratory Investigational New Drug Guidance. This assay measured the activity of an inhibitor of the enzyme poly (ADP-ribose) polymerase (PARP) in patients' blood and tumor biopsy samples [14]. Phase 0 trials have no therapeutic or diagnostic intent—their objective is to establish proof-of-concept of a new agent with minimal risk to participants—to inform decisions about whether the agent holds promise as an anticancer drug [15]. Based on the results from this trial, the NCI decided to continue clinical evaluation of the PARP inhibitor ABT-888 in combination with several different chemotherapeutic drugs to determine whether it potentiates their effects.

LATE-STAGE DEVELOPMENT: THE CTEP CLINICAL TRIALS NETWORK

Clinical evaluation of NExT Program agents is supported by CTEP. Through CTEP, the NCI funds the largest cancer clinical trials network in the world, and the NExT Program provide agents suitable for testing to this network. The primary agents of interest for acceptance into the NExT Program at this stage are those requiring IND-directed toxicology data or agents already in Phase 1 or 2 clinical trials. In addition, CTEP maintains the collaborative aspect of the program wherein industry may submit agents in the late preclinical stage or early clinical stage for CTEP to further develop the clinical program in a niche area that is outside the pharmaceutical industry's scope. As part of this stage of development, the NExT Program provides an array of support services, including regulatory support in the form of filing and maintaining an IND and developing a clinical protocol, while also funding the cooperative group studies. In addition to standard IND support, the NCI has been a pioneer in the development of exploratory IND studies, and can use its expertise to provide support for first-in-human Phase 0 studies [14].

GOVERNANCE

The governance structure of the NExT Program is designed to streamline the decision process within and across projects. The NExT program has a similar application review structure to a grants program but provides nongrant resources to the development community, which require a different mechanism of approval and management. At the core is the project team, which is responsible for day-to-day project operational decisions within the constraints of the allocated resources. Balancing the allocation of the available resources between projects in the portfolio is the responsibility of an internal NExT Senior Advisory Committee (SAC). The SAC is responsible for decisions to initiate projects and allocate initial resources (beginning), to sustain resources throughout project progress based on project prioritization (maintenance), and to close projects by removing resources (close-out). The SAC makes decisions based on scientific input from and analysis of individual projects by the NExT Discovery and Development Committees, as well as on the availability and appropriateness of the resources required. The Discovery and Development Committees are composed of internal NCI staff with scientific expertise in the stage of development a project is assigned. The Discovery Committee provides oversight to discovery projects to ensure that all the scientific objectives have been met before triggering nomination of a clinical candidate. Once a clinical candidate has been nominated by the project team and approved by the Discovery Committee (or if the project entered the NExT Program at a later stage), project scientific review responsibility shifts to the NExT Development Committee.

To ensure a comprehensive scientific evaluation of each project, the internal NExT Committees are advised by a Special Emphasis Panel (SEP) composed of non-NCI staff who have specialized discovery or development expertise. The SEP is dynamic, consisting of members with both industrial and academic experience, and is intended to provide insight into the scientific direction of the NExT Program by bringing diverse perspectives to the conduct of a quarterly project prioritization of the NCI portfolio. SEP members include some of the most productive and experienced leading-edge researchers in the therapeutic development community. NCI relies on these members to share experience, provide guidance on appropriate targets and pathways, advise on medicinal chemistry, and provide strategic insight into what up and coming research is most appropriate for NExT development.

Overall oversight and accountability of the NExT Program is provided by the Senior Management Committee (SMC). The SMC consists of members of NCI's Senior Leadership, including the NCI Director and Directors of DCTD and the Center for Cancer Research; Associate Directors of relevant programs, including DTP; and ad hoc expert government participants, as determined by the NCI Director. The SMC performs evaluative functions and provides guidance, final conflict resolution, and resources for the fiscal stability of the NExT Program. It also has final authority in establishing policies for the operations of the NExT Program.

In addition to the oversight bodies described earlier, the discovery engine of the NExT Program, the CBC, has a separate body whose purpose is to harness the collective expertise of CBC Centers and Participants. The CBC Steering Committee meets quarterly to provide suggestions for improving CBC operations. The CBC Steering Committee consists of participants, most notably principal investigators.

TECHNOLOGY TRANSFER MECHANISMS UTILIZED BY THE NEXT PROGRAM

When the NCI began designing the NExT Program in 2009, a substantial amount of effort was put into developing appropriate models for Technology Transfer. From an intellectual property (IP) standpoint, the goals of the program are straightforward:

1. To promote the discovery and development of novel anticancer agents and ensure they are developed, but retain the option to develop them in the NCI's drug development pipeline.
2. To increase the rate and likelihood of novel cancer therapeutics becoming available to patients.
3. To bundle IP in such a manner as to encourage commercial development by an outside (nongovernment) party.

While the NCI has a significant investment in development activities, the government is neither equipped nor empowered to routinely enter into the business of producing and selling marketed drugs (but may do so under exceptional public health circumstances), so any framework had to include an effective mechanism for transferring commercial development to the private sector.

One of the overriding considerations in developing the NExT Program's IP framework was how programmatic needs would interact with the NIH's responsibilities under federal law. Whenever a recipient's research work is funded either in whole or in part through NIH research grants, contracts, and cooperative agreements, that activity is subject to the requirements of Public Law 96-517, known as the Bayh–Dole Act of 1980 (hereinafter referred to as "Bayh–Dole"). Bayh–Dole was enacted in 1980 and is aimed at turning federally funded research and development into useful patented inventions to benefit American research institutions, industries, and consumers. In general, Bayh–Dole authorizes fund recipients to retain title to inventions resulting from their federally funded research and to license such inventions to commercial entities for development. This created challenges for the NExT Program because the majority of resources the NCI supplies under the program result directly from grants and contracts. Many collaborators, especially those with later-stage technologies, were concerned that the development of IP through the NExT Program would block their commercial development of that technology. Contractors and grantees would retain ownership of any inventions developed under the program, and collaborators feared that if an invention necessary to practice the technology were developed, it might be difficult to negotiate for the rights in the absence of a prearranged licensing option.

Another portion of the Bayh–Dole relevant to the NExT Program is 35 U.S.C. Sec. 202. This section grants to the federal government a "nonexclusive, nontransferable, irrevocable, paid-up license to practice or have practiced for or on behalf of the United States" any subject invention to which a government contractor has elected rights. This section allows the government to use any data or inventions generated by the NExT Program in any internal pipeline, ideally with the consent of the owner or licensee of a lead compound, but if necessary, without consent. This provision was integral to the development of the NExT Program's IP structure because the NCI could utilize this license, especially in early-stage development, to continue work on a project even if diverse interests held IP that was generated as a product of a federal funding agreement.

DISCOVERY TO EARLY-STAGE DEVELOPMENT: THE CHEMICAL BIOLOGY CONSORTIUM

The CBC encompasses several activities, including screening, early hit-to-lead development, and medicinal chemistry activities necessary to advance a promising agent to the clinical candidate phase. The NIH/NCI had previously developed structures to accommodate the initial screening stage of the CBC in the form of the Molecular Libraries Screening Network. The Molecular Libraries Screening Network is a roadmap initiative that has a mechanism for data sharing and IP rights that is both Bayh–Dole compliant and enables the rapid transfer of screening information to other investigators and to the public at large [16]. The language in the Molecular Libraries Screening Network agreement allowed the NCI to utilize the information gathered for its discovery effort, which served as a launching point for the development of modified derivatives that would be essential for lead development.

The medicinal chemistry/lead compound development phase of the CBC required a different mechanism, however, as the IP generated at this stage needed some degree of protection to ensure future development of a successful agent. Such a mechanism had to be stringent enough to protect the lead compound's IP, while at the same time being flexible enough to ensure continued communication between program participants. The NCI had two options for how to structure IP and data rights for this phase of CBC development:

1. The NCI could seek a Decision of Exceptional Circumstances (DEC, similar to the one currently in place at the NCI-Frederick Federally Funded Research and Development Center) in which the NCI would be allowed to deviate from Bayh–Dole and require that all IP be owned by the NCI.
2. The NCI could allow Bayh–Dole to apply and have IP generators negotiate between themselves to bundle all necessary rights for later-stage development.

Option 1 had the advantage of bundling the IP in a single entity, which would avoid significant "anti-commons" problems that often arise when IP related to a technology has diverse ownership; however, there were several drawbacks to this approach. Instituting a DEC is a lengthy process and can be difficult to justify in the absence of a distinct programmatic necessity. While the plain text of the statute states that a DEC is appropriate whenever "it is determined by the agency that restriction or elimination of the right to retain title to any subject invention will better promote the policy and objectives of this chapter," in reality, DECs are only granted when a program can clearly show that the same policy objectives cannot be achieved through a less burdensome method

[17]. More importantly, there was substantial concern that requiring the transfer of all IP rights to the federal government would act as a disincentive for qualified early-stage research institutions to participate in the program.

The primary drawback of Option 2 was the anti-commons problems associated with allowing different parties to own IP in a technology. Unless all groups could come to an agreement on how to manage the IP, it would be difficult for the rights surrounding a promising therapeutic to be sufficiently bundled to allow research and commercial development of the technology.

Ultimately, the NCI decided not to seek a DEC and to allow Bayh–Dole to apply to any inventions generated under the scope of the CBC. While the anti-commons problem was a concern, the risk was mitigated by the application of the government research use license defined in the statutory language. This license allows the NCI to continue noncommercial research and development of a technology created by the CBC in the event that the owners of diverse IP could not come to an agreement on a method of cross-licensing. While this license would not necessarily solve downstream issues when the product was ready to be commercialized (as the license does not extend to commercialization), it does clear a path for noncommercial research and development related to an agent or target. A late-stage therapeutic ready for commercialization is substantially more valuable and less speculative than early-stage technologies, and if a technology was ready to be marketed, it would be in the interest of all parties owning IP related to that technology to effectively cross-license to ensure commercialization. More importantly, allowing CBC participants to own the IP generated in the program served as a substantial incentive for participation in the CBC.

With this in mind, the NCI promulgated a consortium agreement (the CBC Participants Agreement) that all members of the CBC, including government entities, are required to ratify [18]. The consortium agreement includes an overall description of the program and defines the terms by which consortium members interact, exchange material, and maintain confidential information. Membership in the consortium is ratified by the NExT SAC, and members may leave at any time, with the understanding that any materials and data developed through their participation in the CBC up to that point are subject to the terms of the agreement.

MID-STAGE DEVELOPMENT (CLINICAL CANDIDATE PRIOR TO CLINICAL STUDIES)

A clinical candidate, developed through the CBC or submitted by an outside party, that is not yet ready for testing in humans is eligible for mid-stage development resources provided by the NExT Program. At this stage of agent development, the IP for the agent is usually more clearly defined, with a party or parties possessing a composition of matter and/or use patent on the technology entering the program. The primary concern of most development partners at this stage is making certain that appropriate confidentiality is maintained and that no IP is developed that could stand as a barrier for the entry of the agent into clinical trials or for transfer to or codevelopment with another entity for commercialization purposes (many collaborators at this stage are small companies or start-ups interested in partnering with large pharmaceutical companies). The resources that the NCI can offer at this stage in development are primarily contract resources, including GMP scale-up, toxicology, and pharmacology, and these contract resources are channeled through the NCI-Frederick Federally Funded Research and Development Center contractor Leidos Biomedical Research, Inc. (through either prime or subcontracts). The advantage of this structure is that any IP developed by the contractors is subject to the DEC, which has a provision that assign rights in any inventions developed from "third-party" materials to the federal government. Third-party materials are defined as proprietary technology or information provided to NIH by an outside entity that the NCI then transfers to a contractor, including materials generated on the third parties' behalf using data provided by the collaborator. This allows the federal government to control patent prosecution and consolidate licensing activity.

The greater value of the material provided by the collaborator at this stage of development, and the need to align rights for further development, necessitated a different structure than the relatively open model provided by the CBC Participants Agreement. To that end, the NCI developed a comprehensive collaboration agreement, known as the NExT Program Material Transfer Agreement (MTA), that is entered into by the collaborative partner(s) and the federal government [19]. The NExT Program MTA's design was based on the NCI's experience negotiating RAID Program agreements and was intended to be a modular agreement that covers preclinical studies and transfer of GMP-quality material to an outside entity for the purposes of clinical trials with the addition of a clinical addendum. Unlike the CBC Participants Agreement, which focuses on a consortium-based development approach, the NExT Program MTA is intended to be an agreement between the NCI and one or two additional parties. The agreement itself contains more detailed provisions on confidentiality, data use, and publication. In addition, the NCI agrees not to patent any IP it may develop with third-party materials, unless the collaborator and the NCI mutually agree that such patents are necessary for the successful development of the agent.

LATE-STAGE DEVELOPMENT: THE CTEP CLINICAL TRIALS NETWORK

CTEP serves as the clinical development component of the NExT program. CTEP utilizes a cooperative group structure wherein the program funds cooperative groups as well as specific sites that then accrue patients to trials. Sites are free to work directly with outside collaborators for clinical trials, but also receive agents that are routed through CTEP. CTEP provides a centralized structure that holds INDs and manages large multicenter clinical studies.

THE CTEP IP OPTION: THEN AND NOW

One of the driving factors for the size and success of the program was the introduction in 1999 of the CTEP IP Option, a series of standardized terms offered to collaborators, which granted rights to data and inventions that may arise as a result of work conducted under a CTEP study. CTEP funding agreements with the cooperative groups and sites require this language to flow down to all subcontracts and be included as a term in any MTA that transfers proprietary materials and/or data provided for or generated in the course of a CTEP clinical study. The IP option greatly expanded CTEP's ability to work with pharmaceutical partners by creating a concise series of terms that flow down to any CTEP study, and substantially reduced transactional costs associated with negotiating with multiple cooperative groups and clinical sites. Collaborators could enter into a single agreement (a clinical trials agreement, clinical supply agreement, or cooperative research and development agreement) with the NCI and distribute an agent to any one of NCI's funded sites under the same set of terms.

The 1999 version of the IP option covered all inventions that arose from CTEP-sponsored clinical trials with the collaborators' agent and offered collaborators a nonexclusive, royalty-free research use license and a time-limited option to negotiate a royalty-bearing license (commonly known as most-favored nation status).

In 2003, the CTEP IP Option was revised to include language that described the disposition of rights generated from combination studies. This language offered all collaborators providing agents for a combination study a nonexclusive, royalty-free commercialization license on any invention that included their agent that arose from such studies (freedom to operate). This put collaborators on an equal footing with other entities contributing agents to combination studies and was generally viewed as a tremendous success. As of 2011, two-thirds of all cancer-related combination studies listed in ClinicalTrials.gov (over 100 combination studies) are conducted by CTEP under this paradigm. This is tremendously important: cancer therapy has moved more toward personalized medicine combining the best agents to target molecular pathways associated with tumor growth, but this approach has been very difficult to exercise in reality due to the IP constraints of combination studies.

In 2006, the NCI, at the behest of its oversight bodies, undertook a variety of initiatives with the aim of improving the speed and efficiency of the design, launch, and conduct of clinical trials [20–22]. These efforts began with detailed research into the process of clinical trial initiation and culminated in a report issued by the Institute of Medicine of the National Academies (IOM) [23]. This report provided a detailed critique of the NCI Cooperative Group structure and provided recommendations for its improvement. One of the recommendations of the IOM report indicated that the NCI should take steps to facilitate more collaboration among the various stakeholders in cancer clinical trials. The IOM report recommended that the NCI develop standard licensing language and contract templates for material and data transfer and for IP ownership in biospecimen-based studies and clinical trials that combine IP from multiple sources [23].

At the same time the IOM report was being drafted, several other issues evolved that prompted NCI to reevaluate the terms of the CTEP IP Option. The first was an increased interest on the part of CTEP collaborators in the disposition of rights related to inventions generated utilizing agent-treated samples and clinical data generated in the course of CTEP-sponsored studies. At the time the 1999 IP option was promulgated, the study of personalized medicine and biomarker development was in its infancy and collaborators were not particularly interested in what eventually happened to the downstream data and materials generated from CTEP studies. Neither the collaborative agreements nor the IP option addressed rights related to clinical data and agent-treated samples following the completion of the clinical trial. In the late 2000s, there was a growing focus on individualized medicine, rational study design, and diagnostic and prognostic assay development on the part of industry, the NCI, and regulatory authorities. The language in the CTEP collaborative agreements related to the rights in these inventions was ambiguous and contained no reference to inventions generated from agent-treated samples and clinical data from studies conducted after the primary clinical trial was completed. Collaborators asserted that they had rights in inventions generated from these materials and data, while the clinical sites felt that they should be free to utilize these resources as they desired. The ongoing discussions related to the disposition of these rights ultimately culminated in the NCI placing holds on several trial approvals after collaborators refused to sign collaborative agreements [24].

Industry partners also had growing concerns related to inventions generated under the scope of clinical trials related to new indications. The examples most frequently cited were two blockbuster agents: minoxidil (Rogaine) and sildenafil citrate (Viagra). Minoxidil, an antihypertensive vasodilator medication, was first used exclusively as an oral drug to treat high blood pressure. During clinical trials, investigators discovered that one of the side effects was hypertrichosis [25]. As of 2011, minoxidil remains one of the most popular treatments for hair loss [26]. Similarly, sildenafil citrate was initially developed as a treatment for hypertension and angina. Data obtained from clinical trials showed the agent also had utility in the treatment of erectile dysfunction [27]. In both of these cases, these indications not originally conceived by the inventors were not discovered until clinical trials were initiated. Collaborators were concerned about the slight but real possibility that a new use for their proprietary agent might arise during CTEP-sponsored clinical trials and that the language in the 1999 CTEP IP Option to negotiate an exclusive, royalty-bearing license would be insufficient to ensure their ability to market an agent for such a new indication if it did arise.

Discussions related to the IP status of clinical data and agent-treated samples were also occurring within CTEP; the Clinical Trials Working Group of the National Cancer Advisory Board recommended establishing commonly accepted language for clinical trial contracts to facilitate rapid initiation of new clinical trials [28]. To fulfill this charge, the NCI, in conjunction with the CEO Roundtable on Cancer Life Sciences Consortium, commissioned an independent study to determine what terms were most commonly agreed upon in clinical trial agreements between industry and academia. The group then developed a set of common clauses, known as the Standard Terms of Agreement for Research Trial (START) Clauses, which were made available to any party to use as a reference when negotiating an agreement for a clinical trial. These standard clauses provide common language for use as a starting point in the contract agreements that govern clinical trials. The IP portion of these standardized clauses varied substantially from the 1999 CTEP IP Option. Most notably, the study indicated that in the majority of cases evaluated, institutions would offer nonexclusive, royalty-free commercialization licenses on any invention generated as a result of an investigator-initiated clinical trial with a collaborator's proprietary agent. In these studies, the collaborator was providing both the agent and funding to support the clinical trial, and both parties felt this was a fair trade-off.

The lack of clarity on the disposition of IP rights from CTEP-sponsored clinical trials and the variance between the CTEP IP Option and standard language that most industry clinical trial agreements (CTAs) contained was becoming a major impediment in CTEP's ability to obtain proprietary agents from collaborators for use in CTEP-sponsored clinical trials. To address this issue, in 2011, CTEP promulgated a new CTEP IP Option [29].

The 2011 CTEP IP Option revision had three goals:

1. To ensure that the resulting framework incentivizes participation from all stakeholders (industry, academia, government, nonprofit organizations) in CTEP-sponsored clinical trials
2. To ensure that IP rights are managed to promote the development of treatments that reach cancer patients
3. To encourage and support vital ancillary work into the mechanism and biology of cancer treatment, and to make any research tools developed from such work as available as possible to the broader research community.

The 2011 CTEP IP Option addressed these issues by varying from the 1999 iteration in several important respects. Instead of having a single overarching option that described rights to all inventions, inventions were classified into categories based on the type of invention and studies the invention came from. Section A inventions were defined as inventions generated under the scope of a clinical study that used or incorporated the provided agent, otherwise known as "agent inventions." The types of inventions that would fall under this category would be new indications, and in some cases, unique methods of administration and dosing. Section B inventions were those inventions generated under the scope of a clinical study but did not use or incorporate the agent, as well as inventions generated using clinical data and/or agent-treated samples collected under the scope of the clinical study, otherwise known as "biomarker inventions." Section B inventions would include diagnostic, pharmacokinetic, and pharmacodynamic assays, as well as research tools related to specific compounds and broad classes of agents.

The rights the 2011 CTEP IP Option granted collaborators varied substantially between Section A and Section B. As a general rule, Section A covered a narrow scope of inventions but offered broad rights; most importantly, it offered nonexclusive, royalty-free commercialization licenses for Section A inventions. This addressed industry concerns related to freedom to operate in new indications. Section B, in contrast, covered a broad scope of inventions but offered relatively narrow rights, in the form of a research use license and a limited commercial use license for labeling and regulatory purposes only. The intent of Section B is to offer enough rights to collaborators to market an agent they provide to the CTEP program, but provide enough space for investigators to exploit the commercial potential of these inventions, either alone or in partnership with another entity.

THE COLLABORATIVE PROCESS: THEN AND NOW

Most CTEP-sponsored clinical trials were conducted via CTAs until 2004–2005. After 2005, most agreements with industry were established via cooperative research and development agreements (CRADAs). There were several reasons for the shift in agreement type:

1. The NCI was conducting more internal studies with CTEP agents, and a CRADA was necessary to allow the NCI to offer IP rights to any inventions generated under the scope of these studies. The NCI, as a federal agency, is statutorily barred from prenegotiating the disposition of IP rights in the absence of a CRADA [30].
2. CRADAs represent one of the few mechanisms by which federal laboratories and agencies may receive funding from outside sources [30]. The ability of the NCI to share some costs with collaborators became essential as appropriated funding was not sufficient to cover all the costs of these studies, especially the costs of ancillary preclinical scientific work.

While CRADAs offered advantages over the CTA mechanism, there are some limitations. The fact that CRADAs allow the transfer of funds into the federal government increases the chances of conflicts of interest; as such, CRADA approval requires greater scrutiny, including review and approval by NIH committees. As a result, CRADAs take substantially more time to initiate than CTAs. In addition, CRADAs are intended as a mechanism to collaborate with industry and therefore may not be used for collaborations with nonprofit institutions.

In light of these disadvantages, the NCI developed mechanisms for bringing in agents for particular purposes when a full CRADA is unnecessary. The most commonly used alternative mechanism is the clinical supply agreement (CSA). CSAs allow the NCI to bring in agents from outside sources for use in specific extramural clinical trials. In almost every circumstance, a CSA is used to bring in a secondary agent for combination clinical trials, and offers a quick and effective mechanism to promote combination studies without going through the laborious process of negotiating a new CRADA.

The NExT Program continued the policy of negotiating CRADAs for late-stage development projects, and also included provisions for other forms of collaborative agreements (primarily CTAs and CSAs) for the development of agents owned by nonprofit organizations.

SUMMARY

An overview of the NExT IP process is illustrated in Figure 10.3. The NCI has developed a variety of mechanisms to encourage entry into its cancer therapeutics pipeline and facilitate the transfer of technologies throughout the drug discovery and development process. The paradigm that drives this model is that the scope of rights offered to collaborators for technologies developed by the program increases as development

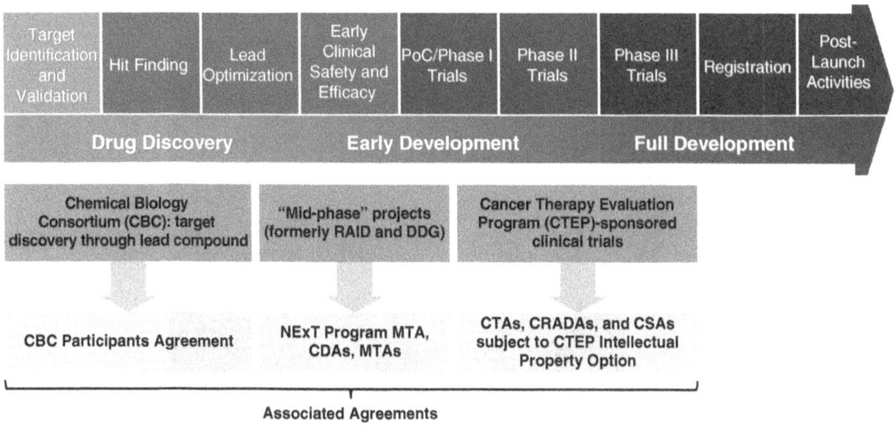

Figure 10.3. Types of intellectual property agreements depending on a project's phase within the National Cancer Institute's Experimental Therapeutics (NExT) Program. These include material transfer agreements (MTAs), confidential disclosure agreements (CDAs), clinical trial agreements (CTAs), cooperative research and development agreements (CRADAs), and clinical supply agreements (CSAs). RAID, Rapid Access to Intervention Development; DDG, Drug Development Group (both previous extramural drug development programs at the National Cancer Institute).

moves forward. The value of a technology is related directly to the stage of development, with early-stage projects having relatively little value and high risk and later-stage projects having higher value and somewhat less risk. Collaborators have difficulty partnering for late-stage development if it exposes the project to the risk of a technology being developed that may hinder commercialization and marketing approval efforts. The NExT Program addresses these concerns by bundling the appropriate degree of IP rights based on the stage of development.

The NCI's ability to create drug development partnerships using this model has been remarkably successful. Over the past 50 years, it is estimated that NCI has played a role in the development of over half the cancer therapeutics currently on the market. As the NCI moves into a new decade, substantial resources are being devoted to ensure that the NCI's leadership role in collaborating with industry, academia, and government for cancer therapeutics development is maintained.

CASE STUDIES

Case Study 1: Halichondrin B (NSC 609395)

Halichondrin B (NSC 609395) is a macrocyclic polyether initially isolated from the sponge *Halichondria okadai* by Japanese investigators in 1986 [31]. In the late 1980s, working with material provided by Dr. Robert Pettit of Arizona State University, NCI

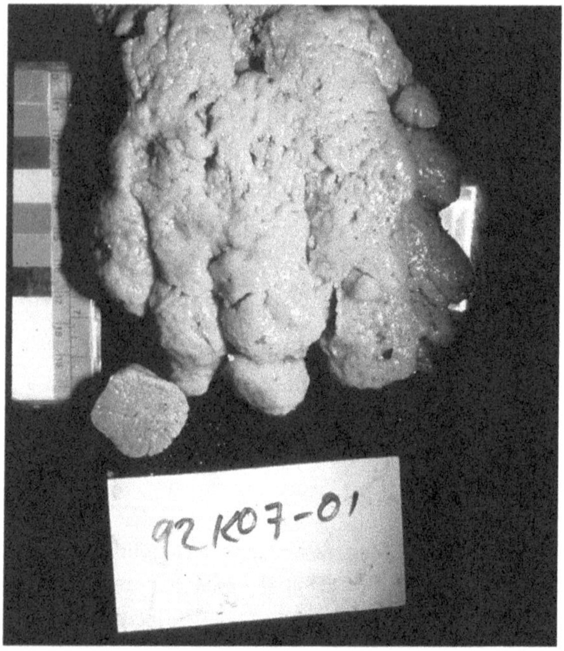

Figure 10.4. *Lissodendoryx* sp., a marine sponge harvested for isolation of halichondrin B, a naturally occurring compound with antitumor properties.

scientists discovered that it had an antitubulin mechanism of action, and in 1992, the NCI approved the molecule for further preclinical development. The primary barrier to development was obtaining sufficient quantities of material for preclinical and clinical work.

The NCI spearheaded two efforts to develop sufficient quantities of halichondrin B. The first was an arrangement with the New Zealand government, NZ academics, and a small NZ chemical company to collect, isolate, and purify halichondrin B from a deep-water collection of 1000 kg of the sponge *Lissodendoryx* sp. (Figure 10.4), following an NCI-sponsored ecological assessment of the sponge bed at a depth of more than 200 m. In addition, the NCI funded the in-sea cultivation of the sponge in mussel farms off the coast of New Zealand (Figure 10.5). The deep-water collection yielded 300 mg of pure halichondrin B from the 1 MT of sponge, a very low yield. However, the aquaculture experiments demonstrated that this method was a viable source if the material continued in development.

In the late 1980s, the NCI funded an investigator-initiated research (R01) grant to Dr. Yoshito Kishi of Harvard University to investigate the utility of a novel chemical synthetic methodology in producing complex molecules such as halichondrin B. During this work, Kishi and the American arm of Eisai Pharmaceuticals (Eisai Research Institute in Massachusetts) tested the intermediates produced during the Kishi synthesis and discovered that its activity resides in the macrocyclic lactone C1-C38 moiety [32, 33].

Figure 10.5. Sponge farms in New Zealand funded by the National Cancer Institute (NCI) to obtain greater quantities of halichondrin B. These efforts were abandoned when a method for synthetic production was discovered—NCI-funded research ultimately resulted in the approved chemotherapy drug eribulin.

The development of a synthetic analogue opened the door for the development of halichondrin B as a cancer therapeutic, as now it was possible to generate sufficient quantities for testing. Harvard University licensed the synthetic technology to the Eisai Research Institute, who then accomplished the synthesis of the resulting drug, E7389 or eribulin (NSC 707389). This compound and its chemical precursor were tested by NCI scientists with halichondrin B in *in vitro* and *in vivo* experiments, demonstrating that eribulin was superior in these models. Subsequently, eribulin was presented to the DDG (the precursor to the NExT Program) for preclinical development in 1998.

Eisai, with the assistance of the NCI through the CRADA mechanism, entered eribulin clinical trials in 2001. A variety of studies ensued, with eribulin showing particular effectiveness in metastatic breast cancer. In 2010, the U.S. Food and Drug Administration (FDA) approved the use of eribulin to treat patients with metastatic breast cancer who have received at least two prior chemotherapy regimens for late-stage disease, including both anthracycline- and taxane-based chemotherapies [34]. Eribulin is currently marketed by Eisai Co. under the trade name Halaven.

Case Study 2: Romidepsin (NSC 630176)

Romidepsin (depsipeptide, NSC 630176) was first isolated from *Chromobacterium voilaceum* by Fujisawa Pharmaceutical Company and was identified by the company in a screen for agents that can reverse the ras-transformed phenotype [35]. At the time the agent was initially reviewed by the NCI, the precise mechanism of action was not known, but based on its structural novelty and interesting activity pattern in human tumor cell-line assays, the NCI Decision Network Committee (the precursor to the DDG) accepted romidepsin for early-stage evaluation in 1991.

Development of the drug by Fujisawa was stopped after cardiotoxicity was observed in dogs. The NCI Decision Network Committee approved further evaluation to determine whether the cardiotoxicity could be overcome through modification of the administration schedule. Subsequent studies showed that intermittent treatment was better tolerated than daily dosing in dogs and mice. In addition to reducing the cardiac risk, DTP resources and expertise contributed to the development of analytical methods and a unique lyophilized powdered clinical formulation.

Romidepsin was eventually found to be a potent histone deacetylase inhibitor by Fujisawa [36]. Once this was discovered, DTP and intramural program researchers, Drs. Susan Bates and Antonio Fojo, using the NCI 60-cell-line screen to determine agents that are substrates for P-glycoprotein (Pgp), identified this compound as an agent that is transported by Pgp, and also discovered that when multidrug-resistant cells were exposed to a Pgp pump inhibitor, sensitivity to Romidepsin increased over 3000-fold, far greater than sensitivity to other Pgp substrates such as the taxanes (400-fold).

A CTA was negotiated between the NCI and Fujisawa. The IND was approved in 1996, and Phase 1 trials were conducted by the NCI intramural program and Georgetown Lombardi Cancer Center. After activity was observed in patients with T-cell lymphoma treated at the NIH Clinical Center, a Phase 2 trial was designed and opened there. Intensive cardiac monitoring was incorporated in this trial, which demonstrated no evidence of myocardial damage, but some evidence of QTc prolongation. This was subsequently noted as a side effect of almost all histone deacetylase inhibitors. In 2009, romidepsin (Istodax, Celgene Corp.) was approved by the FDA for the treatment of cutaneous T-cell lymphoma in patients who have received at least one prior systemic therapy.

Case Study 3: Chimeric 14.18

Chimeric 14.18 (ch14.18) is a mouse/human chimeric monoclonal antibody, and a potent inducer of both antibody-dependent cellular cytotoxicity and complement-dependent cytotoxicity *in vitro*. It was derived from a murine IgG3 monoclonal antibody, 14.18G2a, that had been studied in pilot clinical trials in melanoma and neuroblastoma [37, 38]. The chimeric, humanized version, Ch14.18, is more suitable for clinical development and was thus created in cultured cells using recombinant DNA technology [39].

Since the 1980s, researchers have been testing ch14.18 as an immunotherapy regimen. Dr. Ralph A. Reisfield identified the disialoganglioside GD2, which is expressed on more than 99% of neuroblastoma cells and most melanomas but not most normal cells, as a potential target for immunotherapy in 1985 [40, 41].

Dr. Alice Yu received the ch14.18 antibody from a small pharmaceutical company that was producing the antibody for early-stage trials and was purchased twice by larger companies, the second of which stopped ch14.18 production. In 1996, after Dr. Yu presented her early-phase clinical research on ch14.18 in neuroblastoma, the NCI's Biological Research Branch elected to support future development efforts. In 1997, the ch14.18 clone was transferred to the NCI and the Biopharmaceutical Development Program created master cell and working cell banks and developed a production process utilizing XCELL hollow fiber bioreactors.

Over the next several years, the NCI sponsored seven different clinical trials ranging from Phase 1 to Phase 3. The initial four were Phase 1 trials in which ch14.18 was provided to patients with metastatic melanoma either as a single agent or in combination with IL-2 or granulocyte-microphage colony-stimulating factor (GM-CSF) administered to boost antibody-dependent cellular cytotoxicity; this was followed by a Phase 1 trial of ch14.18 in combination with both IL-2 and GM-CSF. The Phase 2 testing focused on combining ch14.18 with GM-CSF, which resulted in a 1-year event-free survival of $17 \pm 7\%$ and overall survival increase of $33 \pm 9\%$.

On October 18, 2001, the Children's Oncology Group opened a Phase 3 trial, sponsored by NCI, to determine if retinoic acid + immunotherapy (mAb ch14.18 + GM-CSF + IL2) improves survival of children with high-risk neuroblastoma in first response after myeloablative therapy and stem cell transplant, as compared with standard therapy of retinoic acid alone [42]. This trial had a median 2-year follow up, which reported a 20% better disease-free survival than the control. As of 2011, studies were still progressing to collect comprehensive safety data sufficient for FDA approval of this immunotherapy for children with this rare and devastating disease.

Since 1995, the NCI's Biological Development Program manufactured 10 lots of clinical grade ch14.18, and approximately 380 patients have been treated during two NCI-sponsored clinical trials. The final lot from the Biopharmaceutical Development Program is estimated for release in November 2011, at which time United Therapeutics, NCI's development partner, will be responsible for the manufacture of ch14.18 for NCI and for completing the necessary development work for filing a Biologic License Agreement. NCI entered into a CRADA with United Therapeutics on July 1, 2010 to collaborate on late-stage development and commercialization of ch14.18 [43].

ACKNOWLEDGMENTS

The author would like to thank Jena Kidwell for her research assistance (particularly for her diligent efforts in gathering data on Chimeric 14.18) and Heather Gorby and Gina Uhlenbrauck for their help in developing the manuscript.

REFERENCES

1. National Institutes of Health. Drug discovery at the NCI: fact sheet. 2006. Available at http://www.cancer.gov/cancertopics/factsheet/NCI/drugdiscovery. Accessed December 23, 2013.
2. Kola, I., Landis, J. Can the pharmaceutical industry reduce attrition rates? *Nat Rev Drug Discov*. 2004;3(8):711–715.
3. Zubrod, C.G. The national program for cancer chemotherapy. *JAMA*. 1972;222(9):1161–1162.

4. Waud, W., Murine, L. 1210 and P388 leukemia. In: Teicher, B.A., editor. *Anticancer Drug Development Guide: Preclinical Screening, Clinical Trials, and Approval.* Totowa, NJ: Humana Press; 1997, p. 59.
5. Takimoto, C.H. Anticancer drug development at the US National Cancer Institute. *Cancer Chemother Pharmacol.* 2003;52(Suppl. 1):S29–S33.
6. Boyd, M. The NCI in vitro anticancer drug screen: concept, implementation, and operations. In: Teicher, B.A., editor. *Anticancer Drug Development Guide: Preclinical Screening, Clinical Trials, and Approval.* Totowa, NJ: Humana Press; 1997, p. 23.
7. Paull, K.D., Shoemaker, R.H., Hodes, L., Monks, A., Scudiero, D.A., Rubinstein, L., Plowman, J., Boyd, M.R. Display and analysis of patterns of differential activity of drugs against human tumor cell lines: development of mean graph and COMPARE algorithm. *J Natl Cancer Inst.* 1989;81(14):1088–1092.
8. Rixe, O., Ortuzar, W., Alvarez, M., Parker, R., Reed, E., Paull, K., Fojo, T. Oxaliplatin, tetraplatin, cisplatin, and carboplatin: spectrum of activity in drug-resistant cell lines and in the cell lines of the National Cancer Institute's Anticancer Drug Screen panel. *Biochem Pharmacol.* 1996;52(12):1855–1865.
9. National Institutes of Health. Submission of pure compounds for testing in the NCI screens. Available at http://dtp.nci.nih.gov/docs/misc/common_files/submit_compounds.html. Accessed December 23, 2013.
10. Kummar, S., Chen, H.X., Wright, J., Holbeck, S., Millin, M.D., Tomaszewski, J., Zweibel, J., Collins, J., Doroshow, J.H. Utilizing targeted cancer therapeutic agents in combination: novel approaches and urgent requirements. *Nat Rev Drug Discov.* 2010;9(11):843–856.
11. National Institutes of Health. Major ongoing initiatives. 2009. Available at http://dctd.cancer.gov/ProgramPages/dtp/major_drug_development.htm. Accessed December 23, 2013.
12. Mendelsohn, J. Report of the workshop to review NCI's Rapid Access to Development (RAID) program. 2006. Available at http://deainfo.nci.nih.gov/advisory/ncab/139_0906/presentations/MendelsohnRAID.pdf. Accessed December 23, 2013.
13. Cragg, G.M.L., Kingston, D.G.I., Newman, D.J. *Anticancer Agents from Natural Products.* Boca Raton, FL: Taylor & Francis.; 2005.
14. Kummar, S., Kinders, R., Gutierrez, M.E., Rubinstein, L., Parchment, R.E., Phillips, L.R., Ji, J., Monks, A., Low, J.A., Chen, A., Murgo, A.J., Collins, J., Steinberg, S.M., Eliopoulos, H., Giranda, V.L., Gordon, G., Helman, L., Wiltrout, R., Tomaszewski, J.E., Doroshow, J.H. Phase 0 clinical trial of the poly (ADP-ribose) polymerase inhibitor ABT-888 in patients with advanced malignancies. *J Clin Oncol.* 2009;27(16):2705–2711.
15. Murgo, A.J., Kummar, S., Rubinstein, L., Gutierrez, M., Collins, J., Kinders, R., Parchment, R.E., Ji, J., Steinberg, S.M., Yang, S.X., Hollingshead, M., Chen, A., Helman, L., Wiltrout, R., Tomaszewski, J.E., Doroshow, J.H. Designing phase 0 cancer clinical trials. *Clin Cancer Res.* 2008;14(12):3675–3682.
16. National Institutes of Health. MLSCN project team policy on data sharing and IP in the MLSCN program. 2005. Available at http://www.nimh.nih.gov/about/organization/dnbbs/the-mlscn-project-team-policy-on-data-sharing-and-ip-in-the-mlscn-program.pdf. Accessed December 23, 2013.
17. Bayh-Dole Act, U.S. Code 35. 1980, § 202.
18. National Institutes of Health. The National Cancer Institute's Clinical Biology Consortium Participants Agreement. 2009. Available at http://dctd.cancer.gov/CurrentResearch/CBC_Agreement.pdf. Accessed December 23, 2013.

19. National Institutes of Health. National Cancer Institute NExT Program Material Transfer Agreement. 2011. Available at http://next.cancer.gov/pipelineManagement/NEXT_MTA.pdf. Accessed December 23, 2013.

20. Dilts, D.M., Sandler, A.B., Cheng, S.K., Crites, J.S., Ferranti, L.B., Wu, A.Y., Finnigan, S., Friedman, S., Mooney, M., Abrams, J. Steps and time to process clinical trials at the Cancer Therapy Evaluation Program. *J Clin Oncol.* 2009;27(11):1761–1766.

21. Dilts, D.M., Sandler, A.B., Baker, M., Cheng, S.K., George, S.L., Karas, K.S., McGuire, S., Menon, G.S., Reusch, J., Sawyer, D., Scoggins, M., Wu, A., Zhou, K., Schilsky, R.L. Processes to activate phase III clinical trials in a cooperative oncology group: the case of cancer and leukemia group B. *J Clin Oncol.* 2006;24(28):4553–4557.

22. Dilts, D.M., Sandler, A., Cheng, S., Crites, J., Ferranti, L., Wu, A., Gray, R., MacDonald, J., Marinucci, D., Comis, R. Development of clinical trials in a cooperative group setting: the eastern cooperative oncology group. *Clin Cancer Res.* 2008;14(11):3427–3433.

23. Nass, S.J., Moses, H.L., Mendelsohn, J. *Committee on Cancer Clinical Trials and the NCI Cooperative Group Program Board on Health Care Services, A National Cancer Clinical Trials System for the 21st Century: Reinvigorating the NCI Cooperative Group Program.* Washington, DC: National Academies Press; 2010.

24. Goldberg, P. NCI places hold on Genentech drug trials, citing disagreement over IP provisions. *Cancer Lett.* 2010;36(9):8–9.

25. Vanderveen, E.E., Ellis, C.N., Kang, S., Case, P., Headington, J.T., Voorhees, J.J., Swanson, N.A. Topical minoxidil for hair regrowth. *J Am Acad Dermatol.* 1984;11(3): 416–421.

26. Mounsey, A.L., Reed, S.W. Diagnosing and treating hair loss. *Am Fam Physician* 2009;80(4):356–362.

27. Boolell, M., Allen, M.J., Ballard, S.A., Gepi-Attee, S., Muirhead, G.J., Naylor, A.M., Osterloh, I.H., Gingell, C. Sildenafil: an orally active type 5 cyclic GMP-specific phosphodiesterase inhibitor for the treatment of penile erectile dysfunction. *Int J Impot Res.* 1996;8(2):47–52.

28. Clinical Trials Working Group. *Restructuring the National Cancer Clinical Trials Enterprise: Report of the Clinical Trials Working Group of the National Cancer Advisory Board, United States Department of Health and Human Services, National Institutes of Health.* Bethesda, MD: National Cancer Institute.; 2005.

29. United States Department of Health and Human Services. Cancer Therapy Evaluation Program intellectual property option to collaborator. *Fed Regist.* March 2011;76(48): 13404–13410.

30. Cooperative Research and Development Agreements, U.S. Code 15. 1980, § 3710a.

31. Hirata, Y., Uemura, D. Halichondrins—antitumor polyether macrolides from a marine sponge. *Pure Appl Chem.* 1986;58(5):701–710.

32. Towle, M.J., Salvato, K.A., Budrow, J., Wels, B.F., Kuznetsov, G., Aalfs, K.K., Welsh, S., Zheng, W., Seletsky, B.M., Palme, M.H., Habgood, G.J., Singer, L.A., Dipietro, L.V., Wang, Y., Chen, J.J., Quincy, D.A., Davis, A., Yoshimatsu, K., Kishi, Y., Yu, M.J., Littlefield, B.A. In vitro and in vivo anticancer activities of synthetic macrocyclic ketone analogues of halichondrin B. *Cancer Res.* 2001;61(3):1013–1021.

33. Aicher, T.D., Buszek, K.R., Fang, F.G., Forsyth, C.J., Jung, S.H., Kishi, Y., Matelich, M.C., Scola, P.M., Spero, D.M., Yoon, S.K. Total synthesis of halichondrin-B and norhalichondrin-B. *J Am Chem Soc.* 1992;114(8):3162–3164.

34. United States Food and Drug Administration. FDA approves new treatment option for late-stage breast cancer. 2010. Available at http://www.fda.gov/NewsEvents/Newsroom/PressAnnouncements/2010/ucm233863.htm. Accessed December 23, 2013.
35. Ueda, H., Nakajima, H., Hori, Y., Fujita, T., Nishimura, M., Goto, T., Okuhara, M. FR901228, a novel antitumor bicyclic depsipeptide produced by *Chromobacterium violaceum* No. 968: I. Taxonomy, fermentation, isolation, physico-chemical and biological properties, and antitumor activity. *J Antibiot* (Tokyo) 1994;47(3):301–310.
36. Nakajima, H., Kim, Y.B., Terano, H., Yoshida, M., Horinouchi, S. FR901228, a potent antitumor antibiotic, is a novel histone deacetylase inhibitor. *Exp Cell Res.* 1998;241(1): 126–133.
37. Yu, A.L., Uttenreuther-Fischer, M.M., Huang, C.S., Tsui, C.C., Gillies, S.D., Reisfeld, R.A., Kung, F.H. Phase I trial of a human-mouse chimeric anti-disialoganglioside monoclonal antibody ch14.18 in patients with refractory neuroblastoma and osteosarcoma. *J Clin Oncol.* 1998;16(6):2169–2180.
38. Saleh, M.N., Khazaeli, M.B., Wheeler, R.H., Allen, L., Tilden, A.B., Grizzle, W., Reisfeld, R.A., Yu, A.L., Gillies, S.D., LoBuglio, A.F. Phase I trial of the chimeric anti-GD2 monoclonal antibody ch14.18 in patients with malignant melanoma. *Hum Antibodies Hybridomas* 1992;3(1):19–24.
39. Mueller, B.M., Romerdahl, C.A., Gillies, S.D., Reisfeld, R.A. Enhancement of antibody-dependent cytotoxicity with a chimeric anti-GD2 antibody. *J Immunol.* 1990;144(4): 1382–1386.
40. Cheresh, D.A., Honsik, C.J., Staffileno, L.K., Jung, G., Reisfeld, R.A. Disialoganglioside GD3 on human melanoma serves as a relevant target antigen for monoclonal antibody-mediated tumor cytolysis. *Proc Natl Acad Sci U S A.* 1985;82(15):5155–5159.
41. Honsik, C.J., Jung, G., Reisfeld, R.A. Lymphokine-activated killer cells targeted by monoclonal antibodies to the disialogangliosides GD2 and GD3 specifically lyse human tumor cells of neuroectodermal origin. *Proc Natl Acad Sci U S A.* 1986;83(20):7893–7897.
42. Yu, A.L., Gilman, A.L., Ozkaynak, M.F., London, W.B., Kreissman, S.G., Chen, H.X., Smith, M., Anderson, B., Villablanca, J.G., Matthay, K.K., Shimada, H., Grupp, S.A., Seeger, R., Reynolds, C.P., Buxton, A., Reisfeld, R.A., Gillies, S.D., Cohn, S.L., Maris, J.M., Sondel, P.M. Anti-GD2 antibody with GM-CSF, interleukin-2, and isotretinoin for neuroblastoma. *N Engl J Med.* 2010;363(14):1324–1334.
43. United Therapeutics. ch14.18 for Neuroblastoma. Available at http://www.unither.com/ch14-18-for-neuroblastoma. Accessed December 23, 2013.

11

NONINDUSTRIAL PHARMACEUTICAL RESEARCH IN THE BRIC COUNTRIES: LESSONS FOR DRUG DISCOVERY PARTNERSHIPS WITH ACADEMIC AND GOVERNMENTAL INSTITUTIONS

John Watson

Promega Corporation, Madison, WI, USA

Early drug discovery is growing in academic and governmental organizations. The convergence of two powerful drivers has resulted in a shift in early drug discovery from private pharmaceutical companies to academic and governmental organizations. The first is the lack of new chemical entities coming from the existing industrial framework [1], and the second is the desire of governmental agencies to fund translational bioscience research [2]. The outcome has been an explosion in nonindustrial small molecule screening facilities, as well as a number of new government-financed drug discovery programs at public research institutions [3]. These medicinal science resources are often designed to support basic research investigators that have expertise in an important protein target class, but little knowledge of the methods used to develop small molecule modulators of their targets. As these nonindustrial drug discovery programs mature, they have begun to face difficulties related to bringing projects into the drug development pipeline. Challenges related to intellectual property, conflicting organizational missions, transfer of projects to industrial partners, and financial support, are now

Collaborative Innovation in Drug Discovery: Strategies for Public and Private Partnerships, First Edition.
Edited by Rathnam Chaguturu.

becoming bigger issues for the public institutions and the industrial partners that are collaborating to bring this new paradigm of therapeutics development to fruition.

BRIC COUNTRIES CAN PROVIDE A MODEL FOR DRUG DISCOVERY IN NONINDUSTRIAL INSTITUTIONS

While a number of academic laboratories in western industrialized countries have developed new therapeutics, many of the successful drugs were the result of serendipitous discovery by scientists working on more esoteric basic research projects. As public research laboratories take a more deliberate approach in their drug discovery programs lessons can be learned from countries such as Brazil, Russia, India, and China (BRIC) that have a long history of governmentally funded drug programs. Historically, the majority of pharmaceuticals in these countries were produced by government-managed organizations. In the case of Brazil and India, there have been true government-driven drug discovery and development programs since at least the 1950s. In Russia and China, much of the historical investment was in generic drug production; more recently, their interest has shifted to include developing innovative medicines.

The issues faced by the governmental institutions in the BRIC countries trying to develop drugs may be predictive of the ones that will be faced in western countries in the future. These include challenges around managing governmental bureaucracies, developing transition plans to industrial partners, and aligning stakeholder objectives with the expectations of funding sources. The goal of this article is to provide a brief overview of government-financed drug discovery in the BRIC countries in the hope that it provides an understanding of how drug discovery can be performed outside of the industrial process. As western countries look to new models of drug development that include partnerships between industrial and nonindustrial collaborators, the experience of BRIC countries may help model best practices.

COMMON DRUG DEVELOPMENT ACTIVITIES ACROSS THE BRIC COUNTRIES

Before discussing programs in the specific countries, it is useful to understand some of the commonalities across the BRIC countries in terms of their nascent innovative pharmaceutical industries. One key theme in the BRIC countries is that the governments have been the major drivers for developing the pharmaceutical sciences. This is different from western countries, where private companies have driven the evolution of the drug discovery industry. The strategy executed by BRIC countries has resulted in cheap generic drugs being widely available, but has stunted the development of innovative drugs. Recently, all of the BRIC countries have come to understand the importance of having the capability to discover, develop, and produce new pharmaceutical drugs. To provide a competitive opportunity for their programs, the BRIC countries have typically focused their funding efforts on diseases with significant prevalence in their region, or on discovery of drugs from natural products unique to their country. Because the BRIC countries have historically managed their drug development programs within nonindustrial institutions, they are at the forefront of a move by western countries to transfer early discovery research to academic and governmental laboratories.

INDIA

India Was a Pioneer in Nonindustrial Drug Discovery

The first prime minister of independent India, Jawaharlal Nehru, felt that the pharmaceutical industry should be funded by the public sector. To support that strategy, he established the Central Drug Research Institute (CDRI) in Lucknow and Indian Drugs & Pharmaceuticals Limited (IDPL) near New Delhi. Together they were designed to help free India from a dependence on the profit-driven pharmaceutical industry. While the lofty goals set out by Nehru have not been fully accomplished, these two institutions have been successful in laying the groundwork for an indigenous Indian pharmaceutical industry. India has also established a number of other pharmaceutical research institutes designed to provide collaborative opportunities for industrial partners while educating the next generation of drug discovery scientists.

The pharmaceutical industry in India is rooted in its strength in medicinal chemistry related to the manufacture of small molecule generic drugs. This expertise has resulted in India becoming a global center for contract synthesis of new chemical entities with more FDA-approved manufacturing facilities than any country, other than the United States. [4].

IDPL Demonstrates Some of the Challenges Faced by a Publicly Owned Pharmaceutical Company

IDPL was created in the 1960s to provide a publicly owned manufacturing capability for pharmaceuticals important to public health. This has included responding to a plague epidemic in 1994 with the production of tetracycline, and the manufacture of chloroquine to combat a rise in malaria the mid-2000s. In addition, IDPL has been a source of low-cost birth control pills for women and rehydration solutions. Their challenge has been trying to compete in a free market economy against private companies that are not impeded by governmental work rules and political funding decisions. IDPL was envisioned to be financially self-sustaining but has required additional governmental funding support on multiple occasions. An interesting twist is that IDPL created much of its own competition as chemists from its organization moved to private pharmaceutical companies. In fact, IDPL was a key catalyst for the generics industry in India. Today IDPL is operating at significantly reduced capacity relative to its peak when it was India's largest publicly held corporation [5]. The absence of an academic research component created a lack of purpose for IDPL when it struggled to compete in the marketplace with private companies. It also was unable to garner financial support from programs that have supported basic research missions at other public drug discovery institutions. In the end, IDPL was very successful, if unintentionally, in driving success for the private pharmaceutical industry using public funds. It seems likely that the dollars used to create IDPL have been returned many times over to the Indian government through the taxes generated by private companies that have benefited from the pioneering work done at IDPL.

CDRI Shows the Promise of Publicly Financed Drug Discovery

CDRI was established in 1951, soon after Indian independence, with the intention of creating an indigenous drug research institute that could compete with private pharmaceutical companies. While the full vision may not have been realized, CDRI has been one of the most productive innovative drug discovery institutions in India, private or public, in terms of the number of new drugs developed. More importantly, CDRI's mission to drive the education of new drug discovery scientists has fertilized many of India's academic translational research programs and provided the nucleus for innovator drug development in Indian pharmaceutical companies. A number of the scientists trained at CDRI that had done overseas training have now returned to drive early discovery services at contract research organizations (CROs) based in India.

As outlined on the CDRI website, researchers at CDRI have participated in the discovery and/or development of at least 11 existing therapeutic treatments. Drugs developed at CDRI include Centchroman, a nonsteroidal oral contraceptive, Centbucridine, a local anesthetic, and Arteether, an antimalarial. Another outcome of CDRI research has been to bring drug development processes to the study of traditional herbal medicines. This has included the development of a "Standardized Herbal Remedy" termed *Bacosides Enriched Standardized Extract of Bacopa* (BESEB) that will be marketed as a memory aid and sold under the brand name Memory Sure.

The original strategy of CDRI was to develop compounds up to, and including, early-stage clinical trials. The significant upfront resources required to fund clinical trials, along with the high-risk nature of the investment, limited the number of clinical trials that were sponsored by CDRI. With the construction of their new campus in Lucknow, it appears their focus will be on early discovery and preclinical development.

Public funding for CDRI is through the Council of Scientific & Industrial Research (CSIR). Since the mission of CDRI includes investigating the basic sciences of biological processes important to drug discovery, their contribution is not just measured by pharmaceutical products produced. Scientists at CDRI have produced key publications in medicinal chemistry [6], natural products discovery [7], and preclinical toxicology models [8].

Access to compounds developed at CDRI, or various drug discovery services, are available to collaborators through their business development function. Past collaborations have included western drug companies, Indian pharmaceutical companies, the World Health Organization (WHO), and other public research institutions. CDRI represents one of the more mature organizations available in the BRIC countries for managing public–private partnerships. With a history of successful public–private collaborations, CDRI provides a potential model for other nonindustrial institutions trying to commercialize products developed internally.

Other Academic Drug Discovery Institutions in India Include NIPER and IICT

The National Institute of Pharmaceutical Education and Research (NIPER) has evolved into a multisite program focused on research and education in the pharmaceutical sciences. A central 100+ acre campus is located 250 km north of New Delhi in Mohali. In addition, satellite campuses have been built in Hajipur, Hyderabad, Kolkata, Rae Bareli, Guwahati, and Ahmedabad. Together they provide a national infrastructure to ensure that India has the technology and manpower to be competitive in the global pharmaceutical industry. NIPER is an autonomous institute housed within the Department of Pharmaceuticals, which is in the Ministry of Chemicals and Fertilizers of the government of India. Collaborations between the academic scientists of NIPER and external industrial companies are encouraged. The central animal facility can provide services for approved outside collaborators.

The Indian Institute of Chemical Technology (IICT) in Hyderabad has the broad mission of advancing all the chemical sciences, including medicinal chemistry. A number of scientists at the institute are actively involved in internal drug discovery programs or collaborations with outside companies. IICT actively solicits business development opportunities related to the synthesis of drugs or drug intermediates. Previous clients have included Sami Labs, Nicolas Piramal, Civenti, and Dabur Research Foundation.

BRAZIL

The Brazilian Technology System (SIBRATEC) Provides a Framework for Public–Private Partnerships in Brazil

Historically, Brazil's health-care research programs have been heavily invested in improving public health, not developing innovative pharmaceuticals. This has resulted in Brazil becoming a global leader in producing affordable treatments for infectious diseases, especially HIV/AIDS. As per capita income has risen in recent years, the spectrum of health-care issues has shifted to include "western" diseases such as cancer, cardiovascular disease, and diabetes. With these changes has come a realization that the Brazilian government should encourage new drug discovery.

In an attempt to expand Brazil's efforts in translational research, the government has funded a comprehensive science and technology initiative (SIBRATEC), which includes, as one component, a program for drug discovery. The intention is to help fund the development of promising compounds from academic laboratories through Phase 1 clinical trials, thereby bridging "The Valley of Death." While the program is too young to have significant completed projects, it does represent a new direction for Brazilian research, one that could help transition the country into a strong contributor to innovative drug discovery.

Brazil's Drug Discovery Programs Focus on Natural Products Chemistry and Neglected Diseases

Like many emerging market countries, Brazil focuses their drug discovery resources on areas where they have unique issues. In terms of therapeutic areas, most of Brazil's public funding is directed toward infectious disease research. This includes support for the Oswaldo Cruz Foundation (Fiocruz), which has a long history of supporting infectious disease drug development. Brazil also has built significant academic expertise in natural products pharmacology research through the support of rainforest-derived natural products drug discovery.

Outside of Fiocruz, most of Brazil's investment in early drug discovery is directly tied to principal investigators in academic institutions. Brazil's academic framework is based on a western model where the role of governmental funding is often designed to encourage certain areas of investigation while still permitting significant independence of the PI, at the project level. This research paradigm has the advantage of enabling individual academic scientists to drive new projects with some flexibility, but lacks the power of having an integrated drug discovery program at a single academic institution.

Fiocruz Produces Small Molecule Drugs, Biologics, and Vaccines within the Framework of Brazil's National Health System

Fiocruz is the primary public health institute in Brazil, with key roles in public hygiene, vaccine production, biomedical research, small molecule drug manufacturing, drug discovery, clinical trials, and health sciences education. The institute was founded as a hygiene research institute at the turn of the century in response to a bubonic plaque outbreak in Brazil. In 1902, Oswaldo Cruz, a bacteriologist with training from the Pasteur Institute, took over as director and executed on his vision of creating a global center for public health research.

Since its founding, Fiocruz has become an integral part of the health-care system in Brazil, playing the combined roles of a National Institutes of Health, a Center for Disease Control, a pharmaceutical company, and a National School of Public Health. As a governmental institution it has been a champion for public access to vaccines and therapeutics at an affordable cost. Fiocruz includes a publicly managed pharmaceutical entity with clinical trial programs and manufacturing facilities for small molecule pharmaceuticals, biologics, and vaccines [9].

Farmanguinhos and Bio-Manguinhos Produce Many of Brazil's Pharmaceuticals

Farmanguinhos (Institute of Pharmaceutical Technology) is the division within Fiocruz responsible for the production of small molecule drugs. The institute's capacity was greatly expanded upon acquisition of a GlaxoSmithKline (GSK) facility located in Jacarepagua. Farmanguinhos now produces over 1 billion doses of drugs per year and is a major manufacturer of generics in Brazil. In total, there are over a dozen major publicly owned generics manufacturers in Brazil. The ability to manufacture anti-HIV drugs has enabled the Ministry of Health to provide free multidrug therapy regimes to anyone with HIV infection, resulting in a significant reduction in mortality in Brazil due to the disease.

Bio-Manguinhos (Immunobiological Institute of Technology) is an internationally recognized center of excellence for the development and production of vaccines. It, along with the Instituto Butantan in Sao Paulo State, produces a significant percentage of the vaccines needed for public health immunizations in Brazil. Recent efforts have expanded production plans to include biologics drugs.

CTDS Is Designed to Bridge Discovery and Development

Fiocruz has begun the construction of the Center for Technological Development in Health (CTDS), which will take compounds coming from public research institutes and move them from late discovery into early-stage clinical trials. If successful, this program could make Fiocruz one the few publicly managed institutions globally that can take a compound from discovery to full commercialization. The potential opportunity of collaborating with Fiocruz has attracted GSK, Genzyme, iBio, and Pfizer to partner with Fiocruz for codevelopment of drugs. GSK signed a technology transfer agreement with Fiocruz in 2003 that has enabled the Bio-Manguinhos group to produce and distribute mumps, measles, and rubella (MMR) vaccine in Brazil [10].

Brazil's Universities Support Innovative Drug Discovery

A key strength of Brazil's academic sector is the ability to attract Brazilian-born scientists back to country after overseas training. Unlike the other BRIC countries, the majority of Brazilians choose to return to their home country even when offered opportunities in western countries. This has resulted in a number of very productive drug development research programs distributed throughout academic institutions in Brazil. Sao Paulo University has some of the strongest programs in the biomedical sciences in Brazil, which has resulted in a number of collaborations with disease area specialists at the university. For example, Dr. Maria Shikanai-Yasuda recently published that patients with coinfections of HIV and *Trypanosoma cruzi* have higher levels of *T. cruzi* than those without coinfections, suggesting that benznidazole may be the preferred therapy [11]. Dr. Gilberto Schwartsmann manages the South American Office for Anticancer Drug Development Federal University, Porto Alegre. Dr. João Batista Calixto manages the Reference Center for Preclinical Pharmacology and collaborated with Ache Laboratories to develop the anti-inflammatory drug Acheflan. Brazil provides a unique opportunity to build collaborations with world-class research scientists in a country well positioned to bring compounds through the full drug development process.

CHINA

China's Dualistic Approach to Medicinal Therapies Is Reflected in Its Research Programs in the Pharmaceutical Sciences

The *materia medica* field in China has been influenced by the use of both traditional medicines and modern western drugs. An integrated philosophy of medical treatment is evolving that combines a holistic, natural products approach to improving health, with the pharmacological treatment of specific disease states. Institutions in China reflect a range of strategies for developing improved medical therapies, from western-like drug discovery programs, such as the one at the Guangzhou Institute of Biomedicine and Health (GIBH), to more traditional programs, such as those at the Shanghai University of Traditional Chinese Medicine and the Institute of Materia Medica (IMM), Chinese Academy of Medical Sciences & Peking Union Medical College. This diversity of medical treatment philosophy is reflected in the options available to patients, with pharmacies that offer either natural product remedies or western medicines.

GIBH Uses Western Pharmaceutical Methodologies within a Chinese Public Research Institution

Following an outbreak of SARS (severe acute respiratory syndrome) in 2003, the Chinese Academy of Sciences partnered with the Guandong provincial government and the city of Guangzhou (formerly Canton) to establish GIBH as a world-class public drug discovery institute. The structure is similar to that of a U.S. biotech company, with basic science research groups developing projects that feed into biology, chemistry, and preclinical animal core resources (Drug Discovery Pipeline Department). The research groups focus on three strategic areas; Stem Cell Biology and Regenerative Medicine, Chemical Biology, and Infection and Immunology.

Much like a private biotech company, GIBH actively publishes its compound development pipeline to assure continued financial support as well as to communicate licensing opportunities for interested pharmaceutical collaborators. Current collaborations include partnering with the Center for World Health on the identification of plasmepsin V inhibitors for the treatment of malaria, and screening for the next-generation Cox-2 inhibitors with Dr. John Talley, who developed Celebrex while at Pfizer. GIBH also provides contract services for small molecule screening, drug formulation, and analytical chemistry. The business development group at GIBH facilitates industrial collaborations with the Institute.

The IMM Is a National Resource for Drug Discovery, Development, Production, and Distribution in China

The IMM has multiple missions, including internal drug development, acting as a national resource for drug development services, education of pharmaceutical scientists, and commercialization of drugs developed by the institute. This last role is very rare among public research institutions. The institute is home to six academic societies, seven national research centers, a graduate school, and is a WHO Cooperating Center for Traditional Medicine. In total, the institute is credited with developing over 100 therapeutic treatments [12].

IMM researcher You-You Tu won the 2011 Lasker-DeBakey Clinical Medical Research Award for her work isolating, identifying, and developing the drug artemisinin (qinghaosu) for the treatment of malaria. Her original work was carried out in the early 1970s as part of a secret military initiative to develop a malarial treatment from traditional Chinese medicines. After screening hundreds of medicinal extracts, her breakthrough occurred when she came across an ancient text describing a modified preparation method for preparing Qinghao (*Artemisia annua* L., or sweet wormwood) that did not involve heating and was useful for the treatment of malaria. She then performed a lower temperature extraction of artemisia and found an ether-soluble fraction that reduced Plasmodia infection in animal models. Her team went on to identify the active compound, which they named qinghaosu, but is known as artemisinin outside China [13].

As a national center for the pharmaceutical sciences, the IMM is partnered with half a dozen Chinese pharmaceutical companies for commercialization of drugs developed at the institute. In addition, the IMM has collaborations with WHO, the TB Alliance, and the Institute of Cancer Therapeutics at the University of Bradford, United Kingdom. Collaborations with other pharmaceutical companies are an articulated strategy of the IMM. A partnership with HUYA Biosciences provides a route to commercialization in the United States [12]. Given that the IMM has a commercial interest in Chinese pharmaceutical companies, it may be challenged in partnering with pharmaceutical companies developing drugs for the Chinese market.

The Shanghai Institute of Materia Medica (SIMM) Is Home to the Chinese National Compound Library

SIMM is a stand-alone institute within the Chinese Academy of Sciences, which was recently relocated to the Pudong District of Shanghai. The national research centers located at SIMM include the National Center for Drug Screening and the State Key Laboratory of Drug Research. Together they have the capability to identify new hits via high-throughput screening of the Chinese National Compound Library and to develop the drugs through to clinical trials. SIMM also offers screening services to academic or industrial collaborators.

The China State Institute of Pharmaceutical Industry (CSIPI) Transitioned from a Government-Managed Entity to a Division of Sinopharm

In a unique twist on the evolution of a public–private partnership, CSIPI (which was formerly the Shanghai Institute of Pharmaceutical Industry) was merged in 2010 with the existing Sinopharm Group and the China National Biotic Group to form the new Sinopharm. Sinopharm thus expanded from being primarily a small molecule generics manufacturing and distribution operation to one that included biological drugs and innovator drug discovery. Sinopharm is now the largest private pharmaceutical company based in China, with 24,000 employees and a market capitalization of approximately $5,500,000,000 USD.

RUSSIA

The Soviet Legacy Has Created Challenges for the Continuity of a Pharmaceutical Industry in Russia

Prior to the formation of the Soviet Union, Russia's pharmaceutical industry was similar to those of western countries, especially Germany. The private pharmaceutical industry was nationalized soon after the revolution in 1918–1919. As in many "industries" in the Soviet Union, bureaucratic inefficiencies combined with political interference to generate noncompetitive production and distribution methods. In addition, many of the chemistry research programs were under Soviet Ministries that were primarily focused on cold war priorities. The result was a distributed network of semiautonomous production facilities that were spread throughout the Soviet Union. After the fall of the Soviet Union, a privatized generics pharmaceutical industry arose from the former manufacturing facilities that had been located in Russia proper. Manufacturing plants located outside Russia, such as the major biologics facility in Lithuania (which became SICOR Biotech UAB, now part of TEVA), were separated from Russia when that country gained independence. Recently, the Russian government has recognized the need to support innovative drug discovery as an important technology opportunity.

A Strong Academic Tradition in Chemistry Is Driving Chemical Biology Programs in Russia

Academic basic research in the Soviet Union was relatively well funded, especially if the projects had potential military applications. The Research Institute of Organic Intermediates and Dyes of the Ministry of Chemical Industry was home to the founder of ChemDiv, Alexander Ivachtchenko. Like many of the current leaders in applied chemistry in Russia, he performed his graduate work at the Mendeleev Institute of Chemical Technology. One of the key chemical biology programs in Moscow is located at the A.N. Belozersky Institute of Physico-Chemical Biology. The Institute's focus is primarily on basic biological research; however, some of the faculty performs drug discovery related research in affiliation with the Center for Molecular Medicine-Moscow and the Ludwig Institute for Cancer Research.

The Institute of Chemical Biology and Fundamental Medicine in Novosibirsk is one of Russia's leading institutions focused on the interface between chemistry, biology, and human health. The Institute has sponsored international conferences on chemical biology and continues to focus on translational research, including early-stage drug discovery [14]. Russia has not developed significant capacity for high-throughput chemical screening technology in academic research. It is likely the recently announced Biopharmaceutical campus (Moscow Institute of Physics and Technology's Center of

Excellence for Innovations in Drug Development and Life Sciences Technologies) will provide access to this important technology.

The Federal Government Is Initiating a National Effort to Support Drug Discovery and Development in Russia by Partnering with Biotechs

The Russian government has helped to seed venture investment funds designed to provide capital for biotech companies interested in creating innovative drug discovery programs inside Russia. The Maxwell Biotech Venture Fund has funded multiple start-up biotechnology companies including Oncomax, Metamx, Neuromax, Cardionova, and Infectex. Their model involves active management of the companies. Rusnano is another government-supported venture fund that focuses on supporting companies interested in bridging early pharmaceutical research between discovery and late development. They have helped fund the startup of Cleveland BioLab's new subsidiary, Panacela Labs, in Russia.

A core group of Russian pharmaceutical scientists has expanded out of the successful chemistry CRO, ChemDiv. The company, started by Dr. Ivachtchenko, has become one of the largest contract chemistry companies in Europe. Their success in the services business has enabled the creation of the ChemRar Hi-Tech Center in Moscow, which is designed to enable drug discovery and development up to late-stage clinical trials. They currently have a joint program with Novartis to codevelop novel treatments for tuberculosis, as well as a technology sharing agreement with Pfizer. They will likely be benefactors of Russia's federal target program, "Development of the Pharmaceutical and Medical Industry of the Russian Federation until 2020 and Beyond."

BRAZIL AND INDIA PROVIDE RELATIVELY WESTERN MODELS OF COLLABORATION

Opportunities for collaboration with academic scientists in Brazil and India will likely have fewer burdens to entry for western companies than those in China and Russia. There are multiple factors contributing to this, including the fact that China and Russia are still influenced by a historical political ethic based on communism, which may limit their perception of the value of collaborating with private companies. In addition, Brazil and India have evolved academic research systems similar to those of western countries, with open competition for research grants and common expectations of academic productivity. The Indian government has a history of actively supporting collaborations between its major research institutes and private companies, including business development services and contract research offerings in many of their major public research institutes. Brazil has historically been more focused on enabling public institutions, such as Fiocruz, to deliver pharmaceutical products directly to the Brazilian population. More recently however, the Brazilian government has begun to encourage collaboration with private pharmaceutical companies. This has provided a more formal framework for public–private collaborations and should lead to an expansion of Brazil's role in the global drug discovery industry.

COLLABORATION WITH CHINA WILL BE DRIVEN BY THE OPPORTUNITY TO ENTER A MAJOR PHARMACEUTICAL MARKET

Virtually all global pharmaceutical companies are aggressively pursuing strategies that would expand the sales of their products in China. Historically, the Chinese government has often enabled access to their market via joint ventures that were designed, in part, to educate local businesses on the commercial fundamentals of a particular industry. This has resulted in many of the major pharmaceutical companies establishing drug discovery research centers in China, primarily in the Shanghai area. Collaborations with public research institutes in China may provide an alternative route for access to this market.

RUSSIA WILL REMAIN A COMPLEX OPPORTUNITY FOR COLLABORATION

While Russia has articulated a strategy that encourages foreign companies to establish drug discovery programs in the country, execution of the opportunities remains complex. Governmental regulations and bureaucratic structure provide obstacles to setting up a business within Russia. However, many of Russia's top academic scientists are well connected to key governmental decision makers; therefore, collaboration with established public research institutes may provide access to both scientific talent and key governmental officials.

SUCCESSFUL COLLABORATIONS IN BRIC COUNTRIES WILL OFTEN BE DRIVEN BY RELATIONSHIPS WITH THEIR WESTERN-TRAINED SCIENTISTS

A common theme for successful cross-border academic-industry collaborations in the BRIC countries is the role played by researchers that have worked, or trained, outside of their home country. They often provide connections to decision makers and an understanding of the social protocols necessary for building long-term relationships. Since most good collaborations are driven more by soft issues than hard numbers, it is important to have a liaison between the two parties that understands issues relevant to both sides. A project liaison to facilitate the collaboration should be identified early in the partnership process. Proper preparation and a strong understanding of a specific country's translational research priorities will be key to driving successful public–private partnerships in the BRIC countries.

LEVERAGING THE EXPERIENCE OF NONINDUSTRIAL DRUG DISCOVERY IN THE BRIC COUNTRIES WILL HELP INFORM WESTERN STRATEGIES

As the United States and European governments expand funding for chemical biology programs to include high-throughput screening, lead optimization, and preclinical development, it will be useful to utilize the lessons learned in the BRIC countries. One theme is that the creation of "pseudo" private companies will be challenging. IDPL struggled because they were not given full public funding support, nor were they allowed to compete as a fully private company. China has been more successful by cleanly moving CSIPI into Sinopharm while providing key academic research institutes with sufficient funding to continue their missions.

Another lesson from IDPL is that publicity funded programs should have a basic research program and/or a scientific training mission. If an organization intends to secure long-term public funding, it is crucial to have a "noncommercial" function. Institutions such as Fiocruz, IMM, and CDRI have the ability to commercialize new therapeutic products but maintain a responsibility to participate in public health research. A publicly funded drug "company" without a research mission would likely fall victim to the competitive pressures in the pharmaceutical industry. It will be interesting to see how GIBH evolves, given that its mission is heavily weighted toward the successful commercialization of new therapeutics with less focus on basic research funding.

Finally, the BRIC countries are informative on how political priorities can drive nonindustrial drug discovery programs. Both Brazil and India have struggled in the transition to innovative drug discovery because of their history of prioritizing low-cost public health care over supporting the private pharmaceutical industry. A similar issue may arise in Europe as European governments try to find support from private companies for their publicly financed chemical biology programs. If the European governments hesitate to adopt higher cost, innovative medicines due to government-funded health-care cost restraints, there may be little incentive for private companies to partner with academic researchers developing innovative drugs. It is possible that Europe will focus its publicly funded drug development on therapies relevant to neglected diseases where commercialization would be based more on public welfare and less on revenue potential. Russia and China provide another interesting perspective relative to the role of the military in drug discovery research. During the cold war, both countries channeled much of their public funding to the development of therapeutics that were perceived to have potential use by the military. It is likely the United States funded similar research, and continues to do so. One concern is that the trend toward open source drug discovery, where the data on drug development methodology are published, will not benefit from military research. Hopefully, the majority of public funding for drug discovery and development in the United States is channeled into institutions that can publish their data, so that all scientists involved in drug discovery can benefit from the knowledge developed.

REFERENCES

1. Perks, B. UK plans 30% funding bump for translational research. *Nat Med.* 2011;17:394.
2. Paul, S., Mytelka, D., Dunwiddie, C., Persinger, C., Munos, B., Lindborg, S., Schacht, A. How to improve R&D productivity: the pharmaceutical industry's grand challenge. *Nat Rev Drug Discov.* 2010;9:203–214.
3. McDonald, P., Roy, A., Taylor, B., Price, A., Sittapalam, S., Weir, S., Chaguturu, R. High throughput screening in academia: drug discovery initiatives at the University of Kansas. *Drug Discovery World* 2008;Fall:59–74.
4. Gupta, J. Drug Discovery and India: a force to reckon with. Pharma Focus Asia, 2008. Available at http://www.biojobblog.com/uploads/file/www-pharmafocusasia-com_strategy_drug_discovery_india_force_to_reckon-h.pdf. Accessed December 23, 2013.
5. Singh, P. IDPL, India's oldest drug PSU, languishing in wilderness. The Financial Express, 2009. Available at http://www.financialexpress.com/news/idpl-indias-oldest-drug-psu-languishing-in-wilderness/437406/. Accessed December 23, 2013.
6. Bhandari, K., Srinivas, N., Marrapu, V., Verma, A., Srivastava, S., Gupta, S. Synthesis of substituted aryloxy alkyl and aryloxy aryl alkyl imidazoles as antileishmanial agents. *Bioorg Med Chem Lett.* 2010;20(1):291–293.
7. Banga, J., Tripathi, C.K. Purification and characterization of a novel heparin degrading enzyme from *Aspergillus flavus* (MTCC-8654). *Appl Biochem Biotechnol.* 2010;160(4): 1004–1016.
8. Bhadauria, S., Mishra, R., Kanchan, R., Tripathi, C., Srivastava, A., Tiwari, A., Sharma, S. Isoniazid-induced apoptosis in HepG2 cells: generation of oxidative stress and Bcl-2 down-regulation. *Toxicol Mech Methods* 2010;20(5):242–251.
9. Rodriques, D. Brazil: an emerging partner in drug R&D. *Drugs* 2009;12(8):497–502.
10. Esteves, B. Hopes build that new infrastructure can aid drug discovery. *Nat Med.* 2011; 17:1174–1175.
11. Teixeira de Freitas, V., Vicente da Silva, S., Sartori, A., Bezerra, R., Westphalen, E., Molina, T., Teixeira, A., Ibrahim, K., Shikanai-Yasuda, M. Real-time PCR in HIV/*Trypanosoma cruzi* coinfection with and without chagas disease reactivation: association with HIV viral load and CD4+ Level. *PLoS Negl Trop Dis.* 2011;5:e1277.
12. Gingras, M. Drug discovery in China. *Nat Rev Drug Discov.* 2008;7:869.
13. Neill, U. From branch to bedside: You-You Tu is awarded the 2011 Lasker~DeBakey Clinical Medical Research Award for discovering artemisinin as a treatment for malaria. *J Clin Invest.* 2011;121(10):3768–3773.
14. The Institute of Chemical Biology and Fundamental Medicine of the Siberian Division of the Russian Academy of Sciences (until 2003 the Novosibirsk Institute of Bioorganic Chemistry). *Acta Naturae* 2009;2009(2):13–14 (English edition).

12

DEATH OF DRUGS AND REBIRTH OF HEALTH CARE: INDIAN RESPONSE TO DISCOVERY IMPASSE

Bhushan Patwardhan

Interdisciplinary School of Health Sciences, University of Pune, Pune, Maharashtra, India

HEALTH AND DISEASE

The last 50 years have been perhaps very remarkable and unique in the entire history of medical sciences. In fundamental sciences, the discovery of the double helix and completion of the Human Genome Project were achievements that were unparalleled. In diagnostics, technologies like magnetic resonance imaging (MRI) and positron-emission tomography (PET) scan added the brightest feathers ever to our caps. In surgery, bypass, transplants, prosthetics, and robotics changed life expectancies upwards dramatically. In therapeutics, chemotherapy and biologics revolutionized longevity and the quality of life. The Human Genome Project has allowed new understanding of genomics and epigenomics, progressing toward personalized medicine. Developing countries like India lately have come to be identified and distinguished from the developed world through three key characteristics: disproportionate population growth, double burden including infectious and lifestyle diseases, and health care becoming increasingly out of reach for the most of their inhabitants. Leaders, policymakers, and scholars the world over are grappling with possible solutions to these complex, not just complicated, problems—because there is no single magic bullet and no single solution [1].

Collaborative Innovation in Drug Discovery: Strategies for Public and Private Partnerships, First Edition.
Edited by Rathnam Chaguturu.

The World Health Organization defines health as a state of complete physical, mental, and social well-being and not merely the absence of disease or infirmity. However, there is no way to measure "well-being" and so generally absence of disease or infirmity is considered as "health." In many parts, the terms "health" and "medicine" are actually used as synonyms. It is possible to diagnose and treat diseases, but it is very difficult to diagnose or attain true health. Medicines and surgeries exist to treat diseases or infirmity but not for attaining health. Ayurveda, the Indian science of health and wellness, as well as Traditional Chinese Medicine, consider health in a more holistic sense [2]. Ayurveda defines health in a much more positive and broader manner as "Prasanna atmendria mana" where body, mind, and spirit are in the state of happiness or bliss.

The health of an individual depends on four primary determinants, including nutrition, lifestyle, environment, and genetics. Only when any one or more of these foundational pillars of health become weak is health care then required by way of medical interventions. Significant advances in biology and genetics offer good understanding of several genes that may be responsible for many diseases, but our knowledge about the genes involved in health is very limited. The present undue focus on medicine to achieve health is also unsustainable. Health care is not only about treating a disease or disorder, but it is more about dealing with a person in a holistic manner involving mind, body, and spirit. Here, traditional knowledge from Indian Ayurveda and yoga will be of immense importance for health promotion. Therefore, it is important to adopt health promotion and prevention strategies rather than merely focusing on therapeutics. Although this may sound detrimental to pharmaceutical business, as a long-term vision, this change in mind-set will do a lot of good to the industry as well as society and so will emerge as a sustainable solution to present impasse facilitating the "rebirth of health care."

DEATH OF DRUGS

Although sometimes they are portrayed as being evil, toxic, or bad, overall, drugs, when used wisely, are beneficial and good. Modern drug discovery has significantly contributed to improved quality of life and has been responsible for curing, eradicating, and controlling several diseases and disorders. While many drugs may have untoward effects, the benefits from drugs are far reaching. It is unreasonable to aggressively critique modern drugs as killers—in most cases these problems are due to misuse of painkillers, tranquilizers, opioids, and such other drugs [3].

At the same time, we also must accept that the "drugs are dying." This may sound strange but it is reality. This is not a new phenomenon but has been happening for many years. A critical comparative review of monographs from *British Pharmacopoeia* from the first edition in 1864 to the latest edition in 2013 will reveal this fact easily. Very few exceptional drugs (like aspirin) have been able to sustain the real test of time. Many other drugs have actually died for several reasons, making any pharmacopeia an extremely dynamic document.

Drugs fail or die due to several reasons, but most of the time it is because of toxicity, untoward events, or other serious safety-related issues. Drugs also die because of microbes or systems developing resistance or maybe simply because the target on which the drug is acting gets modulated, making that drug redundant. Drugs acting on single targets as agonists or antagonists are likely to produce direct and or indirect cascading effects in the whole physiological system because a body actually constitutes a meta-network of targets interacting with each other to regulate complex biological metabolic processes [4].

The systems biology principles indicate that a seemingly simple drug acting on a specific target can actually trigger a complex reaction somewhere else quite similar to a butterfly effect. In many cases, these are the main reasons of untoward effects of present drugs, which in the long run may lead to death of that drug. Drugs also die if they are ineffective. Interestingly, drugs are also killed by pharmaceutical industries for business reasons. When new-generation drugs are developed, older ones automatically die. As a principle, drug discovery scientists would like to kill a drug as early as possible in the pipeline, as carrying a bad drug toward the end of the pipeline only increases the risks and costs.

Many drugs are getting pulled off the market due to the findings made during postmarketing surveillance mandated by the regulatory authorities like the Food and Drug Administration (FDA). This is more worrisome especially because the regulatory process to approve any drug is indeed a very stringent process, which ensures safety and efficacy of any new drug through series of chemical, biological, and clinical investigations. Still, starting with Thalidomide in the decade of 1950 till today, hundreds of drugs have been killed by the regulators and were withdrawn from the market.

A representative list of such drugs withdrawn from market is also available on Wikipedia (http://en.wikipedia.org/wiki/List_of_withdrawn_drugs). This massacre of drugs has resulted in significant financial loss, putting the whole sector in a panic. Clearly, market-driven approaches to drug discovery have not been successful or sustainable.

INNOVATION DEFICIT

Agreeably, the drugs landscape is not a static picture. It is a highly dynamic terrain where, like waves in the surf, drugs appear and disappear. This dynamism gets shaped by health care, but complementarily, and in turn it also shapes health care. Drug discovery in modern times is closely and strongly connected to huge financial stakes, not only from a profit point of view but also from the perspective of investment in future drugs research. When a drug dies, no matter when or at what stage of its life cycle, it represents a huge financial loss to the pharmaceutical industry. Actually, the longer it survives in the pipeline, the greater is the financial loss. Naturally, the drug pipeline does not get dry or empty since new drugs follow and replace the old ones. This flow keeps and has kept the industry going.

It was hoped that genomics and new high-throughput technologies would help the discovery process. However, in reality, the rate of new chemical entities introduced in the market has actually declined. The discovery pipelines are drying, are becoming inefficient, and are extremely expensive. A recent analysis by the Centre for Medicines Research in the UK has concluded that since 2008, the failure rate for drugs in clinical trials has been rising [5].

At present, the pharmaceutical industry is facing a serious innovation deficit, which has led to many closures, mergers, acquisitions, and strategic alliances witnessed never before in the past. The number of new drugs coming on the market has been considerably reduced despite the availability of newer, more powerful high-throughput technologies. Experts are appealing the need for immediate innovation by asking powerful questions, seeking the outliers, accepting defeat, and populating astutely [6]. The one-disease, one-target, one-drug approach seems to be unsustainable. Moreover, the focus of discovery is rapidly shifting from chemistry to biology. Many thought leaders have started arguing that in a complex biological system, while a quick fix can be best obtained by manipulating single targets, in the long run, one has to closely look at the whole system and examine multiple targets. Scientific advances in systems biology and pharmacogenomics is driving therapeutics toward personalized and integrative medicine.

REDUCTIONIST DRUG DISCOVERY TO HOLISTIC MEDICINE DISCOVERY

Historically, the pharmaceutical sector has witnessed many approaches in drug discovery, including ethnopharmacology, serendipity, chemistry-pharmacology, structure–activity-based synthesis, pharmacophore, and, more recently, chemical biology, genomics, and systems biology bringing more specificity. Since the discovery impasse has become very serious, it is important to take a pause and rethink a few basic concepts and strategies in discovery and therapeutics.

The Oxford dictionary defines a drug as "a medicine or other substance which has a physiological effect when ingested or otherwise introduced into the body." It also extends this to "a substance taken for its narcotic or stimulant effects, often illegally." On other hand, the term "medicine" is defined as "a drug or other preparation for the treatment or prevention of disease." Thus, typically, "drug" means a substance that may have medicinal or poisonous properties. Historically, drugs have been developed from poisonous sources. Thus, some of the early discovered drugs like morphine, cocaine, tubocurarine, codeine, quinine, colchicines, belladonna, and many others actually originated from potentially poisonous plants. This is particularly true because it is often easier to identify the poisonous characteristics of substances than it is to recognize medicinal properties [7]. We are still continuing the same mistake of confusing a drug and a medicine, and (perhaps as a result) our efforts still are in a direction of "drug discovery" and not "medicine discovery."

The traditional knowledge systems like Indian Ayurveda and traditional Chinese medicine will be extremely valuable to identify real "medicinal" resources for bioprospecting, discovery, and development of safer and effective medicines. Such a strategic shift from "drug discovery" to "medicine discovery" might open up entirely new avenues to usher in the age of health and its care and not remain restricted to disease treatment. A thorough examination of Ayurveda may help jump-start the drug discovery engine anew.

INDIAN PERSPECTIVE

India has history of health, wellness, and indigenous medicine that has lasted a few thousand years. However, during the last few centuries, when modern medicine was evolving with the help of advances in biological and chemical sciences, India was left behind the developed world. As a result, until its independence in 1947, the Indian pharmaceutical sector was dominated by western pharmaceutical companies, which controlled more than 80% of the market share. However, post independence, the Indian pharmaceutical sector received due policy attention, especially after the World Trade Organization (WTO) regime initiated major policy changes with an objective of achieving self-reliance for affordable health care to the masses.

As a signatory to WTO, India complied with Trade-Related Intellectual Property agreement and from January 2005, reintroduced protection for product patents and discovery pipelines through exclusive marketing rights. This partially addressed a few concerns of the global pharmaceutical industry. Following this, India has made multiple changes in the policy framework as a part of economic reforms, which include import liberalization and removal of restrictions on foreign firms. This also helped India to facilitate the long-awaited transformation from an image of mere copycat to an emerging knowledge partner in pharmaceutical R&D, innovation, and discovery.

INDIAN GOVERNMENT POLICY

The government of India, through the 12th Five-Year Plan (2012–2017), is planning to boost pharmaceutical R&D and human resource development by setting up national centres of excellence for R&D in phyto-pharmaceuticals, nano-pharmaceuticals, and biosimilars; national facilities for new drug discovery; and national centre for active pharmaceutical ingredients. New schemes for supporting R&D in industry through assistance for setting up Good Laboratory and Good Manufacturing Practice-compliant animal house facilities are being proposed during the plan period.

The Indian government is also in a process of setting up a Pharma Venture Capital Fund to support entrepreneurship and innovations in drug discovery, including incubators for translational research, Pharma Innovation and Infrastructure Development Initiative for R&D infrastructure, and funding to private sector initiatives in private–public partnership mode. The emphasis will be on supporting extramural research in the development of new drugs, novel delivery systems, and dosage forms for mass affliction diseases like Chikungunya, tuberculosis (resistant strains), leishmaniasis, malaria, and lifestyle diseases like diabetes, cardiovascular disorder, and metabolic disorder [8].

PHARMACEUTICAL SECTOR LANDSCAPE

The worldwide sales of medicines are expected to reach 1.5 trillion USD by 2020. Emerging markets like India represent the fastest-growing segment of the global pharma industry. Most of the projected increase in revenues is expected to come from branded generics rather than innovator products. The Indian pharma industry is on a good growth path and is likely to be in the top 10 global markets by value by 2020 [9]. Reputed agencies including McKinsey [10] have predicted that Indian pharmaceutical market will reach 20 billion USD by 2015 with an annual growth rate of 12.3%. Indian drugs are exported to around 200 countries in the world with highly regulated markets, including the United States, the European Union, and the United Kingdom. The major therapeutic categories of export are anti-infective, anti-asthmatic, and antihypertensive. The Department of Pharmaceuticals of the government of India has a vision to make India the largest global provider of quality medicines at reasonable prices. The government is supporting activities to develop human resources for pharmaceutical industry and drug R&D in public–private partnerships to enable availability, accessibility, and affordability, especially of essential and life-saving drugs.

India has over 24,000 pharmaceutical companies, of which around 250 fall under the organized category, which control nearly 70% of the market. About 8,000 small-scale units, including five Central Public Sector Units, together form the core of the pharmaceutical industry. About 75% of the top 20 pharma companies are Indian-owned. Some of the major Indian private companies are Ranbaxy Laboratories, Dr. Reddy's, Nicholas Piramal, Torrent Pharma, Cadila, Cipla, Lupin, Orchid, Sun, Centaur, Wockhardt, Biocon, and Serum Institute, just to name few. The foreign companies in India include Abbott India, Astra Zeneca India, Aventis, Burrough-Wellcome, GlaxoSmithKline, Merck India, Novartis, Pfizer, and Wyeth. Prominent Indian industry associations include the Organization of Pharmaceutical Producers of India (OPPI), the Indian Drug Manufacturers Association (IDMA), and the Association of Biotechnology-Led Entrepreneurs (ABLE), which are focused on drugs, pharma, and the biologics sector, while the Confederation of Indian Industries (CII) and the Federation of Indian Chambers of Commerce & Industries (FICCI) also contribute significantly to keep track of policy and trends in this sector.

SITUATION ANALYSIS

Major consulting firms, including McKinsey, PwC, E&Y, and KPMG, have released well-researched reports highlighting opportunities and policy drivers to boost this sector. There has been valid criticism that the indigenous Indian pharmaceutical industry has not been original and innovative enough. No major drugs or new chemical entities have emerged out of this large sector, despite the availability of valuable traditional knowledge, scientific talent, and infrastructure.

In recent years, Indian companies have started partnering with multinational companies (MNCs) to benefit from the R&D (formulation development) and manufacturing capabilities of the Indian partners and the extensive marketing and distribution footprint of the MNCs in emerging markets. There is also an increasing trend among MNCs for partnering in the domestic market, where the marketing and distribution footprint of Indian companies and the product portfolio of MNCs are being leveraged.

Six biotech parks are functioning in various parts of India under the public–private partnership, and 10 parks are in different stages of development. Several world-class scientific institutions like the National Institute of Pharmaceutical Education and Research, Central Drug Research Institute, Industrial Microbial Technology, National Chemical Laboratory, Indian Institute of Chemical Biology, All Indian Institute of Medical Sciences, Indian Institute of Chemical Technology, National Center for Biological Sciences, National Center for Cell Sciences, the Indian Institute of Science, and the Center for Cellular and Molecular Biology further lend support to the sector. The Council for Scientific and Industrial Research actively collaborates with Indian and foreign companies for contract and collaborative research projects. The Indian government has promoted development of special economic zones (SEZ) for the pharma sector. There are 19 dedicated SEZs in India at various stages of development. Functional pharmaceutical SEZs in India include Jawaharlal Nehru Pharma City (JNPC) in Visakhapatnam (Andhra Pradesh); PHARMEZ (Gujarat), developed by Zydus Infrastructure; and PhaEZ Park (Gujarat), developed by Cadila Pharma. All these developments are expected to give a necessary boost to promote R&D and innovation in the pharmaceutical sector.

INTELLECTUAL PROPERTY

The pre-WTO regimen may have been responsible for Indian pharmaceutical companies emerging as super copycats by developing new cheaper alternative processes to manufacture novel drugs without investing much in innovations, research, and development. This may have also given some temporary financial advantages to Indian companies. However, it also caused significant long-term harm by actually killing the spirit and culture of original research in drug discovery. Therefore, Indian pharmaceutical companies remained more as bulk drug or formulation manufacturers rather than original inventors. In the year 2005, India had to change patent policy in line with the WTO, giving special protections for patents and generic manufacturers. These provisions prohibit the practice of "evergreening," where patents are attempted with minor changes to existing drugs. A WTO agreement also allows countries to force a company to license a patented drug to a generic company. However, frequent use of these provisions is making major global players unhappy.

Affordability has become a major social and political issue, and the governments, especially in poor countries, are under pressure to make exceptions and allow evergreening for a few life-saving drugs. For example, under the compulsory licensing provisions of patent laws, Indian drug manufacturer Natco Pharma was allowed to produce Nexavar (Bayer's kidney and liver cancer drug), making it available at a reasonably affordable price. Bayer's cost of a 1-month supply of Nexavar is about 5600 USD, compared with Natco's generic version, which costs 175 USD a month. Just recently, the Intellectual Property Appellate Board rejected Bayer's appeal to stop manufacturing of the generic version by Natco. Although the patent appeals office has directed Natco to pay 7% royalties on net sales to Bayer, still the economic losses to the discovery companies would be quite substantial. Similarly, in 2006, India denied Novartis a patent for Glivec on the grounds that modification of an existing substance imatinib was not sufficient reason to grant a patent. Such discretionary decisions are making global majors extremely uncomfortable.

At present, India is grappling with finding a reasonable balance between satisfying MNCs on one side and on the other side assuring needy people that life-saving drugs will be made affordable by encouraging indigenous technology. Managing this intellectual property conflict situation is really a tough task. In the long run, the Indian pharmaceutical industry will have to become more original and innovative by developing new discovery models that are inspired by traditional knowledge systems like Ayurveda [11].

NEW PARADIGMS

The Indian pharma sector has now realized the importance of investing for original and innovative R&D. Many industries have strengthened in-house R&D facilities and have started active collaborations with academia as well as other industries. There are increasing incidences where MNCs are either partnering or contracting R&D to India. Indian companies are now working closely with MNCs such as Takeda, BMS, Merck, Pfizer, Abbott, and GlaxoSmithKline for R&D collaboration. This indicates an increasing trend among progressive Indian industry to be a part of global drug discovery programs. Many R&D activities are getting shifted or contracted to Indian companies. The consequences of the "patent cliff" and emergence of opportunities in biologics and biosimilars are bound to create significant opportunities as well as threats to Indian industry.

Obviously, the Indian pharmaceutical industry environment must be conducive and prepared to face these challenges. Probably as a result, inter- and intra-institutional and industrial collaborations, and partnerships with governments and academia are on the rise. During the last decade, several successful partnerships between government, academia, and industry have emerged through national research agencies, universities, and academic institutions.

We present here a broad vision, objectives, and facilitating programs of such important agencies where notable collaborations between academia and industry have been facilitated with support from the government. We also present a few representative case examples.

COUNCIL FOR SCIENTIFIC AND INDUSTRIAL RESEARCH

New Millennium Indian Technology Leadership Initiative (NMITLI)

The New Millennium Indian Technology Leadership Initiative (NMITLI) was launched in 2003 by the Council of Scientific and Industrial Research (CSIR) to foster public–

private partnerships designed to promote innovation-led development and achieve a global technology leadership position across various select science and technology sectors. Now, NMITLI has emerged as the largest public–private partnership effort within the R&D domain in India. Apart from promoting public–private partnerships, NMITLI also aims to fund scientifically and technologically high-risk projects and build "team India" spirit by bringing together the brightest from public institutes, academia, and the private industrial sector. By 2008, the scheme involved 57 projects that networked public and private institutions and enterprises, including 270 R&D groups, 80 industry partners, and 1700 researchers, with a total project investment outlay of more than 80 million USD.

Through NMITLI, public institutions can get a grant-in-aid, while private enterprises are eligible for soft loans. The primary focus of CSIR's funding is to select projects that can give sustainable competitive advantage for India. Under this scheme, CSIR provides funds to a variety of fields, including information technology, metallurgy, pharmaceutical discoveries, agriculture, and biotechnology. Broadly, the projects under NMITLI are under two categories: first, Nation Evolved Projects (NEPs), and second, Industrial Originated Projects (IOPs).

Proposals for IOPs are solicited through advertisement, and after multiple rounds of screening, the meritorious projects are selected for funding. The financial viability and business propositions are thoroughly examined. Indian companies having more than 50% of shareholdings of Indians and/or nonresidents are eligible for proposing under IOP. The financial assistance is distributed as a soft loan at a low interest rate to private sector company partners (Box 12.1 and Box 12.2 are examples of NMITLI projects).

Open Source Drug Discovery

CSIR envisaged that the open source model could be applied to the health-care sector, especially for identifying drug targets for neglected diseases in affordable manner [16]. Open Source Drug Discovery (OSDD) is a consortium conceived and led by CSIR with

Box 12.1. CSIR NMITLI Program: Tuberculosis
NMITLI scheme supported partnerships between experts from 12 institutions and an industry for the development of new target, drug delivery system, bio enhancers, and therapeutics against tuberculosis (TB). This engagement was successful and resulted in a new therapeutic molecule which works through combination therapy and is also less toxic. Moreover, this therapeutic agent (when given in combination with existing drugs) clears the total infection within 2–3 months, cutting short the span for treatment substantially without any episode of relapse as per animal studies. This is the first success achieved in developing new molecule for tuberculosis in the last 40 years, globally. An Investigational New Drug (IND) application for the molecule has been cleared and is currently undergoing Phase 2 clinical trial.

multiple global partnerships. This consortium was launched in 2008 with a vision to provide affordable health care to the developing world. Launched on the three cardinal principles of collaborate, discover, and share, OSDD is a community-driven open innovation global platform for researchers and scientists from multiple streams of sciences such as informatics, chemistry, pharmacy, wet-lab scientists, clinicians, and hospitals to collectively work together and collaborate for discovering novel drug targets, biomarkers, and therapeutics for the neglected diseases such as tuberculosis and malaria. At present, OSDD has more than 6700 registered users from more than 130 countries and has emerged as the largest collaborative effort in the field of drug discovery.

As a philosophy, OSDD believes that the governments and publicly funded institutions of the countries with a high burden of neglected disease need to actively participate

Box 12.2. CSIR NMITLI Program: Arthritis

NMITLI also supports the traditional approach projects which are based on the golden triangle of modern sciences, modern medicines, and traditional sciences and philosophy which assist into the discoveries in less curable diseases like arthritis, diabetes, and hepatic protection. For instance, osteoarthritis herbal drug development project under the NMITLI scheme involved a network of 16 national research institutions, modern medicine hospitals, and pharmaceutical industries from India (Figure 12.1). Following prior art [12] and several rounds of national level consultations with Ayurvedic physicians and scholars, judiciously short-listed botanical drugs entered a parallel track of animal pharmacology and open-label observational studies by clinicians. The project used traditional knowledge-guided platform where the base formulation was optimized with additional ingredients to obtain desired therapeutic activities. All the formulations were manufactured under Good Manufacturing Practices in accordance with U.S. FDA guidance to Industry for botanical drugs. The preclinical evaluation was designed on the basis of systems approach, wherein the assay battery involved targets relevant to inflammation, pain, immunomodulation, and chondroprotection (proteoglycan release, nitric oxide release, aggrecan release, and hyaluronidase inhibition as markers) in human explant model of OA cartilage damage [13]. This led to the design of few variants of synergistic polyherbal formulations that were found to be safe and devoid of any genotoxicity or mutagenic activity. The shortlisted formulations entered a series of randomized clinical trials together with known drugs glucosamine and celecoxib, for comparison. Finally, the best formulation was selected that led to one Indian and one Patent Cooperation Treaty (PCT) applications with a dossier of necessary data required for possible regulatory submissions [14, 15]. Thus, this project was completed in 5 years with an expenditure of about 2 million USD. If successfully introduced in the market, the treatment based on this formulation may cost about 25 USD a month for patients. And remarkably enough, this will be achieved with better therapeutic benefits including chondroprotection that apparently no other modern drug offers. Presently, CSIR is in the process of identifying a suitable industrial partner for further development, optimization, manufacturing, registration, and marketing.

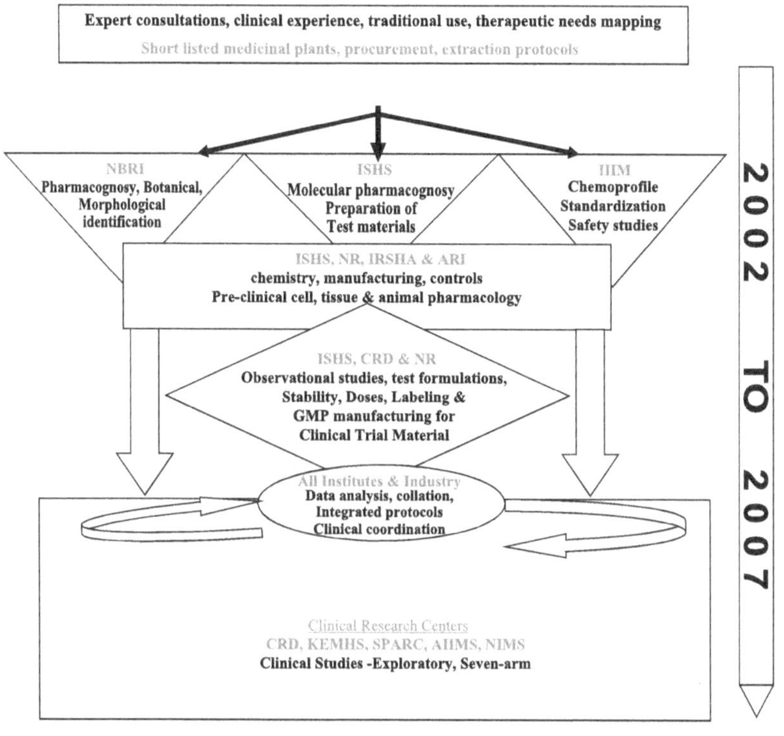

Institutes

Interdisciplinary School of Health Sciences (ISHS), University of Pune
National Botanical Research Institute (NBRI), Lucknow
Indian Institute of Integrated Medicine (IIIM), Jammu (Formerly RRL-J)
Interactive Research School for Health Affairs (IRSHA), Bharati Vidyapeeth University, Pune
Agharkar Research Institute (ARI), Pune
Center for Rheumatic Diseases (CRD), Pune
Medical Research Center, Kasturba Society (MRC), Mumbai (Formerly SPARC)
All India Institute of Medical Sciences (AIIMS), New Delhi
King Edward Hospital & Research Center (KEMRC), Mumbai
Nizam Institute of Medical Sciences (NIMS), Hyderabad
Natural Remedies (NR), Bangalore
Zandu Pharmaceutical Works (ZPW), Mumbai; Arya Vaidya Pharmacy (AVP), Coimbatore; Arya Vaidya Shala (AVS), Kottakal; Dabur India Ltd (DIL), New Delhi; Nicolas Piramal India Ltd. (NIPL), Mumbai

Time: Five years

Figure 12.1. Institutes involved in NMITLI herbal drug development projects.

in the drug development initiative and that trials need to be conducted under internationally accepted standards. More regarding the proposed methodology for clinical trials is available on the OSDD web portal. OSDD aims to de-risk clinical trials for tropical diseases by investing public funds and involving public institutions in the countries where disease occurs. The fundamental principle governing the OSDD model is accessible and affordable health care to the developing world. Anything that is developed in OSDD will be available to the developing world in open source, generic mode, without price monopolies. Therefore, the OSDD approach to drug discovery and development is neutral with regard to intellectual property. Affordability and accessibility remain the core concerns of delivery of drugs for tropical diseases. Once a drug is approved for use by the regulatory agencies, OSDD will depend on the business model of the generic drug industry, which made drugs affordable in the developing countries. Drugs developed by OSDD will be available to any industry player with appropriate manufacturing practices to distribute the drugs to the market. The market competition will ensure accessibility and affordability.

DEPARTMENT OF SCIENCE AND TECHNOLOGY (DST)

Drugs and Pharmaceutical Research Program (DPRP)

Recognizing the profound influence of R&D on the prospects and opportunities for the growth of the Indian drug industry, the Indian government's Department of Science and Technology (DST) mounted a program on drug development promoting collaborative R&D in drugs and the pharmaceuticals sector with specific objectives to synergize the strengths of publicly funded R&D institutions and the Indian pharmaceutical industry. In January 2004, the government of India established the Drugs and Pharmaceutical Research Program and the Drug Development Promotion Board, under the administrative control of DST, to support R&D projects jointly proposed by industry and academic institutions/laboratories and to extend soft loans for R&D to the drug industry. DST has significantly contributed to boost indigenous R&D and innovations in Indian pharmaceutical industry. Since 2004, DST has invested over INR 650 crores by way of collaborative grants and soft loans to industries. Over 300 collaborative research projects have been supported and more than 100 national facilities involving academia and pharma industry have been established. Over 70 institutes have been given soft loans and grants, especially for conducting clinical trials on neglected diseases (Box 12.3 shows a vaccine adjuvant project supported by DPRP).

Innovation Funds and Partnerships

The DST has partnered with a large number of national and international agencies and entities in the promotion of innovations. Some of the lead innovation partnerships of DST are with INTEL, Lockheed Martin, Indian Institute of Management, Ahmadabad. The partnerships have focused on scouting and supporting innovators to become entre-

Box 12.3. DST Vaccine Adjuvant Project

A project supported by the Department of Science and Technology (DST) to develop herbal vaccine adjuvant was undertaken by the Interdisciplinary School of Health Sciences, University of Pune with Serum Institute of India as an industry partner. Despite centuries of vaccine use, still alum salts remain universal vaccine adjuvants licensed for human use. However, they have limitations in engaging cellular immunity. Various immunostimulants from natural and synthetic origin are being studied either to replace or complement alum salts in vaccine formulations [17]. This project generally follows the reverse pharmacology approach based on Ayurvedic knowledge and previous scientific studies [18, 19]. The project used flow cytometry to monitor effects of fractions on antigen-specific protective immunity. Test materials were screened using *in vivo* potency assays for pertussis or diphtheria. These models are known to have relevant correlations with clinically established protective levels and routinely used in potency testing of vaccines [20, 21]. Just within 3 years, the project has resulted in identification of few semi-pure leads that have considerable efficacy against polysaccharide, toxoid, and recombinant group of vaccines. Furthermore, these leads were found to modulate T-helper cell immunity, which is crucial for efficient cellular and humoral immunity. Detailed safety profile of one of the lead has been established, which is over 20 times than its effective preclinical immunomodulatory dose. Thus, in a time span of 3 years, a chemically characterized herbal fraction is ready to enter human clinical trial.

preneurs. Partnerships with INTEL and Lockheed Martin are associated with an award scheme for innovators that is backed up by participation in international fairs and mentoring in the formation of start-up companies.

International R&D Endowment Funds

The DST has a health and pharma focus in many international collaborative programs where Indian academia and industry are collaborating with their counterparts in respective countries. India and the United States have established a joint endowment fund for promoting innovations and joint research and development. The size of the fund is 30 million USD and will be serviced by the Indo US Science and Technology Forum (IUSSTF). The focus of the fund will be on the promotion of industrially relevant innovations to be supported by the private sector for commercialization. The board has identified two topics for focus, namely "healthy individuals" and "citizens."

The EU and India have entered into an agreement to announce coordinated calls for research and development in select priority areas of mutual interest to partners. The EU and India have committed 5 million euro per year for 5 years to support research in priority areas. A dedicated binational Indo-German Science and Technology Centre has now been established at Gurgaon. The center will focus on the promotion of industrial research in mutually gainful areas of interest to both Germany and India. Both

Germany and India have committed 10 million euro each for a period of 5 years. A binational research partnership between India and Australia was launched in 2006. A joint strategic research fund of Australian $30 million was established to support joint R&D; within 2 years, based on the success of the fund, it was increased to Australian $100 million. R&D programs in the areas of biotechnology and other disciplines are supported with 30% of funds each. The remaining 40% of the fund is deployed in addressing grand challenges in the areas of agriculture, energy and environment, health care, and so on.

Good Laboratory Practice (GLP)

India has been making an effort to adopt the Organization for Economic Cooperation and Development protocol for observing Good Laboratory Practice and seeking mutual recognition for certifications. DST is India's national GLP authority. India gained full membership and signed the mutual recognition treaty for GLP in March 2011, and more than 18 laboratories have since gained GLP certification in the country.

DEPARTMENT OF BIOTECHNOLOGY (DBT)

SIBRI and BIRAC

Although over the years biotechnology research contributed considerably to new drug discovery, no scheme was present to promote innovation and commercialization of biotech products through private enterprises. Hence, in 2005–2006, the DBT launched the Small Business Innovation Research Initiative (SBIRI) to support small- and medium-sized enterprises (SMEs) with a grant or soft loan to support the early phase of product development. Under SIBRI, companies with up to 1000 employees are eligible for funding, and the scheme supports pre-proof of concept, early stage innovative research, and provides mentorship.

The Department of Biotechnology also established the Biotechnology Industry Research Assistance Council (BIRAC) as an autonomous, flexible, and futuristic organization for promoting innovations by facilitating start-ups and SMEs. Through the National Biotechnology Development Strategy, DBT has made a commitment of investing up to 30% of its budget in academia–industry collaborations to promote innovation, pre-proof-to-concept research, accelerated technology, and product development in biotechnology related to pharmaceuticals, agriculture, human health, and environment, and so on.

To support high-risk discovery and innovation projects, the new Biotechnology Industry Partnership Programme (BIPP) has been introduced as a government–industries partnership program for supporting R&D on a cost-sharing basis. This new scheme is expected to be an enabling mechanism to promote biotech industry R&D and public–private partnerships. Through this program, the government contributes between 30% and 50% of the cost and in some cases provides grant-in-aid support for evaluation and validation of technologies and products.

Box 12.4. DBT Stanford India Biodesign Program
In 2007, the Stanford-India Biodesign (SIB) program was jointly conceived by DBT and Indo-US Science and Technology Forum (IUSSTF), which encourages multidisciplinary approaches to biology and medicine by training the next-generation innovators in India. The primary objective of this program is to produce leaders in medical technologies through training, fellowships, and one-on-one mentorship. The teams work closely with Stanford University, All India Institute of Medical Sciences (AIIMS), and Indian Institute of Technology (IIT), Delhi. So far, SIB has trained 24 Fellows and 28 interns and the program so far has resulted in the development of 12 devices. Moreover, about 20 provisional patents have been granted and five patent cooperation treaty (PCT) applications have been signed. As per available report, five products are in various stages of trials while two companies have been formed, and one product has been licensed for sale and manufacture by Hindustan Latex. Recently, a national biodesign alliance has been established by DBT with various partners such as the Indian Institute of Technology (IIT), New Delhi; IIT, Madras; AIIMS, New Delhi; Regional Center for Biotechnology (RCB), Faridabad; Translational Health Science and Technology Institute (THSTI), Faridabad; International Center for Genetic Engineering and Biotechnology (ICGEB), Delhi; Indian Institutes of Science (IISc), Bangalore; and Christian Medical College (CMC), Vellore, to coordinate with each other on the biodesign programs in the country.

Global Partnerships and Alliances

New bilateral collaboration schemes were launched with countries like Canada, Denmark, Finland, Australia, Germany, Sweden, Switzerland, the EU, and the UK, and international agencies like the Wellcome Trust and the Bill and Melinda Gates Foundation. A joint Wellcome Trust-DBT program for the development of affordable healthcare products has been initiated with a commitment to invest 165 crores jointly for 5 years. Another example is the Stanford-India Biodesign program (see Box 12.4).

INDIAN COUNCIL OF MEDICAL RESEARCH

The Indian Council of Medical Research (ICMR) is a conglomeration of 32 national institutes and also functions as an apex body in India for the formulation, coordination, and promotion of biomedical research. ICMR also has large extramural funding programs where interdisciplinary collaborative research at national and international levels is encouraged. ICMR research priorities coincide with the national health priorities such as control and management of communicable diseases; fertility control; maternal and child health; control of nutritional disorders; developing alternative strategies for healthcare delivery; containment within safety limits of environmental and occupational health problems; research on major noncommunicable diseases like cancer, cardiovascular diseases, blindness, diabetes, and other metabolic and hematological disorders; mental health research and drug research (including traditional remedies). All these efforts are undertaken with a view to reduce the total burden of disease and to promote health and well-being of the population.

Clinical Trial Registry

The Clinical Trials Registry-India (CTRI) has been established to encourage all clinical trials conducted in India to be prospectively registered before the enrollment of the first participant. The ICMR Bioethics initiative has developed ethical guidelines for the conduct of trials and for ethics committees. Postmarketing surveillance studies as well as BA/BE (bioavailability/bioequivalence) trials are also expected to be registered in the CTRI. The CTRI is working with the WHO ICTRP to ensure that results of all trials registered with the CTRI are adequately reported and publicly available. Being a Primary Register of the International Clinical Trials Registry Platform (ICTRP) (http://www.who.int/ictrp/search/en/), registered trials are freely searchable both from the WHO's search portal, the ICTRP as well as from the CTRI (http://www.ctri.nic.in).

Pharmaceutical and Vaccine Development through Collaborations

The Indian Council of Medical Research has completed a Phase 2 expanded safety and acceptability study of vaginal microbicide 1% Tenofovir Gel with a 100% retention rate. The overall safety profile of the product was found to be good. Five clinical trials with new artemisinin combinations have been successfully completed for malaria, and combination therapies for visceral have been leishmaniasis evaluated (Ambisome, Miltofosine, and Paramomycin). Efficacy and safety testing of 3 and 4 months moxiflox-containing regimens for pulmonary tuberculosis are under way. Efficacy of combined immunotherapy and chemotherapy in pulmonary tuberculosis is being evaluated.

Two Phase 1 studies for HIV vaccines were completed: AAV at NARI, Pune, and MVA-based vaccine at TRC, Chennai. The vaccines were found to be safe. Based on better immunogenicity, MVA is now being used in Pune and Chennai in a prime-boost approach, where priming is done with a DNA-based vaccine. A Phase 3 trial of rotavirus vaccine has also been completed, as has a Phase 2 trial of bivalent killed whole cell oral cholera vaccine VA 1.3. A Phase 1 trial of a three-site measles aerosol vaccine was completed and found to be safe and immunogenic. A recombinant protein-based vaccine for Chandipura has been developed. Combination of this vaccine with DPT was found to be stable in animal experiments. Development of combined Hep E and B vaccine is under progress. Enzyme-linked immunosorbent assay (ELISA)-based hepatitis A diagnostic technology was transferred to BBIL. A recombinant protein-based, highly specific, and sensitive assay for the diagnosis of Hepatitis E was developed.

AYUSH

The acronym "AYUSH" stands for Ayurveda, Yoga, and Naturopathy, Unani, Siddha, and Homoeopathy. The Indian medical system known as Ayurveda is one of the oldest extant health systems in the world with fundamental principles and theory-based practices. Literally, the Sanskrit meaning of *Ayu* is life and *Veda* is knowledge or science.

Therefore, Ayurveda is also generally translated as the Science of Life. India has over 600,000 registered practitioners of Ayurveda and other traditional medicine. Due to better accessibility and affordability, these systems enjoy wide acceptance among large segments of the population, especially in India. Thus, Ayurveda remains a most comprehensive and practical medical science that receives acceptance and support of the public. The Department of AYUSH emphasizes educational standards, quality control, and standardization of drugs; improving the availability of medicinal plant material; research and development; and awareness generation about the efficacy of the systems domestically and internationally. The Traditional Knowledge Digital Library (TKDL) is a collaborative project between CSIR, DST, and AYUSH, which involves documentation of the knowledge available in the public domain on traditional knowledge from the existing literature related to Ayurveda, Unani, and Siddha in digitized format searchable in patent databases in five international languages (English, French, German, Spanish, and Japanese). A lot of work related to evidence-based Ayurveda has been undertaken in collaboration with industry and academia [22].

Advantage India

Traditional knowledge (TK) base from Ayurveda allows drug researchers to start from time tested and safe botanical material. The normal drug discovery course of "laboratory to clinics" in this case actually becomes from "clinics to laboratories"—a true "reverse pharmacology" approach (Figure 12.2). In this process, safety remains the

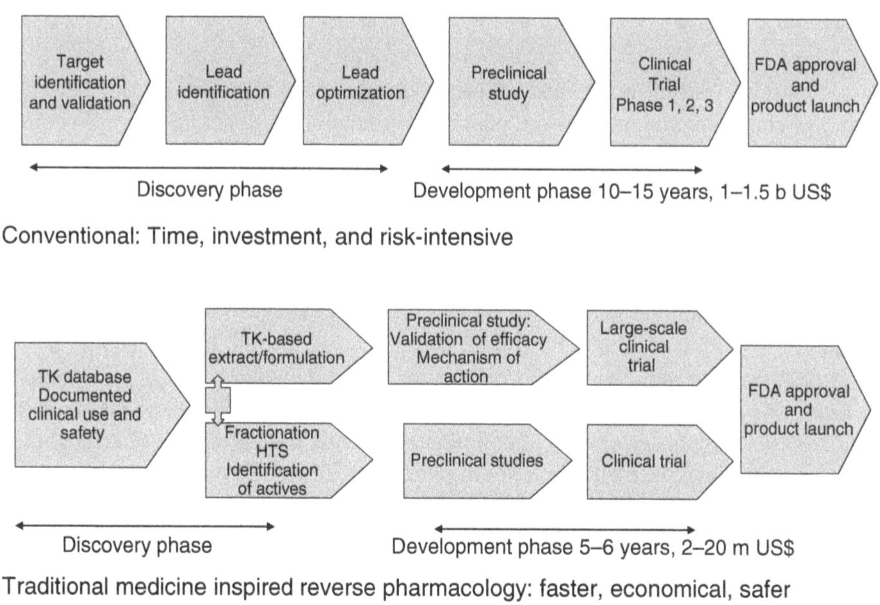

Figure 12.2. Drug discovery and development: conventional versus reverse pharmacology approaches.

TABLE 12.1. Comparison of Traditional (TK)/Ayurveda and Modern Drugs

Attributes	TK/Ayurveda	Modern Drugs
Cost of drug development	Affordable	Expensive
Time taken to design and develop	3–5 years	10–15 years
Risks to fail due to safety	Very low	Very high
Clinical trials for safety and efficacy	Pharmaco-epidemiology and documentation of practices is more relevant since drugs are in active use	Randomized controlled trials are absolutely required in most rigorous mode
Untoward effects	Negligible	Reported for many drugs
Basis of drug designing	Personalized	Pathophysiological pathways
Basic ingredients of the drug	Array of natural substances	Synthetic material
Quality control	Challenging because of complexity of multiple natural ingredients	Comparatively simpler because of easily measurable synthetic substances
Infrastucture and investment	Comparatively less	Very large
Global acceptance and marketing	Challenging because of inadequate scientific evidence	Easier because of sufficient scientific evidence
Acceptability level	Innovation rich	Innovation deficit

most important starting point, and efficacy becomes a matter of validation. In such cases, systematic documentation and pharmacoepidemiology of actual practice is more important than clinical trials. On many attributes, TK-inspired drug discovery is much more efficient and affordable as compared with modern drug discovery (Table 12.1). The best example of bioprospecting using traditional knowledge is Reserpine, the antihypertensive alkaloid from *Rauwolfia serpentina*, which became available as a result of work carried out by CIBA in India in close collaboration with Ayurveda experts. This process of natural product drug discovery was later named "reverse pharmacology" [23, 24].

A large number of molecules have come out of the Ayurvedic clinical base, including *Rauwolfia* alkaloids for hypertension, psoralens for vitiligo, *Holarrhena* alkaloids for amebiasis, guggulsterones as hypolipidemic agents, piperidines as bioavailability enhancers, bacosides in mental retention, picrosides in hepatic protection, curcumins in inflammation, withanolides, and many other steroidal lactones and glycosides as immunomodulators. In the future, modern medicine could well come to be based on such ancient eastern time-tested remedies, developed using advanced technologies from the west. In this process, rationale and science will be key attributes. Traditional medicine-inspired drug discovery and development is therefore considered to be an efficient, faster, and more affordable strategy [25].

Box 12.5. Innovations from Traditional Knowledge: Ayusoft and Ayugenomics
AyuSoft is a collaborative project between the Government of India's Center for Development of Advanced Computing (C-DAC) and the University of Pune. While the Traditional Knowledge Digital Library (TKDL) helps in protecting intellectual property, AyuSoft converts the logic of classical Ayurvedic texts into comprehensive, authentic, intelligent, and interactive knowledge repositories with the help of complex analytical tools (http://ayusoft.cdac.in). The AyuSoft database includes more than five Lakh records, capturing information from nine classic texts. This knowledge base is accessible through a Decision Support System, data mining tool, and digitized searchable texts. The concept of prakriti in Ayurveda has strong genetic connotations that can form base for pharmacogenomics and personalized medicine [24]. Ayugenomics was planned as a platform to undertake the challenge of developing new strategies of drug discovery by integrating the ancient science and knowledge of Ayurveda with modern science, and the technologies of genomics, proteomics, and pharmacogenetics. We demonstrated that different prakritis may possess different drug metabolism rates associated with HLA and CYP polymorphism [26, 27].

Many basic concepts from traditional knowledge systems may give new insights to basic research also. For instance, an informatics platform known as AyuSoft, based on knowledge and logic available from Ayurveda, is a very useful tool for bioprospecting and decision support system. Concept of prakriti or constitution types based on Ayurveda and their genetic basis, known as Ayugenomics, is useful to better understand modern pharmacogenomics and personalized medicine (Box 12.5). Clearly, the vast traditional knowledge base could be efficiently used as a discovery engine or decision support system. India may be able to address the present innovation deficit faced by the global pharmaceutical sector. In the process, it will be very important to build and strengthen academia–industry collaborations not only within India but also those involving global industrial majors and thought leaders.

BUILDING NEW PARTNERSHIPS

At the academic level, an effective network of people with multidisciplinary backgrounds such as clinical pharmacology, cell biology, and genetics is missing [28]. Experts are strongly prescribing transformation of the unsustainable R&D model where many steps are pursued by academia and industry without effective collaboration to bridge the translational and interdisciplinary divides. It is important to recognize that to effectively address the present impasse and overcome the bottleneck in drug discovery we need to strengthen existing partnerships between academia, industry, and the government. It is also important to change our mind-set from drugs to medicine and from a disease-centric to a personalized approach. The national research agencies from India such as CSIR, ICMR, DBT, DST, and AYUSH need to build stronger bonds with Indian and foreign academia for innovative translational research. The Indian National

Science Academy and its counterparts in different countries also can play a major role in facilitating high-quality research in the area of drug discovery and health care. Close interactions with global agencies such as WHO, NIH, UKIERI, EU, and foundations like Wellcome and Gates need to be brought together in a consortium model so that available resources can be focused on creating new alliances and partnerships to address affordability and accessibility of health care.

CONCLUDING REMARKS

While there is increasing consensus that promotion of health must get precedence and preference over the disease-centered model of "health" care, medicine and drugs may continue to remain a part of life. In the present situation when affordability is emerging as major concern, we need to think of new health care that moves beyond diseases, drugs, and medicine in a more holistic and integrative fashion. The TK base in India may serve as a powerful discovery engine to also create innovative models based on reverse pharmacology, systems biology, and personalized approaches. As in the past, the toxic, expensive, inefficient, or ineffective agents may continue to face the "death of drug" phenomenon. However, the new collaborative strategies involving academia, industry, and government, together with the confluence of traditional wisdom, modern science, and technology, can truly facilitate the much-desired emergence of translational research, which may lead to safer and effective "medicine discovery." The present environment in India and the emerging global collaborations between the governments, academia, and industry seem to be conducive for collaborative research not only limited to new drug discovery but also to facilitate much-desired integrative and holistic approaches leading to the rebirth of health care.

ACKNOWLEDGMENTS

I thank Dr. Abhay Jere, Senior Vice President, Persistent Systems Pvt Ltd, Pune, for help in providing useful information. Thanks are also due to Dr. Avinash Patwardhan, Dr. Dada Patil, and Dr. Gururaj Mutalik for offering valuable comments and edits.

REFERENCES

1. Patwardhan, B. Traditional medicine for affordable global health. Commission on Intellectual Property Innovation and Public Health, World Health Organization, Study 9, 2005. Available at http://www.who.int/intellectualproperty/studies/B.Patwardhan2.pdf. Accessed December 23, 2013.
2. Patwardhan, B., Warude, D., Pushpangadan, P., Bhatt, N. Ayurveda and traditional Chinese medicine: a comparative overview. *Evid Based Complement Alternat Med.* 2005;2(4): 465–473.
3. Harmon, K. Prescription drug deaths increase dramatically. *Sci Am.* 2010. Available at http://www.scientificamerican.com/article/prescription-drug-deaths/.

4. Pujol, A., Mosca, R., Farrés, J., Aloy, P. Unveiling the role of network and systems biology in drug discovery. *Trends Pharmacol Sci.* 2010;31(3):115–123.
5. Arrowsmith, J. Trial watch: phase II failures: 2008′2010. *Nat Rev Drug Discov.* 2011;10: 328–329.
6. Bennani, Y.L. Drug discovery in the next decade: innovation needed ASAP. *Drug Discov Today* 2012;17(Suppl.):S31–S44.
7. Patwardhan, B., Vaidya, A.D.B. Natural products drug discovery: accelerating the clinical candidate development using reverse pharmacology approaches. *Indian J Exp Biol.* 2010;48(3):220–227.
8. Report of the Steering Committee on Health for 12th Five Year Plan. Planning Commission, Government of India. 2012. Available at http://planningcommission.nic.in/aboutus/committee/strgrp12/str_health0203.pdf. Accessed December 23, 2013.
9. Indian Pharma Inc. Gearing up for next level of growth. Pharma Summit 2012—a report by CII and PwC. Available at http://www.pwc.in/en_IN/in/assets/pdfs/pharma/pharma-summit-report-31-10-12.pdf.
10. India Pharma 2015. Unlocking the potential of Indian pharmaceutical market. A report by McKinsey & Company. Available at http://agoraprofessional.com/wp-content/uploads/2013/04/India-Pharma-2015.pdf.
11. Patwardhan, B. Traditional medicine for affordable global health. Commission on Intellectual Property Innovation and Public Health, World Health Organization, Geneva. Study report 5, 2005.
12. Chopra, A., et al. A 32-week randomized, placebo-controlled clinical evaluation of RA-11, an Ayurvedic drug, on osteoarthritis of the knees. *J Clin Rheumatol.* 2004;10(5):236–245.
13. Sumantran, V.N., et al. Chondroprotective potential of root extracts of *Withania somnifera* in osteoarthritis. *J Biosci.* 2007;32(2):299–307.
14. Chopra, A., et al. Validating safety & efficacy of Ayurvedic derived botanical formulations: a clinical arthritis model of NMITLI. 5th Oxford International Conference on the Science of Botanicals (ICSB), 2006, University of Mississippi at Oxford, USA.
15. Patwardhan, B., et al. A synergistic herbal composition for treatment of rheumatic and musculoskeletal disorders (RMSDS), 2008; Application No: PCT/IN2008/000462.
16. Bhardwaj, A., Scaria, V., Raghava, G.P., Lynn, A.M., Chandra, N., Banerjee, S., Raghunandanan, M.V., Pandey, V., Taneja, B., Yadav, J., Dash, D., Bhattacharya, J., Misra, A., Kumar, A., Ramachandran, S., Thomas, Z., Open Source Drug Discovery Consortium, Brahmachari, S.K. Open source drug discovery—a new paradigm of collaborative research in tuberculosis drug development. *Tuberculosis (Edinb).* 2011;91(5):479–486.
17. Patwardhan, B., Gautam, M. Botanical immunodrugs: scope and opportunities. *Drug Discov Today* 2005;10(7):495–502.
18. Gautam, M., et al. Ethnopharmacology in vaccine adjuvant discovery. *Vaccine* 2008;26: 5239–5240.
19. Gautam, M. et al. Immunoadjuvant potential of *Asparagus racemosus* aqueous extract in experimental system. *J Ethnopharmacol.* 2004;91(2–3):251–255.
20. Gautam, M., et al. Immune response modulation to DPT vaccine by aqueous extract of *Withania somnifera* in experimental system. *Int Immunopharmacol.* 2004;4(6):841–849.
21. Bani, S., et al. Selective Th1 up-regulating activity of *Withania somnifera* aqueous extract in an experimental system using flow cytometry. *J Ethnopharmacol.* 2006;107(1):107–115.

22. Patwardhan, B. The quest for evidence based Ayurveda: lessons learned. *Curr Sci.* 2012;102(10):1406–1417.
23. Patwardhan, B., Vaidya, A.D.B., Chorghade, M., Joshi, S.P. Reverse pharmacology and systems approaches for drug discovery and development. *Curr Bioact Compd.* 2008;4(4):201–212.
24. Patwardhan, B. Ayurveda: the "designer" medicine: a review of ethnopharmacology and bioprospecting research. *Indian Drugs* 2000;37(5):213–227.
25. Patwardhan, B., Mashelkar, R.A. Traditional medicine-inspired approaches to drug discovery: can Ayurveda show the way forward? *Drug Discov Today* 2009;14(15–16):804–811.
26. Bhushan, P., Kalpana, J., Arvind, C. Classification of human population based on HLA gene polymorphism and the concept of prakriti in ayurveda. *J Altern Complement Med.* 2005;11(2):349–353.
27. Ghodke, Y., Joshi, K., Patwardhan, B. Traditional medicine to modern pharmacogenomics: ayurveda prakriti type and CYP2C19 gene polymorphism associated with the metabolic variability. *Evid Based Complement Alternat Med.* 2011. Article ID 249528. Available at http://dx.doi.org/10.1093/ecam/nep206.
28. Editorial. Where will new drugs come from? *Lancet* 2011;377(9760):97.

Part III

A GAME CHANGER FOR AVERTING FUTURE PHARMA CLIFF

Coming together is a beginning, staying together is progress, and working together is success. —*Henry Ford*

Collaborative Innovation in Drug Discovery: Strategies for Public and Private Partnerships, First Edition.
Edited by Rathnam Chaguturu.

13

ACCELERATING INNOVATION IN THE BIOSCIENCE REVOLUTION

Bernard H. Munos

InnoThink Center for Research in Biomedical Innovation, Indianapolis, IN, USA

A GREAT INDUSTRY FACES A PAINFUL "KODAK MOMENT"

Over the past century, the pharmaceutical industry has contributed enormously to public health and welfare. It conquered major ailments (e.g., diabetes, hypertension, ulcers), tamed infectious diseases (e.g., AIDS and hepatitis), produced psychiatric treatments that emptied "lunatic asylums," as they were called, and turned rDNA technology into a new generation of biopharmaceuticals that help battle cancer and other deadly diseases. The investors who put their faith (and savings) in the industry were well rewarded (Figure 13.1). In the last two decades of the twentieth century, the collective market cap of large U.S. drug companies swelled nearly 30 times [1, 2].

Yet, for the last 15 years, the industry seems to have lost its magic. The number of new drugs brought to market has slumped. Large drug companies that had dominated pharmaceutical innovation now only produce a third of it. The share of prescriptions filled by generic drugs has doubled over the last 10 years and reached 84% in 2012. R&D spending per new drug at major companies has escalated to staggering levels, ranging from $4 billion to $12 billion [3]. Combined, these challenges have thrown into jeopardy the industry's proud legacy and its future.

Collaborative Innovation in Drug Discovery: Strategies for Public and Private Partnerships, First Edition.
Edited by Rathnam Chaguturu.

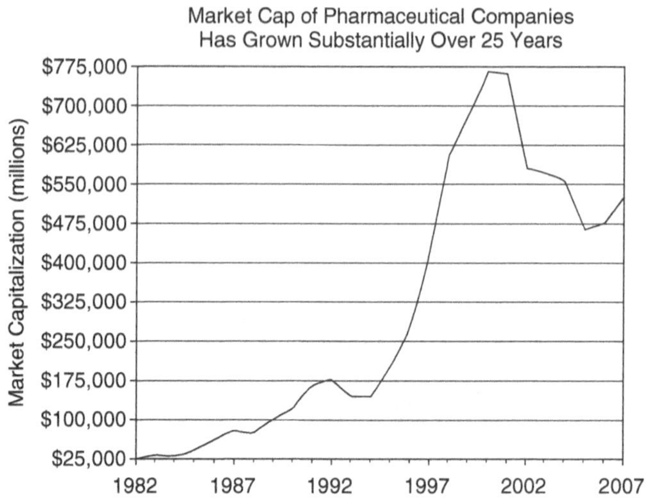

Figure 13.1. Aggregate capitalization of Pfizer, Merck, Eli Lilly, Schering-Plough, Bristol-Myers Squibb, and Abbott (as of August 29, 2007).

If new drugs meet only 15% of the pharmaceutical needs of Americans, does the industry that produces them still matter? If pharmaceutical companies produce drugs that only governments can afford [4], at staggering prices per patient, how long will it be before even governments decide they can no longer foot the bill? If, in addition, many of these drugs are hardly better than generics, when will payers and patients conclude that they are better off with the older medicines? It is already happening. Governments in Germany and England, and insurers worldwide, are pushing back. They increasingly resist paying unless the drugs represent real innovation. Other countries such as India, China, and Brazil have rescinded patents or granted local firms compulsory licenses to copy expensive Western medicines and sell them at prices their patients can afford [5]. The industry is increasingly caught in a bind: it must produce more, better, and affordable innovation, or face severe disruption and the possible demise of some of its best-known companies. Can the mighty pharmaceutical industry, with global annual sales of over half-a-trillion dollars [6], be facing such dire prospects?

As the history of Kodak shows [7], even the mighty can fall [8]. In fact, they do so fairly frequently, and for consumers it is usually a good thing. Disruption may connote painful visions of devastation, but we love the products that cause it, and would never give them up. We would never surrender our iPads®, Kindles®, iPods®, cell phones, or e-commerce, e-publishing, e-trading, GPS, digital photography, email, Internet, and other advances that have disrupted so many industries. For most of us, disruption is a good thing . . . that happens to others. This is likely the way the rest of the world sees the drug industry. For most patients, the status quo means few and mediocre new drugs that they cannot afford, and whatever disruption might bring can hardly be worse. In fact, history suggests that it is unlikely to have negative long-term consequences. It may spell the end of a model that can no longer deliver it, but innovation will survive. The disruptive products mentioned earlier did not mean the end of computers, telephones, cameras, publishing, or securities trading. Someone always found a way to redesign the business model and make it sustainable again, although that someone was seldom an incumbent firm. The same can be expected with drugs. Some pharmaceutical CEOs readily admit that the industry's return on R&D is negative [9, 10]. In 2011, the net present value of the top-20 drug companies' pipeline was $85 billion [6]. This is the value created by past cumulative R&D spending. Paradoxically, it is less than what those companies spent annually on research and development (currently $88 billion) [6]. Drug R&D has become an enterprise where value destruction has reached a massive scale because we spend far too much for what we get. As Josef Schumpeter showed decades ago, industries that experience such misallocation of resources are primed for "creative destruction" [11]. Is this what is in store for the pharmaceutical industry? Can it bounce back like Apple, or will it wither like Kodak?

THE TRANSFORMATION OF A GREAT INDUSTRY

Near-death experiences are common in business, and getting more so. "Nearly all the companies that our grandparents admired have disappeared. Of the top 25 industrial corporations in the United States in 1900, only two remained on that list at the start of the 1960s. And of the top 25 companies on the Fortune 500 in 1961, only six remain there today . . . the average life expectancy of a Fortune 500 company has declined from around 75 years half a century ago to less than 15 years today, and heading towards 5 years [12]."

A common reason for this declining life span is that, in large organizations, short-term priorities always favor marginal improvements at the expense of major change. Companies, focused on meeting near-term goals, tend to concentrate on quick payback projects that produce predictable value. They keep refining their processes to squeeze out inefficiencies and reduce variances. By doing this, they get better at what they do, but also worse at everything else. When the environment changes, because of new technology, shifts in customer behavior, or global economic forces, companies have become so specialized at what they do that they can no longer adjust. Kodak was brought down by the digital photography it invented but could not parlay it into a new business. Xerox invented many of the technologies that power the Internet but remains a troubled copier-maker with a stock price lower in mid-2013 than it was in 1978.

For most of the past century, pharmaceutical companies seemed immune to such problems. They kept bringing to market innovative therapies that drove the industry's remarkable success. At the core of that success was what we now call the translational research model, which simply consists in taking cutting-edge discoveries, usually coming from academia, and turning them into commercial products. The model's focus on breakthrough science largely protected the industry from getting trapped in incremental innovation. That model, however, had a major weakness: the translation process is inherently risky. It entails taking on daunting scientific and engineering challenges that no one has ever solved. When scientists master them, it brings their companies 10–20 years of prosperity, sometimes more. When they fail, their companies can disappear, and many did. Sometimes, the challenges are simply too big for a single company to tackle on its own. Such was the case with Alexander Fleming's discovery of penicillin, which business historians reckon took about 1000 scientists at 39 major labs to translate into commercial products [13]. In fact, had it not been for the war effort, which pressured these labs to join forces to overcome the challenges, we might have witnessed a case of collective industry failure.

The other problem with the translational model is that it tends to be disruptive. This is because translatable discoveries appear at random and not necessarily in the therapeutic areas in which companies are already invested. To launch the new drugs coming from their R&D divisions, companies often had to learn new markets and build infrastructure in different therapeutic areas. At the same time, their existing marketing franchises might wither for lack of successor products to replace those that had become generic. For scientists, this is just the way science works. For marketers, it is a waste that begs to be corrected.

In the 1990s, the marketers got the upper hand and conquered the executive offices at many companies. They went on to remake the industry, in a manner that took it away from the model and the values that had made it great. Up until then, R&D had been the engine that had powered the industry, while marketing was the caboose. These roles were now reversed. The focus was no longer on translating breakthroughs, but on selecting large market opportunities and asking scientists to produce drugs that would address those "unmet needs."

FROM ETHICAL PHARM TO ANIMAL PHARM

The problem with this approach is that innovation is not a linear process. It does not appear predictably on demand, but comes in waves separated by long dry spells. Figure 13.2 illustrates this for several psychiatric and neurological diseases. For schizophrenia, for instance, there was a wave of innovation between 1956 and 1975 with 18 drugs approved, including such breakthroughs as chlorpromazine (Thorazine®), an antipsychotic that is widely credited for bringing down the population of institutionalized patients by as much as 80%. But after that, there were no new approvals for 14 years, and a similar pattern can be observed for every other disease. Companies that ask their scientists to produce drugs in areas that are in the midst of a dry spell will probably get something, but it is unlikely to be a breakthrough.

The pioneers who built the industry, such as Paul Janssen, George Merck, and Eli Lilly, were scientists who understood this, as most scientists intuitively do. They knew that biology and pharmacology are not predictive sciences, and cannot deliver innovation at will. Instead scientists must roam wide and far, and look for opportunities where they can find them. To guarantee that freedom, they set their research divisions apart from the rest of their companies. Thus, Merck Research Labs or Lilly Research labs

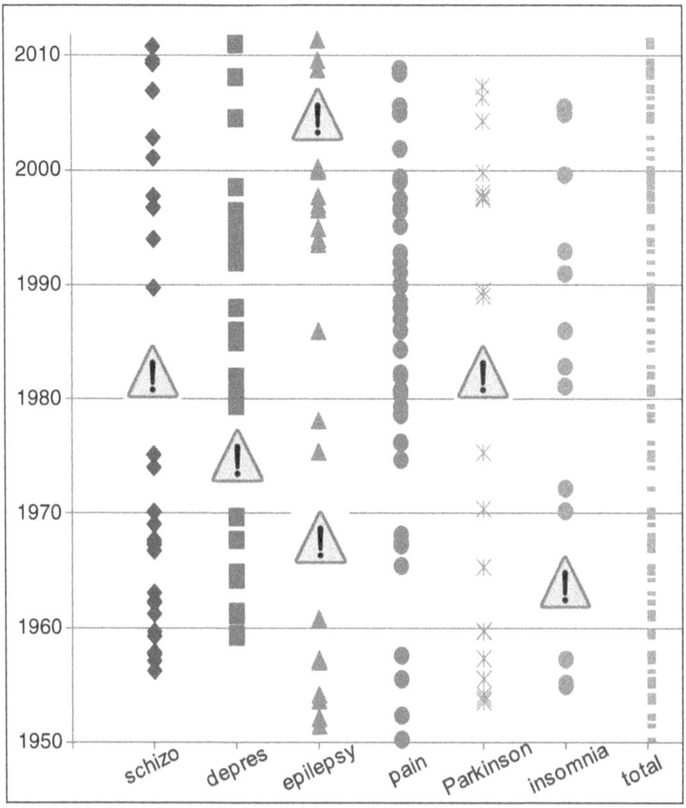

Figure 13.2. Innovation comes in waves separated by long dry spells. Ten-year dry spells are not unusual. See color insert.

were created as independent structures where drug hunters could practice their craft and engage in unfettered translational research. That model worked and brought unprecedented prosperity to the industry. It was that freedom that was largely abrogated in the 1990s as marketing-driven executives strived to make research more "responsive" to the needs of the market.

This was part of an ambitious attempt to redesign the pharmaceutical enterprise in a way that would make it predictable and would deliver innovation when and where companies wanted. It rested on the codification of companies' activities into thousands of standard operating procedures. These were step-by-step recipes that spelled out how to do things like anesthetize an animal, calibrate an instrument, or process an expense report. These "SOPs," as they are known, were then optimized by roving teams of experts, also known as "black-belts" to convey their alleged competence. The idea was to enshrine the "best practices" of the day into a corpus of rules that could be reliably followed by operators to deliver desired outcomes time and again. It was thought this would increase productivity and quality by reducing variances, and cut costs by eliminating waste. The disorderly nature of research would be tamed and replaced by a finely tuned, high-performance machine that would become the new engine of success.

Instead, it brought an unprecedented innovation crisis. There are many reasons for this, most of which could have been anticipated, and indeed were by many scientists.

Processes are not objectionable by themselves. Any organization needs them to function properly. But research has shown that, in innovative biomedical research organizations, processes are fuzzy [14]. They allow for some degree of initiative and flexibility to cope with unexpected situations, which in science are common. Problems crop up when processes become overly rigid and/or when they overreach and turn into micromanagement.

BUT THERE ARE OTHER ISSUES

First, putting best practices into a rule book that must be rigorously followed is often a sure way to ossify a company and kill creativity. This is because technology advances and best practices change, but processes fail to keep up because it is impossible to constantly update them. It would also defeat their purpose. Processes are meant to boost productivity by standardizing tasks. If they keep changing, why bother?

The overreliance on processes disenfranchised many pharmaceutical scientists. It turned them into operators. Creativity became a liability since creative people like to tinker with the status quo. Brilliant minds were no longer required because "excellence" was built-in. Companies could rid themselves of smart, but quirky, scientists, and replace them with docile employees. If they followed processes, performance would follow. SOPs became the answer to every problem, but over time, this promoted a culture of mediocrity. The standards of performance eroded. Positions that for decades had required minimum qualifications, such as a pharmacist degree, became open to less skilled employees.

Along with disenfranchisement came a loss of passion. Scientists no longer had the opportunity to make their mark. Many of them, who had joined the industry to "make a difference," lost this opportunity. Yet, in drug R&D, passion makes a difference. It heightens creativity, shortens timelines, and cuts costs. This is readily seen in patient advocate organizations, such as rare disease foundations created by parents fighting to save their child, or public–private partnerships focused on eliminating neglected diseases.

Ironically, the whole system, which was designed to promote best behaviors, failed to do so in the area that mattered most, that is, ethics. The pharmaceutical industry, which until the 1980s was routinely called the "ethical pharmaceutical industry," saw its reputation tarnished by rampant scandals. The result was a surge in condemnations. Between 1991 and 2012, drug firms paid over $30 billion in penalties, 82% of which occurred after 2005 [15, 16]. Surveys routinely put the industry's reputation toward the bottom of the scale. A problem of this magnitude can hardly be blamed on rogue employees. One must accept that it has become systemic across the industry.

MYTHS AND FACTS ABOUT INNOVATION

Until the turn of the millennium, little was known about what produces innovation. Paradoxically, the drug industry, which spends tens of billions of dollars each year trying to foster it, never bothered to develop the tools it needed to manage this investment effectively. As it turns out, many things done in the name of innovation actually damage it. But it is only in the last 15 years that research has started to shed some light.

Some findings are intuitive; for instance, we now know that regimentation does not foster innovation. Others run against deeply entrenched management practice. Forecasting, for example, is universally seen as an indispensable management tool, yet it is impossible in the pharmaceutical industry. Not merely hard, but impossible, except by chance. This has to do with the statistical distributions that underpin the industry [17, 18]. The "blockbuster" drugs* that dominate pharmaceutical markets are rare events that some authors call "black swans" [19]. They are random and unpredictable, but critically important to all companies. They bring prosperity to firms that have them and misery to those that don't. Understandably, companies strive to produce them, and it is fair to say that no drug candidate moves into clinical development unless it is backed by elaborate forecasts that document its blockbuster potential. Firms employ teams of talented PhDs to develop these projections. Yet 80% of drugs fall short of their target, most of them by very large amounts. The median peak-sales of drugs currently stands at about $400 million [6], far short of the $1.34 billion mark that defines blockbusters in 2013. Drug marketing lore is replete with stories of molecules that were supported by high forecasts, but failed to live up to their billing. Xigris® (drotrecogin alfa), a sepsis drug approved in 2001, was one of them, but its annual sales never rose much above $100 million [20]. It was eventually withdrawn from the market for lack of efficacy. Exubera®, an inhalable insulin, was hailed as a potential $2-billion a year winner. It was withdrawn 1 year after launch with sales languishing at $12 million [21]. Forecasting errors can also work the other way and doom potential winners. Gleevec® (imatinib), a leukemia drug, was forecast to peak at $50 million. Its sales are now nearing $5 billion. It would have been killed, had planners prevailed, but was rescued by an unlikely champion: Novartis' CEO Daniel Vasella, who, perhaps because he was a physician, understood what breakthrough drugs mean to patients and had the courage to challenge the madness of a model that would have killed a cancer cure for alleged lack of commercial potential.

The reliance on forecasts has seriously undermined the industry. Instead of the breakthroughs of the past, it brought into the pipeline compounds that were selected on the basis of well-intended, but flawed, financial projections. But only Novartis has banned the use of forecasts in its labs. Perhaps this is related to its becoming the most innovative company in the industry, with 17 drugs approved in the last 12 years. Other

*When the blockbuster concept became popular in the late 1990s, these products were defined as those that garnered sales in excess of $1 billion. Adjusted for inflation (2.3% p.a. over 13 years) that figure is now $1.34 billion.

companies, such as J&J, Sanofi, and GlaxoSmithKline, are returning to the translational model. A few, such as Vertex, have never abandoned it, but much of the rest of the industry continues to put its faith in a flawed model that has not served it well.

BRINGING ETHICS BACK TO THE INDUSTRY

To rescue innovation, the industry must reconnect with what made it great. It must return to what works and stop doing what does not. The traditional translational research model worked because, although no one can predict blockbusters, scientists are often good at identifying potential breakthroughs, which is what society expects from the industry.

The pharmaceutical industry is unlike most other industries because of its unwritten covenant with society. Drug companies get intellectual property (IP), which is essentially a license to print money, but in return society expects breakthrough drugs. Until the 1990s, that compact benefited both sides. Since then, drug shortages, ethical scandals, extreme pricing, retrenchment from some of the more difficult therapeutic areas (such as infectious, cardiovascular, and psychiatric diseases), as well as declining drug approvals, have seriously damaged the industry's standing. It is problematic because if the industry fails to uphold its side of the bargain, society can scale back IP rights. Producing mediocre drugs and offering them at unaffordable prices may invite such backlash. Some governments are already rescinding patents or granting compulsory licenses to generic manufacturers. To contain this threat, the industry must reconnect with society, re-embrace innovation, and put patients back at the center of what it does.

IT DOES NOT NEED TO COST BILLIONS

A consequence of the innovation crisis has been an extraordinary rise in drug prices. Many new drugs launched nowadays cost tens of thousands of dollars per patient. The typical justification is that drug R&D is expensive, a situation made worse by overzealous regulators who delay approvals. Unfortunately, this does not stand up to scrutiny. Instead, research suggests that questionable management decisions are at the core of drug unaffordability.

Efforts to smooth the flow of innovation and produce it on demand have led to the widespread adoption of "portfolio management" across the industry. This technique involves detailed calculations about how many compounds need to be in the pipeline of a company at various stages of development to yield an uninterrupted flow of new drugs. As they implemented this approach, drug companies created incentives to both encourage the production of such drug candidates and reward their progression toward regulatory approval. Unfortunately, it turned drug innovation into a spreadsheet exercise. Strong pressure to produce the required number of drug candidates, combined with financial incentives, brought mediocre compounds into the pipeline that should never have entered it. Quantity prevailed over quality with disastrous consequences, one of which has been soaring costs and failure rates.

When a drug candidate is only tenuously better than what is already in the market, proving that superiority requires very large and expensive trials. As the number of enrolled patients climbs, the complexity of the trial rises, reflecting the intricate logistics. The result is a trial that is difficult to manage, expensive to run, slow to progress, and likely to fail. Statistics from the FDA and the European Medicine Agency bear this out. Half of Phase 3 trials fail, mostly for lack of efficacy and/or safety, and 40% of the compounds eventually submitted to regulators are never approved for the same reasons [22–24]. This puts the combined failure rate for compounds in Phase 3 and submission at 70%. This is higher than the attrition rate for Phase 2, which is typically thought of as the riskiest phase of drug development [25]. The cost of these failures, which occur after companies have typically invested hundreds of millions of dollars in these molecules, is staggering. Ironically, these compounds fail not because they embody bold risky ideas, but because they were selected as safe bets, with validated targets and proven modes of action [26]. The strategy that was originally designed to produce a steady flow of innovation at low risk ended up being far riskier than originally imagined. Management's risk aversion comes at a steep cost.

Large pharmaceutical firms spend vast amounts of money on R&D, as much as $88 billion annually for the top 20 companies. Much of that is thought to fund the development of badly needed new therapies. In fact, the data show otherwise. More than half of Phase 3 trials run by large companies involve drugs that have long been approved. The FDA requires two successful Phase 3 trials before considering a drug for approval (although it sometimes waives this requirement for breakthrough drugs). Large companies, however, routinely conduct dozens of Phase 3 trials to support their blockbusters, and undertake most of them after the drugs have been approved. Table 13.1 shows the number of postapproval Phase 3 trials for well-known blockbusters. Humira™, an anti-inflammatory drug, was involved in 73 postapproval Phase 3 trials; Seroquel™, an antipsychotic, in 72; and Abilify™, another antipsychotic, in 65. Some of this is justified by the need to "broaden the label." Once a drug is approved, adding indications to avail it to all the patients it can help is a sensible thing to do. But no blockbuster boasts enough indications to justify such high numbers of postapproval trials. It raises the question of the real purpose of this work. Is it research, and is it needed? One may wonder what is really learned from running, say, the fifty-first Phase 3 or Phase 4 trial for a given drug. In all likelihood, such trials hardly represent genuine research, but are rather a means to generate "new data" to feed the companies' vast sales forces. It keeps them from repeating the same pitch, and it works. It does boost sales, but it also curtails the funds available for discovery research, and ultimately imposes a severe penalty on innovation.

There is other evidence to suggest that the actual cost of bringing a drug to market is much lower than often alleged. Over the last 10 years, the FDA has approved 25 drugs for rare diseases. Most of them were developed by small public companies that publish their financial statements. It is possible to analyze them and tease out how much they really spent to bring their drugs to market. Table 13.2 summarizes the data for seven drugs approved between 2002 and 2009. It shows that the out-of-pocket costs for

TABLE 13.1. Big Pharma Routinely Conduct Dozens of Phase III and IV Trials to Support Their Blockbusters

	Number of Trials	
	Ph3 Post Approval	Ph4
Humira	73	25
Seroquel	72	40
Abilify	65	40
Risperdal	60	60
Lyrica	58	44
Rtuxan	55	18
Avastin	54	14
Diovan	53	53
Singulair	45	23
Remicade	43	23
Januvia	40	9
Lipitor	39	40
Lucentis	39	17
Nexium	35	42
Gymbalta	34	49

Source: http://www.clinicaltrials.gov.

TABLE 13.2. Out-of-Pocket Clinical Development Costs for Several Drugs Approved for Rare Diseases

Brand Name	Generic Name	Sponsor	Type	FDA Approved	Indication	R&D Costs ($ million)
Cinryze	C1 esterase inhibitor	Lev	Biological	2008	Hereditary angioma	34
Prolastin-C	Alpha1-proteinase inhibitor	Talecris	Biological	2009	Panacinar emphysema	43
Xyrem	Sodium Oxybate	Orphan Medical	Small mol	2002	Cataplexy	46
Naglazyme	Galsulfase	Biomarin	Biological	2005	Mucopolysaccharidosis VI	84
Kuvan	Sapropterin	Biomarin	Small mol	2007	Phenylketonuria	95
Aldurazyme	Laronidase	Biomarin	Biological	2003	Mucopolysaccharidosis I	98
Kalbitor	Ecallantide	Dyax	Biological	2009	Hereditary angioma	106

Source: 10-K Forms.

clinical development ranged from $34 to $106 million, which is a small fraction of the billions of dollars that many drug companies typically spend to do the same thing [3]. This suggests that eliminating the costs of unnecessary research can go a long way toward making drugs affordable.

NEW MODELS FOR INNOVATION

Returning the pharmaceutical industry to the values that made it great—ethics, risk-taking, breakthrough science, and patient focus—is essential to restore the industry's health. But there are other needed steps. In the last 15 years, a host of new research models have emerged and created new collaborative approaches to research. In the industry's heyday, being a pharmaceutical company meant doing everything from drug discovery through marketing, and excelling at everything. It was difficult. Today, the disaggregation of pharmaceutical research and the wide availability of powerful tools and databases allow companies to concentrate on what they do best—such as assay development, toxicology, or early clinical research—and outsource other activities. A thriving market for innovation has emerged.

THE NEW PLAYERS

- *Academia.* The traditional role of universities has been the production of cutting-edge knowledge, which serves as raw material to the industry's translational research efforts. Academia though has become increasingly involved in translating its own knowledge through incubators such as QB3 at the University of California San Francisco, Stanford's SPARK, or the Deshpande Center at the Massachusetts Institute of Technology. With some variation, the basic idea is the same, that is, provide low-cost facilities, seed financing, and managerial support to academic entrepreneurs with big ideas (those often too big to draw interest from conservative pharmaceutical companies).
- *Government.* The industry's waning interest in translational research, and its retrenchment from key therapeutic areas (e.g., anti-infectives and neuroscience), have prompted NIH to step in and create the National Center for Advancing Translational Sciences (NCATS). NCATS' goal is to facilitate the translation of breakthrough discoveries such as those made with NIH funding. It also plans to encourage drug-repurposing opportunities as a way to speed affordable new therapies to patients.
- *Small Pharma/Biotech.* There are over 4000 small companies engaged in drug R&D, many of them focused on discovery or early development where they explore novel hypotheses. Given their size and limited resources, they must choose their activities carefully, and partner everything else.
- *Contract Research Organizations.* They started out as vendors that provided clinical research services at lower cost, but have increasingly repositioned

themselves as partners that offer high-value services. Some, like Covance, have purchased discovery research facilities from Big Pharma, and manage them for their former owners [27]. Others, like Quintiles, are moving into Big Data analysis or new sophisticated approaches such as helping manage their clients' clinical trial material supply chain.

- *Patient Groups.* Disease foundations are sponsoring growing amounts of early research and clinical trials. Some, like the Michael J. Fox Foundation or the Leukemia & Lymphoma Society, fund projects worth tens of millions of dollars each year. The Cystic Fibrosis Foundation collaborated with Vertex for over 10 years [28] on a program that led to the approval of Kalydeco® (ivacaftor).
- *Venture Capitalists.* Some of them (Atlas, Third Rock) have created incubators where the companies they fund operate quasi virtually, staffed by a handful of people that rely on a network of partners to carry out the work.
- *Everyone.* Some collaborative models are open to nearly all comers. Eli Lilly and Company's Open Innovation Drug Discovery Platform is one of them. It invites scientists to submit compounds that will be assessed against a range of diseases or molecular targets. Merck & Co's CALIBR and GlaxoSmithKline's Discovery Partnerships with Academia are other examples. They offer academic scientists help in translating their biomedical research into novel medicines. India's Open Source Drug Discovery platform is another approach that invites scientists to join forces in developing novel antibiotics for tuberculosis.
- *Crowdsourcing Platforms.* Several platforms have been created to help potential partners connect with each other. Innocentive® is probably the best known. It allows companies with difficult problems to tap the expertise of a community of more than a quarter-million scientists in the hope that some of them will be able to help. That happens about 75% of the time, and when it does it saves over 95% of the costs that would have been incurred with more conventional problem-solving approaches. Transparency Life Sciences® taps the crowd to identify drug repurposing opportunities for approved medicines. Selected projects are then developed using a unique platform for online patient enrollment, and at-home monitoring by wireless biosensors.

THE NEW COLLABORATIVE MODELS

In recent years, a complex pattern of collaboration has emerged linking these players, as illustrated in Table 13.3. It should be stressed that this alliance network is constantly growing and changing, reflecting the pressures upon the industry, which forces companies to consider novel collaboration opportunities with many partners. The result is dense overlapping innovation networks where value is created as ideas are enriched by cross-pollination and turned into therapies through the collaboration of many nodes in the network.

TABLE 13.3. The Ecology of Collaborative Research in the Biomedical Industry

Partners	Collaborators	Examples
Industry–academia	One-to-one	Sanofi-UCSF; Janssen-Vanderbilt; Janssen-Johns Hopkins Bayer-Tsinghua; Novartis-Penn; Shire-Boston Children's
	One-to-many	Pfizer's Centers for Therapeutic Innovation with 20 AMCs; AstraZeneca's A5 Alliance Bristol-Myers' International Immuno-Oncology Network with 10 AMCs
	Incubators	QB3; SPARK; Deshpande Center; von Liebig Center
Industry–industry	One-to-one	Pfizer-GSK; AstraZeneca-Merck; GSK-Abbott
	One-to-many	Virtual pharmas (Zafgen, Hyperion, Synageva, FerroKin, etc.); Chorus; TLS
	Many-to-many (consortia)	Enlight; C-Path; TIPharma; Transcelerate; Arch2POCM Alzheirmer's Diseases Neuroimaging Consortium Asian Cancer Research Group CEO Roundtable on Cancer Life Sciences Consortium University of Dundee Division of Signal Transduction Therapy Consort University of Manchester Collaborative Center for Inflammation Research; Lilly-Merck-J&J
	Public–private Partnerships	IMI, NCATS, Mass. Neuroscience Consortium; Structural Genomics Consortium; TB Drug Accelerator; Neomed Institute
Industry–suppliers	CROs	Lilly-Covance; Lilly-Quintiles; AstraZeneca-Charles River; AZN-Wuxi; Bristol-Myers-Accenture
Industry–patient groups	Disease foundations	CFF; MJ Fox; JDRF; LLS; CFSS; Myelin Repair Foundation; Lupus Foundation; Breast Cancer Alliance
Industry–venture capitalists	Incubators	Velocity, Atlas, Amgen-Kleiner Perkins; Narrow River Management; TLS
Industry–anyone	Crowdsourcing	Lilly's Open Innovation Drug Discovery; Merck's CALIBR GSK's Discovery Partnerships with Academia; Lilly Innovation Fellowship Awards

WHERE HAVE THE LEADERS GONE?

Will all this suffice to rekindle innovation? It will certainly help, but is unlikely to succeed by itself. If we look back at the legendary leaders who built the industry, such as Paul Janssen (Janssen), Roy Vagelos (Merck), Arthur Levinson (Genentech), Daniel Vasella (Novartis), and George Rathmann (Amgen), they shared a singular trait: they were not only the top executives at their firms, but also their de facto Chief Innovation Officers, even when they did not officially carry that title. By acting as tireless advocates of innovation, and more importantly, by being bold innovators themselves, they inspired their employees, and made it safe for them to be audacious thinkers. This is not unique to pharmaceuticals. Steve Jobs (Apple), Jeff Bezos (Amazon), Richard Branson (Virgin), Bill Gates (Microsoft and the Gates Foundation), and Bill Hewlett and Dave Packard (HP) are examples from other industries. Unfortunately, that leadership style is now in short supply in the pharmaceutical industry. This must be corrected. Innovation is about cross-pollination and bold thinking. It does not thrive on law and order. Great innovators care little about the status quo. Just the opposite, they dream of obliterating it with something better, which will give them decisive competitive advantage. Steve Jobs, Jeff Bezos, and the others were not process optimizers. They followed innovation where it took them, even if it meant disrupting their companies. This is no different from the translational research model that made the drug industry great.

PRIORITIES FOR RESTORING INNOVATION

Innovation is a by-product of culture, not a by-product of organization, process, R&D spending, or anything else. Pharmaceutical companies must return to a culture of innovation. This means welcoming back the curious, passionate, and at times quirky scientist; allowing audacity and vision to drive decision making; and making it safe to be bold. It also means restoring the standards of excellence that once made the industry a haven for talent.

The industry must discard practices, such as portfolio management and blockbuster forecasting, which do not work and are deeply flawed forecasts to select the R&D projects that will drive its future. It must refocus instead on translating breakthrough science into innovative therapies irrespective of perceived market potential or fit with existing business.

Firms should also recognize that some scientific challenges are too big for them to tackle alone. They should join their competitors and other stakeholders into ad hoc collaboration networks to understand the science they want to translate before attempting to do so. This will reduce pipeline attrition rates, save vast amounts of money, and allow them to compete more successfully on developing better drugs.

Lastly, drug companies should put ethics and patients back at the center of what they do. This means refraining from decisions and behaviors that adversely impact patients, including unethical behavior, the deprioritization of key therapeutic areas, and extreme prices.

In short, drug companies need to reconnect with the values and models that made them great, and roll back the excesses of the last two decades.

REFERENCES

1. Steiner, M., Bugen, D., Kazanchy, B. The continuing evolution of the pharmaceutical industry: career challenges and opportunities. RegentAtlantic Capital. 2007. Available at http://www.pharmawealthmanager.com. Accessed December 23, 2013.
2. M&A has failed to build value for pharma, Burrill & Company, April 12, 2011.
3. Herper, M. The truly staggering cost of inventing new drugs. *Forbes*, February 10, 2012.
4. Brill, S. Bitter pill: why medical bills are killing us. *Time*, February 20, 2013.
5. Francisco, M. Compulsory license bandwagon gains momentum. *Nat Biotechnol.* 2012;30:814.
6. World Preview 2018—embracing the patent cliff. *EvaluatePharma* 2012.
7. The last Kodak moment? *The Economist*, March 27, 2012.
8. Collins, J. *How The Mighty Fall: And Why Some Companies Never Give In.* Collins Business Book; 2009.
9. Carroll, J. Sanofi's Viehbacher spotlights Big Pharma's new role creating biotechs. *FiercePharma*, February 16, 2012.
10. Witty, A. CEO, GSK, quoted by William Looney in R&D Costs: it's industry's problems. *Pharmaceutical Executive*, November 1, 2011.
11. Schumpeter, J. Capitalism, socialism and democracy (1942).
12. Denning, S. Why did IBM survive? *Forbes*, July 10, 2011.
13. Obituary for Professor John C. Sheehan. *MIT News*, April 1, 1992.
14. Hollingsworth, R. Scientific discoveries: an institutionalist and path-dependent perspective. In: Hannaway, C., editor. *Biomedicine in the Twentieth Century: Practices, Policies, and Politics*. Amsterdam: IOS Press; 2008.
15. Almashat, S., et al. Rapidly increasing criminal and civil monetary penalties against the pharmaceutical industry: 1991 to 2010. Public Citizen's Health Research Group. December 2010.
16. Staton, T., Palmer, E. Pharma's Top 11 marketing settlements. *FiercePharma*, June 26, 2012.
17. Andriani, P., McKelvey, B. Beyond Gaussian averages: redirecting international business and management research toward extreme events and power laws. *J Int Bus Stud.* 2007;38: 1212–1230.
18. Andriani, P., McKelvey, B. From Gaussian to Paretian thinking: causes and implications of power laws in organizations. *Organ Sci.* 2009;20:1053–1071.
19. Taleb, N. *The Black Swan: The Impact of the Highly Improbable*. New York: Random House; 2012.
20. Staton, T. Lilly pulls Xigris on failed septic-shock study. *FiercePharma*, October 25, 2011.
21. Weintraub, A. Pfizer's Exubera flop. *Bloomberg*, October 17, 2007.
22. Mullard, A. 2010 FDA drug approvals. *Nat Rev Drug Discov.* 2011;10:82–85.
23. Arrowsmith, J. Trial watch: Phase III and submission failures: 2007–2010. *Nat Rev Drug Discov.* 2011;10:87.

24. Eichler, H.G. quoted in Gordian, M., Singh, N., Zemmel, R., Elias, T., Singh, N., Zemmel, R., Elias, T. Why products fail in phase III. In Vivo. April 1, 2006. Available at http://sis.windhover.com/buy/abstract.php?id=2006800066. Eichler is the Senior Medical Officer at the European Medicines Agency. Accessed December 23, 2013.
25. Paul, S.M., Mytelka, D.S., Dunwiddie, C.T., Persinger, C.C., Munos, B.H., Lindborg, S.R., Schacht, A.L. How to improve R&D productivity: the pharmaceutical industry's grand challenge. *Nat Rev Drug Discov*. 2010;9(3):203–214.
26. Clinical failures result from predictable business development strategy, invivo blog, October 19, 2012.
27. Lilly sells its Greenfield, Indiana, operations to Covance, Reuters, August 6, 2008.
28. Vertex and cystic fibrosis foundation therapeutics collaboration, Vertex Corporation, 2012.

14

VALUE-DRIVEN DRUG DEVELOPMENT: UNLOCKING THE VALUE OF YOUR PIPELINE

Valentina Sartori,[1] Michael Steinmann,[1] Petra Jantzer,[1] and Matthias Evers[2]

[1]*McKinsey & Company, Zurich, Switzerland*
[2]*McKinsey & Company, Hamburg, Germany*

Pharmaceutical companies hardly need reminding that drug development has become more difficult: even safe and effective drugs struggle to gain regulatory approval and market access. To transform the situation, R&D and commercial teams should adopt a new paradigm: collaborating at the beginning of Phase 2 to keep a laser-like focus on stakeholder value.

In the 1990s, pharmaceutical companies could bank on a successful drug launch if they could prove that their drug was safe and effective. Since then the goalposts have shifted. Regulators want proof that new drugs are safer and more effective than those already on the market [1], and even regulatory approval is no guarantee of success. Health-care providers the world over are struggling with rocketing costs, making them reluctant to pay for drugs that do not deliver significant incremental benefits to patients—particularly if they come with a high price tag.

The result is that many drugs fail to secure broad market access or to earn the developers an acceptable rate of return. Between 1998 and 2008, for example, the UK's National Institute for Health and Care Excellence (NICE) granted restricted or no market access to almost 60% of drugs from the top 10 pharmaceutical companies.

Collaborative Innovation in Drug Discovery: Strategies for Public and Private Partnerships, First Edition. Edited by Rathnam Chaguturu.

Meanwhile, since its inception in 2004, Germany's Institute for Quality and Efficiency in Healthcare has classified 70% of the drugs it has reviewed as "benefit not proven."

The market-access challenge is likely to increase as payors demand ever more value for their money to contain health-care costs, which have risen twice as fast as gross domestic product since 1970. Accordingly, pharmaceutical companies have experimented with new approaches to try to improve their odds of success. Some, like GSK and Novartis, have worked closely with payors in late-stage development; others, like Pfizer and Janssen, have done so after launch, through risk-sharing agreements, for example. In our opinion, however, the only way pharmaceutical companies can consistently launch successful drugs is by working to meet the market's needs much earlier in the development process.

This requires a new paradigm. R&D and commercial teams need to start working together when planning for proof of concept (PoC) in Phase 2. And instead of searching for a gap in the market for the compounds they develop, these cross-functional teams need to design a compound to fill a market gap. That gap will be defined not just by the needs of patients but also by the needs of regulators, health technology assessment bodies (HTAs), and payors.

The drugs that prove successful will be those that demonstrate their value to all these stakeholders, and do so early in development. We call this new paradigm "value-driven drug development." It seeks to maximize the value of a company's current pipeline and to replenish it with new and valuable compounds by steering research in the right direction. In so doing, it helps mitigate three of the main risks in drug development: discontinuation in Phase 3 due to lack of efficacy; commercial disappointment (often because of lack of differentiation); and failure to gain regulatory approval because the compound's risks are deemed to outweigh its benefits (Figure 14.1).

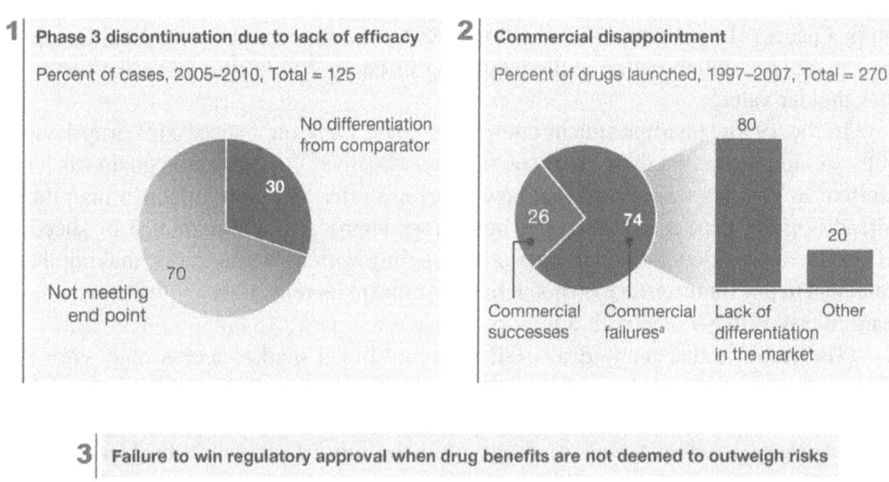

Figure 14.1. Value-driven drug development helps mitigate three key risks. [a]Commercial failure defined as NPV at launch less than average cost of development. *Source*: Pharmaprojects, APM Health Europe, Evaluate, McKinsey analyses. See color insert.

THE FOUR IMPERATIVES OF VALUE-DRIVEN DRUG DEVELOPMENT

Value-driven drug development has four essential components, discussed in detail below:

1. Understand what outcomes matter to patients and other stakeholders at least five years before launch
2. Sharpen the focus of Phase 2 to define value as well as dose
3. Upgrade team and leadership capabilities, and
4. Instill a performance culture that encourages innovation and maximizes value.

1. Understand Which Outcomes Matter to Patients and Other Stakeholders at Least Five Years before Launch

Even as early as 5 years before launch, the patient should be the focus. At this stage, the task is to identify, using real-world evidence, patient needs not yet met by competitors for specific indications, and to understand what profile a new compound should have to satisfy those needs. The next step is to identify a subset of patients who might benefit most from the compound, perhaps because certain genetic variations respond well to it. Admittedly, segmentation in this manner restricts the size of the market for the proposed drug, but it also accentuates the potential differentiation from competitors' compounds.

One example of a successful drug that has been narrowly targeted in this manner is Roche's Herceptin. This drug specifically targets the 25% of breast-cancer patients whose cancer is related to an overexpression of the gene factor HER2. Oncology is the area in which most personalized medicine research has been conducted to date, but we believe other therapeutic areas are suitable too.

Efforts to differentiate a compound and so demonstrate its value can go further still by clearly defining different components of the overall outcome that the subgroup of patients would most value. For example, beyond its efficacy, a compound might also improve a dialysis patient's quality of life by reducing the number of hospital visits required. After patients, the focus turns to other health-care stakeholders that influence registration and reimbursement decisions: governments, regulators, HTAs, and payors. Stakeholders' assessments of a new drug's value will differ, as will the data they require to demonstrate that value.

Regulators are mainly concerned about the risks and benefits compared with the standard of care, and mostly require randomized control trials and "hard" clinical end points directly related to the progression of the disease. Payors care about the total cost impact on their patient population. HTAs want to know whether the incremental benefits of a new drug can justify its cost. They may require observational and experimental studies demonstrating a more subjective assessment by physicians or patients of the drug's impact on symptoms or quality of life. Regulators and payors are aware that their different demands can be hard for pharmaceutical companies to accommodate, and some have started to collaborate to try to reach common ground (see sidebar, "Increasing collaboration").

The development team will need to understand each stakeholder's relative influence. It used to be physicians who decided whether or not a drug was prescribed; now payors and HTAs increasingly hold sway. That said, stakeholders' influence varies by country. For instance, HTAs have little influence over reimbursement decisions in the United States—that is, the remit of insurance companies. In Europe, by contrast, HTAs influence important pricing and reimbursement decisions. For example, NICE rejected the use of Genentech's cancer drug Avastin in two cancer indications (metastatic colorectal cancer and first-line treatment for metastatic renal cell carcinoma) on cost grounds, resulting in sales worth just €10 million in the United Kingdom in 2008. That compared with sales of €300 million in France, where no HTA assessment was made.

Development teams will also need to find an approach that satisfies the main regulatory agencies in the United States and Europe. It is becoming increasingly difficult to submit one registration package that works for both. For example, the European Medicines Agency (EMA, http://www.ema.euroa.eu) always requires a pediatric plan, while the U.S. FDA does not. The EMA always requires a comparator for oncology drugs, while the Food and Drug Administration (FDA) does not. Their assessments differ too. The FDA approved Wyeth's antidepressant drug Pristiq, while the European regulator had concerns about differentiation, prompting Wyeth to withdraw its submission. See Box 14.1 for more details.

Box 14.1. Increasing Collaboration

At the end of 2010, the EMA launched a pilot project with health-care stakeholders from six European countries (France, Germany, Italy, the Netherlands, Switzerland, and the United Kingdom) to assess the therapeutic and economic value of new drugs at an early stage of development, and to share their views with pharmaceutical companies. Astra Zeneca, GlaxoSmithKline, and Johnson & Johnson are involved in the pilot, which currently focuses on drugs to treat type 2 diabetes and breast cancer.

Since early 2011, the EMA has also been collaborating with the European Network for Health Technology Assessment to understand how risk/benefit data contained in European public assessment reports for centrally authorized drugs can be used in HTA assessments.

In addition, regulators and HTAs are collaborating at national level. In the United Kingdom, National Institute for Health and Care Excellence (NICE, http://www.nice.org/UK) and the Medicines and Healthcare Products Regulatory Agency (http://www.mhra.gov.uk) launched a pilot program in 2010 to give pharmaceutical companies independent scientific advice from each agency on how to design drug-development programs that would suit both of them. Although there have been no participants in the program as yet—something NICE attributes to the strict application criteria—many companies have expressed an interest.

In Sweden, the Dental and Pharmaceutical Benefits Agency and the Medical Products Agency also offer joint advice to companies that request it. Since 2009, there have been 20 such joint assessments.

When armed with insights into patients' needs, competitors' strategies, and stakeholders' expectations, development teams are in a position to consider their options strategically. The target product profile (TPP) sought is one that will be clearly differentiated from the future standard of care (as understood at the time of launch); one that delivers maximum value to stakeholders; and one that carries an acceptable development risk profile.

2. Sharpen the Focus of Phase 2 to Define Value as Well as Dose

Having assessed a new compound's safety in Phase 1, most pharmaceutical companies focus Phase 2 on first understanding its efficacy (Phase 2a) and then ascertaining the right dose (Phase 2b). A few companies, such as Novartis and Wyeth, have started to do things differently in an attempt to make the development process more seamless: Novartis's approach is known as Delphi; Wyeth's as Learn and Confirm. We advocate an approach in which Phase 2 homes in as early as possible on where value might lie.

First, Phase 2 is used to identify the subset of patients who have the optimal risk/benefit profile for the compound, as described earlier. AstraZeneca recently received European approval for all lines of therapy for its lung cancer drug Iressa for a subset of patients with a specific biomarker—but only after it withdrew its first EMA submission following a nonconclusive Phase 3 study that targeted the full population of patients and went on to conduct a new, more narrowly targeted study. Its experience underscores the potential benefits of early patient stratification and the use of biomarkers in clinics.

Second, Phase 2 is used not just to test efficacy and dosing, but to start testing the additional questions likely to be raised in Phase 3 by stakeholders seeking value. In this way, the development team can quickly identify compounds that are unlikely to meet stakeholders' needs, stop development, and avoid wasting further costs. Meanwhile, compounds that remain in development have a better chance of gaining regulatory approval and market access.

Interacting with payors, HTAs, and advisory boards at this stage will help the development team test its initial hypothesis about where value lies. Input from these stakeholders will shed light on what a new compound might have to deliver to be judged better than the standard of care, which end points need to be proven, and what data are required. Comparative studies that give an early sense of how the compound differs from the standard of care and how the pivotal Phase 3 study may need to be refined accordingly are also useful. Designing Phase 3 trials to test the compound against the likely future standard of care rather than a placebo is another means of reinforcing the compound's value.

Third, whenever possible, clinical trials in Phase 2 should be designed to optimize costs, time, and data quality, but without sacrificing ethical standards. Take as an example a compound that addresses a well-known and already validated mechanism. Time and costs will be saved by using an adaptive design that combines Phase 2a (proof of efficacy) with Phase 2b (dose ranging), thereby reducing start-up times and improving dose–response estimates. The company can analyze interim results and use modeling and simulation techniques to understand the dose–response curve before continuing the trial and further refining the pharmacodynamic model. If the trial fails to demonstrate that the drug is sufficiently differentiated, the compound can be dropped in the knowledge that only limited resources have been wasted.

On the other hand, should the compound show promise, Phase 3 will be reached more quickly. A good example of innovation in the design of a clinical trial, one that enabled a speedy trial and ultimately faster registration, was Novartis's development of Ilaris, a treatment for Muckle–Wells disease. Novartis used modeling and simulation techniques to select the dose range, which was then confirmed in a seamless Phase 2b/ Phase 3 trial [2].

Fourth, when entering Phase 2, teams need a development strategy for a mechanism of action (MoA) that addresses more than one indication. Even before PoC, a plan is needed that maximizes a drug's potential value, taking into account all the possible indications and respective patient segments. Different indications are likely to have different value profiles. They will meet unmet needs to a greater or lesser extent, carry different risks, require more or less time to develop, be priced differently, and have different interdependencies (e.g., a study for one indication may reveal valuable lessons for another). All this needs to be assessed to understand how best to stagger development.

3. Upgrade Team and Leadership Capabilities

Value-driven drug-development teams require a particular blend of skills and capabilities, as do the governing bodies that oversee them.

The Team. Drug development has tended to be clinicians' turf. But if stakeholder value is the goal, other specialists need to be part of the team too.

Even at the research phase, translational science experts should be present to identify possible biomarkers and develop a biomarker strategy to help patient segmentation. Then, in Phase 1, molecular diagnostics specialists should help develop companion diagnostics to measure in clinics the biomarkers identified. Strategic marketers also have a role in ensuring that market insights—such as what competitors are up to, how other MoAs in development might compete for success, and how the market will have evolved by the time of launch—are incorporated into the development strategy.

When planning for PoC at Phase 2, teams will require still more skills. The strategic-access function seeks to understand where value lies for payors and HTAs. It then works with clinicians to define the data required to satisfy hard and soft end points, comparators, and differentiation requirements. Modeling and simulation will bring in the necessary mathematical skills not only for pharmacokinetic/pharmacodynamic (PK/PD) modeling but also for full drug-to-disease modeling or for decision-analysis support.

The Team Leader. By tradition, a development team is led by a clinician who has little contact with marketing or commercial divisions. But if companies want to promote a value-driven culture and operate effectively within it, a compound in development will need its own CEO: someone capable of managing a cross-functional team, aggregating its members' input, and keeping a balance between clinical excellence and successful commercialization. That calls for drug-development experience, skills in managing projects and teams, and strategic-thinking ability.

In addition, the team leader will need to establish strong knowledge networks with internal and external stakeholders and key opinion formers to stay abreast of research developments, monitor competitors, and be able to react to changing circumstances.

The Governing Body. A similar broad mix of skills and experience needs to be reflected in the governing body that oversees the entire drug portfolio. A governing body that embraces a value-driven approach to drug development will need to have a strategic perspective on the portfolio so that it can assess the relative risk/benefit profile of any single compound within it and decide which compounds to resource and prioritize.

Like the development team, the board will need people with a mix of scientific and business skills and experience, and in particular an understanding of health-care systems in different countries. This mix will help ensure that the board maintains a strong external focus, keeping an eye on what competitors are doing and what the market requires, and providing the right guidance to development teams. In addition, the board should play an important coaching role, challenging teams constructively to ensure that their strategies are robust.

4. Instill a Performance Culture That Encourages Innovation and Maximizes Value

Value-driven drug development has a much broader exploratory remit in Phase 2 than current approaches do. This has repercussions: for example, because Phase 2 seeks to establish how far a compound differs from those that are or will be available, decisions will be taken earlier as to whether to continue or halt development. As a result, the attrition rate of projects may rise in Phase 2, but could decline in Phase 3. In addition, a value-driven approach may shift resources from compounds showing marginal differentiation, even if they are in large indications or segments, from those with greater differentiation but in a narrower segment of the population (e.g., from hypertension to hypertension in the Afro-American population).

This exploratory approach requires a greater degree of transparency, risk-taking, and innovation. To some it may feel liberating; to others, unnerving. To support the new approach, companies need to foster a culture that treats value generation as the key criterion in all-important processes.

New performance measures and incentives will help. Development teams are currently rewarded for meeting milestones on time. In the new paradigm, a team that is frank about the risks of a project or willing to make the tough decision to terminate an unpromising one because of limited differentiation will still be rewarded because it has kept its eyes firmly on the value goal. Similarly, a clinician who fails to show that a compound is different from the future standard of care but uses an innovative cost-saving trial to do so is still congratulated. On the other hand, a clinician who uses a traditional approach to avoid risks and shows mild differentiation compared with a placebo is not.

Cross-functional collaboration is key, but not easy to build. Those accustomed to working in silos tend not to like having their ideas or working practices questioned. A culture that encourages the constructive challenging of ideas and strategies will help break down silos, as will a readiness to dissent and to raise concerns when needed. Those at the top of the organization and in positions of authority—boards and team leaders—will have to show the way. Only when they model the new methods of working will others be likely to adopt them.

CONCLUDING REMARKS

To incorporate a value-driven approach capable of developing innovative drugs with demonstrated value, most companies will need to transform their entire R&D organizations. This will affect the composition of teams, governance, culture, and capabilities.

Our experience suggests that a pragmatic approach is best. Although the architecture of the transformation program needs to be clear, not every detail of the design has to be settled before the company embarks on change. Better to start quickly by piloting different elements of the program so that management can rapidly understand what works and what does not and make any necessary refinements. It also makes sense to stagger program components to avoid overwhelming the organization.

To be sure, such a transformation will stretch executives in R&D and commercial areas as well as their product teams. But done well, it will also unlock the value of the pipeline and deliver a step-change in the organization's performance.

REFERENCES

1. One example is the FDA's tougher scrutiny of data in non-inferiority trials. Report GAO-10-798. July 2010. Available at http://www.gao.gov/new.items/d10798.pdf Accessed December 24, 2013.
2. Orloff, J., Douglas, F., Pinheiro, J., Levinson, S., Branson, M., Chaturvedi, P., Ette, E., Gallo, P., Hirsch, G., Mehta, C., Patel, N., Sabir, S., Springs, S., Stanski, D., Evers, M.R., Fleming, E., Singh, N., Tramontin, T., Golub, H. The future of drug development: advancing clinical trial design. *Nat Rev Drug Discov*. 2009;8:949–957.

15

UNLOCKING THE MARKET POTENTIAL OF ACADEMIC RESEARCH

Assem S. el Baghdady and Yasser M.S. el Baghdady

AlphaBeta Pharma Group, Leatherhead, Surrey, UK

Our world is facing many pressing health-care issues. In developing countries, the basic health components of clean air and water present great technical challenges. Growing populations demand proper nutrition, public health programs, immunization, energy, reproductive health care, and more. Developed countries are seeing more of their health-care resources spent on illnesses caused by an affluent western lifestyle: metabolic syndrome, mental health issues, and the complications of aging. The solutions to these complex issues cut across all borders, and involve many industries and scientific disciplines. Governments need both industry and knowledge institutions to work together to reach tomorrow's technical health-care solutions.

Moreover, in a global economy where conventional manufacturing is dominated by developing economies, the future of industry, including the health industry, in the most advanced economies must depend on its ability to innovate in those high-tech activities that can offer a differential added value, rather than to only improve existing technologies and products.

Universities are at the forefront of knowledge. Most of the research conducted in academic institutions represents small, incremental steps in the advancement of our overall understanding. Academic investigators discover pieces of a vast and complex

Collaborative Innovation in Drug Discovery: Strategies for Public and Private Partnerships, First Edition. Edited by Rathnam Chaguturu.

puzzle, which are eventually made public through journals, conference presentations, and working groups. All of these contribute significantly to the vast body of knowledge in modern society. However, too few of these research programs lead to commercially viable inspiration that can be exploited by entrepreneurs, eventually resulting in a marketable product.

Most basic research that is undertaken in the complete absence of any market insight or commercial valuation is "research for research's sake." Research for research's sake is considered no longer acceptable or justifiable or, some would argue, even ethical. In this era, if research is not beneficial to humanity in any way or shape, then it is a waste: a waste of money, a waste of time, a waste of resources, and a waste of intellectual capital. With the market's fast-paced growth, and the demands of unmet medical needs, we see a wide gap between research produced in academia and the demand in the health-care marketplace. Research should lead to innovation, which in turn should develop into successful commercial exploitation of these ideas. Strategies for adding value and marketability should be planned into the research and development processes so as to bridge the gap between the laboratory and the market, to help ensure the successful commercialization of new technology-based products [1].

We believe that academic research has both ethical and financial obligations. There are ethical obligations toward patients with an unmet medical condition whose lives, against the ticking clock, are depending on the development of efficacious treatments. In addition, we also feel that there is a financial obligation to recoup at least part of the public's research investment in academia, and to recycle it for more research in the future.

Due to diminishing public funding, academia also must attract corporate funding and therefore must align its research with market needs. However, market needs are not always aligned with unmet medical needs or patients' needs. Nevertheless, corporate research generally starts from surveying the current market need (market pull) and takes into account future market, business, and financial projections before embarking on or funding new research. The return on investment for shareholders is the main driver in their decision making: companies conduct research for investors' sake.

Academic institutions no longer have deep pockets, and alternative models are beginning to emerge. In response to these trends, and aware that much university research effort is currently unused, many universities have created Technology Transfer Offices (TTOs). Their purpose is to create a thriving research strategy and to guide academic researchers in aligning their research programs with the requirements of the pharmaceutical industry, to include its accepted research models, regulatory requirements, and a myriad of other aspects that are not within the usual sphere of academic researchers' activity. Many institutions are copying each other's best practices, but the evidence of success is more anecdotal than factual. There are many good reasons for this, an important one being simply the uncertainty and unpredictability of early-stage drug development research.

Many different specialist skills are needed to achieve successful exploitation of research and innovations. Corporate research and development managers, academic researchers, technology transfer officers, intellectual property (IP) specialists, venture capitalists, professional marketers, and policymakers alike have unique and vital contributions to the drug development process. In this chapter, we make an attempt

- To provide an overview of the drug development process to better understand its complexity and the substantial financial and time investment that it requires
- To look closely at the gap between "science for science's sake" versus "science for society's sake" and its causes
- To highlight some of the emerging models that are in place to bridge the gap between the research environment of academia and the commercial world of corporate health care
- To address the emerging role of the academic technology transfer office; discuss strategies and suggestions for improving the effectiveness of these models for the future
- To illustrate some successful initiatives.

WHERE DOES DRUG DISCOVERY FIT INTO THE DRUG DEVELOPMENT PROCESS?

The development of a new drug involves a long, costly journey, hampered by uncertainty. Cost estimation studies, conducted mostly by U.S. entities, provide figures ranging between $500 and $800 million, over an average of 11–15 years, from inception to market for a new drug [2–4]. According to the Pharmaceutical Research and Manufacturers of America (PhRMA), the costs of developing a drug have risen from $138 million in 1975 to $1.3 billion in 2005. The cost to develop a single biologic was estimated at $1.2 billion [5]. Some variations may be attributed to the type and nature of disease being targeted, the type of drug being developed, and the design of the clinical trials that is necessary to have the drug ultimately achieve regulatory approval at local and international levels. However, while these are gross estimations for time and cost, variations notwithstanding, the studies provide concordant conclusions that the process is a lengthy and expensive one (Figure 15.1 and Table 15.1).

In comparison to other industries, the pharmaceutical industry has one of the highest ratios of research and development (R&D) to sales. The PhRMA reports that domestic R&D as a percentage of domestic (U.S.) sales was 20.5% (17.0% for total R&D) in 2010 [5].

In the European Union (EU), the innovative pharmaceutical sector devotes 16% of business expenditure to R&D, the highest of any industrial sector. Pharmaceuticals contribute strongly to a trade surplus, of some €47.8 billion ($50 billion) in 2010 in Europe (Figure 15.2). The strong presence of the pharmaceutical industry in the economy makes it a major employer of skilled labor. In the EU alone, the pharmaceutical industry provides over 640,000 jobs directly, of which 113,000 are highly skilled [6–8]. In the United States, PhRMA reported in 2012 that the biopharmaceutical industry creates approximately 650,000 jobs directly, nationwide, and that each biopharmaceutical job supports nearly five additional jobs. This translates to approximately 4

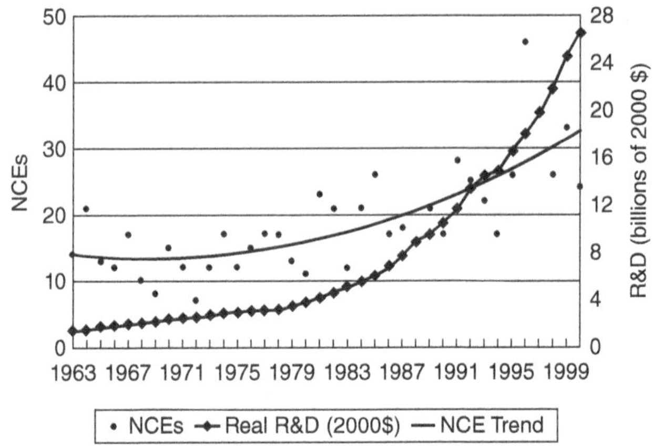

Figure 15.1. Inflation-adjusted industry R&D expenditures (2000 dollars) and U.S. new chemical entity (NCE) approvals from 1963 to 2000. *Source*: PhMRA (2001) and Tufts CSDD Approved NCE Database. DiMasi et al. *J Health Econ.* 2003;22(2):151–185.

TABLE 15.1. Average Out-of-Pocket Clinical Period Costs for Investigational Compounds (in Millions of Dollars)[a]

Testing Phase	Mean Cost	Median Cost	Standard Deviation	N^b	Probability of Entering Phase (%)	Expected Cost
Phase I	15.2	13.9	12.8	66	100.0	15.2
Phase II	23.5	17.0	22.1	53	71.0	16.7
Phase III	86.3	62.0	60.6	33	31.4	27.1
Long-term animal	5.2	31	4.8	20	31.4	1.6
Total						60.6

Source: From DiMasi et al. *J Health Econ.* 2003;22(2):151–185.

[a]All costs were deflated using the GDP Implicit Price Deflator. Weighted values were used in calculating means, medians, and standard deviations.

[b]*N*, number of compounds with full cost data for the phase.

million jobs across the economy, ranging from manufacturing, construction, and services.

Productivity and the rate of increase in investment in drug discovery have slowed. This slowdown is not only attributable to the daunting investment required in time and money, but also to the fact that many of the easier disease targets have been addressed. The many remaining targets are proving extremely difficult to address from several standpoints, including a lack of understanding of disease pathology, limited technological and computational capabilities, and the complex manufacturing processes required to produce a new entity (Figure 15.3).

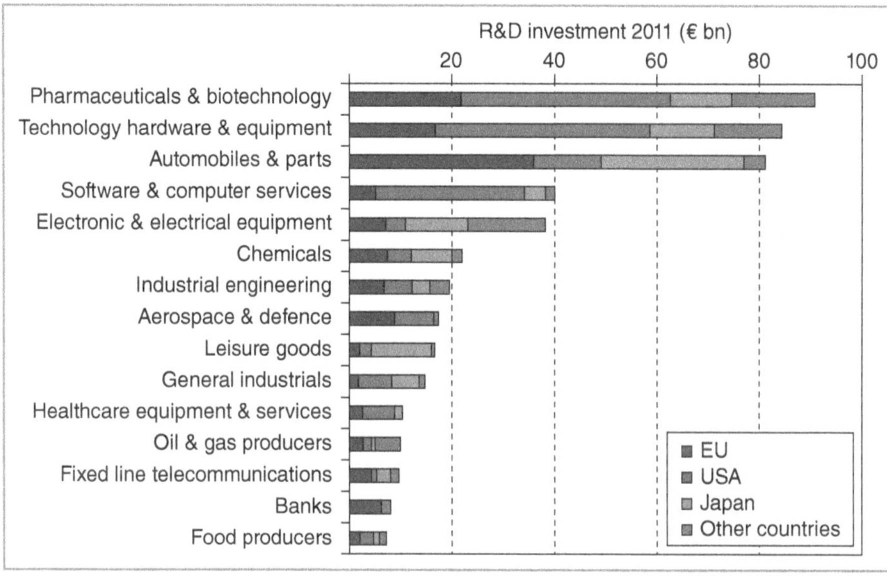

Figure 15.2. R&D ranking of industrial sectors and share of main world regions for the world's top 1500 companies. *Source*: The 2012 EU Industrial R&D Investment Scoreboard European Commission, JRC/DG RTD. See color insert.

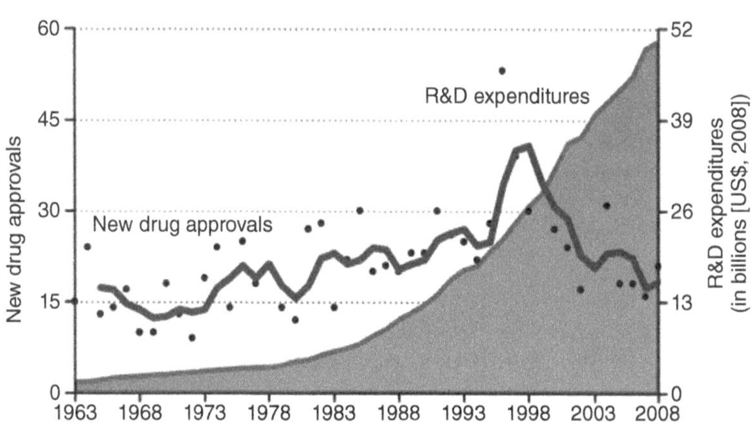

Figure 15.3. New drug approvals and pharmaceutical R&D expenditures. R&D expenditures (shaded) are presented at constant 2008 dollar value. The trend line is a 3-year moving average. Drug approval data from the Tufts Center for the Study of Drug Development (CSDD). R&D expenditure date from PhRMA Industry profile 2009. *Source*: Kaitlin, K.I. Deconstructing the drug development process: The new face of innovation. *Clin Pharm Ther.* 87:356–362.

Another more modern impediment, albeit a necessary one, is that regulatory requirements from safety and tolerability perspectives since the mid-1970s have become far more stringent and thus require further investments in time and money to gather long-term safety data. In 2012, the FDA approved 35 novel drugs.

HOW DOES THE DRUG DISCOVERY PROCESS WORK?

On this long, costly journey from laboratory to marketed drug, the phase of drug discovery is only the beginning. It is the first rung on a very tall, slippery ladder.

While it is well known that the drug discovery process requires a full commitment on the part of researchers and investors, the complexity of the pathway from laboratory to the health-care marketplace will be discussed to gain a greater understanding of its implications for commercialization. Commonly called the development chain or "pipeline," bringing a drug to market occurs in a number of distinct stages, and each may be viewed as a separate cost center.

Step One: Discovery/Basic Research

This stage usually resides within academia, and includes the process of synthesizing, extracting, chemically engineering, or otherwise identifying a molecule, enzyme, hormone, or receptor that has the potential to effect a change in a biological environment, that is, a drug target. It is the drug target that has to be demonstrated to be the modifier of the disease. This is where academic researchers really excel. Validation of the candidate molecule or the intended target is the next step in the discovery stage. It is difficult to say whether the validation step is as important as the identification step, or is even more important! This is where academia is not particularly interested: the primary research has been published already, and no one finds merit in repeating someone else's original work. To be fair, editors of first-tier international journals are also not excited about such work, so that it rarely gets a respected reception, however well deserved. Beyond the extreme difficulty of obtaining funds for a validation project, it can be rather awkward to justify the need for such research work. Validation work is not usually published by the next runner-up team that lost the race to publish first!

Once identification and validation steps are out of the way, another step—no less important than the former steps—called optimization is ready to start. This is optimization of the compound, that is, extraction of an isomer or a metabolite that is much more potent and less toxic, or optimization of the target, that is, identification of a more specific receptor subtype. Optimization can of course occur in parallel without critically impacting the development process and should also involve a validation step. Academic researchers slightly prefer such research work, as it is considered "nearly" original; however, the pharmaceutical industry excels academia in this step.

Once identified, to validate the model, the molecule is then subjected to a battery of *in vitro* biological research and screening investigations, each one designed to explore and clarify the pharmacological activity and therefore therapeutic potential of the agent.

Animal modeling is an essential part of the development process. After the *in vitro* testing stage, an animal model is either identified—if available—or developed for modeling the disease indication. The selection of the animal model is very important and should not be taken lightly. Some animal models of human disease are well recognized within the academic community, but some do not yet exist and therefore require original research and *de novo* development. However, it is crucial that the appropriate animal model is selected in order for a study to be widely accepted by regulatory authorities. This is where academic researchers frequently fall short. Academic researchers often overlook this vital detail, because their focus has traditionally been on the beginning of the development cycle, rather than its end! If it is academia's objective to participate in the drug development process, it is imperative that they align their research program design with the regulatory requirements further down the line in this process. In the pharmaceutical industry, this is part of the responsibility of the regulatory strategy team, to ensure that no problematic issues arise during a discussion/consultation with the regulatory authorities or indeed at the time of submission of a New Drug Application or, even worse, during submission of the Marketing Authorization Application (MAA).

This is an important fundamental role for TTOs to perform; however, frequently at this early stage of discovery, communication on this point does not take place enough between the academic researchers and the TTOs. This is perhaps due to a lack of specialized expertise needed or—may we dare say—a lack of understanding of the entire drug development process (Figure 15.4).

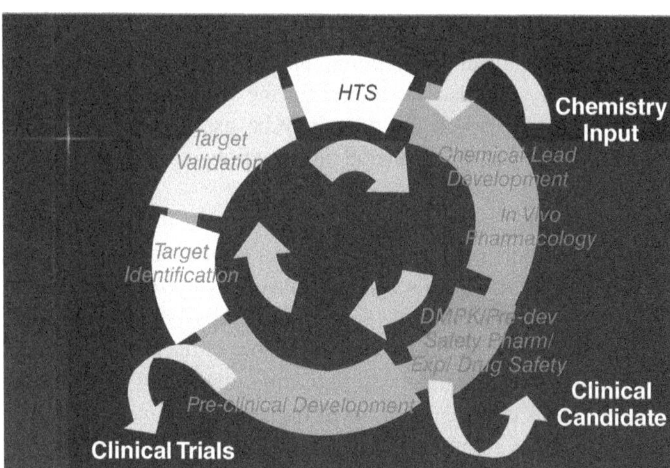

Figure 15.4. Target identification and validation in the drug development process. See color insert.

Step Two: Preclinical Testing

Once a molecule has been defined as having biological potential at the intended target in the disease animal model, preclinical testing for its pharmacodynamic and pharmacokinetic properties takes place to establish allometrically a safe dose to initiate clinical trials in humans. Toxicology tests are done to determine the compound's potential risk to humans, utilizing whatever formulation is available and any potential route of administration to humans.

Typically, preclinical toxicology testing is conducted in two species of animals, most commonly involving murine and canine, although primate and porcine species may also be used. During this stage, work continues in parallel to also focus on turning the active compound into a suitable form and strength for human use in the clinical studies yet to follow. This is called "clinical supply," and it does not have to be the finished item, that is, the end marketed product, either in strength or even formulation, as this will come later on.

An important feature of preclinical testing, and a regulatory requirement, is that each investigation must comply with Good Laboratory Practice (GLP) and associated International Committee of Harmonization (ICH) policies. Needless to say, the majority of the academic laboratories does not hold a GLP certificate, and in the author's experience even argue that they do not need one! At this phase, the compound is referred to as an Investigational New Drug (IND), and the FDA approval for the IND status is based solely on safety data, not efficacy.

Step Three: Phase 1 Clinical Trials ("First in Man")

The IND is now approved for use in humans, and small studies involving 6–20 (though at times reaching 80) healthy volunteers (usually male) will be conducted to determine not only safety and tolerability, first and foremost, but also the basic pharmacological properties in a human subject: absorption, distribution, metabolism, and excretion (ADME). Bioavailability can also be studied for the various formulations. In addition, patients with a disease for which the drug may be useful can be test subjects, such as in oncology, and this can yield useful information regarding the efficacy of the drug.

The start-up dose has always been and will remain the most crucial and most dangerous step in the whole process. The regulatory authorities have issued a number of guidelines that must be followed when calculating (and projecting from animals) the start-up equivalent dose in humans. Typically, single-dose studies are carried out, followed by single ascending dose studies where the dose of the new medicine is gradually increased. This allows the investigator to determine the safety and tolerability and to measure the participant's clinical response to the medicine. Intravenous (IV) formulation and route is the most common in this phase, even if the final formulation is not going to be an IV, and drug metabolism and pharmacokinetics data are collected at all times. This determines whether the medicine is sufficiently absorbed and distributed at the intended site of action, how long the medicine remains active in the system after dosing, and which dosage levels are safe and well tolerated. Phase 1 is sometimes divided into several clinical trials, for instance, into Phase 1a and 1b, with Phase 1a studying a group of healthy subjects, and Phase 1b studying a group of patients with the disease, or specific safety aspects or interactions with other agents.

In recent years, regulators have encouraged a new pre-Phase 1 step called "Phase 0," using a fraction (1/100th) of the proposed Phase 1 starting dose in less than a handful of subjects. This process is very dependent on advanced nanotechnology to detect and trace the minute dose and project the response of the full dose.

Step Four: Phase 2 Clinical Trials

This intermediate phase of testing the investigational drug involves larger groups of human subjects (n = 100–300), usually now including a relevant patient population. While dosing strategies may be investigated, the main aim of Phase 2 clinical trials is to establish effectiveness against the disease or medical condition and to understand any short-term safety risks.

Phase 2 studies are divided into Phases 2a and 2b. Phase 2a studies are generally short, small studies to compare and evaluate different doses of the drug that indicate whether the drug is working in the way that is expected. Phase 2b studies are usually larger (lasting longer and with more patients), with narrow inclusion/exclusion criteria, and are randomized, often blinded, placebo-controlled studies, to investigate the clinical benefit to the patient. These are usually considered the main go/no go decision gate; the outcomes of these trials will be determining factors of whether to proceed to the larger and most expensive part of the process, that is, Phase 3.

Step Five: Phase 3 Clinical Trials

Involving large patient numbers (n = >1000+), Phase 3 trials are designed to mimic clinical real life by adapting a study design that simulates the actual clinical practice in terms of the dose, frequency, and duration of treatment in the intended patient population. A more relaxed inclusion/exclusion criterion includes the targeted patients with that specific disease indication in terms of gender, age, past history, and most common concomitant medication. The primary objective of Phase 3 is to establish efficacy through the most relevant primary efficacy outcome measures, and its secondary objective is to establish the safety profile by further evaluating the safety of the IND through the collection of data on safety, tolerability, and any adverse events. These studies often have a comparative design, either with placebo, or the current clinical gold standard of care, or with a competitor product, or indeed include all of these! In addition, a design that includes more than one formulation, dose regimen, or comparator is not unusual. Upon completing Phase 3, a favorable benefit–risk assessment of the drug should be demonstrated to both the regulator as well as the health-care professionals. Needless to say, the quality of the methodology employed in Phase 3 trials has a great bearing on the later approval by regulatory authorities, as well as on the credence and trust that medical specialists will place in the new drug.

Regulatory Review of Investigational New Drug

Submission of the "dossier" is the biggest event in any pharmaceutical company's diary, perhaps second only to first-in-man dosing in Phase 1. The dialogue between the regulatory authorities and the pharmaceutical company (the "sponsor") does not just start after completing the clinical development or indeed stop at the approval stage! Such discussion ideally starts at the discovery stage and should be ongoing throughout the entire development process; that is, before embarking on the next stage, discussions regarding design, duration, objectives, and so on, should take place. There should be regular updates throughout that stage to discuss progress, difficulties, or any unexpected or unexplained results. Even after obtaining marketing approval, the regulatory dialogue continues on a much more formal basis in the form of periodic safety update reports and risk management plans.

Evaluation of an IND is a lengthy process that could take years. It includes all the data generated by the sponsor (or any other parties) for the intended disease indication (or any other indications) in such patient populations (or any other tested populations). If the regulatory authority does not have the required expertise for this evaluation, a more appropriate external specialist in the field will be engaged for the task. Approval of an IND is dependent on multiple factors besides efficacy and safety, such as the nature of the indication (life-threatening), the target population (vulnerable), the nature of the drug (first in class), the economic environment (budgetary illness prioritization), the political situation (high threat of biological warfare), societal (traditional alternatives), and finally, cultural constraints (stigma).

Postmarketing trials, which are also known as Phase 4 clinical trials, are then conducted to gain understanding and confidence in the drug from safety and efficacy perspectives. In particular, these trials are designed to detect hitherto unseen adverse events, and to further understand the long-term morbidity and mortality profile of the drug in a much larger population. These clinical trials are usually a voluntary exercise sponsored by the MAA holder; however, in some instances (usually due to safety concerns), it is required by the regulatory authority as a condition for the marketing approval and hence becomes a postmarketing commitment by the MAA holder. The regulatory authorities reserve the right to request (i.e., demand) such data as they deem appropriate at any time during the life cycle of the marketed drug (Table 15.2).

Step Six: Manufacturing Process Optimization (Chemistry, Manufacturing, and Control)

Often conducted concurrently throughout the entire drug development process, engineering and manufacturing studies are performed to refine production for efficiency, scale, stability, uniformity, and quality. The results of these tests will often determine the large-scale commercial viability of a compound (Figure 15.5).

The race with the generic drug companies starts as soon as the drug is on the market, and usually creeps over the original drug territories within about 10 years after expiration of IP rights. During that period, the MAA is racing to recoup the cost of

TABLE 15.2. Limitation of Clinical Trials in Establishing Drug Safety of IND[a]

Incidence	Sample size to detect adverse event once with 95% certainty
1:500	1,497
1:5,000	7,488
1:25,000	74,892
1:50,000	149,785
1:100,000	299,572

[a]This is primarily due to the limited number of patients that tested the new drug throughout the clinical phase. Typically; 50 (patients in Phase 1) + 500 (patients in Phase 2) + 1000 (patients in Phase 3) = 1550 patients.

research and development as well as achieve as much profit as possible. Generic drug companies need only to prove that their drug is equivalent to the original drug, which is a far a less costly development process; hence, generic drugs are far cheaper than the originals. This is a different story that could easily fill another book, however, and will not be discussed in detail here.

SCIENCE FOR SCIENCE'S SAKE AND SCIENCE FOR SOCIETY'S SAKE

There is a huge gap that results from the mismatch between academia's discoveries (market/technology push) and industry needs (market pull). Academic research addresses basic, fundamental questions. Universities largely conduct research in the complete absence of "market pull." Research is undertaken from the perspective of the search for new knowledge and understanding, and not necessarily in response to a direct market need. This "science for science's sake" is absolutely appropriate.

Scientists often do not know what the applications of their discoveries could be, or which parties could potentially be interested in their discoveries, and are not familiar with the requirements of research to be accepted within a particular industry.

At a later time, they may try to "push" this new technology or knowledge to a market that they are unfamiliar with. Pushing an unknown technology to an unknown market seldom leads to a success: one unknown in any equation is manageable, but two unknowns are not. Technology push (or market push) is very difficult, and will often fail. Failure occurs not only because of the limitation or the lack of applicability of the technology: it can also be due to unsuitability or the saturation or maturation of the market. Of course, what is deemed "failure" from a commercial standpoint is a natural and normal part of the research process, yielding information of value to the researcher, but not of short-term financial value to any company, especially start-ups (Table 15.3).

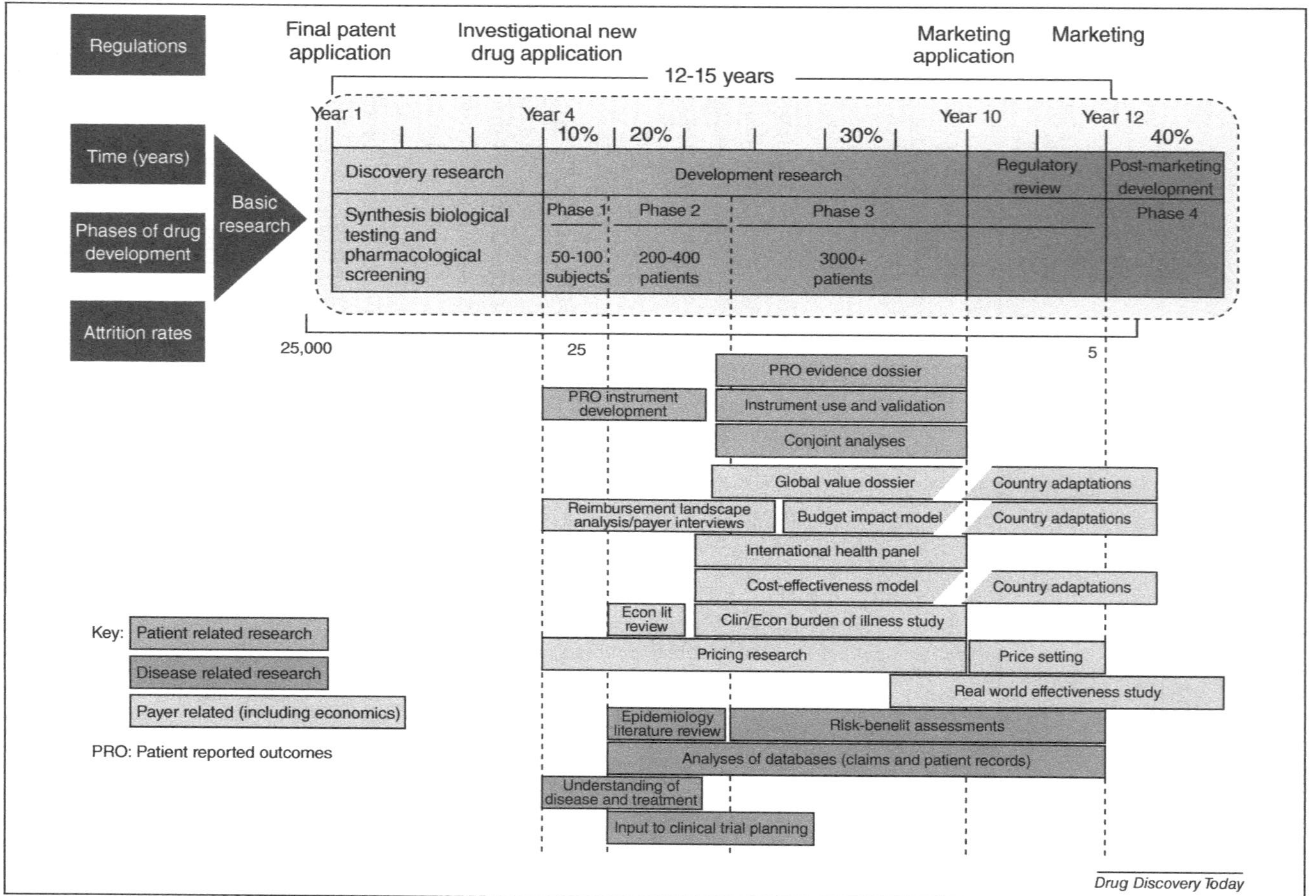

Figure 15.5. Timelines of research phases and attrition rates in drug development. *Source: Drug Discovery Today*. See color insert.

TABLE 15.3. Commercial Contrast between Technology Push and Market Pull

	Technology Push	Market Driven
Business Focus	Funding IP	Customer Requirements
Development Risk	High	Quantified
Market Demand	Unknown	Quantified
Certainty of Returns	Low	High
Business Growth	Unlikely	Projected and Planned

Academia and the pharmaceutical industry have an old relationship, but it has not always been particularly productive or fruitful. As Jose Carlos Gutierrez Ramos of Pfizer said in a recent interview, "In the status quo, money was going from pharma to academia and it was not really being used. You get money from pharma and it's like another R01 (a grant from the National Institutes of Health). You keep doing the same research as before. And therefore pharma was withdrawing. They'd say, 'this is nice, but we aren't getting anything out of it' " [9].

Out of the thousands of medical papers published each year, often one may wonder where the clinical question is among them all. Some studies fail to address the question of genuine interest because of the numbers or ethics involved: "In common with many published studies, the authors allude to just such a question in the final paragraph, clearly stating that their study was unable, and was never designed, to answer such a question, and that other, larger studies might shed more light" [10].

But while scientists may be unaware of what applications their research could address, or which pharmaceutical companies to turn to, at the same time, companies are unaware of what scientific discoveries exist that may be of value to them. "From the time a research group has a cluster of publications to the moment that a pharmaceutical company becomes interested, you could easily expect 5–10 years to elapse. Meanwhile, great research is being done around that pathway of phenomenon, but it's not directional, because that's not the nature of academic research," says Gutierrez Ramos of Pfizer. Adding to the fact that the IP clock continues ticking all the time, 5–10 years means a huge amount of lost profits, which could render the whole project commercially unattractive.

WHAT ARE THE CAUSES OF THE GAP BETWEEN SCIENCE FOR SCIENCE'S SAKE AND SCIENCE FOR INVESTMENT'S SAKE?

Cultural barriers often widen the gap between academic research and market adoption. Many experts with years of experience in both academia and the pharmaceutical industry attest that the cultural barrier is probably the main challenge to bridging this gap. Ambivalent attitudes still hamper monetizing publicly funded research. Particularly in Europe, members of the establishment within many universities and national health services see themselves as public service providers, not profit-making organisations. There is resentment toward the new economic reality, rather than optimism and energy to embrace a dynamic new model.

Industry may not always speak the language of the researchers. Although big pharmaceutical companies have a long history of scouting out and developing products in many therapeutic areas that meet a market need while providing a handsome return on investment for shareholders, it can be difficult to carry a transfer discussion to fruition. Anxiety about their own company's relatively short-term financial goals can lead to impatience, summary rejection of more complex research IP, and a tendency to cut corners. The enormous overheads and colossal expenses put pressure on their profit margins, making them limit their interest to proven technology, rather than the riskier investment necessary to develop early-stage research.

An atmosphere of distrust and lack of understanding often inhibits collaboration. Universities are large bodies whose primary role is teaching and research and do not lend themselves to risk, while entrepreneurs are fast moving, high-risk takers and tend not to comply with rules. This means that it is difficult for both to create a company where they share ownership. In addition, sometimes the researcher leaving the university has been frustrated by his or her experience within the university, and they will do everything possible to avoid the university having any control or influence in their company!

Ironically, misdirected response to public pressure widens the gap. In the past decade, there has been increasing public pressure to "improve the results" of many publicly funded institutions, including universities, by insisting on measurable units and numeric ratings. The previous government in the UK, for example, spent over 10 years laying down such targets and even more time on measures to audit and report these meaningless targets. The brain drain of scientists across Europe in general and in the UK in particular is a direct result of such political policies (albeit coupled with short-sighted taxation laws). Popular articles ranking "The 10 Best Universities" or "The 10 Best Hospitals" are common, despite the lack of methodology, accepted at face value, and university personnel are rewarded in part along these lines. For a university, measurable results primarily include the number of scientific publications, as well as number of patent applications and registrations, and student-generated ratings for teaching excellence. Marketing and selling research findings are far less measurable for universities, and less attributable to individuals. It is hard to quantify and measure, and so receives less structured support, or reward incentive, within the European universities.

Another reason for the gap is that selling research results is simply not part of the academic staff's job. In response to public pressure and slashed budgets, universities have increased the internal workload significantly. Professors must meet tougher requirements for more and better journal publications, generate their own research funds, deal with increased student numbers with fewer support staff; as a result, increasing time is spent on detailed reporting of their activities and expenditures. While many academic researchers are deeply interested in the market potential and the societal benefit of their research, most of them have a hard time meeting their "measurable" objectives within their 24-hour days, let alone guide research findings along toward marketable entities.

With many noted exceptions, most academics lack the entrepreneurial skills that are so important in convincing the pharmaceutical industry to invest in their research results. As mentioned previously, guiding IP along to a marketable product for industry is not included in their academic functions; there are insufficient (or no!) time and resources allotted to it, and academics are not rewarded for it.

What does the process of commercialization entail? What makes it so complicated? The next section will discuss this aspect of unlocking the market potential of academic research.

WHY COMMERCIALIZATION IS IMPORTANT

There are some attitudes and convictions in academia that its proper role is that of a public servant, and that focus on future commercial uses of research could compromise their position of impartiality, thoroughness, integrity, and independence. Universities are concerned that the increasing demand to obtain their funding from private industries will reduce their noble calling to "mere contract research." They may be hostile to the notion of adjusting their research strategy and procedures to align with the needs of potential commercial applications, sometimes even when the potential client is in the science park affiliated with the university.

Commercialization drives research. Yet commercial interests have historically been the drivers of research. Leonardo da Vinci knew that Duke Ludovico Sforza in Milan would be interested in his many discoveries to develop for military use, and he actively solicited the Duke's financial support.

Another familiar example of commercial interests driving research is Edison General Electric, founded by Thomas Alva Edison. The company later became GE. At age 30, Edison built a research laboratory in Menlo Park, New Jersey, to combine the scientists and materials that were required to address complex problems (Figure 15.6). They worked on several projects simultaneously, thinking through subsequent commercial needs and developing solutions. For instance, his power generating and delivery systems made sufficient power available to homes and businesses to ensure the commercial success of his earlier invention, electric light. By creating a critical mass of resources, Edison developed R&D labs that revolutionized the technical research process. He received a record of 1093 U.S. patents [11] (Figure 15.7).

The revenue generated from successful inventions finances future research. In drug development, PhRMA reports that only 2 out of 10 drugs that reach the market return revenues that match or exceed R&D costs [5]. The revenue from the marketed drugs must cover all of the previously conducted R&D for those products, as well as all of the R&D for the "failed" drugs—those that did not reach the market due to any decision factors. However, the "failed" drugs generate much useful research and market knowledge, and have provided employment and income for scientists. These are also noble objectives and do benefit the public. Without commercialization, there are no

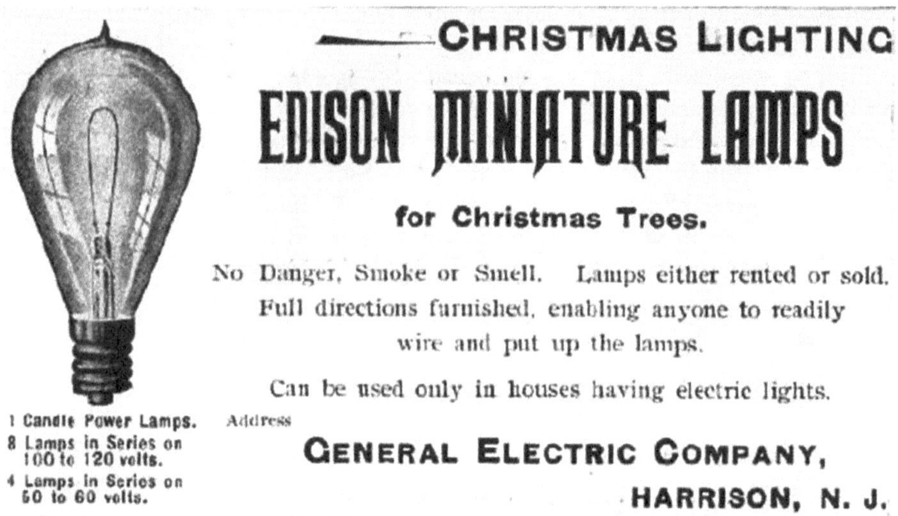

Figure 15.6. One of Edison's many advertisements for electric lighting.

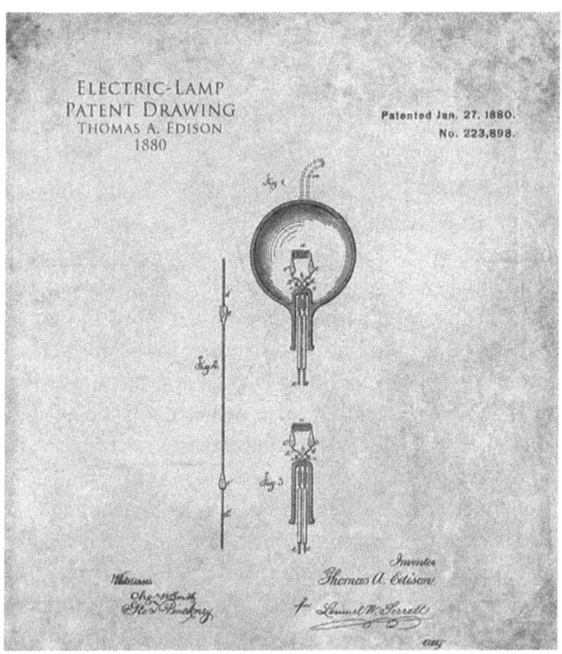

Figure 15.7. A drawing from Edison's patent application for light.

breakthroughs in drug development. One would be hard-pressed to name an innovative pharmaceutical product from any government-owned research organization in any centrally run country.

As for serving the public, the pharmaceutical industry's role cannot be underestimated. Innovative medicines extend life expectancy, treat disease, and increase patients' survival. Although there is criticism of "me-too" products, in many countries, regulatory authorities require that new drugs demonstrate significant improvement compared with medicines already on the market.

Another important public benefit from discovery and innovation in the pharmaceutical industry is that new medicines can reduce the burden of care. Easier routes of administration, fewer administrations, lower dosages, or the possibility for patients to remain at home (or shorter hospitalization periods) translate to significant savings in health care. At present, with all the arguments and debates regarding the price of drugs, both in Europe and in the United States, the share of drugs as a percentage of all health-care expenditure in the United States, which typically has high drug prices, is about 10% [12] (Figure 15.8).

Creating an environment that embraces commercialization will cultivate an attitude in academia that facilitates the development of research into marketable IP. Europe is lagging behind the United States in this area. Many business leaders are calling for policies to address the gap between industry and academia and its consequences: ". . . the patent that goes unexploited, the research report that gets ignored, or the researcher who leaves for richer labs in San Francisco or Singapore" [13].

INTELLECTUAL PROPERTY

Used almost interchangeably with "patents," the term intellectual property (IP) describes a tool for secure exchange of intellectual capital. The patent is the legal registration of the IP, protecting the ownership, and therefore, exploitation, of the IP. A condition for obtaining a patent is that the discovery is new, original, innovative, and has not been made public. Therefore, patent protection should be obtained before a discovery is published. Tension occurs at this stage because rapid publication is of vital importance to academic scientists. However, of course, publications should not be delayed due to patent filing; in practice, this can be achieved by good timing, short decision lines, and the availability of sufficient funds to cover the costs of patent registration [14]. But more important than all that is access to a competent patent attorney, either employed by the research institute or contracted by it—and competent ones do not come cheap! (See Figure 15.9.)

In considering the commercialization of research output, it is important to consider the maturity of the IP. Typically, university IP is very early stage and gets protected via patent well before its equivalent in industry, due to publication pressures. As a result, it frequently becomes superseded during its development phase, when an application is being developed. The formal IP residing in the university is not necessarily required: the research progresses, making the original IP outmoded. If and when the researchers

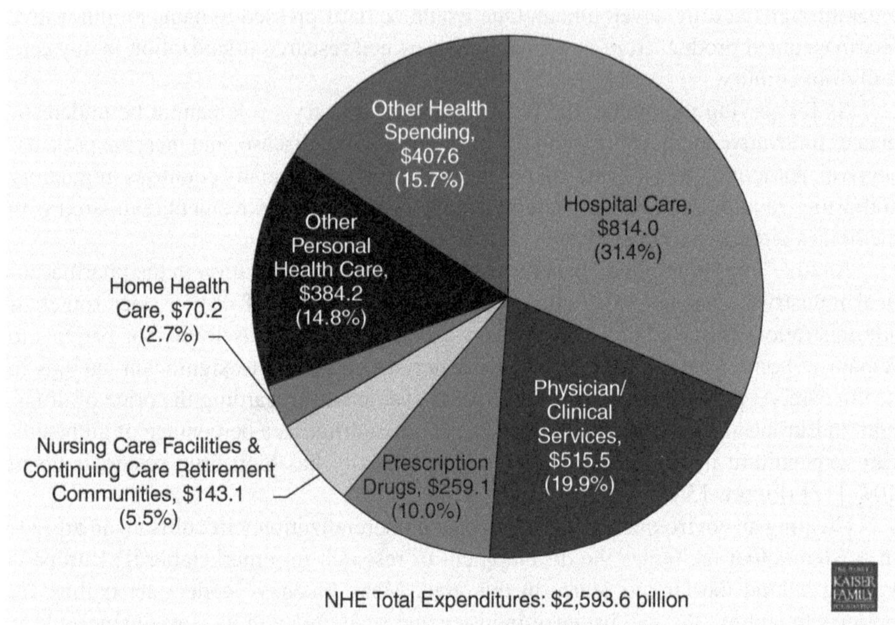

Figure 15.8. Distribution of (U.S.) National Health Expenditures, by type of service (in billions), 2010. *Note*: Other Personal Health Care includes, for example, dental and other professional health service and durable medical equipment. Other Health Spending includes, for example, administration and net cost of private health insurance, public health activity, research, and structures and equipment. *Source*: Kaiser Family Foundation calculations using NHE data from Centers for Medicare and Medicaid Services, Offices of the Actuary, National Health Statistics Group, at http://www.cms.hhs.gov/NationalHealthExpendData/ (see Historical; National Health Expenditures by type of service and source of funds, CY 1960–2010; file nhe2010.zip). See color insert.

leave and start a company, why would they give a shareholding to the university when they do not require the IP it holds? This represents a missed opportunity for the university, both in terms of the lost value of exploitation of the IP, and also a loss in terms of vital contacts and collaboration with the academics who were involved in its discovery, and who can be the very individuals to help bridge this gap between universities and industry.

Possibly because the number of patents held by a university is considered important in its international rankings, the rush to patent precludes thorough "intellectual property due diligence." As one colleague expressed it,

> Unfortunately, says the TTO officer, we have very little money for external small and medium-sized enterprises (SMEs) to review the IP to tell us its commercial value; however, we have the budget to spend $10 K and more to patent it in every country in the world. Also, let's say a great researcher consistently brings in $1 M per year to the university.

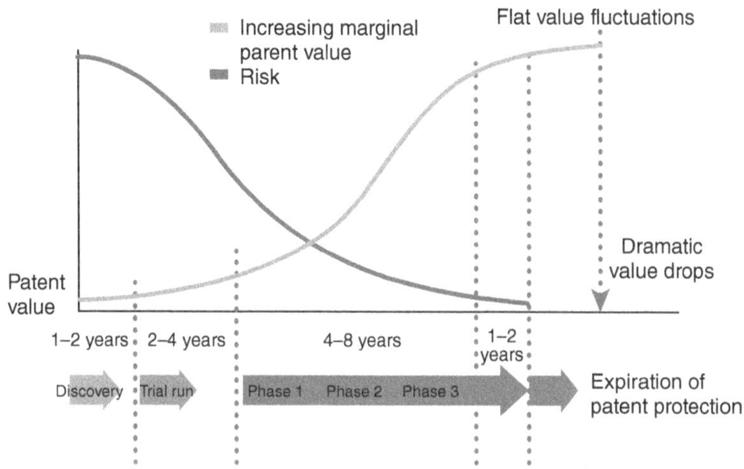

Figure 15.9. Image of pharmaceutical patent life cycle. *Source*: Wu, L. and Wu, L. Pharmaceutical patent evaluation and licensing using a stochastic model and Monte Carlo simulations. *Nat Biotechnol.* 2011;29:798–801.

> This professor is constantly creating IP. Let's say, a pre-patent commercial appraisal is done on a few of his or her IP, with the little money they do have. Is it in the university's best interest not to patent, even if the appraisal comes back negatively? Probably not. It's just how the system seems to work. Thus, if you follow my logic, why spend the commercial appraisal money in the first place? Now, if every university followed a rigorous process to spend patent money only on viable market opportunities, then they'd spend more up front on SME reviewers, but far fewer dollars on patent costs as a result—the author is convinced of that based on my work with university IPs. Of course, this is theoretical. Some universities do a great job in vetting IP before they patent it.

Since the mid-1970s, universities in the United States have greatly increased the number of patents filed. The 1980 Bayh–Dole Act was an important driver for this in the United States.* Patents are a guarantee that the IP generated by the investment is protected against appropriation. While there is evidence of an incentive effect of patents in private sector research and development, it is questionable whether patents serve as incentives within academic research.

Hellmann showed that patent protection increases the returns that scientists can capture from their scientific discoveries [15]. This gives them a greater incentive to seek out pharmaceutical companies that may be able to develop these scientific

*Bayh–Dole gave ownership of inventions back to the universities that created them and gave universities the freedom to negotiate license terms that would encourage development, creating an "institutional ownership model" of academic inventions. The contracting universities may exclusively license the inventions to other parties. The federal government retains rights to license the invention to a third party, without the consent of the patent holder or original licensee, where it deems the invention is not being made available to the public on a reasonable basis.

discoveries. Without patent protection, the only incentive that scientists have is to perhaps gain a contract for doing consulting work for the firm. A patent provides them with a return for their IP [16].

Standard IP agreements facilitate commercialization. It can take 10–20 years to commercialize a patent and to derive substantial value from it. In pharmaceuticals, the profit has to be made during that relatively short period that the drug is on the market, before generic versions are manufactured. What one does with IP, patent or otherwise, depends on the underlying technology, and the future market opportunity, among other ever-changing parameters. Having a patent or patent application is a long way from having a valuable entity, let alone a revenue stream. It is a long road from a scientific discovery, to registering a patent, to producing a marketable commercial product, but successfully following this course generates far more ongoing revenue for a patent holder. But, what is the value of an IP? Or 10 IPs for that matter? It is individuals who create value that has commercial impact. When government research grants are used as an instrument for regional development or social equality rather than to produce commercially viable IP, this is counterproductive. This may lead to some politically unpopular decisions, but it is vital for realizing the commercial and societal benefits of academia's research.

Developing this commercially friendly environment requires a long-term vision. If commercialization is insufficiently supported and rewarded at universities, technology may escape. Entrepreneurial academics will sooner or later chafe at risk-averse culture in universities, which focus largely on education and (basic) research, weighted down by bureaucracy. Business-minded researchers and entrepreneurial inventors may leave academia to start their own companies, actively searching out venture capital.

One cannot start a company or create value simply on the basis of IP. Patents have no intrinsic value; value is created through the transfer and application of knowledge, the application of expertise. This is needed with or without formally registered IP such as patents. Convincing an eminent professor or an institute of this argument generates a great deal of controversy!

COMMERCIALIZING NONPATENTABLE KNOWLEDGE

It is widely believed that the standard route for academics starting their own business is to disclose an invention to the university, get it patented, and venture forth into the spinout world. On the contrary, an extensive survey has found that this is not how the majority of companies are started by U.S. academics, suggesting that government and universities are missing an opportunity in their quest to boost entrepreneurial activity! In what the authors describe as the largest study of its kind, experts on entrepreneurship surveyed 11,572 professors at institutions across the United States. Of the 1948 respondents who had started a business, only 682, or about one-third, had set them up to exploit patents obtained through formal university intellectual-property systems. The remaining 1266 respondents had started businesses—including consultancies and manufacturing and service-based firms—based on nonpatentable knowledge [17]. Social scientists and engineers started the most businesses that were *not* based on patented inventions, but such ventures were also prevalent among biomedical and physical scientists.

"There is a lot of stuff that academic researchers are realizing isn't patentable but they can commercialize for themselves by starting a company," says Scott Shane, an economist at Case Western Reserve University and a coauthor of the study. Surveys of entrepreneurial activity—including government assessments—typically focus on patent activity; they may be significantly underestimating academics' efforts, he notes. Although most commercial ventures from academics who enter industry are in consultancies, there is usually far greater societal benefit from patent-protected IP development than from consultancies. Some experts say that governments and universities are still right to focus on patent-based academic entrepreneurship. Businesses built on inventions tend to need more support to establish themselves, and can potentially generate the greatest returns. It is getting out a new drug, not a new consultancy company, which is going to make the larger contribution to the economy and society.

Unlike in the United States, IP is not well understood in European academic institutions. The cultural difference in attitudes toward monetizing research has hampered interest in the understanding of IP. In Europe, including the five largest markets, the problem is mainly an educational one; that is, industry needs to educate academics on commercial importance, patent registration, and patent protection, while in the United States, the challenge is to meet the commercial expectations of academics and their universities for income generation.

The author's personal experience illustrates this point. The author was discussing the importance of European IP registration with a well-known and successful European professor and the university's legal department. Unfortunately, at the conclusion of their due-diligence exercise, it was discovered that a commercial company in another European country had already exploited the professor's innovative—and published—piece of research for more than 10 years! It is vital that European academic researchers are educated not only about what IP is and its commercial value, but, more importantly about the due-diligence process, the timing and steps of registration, and that there is sufficient specialized legal and technical support to undertake this time-consuming yet absolutely essential work.

BRIDGING THE GAP BETWEEN SCIENCE FOR SCIENCE'S SAKE AND SCIENCE FOR INVESTMENT'S SAKE

Universities know that they have to lessen the gap between their research and industry. Many universities and research institutes have introduced valorization programs. Valorization can be described as a process of exploiting project learning and outcomes (i.e., products, processes, methods) with the purpose of optimizing their value and impact on a target group (e.g., commercial companies). To be effective, the process requires long-term thinking and commitment:

- Focus on the end user's needs from the inception of the project
- Dissemination of results

- Analysis of suitability for transfer to meet these identified needs
- Translation and adaptation to targeted new contexts
- Piloting and experimentation.

Typically, there is a long period of experimentation after the fundamental research that universities conduct, before it becomes clear whether a discovery has value for the pharmaceutical industry. This period has been called the "Valley of Death" because it is fraught with uncertainty, risk, and a high probability of failure. It is a stage at which solid funding is crucial, yet hardest to obtain, and many research projects languish due to lack of sufficient funding, guidance, and support. The early proof of concept phase lasts up to 4 years, and the costs can rise to $50 million. This is only the start of a decade-long development phase that can cost, as we have seen, hundreds of millions of dollars [18] (Figure 15.10).

The most difficult and complex aspect of the commercialization of academic research is funding. At this stage the development is long, uncertain, and prone to becoming derailed. Much experimentation and further development is needed before any statement on applicability can be made with any certainty. At this phase, it is an investment problem: this phase requires solid financial support before there is any

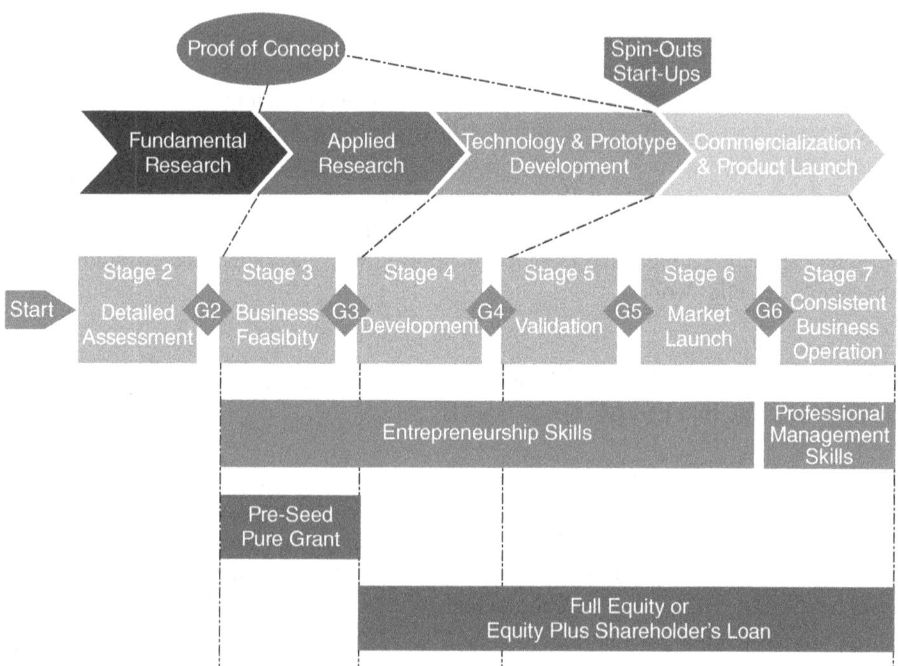

Figure 15.10. Assessment and investment activities throughout the drug development life-cycle. See color insert.

promise of recouping the investment. Where can this be obtained? Industry is not a dependable funding source, because industry is, in most cases, uninterested in an unproven technology, and will hold off on investment until later. Venture capitalists are also not the appropriate party to approach for early-stage funding. Venture capitalists look for rapid returns on investment, and are expecting to recoup their investments within about 2–3 years, selling their share in the venture once its market value has increased and an eager client is found. The U.S. government offers funds for development, and seeding grants are possible. The NIH was established as a vehicle for basic research, not for translating research into marketable medicines. Many European universities have incubators, a cluster of small start-up companies in pursuit of commercial growth of a technology in its early stages. The access afforded by proximity to the university's senior research staff to advise, mentor, and share networks of contacts is invaluable in guiding these projects.

There is no single "one-size-fits-all" business model that fits every situation. Many processes and models exist, and in every situation, the model needs to adapt to the right process, the right product, and the right customer. The single common success factor is adaptability: the willingness and flexibility to adopt a different model efficiently if that is what is required to achieve the best return on investment and customer satisfaction.

Many concepts and operational roadmaps for moving a new-product project from idea to launch have been investigated. One of these is called "Stage-Gate Product Innovation," which is a widely implemented product innovation introduction process that divides product development into distinct stages separated by management decision gates. Cross-functional teams must successfully complete prescribed tasks prior to obtaining management approval to proceed to the next stage of product development decision-making points, go/no-go stages. Better tools to identify early potential of a technology are needed. Many start-up companies and small- and medium-sized organizations have found a more satisfactory niche, and the importance of such experienced individuals and organizations that can help connect academic researchers with the business and public sectors, identify the areas lacking in supporting policy, and introduce programs to encourage interactions. However, this does not happen without considerable effort in identifying suitable drug development partners and staying on course throughout tough, time-consuming negotiations.

TECHNOLOGY TRANSFER OFFICES

Technology transfer is not an easy task; in fact it is a very challenging one. Although universities conduct innovative research leading to patents and original technologies, few of them ever make it to the market or achieve their potential commercial value. This is, of course, not unlike the reality of pharmaceutical research laboratories. Only a very small percentage of Pharma's focused corporate research leads to development; and as we have seen earlier, very few entities ever make it to the market.

Universities have created TTOs in an effort to bridge the research gap. Most large universities have TTOs (also referred to as technology licensing offices, or TLOs).

TTOs function as a third party between the scientist/university and commercial firms, to maximize the university's return on investment in research. The TTO should be more effective in cost and time in searching for the match, because it should have specialized staff that can focus on specific needs in specific industries. Typically, a TTO would have a director, licensing specialist, staff research engineer, staff research scientist, plus administrative staff. It manages the university's patent portfolio, and can also provide intermediate services for the transfer of technology. The intention is that the specialized TTO is able to guide the transfer process efficiently, and that its understanding of potential end markets facilitates the search for a commercial partner or client for the IP. It handles specific administrative steps, such as filing patents. It plays a large role in negotiating contracts with the scientist. Although the university generally owns the patent, the TTO can manage a reward structure with the scientists, in particular if the scientist's involvement is needed for further development with industry alliances.

HAVE TECHNICAL TRANSFER OFFICES FULFILLED THEIR OBJECTIVES?

Universities' TTOs/TLOs are a fairly recent development. There are large differences between UK or European TTOs and those in the United States; their different approaches in dealing with technology transfer issues make it difficult to make accurate direct performance comparisons.

The Situation in the United States

In the United States, business has been a cornerstone of many universities since their inception. There is a long and strong tradition of private universities; indeed, some of the most successful universities in the United States are corporations. By their very nature, they are in the business of selling research. This is such a pervasive argument that they do not have the same type of separate TTOs; researchers see more clearly the path toward a commercially viable IP. For many, it would be unusual not to start an enterprise. The Bayh–Dole Act, giving publicly funded universities ownership of the IP resulting from their research, has accelerated this. Also, in the United States, there is a significant volume of what is effectively technology transfer from academia to industry and through to new ventures, but is termed as consultancy, service, and manufacturing—in other words, often outside the patent-related environment.

The Situation in Europe

European and UK universities have structures, traditions, and cultures that differ from the needs of modern business. Most universities in Europe are government funded; there is no long history of industry investment in universities for early-stage research, which would familiarize the university with working with businesses. It must be remembered that different countries have different policies regarding technology transfer, and the EU is addressing this issue. Similar to the Bayh–Dole Act in the United States, European institutions do retain revenue from their patents and use standard agreements not only to distribute the revenue to the university to fund further research, but also to allocate a portion to the key researchers as an incentive.

Few European TTOs are staffed with teams that have sufficient scientific and business understanding, professional experience from academia and industry, and, most importantly, real market or pharmaceutical industry understanding. Universities and research institutes rarely match the salaries, benefits, and facilities that the pharmaceutical industry can provide, which hamper their success in recruiting and retaining staff with these specialized qualifications and experience. While there are several notable exceptions, most European university TTOs operate according to procedures that almost always delay the process so severely that it obstructs the transfer. Sometimes it is *the* reason that the entire process slides to a halt. As we have seen, research for drug development is a capital-intensive, high-pressure activity, with a race to get a product out before patent expiration. It is a challenge even for mighty Pharma, whose sole focus is drug development. Universities, financed with public funds, traditionally have many masters to serve, in addition to commercialization of IP. Bureaucratic delays and obstructions come about largely by universities mixing priorities: regional development or encouragement of diversity, for instance, often gets added to the objective of commercial transfer, which is quite complicated enough. Too often the result is too many people with different objectives competing for the TTO funding, leading to a bureaucratic quagmire.

In the author's experience, less than 1% of researchers have the ability and desire to be entrepreneurs. Most wish to carry on doing what they do so well, which is research. University administrators have risen through the ranks of academia, and may have little or no experience as R&D managers in an industrial context. This means that they have a different understanding of the technology transfer process.

TTOs can focus unduly on presenting research results rather than effective IP. For instance, in the wave of enthusiasm to establish the TTO and with the pressure to produce results, university administrators can have unrealistically high expectations to monetize an innovative "blockbuster" technology. They may have committed to large financial results within short timelines that cannot be achieved. When these goals are not reached, TTO budgets may be cut, rather than expanded, and the projects lose their momentum. When the TTO budget and its mission are abandoned, the opportunity for learning and lessons about more realistic planning are lost. TTOs are frequently underutilized as a result. Richard Hudson of Science|Business believes that European TTOs need to be reformed. Although TTOs in Europe tend to have more staff, they make only 5% of the revenue of their U.S. counterparts. "University administrators, and their political masters, should set only one goal for the TTO: to make money, as much as possible, as quickly as possible, for the sole benefit of the university. Let the head of the TTO choose the means, rewarding or penalizing that head if the promised results are not achieved. Social goals are right for the rest of the university. A simple financial goal is right for the TTO" [13].

Smaller successes build experience and a market-oriented culture within universities:

- Direct transfer
- (Semi-)exclusive licensing
- Licensing combined with transfer of IP over time
- People transfer
- Licensing in exchange for research collaboration.

The EU is implementing a public–private partnership with the European Federation of Pharmaceutical Industries and Associations that seeks to reinvigorate the biopharmaceuticals sector in Europe. Over the period 2008–2013, €2 billion (approximately $2.9 billion) is budgeted to implement a focused and coherent industrial R&D program that "supports collaborative research projects and builds networks of industrial and academic experts in Europe that will boost innovation in healthcare" [5].

BARRIERS TO SUCCESS OF TECHNOLOGY TRANSFER OFFICES

- Disagreement on the value of the technology
- Value assessment complicated by the uncertainty regarding the market need and unproven fit between the market and the technology
- Complicated IP ownership issues; disagreements in division of revenue to be realized, complicated by different accounting approaches, methods, forms, and timing of payment; insufficient budget and incentive to perform commercial assessment alongside invention assessment.

Because key academic technology researchers usually remain at the university and are not part of the "package" that transfers with the technology, a prospective client must replace that expertise. This makes value assessment and planning for the client complicated and increases risk.

One example of a successful European technology transfer model is the Fraunhofer Institute. It is Europe's largest application-oriented research organization, comprising more than 80 research units, including 60 Fraunhofer institutes at different locations in Germany. Most of their staff of more than 20,000 are qualified scientists and engineers. Of its €1.8 billion annual research budget, €1.5 billion is generated through contract research. The Fraunhofer Institute conducts research in energy, IT, communication, transportation, environmental technologies, as well as pharmaceutical development, regenerative medicine, and nutrition and food technology. This contract research forms about 70% of their total revenue, with 30% coming from the German and regional (Lander) governments in the form of "base funding." Under the Fraunhofer model, each center strives for excellence in one particular area of applied sciences, which enables

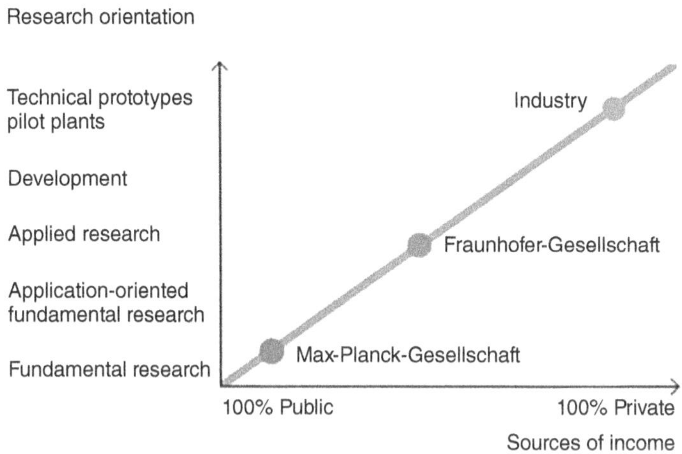

Figure 15.11. Model of the Fraunhofer Institute's funding. The Fraunhofer Institute's research work is oriented toward concrete applications and results. Pure basic research, such as practiced at universities, is funded to nearly 100% by public grants. Industrial R&D, up to prototype level, is largely financed by private enterprise. Approximately 70% of the Fraunhofer Institute's funding is received through contract research earnings, and about 30% of its funding is from the public sector. As a consequence, the Fraunhofer Institute operates in a dynamic equilibrium between application-oriented fundamental research and innovative development projects. *Source*: http://www.Fraunhofer.de.

that center to offer clients focused research in a variety of fields. It also has research centers in the United States, including an alliance for solar energy and building technology with MIT, closely affiliated with at least one of the 60 Fraunhofer institutes in Germany (Figure 15.11).

VENTURE CAPITALISTS

Venture capitalists add value through matching entrepreneurs to technology. Venture capitalists are constantly looking at technologies from universities (more often from North American universities than from European ones), but it is estimated that these represent less than 5% of the companies they fund. While some started out to encourage and capitalize on the high-risk business of helping small businesses grow into medium-sized ones, they too need more stability in their funds, and gravitate toward safer investments. Attracting venture capital to start a company is a difficult and time-consuming task. Another issue is that once the funding is secured, the whole trajectory and direction of how the idea is executed may change, because of the requirement to generate a return on investment within a limited number of years. Venture capitalists invest in individuals, not technologies. They search for the rare entrepreneurs who are able to balance business sense and an understanding of technology, and who can understand the industry for which the technology is being developed. Matching these entrepreneurs with technology wizards is where venture capitalists add value.

NONPROFIT ORGANIZATIONS

An important player in this early phase of drug development is the nonprofit foundation. Some of these are focused on specific diseases or orphan drugs; others may be dedicated to improving older off-patent drugs. Some nonprofits' missions are to address diseases that afflict mainly countries that lack a stable health-care system, such as tuberculosis and malaria. Nonprofit organizations are funded in a variety of ways. They receive funding from disease-focused philanthropies, such as the Bill and Melinda Gates Foundation, or the Cystic Fibrosis Foundation, which invest many millions of dollars in drug development, and from public offerings. The business-like philanthropies of today usually require adherence to strict project deadlines, releasing funds in tranches [19].

For example, SRI International is an independent, nonprofit research institute that conducts client-sponsored research and development for government agencies, commercial companies, foundations, and other organizations. Founded in 1946 by Stanford University, SRI has focused on discovery and the application of science and technology for "knowledge, commerce, prosperity, and peace." It is now independent of the university. SRI's board of directors includes representatives from the nonprofit and commercial sectors. SRI's executive team consists of experts in science, engineering, and technology innovation, as well as policy, growth, and business strategy, and it is they who create and lead the R&D programs. SRI has spun off more than 50 ventures, including Nuance Communications, a provider of speech and imaging solutions for customer service and health care, and Intuitive Surgical, Inc., a leading company in robotic-assisted minimally invasive surgery.

SRI's Biosciences division combines the approaches of a nonprofit research institute, a biotech venture, and a contract research organization, integrating the necessary R&D resources to take drugs from initial discoveries to IND applications. SRI Biosciences has advanced more than 100 drugs to clinical trials, and SRI's own pipeline has yielded several marketed products. In addition to medicines, SRI also conducts bioscience R&D to produce new biomarkers, drug delivery technologies, medical devices, and systems biology tools. It licenses its intellectual properties and creates new ventures. With revenues of around $585 million, SRI International is a successful model.

In the United Kingdom, a network of innovation hubs is being formed to improve the links between universities and businesses. It will receive more than £200 million from the Technology Strategy Board. These Technology and Innovation Centers will be positioned between universities and businesses, and are intended to drive growth by helping to spread knowledge from researchers to entrepreneurs. The Technology and Innovation Centers will connect businesses—large and small, established and start-up—

to potential new technologies, making them aware of funding streams and providing access to skills and equipment. Similar to the Fraunhofer Gesellschaft, each center will focus on a specific technology, one of which has been identified as regenerative medicine.

Funding agencies can play a pivotal part in driving collaboration, which could accelerate development if well managed. They could allow collaboration at the point of sharing data, as well as a multidisciplinary approach to development. Nonprofit disease philanthropies and government agencies that provide small business innovation research grants are starting to bridge this gap between basic research and clinically interesting therapies. New models of providing funding are attracting the right partners to commercialize proof-of-concept findings. Emphasizing early-stage collaboration between these players in drug development spreads the risk in the most difficult stages, so that they can be transferred to pharmaceutical industry for commercial development at a later phase [18].

CENTERS FOR THERAPEUTIC INNOVATION (CTI)

One example of a successful model involving universities and the pharmaceutical industry is the Centers for Therapeutic Innovation (CTI). This model is designed to focus academic research toward translational goals, and to speed the introduction of new scientific research into the pharmaceutical pipeline. A CTI is both a partnership between a very large pharmaceutical company and an academic institution, and a physical location. Close physical proximity benefits collaboration. The people hired for the CTI not only understand both the science and drug development, but also understand the value and the drivers of the investigator, their publication requirements, and other obligations. CTI also forms teams whose focus is to develop the drug for humans and test the hypothesis.

Rather than call for specific areas of interest, CTI's joint scientific committee made up of academic deans and the pharmaceutical company executives vote unanimously whether or not to fund a project. Later on, if the drug works in humans but the pharma company is not interested in pursuing it, it returns to the investigators and the institution with no conditions. To date, the CTI has received 300 proposals, and funded 16 projects. The program is similar to a short-term Howard Hughes grant, in that an investigator can get funding and the possibility to do human research. For 24 of the projects that were not selected, peer review was provided, such as "Extremely interesting, but its translatability lacks 'whatever'."

CTI selects projects based on scientific promise, but must also include a clear path to a proof-of-mechanism or proof-of-concept trials in humans. Publication restrictions are relaxed. Funding for projects selected in a CTI is provided in increments based on research milestones that are shared and agreed upon by the investigators and the pharmaceutical company. The milestone payments are a sharp departure from the "take the money and run" attitude pervasive in most academia–pharma collaborations. The clear focus on translatability, and willingness to end projects that don't meet their marks, represent an industrialization of academia that, while perhaps not welcome everywhere, is crucial to accelerate the translation of innovative academic discovery to actual drugs. For instance, a project may aim to humanize a mouse antibody that shows promise in animal models. An initial grant would fund the work to change the original amino acids to reflect a more human sequence. The investigator would then have to show that the binding and biophysical properties of the molecule were unaffected to unlock more funding. A further milestone might be to demonstrate that the mechanism of action is the same in human cell lines as it was in mouse cells. This type of results-based payment is a sharp departure from typical partnership arrangements, and it aims to provide a clear focus for the project and cut the time to clinic down to 18 months.

What happens to the molecules that meet their milestones, but do not fit within the company's current development goals? Dr. Jose Carlos Gutierrez Ramos explains the three likely outcomes for a molecule from a completed CTI project:

> **One**: If the data are negative, then that's how the science is, and that's a significant risk you take when you go into early science. Maybe the data is negative because you tested the wrong hypothesis, but now you have GMP material to test another hypothesis, and the investigator is completely incentivized to take that next step.
>
> **Two**: If the signal in humans is interesting but perhaps not of sufficient commercial interest to the company, the program stays with the investigator and is equivalent to a series A funding in a biotech. The investigator now has $8 million invested, and IND open, intellectual property around the drug, and half a kilo of GMP antibody. They can go to the lupus foundation or leukemia foundation or company X of a venture capital firm Y. It's equivalent to seed funding.
>
> **Three**: If the molecule is taken fully forward, then there are prenegotiated payments—between $2 and $4 million at proof of mechanism in humans and $15 and $25 million for proof of concept in humans—that are paid to the institution and the investigator.
>
> What is happening more and more often is that the signal is good but the science has evolved. Perhaps there are more experiments to do. CTI is now studying ways to build a hybrid model of funding with investigators, with venture groups, and with biotech, to allow the program to go forward. In this scenario neither the company nor the investigator nor the institution nor the VC takes all the risk. It remains to be seen if more new drugs will result from these projects. But what is true is that it marks a pronounced shift in the way pharmaceutical companies interact with academia. CTIs align incentives and harness, rather than subsume, discovery, one of the core competencies of academic laboratories. Pharmaceutical companies acquire access to innovative science, validated in humans, faster than in the current system. Academic investigators and their institutions get money, resources and the once-in-a lifetime change to ask the truly important questions about their discoveries: "Does it work in people?" [9]

Many European universities are in a transition process. Economies in Europe are all under severe pressure, and public sector budgets are undergoing the deepest cuts in decades. Although universities have commercial awareness, particularly in the technical universities, in practice, universities still treat their IP as a by-product of a research program rather than a core mission. It is a learning process, spanning many years. Models of semipublic contract research collaborating with industry, such as the Fraunhofer Institute, or the Pfizer CTIs, provide examples of successful structures to bridge the gap between research for research's sake and research for society's sake.

RECOMMENDATIONS

The growth of successful technologies is usually led by the founder. If an outside business leader replaces the founder, development often fails. Universities need to support their world-class researchers to remain within their universities *and* develop commercial potential of research. There are several ways to do this, and will probably all need to be individually planned and negotiated depending on the context of the development and the people. Allowing an entrepreneurial academic to take an "industrial leave" to work within another company, or to start his own company, is one approach which has been successful. In addition, universities should create more opportunities for PhDs to take MBA courses. It is important that universities recognize the value and effort of marketing and selling research findings, even though the results are difficult to measure, and individual performance for specific participants are hard to assess.

The project leaders at universities, pharmaceutical companies, and nonprofit organizations must remain involved throughout each phase of the process, rather than hand off legal work to tech transfer offices and IP lawyers [18]. University TTOs should be restructured and develop commercial, independent, sector-specific TTOs: for instance, in the fields of nanotechnology, materials, or life sciences. Recruit university researchers and external industry experts into technical advisory boards, to liaise with corporations on product improvement. These people will have sector-specific knowledge and up-to-date strengths in market appraisal. If the university professors work together with them, there will be tremendous potential synergy. It would be beneficial to survey researchers at universities around the world to see where the best TTOs are, and what it is about them that is so good. Then model their universities' TTOs on them, adopting best practices of the most successful ones. This could be a used as a strong selling point when recruiting new faculty. This would also position the universities to be ready and in a position of strength when the industry approaches them.

From industry's side, there are steps it can take, too, to improve technology transfer. For instance, one large company has five full-time employees out scouting for promising IP opportunities. The company spends around $1 million in salaries and expenses to build strong relationships with many universities that focus on their industry.* That is an effective "pull" strategy, with good return on investment. A common opinion from industry is that the university interface is complex and difficult to deal with. Anything that improves the visibility of the university's capabilities is a good thing.

University PhDs, researchers, and TTO staff need specialized training and education so that they can achieve what is wanted from them in technology transfer. They need to know not only how to file a patent or structure a license or a royalties agreement, but to understand the market and its drivers, learn how to work with investors and industry, and to better understand how to provide what the market and the pharma "customers" need. Although there is a strong cultural resistance to change at universities, scientists can generally be described as open, opportunity-seeking, and willing to accept challenges. These personal characteristics are conducive to entrepreneurial

*From author's personal correspondence.

capabilities and entrepreneurial commitment [20] and so, with the right approach, there is every reason to believe that the effectiveness of technology transfer will improve.

It is important and worthwhile to adopt these recommendations to improve and streamline technology transfer. As one Stanford researcher put it: "The issue of Technology Transfer in our work is crucial, and absorbs much of our research resources, likely more than 50%. Given its importance it is useful to devote some rational discussion and analysis to this topic, since its metrics are today mainly guesses and hopes. If it were science, we would insist on hypotheses and proofs before devoting so many millions to its development and achievement. Let's at least bring the alternatives on a scientific table beyond committee meetings and coffee breaks." [21].

REFERENCES

1. Baker, S. Cameron announces cash to bridge gap between lab and market. *Times Higher Education*. 2010; Oct. 10. Available at http://www.timeshighereducation.co.uk. Accessed December 10, 2013.
2. DiMasi, J.A. New drug innovation and pharmaceutical industry structure: trends in the output of pharmaceutical firms. *Drug Inf J*. 2000;34:1169–1194.
3. DiMasi, J.A., Hansen, R.W., Grabowski, H.G., Lasagna, L. Cost of innovation in the pharmaceutical industry. *J Health Econ*. 1991;10:107–142.
4. Sweeney, K. Technology trends in drug discovery and development. Victoria University of Technology Centre for Strategic Economic Studies Working Paper Series No. 3, 2002.
5. Pharmaceutical Research and Manufacturers of America (PhMRA). 2011. Vibrant Innovation, p. 6. Available at http://www.imi.europe.eu/content/mission . Accessed December 23,2013.
6. European Commission website. Available at http://ec.europa.eu/invest-in-research/policy/ipr_en.htm. Accessed December 23, 2013.
7. European Commission Pharmaceuticals Sector Report December 2011. Available at http://trade.ec.europa.eu/doclib/2012/January/tradoc_148988.pdf. Accessed December 23, 2013.
8. European Commission CORDIS News. Nine ideas to bridge industry-academia gap. Oct. 10 2006. Available at http://cordis.europa.eu. Accessed December 23, 2013.
9. Matafink, J. Innovation in pharma, part II: rethinking academic collaboration. *MIT Entrepreneurship Review*. March 29, 2012. Available at https://miter.mit.edu/articleinnovationnn-pharma-part-ii-rethinking-academic-collaborations. Accessed December 23, 2013.
10. Hutchinson, N. Research for research's sake? *Anaesthesia*. 2003;58(10):1023–1024.
11. Thomas, A. Edison Papers website at Rutgers University 2012. Available at http://edison.rutgers.edu. Accessed December 23, 2013.
12. Kaiser Family Foundation 2012 using NHE data from Centers for Medicare and Medicaid Services. Available at http://www.cms.hhs.gov/NationalHealthExpendData. Accessed December 23, 2013.
13. Hudson, R.L. Science|Business. Available at http://www.science|business.net. Accessed December 23, 2013.
14. Erasmus University Medical Center Department of Knowledge Transfer. Bescherming en exploitatie van de intellectuele eigendom van het Erasmus MC (Protection and exploitation

of Erasmus Medical Center's intellectual property). 2007, Feb. Available at http://erasmusmc.nl/tto-cs. Accessed December 23, 2013.

15. Hellman, T. The role of patents for bridging the science to market gap. *J Econ Behav Organ.* 2007;63(4):624–647.
16. United States Office of Technology Assessment. Patent-term extension and the pharmaceutical industry. Available at http://www.fas.org.ota/repors/8119.pdf. Accessed December 23, 2013.
17. Fini, R., Lacetera, N., Shane, S. Inside or outside the IP-system: business creation in academia. *Res Policy* 2012;39:1060–1069.
18. Ewing Marion Kauffman Foundation 2010. The new role of academia in drug development. White Paper.
19. Munk, S.A. Drug discovery: outside the box. *Drug Dev Res.* 2007;68:145–146.
20. Krabel, S., Mueller, P. What drives scientists to start their own company? An empirical investigation of Max Planck Society scientists. *Research Policy.* 2009;38(6):947–956.
21. Wiederhold, G. Dealing with asynchrony in technology transfer. In: Apers, P., Bouzeghoub, M., Gardarin, G., editors. *Advances in Database Technology.* New York: Springer-Verlag; 1996, pp. 631–634.

16

COLLABORATIVE INNOVATION IN PHARMACEUTICAL INDUSTRY: APPROACHES AND REQUIREMENTS

Monika Lessl and Khusru Asadullah

Bayer HealthCare Pharmaceuticals, Berlin, Germany

The pharmaceutical industry is facing a number of challenges. The model of blockbuster drugs is waning due to patent expiries (influx of generics), "me-too" drugs and increased competition. Expensive late stage failures and increasing value requirements by payors add to these challenges [1]. As a consequence, the pharmaceutical industry needs to consider new approaches to overcome the emerging portfolio gap.

One approach is to promote innovation by joining forces with partners such as biotech companies, other pharmaceutical companies, and academia. Such open innovation concepts are now gaining increasing importance. Whereas classical in-licensing or mergers and acquisitions have been traditionally pursued, early research collaborations aiming to enrich the idea pool and "de-risk" early research projects are gaining interest. Novel models are emerging such as crowdsourcing initiatives or concepts based on the sharing of tools and assets. These endeavors are matched with a rise in interest in academia to contribute to drug discovery, resulting in the establishment of academic screening centers or open access initiatives. This is strengthened by the announcement of the NIH to increase its engagement in drug discovery [2]. This development is on one hand driven by governments to get a

Collaborative Innovation in Drug Discovery: Strategies for Public and Private Partnerships, First Edition.
Edited by Rathnam Chaguturu.

return on their investment in research and on the other hand by the movement of scientists toward more clinically focused research and an interest in bringing their findings to the clinic.

A number of models are currently being tested on how to best promote drug discovery in academic settings that foster the productive and mutually rewarding interaction between pharmaceutical industry and academic research institutes. The challenge remains how the translation of ideas from academic research to novel treatments for the patients can be accomplished and managed best and what models exist to promote the generation of value out of collaborative efforts. Depending on the questions to be addressed, different open innovation models can be applied to foster early drug discovery. These encompass approaches such as crowdsourcing, strategic alliances, incubators, industry on campus concepts, or consortia. For what type of question what model fits best as well as their key characteristics will be discussed in this chapter. Besides novel models for collaborative innovation, the chapter will highlight requirements within the industry to ensure the uptake and further development of innovative ideas from academic partners.

OPEN INNOVATION VERSUS OPEN ACCESS

As the use of the term open innovation varies and often gets mixed up with the term open source or open access definitions—as they will be used throughout this chapter—they have been summarized in Table 16.1. Accordingly, the term open innovation means the flexible use of internal and external paths and ideas to generate value [3]. Thus, this chapter will focus on models of innovation sourcing and not take into account outsourcing of offshoring. To distinguish innovation sourcing from mere outsourcing, definitions are required. Outsourcing subsumes contract research and fee for service contracts, innovation sourcing encompasses collaborative efforts of two or more partners where novel, value-generating ideas are created jointly and translated into products. Thus, innovation is defined as the implementation of novel ideas that create value [4].

OPEN INNOVATION MODELS: KEY CHARACTERISTICS AND SUCCESS FACTORS

To further evaluate different open innovation models, key characteristics as well as key success factors for each model have been determined. Models taken into account in this analysis were crowdsourcing approaches, strategic research alliances between academia and industry (including industry on campus approaches), as well as incubator concepts and precompetitive consortia. Table 16.2 provides an overview on the different collaboration schemes and their key success factors.

TABLE 16.1. Definition of Open Innovation, Open Access, and Open Source

Open innovation

Open innovation is a paradigm that assumes that firms can and should use external and internal ideas as well as internal and external paths to generate value [3].

Open access

Open access is a term used for free access to scientific literature (http://wirtschaftslexikon.gabler.de/Archiv/569867/open-access-v2.html). In the field of drug discovery, this term is discussed for open access to chemical and clinical probes to reduce duplication of efforts and increase productivity [10].

Open source

The term "open source" has originally been used for software whose source code is published and made available to the public, enabling anyone to copy, modify, and redistribute the source code. The term was adapted to the scientific field:

- Research—The "Open Source Science Project" was created to increase the ability for students to participate in the research process by providing them access to microfunding, which, in turn, offers nonresearchers the opportunity to directly invest and follow cutting-edge scientific research. All data and methodology are subsequently published in an openly accessible manner under a Creative Commons fair use license (http://www.theopensourcescienceproject.com/opensourcescience.php).
- Open source drug research and development can help revive the industry, using principles pioneered by the highly successful open-source software movement. There has been another proposal for open-source pharmaceutical development, which led to the establishment of the Tropical Disease Initiative. There are also a number of not-for-profit "virtual pharmas" such as the Institute for One World Health and the Drugs for Neglected Diseases Initiative.
- The term "open source genomics" was coined to describe the combination of rapid release of sequence data (especially raw reads) and crowdsourced analyses from bioinformaticians around the world that characterized the analysis of the 2011 *Escherichia coli* O104:H4 outbreak.

TABLE 16.2. Overview of Different Collaboration Schemes and Their Key Success Factors

Models	Key Characteristics	Key Success Factors
Crowdsourcing (drug discovery)	Funds are provided to researchers to solve a specific question in drug discovery Useful to establish initial contacts Starting point for further collaborations	Easy access to funds Transparency in operational processes and awarding of funds Clear IP regulation required Clear definition of scope and goals of initiative

(*Continued*)

TABLE 16.2. (*Continued*)

Models	Key Characteristics	Key Success Factors
Strategic alliances between industry and academia	Bilateral partnerships encompassing multiple projects Useful to evaluate and translate ideas from research to the clinic Joint teams (industry and academia) to select and promote projects Exchange of personnel or joint labs to promote understanding of each other's goals and to exchange know-how	Identification of partners with complementary skills and competencies as well as matching goals Establishment of trustful relationship as a basis for collaboration Open-minded attitude and receptor in the company required to adopt novel ideas Professional alliance management Sufficient resources (in terms of capacity) at both partners to ensure close interaction and exchange (communication, meetings, etc.) Long-term commitment
Incubators	Physical entity for start-up companies where additional support in terms of know-how in drug development, project management, and/or access to technology platforms is provided	Long-term commitment required Incubators require innovation promoting surroundings (innovation hubs) In case of industry incubators independence of start-ups and influence by Big Pharma have to be balanced
Precompetitive consortia	Association of multiple partners with the goal to develop standards and infrastructure for drug discovery and development Increase knowledge on drug discovery in academia	Professional overall alliance management and consortium leadership Clear responsibilities for subprojects (strong project leaders) Clear definition of joint goals and milestones—recognitions of the value each partner brings into the consortium Infrastructure to share data Long-term commitment Clear communication schedules (regular meetings, communication rules)

CROWDSOURCING APPROACHES ARE USEFUL TO LEVERAGE THE KNOW-HOW OF A LARGE GROUP OF PEOPLE TO ADDRESS SPECIFIC QUESTIONS

While there are a number of examples of industry collaborating with one major partner, there has been a lack of approaches that make use of the expertise of a larger scientific community to address specific questions. The Internet provides an ideal platform for such an approach, being easily accessible from all over the world. Such an innovation model called "crowdsourcing" was first introduced by Jeff Howe [5] and has successfully been used in the Business to Consumer sector. One example is the Procter & Gamble (P&G) "Connect and Develop" portal (http://www.pgconnectdevelop.com), which allows consumers to bring their ideas for product improvements or novel product ideas. According to P&G, around 50% of their product initiatives involve significant collaborations with outside innovators, and they achieved a 70% higher-than-average net present value from connect- and develop-enabled projects [6].

Pharma companies are now piloting this approach in drug discovery [7]. In the original meaning of the term, crowdsourcing is a concept by which specific problems are communicated to an unknown group of potential solvers in the form of an open call, usually via the Internet. The community (the crowd) is asked to provide solutions and the "winners" are rewarded. The first company to introduce this concept in drug discovery was Eli Lilly in 2001, with the establishment of the InnoCentive platform (http://www2.innocentive.com). InnoCentive is now an independent organization with a solver community or more than 200,000 experts in more than 20 countries. Further crowdsourcing initiatives have followed in recent years [7]. In these cases, the classical concept has been extended in a way that companies or institutions seek for novel ideas, which are then further evaluated in a collaborative approach (Figure 16.1). Examples are the Phenotypic Drug Discovery (PD^2) platform by Eli Lilly (http://www.pd2.lilly.com), the Grants4targets (http://www.grants4targets.com) initiative by Bayer HealthCare or the Call for Targets (http://www.callfortargets.org) program by the Medical Research Council Technology UK.

Within its Grants4Targets initiative, Bayer HealthCare is seeking for novel targets in their strategic disease areas, namely oncology, cardiology, and selected gynecologic diseases [8]. Through the platform, scientists can apply for financial support to further evaluate and validate a specific target. To make it attractive, a nonbureaucratic submission process has been established, a fast processing of the proposals set up, and an intellectual property (IP) policy developed where the IP rights remain with the applicants. After the grant period, promising targets could be further pursued via collaborative agreements. Since the beginning of the initiative, 10 calls have been completed, approximately 1000 grant applications received, and more than a hundred grants awarded to academic groups all over the world.

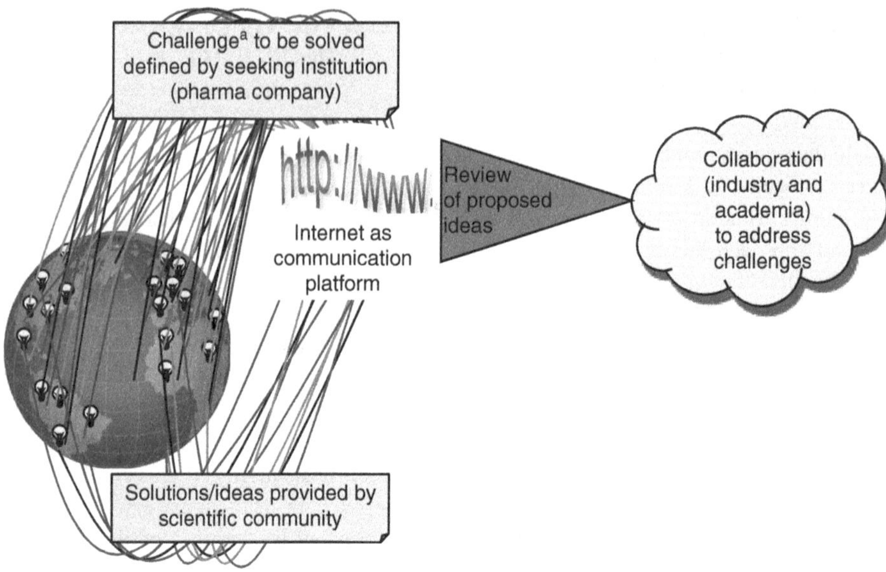

Figure 16.1. Crowdsourcing: an approach to leverage the worldwide know-how to solve a specific question. [a]Such as the search for novel targets, compounds, or indications for known assets.

STRATEGIC ALLIANCES

Strategic alliances require long-term commitment and provide the basis for a close interaction between the alliance partners (Table 16.3). They usually encompass a substantial—in terms of resources and finances—and long-term commitment. To make alliances successful, not one-dimensional technology transfer, but multidimensional interchange of information and know-how should be the basis of such relationships [9]. Strategic alliances are an appropriate model to evaluate and translate ideas from research and jointly develop them along the value chain. Thus, complementary skills and competencies as well as matching goals are a key prerequisite for success. Exchange of personnel or joint labs further promote such partnerships. Examples are the strategic alliance between Bayer HealthCare and the German Cancer Research Center [11] or the alliance between Pfizer and the University of California San Francisco. Joint Committees to select projects, joint project management, and financing additionally promotes real collaborative efforts.

INDUSTRY ON CAMPUS OR INCUBATOR MODEL

Also, industry labs directly located on the university campus provide a good model to promote close interaction and make use of the informal knowledge spillover effect promoted by physical proximity. The increasing importance of such models can be

TABLE 16.3. Examples of Different Collaboration Schemes and Their Main Characteristics

Models to Foster Drug Discovery	Examples	Main Characteristics
Crowdsourcing	Grants4Targets Program (Bayer Healthcare) (http://www.grants4targets.com)	Grants are provided for the validation of novel drug targets in BHC's strategic areas (oncology, gynecology, and cardiology).
Strategic alliances	Strategic alliance Bayer Healthcare German Cancer Research Center Heidelberg (http://www.dkfz.de/de/presse/pressemitteilungen/2013/dkfz_pm_13_22.php)	Strategic alliance to develop novel treatment paradigms for cancer; innovative setup by risk and reward sharing approach (each partner contributes 1 million euros/year to the join budget), joint selection committees, and joint project groups.
Industry on Campus	Bayer Healthcare Innovation Center at Mission Bay (UCSF) (http://www.pharma.bayer.com/scripts/pages/en/news_room/news_room/news_room92.php)	Bayer Innovation Center at Mission Bay (UCSF) represents an innovative concept to locate the industry labs directly on the campus to facilitate the exchange and translation of ideas.
Incubators	Bayer CoLaborator (www.colaborator.bayer.com)	The CoLaborator is an innovative incubator space for young start-ups close to Bayer labs. It provides infrastructure and options to exchange and collaborate with Bayer experts on aspects related to drug discovery.
Consortia	Innovative Medicine Initiative (http://www.imi-europe.org)	IMI is a partnership between the European Union and the European Federation of Pharmaceutical Industries and Associations (EFPIA). The goal is to promote drug discovery by precompetitive research driven by consortia between industry and academia. Focus of activities is to increase efficacy and safety and to promote knowledge and education on drug discovery.

(*Continued*)

TABLE 16.3. (*Continued*)

Models to Foster Drug Discovery	Examples	Main Characteristics
	Enlight Biosciences (http://www.enlightbio.com)	Enlight is a VC company established in partnership with a group of pharmaceutical companies to support the development of breakthrough technologies promoting drug discovery.
	Asian Cancer Research Group (ACRG) (www.asiancancerresearchgroup.org)	ACRG is a nonprofit organization established by Lilly, Merck, and Pfizer with the goal to establish a publicly available genomic database of profiled tumor samples as a basis for the development of novel approaches for cancer treatment.

demonstrated by the successful development of the mission bay campus at the UCSF. In the meantime many biotech companies have established their labs on campus and in January 2011 also, Bayer HealthCare opened its new Innovation Center on the campus. The concept—to create an advantage out of the physical proximity promoting know-how exchange and joint learning—is also a key driver of the incubator model. As an incubator we understand a physical entity that hosts young start-up companies—mostly spinouts from universities. Besides infrastructure, some incubators offer financial support for the companies as well as advice regarding development of their products. Examples of industry incubators are the Biogen Idec bi3, the Bayer CoLaborator, or the Janssen Labs. The QB 3 garage is another example fostered by the academic sector. To make such a model successful, a long-term commitment, appropriate support, and an innovation-promoting environment are key prerequisites. In case of industry incubators, a balance has to be found in creating synergies by close interaction between the biotech partner and the industry to preserve the integrity and independence of the biotech company in the incubator.

PRECOMPETITIVE CONSORTIA

Precompetitive consortia are important models to address general aspects in drug discovery such as tools or standards. Examples are predictive toxicity models, biomarkers, or improved simulation tools for pathway analyses (Table 16.2). Recently, a number of consortia between industry and academia have been established to address these kinds of questions. Examples are the Innovative Medicine Initiative with a total funding of 2 billion euros over 7 years (Table 16.3). The goal is to promote safety and efficacy in drug discovery and development and to foster education and knowledge in these fields. Another example is Enlight Biosciences. Enlight Biosciences has been founded by a group of pharmaceutical companies and a venture capitalist to support the development of novel enabling technologies driving and facilitating the R&D process. The key driver for the establishment was the fear that transformative technologies fostering discovery of novel drugs may remain overlooked if their development is not adequately supported. Similarly, the nonprofit partnership between Lilly, Merck, and Pfizer to accelerate research and improve treatment of lung and gastric cancers in Asia (Table 16.3) is an example of a precompetitive consortium.

PROFESSIONAL ALLIANCE AND COLLABORATION MANAGEMENT

In addition to the selection of appropriate models, professional management of the collaborations and alliances is a prerequisite to generate value out of the collaborative efforts. By a number of interviews with key leaders from industry and academia, the RESOLVE model has been developed highlighting key success factors for collaborative projects (Figure 16.2) [8]. A key aspect of the model is to put effort into relationship

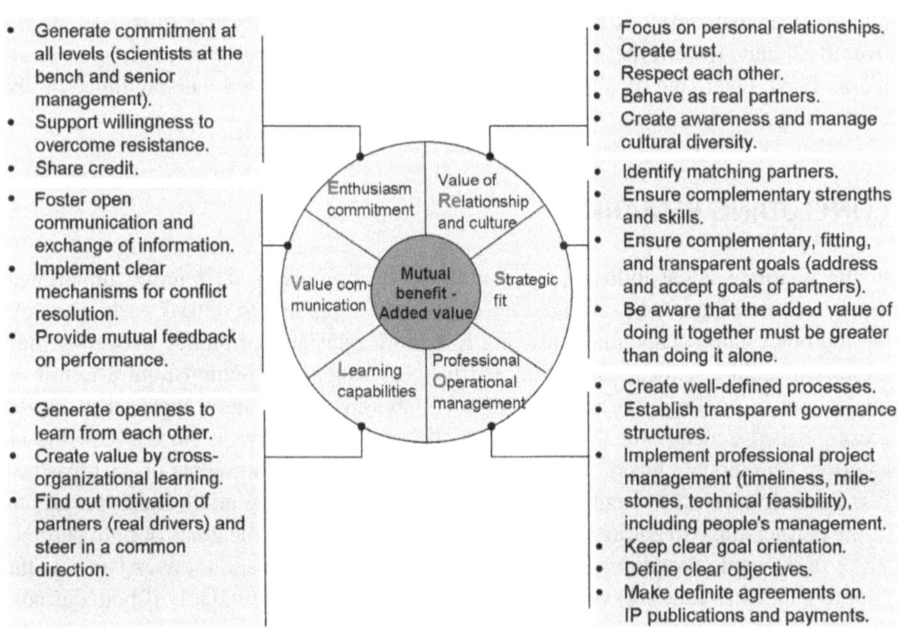

Figure 16.2. RESOLVE Model: Key success factors for industry–academic collaborations.

management. This is often undervalued by scientists, but studies have shown that poor relationships are the main cause of failures in alliances. Personnel contacts, exchange of personnel, respectful behavior built on trust, and integrity are key aspects to provide fertile ground for the collaboration. Another prerequisite for successful collaborations is to ensure a strategic fit between the partners and matching goals. Complementary strengths and competencies are other key aspects as the value of doing things together must be greater than doing it alone. Often this aspect isn't addressed properly prior to the start of an alliance and not enough time is invested to identify the real drivers of the potential partners. This could result in disappointments later on as partners don't strive in the same direction. Professional operational management forms another key pillar to make alliances successful. The outline of the contract and the implemented governance structures have to match and support the goals of the collaboration. These have to be clearly defined and agreed upon. Learning capabilities and openness toward new ideas are key aspects to promote innovation. Partners have to be open to look beyond their own principles and boundaries. Added value has to be created; otherwise, each partner could do the task alone. Furthermore, valued communication as well as enthusiasm and commitment at all levels are key aspects for successful collaborations. Based on the RESOLVE model, a checklist has been developed summarizing key aspects to be considered when initiating a collaboration [9] (Table 16.4). To support the professional management of collaborations, "alliance managers" in industry and academia are important. Whereas industry has recognized the need of dedicated functions and is currently establishing respective positions, only few examples exist in academia so far. Seminal examples can be found at the UCSF and the German Cancer Research Center, where professional alliance management have been established. As alliance management goes beyond mere administrative aspects and aims to guide the overall alliance toward the set goals, respective positions have to be at an appropriate hierarchical level and position holders have to be respected personalities in the community.

CONCLUDING REMARKS

In the pharmaceutical industry, open innovation approaches are gaining importance due to the pressing need to overcome the innovation gap and to de-risk early research. On the other hand, academia is moving to a more translational model, driven by their interest to field-test ideas and promoted by governments to demonstrate a return on investment in public research. How such collaborative innovation approaches can be accomplished depends on the question to be answered. There is no one-size-fits-all solution. The model chosen has to match the needs and requirements of each partner. It is critical to invest enough thoughts in the beginning on the goals and expectations of the partnerships. In addition, it has to be evaluated whether the goals of both partners strive in the same direction. Partnerships driven by opportunistic motives (such as the goal to gain extra funding) will not be successful in the long run. Only if both partners are willing to invest time, commitment, and resources can a long-term success can be

TABLE 16.4. Checklist for Initiation of Joint Projects between Pharmaceutical Industry and Academia

1. *Identifying matching partners*
   Do the partners fit together?
   Do they share the same values? Is there a cultural fit?
   Is there a willingness to respect and support each other?
2. *Strategic fit*
   What are the expectations of the partners and the respective goals?
   Can the expectations with respect to scientific quality and expertise be provided by the partner?
   What do the partners want to get out of the collaboration?
   Will the goals fit together?
   Do the partners provide complementary skills and expertise?
3. *Operational management*
   Who is in the driver's seat for the collaboration, on a management level and on an operational level?
   Will there be a single interface for effective collaboration management?
   Have clear agreements been made before the start of the collaboration regarding objectives at each site, timelines, publication strategy, IP rights (who owns what?), budget, decision processes, conflict management?
   Have all stakeholders been involved in the agreement and do they support it?
4. *Learning capability*
   Is there a willingness and interest to learn from each other?
   What programs can be initiated to foster cross-organizational learning (e.g., exchange programs, seminars)?
   Can specific training be offered for members of the alliances (skill and competency building)?
5. *Communication*
   How is the communication culture of the partner?
   How will timely communication be ensured?
   How can an efficient and satisfactory exchange of information be ensured (e.g., regular meetings, communication platforms)?
   Are conflict management and conflict escalation measurements in place?
6. *Commitment*
   Are all stakeholders committed to the collaboration (from scientist at the bench to senior management)?
   What can be done to overcome resistance and to motivate the people involved?
   Will the credit of the collaboration be shared? What's in for each partner?

achieved. To guide the collaboration, alliance management is key, which needs to be implemented and professionalized on both sides (industry and academia). The partnership should be built on the strengths of each partner to generate added value. Only if the value in doing it together is greater than doing it alone, can a partnership be successful.

ACKNOWLEDGMENT

The authors thank Stefanie Schoepe for her help and fruitful contributions in preparing the manuscript.

REFERENCES

1. Paul, S.M., Mytelka, D.S., Dunwiddie, C.T., Persinger, C.C., Munos, B.H., Lindborg, S.R., Schacht, A.L. How to improve R&D productivity: the pharmaceutical industry's grand challenge. *Nat Rev Drug Discov*. 2010;9:203–214.
2. Collins, F. An audience with . . . Francis Collins. Interviewed by Asher Mullard. *Nat Rev Drug Discov*. 2011;10(1):14.
3. Chesbrough, H.W. *Open Innovation: The New Imperative for Creating and Profiting from Technology*. Boston: Harvard Business School Press; 2003.
4. Lindner, J.C., Jarvenpaa, S., Davenport, T.H. Towards an innovation sourcing strategy. *MIT Sloan Manage Rev*. 2003;Summer:43–49.
5. Howe, J., editor. *Crowdsourcing: Why the Power of the Crowd Is Driving the Future of Business*: New York: Crown Publishing Group; 2008.
6. Neville, H. Interactive added value. Stifterverband für die Deutsche Industrie. 2011. Available at http://www.stifterverband.info/. Accessed December 23, 2013.
7. Lessl, M., Bryans, J.S., Richards, D., Asadullah, K. Crowd sourcing in drug discovery. *Nat Rev Drug Discov*. 2011;10:241–242.
8. Lessl, M., Schoepe, S., Sommer, A., Schneider, M., Asadullah, K. Grants4Targets—an innovative approach to translate ideas from basic research into novel drugs. *Drug Discov Today*. 2011;16(7–8):288–292.
9. Lessl, M., Douglas, F. From technology-transfer to know-how interchange: the role of academic-industrial collaborations in innovative drug discovery. *Wissenschaftsmanagement*. 2010;2:34–40.
10. Edwards, A. M., Bountra, C., Kerr, D.J., Wilson, T.M. Open access chemical and clinical probes to support drug discovery. *Nat Chem Biol*. 2009;5:436–440.
11. Wellenreuther, R., Keppler, D., Mumberg, D., Ziegelbauer, K., Lessl, M. Promoting drug discovery by collaborative innovation: a novel risk- and reward-sharing partnerships between the German Cancer Research Center and Bayer Healthcare. *Drug Discovery Today* 2012; 17(21/22):1242–1248.

17

CLOSE CONTACT: A COLOCATION MODEL FOR ACADEMIC–INDUSTRIAL PARTNERSHIPS IN DRUG DISCOVERY

Peter A. Covitz[1] and Terrence D. Ruddy[2]

[1]*Covitz Life Science, Ottawa, Ontario, Canada, and Somerville, Massachusetts, USA*
[2]*University of Ottawa Heart Institute, Ottawa, Ontario, Canada*

The translation of biomedical research findings into commercially relevant medical products is fraught with obstacles and challenges. Seemingly promising findings often languish for lack of a commercial path forward. A number of issues can prevent a concept from proceeding from discovery to development and commercialization: incomplete access to scientific and medical expertise, funding shortfalls, inadequate attention to intellectual property (IP) protection, lack of knowledge of market dynamics, inattention to competition, unrealistic valuation in licensing discussions, lack of familiarity with quality and regulatory requirements, lack of access to preclinical testing facilities, manufacturing challenges, and difficulties in launching a clinical program. We sought to overcome these obstacles by creating a novel collaborative structure for an academic–industrial research and development (R&D) partnership focused on nuclear cardiology. The two partners prenegotiated terms for the handling of innovations that might emerge from the collaboration, and each entity provided complementary expertise and resources. The program has been a success thus far, having attracted significant grant funding from government agencies and produced a promising novel imaging tracer for evaluating myocardial perfusion. Key novel features of the

Collaborative Innovation in Drug Discovery: Strategies for Public and Private Partnerships, First Edition. Edited by Rathnam Chaguturu.

collaborative arrangement, especially the close contact between the parties at a working level, will be described and discussed.

STAKEHOLDERS AND THEIR INTERESTS

Academic Investigators

The focus of researchers working at academic university medical centers ranges from traditional basic science to clinical research, including human mechanistic studies and clinical trials. Many have recognized the possible benefits of translational research with the transfer of basic science discoveries into the realm of clinical care. However, success with "bench to bedside" implementations of biomedical discoveries has been limited. Collaboration between basic science and clinical researchers is discouraged by the traditional academic culture. Researchers from one camp have difficulty appreciating the value of the research contributions of those from the other. Frequently, basic science and clinical researchers are physically separated and have few common educational or research rounds. Team participation for both basic science and clinical researchers is less helpful for academic promotion, as recognition of individual success is more valued. Success as a principal investigator on a peer-reviewed grant has greater merit attributed to it than success as an essential collaborator on a team grant.

The culture of industry differs, offering more acceptance of a team approach. Emphasis is placed on common goals, and success is measured by defined deliverables. Introduction of the philosophy of industry into an academic center may alter perceptions of value and facilitate development of teams that are more effective for translational research. The added resources often provided by industry may result in more rapid and greater project success. The academic gains associated with publications and grants may offset perceived losses related to the team approach and facilitate acceptance.

Academic researchers strive for peer-reviewed publications and grants to achieve peer recognition and promotion through the ranks to associate or full professor, with associated personal, financial, and institutional rewards. However, the work tempo, time urgency, and intensity vary among staff. Young researchers need significant publications early in their careers to obtain grants, which are essential to their initial job security (described as "publish or perish"). More established senior staff, with previous academic promotions and an adequate number of publications, frequently pursues a slow and steady approach, with "safe" research that continues with previous models and concepts and introduces little innovation. For many, the time horizon is extended from months to years, with adverse effects on productivity. In contrast, the industry researcher typically has a shorter time focus and needs to submit quarterly reports and generate results that rapidly turn into value for shareholders. The industry appreciation of the importance of IP and patent rights leads to more emphasis on innovation. Infusion of the industry mind-set into the academic center may speed up the metronome and lead to greater and timelier academic successes.

Industry R&D Executives

At its core, the practice of industrial science is not appreciably different from that of academic science. Application of the scientific method through hypothesis-driven experimentation is fundamental in both cases. The primary differences arise from the incentives provided to scientists in the context of their employing institution's goals and funding mechanisms. Academic performance is largely measured by successful publication in peer-reviewed journals and attraction of grant funding. Industrial R&D performance is measured by achievement of project milestones within a specified budget and timeframe, and in accordance with all relevant regulatory requirements.

Notwithstanding the emphasis on budget and time frame, industrial R&D is still subject to all the vagaries and unpredictable outcomes of experimental science. It is still research, after all. One of the primary tasks, then, of the industrial R&D executive is to assemble as competent a team as possible and to mitigate the inherent risks. While an academic environment is—and should be—tolerant of a learn-as-you-go approach, the industrial setting demands the avoidance of costly or time-consuming mistakes. The executive's job is to pull in all the needed expertise at inception phase of a program and ensure the team is communicating effectively to minimize the chances of wasting time chasing dead-end lines of inquiry.

In many small- and medium-sized life science companies, the R&D department is constrained by the financial wherewithal of the company. Even in a profitable company, shareholder demands for short-term returns on investment often preclude lavish spending on R&D. This situation presents two funding challenges for the R&D executive. First, given the interdisciplinary nature of contemporary biomedical research, he may not be able to hire staff or retain consultants in all the technical areas that need attention to avoid the risks noted earlier. Second, even with staff on board, the costs associated with doing industrial-quality R&D, which often needs to conform to Good Laboratory Practice (GLP) guidelines, may limit the ability to conduct a full program. Finally, the facilities infrastructure, especially for animal and clinical studies, can be cost-prohibitive to either build internally or even to outsource to commercial service providers. Given this predicament, the industry R&D executive must seek creative ways to acquire the needed expertise, gain access to needed facilities, and to supplement his internal budget with outside funding.

Beyond funding issues, the industrial R&D executive must prove his team's credibility to the biomedical community. There are outstanding scientists in industry as well as academia, and industry scientists can, in many cases, publish their work. Nonetheless, the public prestige associated with successful academic careers has significant weight in the community. Affiliation with highly regarded academic scientists and clinicians who are known to be discriminating in their selection of partners can be an effective means to demonstrate credibility.

Intellectual property (IP) is a key business deliverable for any industry R&D activity in biomedical science. Universities vary widely in their perceptions of how IP is valued, and in their ability to professionally negotiate licensing terms. When a university technology transfer officer calculates an inflated valuation, or has little familiarity with business risks and drivers, the result can be a breakdown in negotiations and wasted time and effort spent on due diligence. An R&D program manager must therefore carefully balance the time and energy spent conducting technical evaluation with that spent on business and licensing negotiation. If the discussions and evaluations on either track run into a significant barrier, then activities on the other track need to be paused. Such balancing requires a significant amount of management attention.

RELATIONSHIP MODELS BETWEEN INDUSTRY AND THE ACADEMY

Arm's Length

In the *arm's length* model, industry generally provides financial assistance to an academic science program, but no staff or expertise of its own. The size of such a contribution can vary from a small grant to partially subsidize a graduate student all the way to a major center grant to create a new facility. The subject matter focus of the award is typically of interest to the industry partner, but the contributor may not directly influence the specific research program that emerges. In many cases, the industry partner will negotiate rights to whatever discoveries that may emerge as part of the initiative. In other situations, the industry contributor may choose to simply seed the field with funding to see what the result may be.

Such arm's length partnership is beneficial to both parties. While the academic recipient may appear to be deriving the greater benefit, the industry partner will also gain significant indirect benefits in the form of enhanced prestige and improved public relations. Moreover, such contributions can often form the basis of a relationship that can later lead to additional collaborations that tie more directly into the industry partner's own R&D initiatives.

Separate but Equal

The *separate but equal* model has partners working together using their respective resources, but without frequent close interaction or resource sharing. A governance process for the collaboration will identify some common goals and allocate tasks to each respective partner's team. Periodic data reviews and sharing of results are scheduled as needed, but typically a minimalist approach to communication is taken. This model requires that key (limited) points of contact between the partners are identified, and these personnel are empowered to manage their respective sides of the program.

Open Door

In the *open door* model, staff and facilities are shared as needed, but the primary research location of each partner generally remains separate. Research staff at any level arrange to spend time in the partner's lab as needed. Communication between the teams is more frequent and often informal. This model is most frequently seen in collaborations between academics. However, it can be a fruitful way for industry–academy partners to give their staff exposure to a wider array of ideas and facilities. The key risk to an industry partner is that the relationship will evolve to be too informal and "academic" in nature. The agreed-to terms for confidentiality, data sharing, scope of research, and IP handling must be clearly communicated to staff and monitored by management.

Close Contact: Staff, Facilities, and Equipment Colocated at Common Site

The *close contact* scenario is the focus of the this chapter. In this case, the industry and academic partners agree on details for a comprehensive joint R&D initiative that will involve a variety of contributions by both parties. Either or both partners may contribute financial resources, staff, facilities, and equipment, all of which are located at a single primary site. The partners bring complementary expertise to the relationship, and agree to work intensively on R&D subjects of common interest. Each partner brings strengths to the partnership that match the weaknesses of the other, with the result being a vertically integrated team that is more than simply the sum of the parts.

The model does require a substantial amount of time to negotiate numerous issues prior to initiating the collaboration. However, a close contact relationship has numerous benefits, and the time spent dealing with issues up front can substantially reduce research program barriers later. In fact, it has been the experience of the authors that a close contact relationship can dramatically accelerate the pace of translational research from the laboratory to the clinic.

CLOSE CONTACT CASE STUDY: NORDION AND UNIVERSITY OF OTTAWA HEART INSTITUTE

The University of Ottawa Heart Institute (UOHI) is Canada's largest and foremost cardiovascular center for clinical care, teaching, and research. UOHI researchers have made many contributions that have impacted the way cardiovascular disease is diagnosed and treated. As a cardiovascular health-care center, UOHI serves as a center of excellence that draws patients from all over Ontario and across Canada. The Institute is particularly strong in the field of nuclear cardiology, where radioisotope tracers and nuclear medicine imaging techniques are used to diagnose disease and monitor therapy. Faculty is deeply involved in the study of new applications for existing tracers, and in the development of new diagnostic tracers and imaging methods.

Nordion Inc. is a global provider of radioisotope technologies for health care. The company is a leading manufacturer of active pharmaceutical ingredients and finished radiopharmaceuticals for nuclear medicine imaging and therapy. Nordion has extensive knowledge and infrastructure to satisfy the unique combination of nuclear and pharmaceutical safety and regulatory requirements that must be met for products that are both radioactive and used in humans. The company's R&D capability includes depth in isotope production, radiochemistry, and pharmaceutical formulation of radioactive compounds. The firm's activities span from discovery research to clinical development to commercial manufacturing and distribution. Nordion is therefore well versed in the details of current GLP, Good Manufacturing Practice, and Good Clinical Practice.

Leaders at UOHI and Nordion took a close look at each institution's strengths and identified areas of complementary capability. UOHI has world-leading clinical and research expertise in cardiology, medical physics, and nuclear medicine. The Institute is located on the University of Ottawa Civic Hospital campus, and has research laboratories in the same building complex as the hospital. Facilities include both small animal and human-scale imaging cameras as well as a research animal care facility. The UOHI academic culture is strongly biased toward cooperation and collaboration.

Nordion, in turn, has talented scientific staff in radiochemistry and biology, as well as quality and regulatory staff who provide guidance and oversight on compliance issues. Nordion's engineering team is experienced in the design and construction of nuclear research laboratories, and in radiopharmaceutical process development.

Combining the talents and expertise of the two entities seemed like a compelling way to foster innovation and to translate discoveries more quickly into product candidates that could be tested in the clinic. But what collaboration model to use? The various models described previously were all considered. In the end, the close contact model was chosen as the surest way to create a framework that, once instantiated, would be conducive to the most rapid progress through the stages of the drug/diagnostic development process.

The close contact implementation between Nordion and UOHI includes the following key provisions: (1) research agreement that describes areas of focus and staffing; (2) IP agreement that details the ownership and licensing rights to technologies developed under the research agreement; and (3) facilities agreement that describes the space, laboratory, and equipment contributions from each party. The details of each provision, and how they contribute to the improved efficiency in the drug development process, are given later.

RESEARCH AGREEMENT

Larger scale academic collaborations often follow from a successful joint or team grant submission. The process of preparing the structured grant application drives the definition and planning of the anticipated program. In the case of an industry-academic collaboration that is not directly tied to a grant submission, there is more freedom to define the program at whatever level of specificity the partners require. This flexibility can be very advantageous, especially when the interests of each partner are well understood and considered to be compatible. However, the lack of structure from an externally imposed process can also be a drawback, as the uncertainty about what level of specification is needed can result in caution during negotiations and potentially slow down the process of reaching agreement.

The research agreement between Nordion and UOHI lays the foundation for collaboration in several areas. To reach agreement, each of these areas needed to be specified with sufficient clarity to make concrete implementation possible, but with enough flexibility for the collaboration to evolve. The first area of specification was the medical field of interest. While cardiology was of course the starting point for that discussion, the agreement needed to further narrow the field. Both Nordion and UOHI have common, overlapping interest in imaging, so it was decided to focus the research plans on molecular imaging related to cardiovascular disease. Given Nordion's commercial focus on radioisotopes for nuclear cardiology and UOHI's leadership in radiotracer development, the program coalesced around an agenda to pursue tracer development.

The focus on radiotracer development provided the necessary starting point for the next areas of specification: the technology platforms and laboratory requirements. To conduct research that was both novel and commercially relevant, the program needed to be properly equipped to take advantage of the latest imaging modalities in a preclinical setting. Since discovery-stage work was expected, the facility needed to be equipped for biological tissue culture, synthetic chemistry, and radiochemistry. Imaging studies in animals require handling of radioactivity on the order of tens of millicuries, and both single photon and higher energy positron-emission isotopes are used. These activity levels are common in radiopharmacies, but less common in discovery-stage research labs, where microcurie-level quantities are more often seen.

To be able to move candidates from discovery into preclinical testing, the facility must have the capacity to operate under GLP. GLP is commonly understood in industry, but generally not followed in academic labs. As a result, academic work often needs to be recapitulated in a GLP setting to verify results are accurate and can be used in a regulatory filing. This repetition of previous work is often responsible for considerable additional cost and time in the pharmaceutical development process. We decided to eliminate the gap between academic research and subsequent GLP lab validation by incorporating GLP capability into the collaboration. This element of the agreement has probably been one of the greatest accelerators of the drug discovery process.

The governance structure of the collaboration is one of the more important aspects of the relationship. We have implemented a two-tiered hierarchy for managing the activities of the relationship: a Senior Leadership Committee and a Project Team. The Senior Leadership Committee includes the executive sponsor from UOHI and from Nordion, as well as the faculty and senior scientists involved in the collaboration. The Senior Leadership Committee meets once per month to review program status and make decisions about key issues escalated from the Project Team. The Project Team meets approximately biweekly or as needed, and is made up of the staff performing the hands-on work as well as a project manager. Notably, the Project Team also includes representatives from supporting resources such as the imaging facility and quality assurance in addition to the core members of the laboratory.

INTELLECTUAL PROPERTY AGREEMENT

Commercialization of new discoveries is of great importance to the academic and industry partners. In addition to a potentially significant revenue stream, discoveries leading to improved health care will enhance the profiles of both partners. Industry typically has greater awareness of the importance of IP and has greater resources available for evaluation and patenting of new inventions. Similarly, industry usually has greater expertise in manufacturing and overcoming the many the regulatory hurdles necessary to successful marketing of a new product or therapy.

The ownership of IP needs to be categorized into technology previously developed by either partner ("background IP") and new, jointly developed technology. Patents need to be filed to protect ownership of new IP. Terms for ownership of the property and assignment of the patents should be determined by agreement at the initiation of the partnership. In this case, the partners agreed to share revenue from jointly owned IP according to a predefined schedule. It was also agreed that ownership of each partner's background IP would stay with the partner and not be shared.

FACILITY AGREEMENT

After assessing the options for implementing a laboratory with the capabilities outlined earlier, we decided upon a novel construct that fully leverages the strengths of each partner: close contact. The UOHI has a number of resources that were important to include, and the best way to do that was to locate the laboratory within the Institute. It was therefore agreed that the laboratory would be instantiated in newly dedicated space on the UOHI campus. At the same time, we recognized that the laboratory could not operate entirely as an academic laboratory and would need the discipline and rigor required for GLP implementation. For that reason, the lab is managed and staffed by Nordion. In this way we bring the best of both partners to the collaboration: the spirit, culture, faculty, students, and facilities of an internationally acclaimed academic research institute, combined with the experience, rigor, discipline, and entrepreneurial culture of industry.

A lease agreement between Nordion and UOHI was crafted and signed, with Nordion agreeing to equip the laboratory space and staff the laboratory. Terms for occupancy and equipment ownership were worked out and provide for the usage of the key resources. UOHI contributes space for the radiochemistry laboratory and space, equipment, and technician time from its imaging and animal care facilities. In this way the investment that each partner makes in the resources they are responsible for is leveraged and amplified through the investment from the other partner. Since the facilities and personnel are all colocated on the UOHI campus, access and usage are equitable.

BENEFITS OF CLOSE CONTACT

The research laboratories at the Institute were renovated and equipped over several months. Staff were recruited with team interviews. Communication was facilitated by on-site meetings, scheduled weekly and as necessary for urgent issues. Appropriate expertise could be rapidly accessed on-site from either partner, or off-site via email or telephone. Research enthusiasm and momentum were high and led to pilot data for a successful peer-reviewed grant application. The grant application was multidisciplinary, with aspects including synthetic chemistry, radiochemistry, physiology, imaging physics, and small and large animal imaging, as well as a Phase 1 clinical trial. The majority of the team was on-site, and this greatly facilitated the development of the grant application. Once funded, the team spirit grew, and the work intensity and capacity grew as well. Long hours and weekend work were necessary to meet deadlines and were made tolerable by the camaraderie. Success with a patent filing was achieved in 2 years, and the team is now in the midst of preparations for a Phase 1 clinical study on a novel myocardial perfusion imaging tracer (Figure 17.1). Use of the close contact model contributed to a common sense of purpose and greater productivity than expected.

The academic culture benefited from the infusion of industry emphasis on time lines and deliverables. The industry partner brought in project management leadership employing Gantt charts and time management skills. The industry scientists appreciated the interaction with their academic associates at research meetings and rounds. Equipment was made available for shared use by other laboratories in the Institute. Meetings with other UOHI scientists led to concepts for team grant proposals and plans for pilot data studies. Synergy with other research programs within the Institute will lead to more synergy and more grant applications. Industry has also gained, with greater access to peer-reviewed funding opportunities, more awareness, more highly qualified collaborators, and a greater variety of funding sources.

The expertise of the industry partner in regulatory issues, GLP procedures, Good Manufacturing Practice, and preclinical toxicology has been extremely important in the effort to translate the radiochemistry discovery and preclinical work into a Phase 1 trial. This industry "know-how" combined well with the clinical and research strengths of the academic partner and resulted in a stronger platform for approaching regulators.

Prenegotiation of the IP issues concurrently with the partnership agreement accelerated progress. The usual academic approach to discovery and subsequent efforts to find an industry partner would have required more time for the development of a legal agreement as well as a working relationship.

RISKS AND CHALLENGES OF CLOSE CONTACT

Close contact can result in issues for the partnership. The "time-consciousness" of industry can be demanding for the academic partner who has other duties in the clinical realm or teaching responsibilities. Research agendas may differ, with the academic partner seeing greater value in projects less likely to lead to important IP for commercialization, or requiring a longer time for fruition. Funding success for either partner may vary due to changes in the economy that affect one more than the other or simply due to failed grant applications. As with any successful partnership, a degree of flexibility and tolerance is required. Our governance structure with equal decision making and

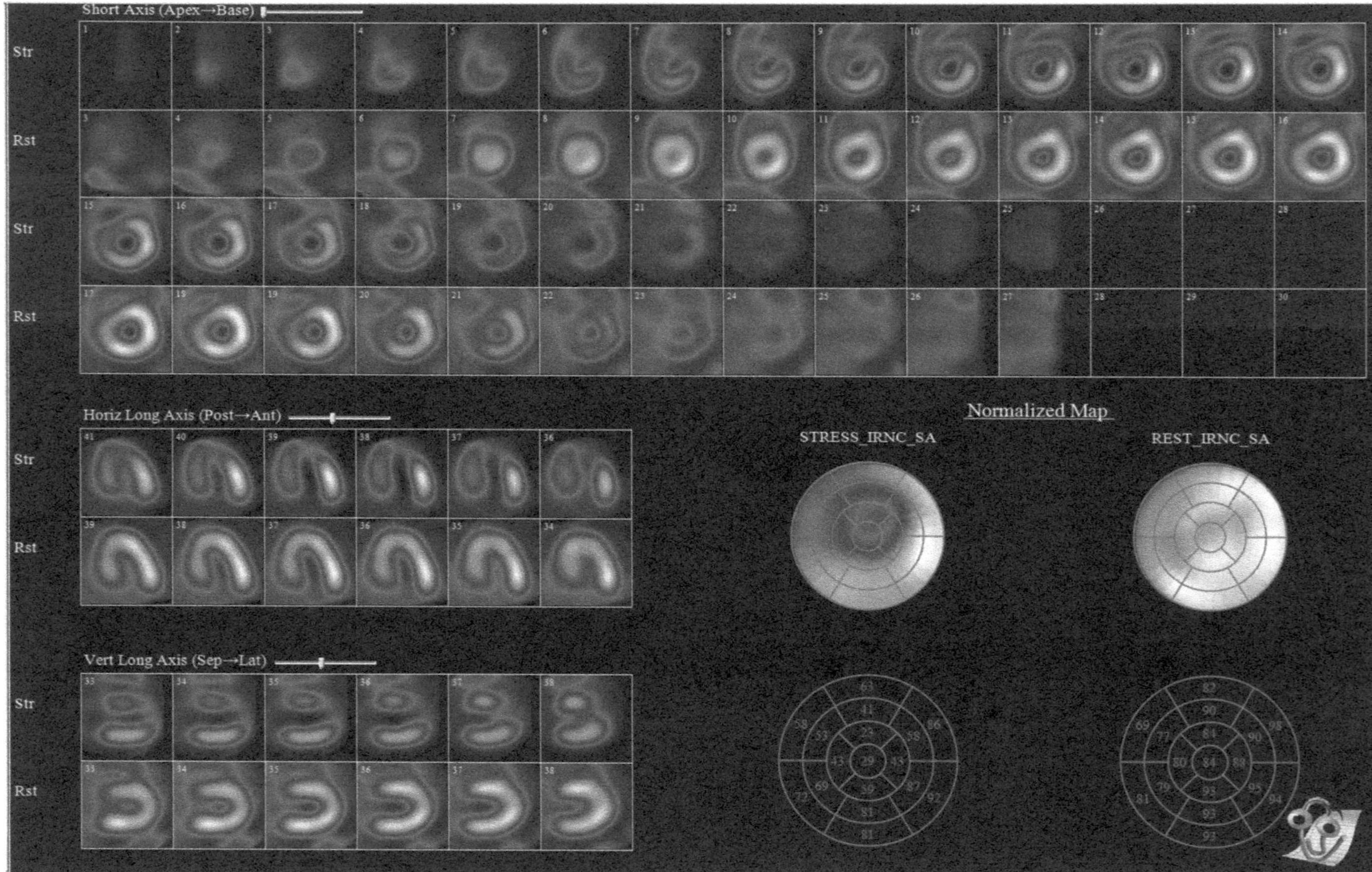

Figure 17.1. Single photon emission computed tomographic (SPECT) myocardial perfusion images demonstrating ischemia during dipyridamole stress (Str) and normal perfusion at rest (Rst) in an experimental porcine model. Ischemia was induced with a 2.5-minute occlusion of the left anterior descending coronary artery and produced a large area of reduced uptake in the anterior region on the stress images.

opportunity for easy and open input from both partners was beneficial for dealing with issues as they arose.

CONCLUSIONS

As always, people make the difference in any effort, and our partnership has developed from a group of outstanding individuals into a highly motivated team working together to achieve common goals. The close contact model has been very helpful in expediting the process and catalyzing the formative stages. Our partnership has been very successful in the short term, winning more funding through a major peer-reviewed grant and filing a patent for a compound with significant commercial value and the potential to improve health care. Time will tell as we move ahead and face the challenges of uncertain funding and less successful projects. Close contact should continue to help, as the partnership should be able to deal more quickly and efficiently with adversity and change direction as needed for eventual success.

ACKNOWLEDGMENTS

The authors would like to thank the senior scientific team of the collaboration, Drs. Corinne Bensimon, Glenn Wells, Lihui Wei, and Pasan Fernando, for their dedication and innovation. Russell Redshaw and Marion Fraser provided crucial support for intellectual property management and other joint administrative activities. Richard DeCaire and Michèle Légaré-Vézina have been diligent in their establishment and monitoring of safety requirements. Melanie Mintz, Marlie Poirier, and Janet Bukovcan ensure the program remains compliant with quality and regulatory requirements. The joint research described herein has been financially supported by Nordion, UOHI, and grants from the Canadian Institutes of Health Research and the Natural Sciences and Engineering Research Council of Canada.

18

SUCCESS FACTORS AND OBSTACLES IN ACADEMIA–INDUSTRY PARTNERSHIPS: A CASE STUDY OF A GRADUATE PROGRAM WITHIN THE BAYER–UNIVERSITY OF COLOGNE "PRIVILEGED PARTNERSHIP"

Stefan Herzig,[1] Marion Rozowski,[1] and Ingo Flamme[2]

[1]*Department of Pharmacology, University of Cologne, Cologne, Germany*
[2]*Bayer HealthCare AG, Wuppertal, Germany*

The development of new active principles for medication has been the preserve of industrial research and development on the one hand, and of academic discovery and enhancement on the other hand. Even though small biotechnology companies have successfully developed biologicals, the invention or discovery of small molecules has remained the domain of large pharmaceutical companies. Over the past decades, academic pharmacological research has increasingly focussed on the discovery and investigation of the drug targets of small molecules. In this respect, university research plays a key role in the discovery of new therapeutic principles.

Over the past 10 years there was a marked decrease in the number of new chemical entities with novel active principles introduced into the market. This development combined with the ever-rising costs of every single innovation has called into question

Collaborative Innovation in Drug Discovery: Strategies for Public and Private Partnerships, First Edition.
Edited by Rathnam Chaguturu.

the commercial sense of pharmaceutical research in industry. Thus, it does not come as a surprise that over the past years, there have been intense discussions in which way the development of new low molecular weight therapeutic principles can be sped up. Uniting the expertise of academia and of the research sector of the pharmaceutical industry might be a successful approach.

As an example for the possible rapprochement of both areas, we will describe and critically analyze here an initiative, which constitutes the framework in which the University of Cologne and Bayer foster and strategically plan their collaboration. In 2008, Bayer and the University of Cologne agreed on establishing a privileged partnership. On the basis of a framework contract, this partnership comprised collaboration in the field of clinical studies on Bayer substances as well as the establishment of a graduate program in pharmacology and experimental therapeutics. We have published the concept and the development of the graduate program elsewhere [1].

Here we want to emphasize the conception of the structure and organization of the program following an analysis of the challenges such a program faces and the aims it wants to pursue. Furthermore, we will analyze typical difficulties encountered in implementing the program in the light of 3 years of experience with the program and based on a first external evaluation. We will report on how we could address these problems in the further development of our concept.

We propose that collaboration in the area of fundamental research without a specific or immediate development or application goal in mind constitutes an excellent foundation for a long-term strategic and effective partnership between an academic and an industrial partner.

SWOT ANALYSIS OF AN INDUSTRY–ACADEMIA PARTNERSHIP

For the planning of the structure and organization of a joint postgraduate program in pharmacology and experimental therapeutics, the leading pharmacological research staffs of both University and Bayer undertook a process analogous to a strengths, weaknesses, opportunities, and threats (SWOT) analysis and identified risks and opportunities of a joint initiative in the area of pharmacological research. Before we proceed to this critical self-analysis we want to summarize some important points from the literature concerning industry–university collaborations. We have consulted a review by Blumenthal from the 1990s [2] from the area of life science research in the United States, a comprehensive analysis of industry–academia collaboration in the United Kingdom [3], and some further important reviews [4–6]. We will include these points in our overall SWOT analysis of an academic–industrial partnership.

Overall, the financial investment of the pharmaceutical industry in collaborations with universities is increasing. However, the total investment into R&D collaborations with universities compared with other industries is still rather small. It has been criticized that most activities have a small financial volume and are of limited duration [2, 3, 5]. The funding of specific projects in the area of clinical research and the recruitment of university professors as independent advisors plays an important role in collaborations of the pharmaceutical industry with universities.

BAYER–UNIVERSITY OF COLOGNE COLLABORATION

Opportunities

The research staff considered the collaboration as a specific opportunity to be able to recruit and educate particularly talented scientists for pharmacological research in academia or industry. By combining the expertise and technological abilities of both partners, they were further expecting a competitive advantage both in the area of fundamental research and in the area of drug development. Finally, both partners perceived a competitive advantage due to the sponsorship. Bayer hoped to be able to advertise its sponsorship activity in a positive way and the University welcomed the extra income of federal state funds coupled to the acquisition of third-party funds.

Risks

Potential conflicts of interests were considered as risks to the collaboration. The financial support by industry of university scientists could lead to public allegations of loss of neutrality on scientific issues. The freedom and independence of university research, which receives primarily public funding, could be brought into disrepute. Other perceived risks were that the organizational and temporal investment for the establishment of the new collaboration could potentially impede the efficiency of the core R&D activities and therefore yield a critical advantage to competitors in decisive areas.

Strengths

We considered the following to be strengths of the collaboration: comparatively fast availability of funding and resources for carrying out projects through an industry sponsorship; the combination of know-how from both areas; the existence of patentable substance libraries at the industry partner; the University's ability and determination to use innovative chemical processes; the University's readiness to undertake high-risk projects; the ability of the industrial partner to formulate clear goals and to implement mile stone planning and overall professional project management. The university considered as a strong point that outstanding fundamental research could serve as a source for economically attractive innovations and developments.

Weaknesses

The obligation to conduct socially responsible research that is not primarily profit-oriented was considered a weakness of an industry-university collaboration. With regard to planning periods in the case of industry sponsorship, one had to expect that they would correspond to the short-term, project, and milestone oriented procedures of industry. The scientists also considered as problematic the requirement to match their projects with specific project areas as defined by the upper management especially with regard to the possibility of unexpected research results. This potential source of conflict was resolved to some degree by the very good match between the university and the industrial partner in several areas of research, namely cardiovascular science and oncology. One could foresee that information and communication deficits between and within institutions could pose obstacles to this new form of collaboration. Also bureaucratic procedures could potentially impede research. Joint projects would have to follow both industry and university regulations. Likewise, there was the possibility that the decision and approval paths typical of each institution could obstruct innovative procedures.

It was important to find ways to avoid incompatibilities between the different decision processes.

We have summarized the points made above in a SWOT analysis in the form of a table (Table 18.1). Where risks or internal weaknesses threatening opportunities were identified, we tried to address these problems by defining the aims and planning the organization and structure of the graduate program (http://pharmakologie.uk-koeln.de/graduiertenprogramm) accordingly. In Table 18.1 we have indicated our measures to address problems by black arrows.

ORIGINAL CONCEPT AND IMPLEMENTATION

The history and concept of the graduate program "pharmacology and experimental therapeutics" has been published in detail elsewhere [1]. We want to describe and discuss the program concept here only in so far as it can be derived from and is motivated by the above analysis of strengths and weaknesses. The aims given in the following headings refer directly to the aims given in Table 18.1, which have been derived from the SWOT analysis.

Connecting, Binding, and Simple Governance Structures

For the planning, organization, and realization of the graduate program, we wanted to implement structures, which were already familiar to most of the people involved. The structures should be simple, transparent, and quickly allow the operation of the program. Therefore, the internal structure of the graduate program was modeled on similar programs, for instance those funded by the German Research Foundation (DFG). All principal investigators whose projects are accepted into the program form the so-called "Programmrat" (program council). The program council meets once yearly, preferably at the occasion of the Annual Retreat to exchange information and to discuss strategic questions.

In addition to the program council, the graduate program has further supervisory bodies, which are supposed to ensure that all stakeholders of the program can participate in and agree to its development. The program council elects from among its members the speaker of the graduate program and two additional representatives to the program committee. Bayer, the dean of the faculty of natural sciences, and the dean of the faculty of medicine of the University of Cologne nominate additional members of the program committee (one each). The speaker is responsible for directing the operation of the program and is supported by a program coordinator. The program committee designs

TABLE 18.1. Consequences for Structure and Organization Resulting from a **SWOT** Analysis

		Internal Analysis	
		Strengths	Weaknesses
External Analysis	*Opportunities*	***O*** *Competitive advantage due to sponsorship* ***O*** *Recruitment of particularly talented staff* **S** Specific know-how **S** Fast provision of funding and resources **S** Fundamental research as a source for innovations **S** Willingness to undertake high-risk projects ***O*** *Competitive advantage by the combination of complementary expertises* **S** Specific know-how **S** Fast provision of funding and resources **S** Patentable substance libraries **S** Innovative chemistry ***O*** *Competitive advantage by combination of complementary technologies* **S** Willingness to undertake high-risk projects **S** Clear goals, professional project management **S** Fundamental research as a source for innovations	***O*** *Competitive advantage due to sponsorship* **W** Obligation to conduct socially responsible research ***O*** *Competitive advantage by the combination of complementary expertises* **W** Short (financial and temporal) planning periods **W** Focus on a specific project portfolio **W** Insufficient communication between the institutions → **Concept designed for sustainability** → **Openness concerning project proposals** → **High emphasis on networking to foster inter-institutional communication and lasting personal relationships** → **Curriculum offering "core scientific"** ***and*** **"soft" skills**
	Threats	***T*** *Competition* **S** Patentable substance libraries **S** Innovative chemistry **S** Willingness to undertake high-risk projects **S** Clear goals, professional project management **S** Fundamental research as source of innovations ***T*** *Vulnerability and loss of reputation by conflict of interest allegations* **S** Clear goals **S** Professional project management → **Focus on fundamental research instead of R&D** → **Maximal transparency**	***T*** *Competition* **W** Obligation to conduct socially responsible research **W** Bureaucratic obstacles **W** Incompatibilities of bureaucratic procedures between institutions **W** Complicated and potentially incompatible governance structures → **Connecting, binding and simple governance structures** → **Familiar funding structures but with faster operating sequences** ***T*** *Vulnerability and loss of reputation by conflict of interest allegations* **W** Danger of unclear regulation of intellectual property rights **W** Delay, insecurity of publication **W** Insufficient communication within institutions → **Clear and binding agreements on confidentiality, IP, and publication rights**

the curriculum and develops it further, controls the financial planning, referees new projects resulting from project calls, and decides on the acceptance and funding of projects within the program. The committee also supports the recruitment of new graduate students by participating in applicant job talks and interviews. The program committee, as represented by the speaker and a vice speaker (one will be member of the University one will be at Bayer), reports to a supervisory body of the graduate program—the steering committee. This regulatory body consists of representatives of the University, the University Hospital, and Bayer. Members of the board of directors of all three institutions as well as representatives of the financial boards form the steering committee. The steering committee has the responsibility to supervise and develop the privileged partnership. This includes both development and support of the graduate program and of the area of clinical studies.

The activities of the graduate program and the state of all cooperative studies between Bayer and the University under the framework of the privileged partnership are reported at the twice-yearly meetings of the steering committee. These meetings also serve to discuss the strategic concept and future of the program. Within Bayer, the steering committee is directly subordinate to the Bayer management board. Within the University, it prepares the development of the cooperation for the rector (Figure 18.1).

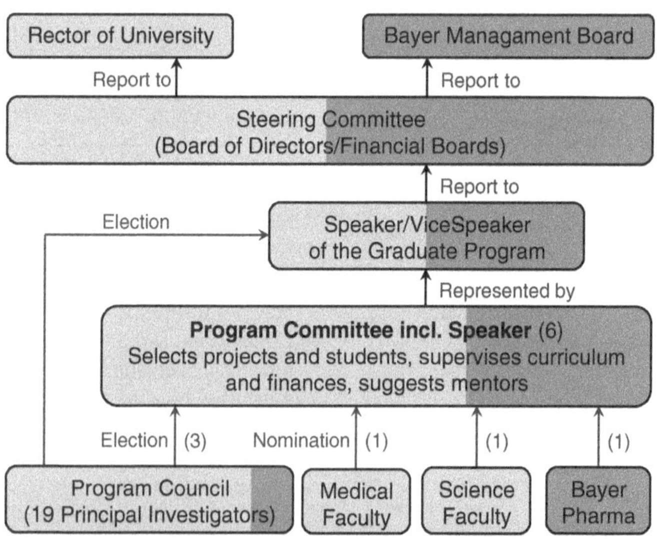

Figure 18.1. Governance structure of the graduate program in pharmacology and experimental therapeutics at the University of Cologne. Several layers of supervisory bodies allow different stakeholders at the University (light gray) and Bayer (dark gray) to follow and participate in the development of the program.

Familiar Funding Structures but with Faster Operating Sequences

To determine in more detail what kind of scientific projects the future graduates should carry out and to recruit such projects, we decided on a funding structure closely resembling established procedures within the University for other third-party funding. After several initial conversations between scientists from the pharmacological departments of the University and of Bayer with representatives of thesis committees and the Graduate School of Biological Sciences, we invited interested scientists of the medical faculty to discuss about the future graduate program. We had two rounds of discussions with about 25 participants each. In these discussions we reflected upon the common research areas of both partner institutions (cardiovascular sciences, oncology, neurosciences, defense, and intensive care) and narrowed down these areas more closely. We outlined aspects of the University's methodological and thematic repertoire, which would be especially useful for the education of future pharmacologists: *in vivo* and *in vitro* studies for the establishment or generation of disease-relevant animal models reaching from mouse models to large animal models, methods of molecular image generation *in vivo*, telemetrics, the analysis of biomarkers, and finally, clinical studies. These discussions led to the focus on the scientific areas included in the first (2008) and second (2009) project calls, which offered funding for a period of 3 and 2 years, respectively.

The following additional criteria for the inclusion of projects were determined: scientific excellence of the project, scientific track record of the principal investigator, and his/her experience as a supervisor of natural science and medical graduate students. For the project calls we only allowed 1 month from launching of the project call to submission deadline, from submission deadline to peer review, and from peer review to the decision which projects to accept. We received following the first and second project calls 23 and 9 applications, respectively. The acceptance rate of projects was around 50% (19 out of 35). For three of the accepted projects, intramural funds from Bayer were provided. In the first funding period, seven doctoral theses (PhD) were funded by Bayer sponsorship at the University of Cologne. The funding of seven instead of the intended five to six doctoral students at the University through Bayer was due to the excellent quality of the received project proposals. The medical faculty of the University of Cologne funded three further medical theses (Dr. med. and Dr. med. dent.) over a 2-year period with a less well-endowed stipend. All other projects were exclusively funded by the project PIs. In the end, the percentage of intramural projects at Bayer, which was intended to be one-third of the size of the graduate program, was smaller than planned. While the project call resembled calls for public funding, the proposals needed to be restricted to 10 pages including abstract, introduction, and proposed scientific investigation, planned methods and scientific aims, timetable, publication list, funding track record, and list of previously and currently supervised students. This is about half of the volume of a grant application to the DFG.

This format and the temporal sequence could be implemented without any problems. As expected, some applicants withdrew their application after it became clear that their project was accepted into the program based on program match and quality

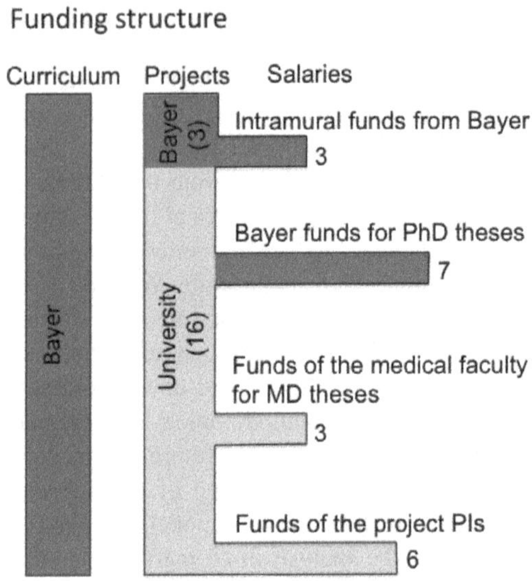

Figure 18.2. Funding structure. Bayer (dark gray) funds the administration and the curricular activities of the graduate program; the University of Cologne (light gray) and Bayer equally fund the PhD student positions.

standards but would not be sponsored by Bayer. However, in the course of the program, we succeeded in winning over PIs to include already funded projects into the program, so that we could achieve a balanced proportion of Bayer sponsored and University-financed doctoral students at the University of Cologne (Figure 18.2).

Focus on Fundamental Research Instead of R&D

In spite of the fact that the first project call had included clinical studies, already the first round of proposals showed that the available funding volume was insufficient for any meaningful clinical studies. As a consequence, only a small number of clinical studies was proposed, none of which was recommended by the reviewers to be accepted into the program. In hindsight, we are able to say that concentrating the doctoral projects on "preclinical" scientific questions warrants the inclusion of projects which have clear scientific milestones and which can be tackled within the framework of a three-year PhD-equivalent doctoral thesis. Concentrating on this type of project also allows avoiding conflicts of interest whose financial background would exceed the mere making available of project funding. These observations are in agreement with similar experiences reported in the literature (see previous discussion). In addition, we learned that projects, which dealt with a basic science question and not an immediately therapeutic objective, were more suited to lead to PhD titles conferred by the natural sciences faculty.

Maximum Transparency

From the start we paid attention to a high degree of transparency in our external communications as we were confronted with allegations from the public that the financial support of the Bayer company would endanger the freedom of scientific research at the University of Cologne. However, in the initial phase of the graduate program, public communication was only possible in a very limited way, for instance, via a publicly available homepage. We faced a number of trivial but not immediately resolvable uncertainties and a certain legal insecurity. For instance, we could not clarify whether a webpage describing the joint program would have to be published in the corporate design of the University, of Bayer, or a newly designed "joint" layout.

Therefore, the homepage was in the first instance designed in a very basic way, foregoing the placement of a Bayer or a University Hospital Logo. Only following the external evaluation of the graduate program, we could clarify within the steering committee how the webpage should be designed. In this context, it is also interesting to note that some scientists from the University of Cologne were not prepared to make the abstract of their research project publicly available. They did not want to endanger their lead over their scientific competitors outside the collaboration by giving away information on their research project.

Clear and Binding Agreements on Confidentiality, IP, and Publication Rights

On the level of the steering committee, it was made clear from the beginning that the publication of doctoral thesis projects should not be delayed or otherwise restricted. Restrictions on publication have been a prevalent problem in industry–academia collaborations [2]. The right to publish was granted to researchers at the University of Cologne, and the ownership of IP rights was contractually regulated.

Concept Designed for Sustainability

Already during the initial discussions, all stakeholders realized that to establish a sustainable collaboration between the University of Cologne and Bayer, a concept aiming for longevity was needed for the graduate program. In the planning of the program, which was in the first instance funded for a 3-year period, we took into account that in case of a positive evaluation by external reviewers after 2 years, we would want to continue the program. Therefore, we decided on two time points of student recruitment for the first funding period so that in the case of a continuation of the program, about the same number of doctoral students would be engaged in different phases of their PhD projects. At the start of the cooperation, we could not predict to which extent and at what pace we could reduce the sponsoring of doctoral positions by Bayer without reducing the volume of the program or its scientific quality. In the meantime, it has become clear that the other advantages of the program (see later discussion) provide enough incentives for many scientists to join the program with funded projects. We may hope that in the long run, the financial support for the administrative overheads and the curriculum of the program alone will be sufficient to recruit excellent scientists and doctoral students into the program.

Openness Concerning Project Proposals

This positive development has certainly been possible because of the extensive overlap in research interests between Bayer and the Medical Faculty of the University of Cologne in the areas of cardiovascular science and oncology. This is reflected in 12 and 9, respectively, project proposals and 8 and 3, respectively, projects accepted into the program in the areas of cardiovascular science and oncology, which constitute the majority of applications and accepted projects. The envisioned methodological focus on animal experiments, imaging, or biomarker research was, however, not reflected in the received project proposals. Most of the accepted projects were using the usual molecular biology techniques and *in vitro* systems. Only a minority of projects was at all intending the pharmacological characterization of small molecules. In the course of project recruitment we had to realize that a substantial number of PIs did not hand in an application because they wrongly believed that only projects using Bayer substances would be accepted into the program. To the contrary, we believe that a great variety of scientific questions and methodological approaches is appropriate to prepare junior scientists for a scientific career in the pharmaceutical industry as well as academia. For this reason, the selection criteria were revised for the second funding period.

Interinstitutional Communication

It was decisive for the success of this new form of collaboration to improve and intensify the communication between the partner institutions. Therefore, we emphasized opportunities for personal interactions in planning the graduate program. We wanted to encourage networking among the graduate students, facilitate contact between the graduates and scientists, and create opportunities for contacts between scientists within and between all partnering institutions. To serve all these purposes, we designed an obligatory curriculum for the students, which at the same time corresponds to the core graduate program of the umbrella organization Graduate School for Biological Sciences (GSfBS) (Figure 18.3). It contains a 3-year program of curricular activities with a total workload of 900 hours. To monitor the progress of the students, the European Credit Transfer and Accumulation System (ECTS) of awarding credits has been adopted. The graduate program is *not* an "ECTS Label Program"; it merely uses the European Commission ECTS guidelines to achieve an objective assessment of the students' learning within the program. The intention of using ECTS points rather than asking for participation in one mandatory training program is to strike a balance between providing students with a curriculum and allowing them to decide for themselves which learning activities are best for them.

Part of the core curriculum is a monthly 2-hour journal club where the graduates report about their own projects or about publications from the literature and discuss

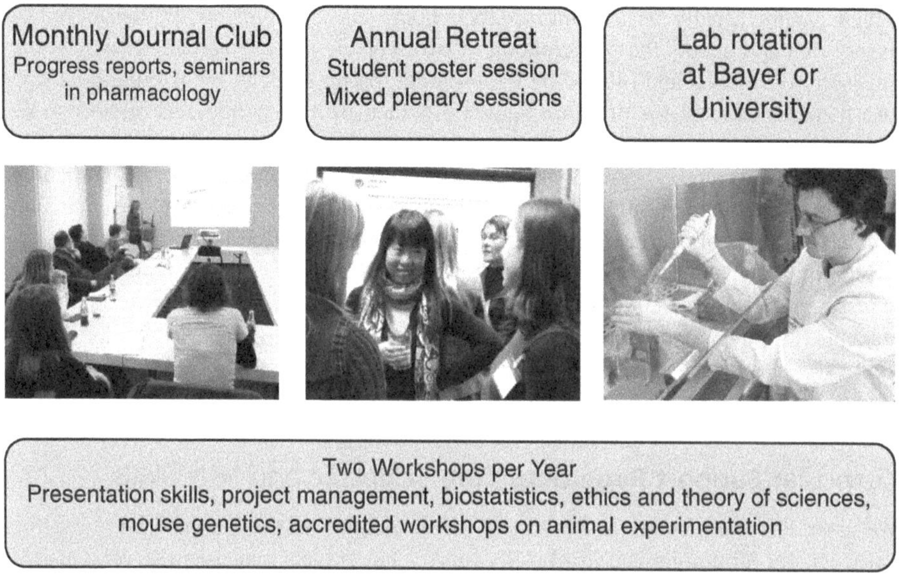

Figure 18.3. Core curriculum. Obligatory curriculum of the graduate program including scientific and soft skills elements.

under the guidance of moderators from the circle of PIs. These meetings take place alternately at Cologne and Wuppertal. We expect that our students also take part in weekly group meetings at their own institutions.

Once a year we organize a meeting of all principal investigators and all students and additional invited speakers and, on occasion, external reviewers in the form of an Annual Retreat. We spend one and a half days at a retreat location in a pleasant environment with good logistics.

In addition to their supervisor, each of our students has two additional mentors who are updated on the progress of the work at regular intervals and who give recommendations on how the investigation could be continued from their perspective. They also recommend curricular activities to the students. When choosing mentors, we make sure that all three institutions contributing to the program are represented: Bayer, the medical faculty, and the natural sciences faculty. Furthermore, every student in the 3-year program has to undertake a minimum lab rotation of 4 weeks full-time in the partnering institution. In the course of this lab rotation, the student can either gain an overview over the scientific activities of the partnering institution or carry out an experiment following careful planning. This makes particular sense when the methodological expertise is located exclusively or in a particularly advanced way at the partnering institution. Especially these lab rotations have been very useful in establishing personal contacts between the graduates and the scientists of those institutions with which they were not originally affiliated. In this context, the students can develop skills, which cannot usually be acquired in their home lab environment. Furthermore, we expect of our students the participation in two workshops per year. These workshops are partly developed and offered by the program, partly realized by external providers and financed through the program budget provided by Bayer. Subjects offered so far contain biostatistics, project management, mouse genetics courses, a workshop on the ethics of clinical studies, a workshop on presentation skills, and accredited workshops on animal experimentation. The originally envisioned participation of students in undergraduate lectures of, for instance, medicine or chemistry to complement the knowledge base of the graduates according to their individual degrees was not taken up. An interim assessment of the curricular activities of our graduates after 2 years showed that most students fulfilled or exceeded the expected curricular workload. In particular, the opportunities created by the curriculum to get to know one another allowed networking among the students and between the involved scientists of both institutions.

Curricular Support Regarding Core Scientific and Soft Skills

We have described above the content of the curriculum of the graduate program. Initially, we put the emphasis on soft skills courses like project management, presentation skills, or on qualifying students to carry out animal experiments. The acquisition of core scientific skills was expected from working on the thesis project and from project-overarching activities of scientific information exchange (e.g., journal club). Only at the occasion of the first external evaluation of the program did we discuss that due to the small number of pharmacological projects within the program, the targeted communication of specific pharmacological knowledge should in the future receive more emphasis. We will report about this in the section on Curriculum Assessment.

SYNERGIES IN GRADUATE EDUCATION WITHIN THE UNIVERSITY

At the University of Cologne there are currently two other programs for young scientists with a similar focus on fundamental medical research and which are also collaborations of the medical and the natural sciences faculties.

With the introduction of the interdisciplinary graduate program Molecular Medicine (IPMM) at the University of Cologne in 2003, a 3-year degree conferred by the medical faculty in cooperation with the natural sciences faculty was first established. The IPMM program has developed a particular curriculum for graduate courses and runs a number of soft skill courses on a regular basis. With the funding of the cluster of excellence CECAD (Cluster of Excellence "Cellular Stress Responses in Aging Associated Diseases"), another strong program for young scientists could be founded at the University. The IPMM program, the CECAD program, and our pharmacology program have shared aims and areas of research. This is also manifest in a considerable overlap of PIs.

Thus, our specialized program forms an additional network with a new focus in an already highly networked environment. In addition, all these programs are under the umbrella program of the GSfBS, organized by the biology group. Also, the GSfBS offers training to students. It has provided a number of high-quality soft skills courses our students could take advantage of. Thus, the cumulative efforts of at least three specialized programs and a more general soft skills umbrella program allow our students to assemble their individual educational programs parallel to their thesis work from the many educational activities on offer. This also allows the critical mass of students to offer courses to students when they most need them. Similarly, this gives the degree problems enough clout to force the University to find solutions, for example, with the introduction of a three-year PhD degree at the medical faculty.

SYNERGIES OF INDUSTRY AND ACADEMIA IN VOCATIONAL FIELD EXPLORATION

Our graduate program fosters personal communication between the graduate students and PIs at Bayer and the University, especially their mentors and lab rotation supervisors. This allows the students to find out about individual careers in academia and the pharmaceutical industry. In addition, we were able to recruit Bayer Hochschulmarketing to give our students an overview of career options at Bayer and to arrange a meeting with recent graduates who now have a position at Bayer.

OPPOSING STRUCTURES AND OTHER OBSTACLES WITHIN THE UNIVERSITY

Value Recognition

In the beginning, the program received a mixed reception as reflected in the spontaneous or invited commentaries of University PIs and their reactions. Naturally, PIs liked the financial support for their research by industry. However, the funding provided by the program did not entail all the benefits of third-party funding from other sources. The acquisition of third-party funds is one of the factors in the achievement-oriented remuneration scheme (leistungorientierte Mittelvergabe [LOM]) by the federal state NRW (North Rhine Westphalia). The seven medical faculties of NRW have decided that the "bonification" (betterment) factor for public funding should equal 1.0, whereas the bonification factor for third-party funds from industry should equal 0.3. This corresponds to a recommendation made by the DFG in 2005 [8]. The background for this decision is that in the past, industry funding of medical faculties was primarily in the context of clinical studies with a large financial volume and a low level of peer review scrutiny. For our program, this big difference in bonification constitutes an important disadvantage for attracting the most scientifically successful PIs. In addition to the remuneration scheme by the federal state, there is a bonification scheme in place at the UK Köln—Köln Fortune—which automatically provides 10% extra funds for DFG-funded research and similar percentages for charity-funded research but no additional funds for industry-sponsored research. Thus, there is a discrepancy between the demands on the program projects concerning quality (comparable to the criteria of government funding or funding by charities) and the appreciation the projects receive with regard to the inner-institutional bonification of the acquired third-party funds.

In contrast, the students, which the program could attract, were seen unanimously in a very positive light. Not only the large number of applications (70 after the first advertisement) but also the specific motivation of many selected candidates to work within the framework of a university–industry collaboration was regarded by the principal investigators as a positive selection attribute. We provide some PI comments, as follows:

> I think that the student enrolled in the program belongs to one of my better students and this selection bias by the program makes the selected students those who will likely flourish in academia or the corporate sector > BOTH, due to the selection FOR the program AND due to the PROGRAM itself, which provides a very good platform for exchange, a good formal curriculum, and added tasks requiring well managing students surviving in that program.

> Students in the graduate program may have undergone somewhat harsher selection than students outside similar programs and have agreed to devote time and effort to the framework program. Thus they have taken a decision for widening their horizons without immediate effect on the progress of their thesis projects.

Degree Procedures and Diversity of Academic Titles

The program offers all of the academic titles conferred by the medical faculty and the natural sciences faculty (3-year program: Dr. rer. nat., Dr. nat. med.; 2-year program: Dr. rer. medic. and Dr. med. (dent.)). We now believe that it might have been better to establish a new degree with its own degree regulation. As a matter of fact, students of our program have pursued all of these titles. Students who have a degree in pharmacy or natural sciences have aimed for the first three titles, the medical students enrolled within the program aim for the Dr. med. (dent.). Among the first three titles, the internationally best known Dr. rer. nat. title is clearly preferred by our students. As the natural sciences faculty confers the Dr. rer. nat., the acquisition of this title by students working in the medical faculty is obstructed what concerns thesis content and administrative procedures. The Dr. rer. nat. degree regulations have been specifically written for natural science students and for instance do not mention pharmacy graduates. In addition, natural science faculty members within the joint degree committee of medical and natural sciences faculty, which decides on title recommendations for students within the medical faculty, restrict the possible content of a thesis suitable for the Dr. rer. nat. degree to fundamental research questions.

The obstacles toward obtaining the Dr. rer. nat. degree are the most criticized aspect of the program. The registration procedure has been criticized by the graduates, their supervisors, and the external referees.

> Registration procedures for the Graduate School of Biological Sciences were supported by the program, but the process in my view still remains unnecessarily complicated and hardly transparent.

> The variety of doctoral degrees is confusing. There is a need for university-wide policies on (PhD) degree clearance. Awarding the degree through colleges/schools (i.e., Fakultaeten) and not through the University of Cologne creates "turf wars" and disadvantages the students. It is not clear to the students if the Dr.rer.medic. receives sufficient recognition abroad.

The medical students in the program welcome the long-term participation in our program, which allows them to carry out a quality scientific project with sufficient financial and temporal resources. However, they feel that the title they can acquire (Dr. med. (dent.)) does not reflect the quality of their work. The low appreciation the title Dr. med. receives nationally and internationally [7, 8] explains this disappointment. Furthermore, the 2-year degree Dr. rer. medic., which natural science students can acquire at the medical faculty, is unknown outside the University of Cologne and is possibly considered as less valuable and not sufficiently preparing for a career in research. Consistent with this evaluation, the medical faculty is now planning to abolish this title and to introduce a 3-year PhD-equivalent degree, focused on projects related to health sciences (e.g., clinical research). This PhD can be undertaken by natural scientists (and also postgraduates holding a degree from a variety of disciplines within the research area of health sciences) with a master or diploma degree or similar academic degrees preparing for health sciences research and those with a state board examination in medicine.

The most poignant critical remarks from our external reviewers about doctoral degrees and degree regulations fall into the responsibility of the University:

> A point of contention for many doctoral students is the type of conferred doctoral degree. The fact that according to the knowledge of the reviewers no German university has found an in every respect satisfying solution to this issue indicates the complexity of the problem. The reviewers sympathise with the differing objective interests of the different parties: the students aim for especially in the case of a first degree in the natural sciences in almost all cases for the Dr. rer. nat., as this degree is held in higher esteem nationally and internationally than the Dr. rer. medic or the Dr. nat. med.. The natural sciences faculties are in general financially less well endowed than the medical faculties, but have to provide for Dr. rer. nat. degree procedures a disproportionately high share of examination workload, also for candidates whose thesis project is performed within the medical faculty. Therefore ways of compensating the natural sciences faculty should be explored. In addition to this there is a general partly justified scepticism toward the scientific quality and the extent of dissertations, which have been produced under the guidance of medical colleagues. Similar conflicts exist in many other graduate programs where natural sciences and medical faculties are collaborating. The reviewers understand the different points of view of all involved but are of the firm opinion that in this case the interest of the doctoral students has priority. It appears essential that already at the point of acceptance into the graduate program there are clear rules which are the prerequisites for the acquisition of the Dr. rer. nat.. Clarity means here that every single graduate should learn at the start of the project if the project and his previous education would lead to this title. If this cannot be achieved there will not only be disappointment from current participants but a diminished attractiveness of the graduate program for future applicants. A more strongly centralised, faculty overarching and if necessary university wide regulation of awarding doctoral degrees is desirable and could avoid conflicts between particular faculties at the expense of the students. The reviewers also want to point out that in the current degree

regulations the main advisor of a thesis is also the primary examiner at the occasion of the final thesis examination. While this is common practice at many German universities it can lead to a non-negligible conflict of interest.

OPPOSING STRUCTURES AND OTHER OBSTACLES WITHIN INDUSTRY

According to the original plan, Bayer would have offered five to seven doctoral thesis projects within the program. However, it soon became apparent that the scientists at Bayer did not exhaust the quota of five to seven students. They could only offer half the envisaged doctoral positions. Most of the doctoral thesis supervisors at Bayer are industrial researchers, who have previously had an extended career as university researchers and who command a high international reputation. Younger colleagues working in industry hardly have the time needed for the supervision of doctoral students. Also, the necessary long-term planning for a three-year research project is not easily compatible with the reality of working as a junior researcher in industry. Company goals may change, the junior group leader might move to a different position and leave behind graduate students who then find it difficult to finish their thesis work.

Individual researchers at Bayer showed a high level of interest in the graduate program and were clearly willing to participate in joint events and to make contributions. In this way, the very important added value in terms of the establishment of personal contacts could be realized.

COMMUNICATION DEFICITS BETWEEN AND WITHIN INSTITUTIONS AND THE GRADUATE PROGRAMS ROLE IN OVERCOMING THEM

It surprised us how little many graduates knew about the tasks and procedures of pharmaceutical research. The exceptions in this respect were graduates in pharmacy who had previously done a lab rotation in industry. Offering insight into the pharmaceutical industry and dealing with information deficits is an important task of our graduate program. Even more surprisingly, quite a few researchers in industry who all have a background in academic science voiced judgments of university research that are in clear need of revision. We will devote a high number of original citations to this important aspect of our collaboration in postgraduate education. These citations will illustrate information needs and how they could be met by specific curricular activities:

> Generally it would be of great interest how one can step into a company like Bayer. If there is any piece of advice people from Bayer can give. Is it of use to apply for an industrial position right after finishing the PhD? Is working abroad (US?) of any benefit for getting a job at Bayer? Where do the organizers of the program see our chances to start working at Bayer or in industry in general? In which positions? How did different people working there get into the company? How much do you earn when you start working at Bayer or smaller companies (at one point we will have to add "desired salary" to our job application)? At that point it would also be interesting to talk to people from the human

resource management. What are the things they are looking on? What draws their attention when they see an application? What are No-go's? Etc.

The tour of cardiovascular research was very interesting for me. They have tried to introduce their different models in different areas to us. For pharmaceutical industry the pharmacological screening is one of the most important parts in the preclinical research. The pharmaceutical screening requires the models as precise and reproducible as possible. This is the big difference, which I have gotten through this tour.

I enjoyed the friendly work atmosphere and the approach of result-driven, focussed work that is done in the Bayer lab. I think in comparison to university, the industry asked more precise questions and expects a certain "answer" with certain standards like reproducibility, significance, etc. In university those research questions are broader and the "proof of principle" is often much more important. I also liked the style how each experiment has its own protocol, result sheet etc. It made one work more precisely and finish the protocols and result section of each experiment faster. It also leads to a more structured work approach, which I would have really liked at the beginning of my thesis. Despite the fact that I dealt with a totally different topic than in my PhD project at university, I enjoyed working on the new topic very much. I felt nice to see that I am able to get into a new field in a rather short time. Aside I was also allowed to perform some experiments for my PhD project, too which worked out quite well. The chance to use equipment from the Bayer labs, which is not present in university, was also a big advantage.

Several conversations with lab heads touched on differences between academia and industry, and I was struck by the impression that many people seemed to think that academic work is somehow not as rigorous. I do think several aspects in industry are more efficient: obviously the machine automation, but also the business side ensuring that the most productive projects continue, and many labs shared resources which was also nice to see.

CURRICULUM ASSESSMENT

Curricular Content

The main initial focus of our curricular concept was derived from the example of similar structured graduate programs. We offered on the one hand generic soft skills courses, for instance, a workshop on presentation skills or project management, and on the other hand asked the graduates to undertake a lab rotation at the partnering institution. Both measures shall be evaluated here drawing on typical feedback from students.

Feedback on Project Management Course:

Considering the timing of the project management workshop I would say that 6 months after the start of the PhD would be ideal. The course addresses rather intimate questions, so that it will be important for most participants to be familiar with the other people in the course before being able to talk about fears and worries and to assess oneself in certain respects.

What I liked about the workshop was hearing about the problems and especially their solutions in other labs.

> Which aspects of the workshop were the most relevant/helpful to your PhD?< How can I manage my time and get along better with my supervisor.> Would you recommend this workshop to other PhD students? Why/why not? <Yes, because everyone had the possibility to speak about his/her own problems and together the group tried to solve it.
>
> Well, in a PhD program of this kind, some aspects like ego management of executive Profs is required.

It was interesting to see that the joint participation in workshops, some of which lasted several days, led to networking among the graduates and also to personal relationships between the two research locations, Cologne and Wuppertal. At least as important as the original learning outcomes was the achievement of implicit tasks by the students like the support of fellow students in situations of conflict in the laboratory team or with the supervisor and the development of a positive feedback culture among the student peers.

Lab Rotation: A Useful Experience?

All lab rotations so far took place at Bayer, as the few students performing their thesis at Bayer either joined the program relatively late or took a break (maternity leave). The feedback from the lab rotation experience had positive and negative components:

> Beforehand I was not sure whether these 6 weeks could be connected to my project in Cologne. One difficult point for me was the way from my home to Bayer; with the train it would have been 2h in the morning and 2h in the evening. Therefore I decided to rent a room in Wuppertal. In a beforehand discussion with Dr. XXX, we found some overlap of our research interests. So I could perform experiments with "our" mice. I learned several methods that will be really useful for my further work at Cologne.
>
> I had several discussions with XXXX, who agreed that there was not much relevance between work done in his lab and my PhD thesis. I did have the opportunity to observe several surgeries, but these were not relevant in any way to my PhD. We did spend some time discussing project design, which—although I do not currently work with animals or pharmaceutical kinetics—might someday be relevant to future work.
>
> My supervisor at Bayer thought of designing this internship in such a way, that I do not only have the chance to learn new methods which might be useful to me but also to explain to me the structure of Bayer, the designing of projects in commercial background, the different steps which a project has to pass, the different meetings with the different hierarchical levels and a lot of other things.
>
> In my case it was very useful to contact my host lab in time before my internship started. We had enough time to think about what I could do during my internship. We planned a little project and I had the chance to send my own mice to Bayer. . . . My only "complaint" is that it really takes a lot of time each day to get to Wuppertal and back (both by car or train)!!

We conclude from the student feedback that the decision to have the lab rotation at an advanced point of the thesis work was right and that it is worth the effort to take time to discuss and plan in advance what could be the potential project or activity of the student during the lab rotation. We agree that the students should be able to stay in Wuppertal during their lab rotation and we reimburse them for their travel or accommodation expenses. However, there is no accommodation available from Bayer Wuppertal so that the students have to find individual solutions for accommodation.

External Reviewer Advice on the Curriculum

Our external referees welcomed the curricular activities we offered in the area of soft skills development. They criticized, however, that we did not offer a core scientific training in pharmacology. As the projects within the graduate program are thematically diverse and the students have different scientific backgrounds, it would be especially important to offer a graduate-level curriculum in pharmacology.

> There were consistently positive student comments on the workshops related to "transferable skills" (time management, scientific writing, poster presentation, communication). The "journal club" was highly appreciated by all students I talked too. The students with a background in biochemistry or biology asked for more disciplinary coursework in pharmacology. They felt that going to "undergraduate" lectures did not make much sense. They would prefer graduate level courses or training (more custom-tailored to their needs).
>
> The common scientific thread of the graduate program has to be more emphasised in the curriculum as well as in the external presentation of the program. Central elements of this should be an emphasis on the process of pharmaceutical research within the projects as well as reinforced curricular units in this area; however, the overall curricular load of the students should not be extended. Here as in other aspects of the graduate program the students should be able to have a say about the development of the program.
>
> The students perceive a general added value of performing their thesis within a graduate program in comparison to a thesis under the supervision of just one PI. As reasons for the added value are given primarily:
>
> A more extensive exposure to subjects outside their own thesis project, interdisciplinary experiences, and being part of an intellectual community. Also the workshops on transferrable skills like time management, scientific writing, poster presentations, or communication techniques received a consistently positive appreciation. These responses of the students are in agreement with the overall experiences in the education of doctoral students in different countries. Such experiences have led in Germany and many other countries to the recommendation to introduce graduate programs for all doctoral students.
>
> The students are reporting unanimously that the cooperation between the university and an industrial partner was a decisive characteristic of this graduate program, which made it more attractive to them compared with other graduate programs. They regarded as advantageous the greater interdisciplinarity and the greater emphasis on medical application. In addition to this, they were expecting that the education in a graduate program in

which a company was involved would facilitate their access to different occupational fields. Here they are not just thinking of a potential appointment at Bayer but also of the general qualification they gain by being exposed to industrial research.

The most important scientific criticism the reviewers want to raise is the great thematic heterogeneity of the graduate program. This does not necessarily mean that individual projects are outside a desirable framework. To the contrary, many of the projects could profit from a clarification of the common thread of target research and validation and pharmaceutical development even within the projects. Apart from this, this common thread should be more explicitly reflected in the courses offered by the program.

Tailored Seminars on Pharmacology

As a first and quite immediate response to these comments by the reviewers we started to invite speakers for our journal club events. For this purpose we could recruit experts from Bayer on subjects like patents and clinical trials. We were also able to recruit high-ranking experts from academia, who were working in basic science as well as substance-oriented pharmacological research. The successful recruitment of guest speakers was certainly not due to the fact that we could offer remuneration to the speakers. The invited colleagues showed a genuine interest in and a high appreciation of our program initiative. It is still too early for a critical evaluation of this new curricular element. However, we already received a very positive feedback on the exposure of our graduates to scientific role models at the occasion of the two preceding annual retreats.

Finally, we followed the advice of the reviewers to let two student representatives take part in the program committee meetings and thus participate in the decisions related to the future development of the curriculum.

REVISED CONCEPT AND FUTURE PROSPECTS

In the following we give a summary of the changes to the concept of the graduate program in collaboration between the University of Cologne and Bayer, which we implemented after the first external review of the program. The involvement of independent, external reviewers has proved useful in several respects.

First, in the new version of the concept, we were able to specify in which way the decision not just for the acceptance of projects into the program but also the decision on which of the accepted projects would receive financial support should be taken. This is now unambiguously a task of the program committee, in which Bayer is represented with a one-third minority. This regulation invalidates any allegation of industry influence on university research.

Second, through an extended discussion process we could reach the decision to make the concept of the graduate program and all related further information publicly available via a graduate program homepage. Further activities aiming at maximal transparency were the presentation of the graduate program at a large international conference [9] and the writing of this book chapter.

Furthermore, the reviewers raised attention to the fact that with regard to degree procedures at the University of Cologne, the situation needs to be improved. Other universities who offer studies in pharmacy may have more workable arrangements. In Cologne, many pharmacological projects will have clear affinities to medical as well as chemical research or other natural sciences subjects. This double affinity means that the projects are thematically at the interface between two faculties. Such thesis projects at the interface of several faculties are not classically regulated by the degree regulations of a single faculty. The University of Cologne has undertaken first steps with the introduction of an MD-PhD program, but there is further need for improvement. We cannot accept that students of our program have to take more and higher bureaucratic hurdles than students with comparable doctoral theses who work in a different thematic or organizational context.

And finally, the reviewers drew our attention to the fact that there was a need to develop the content of the scientific education we offered to the graduates. In agreement with the literature [3], the sharpening of the profile of qualification of the future pharmacologist working either in industry or at the university is highly desirable, even if its definition will have partially prophetic character. We are planning further steps in this direction after the successful recruitment of pharmacological experts for lectures within our curriculum, a measure that was highly welcomed by our students.

However, to address the problem of what the thematic focus of a successful pharmacologist for the future should be, the establishment of a master course in pharmacology would be better suited than the optimization of a PhD program. An intensification of the curricular content of a PhD program is hindered by the heterogeneous previous education of the students, so that many of the curricular activities of the students need to be tailored to their individual needs and there should not be an obligation to participate in a fixed curriculum. A further expansion of the curriculum would be in constant conflict with the interest of the PIs to maximize the scientific output of their students. And finally, within the German system, the establishment of a master course allows to offer courses, seminars, and lectures in pharmacology and neighboring disciplines with the support of public funding. In the form of a master course, the scientific education in pharmacology is justified as a core duty of the university according to its public mission.

As a positive outcome of the first 3 years of the graduate program, we can fully underwrite the hypothesis put forward by Blumenthal and Gray [5, 6] that the education at PhD level on the basis of fundamental science-oriented research projects constitutes an ideal area of overlap of interests of the university and an industrial partner. This focus does not only make use of the central competence of the university of teaching science but also addresses the interest of the research-oriented pharma industry in the education of qualified researchers. In a collaboration of this form, it is also relatively easy to find solutions to the critical questions surrounding industry–

university collaborations (intellectual property, publication rights) and to achieve transparency with regard to a critical public. In our experience, a decisive effect of our collaboration in graduate education has been the development of personal contacts between the PIs at both institutions in addition to the contacts among the graduates and between the graduates and the PIs. These personal contacts will be important for future scientific collaborations.

We believe that only on the basis of a greater number of such stable personal contacts and the mutual trust they entail the domains of added value of such a collaboration can be conquered. These are particularly attractive for both sides but can only rarely be realized. One of the domains of added value is certainly that industrial researchers are able to offer their students a high-quality university degree, an undisputed privilege of the university. Conversely, from the perspective of the university researchers, it would be desirable for the future to gain access to the industry-owned extensive libraries of chemicals and the industry's ability to chemically optimize new interesting molecules with regard to their medical properties. In this respect we need to recognize the sovereign rights of industry, as the arsenal of extensive chemical libraries can be considered the commercially decisive, intellectually and substantially genuine property of the pharmaceutical industry.

It remains to be investigated whether general regulations of university–industry partnerships promote a systematic access to such added value. An extensive discussion involving different stakeholders of public interests seems to be inevitable in the question of how to regulate industry–university collaborations.

ACKNOWLEDGMENTS

The authors thank the graduates, the PIs, the program committee, and the steering committee of the preferred partnership. Special thanks to the external reviewers Prof. Dr. Henning Schröder and Prof. Dr. Martin Michel for their valuable comments. The authors also gratefully acknowledge the great support from administrative, laboratory technical, and IT staff at both institutions.

REFERENCES

1. Herzig, S., Rozowski, M., Flamme, I. Graduate program in pharmacology and experimental therapeutics—a joint endeavor of the University of Cologne and Bayer Schering. *Open Conf Proc J.* 2011;2:73–79.
2. Blumenthal, D., Causino, N., Campbell, E., Louis, K.S. Relationships between academic institutions and industry in the life sciences—an industry survey. *N Engl J Med.* 1996 February 8;334(6):368–373.
3. Lambert, R. Lambert review of Business-University Collaboration. Final report. 2003. Available at http://www.hm-treasury.gov.uk/lambert_review_business_university_collab.htm. Accessed December 24, 2013.
4. Campbell, E.G., Blumenthal, D. Industrialization of academic science and threats to scientific integrity. In: Emanuel, E.J., Grady, C.C., Crouch, R.A., Lie, R.K., Miller, F.G., Wendle, D.D.,

editors. *The Oxford Textbook of Clinical Research Ethics*. New York: Oxford University Press; 2008, pp. 780–786.

5. Gray, N.S. Drug discovery through industry-academic partnerships. *Nat Chem Biol*. 2006;2(12):649–653.
6. Blumenthal, D. Academic-industrial relationships in the life sciences. *N Engl J Med*. 2003 December 18;349:2452–2459.
7. Deutsche Forschungsgemeinschaft 2010. Empfehlungen der Senatskommission für Klinische Forschung zur Strukturierung der wissenschaftlichen Ausbildung für Medizinerinnen und Mediziner. Available at http://www.dfg.de/download/pdf/dfg_im_profil/geschaeftsstelle/publikationen/medizinausbildung_senat_klinische_forschung.pdf. Accessed December 24, 2013.
8. Deutsche Forschungsgemeinschaft. Deutsche Forschungsgemeinschaft Empfehlungen zu einer "Leistungsorientierten Mittelvergabe" (LOM) an den Medizinischen Fakultäten Stellungnahme der Senatskommission für Klinische Forschung der Deutschen Forschungsgemeinschaft, Stand Juli 2004. Available at http://www.dfg.de/dfg_profil/gremien/senat/klinische_forschung/index.html. Accessed December 24, 2013.
9. Herzig, S. Graduate program in pharmacology and experimental therapeutics—a joint endeavor of the University of Cologne and Bayer Schering. International Conference on Drug Discovery and Therapy. Dubai, UAE, February 7–10, 2011.

19

ACADEMIC, COMMERCIAL, AND BIODEFENSE CASE STUDIES FOR COLLABORATIVE DRUG DISCOVERY: POTENTIAL FOR DISRUPTING DRUG DISCOVERY

Barry A. Bunin and Sean Ekins

Collaborative Drug Discovery, Inc., Burlingame, CA, USA

Historically, drug discovery has evolved at fairly linear rates, with modest improvements as new technologies were developed; at worst, some have even correlated the additions of new technologies with the decrease in pharmaceutical productivity [1, 2]. This evolution at linear rates may be because (as many people believe) the easier problems were solved first and/or because of additional regulatory requirements that have been added as R&D has progressed. Even the 4.9% increase in scientific publications [3] seen each year, much like compounded interest, should be paying more handsome dividends than it does, at least for drug discovery. We know that other fields can improve dramatically when they evolve around a common standard. For example, in computer chip R&D, Moore's law states, "The number of transistors that can be placed inexpensively on an integrated circuit doubles approximately every two years" [4]. In contrast, network-based phenomena evolve much faster per Metcalfe's law, which states, "The value of a telecommunications network is proportional to the square of the number of connected users of the system (n^2)" [5]; hence, we see astonishing exponential rates of evolution in adoption of Internet tools like Facebook [6].

Collaborative Innovation in Drug Discovery: Strategies for Public and Private Partnerships, First Edition. Edited by Rathnam Chaguturu.

We have previously described a brief history of Collaborative Drug Discovery, Inc. (CDD [7]) and our role as a provider of the first collaborative drug discovery data-hosting platform in the cloud [8]. Our years of experience developing software and efforts to promote collaboration to speed drug discovery provide a valuable perspective, which we will share through the course of this chapter. We will demonstrate how scientific teams and their projects mutually benefit from the secure exchanges of information. Both commercial and altruistic case studies clearly demonstrate how a collaborative software platform can accelerate the efficiency of drug discovery and reduce redundancy. The key question today is how these economic efficiencies can be scaled to the entire drug discovery ecosystem.

We have shown how a web-based database technology can allow anyone to collaboratively leverage others' drug discovery results [9–13]. This enables the scientists within their natural workflow to effectively collaborate while maximizing the intellectual property rights required for commercialization. The first important prerequisite is having a scientist with a collaborative mind-set, who appreciates the value gained by collaborating with others. Efficiency is possible with the right attitude: a recognition and nuanced appreciation of the additional opportunity and challenges inherent in initiating and executing successful collaborations. A second prerequisite is the development and utilization of technologies that make collaborations work better [14]—of which there are many. Fortunately, there has been rapid progress in this field. With the enlightened use of collaborative technologies, more opportunities are realized because everyone with complementary capabilities is on the same page. As we will illustrate, complex collaborations can become manageable and scalable using technologies such as CDD and others.

Collaborative drug discovery platforms like CDD are broadly applicable because drug discovery projects have similar technical and commercialization hurdles. Apart from the perplexing, idiosyncratic features of individual molecules, targets, and assays, there are common, systematic technical challenges associated with lead identification via screening (whether high throughput, computational, or a combination of both) and lead optimization. Despite the great variation in scientific projects and assay methods at a data level for the vast majority of cases, the common standards have fortuitously become self-evident (sdf, csv, SMILES, xls, controls, IC50's, hyperlinks, etc.), which has in turn led to the development of early bioassay ontologies [15]. These common standards developed over time, coupled with conserved drug optimization process workflows, suggest that a finite set of collaborative technologies can meet the majority of drug discovery informatics requirements. By being first movers, these collaborative tools set the standards for de facto collaboration and have formed a common foundation for collaborative science.

While in the past discovery of new drugs could be attributed to individual scientists [16], it is now incredibly rare for a scientist working in isolation to discover and develop a drug alone. Drug discovery work is collaborative, involving massive teams doing interdisciplinary preclinical and clinical research. We are also seeing not only companies outsourcing research that traditionally would have been performed in-house, but also companies collaborating with academics and other organizations, placing an extra emphasis on the need for collaborative technologies that effectively bring researchers together.

From our experiences with leading commercial, academic, and government laboratories [7], a recurrent theme has been the importance of giving each researcher control over whom they collaborate with (if indeed they need or want to). Providing facile control of each collaborator's data access within natural workflows (with minimal activation barriers) has been a key for more effective collaborations and is therefore important for any technology in this collaborative domain. Providing fine-grained control in a secure environment for selective data sharing without multiple data uploads for prepatent and prepublication data (i.e., intellectual property and data-sharing capabilities as the data are being created) is a driver for effective collaborations. Researchers want the freedom to do what is necessary within their regular workflows or as demanded by their organizations. We should provide researchers with a convenient way to share data any time but always with the default being maximum privacy.

We believe there are at least two major "activation barriers" that need to be surmounted for facile collaboration around data. The first is reliably, accurately, and meaningfully capturing the data from the experiments. This is not as trivial as it sounds, and it is generally underappreciated. The second barrier is assuming the need to keep data private for commercialization reasons, which is generally overemphasized. In this context, it is worth remembering that, upon patenting, information is shared, albeit not directly in a reusable database format. By lowering barriers for both data archiving and selective sharing, we can begin to reap the benefits of Internet-catalyzed collaborations across the entire academic, industry, nonprofit, and government drug discovery ecosystem [17]. Technical innovation allows researchers to decide if, when, and with whom to share data. The realization of our vision to develop software for sharing data is seen in concrete examples like a leading academic high-throughput screening center simultaneously managing 26 separate, secure projects in a single collaborative "vault" (with appropriate access permissions as projects roll in and out of the screening queues).

We are clearly at the start of a new era for drug discovery, one of specialization, integration, and collaboration. It is an era driven by technologies that provide data availability, such as PubChem [18] and ChemSpider [19, 20] (for all open data), and better collaborative technologies like HEOS® (for all private data) [21, 22] and CDD (for private, collaborative, and open data) [9, 12, 13]. These technologies have become widely adopted, supporting thousands of researchers around the world who routinely trust these hosted systems, which in turn sets the foundation for a more effective paradigm. Drug discovery research can directly benefit from previous and complementary collaborative efforts, while collaborative informatics provides cohesiveness when scaling these efforts. Researchers can now potentially balance appropriate ownership of intellectual property (IP) with maximum collective learning. We should see new discoveries in turn occurring more rapidly as scientists collaborate, from which we can all benefit. The various case studies detailed later represent a wide array from the commercial, academic, government, and biodefense arenas to demonstrate how researchers across all types of organizations can form a collaborative team. We hope that the reader will join us in collaborative drug discovery soon and help us in our mission to positively disrupt drug discovery.

ACADEMIC COLLABORATIVE DRUG DISCOVERY CASE STUDIES

Antimalarial Resistance Reversal

This was our first published case study [13] and involved a four-way collaboration. Academic and ex-industry scientists with complementary capabilities in the United States and Africa came together for the purpose of identifying antimalarial resistance reversal agents in combination with chloroquine [13]. Professor Kelly Chibale (University of Cape Town, South Africa) and his existing collaborators (including Dr. Peter Smith at University of Cape Town) had finite resources, budgets, and time. Dr. Chibale had done his postdoctoral work at the University of California, San Francisco (UCSF), and he knew other members of the CDD community at UCSF (e.g., Professors Phillip Rosenthal and James McKerrow) who also were working on small molecule drug discovery against malaria and other neglected infectious diseases like Chagas disease and African sleeping sickness. Given their personal and professional connections, they were willing to pool and share existing resources and compounds. Selected samples were shipped on dry ice from UCSF to Dr. Peter Smith's laboratory at the University of Cape Town to screen for reversal of chloroquine resistance activity. New compounds were identified that almost completely reversed the chloroquine resistance. There was another discovery to come, however, using a set of 1720 known drugs and their therapeutic indications kindly provided by Dr. Christopher Lipinski (retired, Pfizer) and available in CDD Public for the use and benefit of researchers throughout the world. Within this data set, a further 18 matches were found with the same substructure that Prof. Chibale had identified as being responsible for reversal of chloroquine resistance activity. These known drugs were then purchased, shipped, and tested by Dr. Smith for antimalarial resistance reversal activity in combination with chloroquine. Several compounds almost completely reversed chloroquine resistance *in vitro* (sevenfold); these included the FDA-approved drugs pimozide, vinblastine, sertraline, and dihydroergotamine mesylate. The discovery of effective compounds that already have known human safety profiles and FDA approval can save years or even decades in the race to overcome malaria drug resistance, relative to starting a new project from scratch. This represents a potential *in silico* drug repurposing strategy, which we have discussed elsewhere as a way to accelerate drug discovery [23, 24]. It is worth highlighting that these public–private and private–private data sharing collaborations required multiple parties all providing complementary pieces, as well as the CDD platform that brought them together and facilitated it.

Tuberculosis Research

This second academic case study deals with another neglected disease and involved building a new community of Tuberculosis (TB) Drug Discovery Researchers funded by the Bill & Melinda Gates Foundation. With finite resources and software capabilities, CDD needed to balance the ability to support anyone working on TB drug discovery while accelerating the progress of the most promising hits and leads within a pipeline drawn from academic, nonprofit, government, and corporate laboratories distributed across the globe. The scaling challenges inherent in this project led to CDD's current policy of freely hosting any data that researchers are willing to share publicly (via CDD Public) while charging for privacy, collaborative capabilities, import, and export. While several respected information technology leaders encouraged us to make CDD open source and offer services (following a Linux type model), a credible alternative business model has emerged that ensures sustainability of an increasingly valuable commercial software offering, balanced with an equally valuable free content offering for the general public.

Within the TB arena in the first 2 years of this project, *250 users from 58 labs used the CDD collaborative platform.* Perhaps somewhat surprisingly, big pharmaceutical corporations have also been among the more forward-thinking participants in this project in terms of actively and publicly sharing private neglected disease structure–activity relationship (SAR) data. The project led to CDD publicly hosting GlaxoSmithKline malaria and Novartis TB SAR data—harbingers of the winds of change impacting our field.

The CDD TB database became an integral part of the workflow for most of these TB groups, enabling them to better understand their data and exploit their finite resources more effectively. The technologies and support enabled them to selectively compare their results, avoid duplicate work, and formulate better future research priorities. These collaborations allowed previously disjointed efforts to begin to coalesce into a "virtual pharmaceutical organization." The participation of the largest global pharmaceutical companies and their willingness to facilitate the exchange of data between themselves and the academic and other nonprofit research groups through both open and secure private channels is an important, and historically irreversible, collaborative drug discovery development.

Some of CDD's most advanced features have debuted within this TB community. For example, one laboratory used CDD's software to manage a portfolio of distributed projects, much as a major pharmaceutical company manages internal or external projects. CDD provided this "Projects" functionality to all users. Projects allows for secure, selective data sharing or data partitioning with fine-grained controls of individual objects within secure vaults. Each vault is a separate database with customizable business rules for registration, field definitions, protocol definitions, assay readouts, assay-specific normalization options, and so on. Moreover, data within each secure vault are further partitioned into separate projects with carefully controlled personal access permissions for an extra level of data control.

In an attempt to obtain a greater understanding of the chemical space of molecules tested against Mycobacterium tuberculosis (Mtb), we created a Collaborative Drug Discovery Tuberculosis Database (CDD TB) for related molecular libraries of compounds from the literature [9]. In CDD Public more than 300,000 unique molecules and well over 300,000 TB assay data points with minimum inhibitory concentration, IC_{50}, and/or toxicology data from published and patented sources are available for anyone to view and mine. To the best of our knowledge, this is the largest, most comprehensive source of TB drug discovery SAR data available in the world, and it is hosted by CDD for the benefit of the entire research community [8]. Our analysis of the public data sets [9] provided insights into molecular properties and features that are determinants of activity in whole cells [9, 12]. We have recently published several studies using computational approaches with these TB data sets to produce computational machine learning models that in turn have been used to identify new hit compounds. The first study involved two large publicly available Mtb data sets with over 200,000 molecules and a second with over 100,000 molecules. We used Bayesian classifier models and predicted data sets from additional laboratories and identified 4-fold and 10-fold enrichment factors [9, 12, 25]. These models were then used as part of an in-drug discovery program with collaborators at SRI, International, who identified two new hits for TB [26]. A second approach we have taken with collaborators at Johns Hopkins and the University of Medicine and Dentistry of New Jersey is to use the 3-D pharmacophore around a molecule of interest [9] to find molecules with a similar shape, while using a molecular shape restriction to search large multiconformer libraries of compounds. The 3-D pharmacophore searching was used to create a pharmacophore model for L-glutamate. Azaserine favorably fits this pharmacophore model, and after searching several vendor databases with this model, we purchased and tested 12 molecules. Among these compounds, L-methionine-S-sulfoximine showed activity against Mtb with a minimum inhibitory concentration in the range of 8–16 μg/mL. The examples of azaserine and L-methionine-S-sulfoximine provided a proof of principle [27] that mimics of essential metabolites, as identified through cheminformatics methods, and represent effective molecules with anti-TB activity *in vitro*.

More recently we have annotated TB molecules with putative targets as part of a study with SRI funded by a Small Business Technology Transfer award from the National Institute of Allergy and Infectious Diseases. We have also developed a unique community with over 20 groups collaborating in the European Union-funded More Medicines for Tuberculosis (MM4TB) project [28, 29], which includes major pharmaceutical companies such as Sanofi and AstraZeneca alongside academic groups and smaller companies. MM4TB works as if it is one organization within a single CDD Vault using the full projects and structure masking capabilities (as needed when dealing with big pharma screening libraries).

The many computational analyses described previously were performed outside of CDD and using best-in-class computational modeling technologies in Discovery Studio from Accelrys [30] and ChemAxon [31]. These enable identifying potential target mechanisms by computational methods; searching for potential compounds to test with pharmacophore models; focusing priorities on desirable molecular properties, including ADME (absorption, distribution, metabolism, and excretion); rendering statistical analyses of data sets; and providing advanced visualization capabilities. It is therefore important that collaborative tools can integrate with third-party gold-standard tools that add considerable analysis capabilities for data interpretation.

CDD continues to support pharma-academic TB collaborations today. The CDD TB database project was recently extended from a 2-year to a 5-year project with the

Bill & Melinda Gates Foundation, based on the initial success. The collaborative approach has certainly become a fixture of the TB drug discovery researcher community. The continued collaborative relationships between academic and research institute TB research groups and big pharma are important, and it remains to be seen how they will be impacted by the various company mergers, downsizing, and refocusing.

COMMERCIAL COLLABORATIVE DRUG DISCOVERY CASE STUDIES

Although our initial applications for collaborative technologies started with academic researchers doing drug discovery for neglected disease, today collaborations are just as applicable to commercial targets and therapeutic areas. The involvement of more commercial companies comes with the realization that the cloud can be used for secure molecule and data transactions in the same way that it is used for other types of business transactions.

Acetylon Pharmaceuticals

Acetylon is a Harvard University spinout company which has worked with CDD to manage academic–industry and China-based contract research organization (CRO) collaborations. In 18 months, Acetylon filed an investigational new drug application with the U.S. FDA for their selective Class I histone deactylase inhibitor drug and raised $27 million in Series-B financing [32].

CDD's role was to provide the informatics support for their academic–industry–CRO collaborations from the initial funding stages through filing of the investigational new drug application and beyond. This is a quite common use case for budget-conscious organizations that want to bring together complementary capabilities around a single mission. From a process perspective, this collaboration taught us that when sensitive data are shared, it is important to have permission from both the sharing and receiving parties. We were initially concerned with permissions and paper trails from the sharing party prior to the sharing of outgoing data, but we soon realized that it makes sense for the recipient to formally approve receipt of incoming data too.

Semafore Pharmaceuticals

Semaphore developed clinical stage selective and pan phosphoinositide-3-kinase (PI3K) pathway compounds [33]. This is another example of a virtual drug discovery organization where a small team of internal chemists and biologists used CDD as they worked with CROs to advance oncology compounds targeting the PI3K pathway space. Today the company is focused on advancing its clinical pipeline, but continues to rely on CDD to host the IP behind their clinical candidates, which include a novel small molecule targeting the PI3K pathway and a dual PI3K-mTOR inhibitor that is a prodrug targeting integrins expressed on the surface of new tumor vasculature and within the tumor compartment.

Bioseek (Asterand)

Bioseek has developed customized, complex, *ex vivo* high-throughput cell-based screening assays. These assays provide statistically superior correlations to human conditions in live tissues. The company provides services using these technologies in a number of collaborations with major pharmaceutical and government agencies, including an ongoing effort at improved toxicity metrics with the U.S. Environmental Protection Agency (EPA). CDD has been used to privately share data generated as part of the ToxCast project with the collaborators at the EPA [34] and with their other collaborators in a conventional manner.

Indel Therapeutics

Indel is a biopharmaceutical company dedicated to developing new drugs to address antibiotic resistance, focused on curing difficult-to-treat and hospital-acquired infections. As a University of British Columbia spinout company, it has technology licensed from academia that helps it exploit divergence in proteins required for both eukaryotes and prokaryotes. More specifically, it identifies and targets discrete amino acid insertion/deletion differences ("indels") between essential, homologous proteins that are evolutionarily conserved across bacterial pathogens and humans. Indel works with consultants, academics, foundations, and industry researchers across multiple time zones. Together, they focus on the early, innovative stages of the drug discovery pipeline and rely on CDD technologies for all of their chemical registration, SAR analyses, and to manage ongoing collaborations with academics, consultants, foundations, and companies.

BIODEFENSE COLLABORATIVE RESEARCH CASE STUDY

In many regards, the challenges with biodefense parallel the challenges that researchers face working on rare and neglected diseases. In both cases, a more economical model is mandatory, not simply desirable, given the critical need combined with the lack of a natural commercial market. Radiation reversal agents are feasible, and as the Fukushima Daiichi nuclear disaster in Japan recently demonstrated, their further development would be useful in the event of an emergency that is not caused by terrorists as well. The University of California, Los Angeles (UCLA) Center for Medical Countermeasures against Radiation (CMCR) discovers and develops novel small molecule drug leads that serve as radiation mitigators, in addition to advancing our understanding of radiation toxicity and lethality in humans. CDD is used in the UCLA Radiation Protection Screening Core as part of that effort. Within this core (Figure 19.1), phenotypic and target-based screening results are entered into the CDD database, linking a biological result to a chemical structure. The data are mined for trends such as SARs, substructure analysis, and other determining information that may be inside or outside the UCLA program. Thoroughly tested chemical entities (leads) are synthesized, and this is then further explored via proteomics, chemical informatics, and analog synthesis as

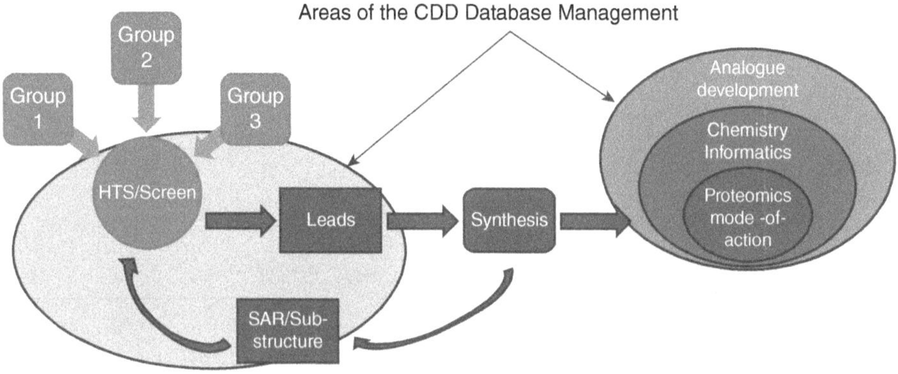

Figure 19.1. A schematic of the overall strategic approach of the UCLA-CMCR screening program, where CDD plays an integral part in the storage, management, processing, and social networking efforts.

a platform to support the management of the screening information as well as the discovery of mechanism of action. Multiple lead candidates have been tested in animals, and these data are all compiled back into the database. Over the years, the breadth of data associated with the compounds has become significant, and several successful lead molecules have been produced. The CDD platform is able to network the multiple research groups from entirely different locations and disciplines with the small molecules being the common language between them all. The database serves to distinguish what aspects of the small molecules analyzed are most effective or responsible for mitigating cellular damage or death from ionizing radiation. To date, over 15 research groups from different backgrounds are networked through the database, and over 100,000 compounds have been investigated for activity [35].

DISCUSSION

As the academic, commercial, and biodefense case studies showed, collaboration and collaborative technologies are target and therapeutic area "agnostic." Technical capabilities for data mining, data partitioning, and collaborative communication are equally applicable to almost any research area. In other words, the platform for bringing together scientists can be independent of the specific scientific details brought to and used within the collaborative platforms. Collaborative technologies generate better and faster results than do the old paradigms, where organizations worked separately using traditional databases. An interesting aspect of collaborative technologies is that their value accrues as a function of the number of users, not merely as a function of the number of researchers collaborating around a set of data. Another virtue of collaboration is that the high costs of technology development are spread across many researchers. Innovative functionality developed for one group benefits other groups that use

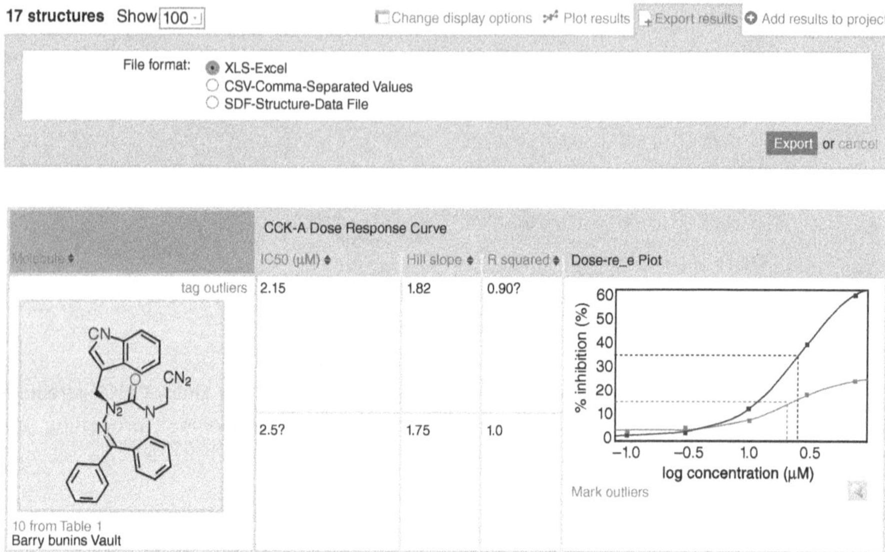

Figure 19.2. New collaborative capabilities for efficient drug discovery.

the collaborative platform. Substantial capabilities were needed in CDD to optimize collaborative drug discovery within a researcher's natural workflow, including intuitive scientific data handling capabilities [13] and innovative collaborative capabilities required for widespread adoption. Core drug discovery capabilities require a graphical user interface (GUI) that is much more intuitive than is found in the traditional chemical registration and SAR technologies [36]. An optimal technology for widespread adoption should be more like common, intuitive social networking technologies such as LinkedIn or Facebook showing the molecule structure, the numbers, and the curves (Figure 19.2).

When people think of collaboration, it is sometimes thought of as only for 100% open collaborations that do not involve any IP ownership, but that is a misconception. Collaboration can involve IP. Collaboration can be between a single chemist and a biologist. Collaboration can involve prepatent and prepublication, data. Whoever owns data now has the power to decide if, when, and with whom to collaborate. Collaboration even occurs within a single group: any communication from a biologist to a chemist, from a professor to a student, from a boss to a subordinate, is a form of collaboration. It is especially important to have collaborative data sharing before a student, postdoctoral fellow, or employee leaves a group, so others benefit from the detailed data he or she has generated. Beyond being just possible, the technical solution needs to be easy and convenient, given the natural workflow of already busy researchers. For example, academic screening centers often wish to batch import chemical library structures and master plates once, and then need to share the molecules and child plates with multiple

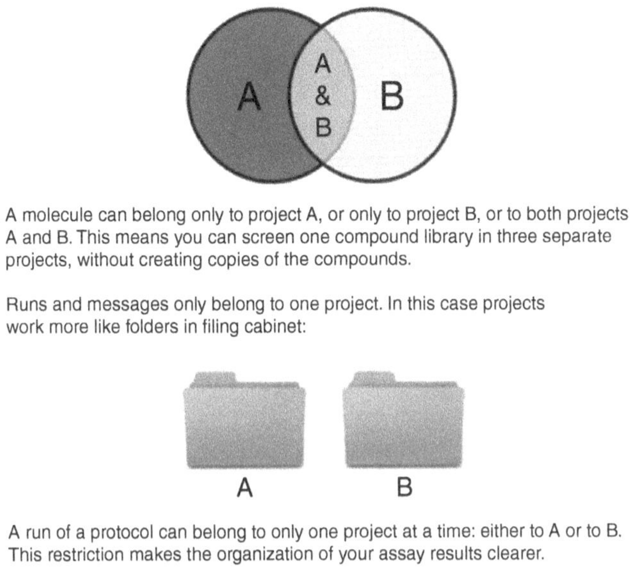

Figure 19.3. Data can be shared between projects or kept entirely separate in CDD.

collaborators; they want to do this without having to upload the data multiple times. Similarly, if a biologist wants to double check or confirm assay results before sharing them with a chemist, it is most natural for that biologist to simply start with the data in a private project and then move it to another collaborative project when appropriate. Moreover, researchers may tag data for one or more collaborative, private, or public projects. Use cases for individuals and organizations include Temporal Control (1 month for my eyes only and then I want to share the results with others), Spatial Control (I want project 1 for internal-only access, and I want to share project 2 with a specific CRO), and by Data Type Control (I want to just share this IC_{50} or a specific batch of a molecule) (Figure 19.3 and Figure 19.4).

There are very good reasons for collaborating via a secure collaborative platform hosted in the cloud. Data standards for chemical and biological data are embraced, there is no hardware required, no software to download, and collaborative platforms are accessible from any PC, MAC, or Linux computer with Internet access. Nonetheless, we realize there are other software platforms and databases that researchers will continue to use for legacy reasons or personal choice. We also have a glimpse of a mechanism to extend the collaborative capabilities beyond data sharing to any researchers using any database, whether they wish to upload molecules or biodata to the cloud or not. Building upon existing selective and secure chemical registration and SAR data-sharing capabilities is a new research initiative for selective and secure computational model sharing. Initial published results by CDD with Pfizer [37] have demonstrated that free, open source descriptors and models are comparable. While our initial efforts

User Roles and Privileges in CDD

Privilege/Role		Read Only	Read Down load	Read Export	Read Add - Biology	Read Add	Full Access - Biology	Full Access	Vault Admin
View and search data		✓	✓	✓	✓	✓	✓	✓	✓
Post and reply to messages		✓	✓	✓	✓	✓	✓	✓	✓
Download documents			✓		✓	✓	✓	✓	✓
Export search reasults				✓	✓	✓	✓	✓	✓
Protocols	Create				✓	✓	✓	✓	✓
	Edit others'						✓	✓	✓
	Add data				✓	✓	✓	✓	✓
Plates / References	Create				✓	✓	✓	✓	✓
	Edit others'						✓	✓	✓
Molecules / Batches	Create					✓		✓	✓
	Edit others'							✓	✓
Data / Publication	Request				✓	✓	✓	✓	✓
	Approve								✓
Edit vault settings									✓

Figure 19.4. Different per project user privileges such as read and write, read-add only, structure masking, and read-only in CDD.

at CDD have been focused on ADME/Toxicology and tuberculosis quantitative structure–activity relationship (QSAR) models, the approach is potentially extensible to any SAR/QSAR data sets. A reasonable argument can be made that a big pharmaceutical company can only gain from this approach, as it could profile external potential drug candidates for partnering. If the predictions look promising, what better place to test the candidate than the facility with the data upon which the model was built? We hope companies will join us to demonstrate how distributed data sets and corresponding computational models can accelerate collaborative drug discovery and improve their predictions and coverage of chemical space. We may also see such approaches result in an enhanced usage of computational ADME/Tox models, decreasing *in vitro* testing and perhaps limiting it to confirmation of predictions.

In summary, what is implemented in software guides common business rules and creates efficiencies. The value of a collaborative platform accrues via aggregated features, data, and users. This is as true for collaborative drug discovery platforms like CDD as it is for Facebook: the underlying trends and drivers are the same. These new collaborative drug discovery technologies have lowered the barriers for data archiving and intuitive collaboration. Those who wish to survive as pharmaceutical companies will need to be efficient in drug discovery; they will need to focus on the area where they are best in the world. This will involve identifying partners with complementary world-class capabilities and finding ways to effectively collaborate around the globe as if researchers were around the corner, down the hall, or in the same room. The era of collaborative drug discovery is under way and, as our case studies show, there is potential for it to have a broad impact.

ACKNOWLEDGMENTS

We wish to acknowledge our colleagues for assisting in developing CDD, as well as the many research collaborators and those whose case studies are included here, including Dr. Ellen Berg, Dr. Kelly Chibale, Dr. Chris Lipinski, Dr. Peter Smith, Prof. Phillip Rosenthal, Prof. James McKerrow, Dr. Malcolm Kendall, Dr. Andrew Norris Dr. Joel Freundlich, Dr. Gyanu Lamichhane, Dr. Rishi Gupta, Dr. Eric Gifford, Dr. Ted Liston and Dr. Chris Waller, as well as Dr. Deborah Bunin for proofreading the chapter. We gratefully acknowledge The Bill and Melinda Gates Foundation Grant#49852 "Collaborative Drug Discovery for TB through a Novel Database of SAR Data Optimized to Promote Data Archiving and Sharing." Some of our work described herein are also supported by Award Number R41AI088893 from the National Institute of Allergy and Infectious Diseases. The content is solely the responsibility of the authors and does not necessarily represent the official views of the National Institute of Allergy and Infectious Diseases or the National Institutes of Health.

REFERENCES

1. Munos, B. Lessons from 60 years of pharmaceutical innovation. *Nat Rev Drug Discov.* 2009;8(12):959–968.
2. Paul, S.M., et al. How to improve R&D productivity: the pharmaceutical industry's grand challenge. *Nat Rev Drug Discov.* 2010;9(3):203–214.
3. Larsen, P.O., von Ins, M. The rate of growth in scientific publication and the decline in coverage provided by Science Citation Index. *Scientometrics* 2010;84(3):575–603.
4. Moore, G.E. Moore's law. Available at http://en.wikipedia.org/wiki/Moore%27s_law. Accessed December 24, 2013.
5. Metcalfe, R. Metcalfe's law. Available at http://en.wikipedia.org/wiki/Metcalfe%27s_law. Accessed December 24, 2013.
6. Foster, B. Number of Facebook users. Available at http://www.benphoster.com/facebook-user-growth-chart-2004-2010/. Accessed December 24, 2013.
7. Collaborative Drug Discovery. Available at http://www.collaborativedrug.com/. Accessed December 24, 2013.
8. Ekins, S., et al. Pioneering use of the cloud for development of the collaborative drug discovery (CDD) database. In: Ekins, S., et al., editors. *Collaborative Computational Technologies for Biomedical Research*, Vol. 335–361. Wiley and Sons; 2011.
9. Ekins, S., et al. A collaborative database and computational models for tuberculosis drug discovery. *Mol Biosyst.* 2010;6:840–851.
10. Ekins, S., Williams, A.J. Meta-analysis of molecular property patterns and filtering of public datasets of antimalarial "hits" and drugs. *MedChemComm* 2010;1:325–330.

11. Ekins, S., Williams, A.J. When pharmaceutical companies publish large datasets: an abundance of riches or fool's gold? *Drug Discov Today* 2010;15:812–815.
12. Ekins, S., et al. Analysis and hit filtering of a very large library of compounds screened against *Mycobacterium tuberculosis. Mol Biosyst.* 2010;6:2316–2324.
13. Hohman, M., et al. Novel web-based tools combining chemistry informatics, biology and social networks for drug discovery. *Drug Discov Today* 2009;14:261–270.
14. Ekins, S., et al., editors. *Collaborative Computational Technologies for Biomedical Research*, Vol. 10 of Wiley Series on Technologies for the Pharmaceutical Industry. John Wiley & Sons; 2011.
15. Williams, A.J., et al. Current and future challenges for the collaborative computational technologies for the life sciences. In: Ekins, S., et al., editors. *Collaborative Computational Technologies for Biomedical Research.* Wiley and Sons; 2011, pp. 491–517. See also http://bioassayontology.org/. Accessed December 24, 2013.
16. Sneader, W. *Drug Discovery: A History*. John Wiley & Sons; 2005.
17. Bunin, B.A., Ekins, S. Alternative business models for drug discovery. *Drug Discov Today* 2011;16:643–645.
18. PubChem. Available at http://pubchem.ncbi.nlm.nih.gov/.
19. ChemSpider: a free chemical database. Available at www.chemspider.com. Accessed December 24, 2013.
20. Williams, A.J. Internet-based tools for communication and collaboration in chemistry. *Drug Discov Today* 2008;13(11–12):502–506.
21. HEOS®. Available at http://www.scynexis.com/heos-collaboration-software-suite/product/. Accessed December 24, 2013.
22. Bost, F., et al. Informatics for neglected diseases collaborations. *Curr Opin Drug Discov Dev.* 2010;13(3):286–296.
23. Ekins, S., et al. In silico repositioning of approved drugs for rare and neglected diseases. *Drug Discov Today* 2011;16:298–310.
24. Ekins, S., Williams, A.J. Finding promiscuous old drugs for new uses. *Pharm Res.* 2011;28:1786–1791.
25. Ekins, S., Freundlich, J.S. Validating new tuberculosis computational models with public whole cell screening aerobic activity datasets. *Pharm Res.* 2011;28:1859–1869.
26. Sarker, M., et al. Identification of novel molecules for tuberculosis combining cheminformatics, diverse databases and logic-based pathway analysis. Submitted, 2011.
27. Lamichhane, G., et al. Essential metabolites of *M. tuberculosis* and their mimics. *MBio* 2011;2:e301–e310.
28. More Medicines for Tuberculosis. Available at http://www.mm4tb.org/. Accessed December 24, 2013.
29. Makarov, V., et al. Benzothiazinones kill *Mycobacterium tuberculosis* by blocking arabinan synthesis. *Science.* 2009;324(5928):801–804.
30. Accelrys. Available at http://www.accelrys.com. Accessed December 24, 2013.
31. ChemAxon. Available at http://www.chemaxon.com/. Accessed December 24, 2013.
32. Acetylon. Available at http://acetylon.com/. Accessed December 24, 2013.
33. Semaphore Pharmaceuticals. Available at http://www.semaforepharma.com/index.html. Accessed December 24, 2013.

34. Houck, K.A., et al. Profiling bioactivity of the ToxCast chemical library using BioMAP primary human cell systems. *J Biomol Screen*. 2009;14(9):1054–1066.
35. Collaborative Drug Discovery at the UCLA Center for Medical Countermeasures against Radiation (CMCR). Available at https://www.collaborativedrug.com/buzz/2011/10/05/cdd-implementation-at-the-ucla-cmcr/. Accessed December 24, 2013.
36. Bunin, B.A., et al. *Cheminformatics: Theory, Practice, and Products*. Springer-Verlag; 2006.
37. Gupta, R.R., et al. Using open source computational tools for predicting human metabolic stability and additional ADME/TOX properties. *Drug Metab Dispos*. 2010;38:2083–2090.

20

ACCESS PLATFORM: A STREAMLINED INTEGRATIVE PARTNERING PROCESS AT SANOFI TO COMMERCIALIZE UNIVERSITY-BASED INTELLECTUAL PROPERTY

Paul R. Eynott[1] and Carole Fages[2]

[1]*Tech Launch Arizona, Dean's Office, College of Science Administration, University of Arizona, AZ, USA*

[2]*Early to Candidate Unit, SANOFI Group, Montpellier, France*

Innovation has become a widely used concept over the last few decades. It is frequently associated with enhanced opportunity for success and scientific excellence. The current rules-based, multilateral trading system presents innovation as a global approach to increase productivity and enhance perceived value. It is often stated that encouraging innovation is key to economic growth and sustainable development [1]. Taking this perspective into account, the present chapter outlines the importance of surrounding the concept of innovation and how organizations aim to leverage this for the pharmaceutical industry. The objective of this chapter is to highlight and present the importance of collaborative, "integrated" drug discovery and outline some of the approaches to academia, industry, and/or government partnerships that have been implemented. In particular, this chapter will look at the strengths and weaknesses of existing partnering models and in this context introduce The Access Platform, an innovative partnering model developed by SANOFI.

Collaborative Innovation in Drug Discovery: Strategies for Public and Private Partnerships, First Edition. Edited by Rathnam Chaguturu.

This chapter provides an in-depth description of collaborative platforms that exist within the pharmaceutical industry and, in particular, will focus on the development of new drugs, bringing together first-class public labs and major pharmaceutical industry players like SANOFI through its Early-to-Candidate (E2C) unit. Starting from targets, novel phenotypic assays, or tool compounds from internal activities and/or academic laboratory sources, the pipeline leads to the identification and evaluation of novel tool compounds and/or biologics in the E2C Access Platform. The specific objectives of E2C within SANOFI are to capture external innovation as early as possible, and to diversify its current scientific activities to deliver first-in-class, highly innovative compounds (mostly development candidates). To achieve these goals, E2C needs to develop and extend its collaborative activities. These dramatic steps are taken to address the industry's core underperformance. A decade of declining research and development (R&D) returns has resulted in a spur to move or to externalize research, with deep cuts in in-house early-stage research [2]. The Access Platform model will be discussed in detail, as our analysis suggests this model could create material value for shareholders through higher earnings and multiples.

INTRODUCTION TO THE EARLY-TO-CANDIDATE (E2C) UNIT WITH THE SANOFI GROUP

SANOFI, one of the world's largest pharmaceutical groups, is a global leader in healthcare focused on patients' needs. SANOFI has 110,000 employees in 100 countries [3]. Between 2009 and 2011, the company evolved through the acquisition of 23 companies, including Genzyme, negotiation of ~61 in-licensing agreements and two joint ventures, and investment of a total of ~€23 billion in external growth. In 2009, SANOFI pursued 33 acquisitions and partnerships, and 37 acquisitions and partnerships in 2010 [4]. SANOFI is a diversified health-care company that collaborates with external partners to improve health, enhance life, and respond to the future health-care needs of the people around the world. The SANOFI group demonstrates leadership both in business achievements and in communities or local innovation ecosystems, and incorporates SANOFI R&D as an instrumental part of the organization. E2C is an entrepreneurial unit within SANOFI R&D that is dedicated to transforming scientific innovations into therapeutic solutions for patients. This unit is entrepreneurial in nature, being charged with in-licensing and out-licensing discovery at early stages in SANOFI R&D across the many therapeutic areas and various approaches, including devices, drugs, and biologics. The E2C unit will contribute to the global strategy of the SANOFI group, which is to (1) increase innovation in R&D, (2) seize external growth opportunities, and (3) adapt the group to future challenges and opportunities in the health-care field. In an effort to achieve these goals, a new R&D model was created to emphasize simplicity, openness, and partnerships. The E2C mission is to adopt "validated concepts" and "early-stage leads" from both within the company and externally through academia, other biotechs, or other divisions within SANOFI R&D that have core competencies that will enable the E2C unit to more rapidly convert them into well-defined tools with established therapeutic orientations. In particular, E2C brings in candidates for clinical proof-of-concept in collaboration with other SANOFI R&D divisions/units or external partners. The E2C unit aims to complement and support innovative internal research through developing external partnerships. Successful innovation depends on the development and integration of new knowledge in the innovation process. To innovate successfully, E2C combines different innovation activities. In addition to doing its own research, E2C is engaged in the acquisition of knowledge and will cooperate actively in R&D with biotechs and research organizations. The strategy combines different ways of partnering to maximize the capture of innovation. Because partnering is dependent on different parameters such as maturation, new ways of capturing external innovation are engaged in addition to typically sponsored research.

Biomedical Innovation Ecosystems

Innovation in the development of therapeutics is critical to the survival of the pharmaceutical industry innovation engine. From the initial serendipitous discovery of penicillin inhibition of bacterial growth, to the discovery that extracts from the Pacific Yew tree can kill cancer cells, innovation has been central to the growth and sustainability of the pharmaceutical industry [5]. It follows that innovation in drug development is a fundamental requirement for a company to establish and maintain a competitive advantage in an increasingly complex world [6]. Pharmaceutical innovation is an ongoing process with numerous features. The scope of innovation generally ranges from developing, manufacturing, and marketing new drugs, about which relatively little is known at the time of their discovery, to enhancing existing drugs that have been on the market for some time by making minor chemical modifications to them. Between these activities, manufacturers also try to find new methods to increase the safety, effectiveness, and convenience of their products. Given the relationship between R&D, industrial innovation, and growth, the analysis of the pharmaceutical industry undertaken here demonstrates an extremely important part of development economics. Cooperation across government, universities, and industry often combines to provide a high level of expertise, which accelerates the process of bringing new products to the market [7]. Often, colocation of academia, industry, and government organizations results in a "localized innovation ecosystem." Examples of biomedical innovation ecosystems can be found, in the United States in Boston, San Diego, and San Francisco [8]. University research has formed the foundation for a majority of the significant technological and medical advancements in this area. Despite this fact, it is well known that many additional innovative concepts and biological assays developed by academic research remain in the laboratories of academia and are waiting to be disclosed and leveraged by industrial partners for drug discovery purposes [9]. On the contrary, the pharmaceutical industry needs to be more open in what has been tried and tested in early discovery to provide a perspective to academia and help avoid duplicating efforts. In this context, there is a need to build a strong bridge to close the transfer gap between basic "academic" research and market exploitation.

Innovation versus Invention and the Pharmaceutical Conundrum

Ensuring the successful development and launch of new drugs is considered to be an enormous challenge and expense for the pharmaceutical industry [2, 9]. Therefore, collaboration between key external stakeholders is now deemed essential to innovate for the future. High costs of discovery coupled with the lengthy development process constitute significant obstacles for innovation in the pharmaceutical sector. According to the Pharmaceutical Research and Manufacturers of America, the average time required to bring a new chemical entity or new biological entity to market is estimated to be about 10–15 years in the United States. The proportion of substances that are successful and result in marketable products is 1 out of 10,000 [10]. In addition, only 3 out of every 10 successful drugs bring in enough revenues to meet or exceed average R&D costs [9, 11]. Further to this, an economic analysis of the industry reveals that apart from the high cost of R&D activities, it is also necessary for companies to control other important complementary assets. The key complementary assets required for the innovation process to succeed include possessing the competencies developed in the management of large-scale clinical trials, obtaining regulatory approval, and managing the costs of marketing and distribution. Thus, it can be said that the costs of the development of a drug also depend on these factors.

It is widely recognized that innovation is central to the progression of society because the social returns from innovation far exceed the private investments. In other words, the benefits of innovation to society as a whole greatly exceed the benefits to the firms that develop the innovative processes and products. Innovation can be simply defined as making changes to what already exists. Innovation can, therefore, be considered an interactive process in which later steps in the process are linked back to the earlier ones [12]. It encompasses a new idea brought to the market or into the production strategy. Successful innovation makes a significant contribution to economic growth [13]. To a large extent, the possession of technological knowledge regarding potential production arrangements in conjunction with knowledge of the technology in use constitutes the given state of the arts. In this context, invention is described as producing technological changes in the knowledge available. However, innovation is defined as the application of existing knowledge to changes in the actual technological arrangements. Today, it is assumed that the individual firm invents and innovates as a normal part of business activity [14].

In alignment with this revolution in the U.S. research landscape, and as part of the E2C Access Platform initiative, we propose to develop integrated partnerships with the premier drug discovery centers in an effort to rapidly augment SANOFI's discovery pipeline. The plethora of projects ongoing at the discovery centers will facilitate the selection of the most appropriate and best opportunities that fit with the objectives of SANOFI R&D. In addition, the discovery centers can provide access to the expertise and knowledge in major therapeutic areas of interest to avoid redundant parallel research at SANOFI. The principal investigators associated with each project could provide SANOFI with a network of established, recognized, scientific experts to facilitate the joint discovery efforts and increase the probability of project success. This type of partnership has the ability to not only improve the productivity of SANOFI R&D but also increase the scientific respect and credibility with both scientific professionals and the investor community. The E2C unit has spent a year exploring this strategy by directly engaging multiple drug discovery centers in the United States and discussing potential partnering models. We believe that we have identified a partnering approach that is mutually beneficial to SANOFI and the discovery center partners, and will deliver significant value for SANOFI R&D with essentially no upfront financial investment.

WHY DOES COLLABORATION IN DRUG DISCOVERY MATTER?

The drug discovery process is incredibly complex and requires insights from investigators with many different backgrounds. Investigators must first develop a deep understanding of disease biology and the pathways involved. As the biology of disease is elucidated, protein targets can be identified. Once targets are validated, a wide variety of techniques are used to identify molecules that can modulate their activity. These techniques include crystallization, bioinformatics, genomics, proteomics, assay development, high-throughput screening, lead optimization, cellular and molecular pharmacology, cellular disease models, pharmacokinetics, and toxicology [15]. Due to the interdisciplinary nature of the drug discovery process, effective collaboration among investigators is essential. Furthermore, to make the drug discovery process more efficient, organizations such as large pharmaceutical companies should seek input from external investigators. At SANOFI, the Access Platform will serve as an organized model to access this valuable external knowledge.

Collaboration is defined as two or more people or organizations working together to achieve a common goal. In general, the primary goal of an academic researcher is to conduct basic research and publish findings. In contrast, the primary goal of the pharmaceutical industry is to translate basic research into a commercial product that treats an unmet medical need. This fundamental difference in core competencies and mission has in many instances hindered collaboration. More recently, the realization for academia, industry, and government health organizations is that the overall mission is essentially the same: improve the quality of life for people around the world through scientific innovation and discovery. This realization has the potential to pressure change from the position of investment bankers, patient advocacy groups, and, most importantly, shareholders. Few people from government, academia, or advocacy groups will argue that the pharmaceutical industry creates value but instead hyperinflates derivatives without a clear basis in reality. It is evident that the various players in the biomedical ecosystem must make an effort to understand each other, identify shared goals, and work together. It is collaboration among the important players in the biomedical ecosystem that will reduce global R&D costs and stimulate true innovation in the discovery of new therapeutics.

Why Are Public–Private Partnerships Critical to Drug Discovery?

It is indisputable that drug discovery is a time-consuming, risky, and expensive process with high potential for failure at multiple points. Pharmaceutical industry R&D

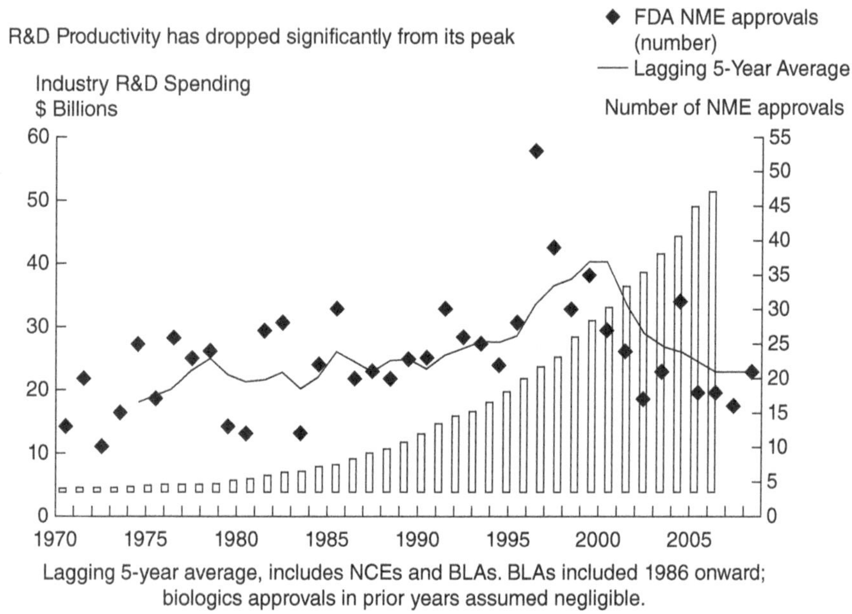

Figure 20.1. NMEs per year versus total R&D spent in billions of 2008 U.S. dollars (prior to 1984 first new chemical entity, or NCE, approvals). The graph shows the number of NMEs per year was trending upward to 1996, but in recent years, the number appears to be going down. Increasing the number of NMEs per year would be beneficial, but dramatic increases in R&D spending have not improved NME statistics. The increasing distance between NMEs and pharmaceutical R&D spending in is often referred to as the "Innovation Gap."

expenditures have been steadily increasing since the 1970s (see Figure 20.1) [16, 17]. Alarmingly few new molecular entities (NMEs) were approved by the FDA in the past decade [18–20]. In 2010, the pharmaceutical industry spent $50 billion on drug development, with an additional $31.2 billion spent by the National Institutes of Health (NIH) on biomedical research [5]. Despite this enormous investment, only 21 applications for NMEs were approved by the FDA, which is less than half of the peak level in 1997 [19]. Although there are many factors responsible for the low approval rate, with increased risk and complexity of new targets as one aspect, it is disturbingly clear that R&D productivity in the pharmaceutical industry is declining.

In the private sector, the decline in research productivity can be attributed to the conservative portfolio management approaches adopted by large pharmaceutical companies in the 1990s. In an attempt to mitigate risk, pharmaceutical companies chose to focus on late-stage research on validated targets, proven modes of action, and well-known drug classes [5]. This strategy inhibits innovation and results in the development of drugs that are marginally better than or not as effective as drugs already on the market. In 2009, less than 7% of industry sales could be attributed to drugs launched within the past 5 years [21]. Therefore, it is evident that industry has suffered a deficit

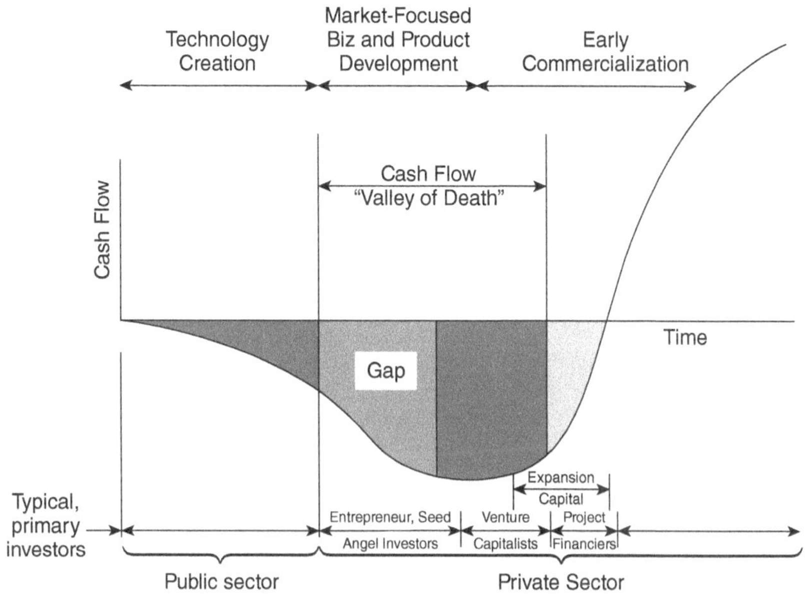

Figure 20.2. The Cash Flow Valley of Death as a function of development stage (time), with typical investors shown for the various stages. The transition from research and preclinical development to the clinic is so perilous that it is frequently referred to as the "valley of death."

in both productivity and innovation. In the public sector, the deficit in R&D productivity is due to the lack of resources required for academic entrepreneurs to advance their technologies to a point where they can be licensed or used to form start-up companies. Since many entrepreneurs are often unable to find the amount of capital needed to effectively commercialize their technologies, their projects are often trapped in a cash flow "valley of death" (see Figure 20.2) [22]. There are four critical resources needed to execute successful commercialization efforts: (1) capital, (2) scientific or technical expertise and qualified intellectual property (IP), (3) business expertise, and (4) facilities. There is interdependency between these resources, whereby not having one can make it more difficult to obtain the others. Nevertheless, an entrepreneur contemplating an early-stage technology business must obtain access to all of these resources to achieve success.

In an effort to address the decline in research productivity and innovation, government health organizations, academic institutions, and pharmaceutical companies have begun to form public–private partnerships [7]. These partnerships will allow the public and private sectors to share the costs and risk of translational research. The public (industry) partners benefit by gaining access to knowledge and novel ideas possessed by academic researchers. In turn, the researchers will gain access to the funding and/or business expertise necessary to turn their ideas into innovative technologies. Essentially, public–private partnerships are an effective way to facilitate the discovery and

TABLE 20.1. Key Ingredients for Evaluating Technology

Value proposition
Problem–solution
Barriers to entry (market penetration)
Regulatory pathway
Defendable IP
Cost of acquiring a license

the commercialization of new technologies. Public–private partnerships can take on many forms, which will be discussed later in the chapter.

BRIDGING THE VALLEY OF DEATH: EXAMPLES OF CURRENT MODELS

As referenced earlier, there has been a lack of focus (and resources) on several of the early steps that must be taken to transform a novel technology into a marketable product or service. There appears to be an assumption that once a discovery is made, it will become a product or service in the marketplace through some natural evolution. Unfortunately, this scenario is becoming more difficult. Two critical steps required for successful technology commercialization include proper evaluation of the technology (see Table 20.1) and creating an integrated ecosystem that supports the ongoing development of the technology. Relative to the resources devoted to building and supporting research infrastructure, the resources dedicated to commercialization efforts are sorely lacking. Resources include funding and the necessary business and technical advice required to move research discoveries along the commercialization continuum. Given this imbalance, it is not surprising that the number and pace of novel innovations and new commercial enterprises in the United States are frustratingly low [22]. Additionally, the process to acquire the necessary resources is time-consuming, which contributes to the low outcomes as great ideas fall away due to the burden of translation.

The need for science and innovation to drive economic growth in the United States has never been greater. The United States has invested hundreds of billions of dollars in research and research infrastructure over the last 25 years. In fact, the NIH provided over $31 billion in funding for the 2010 fiscal year [23]. These investments have resulted in many advances that have benefited society; however, there is concern that the size of investment is not yielding new therapies to the extent proposed. Despite this concern, the NIH budget rose $1.0 billion (3.2%) to $32.2 billion in FY 2011 from the FY 2010 budget. New Director Francis Collins named five priority areas: high-throughput technologies, translational medicine, informing health-care reform, global health, and reinvigorating the biomedical research community. Transformation of various research discoveries into new products and services is not as productive as it could be, and the public's concern is being articulated in various media. For example, a recent *Newsweek* article focusing on the large public investment in research stated that, “More and more policymakers and patients are asking, ‘Where are the cures?’ ” The answer is that potential cures, or at least treatments, are stuck in the chasm between a scientific discovery and the doctor’s office known as the valley of death. Clearly, a focus on innovative commercialization efforts would help address this concern [24]. To this end, HR3590, the Patient Protection and Affordable Care Act, authorizes the creation of the Cures Acceleration Network, which focuses on accelerating the development of high-need cures, including the development of medical products and behavioral therapies. Other efforts are also being made to provide resources to support economic development through the creation of more small businesses across the United States.

Currently, there are many models in the public and private sector that address the cash flow valley of death faced by entrepreneurs. In addition to traditional entrepreneurial funding such as venture capital firms and angel investors, there are proof of concept centers and specialized federal, academic, and pharmaceutical industry programs focused on developing early-stage technologies.

Venture Capital and Angel Investments: The first “valley” presenting a challenge to commercializing a technology can be found at the early-stage transition point between discovery and development. Access to early-stage capital to finance pre-seed and seed stages has changed significantly over the past 10 years. “Changed significantly” means that it is increasingly difficult to raise the \$100K–\$2M necessary to advance beyond the pre-seed stage. Only 4% of venture capital investment went into early-stage companies last year. In fact, after the 2008 financial crisis, it became much more difficult to even raise the pre-seed funds (\$25K–\$100K) necessary to get beyond proof of concept. Technologies with uncertain commercial potential represent significant risks for investors. Therefore, to fund innovation development through this technical risk phase, sources of early-stage capital typically include friends, family, and angel investors. In this scenario, demand far outweighs supply, and demand is further augmented by the fact that funding biotechs is much more expensive than developing software. Furthermore, in addition to the shifting support of venture capital funding to later stage investments, early data are showing that angel investors may also be shifting their investment strategy. Angel investment groups, and some individual angels, are now co-investing in larger deals or later stage companies. According to the Angel Capital Association, in 2008, the majority of angel groups or networks looked to invest between \$250K–\$500K, with \$25K–\$250K running a close second [25]. Now, 3 years later, rather than one angel group putting \$250K in an early-stage company and running the risk that the company would not be able to raise additional funding, angel groups are now co-investing with one another or with venture funds so that the total is closer to \$1M or \$2M. This early-stage funding gap is further expanding the size of the valley of death and thus, new solutions are required to bridge the divide. The longer this early-stage funding gap continues without a workable solution, the greater the chance that institutional venture funds and growth capital investors will see a widening gap in the number of companies that reach their respective stage of investment [26].

Federal Programs: The federal government has implemented various initiatives and programs in an attempt to address some of the valley of death concerns. Over the past 25 years the federal government has promoted the Small Business Innovation Research (SBIR) and the Small Business Technology Transfer (STTR) programs. The programs' legislative objectives are to stimulate technological innovation and use small businesses to meet federal research and development needs. According to a recent report, the NIH has had some success in focusing its SBIR/STTR program on commercialization, but room for improvement remains [27]. The NIH is beginning to recognize that a greater emphasis at the pre-SBIR (proof-of-concept) stage could significantly increase the effectiveness of the SBIR program and assist in furthering important technology. For example, many applicants do poorly or fail because of prematurely formed companies (necessary to qualify for the program), insufficiently developed technology, and an inadequate assessment of commercial feasibility. Currently, there are few federal programs or mechanisms to address these issues.

Proof of Concept Centers: Early stress testing of a new technology's commercial feasibility is critical for the development of a successful product. Early indications show that Proof of Concept Centers (POCCs) can serve a very effective role in facilitating technology commercialization and economic development. According to the Kauffman Foundation, "Proof of Concept Centers are a new model of support at some universities that provide seed funding and expert assistance to help entrepreneurs prepare for the strongest market entry possible." Recognizing that the POCCs are an effective method for launching the commercialization of academic innovation and to fill the seed-stage funding gap for new technologies, the Kauffman Foundation is facilitating networking among these centers, studying best practices, and establishing metrics for measuring outcomes. Kauffman has reviewed two examples of POCCs, the William J. von Liebig Center at the University of California, San Diego, founded in 2001, and the Deshpande Center at MIT, founded in 2002. Results thus far suggest that the POCC is a good model. By early 2008, the two centers combined had provided nearly $10M in grants, producing 26 spinout companies that raised an additional $159M in private investment. As the Foundation states, "the process is useful even when it demonstrates that a research idea will not be viable. The researcher can move on quickly to other work, better informed about what could help make the next idea a winner" [28]. Failing quickly is a good outcome in the case of technology commercialization so that limited resources are not wasted.

The number of states forming and supporting bioscience POCCs has increased from 23 to 33 since 2006 [29]. This increase is due in large part to the broader realization of the effectiveness of these Centers in technology commercialization and the value they can bring to the state. However, technology "deal flow" is also an essential element to success; creating POCCs at every research institute is therefore not an effective solution.

Academic Programs: Most academic institutions in the United States have some form of "technology transfer office," which is responsible for generating IP around technologies developed on campus. These offices generally focus on licensing the technology rather than commercializing it through the creation of a new entity. Therefore, there are often few resources available for academic entrepreneurs seeking funding and commercialization advice. However, some universities are making an effort to do more to address technology commercialization issues faced by entrepreneurs. Programs such as the Biomedical Accelerator Fund at Harvard University provide funding for academics pursuing translational research. The Bioscience Discovery Evaluation Grant Program at the University of Colorado provides technology commercialization resources to students, faculty, and research staff who want to form start-ups.

Industry Programs: In addition to traditional sponsor research agreements or grant programs, large pharmaceutical companies such as Lilly, Pfizer, GSK (and others) have implemented new public–private partnership models that can be divided according to their organizational structure into either centralized or virtual organizations. For example, a central organization has a lean organization at its core. Pfizer has recently announced the creation, together with the University of California, San Francisco, of the first Global Centers for Therapeutic Innovation (CTI) focused on the development of biotherapeutics. The CTI will be staffed with 20 or so Pfizer employees, with each project being led by local academic researchers and CTI Pfizer employees. Pfizer plans to create several CTIs with major academic institutes around the world. The budget committed to each center will be around $90 million. Virtual organizations, on the other hand, include the model announced by GSK. This new initiative aims to create up to 10 relationships with individual researchers throughout the world, forming a virtual project team with each of them to provide immediate access to GSK resources for developing new therapeutics. Similarly, Lilly has launched the Open Innovation Drug Discovery Program, which includes the Lilly Phenotypic Drug Discovery Initiative (PD^2) and the Lilly Target Drug Discovery Initiative ($TargetD^2$) Platforms. This program allows academic researchers to submit tool compounds to be tested in Lilly's internal disease-relevant and target-based assay panels, respectfully, but neither approach provides opportunities for academic researchers to be acquainted to industrial needs. In Table 20.2, we have highlighted example models that are aimed at commercializing technologies through partnerships from within industry and outlined some of the pros and cons of the various models.

NEW STRATEGIES FOR IMPROVING R&D PRODUCTIVITY: THE ACCESS PLATFORM

Each year, investigators at universities throughout the world are developing innovative assays and finding tool compounds and biologics that are never fully evaluated as potential drug candidates. One of the challenges for academic investigators is to have

TABLE 20.2. Industry Models for Partnerships

Lilly—Open Innovation Drug Discovery Provides academic researchers access to their assays and computational tools and maintain first right to negotiate	**Pros:** Clear timeline and exit, IP belongs to investigator, requires minimal investment **Cons:** Not very proactive, minimal communication
Pfizer—Global Center for Therapeutic Innovation Laboratories on or near campuses staffed with Pfizer and academic researchers	**Pros:** Integrated process, monetary incentive for investigators, clear expectations surrounding IP, designated facility promotes open communication, local proximity, embedded scientists together **Cons:** Requires significant investment
Johnson & Johnson Development Corporation Venture Capital subsidiary that invests in emerging health-care businesses	**Pros:** Proactive, clearly defined focus areas, rapid review of technology, early partnerships possible **Cons:** Unclear timeline and expectations regarding exit/IP
Johnson & Johnson COSAT Team of researchers that identify and aid high-risk projects	
Abbott—Global Licensing and New Business Development Team	**Pros:** Promotes open communication **Cons:** Unclear timeline and IP expectations, not proactive in finding entrepreneurs
AstraZeneca—Innovative Medicine Units Teams focused on an area of study identify and evaluate	**Pros:** Promise to be prompt, unclear IP expectations **Cons:** No clear timeline, they prioritize late-stage projects
Novartis	**Pros:** Defined areas of interest, single database to organize opportunities **Cons:** Not proactive, no clear timeline
Amgen Extramural Research Alliances and Licensing Provides Amgen products and related materials to academic investigators at no cost	**Pros:** Actively seeks new projects, integrated and interdisciplinary team handles valuation and communicates with entrepreneur **Cons:** No clear timeline, focuses on preclinical and clinical projects
Merck Partnering Worldwide scouts seek out external opportunities. Projects are evaluated and collaborations established in a four-step partnership process	**Pros:** Flexible and clear process, definitive agreements drafted before collaborative work begins **Cons:** No clear timeline, IP expectations unclear
Roche—Expanding the Innovation Network Provides strategic questions, funding, and guidance for academic projects	**Pros:** Specific needs of the company are made clear, first right to negotiate for licensing or partnering **Cons:** IP expectations unclear
GSK—Open Innovation Team http://innovation.gsk.com/ Entrepreneurs can form virtual project teams with them	**Pros:** Relatively clear timeline, interdisciplinary team dedicated to communicating with entrepreneurs **Cons:** Not very proactive, very broad input expectations

ready access to know-how and expertise in drug discovery and development. Therefore, E2C aims to become part of the valorization process of academic institutes by providing access to its expertise (in assay development, screening, and hit identification/optimization) through a collaborative platform model known as the Access Platform. If promising, results can further serve as a basis for requesting options or license agreements. Equally important, the Access Platform model will create a 50/50 joint relationship where the goals of the academic investigators (or biotechs) and E2C are aligned and both sides are empowered to succeed.

A key part of this initiative is to expand the level of collaborative research with selected academic institutes to fuel innovative discovery in the E2C key domains while also providing a means for E2C to rapidly acquire and exploit new expertise in various scientific domains. At steady state, the E2C Access Platform is designed to deliver optimized tools and leads (small molecules and biologics).

University research has formed the foundation for a majority of the significant technological and medical advancements. Despite this fact, it is well known that many additional innovative ideas and assays, developed by academic research, remain in the laboratories of academia and are waiting to be disclosed and used by industrial partners (mostly big pharma) for drug discovery purposes. In this context, there is a need to build a strong bridge and close the transfer gap between basic "academic" research and market exploitation.

The strategic aim of the Access Platform is to create a collaborative platform for the development of new drugs, bringing together first-class public labs and a major pharma player like SANOFI through its E2C unit (see Figure 20.3). Starting from

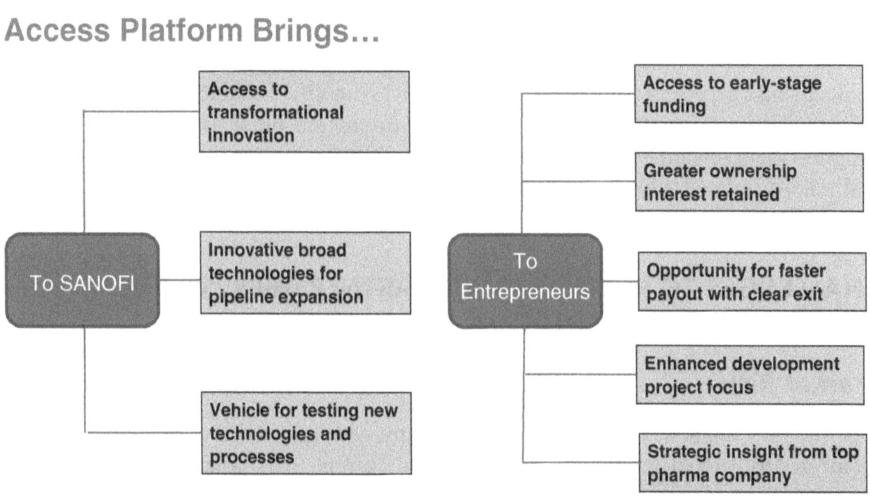

Figure 20.3. The schematic highlights the strategic aim of the Access Platform, which is to create a collaborative platform for first-class public laboratories to partner with a major pharma player like SANOFI to develop new drugs. This unique early partnering concept operates in a mutual shared-risk framework with benefits for all stakeholders, in particular embracing the key drivers of both industry and of entrepreneurs.

targets, novel phenotypic assays, or tool compounds from public lab sources, the pipeline leads to the identification and evaluation of novel tool compounds and/or biologics in the E2C Access Platform to accelerate discoveries in the several scientific domains. This Access Platform is a unique early partnering concept operating in a mutual shared-risk framework with several benefits for both partners:

- SANOFI is able to access early external innovation requiring maturation or de-risking steps (no additional costs except operational cost of the Access Platform will be required for the de-risking step).
- Academic partners are able to access know-how and skills in pharmaceutical drug discovery through a fully operational and integrated platform.
- A win–win situation is created: E2C is afforded the opportunity to enter into in-licensing agreements on "validated" projects, and external partners are able to valorize their project according to "pharmaceutical" standards.

The E2C Access Platform model creates an efficient, cost-effective mechanism to open innovation portals for SANOFI R&D. An adaptation of the current E2C Access Platform that exists in Europe and China is being launched to leverage efforts at academic drug discovery centers throughout the United States. This strategy permits access to mature discovery projects (assays, tools, leads) where significant investments and validation has already been accomplished. The Access Platform DC^2 partnering model (U.S. arm of the Access Platform) only relies upon the investment of internal SANOFI resources initially and postpones the commitment of any R&D funds until after key project milestones have been achieved, thereby decreasing the financial risk associated with external partnering. In addition, since the projects engage partnered resources within SANOFI, the quality will be higher and the investment will be lower than in-licensing the project from the discovery centers directly. Overall, the Access Platform DC^2 model increases the ability of SANOFI to capture innovative opportunities and progress them toward development candidates while decreasing the scientific and financial risk associated with early drug discovery.

CHALLENGES TO PUBLIC–PRIVATE PARTNERSHIPS: SEIZING THE OPPORTUNITY

Market Risk

Pipeline development comes from relationships with universities, economic development groups, global relationships, and, in the current marketplace, terminated R&D staff from competitors. The challenge to university relationships, as noted by Pertuzé et al., is that "There is an outcome-impact gap in university collaborations: Promising outcomes of university projects often fail to translate into tangible impacts for the companies involved" [30]. Universities are also tracking research dollars associated with professors, but whether this is a factor for businesses in evaluating longer term relationships is up for review. Technologies in the proof-of-concept stage have the lowest cost entry for outside companies and also the lowest cost exit. Many of the innovators at this point need assistance moving into the patent or testing phase, where the outside corporation can provide the highest level of service.

Intellectual Property

This section explores the economic rationale of patent rights and discusses how this rationale applies to the context of innovation. The familiar argument is that patents create incentives to innovate, yet little is known about how those incentives work in practice. Whether patents actually encourage and promote innovation is still an open question in the literature. Although the empirical evidence is to some extent ambiguous, it does indicate a positive correlation in favor of the pharmaceutical industry. Thus, this chapter describes and critically appraises a number of studies reporting on (1) innovation, (2) innovation and patents, and (3) the economic effects of patents and how efficiently they work creating optimal incentives. Furthermore, the present chapter details the research methodology undertaken to identify and assess the relevant economic and legal literature. Cooperation with universities and smaller organizations has traditionally been challenging, especially in the area of patent support and market development, which are the vulnerabilities of these two groups. In working with innovators, the period after "proof of concept" is particularly treacherous—the famous chasm or "valley of debt." In working with innovators in this stage of development, seeker can enter into relationships at the point of lowest cost and lowest risk. At this stage, patent and testing resources are critical to the growth of the innovators company. By working with innovators at this stage, a portfolio of innovations can be managed at a lower cost. The return on investment (ROI) expectations to the company are more easily managed with a 2% success rate over a larger group of technologies. Recent changes in the patent law open new opportunities for corporations to work with universities and smaller organizations. In the past, the patent applicant was required to demonstrate work on a project to justify a claim of being first to the marketplace. Under the new law, the patent applicant is required to be first to file. As universities and smaller companies do not have the resources to manage a large portfolio of technologies that are eligible for patent and the associated expense of developing these applications, they are seeking alternate strategies to manage the costs. These strategies include filing provisional patents while investigating the current market for the technology. This challenge represents an opportunity for larger organizations who seek to develop a portfolio of early-stage technologies.

Budget

Although the collaborative platform does not aim to finance the innovation process of the academic lab or the E2C Access Platform, but rather valorize academic innovation through E2C drug discovery expertise, E2C is considering several scenarios for additional income to facilitate project transfer between partners or/and to cofinance different projects. Potential sources of additional financing include the NIH, government, and private equity funding (copartner with venture capitalists, angel investors, or biotechs such as Audion Therapeutics for selected projects). It is noteworthy to mention the possibility of valorizing the output of Access Platform projects outside SANOFI (when not entering the SANOFI pipeline) through out-licensing, spin-off, or joint venture (opportunities to create additional value). With this unique early partnering model, E2C aims to accelerate innovation originating from academic projects and translate them into therapeutic applications to the benefits of patients all over the world. Through its positioning at SANOFI R&D and its scientific expertise, E2C is the best partner to make the Access Platform a reality. The role of the university, small company, capital organization, government agency, and major corporation will change, and each will play a role in bringing new technologies and solutions forward. Technology, like money and water, will seek its own level and most efficient path to the marketplace. To be successful, we need to become agile enough react to these changes and our roles in this process.

Legal Environment

The local pharmaceutical industries in developing countries owe their beginnings to the absence of patent protection for pharmaceutical products within these countries. Increasing health costs as well as unfair trade practices, coupled with the market disturbances of multinational pharmaceutical companies, led to the abolishment or nonintroduction of patent protection for any type of pharmaceutical or chemical product or process in developing countries. Over the years, this trend spread and became common practice in many developing countries. Nevertheless, the Agreement on Trade-Related Aspects of Intellectual Property Rights (TRIPS) changed the global landscape with respect to patent protection and enforcement. More importantly, TRIPS granted patent protection to any invention, whether it was a product or process, in all fields of the technology, without any discrimination. As a part of the TRIPS agreement, developing countries were given transition periods in which to bring national legislation and practices into line with the TRIPS provisions. The transition periods expired in 2005 for most developing countries, including India and China. Since then, the new and strengthened patent regime is applicable.

GUIDING PRINCIPLES FOR PARTNERING: WHAT HAVE WE LEARNED TO DATE?

Drug Discovery in Pharma

The declining productivity of R&D combined with the sobering realization of the complexity of biology emphasizes the need for a new model. This new model could include organizations working together to fund a project, researchers from different organizations working on the same project in a common space, organized data sharing across organizations, and organized transfer of technology from public or academic research labs to industry. The key to the process is to develop a pipeline that the organization will use to manage the risk and expectations for the marketplace. As time passes, the corporation must find new products and technologies to insert into the pipeline gaps. Beginning 4–5 years in advance of the marketplace, the organization must demonstrate a pool of up to 5000 compounds available in the initial pool of technologies. To develop an effective pipeline, the organization must continually turn over the batch of initial compounds, retesting for new indications where appropriate, and dropping compounds from the pool when adverse reactions are detected. At the end of the evaluation phase, the organization must decide how to proceed to the next step and how many resources to dedicate. The organization may choose to keep development internal, utilize external resources or partnerships, or release the compound to another organization for development.

Changing Landscape of Drug Discovery Paradigm: A Call to Action

All of us are aware that the process, as it exists today, has become cumbersome, with many well-intentioned participants wanting to monetize technological advances and generate royalties for their organization in a nimble manner. Today, we must embrace with a sense of urgency that the model is now about throughput, advancement, and velocity. A new streamlined integrative partnering process is imperative to achieve success. Smaller, smarter, and more agile behaviors and processes must take precedent to pave the road to success for all stakeholders. Collaboration is critical, and education is paramount around the resources required to actually develop new therapeutics. Society as a whole must dedicate sufficient resources and focus the mechanisms of insurance, patents, and governments, to accelerate inventions into innovations for patients.

EXPANDING THE ACCESS PLATFORM TO EVOLVE WITH GLOBAL CHANGE: OPEN INNOVATION VERSUS CROWDSOURCING

"Open innovation" has gone from theory to case studies and has now moved into corporate strategic planning as a recognized tool and current hot topic in management circles. As such, organizations must review the decisions associated with the process as part of their ongoing financial strategies. To be effective, financial groups need to review decisions in terms of impact to the product pipeline and the development of new cash flow sources on an ongoing basis. Technology managers plan around the completion of a project; financial organizations seek success of the overall program and manage risk across multiple events. The challenge is to find the ROI in the process. Case studies have many examples of the importance of open innovation, but roll back to high technology-centered corporations in markets with rapid turnover (Intel), global product development groups (Procter and Gamble), or companies that have made R&D part of their culture (3M). For example, at 3M, each associate had 10% of their time to work on a back-burner project with other associates and at Raytheon, chosen (up to 20 per year) associates were given a 1-year period to create a team of their choosing on a project of their choosing with full authority. What we do not see are the case studies on the failure of innovation (Mark Hurd's cutbacks at HP) or where the longer term payback was for companies who run extensive programs (GSK).

The existing open innovation model tends to focus on the value-generating properties of collaboration with external partners. In particular, it seems to assume that collaboration with external partners improves product innovation performance, which in turn has a positive impact on financial performance. At the same time, the open innovation model remains relatively silent on the cost implications of collaborative strategies. In this way, we were able to show that greater technology alliance portfolio diversity indirectly (i.e., via increased internal innovation efforts and product innovation performance) decreases the share of personnel costs in the added value. At the same time, though, we also found greater technology alliance portfolio diversity directly increases the share of personnel costs in the added value. Moreover, our analyses (data not shown) indicated that the direct cost-increasing effect supersedes the indirect value-enhancing effect. As a result, the total effect of the technology alliance portfolio on the profit margin of the firm turned out to be negative. The development of the collaboration pipeline must be part of a strategic plan for company growth. On steady earnings, the stock price remains unchanged. To grow the stock price, growth in new products and revenue must occur on a consistent and regular basis. Innovation projects, whether new technologies or revenue generation, are key to maintaining long-term corporate growth and stock price.

SUMMARY, CONCLUSIONS, AND RECOMMENDATIONS

Finding Common Ground through a Common Understanding

Organizations must develop a balanced strategy around innovation, provide a mix of risk and nonrisk investments, and allow the organization to work simultaneously with sustaining and disruptive technologies. Many innovation consulting groups focus on the value and strategies around the importance of innovation, open innovation, crowdsourcing, and other ideas, and do not focus on maintaining the ROI to the corporation, allowing them to manage risk and support innovation programs. Predictive systems based upon risk assumptions have their own challenges. Financial models will be used to understand the risk points and allow the organization to make decisions earlier in the development process to proceed or abort a project based upon criteria developed in advance. What is not critical is the guarantee of success of an individual project, but the more successful management of the risk across the system:

- Development of a process of innovation
- Development of financial models that can be verified with the marketplace for acceptance, rather than an inward focus
- Development of a portfolio that shows managed risk over an extended period of time
- Aggressive pipeline management to ensure new compounds meet certain completion steps or are terminated from the pipeline

- Management of the portfolio to ensure that the gaps in the annual pipeline are filled by new compounds, licensing of other technologies, purchase of companies, or development of cooperative agreements
- Management of internal R&D into a structure of joint ventures with scientists and external financial organizations
- Cooperation with universities and smaller organizations especially in the area of patent support and market development, which are the vulnerabilities of these two groups
- Developing long-term interests in new markets and using older products to enter these markets and create relationships which can be leveraged when newer products are introduced to these markets.

Corporations need to manage collaborative relationships. These relationships are critical to growth, process, and risk management. Successful corporations look extensively outside for new ideas and technologies and the ability to supplement current product offerings with external ideas. The execution of this process is what differentiates the winners from the losers. In "The Road to Positive R&D Returns," the authors note several strategies for managing innovation as it relates to generating bottom-line results. In their Exhibit 2, these strategies include moving testing to earlier phases, reducing launch time for drug development, and reviewing cost models for the development of the compound. The authors state,

> A consistent, aggressive, and simultaneous focus on costs, speed, and decision making could raise an average small molecule's IRR [internal rate of return] to about 13 percent, from 7.5 percent. Let's say that a leading pharmaceutical company has a typical R&D portfolio split 75 percent to 25 percent between small molecules and biologics, respectively, and distributed across various phases of development. Our approach would raise the portfolio's returns to between 14 and 15 percent, from 9 to 10 percent currently. [31]

This is an interesting claim and effective to readjust focus on the execution of innovation strategies, which will be different in each situation. The key is execution. Execution is based upon the development of research schedules with milestones. The failure to achieve a milestone is the justification to cease funding the project. By moving projects external to the corporation, organizations will find it easier to terminate external resources rather than manage the political challenges of reducing internal staff.

An Exciting Time for Collaborative Drug Discovery?

The focus of this problem is the development of an effective pipeline for pharmaceutical development. Many consultants in this area focus inward to the corporation with the idea of developing a financially lucrative solution with individual technologies, when their focus should be on the development of a portfolio of solutions and the creation of a value companies must evaluate capital versus capacity, while also evaluating their strategic R&D skill set: less focus on older product lines, more focus on new resources to help with new strategic goals. Companies should evaluate their R&D skill set over an extended period of time, comparing internal and outsourced resources [32].

Several strategies—partnerships with organizations having technologies for commercialization—have their own challenges. Additionally, organizations can group multiple similar technologies and learn which idea has the most effective solution. Testing strategies can bring end to all similar projects by effectively providing more thorough testing process. The February 2010 McKinsey report notes biotech expected returns are usually negative without a risk management strategy. Return to organization is lower costs, managed risks, more options in the early stages of the product funnel, leaving higher odds on successfully completing project at the end. In the February 2010 McKinsey article, Symphony Capital is a good example of portfolio management. By financing R&D and proof of concept for major companies, Symphony can make money for their services or in selling product back to the company. Failing products are funded by the successes. They provide finance and staffing support and allow the companies to sell the early-stage technology or purchase it on the market for an agreed-upon price if it has passed certain tests. Financially, they make a small profit past the proof-of-concept stage and purchase back the winning techs later in the process. This is another example of a strategy that the organization may use to manage risk on larger portfolios.

CONCLUSION

With more structured execution of product entry, an organization can add revenue-enhancing projects on a regular basis, which is critical to the growth of the company in the financial markets. Revenue enhancement models include comarketing or cobranding products, in which two large organizations can share risk of product entry and increase the immediate visibility of a new product in the marketplace. Disruptive technologies must be managed at arm's length. It is difficult for companies to manage competing technologies within their own space, as there is a concern over "cannibalizing" their own offerings. This is the marketplace, and, through branding and other steps, this can and must be managed as separate activities. Within the corporate model, hirings had a focus of "team players," rather than risk takers. Groups who will react to risk must manage disruptive technologies separately. Internally, most corporate staff members will risk their department with new initiatives, but not their personal careers. The risk/reward structure does not fit the model of the corporate employee, especially senior management.

REFERENCES

1. Hall, B.H., Rosenberg, N. *The Handbook of the Economics of Innovation*. Amsterdam: North-Holland; 2010.
2. Munos, B. Can open-source R&D reinvigorate drug research? *Nat Rev Drug Discov*. 2006;5:723–729.

3. 20F Form 2010. Employees Estimation March 2011, Annual Review 2010.
4. Christopher, A., Viehbacher, C.A. Strategy and outlook IR thematic seminar, September 6, 2011.
5. Munos, B.H., Chin, W.W. How to revive breakthrough innovation in the pharmaceutical industry. *Sci Transl Med.* 2011;3:89cm16.
6. Loftus, P. Spending on research crucial for long-term success. *Wall Street Journal* 2011; July 15.
7. Mullard, A. Partnering between pharma peers on the rise. *Nat Rev Drug Discov.* 2011;10:561–562.
8. Cortright, J., Mayer, H. *Signs of Life: The Growth of Biotechnology Centers in the U.S.* Washington, DC: The Brookings Institution; 2002.
9. DiMasi, J.A., Hansen, R.W., Grabowski, H.G. The price of innovation: new estimates of drug development costs. *J Health Econ.* 2003;22:151–185.
10. Gassmann, O., Reepmeyer, G., Von Zedtwitz, M. *Leading Pharmaceutical Innovation*, 2nd edn. Heidelberg: Springer; 2008.
11. Collier, R. Drug development cost estimates hard to swallow. *CMAJ.* 2009;180(3): 279.
12. Lundvall, B. *Innovation, Growth and Social Cohesion.* Cheltenham, UK: Edward Elgar; 2002.
13. Solo, C.S. Innovation in the capitalist process: a critique of the Schumpeterian theory. *Q J Econ.* 1951;65(3):417.
14. Dasgupta, P. The theory of technological competition: new developments in the analysis of market structure. London School of Economics and Political Science; 1982.
15. Drews, J. Drug discovery: a historical perspective. *Science* 2007;287:1960–1964.
16. Peltzman, S. An evaluation of consumer protection legislation: the 1962 drug amendments. *J Polit Econ.* 1973;81(5):1049–1091.
17. Huntt, P.B. The importance of patent term restoration to pharmaceutical innovation. *Health Aff (Millwood)* 1982;1(2):6–24. Spring.
18. Mathieu, M.P. *PAREXEL's Pharmaceutical R&D Statistical Sourcebook.* Waltham, MA: PAREXEL International Corp; 2012.
19. FDA, New Molecular Entities, 2013. Available at http://www.fda.gov/downloads/AboutFDA/Transparency/Basics/UCM247465.pdf. Accessed December 24, 2013.
20. Munos, B.H. Lessons from 60 years of pharmaceutical innovation. *Nat Rev Drug Discov.* 2009;8(12):959–968.
21. CMR International. Global pharmaceutical R&D productivity declining according to Thomson Reuters. 28 June 2010. Available at http://thomsonreuters.com/content/press_room/science/RandD-Productivity-Declines. Accessed December 24, 2013.
22. Murphy, L.M., Jerde, P., Rutherford, L., Barone, R. Enhancing commercial outcomes from R&D: a framework for a public–private partnership to increase the yield of federally funded R&D investments and promote economic development. Technical Report NREL/TP-110-40463; National Renewable Energy Laboratory. May 2007. Available at www.nrel.gov/docs/fy07osti/40463.pdf. Accessed December 24, 2013.
23. NIH. NIH Budget. 2010. Available at http://www.nih.gov/about/budget.htm. Accessed December 24, 2013.
24. Begley, S., Carmichael, M. Desperately seeking cures. *Newsweek* 2010;155(22):38–42.

25. Angel Capital Association. ACA member landscape—2009 different than 2008 (and not). 2009 April 16. Available at http://www.angelcapitalassociation.org/data/Documents/Press%20Center/ACA%20Statistics%202009.pdf. Accessed December 24, 2013.
26. McDemus, C. Are true early stage investors an endangered species? 2010. Available at http://www.vcdeallawyer.com/2010/06/15/are-true-early-stage-investors-an-endangered-species/. Accessed December 24, 2013.
27. Wessner, C.W. *An Assessment of the Small Business Innovation Research Program at the National Institutes of Health [National Research Council (US) Committee for Capitalizing on Science, Technology, and Innovation: An Assessment of the Small Business Innovation Research Program]*, Washington, DC: National Academies Press; 2009.
28. Gulbranson, C.A., Audretsch, D.B. Ewing Marion Kauffman Foundation: proof of concept centers. 2008. Available at http://www.kauffman.org/advancing-innovation/proof-of-concept-centers.aspx. Accessed December 24, 2013.
29. Goodno, G. *State Bioscience Initiative 2008*. Columbus, OH: Battelle; 2008.
30. Pertuze, J.A., Calder, E.S., Greitzer, E.M., Lucas, W.A. Best practices for industry-university collaboration. *MIT Sloan Manage Rev*. 2010;51(4):2.
31. David, E., Tramontin, T., Zemmel, R. The road to positive R&D returns. *McKinsey Quarterly* 2010; February. Available at http://www.mckinseyquarterly.com/Health_Care/Pharmaceuticals/The_road_to_positive_R_and_D_returns_2528. Accessed December 24, 2013.
32. David, E., Mehta, A., Norris, T., Singh, N., Tramontin, T. New frontiers in pharma R&D investment. *McKinsey Quarterly* 2010; February. Available at http://www.mckinseyquarterly.com/New_frontiers_in_pharma_RD_investment_2530. Accessed December 24, 2013.

21

ENTREPRENEURSHIP: DRUG DISCOVERY INNOVATION AT START-UP AND MEDIUM-SIZED BIOTECHNOLOGY COMPANIES

Allen B. Reitz and Kathleen M. Czupich

Pennsylvania Drug Discovery Institute, Pennsylvania Biotechnology Center, Doylestown, PA, USA

Drug discovery research has changed tremendously over the past 20–30 years. A restructuring in the way biomedical research is done in the private sector is currently under way, including substantial net downsizing of personnel in the pharmaceutical industry particularly in the United States and Europe, with growth in China, India, and certain other developing countries. For example, it is estimated that more than 110,000 pharmaceutical industry employees have lost their jobs in the calendar years 2009 and 2010. However, in terms of perspective, the modern pharmaceutical industry grew rapidly during the period 1970–2000, and we are now returning to a scale and level of productivity observed in the 1950s and 1960s [1–3].

Other chapters in this book will address the "why" for the megatrends involving change in the pharmaceutical industry. Here we present our own example of "what's next"—a focused, regional initiative to promote entrepreneurship and proof of concept product development, using a model of focused, smaller work groups that is becoming more prevalent and is more efficient than the larger, cumbersome, legacy research organizations of the recent past. This chapter is a personal account of our own experience at the Pennsylvania Drug Discovery Institute (PDDI, http://www.padrugdiscovery.org) located at a Proof of Concept Center (POCC) called the

Collaborative Innovation in Drug Discovery: Strategies for Public and Private Partnerships, First Edition. Edited by Rathnam Chaguturu.

Pennsylvania Biotechnology Center (PBC) in Doylestown, PA, promoting the thesis that innovation is regional and empowered one Principal Investigator at a time through innovation, appropriate team building, and creative access to early funding options.

THE REAL TRAGEDY

Tens of thousands of highly trained drug discovery professionals are out of work with only limited hope of working in their field again. The work of the commercialization of university technology should include programs to integrate these highly trained former pharmaceutical company workers into the small business discovery community [4]. While not all scientists are amenable to this shift as many just want their old job back, those that wish to continue working in science are often flexible to the new scenarios facing them and their coworkers.

CONTEXT

The modern era of science has seen tremendous advances in our understanding both disease pathology and drug toxicology, to a greater extent than ever before. In spite of this, there are fewer new molecular entities (NMEs) approved by the United States per year than 10–15 years ago. The pharmaceutical industry is experiencing unexpected stagnation as measured by the approval of NMEs by the U.S. Food and Drug Administration (FDA). Several of the factors for this phenomenon include major losses of revenue due to patent expirations, increasingly cost-restrained health-care reimbursement programs, and more demanding regulatory requirements. In 2007, for example, only 19 NMEs were approved by the FDA, including both small-molecule and protein-based biologic therapeutics, the lowest number since 1983, rising slightly to 21 NMEs in 2008 and 24 NMEs in 2009, out of which fewer than 20% are considered first-in-class [1]. The combination of fewer truly innovative NMEs, coupled with the loss of revenue due to conversion to generic drug status, has resulted in substantial contraction of the pharmaceutical industry.

The 1980s and 1990s represented an era of increasingly productive pharmaceutical company research, characterized by record approvals and introductions of new drugs into the marketplace. A part of this boom has been attributed to the Bayh–Dole Act of 1980, which was a major change in U.S policy with respect to the ownership of intellectual property rights for federally funded research. The Bayh–Dole Act promoted technology transfer by allowing federally funded grant recipients to retain ownership of patent rights resulting from such research in the absence of special circumstances, including a requirement that universities and other nonprofit research institutions share royalties with the inventors. Before Bayh–Dole, many inventions resulting from federally funded scientific research were not commercialized. The personal incentives thus created promoted technology transfer and commercialization, resulting in economic benefit for the U.S. economy as a whole [5, 6]. The importance of Bayh–Dole in stimulating university patenting and licensing activities can be seen from the fact that only

25 universities received just 96 U.S. issued patents in 1965, and 25 universities received just 150 U.S. issued patents in 1980. However, 150 universities received 1500 U.S. issued patents in 1992, an increase of more than 1500% during a period of time where overall U.S. patenting rose by less than 50%. In the 5 years between 2005 and 2010, the number of patent from universities has increased by threefold, whereas industry-generated patents only doubled in that period of time.

Small is good. Large pharmaceutical companies can no longer sustain their basic research efforts. With only ~25 NMEs approved every year, the great fortunes spent on drug development have turned out to be an economic lottery in which the cost per ticket far exceeded the expected winnings. With the end of patent protection of the leading cash producers, the pharmaceutical industry has had no other viable option but to reduce expensive internal drug discovery efforts. Driven by the invisible hand of economic realities, the future of early-stage drug discovery has switched to research done in university settings and smaller biotechnology companies. Luckily, drug discovery professionals with vast industry experience are available to assist such efforts. Unluckily for these researchers, the funding available does not result in the same numbers of jobs lost or establish the earlier large multidisciplinary project teams characteristic of the pharmaceutical industry.

Innovation in the area of drug discovery research is now most appropriately being carried out in smaller, more-focused research groups, as opposed to the monolithic organizations of the past. Whereas 80% of NMEs have originated in the largest pharmaceutical companies since 1950, in the past several years an equal number have come from smaller organizations [1]. Small companies are able to make decisions and advance research quickly, as they are not restricted by the layers of approval and interminable review committees found elsewhere. Larger organizations are keenly aware of changes in the pharmaceutical industry, but have demonstrated a relative inability to adapt as seen by fact that over the past decade the biophysical properties of new drug candidates introduced into preclinical and clinical development are the same as before even given new understanding about the ADME (absorption, distribution, metabolism, and excretion) and other factors that promote success in drug discovery [7]. Lack of resources in a small company environment are compensated for by the lack of multilayer reporting requirements and the elimination of nonessential work.

The ability to transfer information quickly via the Internet has increased the efficiency of smaller organizations, a development entirely unanticipated 50 years ago. Knowledge itself, and the transfer of knowledge and information, has become more of a commodity, which allows for rapid access to the most recent and relevant information on any particular topic of interest. The Internet has facilitated broad dissemination of information such as by the use of SciFinder, PubMed, and PubChem. Use of the Internet has also promoted collaborations across the world and has discounted the regional advantages of locating in any particular area with highly trained research staff and colleagues, as these are now available instantly by computer and e-mail. Small companies can only provide some of the expertise needed for drug discovery, and thus need to actively seek external review and collaboration. Previously, the drug discoverer would "throw the discovery over the wall" internally when their area of expertise was expended. By establishing discipline-based research networks worldwide, small companies can have access to the needed expertise instantly and make decisions as to interpretation and strategy on their own quickly.

BIRTH OF AN INDUSTRY

The implosion of the pharmaceutical industry has created a new cottage industry where small companies are at least as productive on aggregate as larger ones, in early-stage drug discovery. At a suitable POCC, state-of-the-art facilities can be made available to small businesses to conduct the product development that is critical to being able to attract additional investment. Researchers as entrepreneurs are willing to work for less when they have a share in the upside of a discovery, significantly reducing costs to a small business. Considering the scarcity of high-paying jobs in the pharmaceutical industry, entrepreneurship is often the only alternative to permanent unemployment, early retirement or retraining in areas outside of science. Facilities and equipment are often cash burdens to a small business, but these costs can be leveraged substantially such as by purchasing used, quality equipment, which is far less expensive than similar capital expenses at larger companies. Administrative expenses at the start-up biotechnology company are also low when compared with larger organizations. By keeping costs down to a bare minimum, the runway for product development or the time to do research with currently available funds is extended.

The Pennsylvania Drug Discovery Institute

The PDDI is a 501(c) (3) nonprofit organization located at the Pennsylvania Biotechnology Center (PBC) in Doylestown, Pennsylvania, about 30 mi north of Philadelphia. The PBC was founded in 2006 by Prof. Timothy Block of the Drexel University College of Medicine and the Hepatitis B Foundation, along with Delaware Valley College, and is a mixed-use academic–industrial POCC facility. There are more than 220 researchers located at the PBC, including Drexel University College of Medicine faculty, postdoctoral fellows and graduate students, several nonprofit foundations including the Blumberg Institute, and more than 30 small biotechnology start-up companies.

Although the conceptualization of research is promoted by wide access to publically available tools, on-ground implementation is best promoted by face-to-face collaborations with a traditional project team operational structure. It is no surprise that innovation in drug discovery research is "one PI (Principal Investigator) at a time," critically dependent upon a high level of innovation and a complementary, qualified research team. It is essential for the development of nascent platform technologies that they are nurtured in an environment of shared services, which characterizes a functioning POCC, including a supportive physical environment and the accompanying knowledge-based community and culture. POCCs are more effective at promoting product development than are more traditional research incubators, which offer limited shared services or on-site collaboration opportunities.

The PDDI is currently focusing on service and outreach activities in the area of workforce reentry assistance for highly qualified researchers who have been laid off elsewhere, given the geographic location of our facility as being close to several major pharmaceutical companies who have downsized in the recent past. Although we see this type of function as having constant value in the future, the fact is that the mass layoffs experienced so far will not continue indefinitely simply because sooner or later there will be no one left to lay off, in the most extreme case. Therefore, our activities will shift into serving as a means to promote the commercialization of university-based technologies by fostering independent research in the start-up biotechnology area, and as a think-tank to consider ways in which innovation occurs, including best practice examples. We recently initiated on-site research activities headed by Prof. Doron Greenbaum based at the Science Center in Philadelphia, PA.

FORWARD TRACK™ PROGRAM

The Forward Track Program is the primary means by which we provide workforce reentry assistance to qualified biomedical researchers in the tri-state area of Pennsylvania, New Jersey, and Delaware. Many highly trained scientists with advanced academic degrees and impressive scientific credentials have been displaced recently. Through Forward Track, qualified senior-level research scientists locate at the PDDI and are mentored in developing research plans and teams. Researchers are given an affiliation with the PDDI as well as career development advice and assistance, especially regarding forming collaborations, accessing federal and other funding, and in creating their own start-up biotechnology companies. Program members are encouraged to participate in the seminars, journal clubs, research rounds, and ad hoc interactions that are hallmark features of the collaborative spirit fostered by the PBC. Once per month, we sponsor the Outreach Seminar Series, in which one to three program participants come to the PBC and give oral presentation on one of their particularly significant research achievements to an audience consisting of senior members of the facility.

INDUSTRY-STANDARD DUE DILIGENCE REVIEW

University research was historically more basic in nature, contrasted with the applied focus found in the pharmaceutical industry. Now, more translational research with an applied focus is done at universities and nonprofit research institutions. To be properly developed further, public–private partnerships are generally needed in which researchers focus on product development derisking activities, such as ADME profiling, to warrant further investment. Several criteria for evaluation of any nascent technology are as follows: Is it innovative? Is it a substantial improvement over current standard of care? Is the preliminary data of sufficient merit to justify further investment of funding, time, and personnel allocation?

At the PDDI, we take advantage of the immense talent pool in our region of individuals with industry experience, and have created an expert's list of highly qualified reviewers and mentors separated by discipline to evaluate basic discoveries at academic and nonprofit institutions using an industry-standard due diligence review process. We use this list to match experts in appropriate disciplines with others as required to create project teams that are most likely to succeed. Many professors and principal investigators are conducting important basic research but do not have the expertise required for the applied value-added experiments and studies needed prior to generating commercial interest. We provide advice on project flowchart design and the interpretation of data and may recommend the creation of one or more associated spin-off companies, which could add value to the technology and create new jobs and expertise.

PDDI KNOWLEDGE-BASED COMMUNITY

One of the strengths of the PBC is a knowledge-based community consisting of an active seminar program, frequent journal clubs, and sharing of information and equipment. The PDDI extends this spirit of collaboration into the applied arena of drug discovery research. We grant faculty appointments in the PDDI to those with industrial or biotechnology company experience based on academic credentials and demonstrated accomplishments in relevant and complementary fields of study.

Processing of In-Kind Assets and Donations

The PDDI is a 501(c) (3)-designated charity, meaning that donations may be tax deductible to the donor. The largest donations we receive are in-kind, such as consumables and equipment that organizations donate as their facilities are downsized or closed. We have established a PDDI Board-approved process whereby we can accept and process such donations after review of their suitability, and then provide a physical location for their archiving, curation, and distribution to others. Small biotechnology companies, nonprofit research institutions, and universities can come to the PBC and take these in kind assets for their own use, assuming all ownership and liability, following common-sense guidelines. This program has proven to be tremendously popular with research institutions located both at the PBC and regionally. We have received two separate donations of ~6000 chemical starting materials and reagents from pharmaceutical companies, bar-coded for easy access and searching via Excel-based programs. In addition, we have received more than 60 boxes of glassware and other consumables that would be valued at more than $1 million if purchased new. At many companies experiencing downsizing, these assets are discarded or sent to an incinerator. While not illegal, such disposition of assets is to some extent an unethical waste of the resources of society. However, the reuse of such assets fits into the mission of the PDDI as a nonprofit organization promoting entrepreneurship and biomedical research for the public good.

STUDY OF BEST PRACTICES IN THE COMMERCIALIZATION OF UNIVERSITY RESEARCH

We are studying the changing face of innovation in biomedical research, not because we are expert in public health policy, but because we can thus better adapt ourselves to the changing landscape of early proof of concept research and funding opportunities. We have engaged several groups composed of senior postdoctoral fellows and other researchers from the Philadelphia area as part of the Penn Biotech Group to assist the study of technology transfer. The technology transfer process is a key factor in the translation of basic academic discoveries into commercial opportunities. Academic research supported by the National Institutes of Health (NIH) and the National Science Foundation (NSF) has resulted in many basic biomedical discoveries. Efficient technology transfer has a positive impact on society by creating new businesses and assisting in the success of existing businesses, which in turn generates high-impact jobs and drives economic growth. In our recent study of best practices in biomedical research, we studied Harvard University, the University of Wisconsin-Madison, Emory University, and The Scripps Research Institute, each of which represents of a particular type of institution [6].

The process of technology transfer involves the academic institution and a biotechnology or industrial partner. The industrial partner seeks financial return on investment, is willing to take measured risks, and strongly protects patent rights while attempting to respond to challenges and opportunities. Universities have an educational focus as well as research, and tend to be more risk-adverse, often making decisions carefully through lengthy approval processes. This culture gap can create a major bottleneck in the process of technology transfer. At universities, this process is handled by the Technology Transfer Office, which is responsible for identifying, protecting, and licensing new technologies discovered internally. A first step toward possible commercialization of an invention is obtaining intellectual property or patent protection sufficient to interest a potential partner in a possible commercial return on investment.

A primary means for the commercialization of an invention is licensing, which grants rights to the use of property without transferring ownership. Another outcome of the technology transfer process is the creation of a start-up biotechnology company. In 2008, 595 new companies whose licenses were based on academic discoveries were formed, a 30% increase from 2006. Over 70% of these companies remained in their home states, indicating that university spin-offs are an important vehicle for economic growth within a particular geographic region. The Economic Development Administration of the U.S. Federal Department of Commerce put forward a Request for Information in 2010 on the topic of "Commercialization of University Research." A total of 205 responses were received as posted at this website: http://www.eda.gov/commrfi-responses. Taken together, these documents establish a high level of interest in the topic of technology transfer and commercialization of university-derived technologies throughout the United States.

University involvement in regional economic development initiatives are to be encouraged, as exemplified by the PBC in which academic and biotechnology researchers are colocated, fostering and promoting collaborations. Such initiatives actively seek to provide value-added guidance and support such as matching the university technologies with researchers and companies that offer suitable applied research capabilities and expertise. In general, regional economic development initiatives are the most effective when they offer strategic advice and project management to increase the probability of success of any new start-up venture.

THE VALLEY OF DEATH

"Though I walk through the valley of the shadow of death, I will fear no evil" (Psalm 23.4). The "valley of death" in biomedical research refers to the high attrition rate between promising, innovative ideas and the transition into the expensive development functions required for evaluation in the clinic. There are many causes for and suggestions to address the valley of death that have been expertly described elsewhere [8]. Some of the factors involved in the transition between conception and demonstration of proof of concept sufficient to obtain additional funding from our perspective are shown graphically in Figure 21.1. The technology must represent a platform and merely only a particular product. In the context of small-molecule drug discovery, this means a patent portfolio with backups identified in case the lead falls out of consideration due to unforeseen limitations. However, a bright idea or even a fantastic early-stage data package is not sufficient to guarantee success. An appropriate research team with the necessary disciplines represented and who are willing to work together to adjust to changing opportunities and challenges is also required. In addition, a supportive physical location with nurturing infrastructure as well as early-stage funding including self-funding is required on the part of the founders. It is key that scientists involved in early-stage research operations have access to electronic library resources and

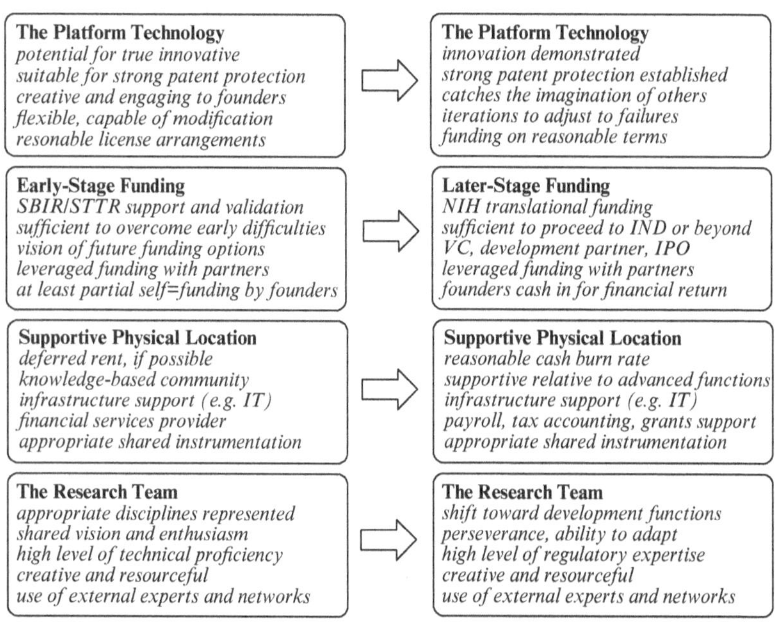

Figure 21.1. The "valley of death," represented as the transition between the initiation of proof of concept research operations and the further validation required for continued product development.

important literature searching tools such as SciFinder from the American Chemical Society, which is now being made available on more reasonable terms to smaller companies.

The following are specific recommendations to promote technology transfer and entrepreneurship in biomedical research.

Consider Orphan Indications and Areas Poorly Treated by Current Therapies

The Orphan Drug Act of 1983 has promoted the development of new drugs for the treatment of conditions with a prevalence of less than 200,000 in the United States. As of 2008, 326 orphan-designated products have been approved by the FDA, primarily for cancer and with an average patient population of less than 10,000 [9, 10]. Many drugs will eventually be demonstrated in clinical practice to be useful for treating multiple indications. By focusing initially on an orphan indication, the developer is often able to minimize the time to market as well as reduce the cost of the clinical trials followed by use for additional indications as appropriate in larger markets. Poorly treated diseases have a low bar to overcome, increasing the chances of acceptance and commercial return.

Orphan drug designation granted by the U.S. FDA provides 7 years of market exclusivity independent of patent protection, as well as other price and fee advantages.

Promote Supportive University Policies and Technology Transfer Offices

A supportive TTO at any university is to be encouraged, and is an incentive for promising junior faculty to join any particular university. Such support includes the presence of associated incubator space or liberal policies regarding the use of university space to conduct start-up proof of concept research. Templates for standard agreements are a great idea, such as at certain universities where standard forms such as confidential disclosure agreements are freely available on their website. External collaborators can download, sign, and submit these documents for immediate acceptance unless they require further negotiation.

It is critical that technology transfer offices (TTOs) respond in a timely fashion to facilitate potential collaborations with external parties. There are many anecdotal stories concerning interminable delays in working with academic TTOs for even simple documentation such as confidential disclosure and material transfer agreements. Academic TTOs should be staffed by individuals with at least some research and industrial background.

Keep the Inventor Involved with the Technology

The longer an inventor is involved with their technology, the greater the chances of success. Often technologies have to be changed or refined based on new data, and the originating inventor is most likely to make the necessary changes so the product can advance.

Focus on Product Development

At the early product stage, many will advise to put together a slide presentation and a "road show." In addition, business development professionals can be brought in to identify targets and raise money. However, much of these business development activities if conducted too early are counterproductive; furthermore, companies that focus on product development as opposed to raising unwarranted expectations are most likely to increase the underlying value of their technology. Research funding is hard to come by, and should be used to support the critical functions of a start-up and not superfluous travel, furniture, or office party expenses.

It is important to focus product development in areas of greatest unmet medical need in the future, using methods most likely to succeed. Chronic diseases such as diabetes and cancer will be a more prevalent fraction of the disease burden for the growing human population on earth over the next 20 years [11]. Drugs have historically been discovered to a large extent by phenotypic screening or pathway modulation methods, which should still be considered as a complement to purely mechanism-based approaches [12].

Laboratory Space and Nurturing Culture of Innovation

A successful POCC adds value to and accelerates the innovation process, and offers increased value when compared with an incubator that provides exclusively laboratory space for rent. The technologies embedded in start-up companies at a successful POCC move more quickly past early milestones by taking advantage of the capabilities, guidance, and assistance provided by the POCC. Associated support disciplines should be represented at the POCC, in the best-case scenario. High-quality incubator research space, access to experts in the area, and access to research journals are all necessary components of success. The key is an open and flexible environment in which the different disciplines and activities required for early entrepreneurial activities are represented. We have found that a flexible renting and leasing model is key: some inventors are willing to self-finance by not taking a salary but need a deferred payment schedule for renting office and/or laboratory space to be viable early on. Also, limited seed grants or convertible debt in the range of $15,000–50,000 help to advance early-stage research, to provide entrepreneurs the opportunity to obtain preliminary data to support grant applications or obtain further funding.

Business Acumen and Guidance

Specific knowledge of grants administration and early-stage funding opportunities are essential. In addition, support to manage the mechanics of a small business such as dealing with regulatory and legal filings, accounting, and tax preparation should be on-ssite or closely nearby.

Appropriate Scientific Expertise and Team

Experts in scientific subject matter are essential to the success of any grant. Many highly trained people are available through redeployment from the large pharmaceutical industry. It is critical to establish a compelling scientific team to support any application for grant or other funding.

SBIR Funding

The federal government has a special "set aside" program to fund Small Business Innovation Research (SBIR) grants. The SBIR and related STTR (Small Business Technology Transfer) programs have just been reauthorized for six additional years starting in early 2012. The SBIR program allocation is being increased from 2.5 to 3.2% of the total budget of the National Institutes of Health, and the STTR allocation from 0.3% to 0.45%, giving small businesses an increased role in the federal research and development effort, while preserving the bulk of federal funding for basic research. SBIR funding is a critical component of the success of our program, and provides up to $225,000 for a Phase 1 award and $1.5 million for a Phase 2 award. Such research support is nondilutive capital, meaning that no share of ownership is exchanged to the government for the funds provided.

PERSONAL STORY OF PROMOTING ENTREPRENEURSHIP: KATHLEEN CZUPICH

In 2007, the reality of the paradigm shifts in the pharmaceutical industry became obvious to one of the authors (K.C.) in her work as the founding administrator for the PBC. The PBC was funded by a grant by the Commonwealth of Pennsylvania to establish a biotechnology center in Bucks County, Pennsylvania, and is also known as the Biotechnology Center of Bucks County. The vision for the PBC was to be a business incubator that would colocate with a division of Drexel University's College of Medicine, and the Hepatitis B Foundation and the associated Institute of Hepatitis and Virus Research. Eighteen months after the PBC opened, she left her role as the chief financial officer (CFO) of the nonprofits at the PBC to create a company called Artemis Solutions Inc., specifically to assist entrepreneurs start their own companies and obtain the required early-stage funding. The purpose was to create start-up companies to do biomedical research, largely through collaborations with universities and funded by federal small business grants. In her previous role as CFO, she was responsible for the application and administration of $5 million in federal and state grants, including maintaining the books of the organization and completing the necessary audit requirements. All of these skills were directly applied to the small businesses that were collaborating with universities funded by federal grants. Since 2008, she has founded ~30 small companies, 11 of which have been successful in obtaining and maintaining federally funded research programs. The experienced researchers who are the founders and owners of these companies had been laid off from the pharmaceutical industry after careers filled with substantial professional accomplishments. Those who had been employees their whole lives instantly became business owners and entrepreneurs, and K.C.'s role was to show them how to do it. Pennsylvania has a program to assist small business with obtaining money through the SBIR program, which provides funds to preferred providers to impart expert assistance for grant applications. As a preferred provider, K.C. received funding to assist companies in crafting and filing grant applications. Once funded, she administered the grants and ran the business aspects of the companies, including obtaining laboratory space, payroll, insurance, bookkeeping, and grant administration. In short, she would fill the CFO role of the company, without their having to hire or fund her on a full-time basis. Furthermore, being on-site at the same facility with the entrepreneurs is a great advantage relative to alternate off-site arrangements. Due to the complexity of grant requirements, a high level of experience and knowledge of government accounting standards are required. Small businesses are guided by the same rules regarding grant administration as nonprofit research institutions. Expertise in grant application and administration is not widely held but is necessary for the success of the small business, and is an essential component of any successful POCC.

PERSONAL STORY OF BEING AN ENTREPRENEUR: ALLEN REITZ

After nearly 26 years with Johnson & Johnson, one of the authors (A.R.) was a free agent who could take on many different professional activities that were previously not allowed. For example, he was a consultant for various organizations, including the World Health Organization and an expert witness in a patent interference case testifying under oath. He had earned an Executive Masters at the University of Pennsylvania while at Johnson & Johnson (M.S. Management of Technology), so he had considered and planned for the possibility of starting an independent small business biotechnology operation in advance. In 2008 and 2009, A.R. formed two small companies (http://www.alsbiopharma.com and http://www.fc-cdci.com) with on-site laboratory operations at the Center, focusing on early drug discovery with a core competency in medicinal chemistry. For 8 months, he self-financed the operation without salary and purchased a 300-MHz NMR and LC/MS instruments using a home-equity line of credit. Since then, the instrumentation has been completely reimbursed, and the company operations have 15 employees total with multiple funding streams and ongoing collaborations. ALS Biopharma, LLC addresses the orphan indication of amyotrophic lateral sclerosis and was cofounded with an angel investor who had ALS in his family, and Fox Chase Chemical Diversity Center, Inc. was so named because the company obtained the first-ever SBIR grant in combination with any Principal Investigator at the Fox Chase Cancer Center in Philadelphia. Both companies are staffed with individuals with extensive prior industrial experience. The key observation is that the secret of start-up biotech research is not how to spend money, but how to not spend it and get the same results. Used equipment and furniture is now available at a very steep discount from new prices, and should be exploited whenever possible. The core business strategy is to create and maintain viability through grant and investment support to create an opportunity for true value creation by establishing platforms based on composition of matter intellectual property.

RECOMMENDATIONS FOR CHANGES IN PUBLIC POLICY AND FUNDING

The political will required to fund initiatives in the area of entrepreneurship and product development is the most critical factor for success, taken together with a robust program of in-kind support and advice. The following are two specific recommendations in this area:

1. We encourage the Department of Commerce to create and fund program grants specifically to sponsor the creation and nurturing of POCCs for technology development, modeled after large consortium grants such as the P01 grants on neuroAIDs administered by the NIMH of the NIH.
2. We encourage private companies to donate legacy intellectual property that currently lies fallow due to portfolio reduction to donate this to regional due diligence panels and organizations such as the PDDI, to facilitate the development of the intellectual property while the parent company retains an equity position. An enhanced federal research and development tax credit could be created to promote such transfer of assets. A probable outcome is to match this technology with experienced researchers, such as those involved with it originally at the originating institution, who would be willing to develop it further via a separate new or existing small business entity, accelerated by self-financing or seed funding from regional POCCs on a competitive bid basis.

METRICS FOR SUCCESS

It is difficult to accurately quantify the relevant metrics associated with an increased net present value of early stage proof of concept biomedical research. Relevant top-level metrics for success include the following over the past three years, with values for the PBC and the PDDI in parenthesis.

a. Technologies licensed (~5)
b. Number of new companies created (~25, 11 with funding)
c. Number of new jobs created (~60)
d. Number of U.S. patent applications (~15)
e. Number of SBIR and other grants obtained (>15)
f. Total amount of financial support such as from NIH funding (~$20 million).

The metric categories would generally be the same for university-associated versus independent POCCs, with the exception that any associated university should be specifically benefited by the adjacent POCC to a measurable extent. The key is to promote the applied as well as basic research required to achieve proof of concept milestones, and warrant further investment and development prior to commercialization.

CONCLUDING REMARKS

It is clear that innovation is now occurring to a greater extent than ever before in small organizations, often in collaboration with inventors at universities and entrepreneurs with experience in the pharmaceutical industry. For the past 2 years, more NMEs have been approved for therapy by the FDA from small than from large companies. The PDDI represents a support function located within a POCC that is attempting to provide leadership in establishing public–private partnerships and accelerate biomedical research in measurable ways.

ACKNOWLEDGMENTS

We thank the Board of Directors of the PDDI for their support and encouragement. We acknowledge and appreciate the assistance of Dr. Bill Kinney, Keystone Innovation Coordinator of the PBC, in tabulating the success metrics shown here. We thank Dr. Tim Block for his vision, energy, and enthusiasm.

REFERENCES

1. Munos, B. Lessons from 60 years of pharmaceutical innovation. *Nat Rev Drug Discov.* 2009;8:959–968.
2. Pammolli, F., Magazzini, L., Riccaboni, M. The productivity crisis in pharmaceutical R & D. *Nat Rev Drug Discov.* 2011;10:428–438.
3. Thomson Reuters International Year of Chemistry (IYC 2011) Report. The changing role of chemistry in drug discovery. 2011.
4. Ahn, M.J., Meeks, M. Building a conducive environment for life science-based entrepreneurship and industry clusters. *J Commer Biotechnol.* 2008;14:20–30.
5. Renault, C. Academic capitalism and university incentives for faculty entrepreneurship. *J Prod Innov Manag.* 2006;31:227–239.
6. Vaidya, V.Y., Kadaba, A.P., Nieves, A., Shi, F., Wang, L., Chen, Y.-L., Yu, S., Gao, L., Moritz, K.P., Czupich, K.M., Reitz, A.B. Emerging metrics in technology transfer I. Case studies in the life sciences. *Int J Technol Transfer Commer.* 2012;11:110–136.
7. Leeson, P.D., St-Gallay, S.A. The influence of the "organizational factor" on compound quality in drug discovery. *Nat Rev Drug Discov.* 2011;10:749–765.
8. Coller, B.S., Califf, R.M. Traversing the valley of death: a guide to assessing prospects for translational success. *Sci Transl Med.* 2009;1(10):10cm9.
9. Braun, M.M., Farag-El-Massah, S., Xu, K., Coté, T.R. Emergence of orphan drugs in the United States: a quantitative assessment of the first 25 years. *Nat Rev Drug Discov.* 2011;10:519–522.
10. Seoane-Vazquez, E., Rodriguez-Monguio, R., Szeinbach, S.L., Visaria, J. Incentives for orphan drug research and development in the United States. *Orphanet J Rare Dis.* 2008;3:33.
11. Reardon, S. A world of chronic disease. *Science* 2011;333:558–559.
12. Swinney, D.C., Anthony, J. How were new medicines discovered? *Nat Rev Drug Discov.* 2011;10:507–519.

22

CHEMICAL CONSULTING

Lester A. Mitscher

Department of Medicinal Chemistry, University of Kansas, Lawrence, KS, USA

Although many academic chemists consult with firms and agencies, virtually none have formal training in this useful art. Furthermore, there is little in the way of formal literature on the topic other than a few books [1] and some trade magazine articles directed toward industrial managers, engineers, and clinicians. Practices vary considerably depending upon the skills and experience possessed by the consulting chemist and the environment in which he/she practices as well as the needs of the client. In this chapter, the author offers several observations based on over 40 years of continuous experience consulting at various levels with small and large firms in the pharmaceutical industry and with governmental, academic, and private agencies. The reader should not suppose therefore that this chapter is comprehensive or that it deals with every aspect of chemical consulting, given the intrinsically multifaceted nature of this activity. Furthermore, the proprietary nature of this activity restricts one's ability to illustrate points by specific examples. Nonetheless, the author hopes that the following account contains interesting and helpful matter to those who are thinking of consulting or to those who may be thinking of engaging a consultant.

Collaborative Innovation in Drug Discovery: Strategies for Public and Private Partnerships, First Edition. Edited by Rathnam Chaguturu.

NEED FOR CHEMICAL CONSULTANTS IN DRUG-SEEKING CAMPAIGNS

One of the particular virtues of consultants is that they are cheaper than hiring a collaborator and do not receive fringe benefits. Furthermore, the relationship is regarded as temporary, so parting company is not normally traumatic to either party. The client usually defines the issue and provides the necessary internal background and the expectations. The consultant asks strategic questions and adds his/her independent thoughts. This relationship requires mutual trust.

In a drug-seeking campaign as in most scientific undertakings, each answer uncovered raises several alluring questions. This can all too easily lead to project drift. One important contribution that a consultant can make is to help keep the project on course.

Among the types of projects that an experienced chemical consultant can often help with are, for example, in assessing the degree of novelty and utility in heritage or potential chemical compound collections; selection of useful hits for further work from the results of screening assays or in open source data; anticipating and avoiding problems with toxicity and pharmacokinetic factors; design of potentially useful ligands for the targets being addressed; improvement of the desirable properties of early leads through efficient analoging; synthetic design; alteration of chemical pathways to avoid development problems later; redesign of synthetic routes to facilitate the preparation of bulk quantities at an attractive cost; interpreting apparently anomalous reaction results and devising schemes to avoid them; devising suitable chemical methods of analysis; designing suitable backup alternatives; and constructing patents that provide adequate patent coverage.

The amount of effort and money involved in introducing a novel pharmaceutical agent into the market place can be very substantial. A drug-seeking campaign can consume a decade or more and require the expenditure of many millions of dollars when there is no obvious structural guidance from which to start. Unfortunately, people are notoriously poor at making decisions in the face of complex and often uncertain or conflicting data when the stakes are high. A poor decision can result either in significant wastage in precious resources through synthesis and evaluation of compounds that ultimately fail or in failure to proceed with compounds that ultimately would have succeeded. The costs of drug-seeking campaigns usually escalate rapidly as the campaign goes on. If a failure is to be the ultimate outcome, it should happen as early as possible and be based on solid scientific evidence. If the properties of a lead compound or series are truly outstanding, or, conversely, are appalling, then decisions are easy. Decision is vexingly difficult when properties lie midway between these extremes! As the work proceeds, one quite often becomes seduced by the belief that success can lie just ahead. Chemists are by nature optimists; otherwise, they would not be in this field. One must guard against the tendency to behave like a compulsive gambler believing that by making just one more analog, success will be at hand! Overestimation of the prospects for success is endemic. Receiving credible advice from a skilled practitioner allows one to cope better with these pressures. This is particularly true if the ultimate decision is that the project should be terminated. The consultant can be of great help in making these vexing decisions.

Advice from an experienced chemist who has successfully navigated these troubled waters is no guarantee of success but does considerably enhance the comfort level of those with responsibility for the effort. Under the best of circumstances, many more molecules fail than not.

Failure stems from many causes, not all of which are easily anticipated. A faulty theory of the molecular cause of a disease and therefore of a posited molecular treatment, a biological test system that fails to address the key characteristic required, an unanticipated difference in species response to treatment, an unavoidable adverse side effect, poor water solubility, adverse pharmacokinetic and pharmacodynamic features, idiosyncratic toxicity, unexpected failure to achieve adequate and defensible patent protection, and so on, are all challenges that must be recognized early and dealt with. Some of these are unavoidable when they happen, but the scientist, assisted by the consultant, has the responsibility of anticipating likely problems and devising means of dealing successfully with them or deciding to change directions.

Temporary problems blocking further progress almost always occur in drug seeking. These must be dealt with expeditiously since time loss is an implacable enemy. When a workable solution does not readily present itself to the researcher, consultation can be extremely useful. In this context, the consultant can supply specialized knowledge or capacities not present in a firm at that time. If problems of this type occur repeatedly, a longer term relationship may be required until this problem disappears. When such problems are found to be widespread in a firm, the consultant can deal with this in an efficient manner by developing and presenting in-house an interactive short course on specialized topics such as the relationship between pharmacokinetic and pharmacodynamic characteristics and the structure of molecules, on the strengths and weaknesses of fragment-based drug discovery, on newer methods of synthesis and novel reagents, and the like.

The consultant will usually be working with very intelligent and highly trained persons who have already given a problem thoughtful consideration without success or not be confident that their solution will be the best approach. Clients often know their individual project very well and are enthusiastic about it. They will resent trivial or obvious suggestions. Giving cogent reasons for suggestions goes a long way toward developing a productive relationship. Consulting, like water, does not flow uphill. Credibility develops naturally when advice works more often than it fails.

The need or desire for consultation often develops due to the intense specialization that can develop due to an individual's faithfulness in carrying out his/her assigned role in the firm. Thus, scholarship and experience may over time become narrow. In other cases, the need may be associated with newness on the job or significant changes in the firm's direction. Thus, consultation is especially useful when a laboratory is planning to or has decided to enter a significantly new area since internal prior experience may not be especially helpful. Sooner or later it will happen that the internal knowledge outgrows the contribution that a consultant can make. In long-term relationships, the client will know the consultant's manner of thinking rather well and can anticipate the advice most likely to be proffered. When this occurs, it is time to move on.

ROLE FOR CHEMICAL CONSULTATION IN OPEN DRUG DISCOVERY PROGRAMS

Open drug discovery is a comparatively new approach to drug seeking as is apparent from the information elsewhere in this work. Traditionally, drug seeking has been a highly proprietary enterprise carried out behind closed doors. This characteristic still largely characterizes discovery research in industry. Working in this environment requires some adjustment for academics who consult because the essence of academic work is the wide sharing of new science at the earliest opportunity. Open drug discovery lies somewhat in between these extremes. Some companies, however, are beginning to open a portion of their books to outsiders as a public service. Often these laudable activities involve releasing potentially useful data and compounds to be applied to "orphan diseases" to be pursued in other hands. These are severe health problems such as malaria, tuberculosis, or various tropical diseases that the company, for economic reasons, cannot pursue with the realistic expectation of providing their shareholders with a satisfactory return on their investment. This is, after all, the essence of capitalism. Companies are not charities.

Unfortunately, by and large, the places where these public health problems are widespread and endemic do not have the human and economic resources to attack them successfully. Historically, these problems have been attacked inefficiently. In an attempt to deal with this otherwise intractable problem, some forward-thinking firms are placing compounds and data in the public domain in the hope that outside individuals will build on this information and develop therapy for these prevalent but neglected diseases.

Thus, both experienced and comparatively naive scientists are beginning to have access to masses of data from which to pursue potential cures. A useful parallel can be seen in the availability of commercial and academic compound libraries stemming from combinatorial chemistry mated with high-throughput screening efforts especially as occurs with the NIH roadmap. Both of these efforts produce "hits" that require further refinement for new purposes. Experience shows that many of the hits uncovered fail to have the intrinsically desirable characteristics likely to elevate them to the "lead" category and thence onward to candidate drug and actual drug status. Selecting among these is an art requiring experience and judgment. This is largely a multifactorial exercise in pattern recognition in which experience is especially helpful. The more one knows in such exercises, the luckier one is likely to be. Characteristics particularly important for the treatment of orphan diseases in impoverished and/or remote populations are that the agents be cheap, orally active, be convenient to consume, require once daily dosing, be effective rapidly, and require short-term treatment schedules. This is not easy to accomplish when added to all of the other characteristics that must be gotten right. Chemists new to drug seeking can profit significantly from the advice of experienced consultants in making these decisions.

Many organic chemists have an excellent background and significant experience in synthesis but have little formal training in human biology. This is a potential handicap in drug seeking. Experienced medicinal chemists have a background that blends these two disciplines in a practical way. This is especially useful when faced with the necessity of choosing a few compounds for further work from a series of hits from a high-throughput screening program. Selecting advancable compounds requires a judicious balancing of a couple of dozen significant chemical and biological characteristics. Furthermore, these characteristics differ individually in importance from series to series. Thus, the problem is anything but trivial. Organic chemists know how to make compounds. Medicinal chemists know what is worth making for medicinal purposes.

For those who may find these concepts novel, a hit may be defined as a compound that produces a response in a screening program designed to reveal compounds with promising properties. A lead is a hit that upon further testing, usually in intact cells or, better, in normal or diseased whole animals, produces a positive effect and which has other attractive physicochemical properties. A candidate drug is a compound that has sufficient novelty, potency, selectivity, lack of toxicity and chemical tractability that it is worth a significant further investment in time and treasure. A drug is a compound that has useful therapeutic and commercial properties in diseased individuals that it can be marketed.

As hits progress in an iterative manner to a lead and then a candidate drug and finally a drug, problems emerge that require a chemical solution. Having a knowledge of the type of problems that have come up before in other campaigns and remembering the means that were found effective in dealing with these or, better, having imagination that can lead to novel means of addressing new situations are priceless characteristics of the best consultants.

NECESSARY BACKGROUND AND EXPERIENCE FOR SUCCESSFUL CONSULTING

Most university professors who consult are comparatively narrow specialists and consult as individuals. Call for their service is sporadic. This kind of activity is more suitable as a sideline than as a principal source of income. On the other hand, there are a significant number of consulting firms that have stables of consultants available to fit a variety of needs. These firms are more visible and are potentially able to offer more opportunities.

For those new to the field or those seeking to expand their client list and wishing to work independently of these organizations, there is, a web site jointly sponsored by the American Institute of Chemical Engineers and The American Chemical Society by use of which consultants and clients can find each other [2]. Also the Association of Consulting Chemists and Chemical Engineers is a nonprofit organization that provides a similar web-based service [3].

The particular value of consultants is that they can offer an independent viewpoint and have no particularly vested interest in the outcome of the matter over which they opine. They do not stand to profit much, gain internal status, or become promoted. The favorable outcome from their standpoint is the satisfaction of being helpful and, particularly, the likelihood that success will be rewarded by an ongoing consulting relationship. Thus, their impartial opinion is of special value. They can also serve as arbiters when there is an unresolved dispute. Their lack of a stake in the outcome enhances the likelihood that the results of their analysis would be accepted by both sides.

Obviously current or recent visible success in performing the sort of work involved in the consultation greatly enhances the comfort level of those to whom advice is to be given. A consultant with a distinguished track record in performing the type of work in question will be in demand and is likely to give satisfaction. Openness and a willingness to become an intellectual partner in the program are also important. If the client perceives a tone of selfishness or believes that the consultant is learning more than he/she is contributing, the consultant will find his/her availability for appointments significantly undersubscribed. The successful consultant brings more to the table than an appetite! Energy, inventiveness, scholarship, intelligence, logical thinking, honesty, and diplomacy are clearly highly desirable. The consultant must understand and make clear the difference between opinion and fact.

The ability to handle technical questions in a practical way is of course paramount, but the ability to counsel a discouraged employee is a useful skill as well. When a problem appears intractable to a scientist, energy can fade. Drug seeking is not an activity for the easily discouraged. Reinforcing the desirability of going on or devising a decisive experiment can revitalize enthusiasm in a positive way. Alternatively, joining in conversations with higher authority when it is apparent that no reasonable outcome can be anticipated from work is occasionally necessary. Clear and logical presentation of factual information is vital here and is most effective when put in writing.

TYPICAL CLAUSES IN CONSULTING CONTRACTS

The consultant is usually not considered to be an employee for fringe benefit and tax purposes and a client–consultant relationship is usually not to be disclosed to third parties.

Consulting contracts usually define in a general way the area in which consultation is to be performed, where, the anticipated frequency of visits, and the mode of interactions. They frequently also include the name of the internal person in charge of monitoring and hosting the activity.

Information developed in the course of consultation is proprietary and thus confidential. There are significant penalties to which the consultant is required to agree in advance in the unlikely event that breaches of confidence take place that damage the client. As a direct consequence, the consultant rarely has publication rights. By mutual agreement when the consultant has made a seminal or pervasive contribution, joint publication may result. When the consultant is experimentally or conceptually active in the same field and there may be overlap of ideas, manuscripts are best published with the client as coauthors so that the paper can be vetted within the firm before being submitted. When this is not appropriate, protection for the consultant can be achieved by having the manuscript reviewed within the firm for clearance. The consultant must also take particular care not to transmit to the client proprietary information belonging to third parties.

The consultant must be prepared for the circumstance that some ideas stemming from their contribution may appear in print without attribution when these, however useful, did not play a decisive role in the outcome of the work. The consultant's reward may well be fiscal rather than ego satisfying.

Patent rights are more complex. Legally, the idea is more important than the specific reduction to practice. In consulting, useful solutions often are the result of group discussions, and it can be rather difficult after the fact to determine just when and by whom key concepts were advanced. Multiple rather than individual inventorships are more common in drug seeking. It is legally crucial that the real inventor be included in an application, and this decision should best be made by the firm's lawyers. If the patent is particularly valuable, this can lead to significant trouble for both parties that is best avoided by prior agreement than by trying to achieve mutual satisfaction after a problem has developed.

Given the potential for significant penalties for inadvertent or knowing disclosure to third parties of proprietary information or the possibility of omitting relevant information that might have avoided liability to the client firm, the consultant should seriously consider arranging for professional liability insurance. The American Chemical Society, for example, offers at a reasonable cost such so-called errors and omissions insurance, giving protection against penalties by third parties (clients, patients, customers, etc.) alleging negligence in the rendering of or the failure to provide professional services. If one is providing consultant services through an employer offering such services rather than as an individual agent, the employer will have some coverage, but this will not likely cover you for any services that you may provide outside of the scope of your official position. This aspect will not be covered in a consulting contract but is placed here for the guidance of interested parties.

In the unlikely event that a lawsuit results from the consultant's failure to live up to the strictures in the contract, the contract will usually indicate the state whose laws govern the activity (and where suit will likely take place).

Since consulting quite often involves short-term relationships with a variety of firms, the consultant must have a clear memory of where and when he/she came into possession of novel information and exercise special care not to present third parties with proprietary information that belongs to others. One can use general concepts freely, but specific information cannot be divulged. The consultant will usually be required to certify at the outset that he/she has no relationship with a third party that would fairly be judged to involve a conflict of interest. Furthermore, he/she may be required to certify that no such conflicting relationship will be entered into during the course of the consultantship or for a specific time after the conclusion of the relationship.

Consulting contracts usually specify a specific time frame in which the consultancy is to take place and will usually further specify the length of time in which matter specifically developed as a consequence of the consultancy must be kept secret. It is common for the consultant to agree to return or provide evidence of destruction of materials specific to the consulting activity once the relationship has run its course. Provisions are given for termination by either party if mutual satisfaction is not achieved.

Fees are specifically spelled out in consultancy contracts. The consultant should be realistic in deciding the magnitude of the fees his/her activities warrant while at the same time valuing his/her work. Time is after all a nonrenewable resource. Usually, fees for time expended in short-term consultantships are assessed by the hour and are somewhat higher than those associated with longer time associations, which are usually compensated on a retainer basis specifying the term involved and a minimum or perhaps a number of hours to be provided to the activity. A monthly invoice is usually provided to the firm detailing the time expended and the purpose for which the time was used. The consultant will find it useful to have specified the time period by which the invoice should have been satisfied by reimbursement. Longer term relationships are usually renewable by mutual agreement at a given time, often annually. Out-of-pocket costs incurred at the request of the client are usually reimbursed monthly following the provision of available receipts. The consultant may only bill for hours actually expended on the project at the behest of the client.

Unless specifically indicated otherwise, ownership of technologies developed as a consequence of a consultancy belong to the firm. The most common exception is where the consultant and/or his associates contribute significant independent experimental material to the final product. In that case, a sharing arrangement should be negotiated in advance.

Most consulting contracts forbid the divulgence of proprietary information to third parties unless specific permission is obtained from the firm in writing. An exception is usually made when such information is given to the consultant's coworkers who have a need to know to assist in carrying out the consultant's assignment. In this case, the consultant is liable for any malfeasance of his assistants or associates who have gained proprietary information in this manner.

If the consultant has an active research program in the same area in which he is giving advice, these strictures can be quite inhibitory. The consultant must have a very good memory as to where specific information was gained to avoid accidental release of proprietary information. Most contracts deal with this in part by exempting information that the consultant can document as having been already in his or other's possession at the time independent of the consulting arrangement.

SUBLIMATION OF EGO

Advice must be freely given but need not be acknowledged or followed by the recipient. This can be frustrating to the consultant. This avoids, however, the danger that the consultant might gain authority without responsibility.

Knowledge and technology developed in conjunction with a consultant that solves a problem may be published by the recipient without attribution. This can also be frustrating to the consultant. An intermediate practice is to acknowledge generally the help of the consultant without giving specifics. The consultant's reward is often limited to the internal satisfaction of having done a good job and having profited through fees.

ETHICS OF CONSULTING

A repeated theme in the foregoing is the requirement for the consultant to behave in an ethical manner. Honesty is not only the best policy, it is the only policy. The consultant is well advised to make careful dated notes of consulting activities and keep these away from other eyes. Memories are notoriously unreliable when significant economic value is involved. A good memory of where ideas originated is helpful, but documentation in disputes is often crucial.

When a major contribution has been made to an invention or product by the consultant, he/she may be listed as an inventor on subsequent patents and/or may be included as a coauthor in publications. When the consultant is active independently in the same field, papers on the consultant's own work must be written with special care to avoid disclosing material that arose in the course of consulting activities. When there is possible doubt as to the specific origin of concepts or data, one can let the client see the paper before submission and/or if warranted include one or more of the clients as coauthors. This has the practical result that the paper will then be circulated within the client's workplace so that any potentially troublesome information can be identified and dealt with to mutual satisfaction. It is almost always better to anticipate and to avoid trouble than to try to extricate one's self after a dispute has arisen.

Establishing inventorship can be complex. Consultants occasionally find themselves in group discussions in which most participants are actively contributing ideas. It can be difficult after the fact to attribute accurately each significant idea to particular individuals. Leaving off an inventor from a patent application can lead to significant trouble down the road. Deciding just who is an inventor is a legal matter that is best decided by those trained in patent law rather than by chemists or consultants. This is particularly important when it appears likely that a patent will be commercialized and result in significant royalties, sales, and profits.

QUESTIONS THAT FREQUENTLY ARISE IN CONSULTING

These questions can only be discussed in general terms as the material is privileged unless or until they are revealed publically by the client. They are also idiosyncratic to particular consultations.

Evaluation of the progress and value of specific projects is mainly the responsibility of middle and higher management who depend in part on the judgment of the consultant in reaching decisions. When the need occurs, the desirability of a change in direction is a critical juncture in a program and is not reached lightly. Specific reasons and justifications must be advanced and justified.

Specific questions such as devising alternative means of constructing a given molecule or drug class occur very frequently and are usually dealt with in small group settings. The breadth of command of the relevant literature by the consultant is of particular importance here.

Similarly, the ability to solve absorption, distribution, metabolism, and excretion (ADME)/toxicity defects as a project matures is of crucial importance. Dealing with these problems requires an in-depth understanding of chemistry, physiology, biochemistry, and pharmacology. Available textbooks of medicinal chemistry catalog examples of what can go wrong but are less forthcoming about how to avoid or overcome such problems. The experience and judgment of the consultant is particularly important here.

Potency and specificity of hits and leads are fairly straightforward but not always easy to optimize. Particularly early on in a project, apparently incongruous results are encountered, reflecting the complexity underlying even apparently simple biological test systems. Devising practical and decisive experiments that reveal the key variables involved is a particular skill that can dramatically speed progress. Practical experience of the consultant is very helpful here. An apparently divergent test result or property is often a very important clue. The commonly misquoted saying that it is the exception that proves the rule is actually a verbal shrug suggesting that one should ignore such data. The saying should state that it is the exception that probes the rule. An unexceptional observation is cumulative. An exceptional observation either means that the experiment was done wrong, or that the observation indicates either a poor understanding of the underlying science or that there is a significant reason why the point must diverge. In the latter case, developing an understanding of this phenomenon can lead to a breakthrough of great importance. Technology is accretive, but discovery is transformative!

Among the general types of questions that arise are the following nonexhaustive examples: "How do I enhance or limit the kind of metabolic reaction that is causing a problem in refining this lead?" "This lead is satisfactory when given orally but is insufficiently water-soluble for convenient injection. How can I overcome this molecular defect without starting all over again?" "Protein binding appears to be excessive. How do I decrease this and yet retain activity?" "This lead gives quite interesting activity when injected but performs poorly when taken orally. What might be the kinds of causes for this phenomenon and how can I overcome this without starting all over again? Would making a prodrug overcome this problem?" "High-throughput screening has generated a couple of dozen hits. I cannot pursue them all. Can you help me prioritize them?" "This hit is about as soluble as ground up paving blocks. How can I enhance its water solubility?" "This candidate drug has a terrible taste. Can we overcome this without loading up a dosage form with massive quantities of sweeteners?" "This lead causes 'X,' which we like but also causes 'Y' which we do not like. Can these effects be separated by analoging?" "The receptors for this drug are in the central nervous system. What molecular characteristics would enhance CNS penetration?" "What are the structural requirements for taking advantage of active uptake mechanisms?" "How can we distinguish experimentally between active and passive uptake?"

The list of such questions is potentially endless, but these examples chosen from recent real-life questions illustrate some of the things that the chemical consultant must be prepared to help with in drug-seeking campaigns.

CONSULTANT, MANAGEMENT, AND THE EMPLOYEES

When the consultant's activities result in naming to the firm's board of advisors, he/she is often asked to comment on the relative abilities and accomplishments of key personnel. This is easy when a positive response is called for but can be an uncomfortable situation requiring both honesty and diplomacy when a less enthusiastic response is warranted. Negative assessments should be accompanied by proposals of a course of action that will result in improvement if followed.

Employees talk freely to consultants once a trusting relationship is established. Morale is good when consultations consist predominantly about technical problems. Morale is not good when consultations consist predominantly about interpersonal relationships. Management is not always aware that such problems exist. The consultant can fill a valuable role in helping address these situations provided that it is done carefully so as not to destroy trusting relationships or further exacerbate problems.

The consultant is surprisingly often asked to convey messages to management or, conversely, to reinforce communications from management to employees. Employees are frustrated when repeated advice to management is ignored or an adverse response is forthcoming. In these situations, the consultant is often asked to convey a reinforcing message either to the employee or to management or to both. Interestingly, this often is given more weight than advice originating internally. This circumstance is an embodiment of the "NIH" principle. That is, if an idea was *not invented here*, there might be something to it! Of course the consultant must believe in the message being carried if one is going to do this.

PURPOSEFUL WORK IS THE PROPER MEASURE

Hard work is commendable, but useful accomplishments are of greater utility. The output of purposeful work is the proper measure of the contribution of a scientist or of a program/project. This concept is not always recognized. Peter F. Drucker, the famous business guru, put it most succinctly with the dictum that "efficiency is doing things right: effectiveness is doing right things" [4]. Recognizing this and applying it to specific situations is an important function of effective consulting.

POINTS TO PONDER

An appropriate joke makes a concept easier to remember and, perhaps, to repeat. A joke is not really funny unless there is an underlying truth or a revelatory insight encoded in it. The following examples use sarcastic exaggeration to make important points.

"A consultant knows and can describe 40 ways of making love but has no sexual partners." What this is saying is that advice is more readily accepted and acted upon when it comes from someone who has had success and is speaking from experience. In consulting, success builds upon success. Obviously, if advice results sufficiently often in success, then the recipients are much more open to advice.

"A consultant borrows your watch and then proceeds to tell you what time it is." What this means is that posing an obvious solution is not impressive and does not build confidence. One is usually talking to serious and competent scientists who are strongly invested in their work. They are looking for useful suggestions that have not yet occurred to them. Perhaps they are too close to the problem and do not yet see the potential applicability of alternative technologies residing in literature outside of their normal reading.

One can hear a variation on the old line: "those who can—do; those who can't—consult." This author knows of a situation where a distinguished academic chemist was perceived as carrying away more than he brought in his consultantship with a major drug house and spent much of his time telling what his own research group was doing rather than listening to the problems before his audience. Consequently, management had to compel chemists to talk with him. The contract was soon allowed to lapse.

CONCLUSION

Effective consulting is useful and a pleasure both to the consultant and to the client. It is not easy to do it well. With respect to open access drug-seeking programs, one is receiving information that neither party has generated themselves, so comparatively little prehistory is available to draw upon. A consultant with broad relevant experience in developing hits to leads and leads to candidate drugs can be a priceless addition to the team, particularly when the work is being taken up by persons with little or no prior experience in discovering drugs.

REFERENCES

1. Sodano, C.S., Sturmer, D.M. *Trends in Chemical Consulting*. Washington, DC.: American Chemical Society; 1991.
2. Chemical Consultants Network home page. Available at http://www.chemconsultants.org/. Accessed January 26, 2014.
3. Association of Consulting Chemists and Consulting Engineers. Available at http://www.chemconsult.org/. Accessed January 26, 2014.
4. Drucker, P.F. *The Effective Executive: The Definitive Guide to Getting Things Done*. Classic Drucker Collection ed. Burlington, MA: Elsevier; 2007.

Part IV

NONPROFITS DRIVE BENCH-TO-BEDSIDE INNOVATION

Can't afford to innovate? Open up! —*Henry Chesbrough*

Collaborative Innovation in Drug Discovery: Strategies for Public and Private Partnerships, First Edition.
Edited by Rathnam Chaguturu.

23

OPEN SOURCE DRUG DISCOVERY FOR NEGLECTED DISEASES

Tonny Johnson and Sanchayita Kar

Sciclips, Madison, WI, USA

In the recent years, there has been a steady decline in the number of drugs approved by the U.S. FDA, which has been a major concern in the drug discovery research industry. Pharmaceutical companies are facing major challenges such as lack of innovation in the research and development (R&D) sector and pressure for production of cheaper medicines due to global competition. Although pharmaceutical companies are increasing their R&D budget every year to overcome this problem, they are losing billions in cash flow as many drugs become generic. Today, pharma companies are constantly looking for innovative ways to reduce R&D cost by adopting new approaches like open source, collaboration, and IPsharing in their business model. Open source and open science are new avenues where companies look for talents, resources, tools, and technologies from outside the company. In biomedical science, the open source model was used for the first time in bioinformatics to produce tools to understand the genome and the proteome. The Human Genome Project (HGP) was the first initiative in an open source model where thousands of researchers worldwide worked together with all the necessary tools and resources to sequence the entire human genome and make it openly accessible to researchers worldwide. This opened up new avenues to accelerate drug discovery research. Because of its close and secretive nature, the traditional drug

Collaborative Innovation in Drug Discovery: Strategies for Public and Private Partnerships, First Edition. Edited by Rathnam Chaguturu.

development process could not embrace the open source model. The entire drug development process, from the early discovery phase to the actual manufacturing of the drug, is covered by intellectual property (IP) rights. These rights give the pharmaceutical companies a legal monopoly and the ability to exclude others from using the invention. IP protection is one of the main causes for the high cost of medicines. Millions of people die every year in developing countries because they cannot afford the high price of drugs set by pharmaceutical companies. For the last 8–9 years, there have been collaborative efforts by various nonprofit initiatives, public–private partnerships, product development partnerships, government institutions, and philanthropic consortia to adopt the open source model through the voluntary sharing of knowledge, data, tools, and infrastructure for developing cheaper drugs for the poor people in developing countries who are suffering from neglected tropical diseases (Figure 23.1). Recently, major pharmaceutical companies like GlaxoSmithKline (GSK), Novartis, Merck, Lupin, Bayer, Sanofi-Aventis, and Pfizer are showing interest in the open source model by donating compounds for target drug candidates, and sharing IP rights and high-throughput screening technologies in the "research pool" to speed up the drug development process for neglected diseases. This chapter will review the open source initiatives in drug discovery with special emphasis toward neglected diseases.

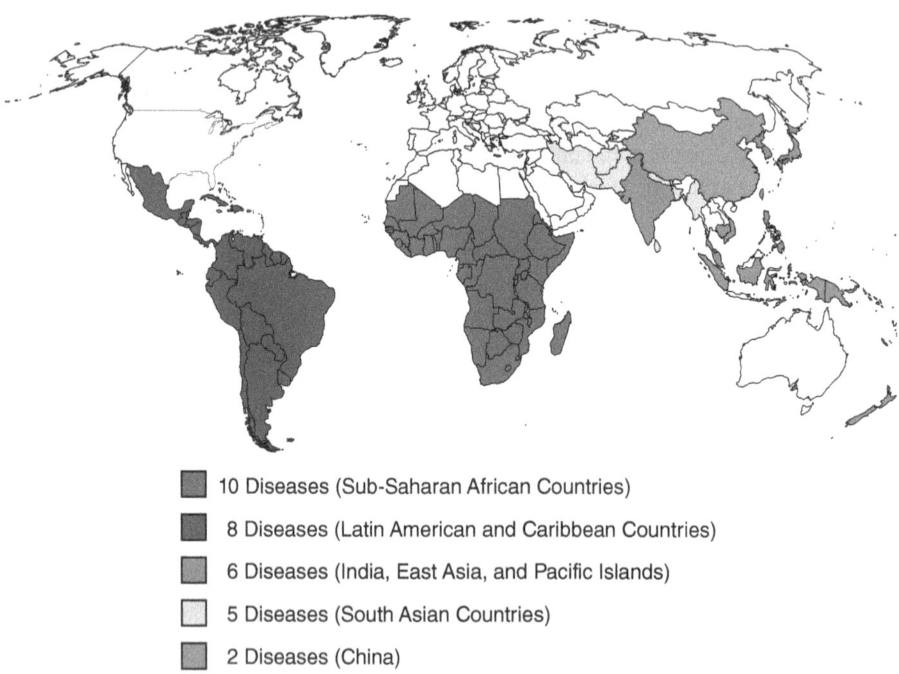

Figure 23.1. Geographical distribution of the 13 tropical neglected diseases ranked by their prevalence. The data have been retrieved from Table 23.1. *Source:* http://www.nejm.org/doi/full/10.1056/NEJMra064142. See color insert.

OPEN SOURCE

The term "open source" came into the limelight in the early 1990s with the development of Linux (a computer operating system, like Microsoft Windows or Apple Mac); however, unlike Windows or Mac, Linux is assembled under the free and open source software development and distribution model [1]. Linus Torvalds, a Finnish American software engineer, adopted the path of free software advocates of the early 1980s like Richard Stallman (mastermind behind the GNU Project), who believed that people have the right to have full control over information and the freedom to share and change software. In 1991, Torvalds launched Linux kernel, which is the most prominent example of free and open source software collaboration to date [2]. Since the source code was made available to the public, Torvalds' approach became known as the "open source" model. Free and open source software allows anybody the freedom to study, use, change, participate in the development of, and even distribute the software. This open source model of product development contrasts with the proprietary or "closed" product development model. Advocates of the free software model came up with a new copyright license, known as the General Public License or "copyleft," which allowed them to make the source code available to the public. In the open system, the transparency of the source code created a platform where skilled people around the world could scrutinize the code and make improvements, using their imagination and creativity to come up with new innovative products [3]. The impact of the open source model adopted in the software industry in the early 1980s is huge. Currently, there are over 30 million users of Linux, which powers over 85% of the top 500 supercomputers in the world and runs on one quarter of all new smartphones. Popular web browsers like Mozilla Firefox, Chrome, and Konqueror are examples of open source software. In fact, anytime we are using Google, Yahoo, Youtube, Facebook, Wikipedia, or any other website, we are communicating with computers running on free open source software [4].

OPEN SOURCE MODEL IN BIOMEDICAL SCIENCE

The unequivocal success of the open source model in software industries caused it to gain momentum over the years and successfully fill the niche neglected by the private sector. The HGP is an example of the emergence of an early open source model in the biopharmaceutical sector. It was the collaborative effort of thousands of scientists all over the world (although primarily in the United States, United Kingdom, Japan, France, and Germany) to map and sequence the entire human genome. The project began in 1990 with the help of the U.S. Department of Energy and National Institutes of Health, and the Wellcome Trust in the United Kingdom; later the National Human Genome Research Institute of Health also joined in the venture. The main objective of the project was to identify disease-causing genes, determine the proteins coding the genes and their function, and use that information for drug development. In 1996, at the summit in Bermuda, all the HGP advocates (scientists as well as policy makers) agreed on a set of rules and principles, now known as the Bermuda Principles, which stated that the DNA sequencing information generated from the project should be open to public within 24 hours of its generation. This rapid release of genomic data from the HGP and making it available to the public revolutionized the scientific approach for data transparency in the biopharmaceutical sector [5]. The Bermuda Principles supported minimizing patent rights for the data generated through HGP. The "First Draft" of the human genome sequence was completed in 2000, and the sequence was published in the public Genbank database in 2001: a 90% complete sequence of all 3 billion base pairs in the human genome.

The open source model of HGP had a huge impact on biomedical research in subsequent years, especially in clinical medicine. Genomic mapping helped scientists understand the genetic basis of common hereditary diseases and assisted with the development of a wide variety of genetic tests for many common diseases. Related fields such as bioinformatics (a relatively new discipline at the convergence of biology and computing that is focused on capturing, storing, analyzing, and integrating biological and genetic information) and proteomics (the study of protein expression throughout an organism) evolved and benefited from the technological and scientific advances made possible by the HGP [6]. It is hardly surprising that in the post-HGP era, more and more biological software and bioinformatics programs and databases surfaced with an open source model: this may be due to the very fact that they are primarily written in codes and can be subject to IP coverage by copyright [7]. Examples of open source include Biojava [8], BioPython [9], Bio-SPICE [10], BioRuby [11], Simple Molecular Mechanics for Proteins [12], and Genetic components for Model Organism Databases [13]. Another major example of open source after the HGP is the HapMap Project. Researchers from six countries combined their efforts to identify and catalog genetic differences in humans and create a haplotype map of the human genome. Data from the HapMap project, which include information about areas known as single nucleotide polymorphisms (where DNA sequences vary among individuals), are open to the public [14].

In 2002, the Swiss Institute of Bioinformatics, the European Bioinformatics Institute, and the Protein Information Resource joined forces to create the UniProt (Universal Protein Resource) consortium [15] to provide a high-quality protein sequence database and functional information (much of it derived from the genome-sequencing project and freely accessible to the scientific community). The Protein Drug Target Database, a freely accessible protein database for *in silico* (computer simulated) drug target identification, currently contains more than 1100 protein entries with 3-D structures presented in the Protein Data Bank [16]. The data included in the open access database launched by Sciclips in 2010, ddTargets (Drug Targets/Disease Target) [17], covers more than 4500 drug targets from U.S. patents, U.S. and international patent applications, and published research articles. Detailed information regarding drug target type, assays, and methods to characterize these drug targets are freely accessible to researchers.

Various biological software programs that evolved after the HGP helped immensely in identification of lead compounds through *in silico* drug design [18]. Currently, various methods are used to identify lead compounds, including computational techniques to study the binding of different drugs or ligands to the target, and computational chemistry, cheminformatics, and bioinformatics techniques for identifying drug targets from genome sequences and databases. The major difference between software development and drug development is that in drug development all these computational techniques and programming could only speed up the process of identifying the lead compound or target molecule, which is not the final product of drug development. However, early identification of lead compounds and target molecules may provide a stepping-stone in the process of open source drug discovery.

WHY IS THE PROGRESS OF OPEN SOURCE DRUG DISCOVERY IN BIOMEDICAL RESEARCH SO SLOW?

It has been more than a decade since the completion of the HGP, and despite the resultant breakthrough technologies and resources there has been very little progress in open source drug discovery research. There are huge differences between software and heath care products, including differences in IP protection, the complexity of the development process and regulatory requirements, the total cost of the development process, and in the time and resources required.

With respect to IP protection, one of the most significant differences between software development and drug development is the licensing system. Software is written in "source codes" and can be protected by copyright licenses, whereas scientific ideas (e.g., new assay development or a new method for a diagnostic test needed for the drug development process) are covered by patents. Patents, which exclude everyone other than the inventor from practicing the invention, are very expensive to obtain and also require a high maintenance cost [19].

The drug development process is also very complex, and is subject to government regulatory requirements. The process of bringing a new drug into the market begins with the identification of thousands of chemical compounds that have the potential to alter the action of the drug target. Out of these, about 250 will be screened by preclinical testing to evaluate the compound's toxic and pharmacologic effects through various *in vitro* and *in vivo* assays in mice and other test animals. Once the candidate drug has gone through a set of rigid tests to prove its safety, compounds are then tested for their pharmacology (absorption, distribution, metabolism, excretion, and toxicology) and sometimes chemically modified to improve their tolerability by the human body. Typically, about 10 of these compounds will be screened for tests on humans, and ultimately only one compound will be selected as a drug for sale after the rigorous process of clinical studies and approval from regulatory bodies like the U.S. FDA or the European Medicines Agency [20]. The approvals required by government regulatory bodies for biomedical products in development are in stark contrast to the development process for software products.

The full cost of bringing a new drug to the market, from discovery through clinical trials to approval, is enormous. The estimated total cost is somewhat controversial, as different studies have calculated different numbers. In 2003, DiMasi et al. reported that bringing a new drug to the market may cost approximately \$800 million [21]. Some health economists estimate the cost between \$500 million and \$2 billion, depending on the therapy or the developing firm [22]. In 2010, in a study published in the journal

Health Economics, one of the authors who criticized the methods used by DiMasi estimated the final cost to be ~$1.2 billion [23]. Critic Marcia Angell, MD, a former editor of the *New England Journal of Medicine*, has called that number grossly inflated and estimates that the total is closer to $100 million. A 2011 study, also critical of the DiMasi methods, puts average costs at $55 million [24].

It takes anywhere between 8 and 14 years to complete the entire drug development process from start to finish. Unlike software products, health-care products require specialized equipment, laboratory space, access to biological reagents, and research tools. All the work at various stages of drug development must be conducted according to stringent Good Laboratory Practices, Good Manufacturing Practices, and Good Clinical Practices [19]. Pharmaceutical companies, like any other for-profit organization, manufacture products that must be sold for a profit in order for the company to grow. Pharma companies have to invest heavily at various phases of drug development, including R&D and preclinical and clinical trials, as well as to protect IP, comply with various organizational requirements, gain approval from regulatory agencies, market the drug, and conduct postmarket monitoring to ensure that IP rights are protected. In the conventional drug discovery model, drug companies usually protect the active ingredients of the drug, the nature and the source of the compound or biological substance, the manufacturing process, and other knowledge-based information through securing patents, which provide protection for a particular exclusivity period during which no one can market the drug without infringing the patent. Because of patent exclusivity, pharmaceutical companies can earn a significant return on their investment.

By contrast, open source drug development has features that won't allow the pharmaceutical companies to impose a monopoly through patent exclusivity. One of the major goals of open sourcing is to promote the free flow of information to the public; this free use and access to information while the drug is in the development process will create further innovation and competition. Innovation and competition will, in turn, significantly lower the price of the drug. It is very unlikely that pharmaceutical companies will ever adopt an open source model for future drug development endeavors, since they cannot recover money invested for the R&D and other drug development processes.

There is no doubt that in the coming years biomedical science will continue to make breakthrough innovations, and pharmaceutical companies will protect these inventions with IP rights and bring new or "me-too" drugs (those that are nearly identical to current medications) to the market. But the question is whether these expensive drugs or high-end treatments can reach the billions of people in developing countries who need them the most. Will pharmaceutical companies invest time and money in conducting R&D to bring new drugs for neglected tropical diseases to the market? The open source model can perfectly fit into drug development for neglected disease, and it could play a vital role in filling this huge unprofitable niche area. Philanthropic consortia, public–private partnerships, and other nonprofit organizations all over the world can come forward and promote the open source model with a common goal to develop and provide low-cost drugs to millions of people suffering from these diseases around the world.

OPEN SOURCE AND PLAYERS IN NEGLECTED DISEASES

Neglected diseases are a group of tropical infections that affect world's poorest populations. These are known as "neglected" since pharmaceutical companies and government agencies do not find these tropical diseases attractive enough to invest in, due to smaller profits and market size. According to the reports, only 10% of global R&D funding goes toward research on neglected diseases that affect 90% of the world's population, a phenomenon known as the 10/90 gap [25]. According to TB Alliance, a nonprofit organization dedicated to fighting tuberculosis (TB), R&D for neglected tropical diseases receives only $1 out of every $100,000 spent worldwide on biomedical research and product development.* The R&D system remains very much focused on developing new drugs for the diseases of the affluent countries (arthritis, acid reflux, bipolar syndrome, diabetes, depression, etc.). There are 13 primary neglected diseases caused by various parasitic, bacterial, and soilborne pathogens. Table 23.1 lists the neglected diseases ranked by their prevelance [26]. HIV/AIDS, malaria, and TB are also considered neglected diseases, as millions of people in developing countries die each year due to these diseases. In addition, the World Health Organization (WHO) also adds the following to the list of neglected diseases: cysticercosis,† dengue or dengue hemorrhagic fever,‡ echinococcosis,§ fasciolosis,¶ rabies,** and yaws;†† and includes two other "neglected conditions": podoconiosis‡‡ (also known as elephantiasis) and snakebite.§§

Each year, millions of people, especially children in the low-income populations in the developing regions of Sub-Saharan Africa, countries of South Asia and East Asia, and the Pacific Islands, Latin America, and Caribbean Islands become silent victims of malaria, tuberculosis, and many other infectious diseases (Table 23.1). These diseases are symptomatic of poor people living in rural areas, city slums, in camps of conflicted zones, in areas with unsafe drinking water and poor sanitary conditions, and with little or no access to health care. Many of these infectious diseases lead to long-term deformities in the patient, cause chronic disabilities, hinder childhood development, and result in slow death. Neglected disease can be categorized into three groups based on the availability of drug/treatment:

- For diseases such as HIV/AIDS, malaria, and tuberculosis, drugs and treatments are available but need to be delivered and made more accessible and affordable to poor people in developing countries.

*See http://www.tballiance.org.

†See http://en.wikipedia.org/wiki/Cysticercosis.

‡See http://en.wikipedia.org/wiki/Dengue.

§See http://en.wikipedia.org/wiki/Echinococcosis.

¶See http://en.wikipedia.org/wiki/Fasciolosis.

**See http://en.wikipedia.org/wiki/Rabies.

††See http://en.wikipedia.org/wiki/Yaws.

‡‡See http://en.wikipedia.org/wiki/Elephantiasis.

§§See http://en.wikipedia.org/wiki/Snakebite.

TABLE 23.1. The Major Neglected Tropical Disease Ranked by Prevalence

Diseases	Global Prevalence (Million)	Population at Risk	Region of Highest Prevalence
Ascariasis (roundworms)	807	4.2 billion	East Asia and Pacific Islands, Sub-Saharan Africa, India, South Asia, China, Latin America, and Caribbean
Trichuriasis (whipworm)	604	3.2 billion	East Asia and Pacific Islands, Sub-Saharan Africa, India, South Asia, China, Latin America, and Caribbean
Hookworm infection	576	3.2 billion	East Asia and Pacific Islands, Sub-Saharan Africa, India, South Asia, Latin America, and Caribbean
Schistosomiasis (snail fever)	207	779 million	Sub-Saharan Africa, Latin America, and Caribbean
Lymphatic filariasis (elephantiasis)	120	1.3 billion	India, South Asia, East Asia and Pacific Islands, Sub-Saharan Africa
Trachoma	84	590 million	Sub-Saharan Africa, Middle East, and North Africa
Onchoceciasis (river blindness)	37	90 million	Sub-Saharan Africa, Latin America, and Caribbean
Leishmaniasis (kala-azar/ black fever)	12	350 million	India, South Asia, Sub-Saharan Africa, Latin America, and Caribbean
Chagas disease	8–9	25 million	Latin America and Caribbean
Leprosy	0.4	ND	India, Sub-Saharan Africa, Latin America, and Caribbean
Human African trypanosomiasis (African sleeping sickness)	0.3	60 million	Sub-Saharan Africa
Dracunculiasis (Guinea worm)	0.01	ND	Sub-Saharan Africa
Buruli ulcer	ND	ND	Sub-Saharan Africa

Source: http://www.nejm.org/doi/full/10.1056/NEJMra064142.
ND, not determined.

- Treatments for some neglected diseases, such as tuberculosis and leishmaniasis, are outdated, and the pathogens have become drug resistant, or the drugs have severe toxicity, poor efficacy, or a complex mode of administration. In spite of huge funding from different nonprofit organizations and government agencies, there has been no approval for any new tuberculosis drug in the last 11–12 years.

The last FDA-approved drugs for tuberculosis (rifabutin and moxifloxacin) came onto the market in 1999 [27].

- For some neglected diseases (dengue fever, Buruli ulcer), no good treatment exists, and effective tools like vaccines, diagnostics, drugs, and other preventive measures are either lacking or highly inadequate [27].

In the past 5–7 years we have seen a tremendous awareness and efforts among global governments, stakeholders, government sponsored public–private partnerships, nonprofit drug R&D organizations, and even multinational pharmaceutical companies worldwide to bring new drugs and treatments for the neglected tropical diseases [28, 29]. Product development partnerships (PDP), including nonprofit organizations such as Drugs for Neglected Diseases initiative (DNDi), Foundation for New Innovative Diagnostics, Institute for OneWorld Health, Médecins Sans Frontières (also known as Doctors Without Borders), Medicine for Malaria Venture, and Global Alliance for TB Drug Development (TB Alliance), have played a major role in raising awareness through advocacy for R&D on developing drugs for neglected diseases. They work with pharmaceutical companies, government agencies, the World Bank, and various philantropic organizations to reduce the cost of R&D through the voluntary sharing of knowledge, data, tools, and infrastructure.

Many pharmaceutical companies over the last few years have showed interest and begun to collaborate with various nonprofit organizations to develop drugs for neglected tropical diseases [30, 31]. A series of events happening in the early 2000s may lead to an environment conducive to the establishment of organizations like PDPs and early involvement of pharmaceutical companies in the open source model.

In 2001, major multinational pharmaceutical companies attempted to prosecute the South African government for allowing production of generic drugs. Various activist groups like the William J. Clinton Foundation and the International Community Foundation, along with the South African government, the European Parliament, and 300,000 people from over 130 countries signed a petition against their action, and finally the pharma companies were forced to back down from their case. In 2000 Cipla, an Indian pharmaceutical company, started to produce generic antiretroviral drugs for a significantly cheaper price. This led to tremendous competition between the brand name, proprietary drugs, and the generic drugs, as the price difference was remarkable. At a cost of $10,000–$15,000 per person per year, the price charged by pharmaceutical companies was practically unaffordable for poor people in developing countries. By the middle of 2001, a triple combination cocktail therapy was available from Ranbaxy, an Indian generic manufacturer, for as little as $295 per person per year [32].

Tremendous social pressure from the public sector, government agencies, and the media, along with increasing awareness created by nonprofit organizations for the neglected diseases during that time, helped to change the dynamics and led many pharmaceutical companies like GSK, Bayer, and Novartis to start collaboration and partnership with TB Alliance and other PDPs.

In the last few years, we have seen tremendous willingness among pharmaceutical companies around the world to open up and give away many of their resources, including patent rights and IP, information about compounds, databases (genome sequence, proteomic database), screening techniques, and diagnostic assays, which will be very helpful for drug development for malaria, TB, and other neglected diseases. The International Federation of Pharmaceutical Manufacturers and Associations found that collaborative efforts between pharmaceutical companies and nonprofit organizations focusing on drug development for neglected diseases have increased fivefold since 2005, and about 80% of neglected disease research efforts are now done through collaborations [33].

In 2008, DNDi partnered with GSK to develop drugs against visceral leishmaniasis (kala azar), human African trypanosomiasis (sleeping sickness), and Chagas disease. In 2009, Sanofi-Aventis started to collaborate with DNDi in the development, manufacturing, and distribution of fexinidazole, a promising antiparasitic drug for safer and more effective treatment of human African trypanosomiasis, a fatal disease that threatens 60 million people in Sub-Saharan Africa. Currently, fexinidazole is in Phase 1 clinical trials and recently researchers have identified fexinidazole as a new potential drug for visceral leishmaniasis [34].

In 2008, the partnership between Sanofi-Aventis and DNDi resulted in the development and distribution of an antimalarial treatment, Coarsucam™ or Artesunate Amodiaquine Winthrop® (ASAQ) in just 3 years. The drug has been marketed for less than $1 for adult treatments and for less than $0.50 for childern in sub-Saharan Africa. ASAQ is made available to governments and nonprofit organizations at a cheaper cost [35]. In 2009, collaboration between Novartis and the Institute for Microbiology and Epidemiology in Beijing resulted in the development and distribution of the antimalarial drug Coartem®. This combination of traditional Chinese plant-based malarial remedy with a synthetic compound is highly effective and carries low risk of resistance when used in combination with other antimalarial drugs [36]. Pfizer partnered with the International Trachoma Initiative (ITI), a nonprofit organization formed to fight trachoma, the leading cause of blindness in developing countries. ITI distributes the antibiobic Zithromax, donated by Pfizer to treat trachoma [37].

In 2008, Cambia, a nonprofit group in Australia, backed by a $3M grant from the Bill and Melinda Gates Foundation, partnered with Queensland University of Technology and established a global open source initiative promoting patent system transparency worldwide. This was to be a giant step forward for making the IP system more open and effective for developing new drugs for the poor countries [38].

In 2009, Novartis, a Swiss drug company, collaborated with the Institute for OneWorld Health, a nonprofit organization that develops drugs for neglected diseases, and gave access to its proprietary research data to develop a drug for diarrhea, a leading cause of death among children in poor countries due to unhygienic living conditions [39].

In January 2010, GSK provided the public domain with 13,500 (potential lead) antimalarial compounds screened out of 2 million compounds. GSK has also created a patent-pool (pool for open source innovation), which aims to remove restrictive IP rights over drugs for neglected diseases. The main aim of the open source patent pool was to accelerate innovation and drug discovery efficiency of biotechnology-based drugs, vaccines, and diagnostics for neglected diseases and to save lives in developing countries. In another collaborative effort, GSK teamed up with Alnylam pharmaceuticals, a leading RNAi Therapeutics company, that donated another 1500 patents into the pool. GSK contributed $8M seed money for "Open Lab" and other new collaborations to share IP rights for neglected diseases [32]. In May 2010, the Massachusetts Institute of Technology (MIT), a leading research institution, announced its intent to join the open innovation patent pool started by GSK, by contributing its IP rights and know-how in neglected tropical disease research. MIT will be the first academic institution to join the group for this noble cause [40]. Now the patent pool has more than 2300 patents, which are freely accessible to the public.

Indian scientists at Open Source Drug Discovery and the Council of Scientific and Industrial Research have mapped the entire *Mycobacterium tuberculosis* genome and posted the information to the web* for free access [41].

The African Network for Drugs and Diagnostics Innovation, which promotes and sustains African-led product R&D activities, discovery, development, and delivery of new drugs, is a good example of an open access initiative that supports scientific networking and the exchange of information between African scientists, policymakers, and their peers on other continents [42].

FUTURE OF OPEN SOURCE DRUG DEVELOPMENT

The poor people of developing countries do not have access to proper health-care services and are often diagnosed when diseases are in their final stages. They frequently cannot afford drugs, even if they are diagnosed in proper time. For this reason, it is absolutely necessary that rapid, low-cost diagnostic tests be made available in these countries. These diagnostic tests are assays designed for use at health facilities with minimal infrastructure. Many of the facilities lack the technology, equipment, and trained personnel required for complicated tests.

Next-generation drugs in malaria, tuberculosis, and other neglected diseases need to be developed through open source innovation to break the current business monopoly of pharmaceutical companies worldwide. To achieve this, it is necessary to scale up the global R&D capacity by promoting newer technologies in drug development for neglected diseases. Open source models should emphasize cutting-edge technologies to develop antibody-, protein-, and peptide-based drugs, rapid diagnostics tests, vaccines, new therapeutic solutions for neglected diseases, and a more efficient and functional open IP system for the discovery of new drugs.

Future open source drug development for neglected diseases needs openness in all fronts. An ideal open source model should have open exchange of scientific information and data sharing not only during the discovery phase but also during drug development and beyond. It will take time to change the traditional "closed" system of drug discovery to an open source model. Open source models need new ideas and new tools to make decisions on effective collaborations and "where" and "how" to introduce opportunities for openness in the drug discovery process (see Figure 23.2). At present, the open source model looks promising for neglected diseases, as the organizations that are investing

*See http://www.osdd.net.

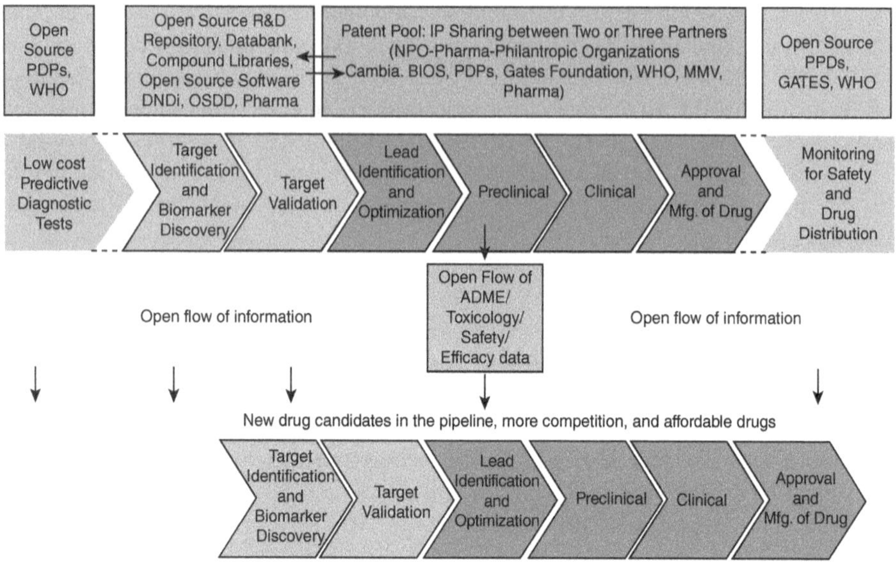

Figure 23.2. An open source model for neglected tropical diseases. The model describes the open flow of information at various levels of the drug discovery pipeline for neglected tropical diseases.

money are nonprofit or philanthropic consortia, or nonprofit organizations collaborating with pharma and biotech companies. The open flow of information during the drug development process will lead to more innovation, parallel research, and more competion. This will result in production of cheaper and more accessible drugs, which are the two major goals for open source drug discovery.

One of the major breakthroughs for neglected diseases took place in 2007, when the U.S. Congress enacted legislation to increase development of new drugs for neglected tropical diseases. A Priority Review Voucher created by the U.S. FDA provided a powerful incentive for companies who are willing to invest in the development of new drugs and vaccines for neglected tropical diseases. The bearer of the voucher is entitled to get priority review rather than standard review, which can cut the FDA review process from an average of 18 months down to 6 months (important for fast release of neglected drugs in the market) [43].

Most recently, in January 2012, the Bill and Melinda Gates Foundation, along with 13 major pharmaceutical companies, the governments of the United States, the United Kingdom and the United Atab Emirates, the World Bank, and other global health organizations, announced that they would work together to eradicate neglected diseases by 2020. This is one of the largest collaborative efforts undertaken to date to combat neglected diseases [44]. They will provide more than $785 million to support R&D programs and take an active role in the efficient distribution of existing drugs so that they are easily accessible. Most of the pharmaceutical companies have agreed to extend their drug donation programs through 2020 and share their expertise and libraries of compounds to support this R&D. The U.S. government, the UK government, and the Bill and Melinda Gates Foundation have collectively pledged to spend over $200 million on prevention and treatment of neglected tropical diseases by 2013. The money will be invested in R&D, sponsoring awards for innovative ideas for low-cost diagnostics and detection tools, and ideas for preventive measures for the neglected diseases.

Until now, most pharma companies extended their help by donating drugs for neglected diseases (which is very important, since these drugs are donated at low cost to the patients), but drug donation is not the only solution. More pharma companies should participate in R&D for development of new and improved drugs and vaccines using cutting-edge technologies and apply their "know-how" in drug development to neglected diseases.

In the past, we have seen uneven allocation of funds within the neglected diseases, with funding highly concentrated on some diseases, like HIV, malaria, and TB. According to the Tufts Center for Drug Development, HIV, malaria, and TB were jointly funded at $2.5 billion in 2007. Though TB and malaria received similar funding, no new drug has been developed for tuberculosis for the last 9–10 years. Expenditure on R&D for HIV/AIDS was $1.8 billion (out of a total $2.5 billion), but no vaccines or new drugs have been approved [45]. At the same time, there are neglected diseases that do not have any treatment or cure and suffer from lack of funding. Governments should take an active role in encouraging new drug development in neglected diseases, take appropriate measures to improve general health care by educating the people of developing countries regarding cleanliness, and provide safe drinking water and proper sanitation to prevent these neglected diseases. Nonprofit organizations and private–public partnerships should consider developing health facilities with adequate infrastructure to aid in proper distribution and accessibility of drug treatments to patients.

REFERENCES

1. Raymond, E.S. *The Cathedral and the Bazaar: Musings on Linux and Open Source By an Accidental Revolutionary*. O'Reilly Media, Inc.; 2001.
2. DiBona, C., Ockman, S. *Open Sources: Voices from the Open Source Revolution*. O'Reilly Media, Inc.; 1999.
3. McGowan, D. Legal Implications of Open-Source Software. Available at http://ssrn.com/abstract=243237 or http://dx.doi.org/10.2139/ssrn.243237. Accessed December 13, 2013.
4. Introduction to Linux, Free Software and Open Source. Available at http://www.linux.org.au/linux. Accessed December 13, 2013.
5. Contreras, J.L. Bermuda's legacy: policy, patents, and the design of the genome commons. *Minn J Law Sci Technol*. 2011;12(1):61–125.
6. Cooper, N.G., editor. *The Human Genome Project: Deciphering the Blueprint of Heredity*. Mill Valley, CA: University Science Books; 1994.
7. Maurer, S.M. Open Source drug discovery: finding a niche (or maybe several). 2007. Available at SSRN http://ssrn.com/abstract=1114371 or http://dx.doi.org/10.2139/ssrn.1114371. Accessed December 13, 2013.

8. Holland, R.C., Down, T.A., Pocock, M., Prlić, A., Huen, D., James, K., Foisy, S., Dräger, A., Yates, A., Heuer, M., Schreiber, M.J. BioJava: an open-source framework for bioinformatics. *Bioinformatics* 2008;24(18):2096–2097.
9. Cock, P.J., Antao, T., Chang, J.T., Chapman, B.A., Cox, C.J., Dalke, A., Friedberg, I., Hamelryck, T., Kauff, F., Wilczynski, B., de Hoon, M.J. Biopython: freely available Python tools for computational molecular biology and bioinformatics. *Bioinformatics* 2009;25(11): 1422–1423.
10. Garvey, T.D., Lincoln, P., Pedersen, C.J., Martin, D., Johnson, M. BioSPICE: access to the most current computational tools for biologists. *OMICS.* 2003;7(4):411–420.
11. Goto, N., Prins, P., Nakao, M., Bonnal, R., Aerts, J., Katayama, T. BioRuby: bioinformatics software for the Ruby programming language. *Bioinformatics* 2010;26(20):2617–2619.
12. Eisenmenger, F., Hansmann, U.H.E., Hayryan, S., Hu, C.K. An enhanced version of SMMP—open source software package for simulation of proteins. *Comput Phys Commun.* 2006; 174:422.
13. O'Connor, B.D., Day, A., Cain, S., Arnaiz, O., Sperling, L., Stein, L.D. GMODWeb: a web framework for the Generic Model Organism Database. *Genome Biol.* 2008;9(6):R102.
14. Thorisson, G.A., Smith, A.V., Krishnan, L., Stein, L.D. The International HapMap Project Web site. *Genome Res.* 2005;15(11):1592–1593.
15. Apweiler, R., Bairoch, A., Wu, C.H., Barker, W.C., Boeckmann, B., Ferro, S., Gasteiger, E., Huang, H., Lopez, R., Magrane, M., Martin, M.J., Natale, D.A., O'Donovan, C., Redaschi, N., Yeh, L.S. UniProt: the Universal Protein knowledgebase. *Nucleic Acids Res.* 2004; 32:D115–D119.
16. Zhenting, G., Li, H., Zhang, H., Liu, X., Kang, L., Luo, X., Zhu, W., Chen, K., Wang, X., Jiang, H. PDTD: a web-accessible protein database for drug target identification. *BMC Bioinformatics* 2008;9:104.
17. Open access database for Therapeutic Drug Targets/Disease Target and Combination Therapy. Available at http://www.sciclips.com/sciclips/drug-targets-main.do and http://www.sciclips.com/sciclips/novel-potential-drug-targets.do. Accessed December 13, 2013.
18. Ekins, S., Hupcey, M.A.Z., Williams, A.J., Bingham, A., editors. *Collaborative Computational Technologies for Biomedical Research.* Wiley and Sons; 2011.
19. Marden, E. Open source drug development: a path to more accessible drugs and diagnostics? *Minn J Law Sci Technol.* 2010;11(1):217–266.
20. Lipsky, M.S., Sharp, L.K. From idea to market: the drug approval process. *J Am Board Fam Pract.* 2001;14(5):362–367.
21. DiMasi, J.A., Hansen, R.W., Grabowski, H.G. The price of innovation: new estimates of drug development costs. *J Health Econ.* 2003;22(2):151–185.
22. Adams, C.P., Brantner, V.V. Estimating the cost of new drug development: is it really $802 million? *Health Aff.* 2006;25(2):420–428.
23. Adams, C.P., Brantner, V.V. Spending on new drug development. *Health Econ.* 2010; 19(2):130–141.
24. Timothy, N. The Make-Believe Billion: how drug companies exaggerate research costs to justify absurd profits. *Slate.* Available at http://www.slate.com/articles/business/the_customer/2011/03/the_makebelieve_billion.html. Accessed December 13, 2013.
25. Kilama, W.L. The 10/90 gap in Sub-Saharan Africa: resolving inequities in health research. *Acta Trop.* 2009;112:S8–S15.

26. Hotez, P.J., Molyneux, D.H., Fenwick, A., Kumaresan, J., Sonia, P.H., Sachs, E., Sachs, J.D., Savioli, L. Control of neglected tropical diseases. *N Engl J Med.* 2007;357:1018–1027.
27. Cohen, J., Dibner, M.S., Wilson, A. Development of and access to products for neglected diseases. *PLoS ONE* 2010;5(5):e10610.
28. Talaga, P. Open innovation share or die. *Drug Discov Today* 2009;14:21–22.
29. Gustavsen, K., Hanson, C. Progress in public–private partnerships to fight neglected diseases. *Health Aff.* 2009;28:1745–1749.
30. Munos, B. Can open-source R&D reinvigorate drug research? *Nat Rev Drug Discov.* 2006;5:723–729.
31. Mullin, R. Paying attention to neglected diseases: the drugs for neglected diseases initiative is mobilizing public/private partnerships. *Chem Eng News* 2009;87(16):21–24.
32. GlaxoSmithKline announces open innovation plans. 2010. Available at http://opensource.com/business/10/1/glaxosmithkline-announces-open-innovation-plans. Accessed December 13, 2013.
33. Bansal, S. Top five ways Big Pharma can address neglected diseases. Available at http://www.forbes.com/sites/sarikabansal/2011/11/22/top-five-ways-big-pharma-neglected-diseases/. Accessed December 13, 2013.
34. Researchers identify fexinidazole as potential new therapy for visceral leishmaniasis. 2012. Available at http://www.news-medical.net/news/20120203/Researchers-identify-fexinidazole-as-potential-new-therapy-for-visceral-leishmaniasis.aspx. Accessed December 13, 2013.
35. Sanofi-Aventis press release. 2010. Available at http://en.sanofi.com/Images/13744_20100525_GBC%20AWARD_en.pdf. Accessed December 13, 2013.
36. Scientists who developed Novartis Coartem® and Glivec® receive 2009 "European Inventor of the Year" awards. 2009. Available at http://www.malaria.novartis.com/downloads/press-releases/2009-05-european-inventor-of-the-year.pdf. Accessed December 13, 2013.
37. Partnership to end blinding trachoma through the International Trachoma Initiative. Available at http://www.pfizer.com/responsibility/global_health/international_trachoma_initiative.jsp. Accessed December 13, 2013.
38. Grose, S. Online resource aims to smooth the biomed patent search. *Nat Med.* 2009;15:1242.
39. Novartis to partner with OneWorld Health to develop diarrhea drug. 2009. Available at http://www.medicalnewstoday.com/articles/157581. Accessed December 13, 2013.
40. Massachusetts Institute of Technology (MIT) joins pool for open innovation against neglected tropical diseases. 2010. Available at http://www.businesswire.com/portal/site/home/permalink/?ndmViewId=news_view&newsId=20100505006034&newsLang=en. Accessed December 13, 2013.
41. Indian scientists refuse to patent tuberculosis genome, encourage anyone to make the drugs. 2010. Available at http://www.techdirt.com/blog/itinnovation/articles/20100412/0118378969.shtml. Accessed December 13, 2013.
42. Nwaka, S. Advancing drug innovation for neglected diseases—criteria for lead progression. *PLoS Negl Trop Dis.* 2009;3(8):e440.
43. Grabowski, H.G., Ridley, D.B., Moe, J.L. Priority review vouchers to encourage innovation for neglected diseases. Available at http://www.law.harvard.edu/programs/petrie-flom/workshops_conferences/2008_workshops/Grabowski.pdf. Accessed December 13, 2013.
44. Bill & Melinda Gates Foundation, 13 pharma companies come together to eliminate tropical diseases. *The Economic Times* 2012. Available at http://articles.economictimes

.indiatimes.com/2012-01-31/news/31010344_1_tropical-diseases-river-blindness-research-work. Accessed December 13, 2013.

45. Progress in neglected disease drug development. Boston Biotech Clinical Research. 2010. Available at http://bostonclinicalresearch.com/blog/bid/45091/Progress-in-Neglected-Disease-Drug-Development. Accessed December 13, 2013.

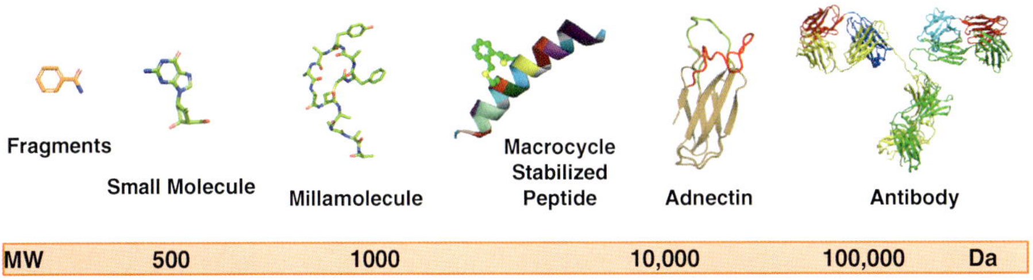

Figure 4.8. The evolution of the chemical space is essential to expand the druggable genome and identify potential therapeutic solutions.

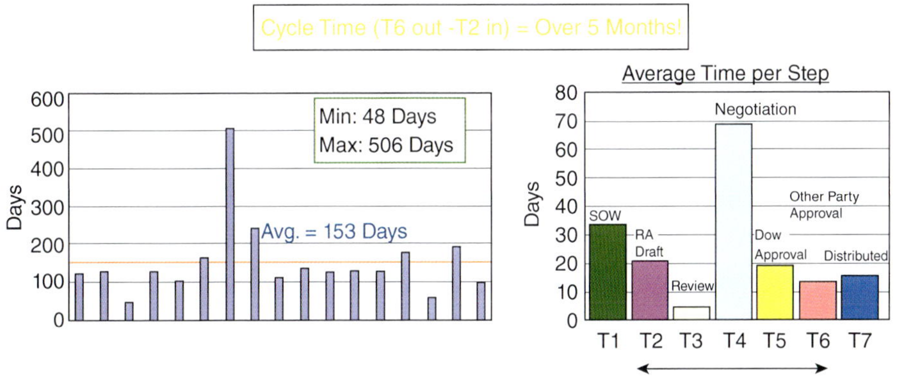

Figure 7.1. Dow Six Sigma study of cycle time for sponsored research agreements in the United States. Courtesy of Susan Butts, Dow Chemical Co.

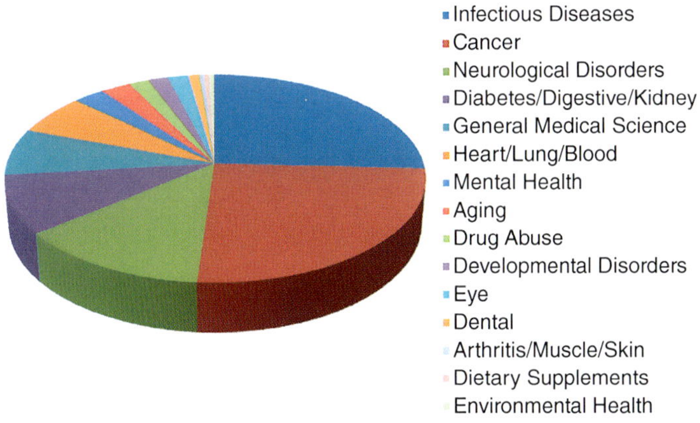

Figure 8.3. NIH Institute representation in the *Assay Development for HTS* Awards.

Collaborative Innovation in Drug Discovery: Strategies for Public and Private Partnerships, First Edition.
Edited by Rathnam Chaguturu

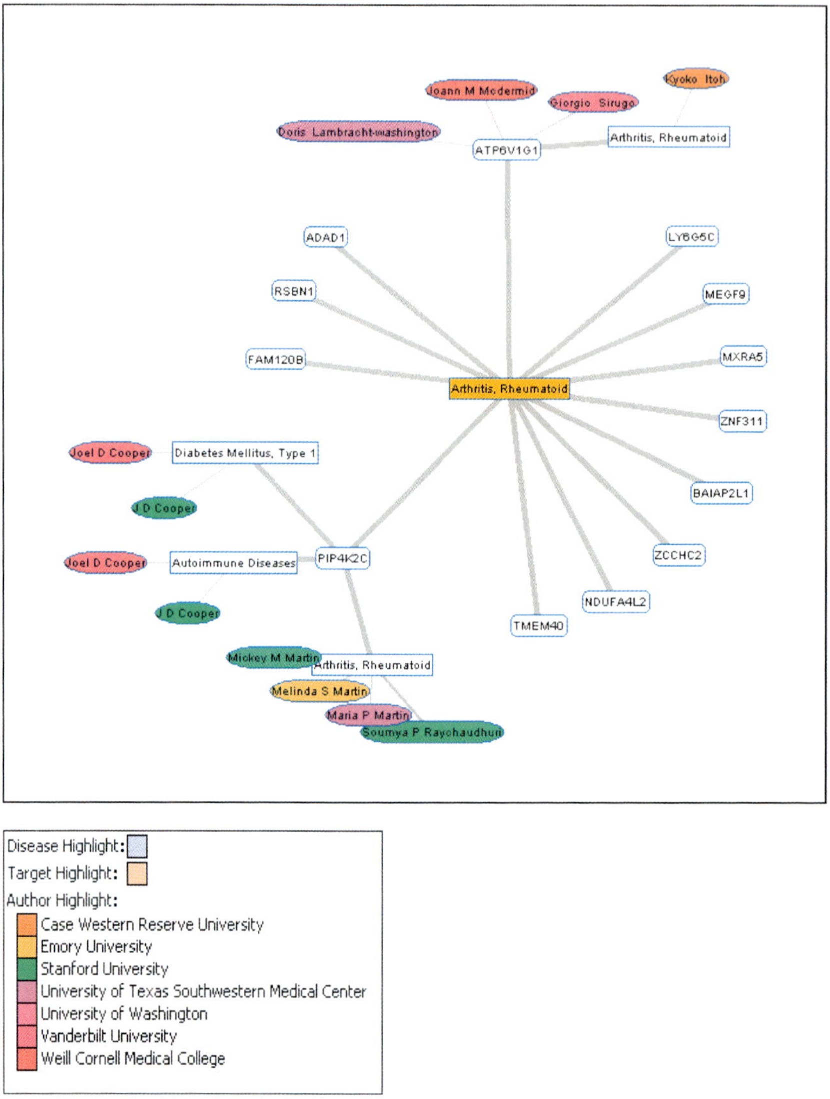

Figure 9.4. Shown are authors working on PIP4K2C and ATPV1G1 in the context of RA and other diseases.

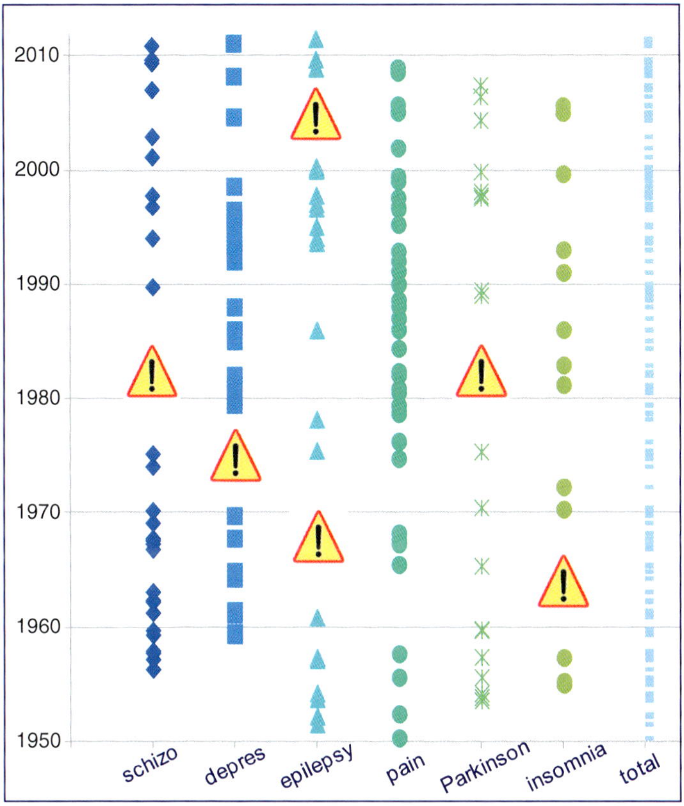

Figure 13.2. Innovation comes in waves separated by long dry spells. Ten-year dry spells are not unusual.

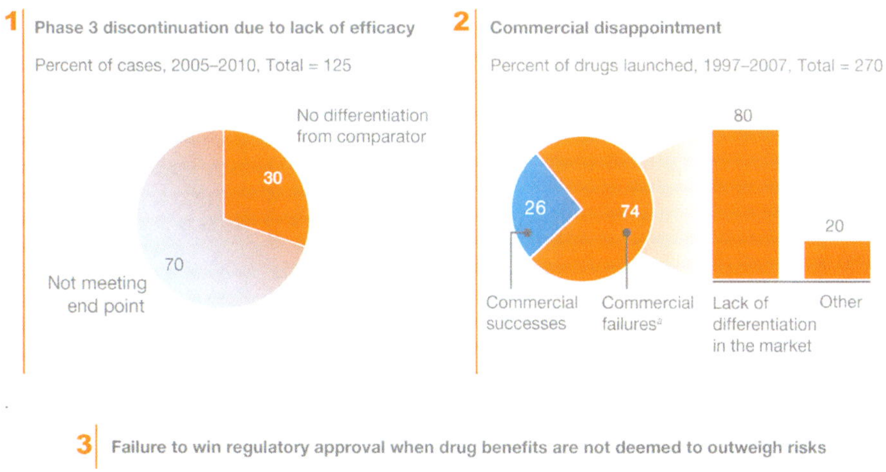

Figure 14.1. Value-driven drug development helps mitigate three key risks. [a]Commercial failure defined as NPV at launch less than average cost of development. *Source*: Pharmaprojects, APM Health Europe, Evaluate, McKinsey analyses.

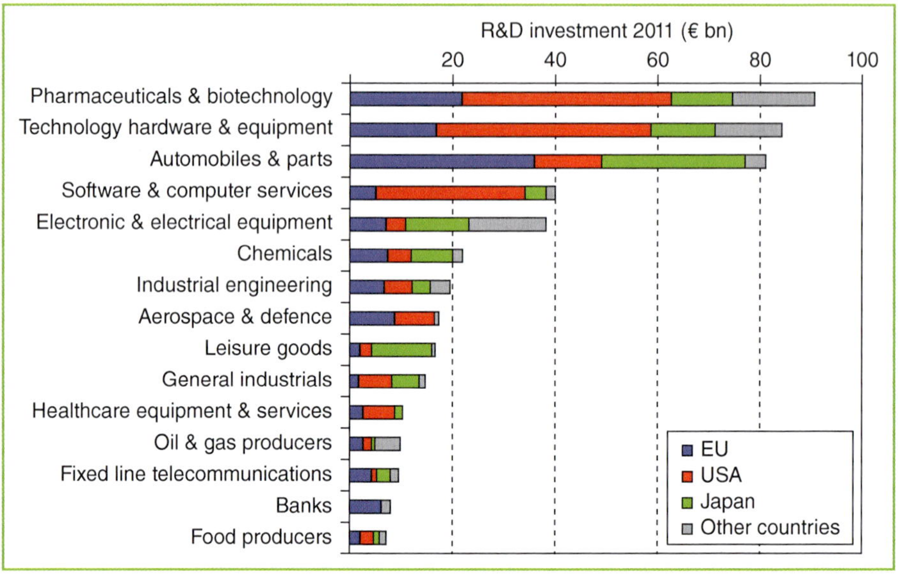

Figure 15.2. R&D ranking of industrial sectors and share of main world regions for the world's top 1500 companies. *Source*: The 2012 EU Industrial R&D Investment Scoreboard European Commission, JRC/DG RTD.

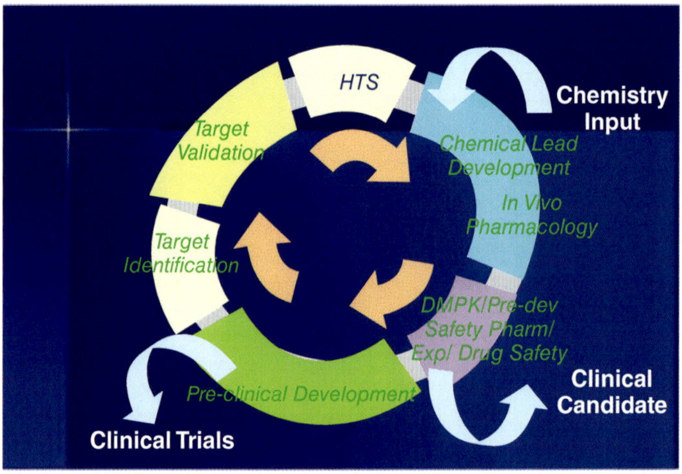

Figure 15.4. Target identification and validation in the drug development process.

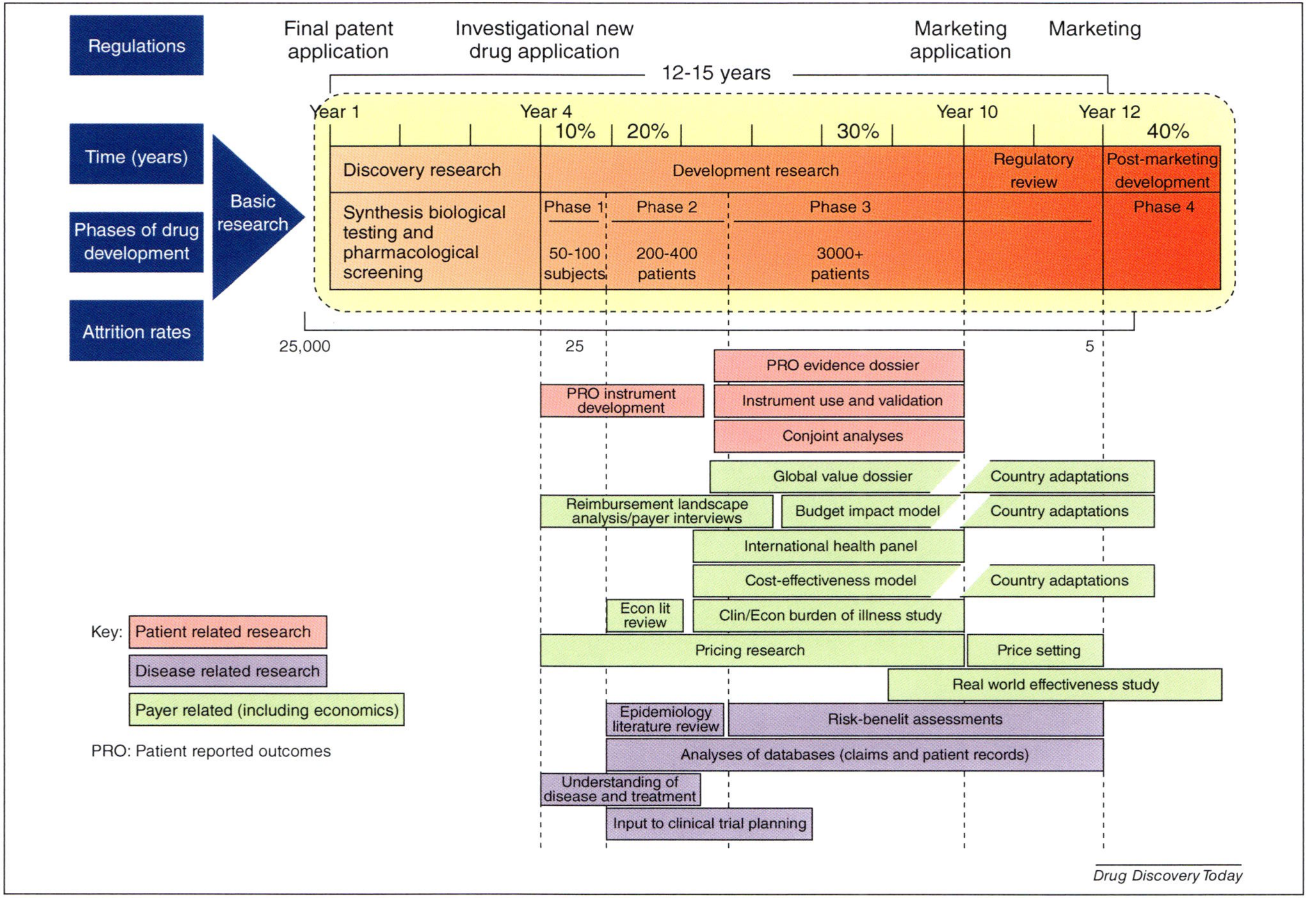

Figure 15.5. Timelines of research phases and attrition rates in drug development. *Source: Drug Discovery Today*.

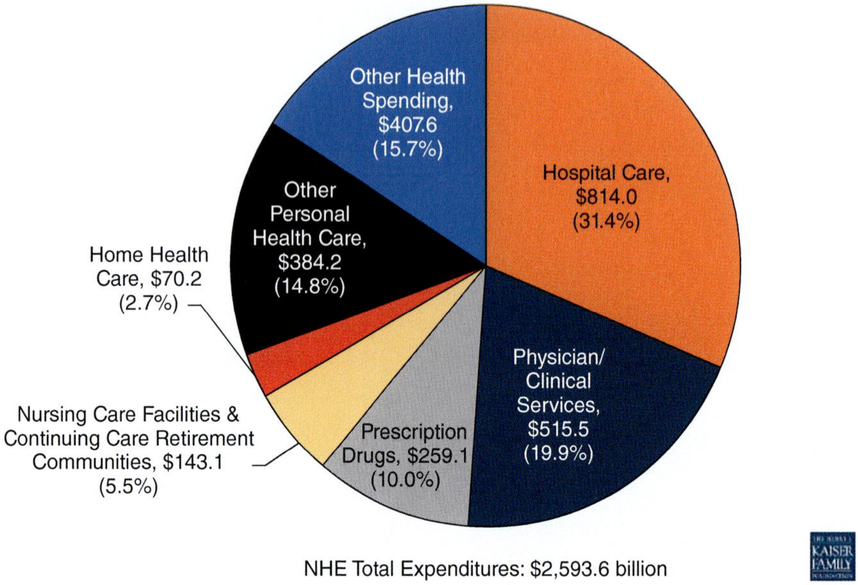

Figure 15.8. Distribution of (U.S.) National Health Expenditures, by type of service (in billions), 2010. (See text for full caption.)

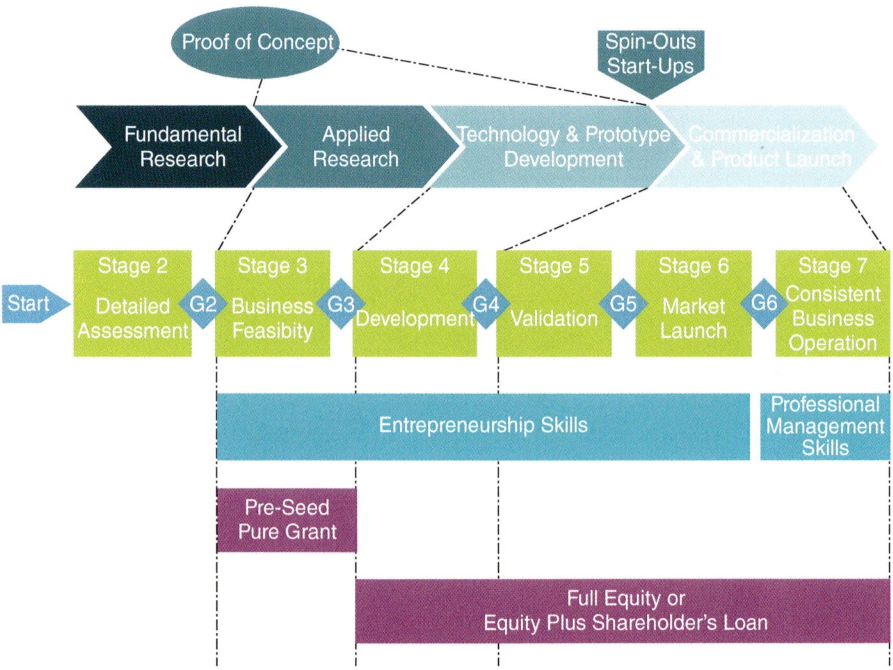

Figure 15.10. Assessment and investment activities throughout the drug development lifecycle.

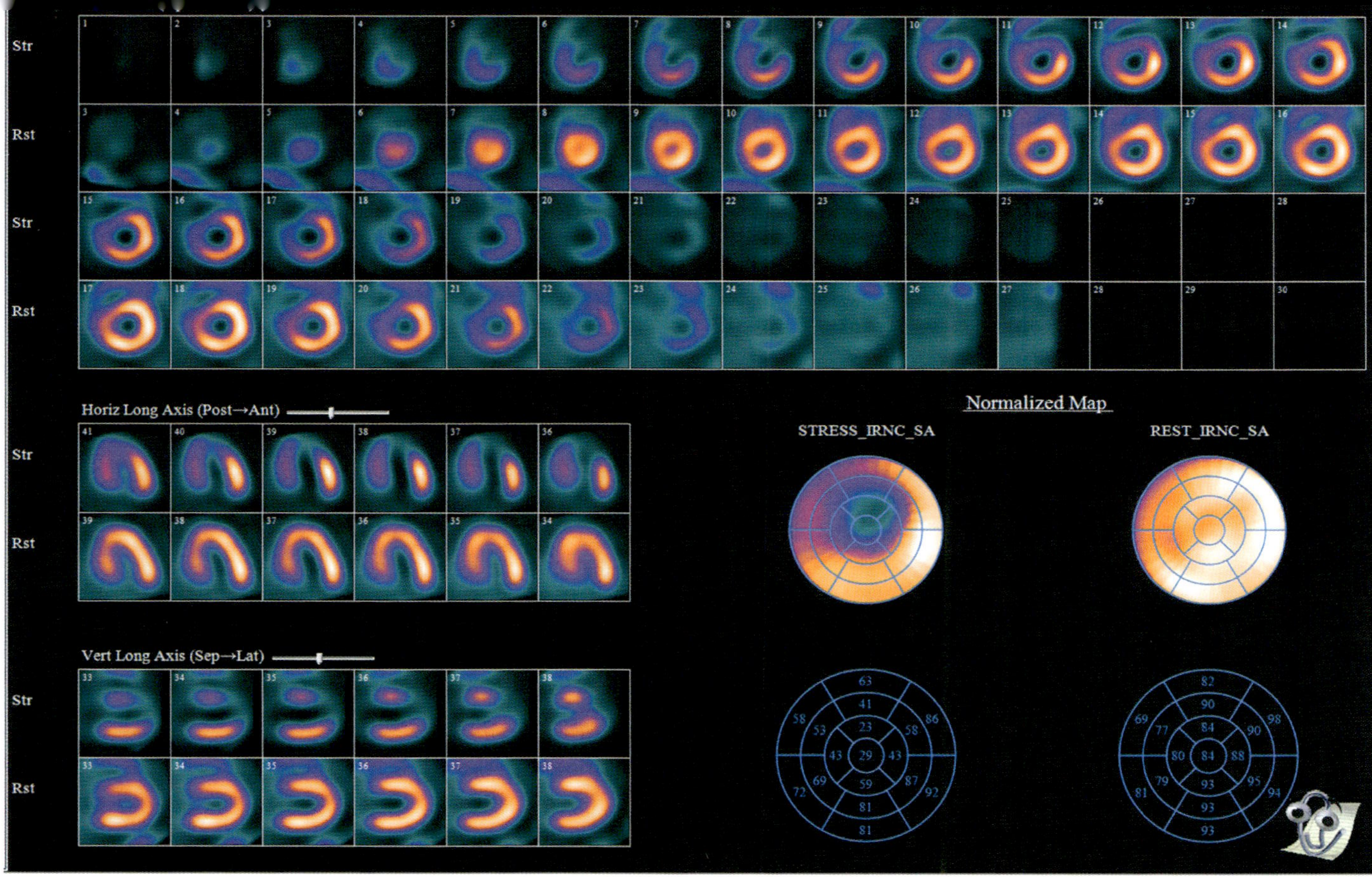

Figure 17.1. Single photon emission computed tomographic (SPECT) myocardial perfusion images demonstrating ischemia during dipyridamole stress (Str) and normal perfusion at rest (Rst) in an experimental porcine model. Ischemia was induced with a 2.5-minute occlusion of the left anterior descending coronary artery and produced a large area of reduced uptake in the anterior region on the stress images. The radiotracer is a new I-123 iodinated derivative of rotenone designated as CMICE-013.

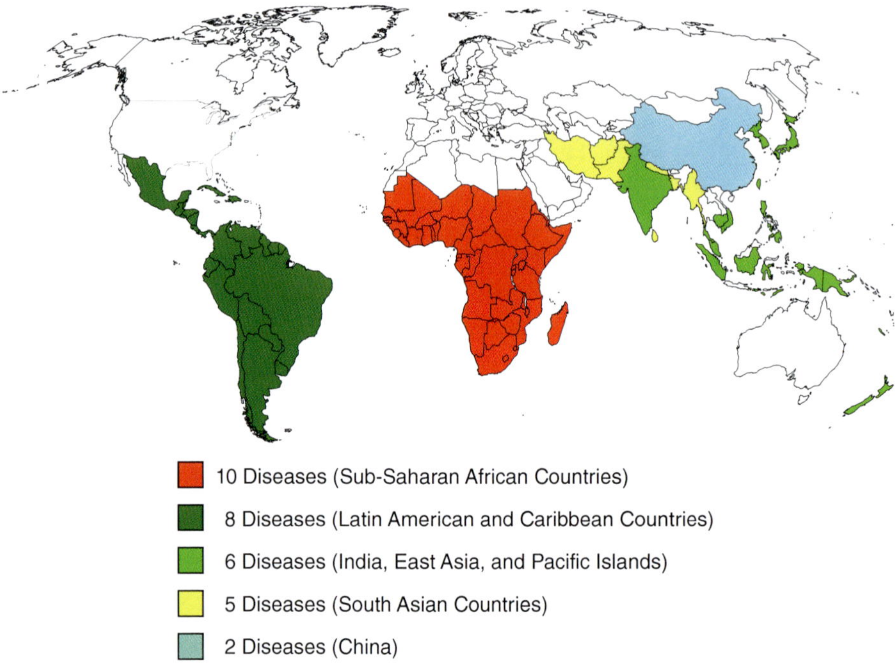

Figure 23.1. Geographical distribution of the 13 tropical neglected diseases ranked by their prevalence. The data have been retrieved from Table 23.1. *Source:* http://www.nejm.org/doi/full/10.1056/NEJMra064142).

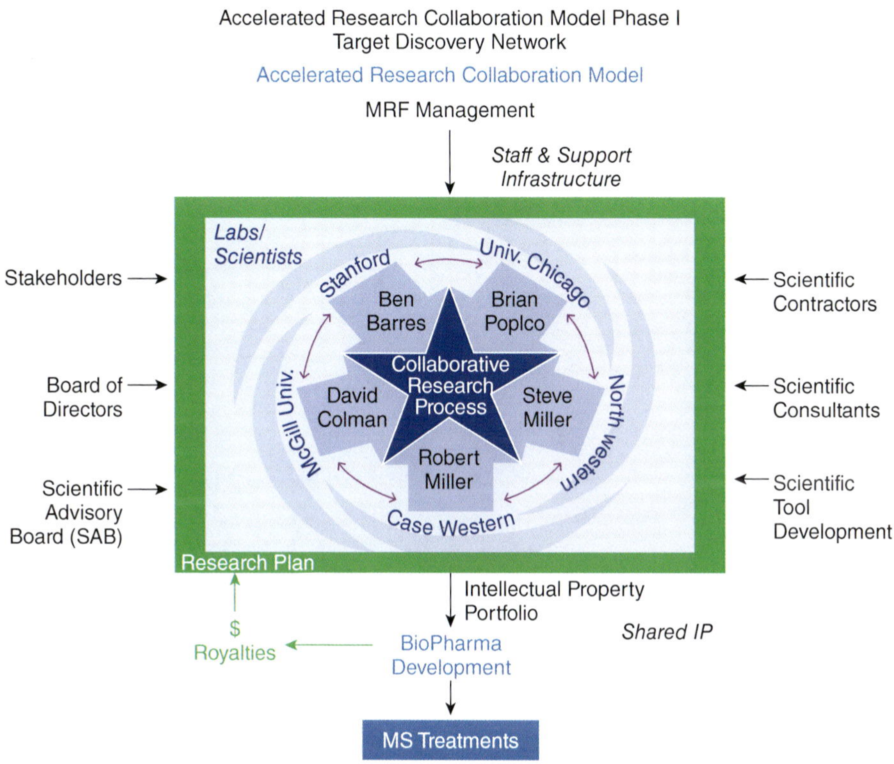

Figure 24.1. 2003 ARC model graphic.

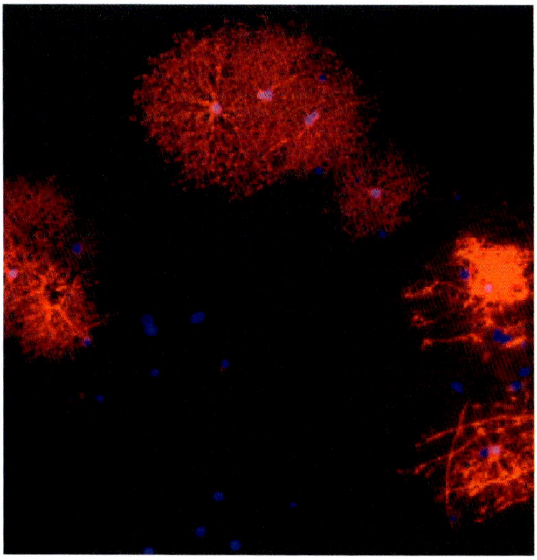

Figure 24.3. *In vitro* myelin image.

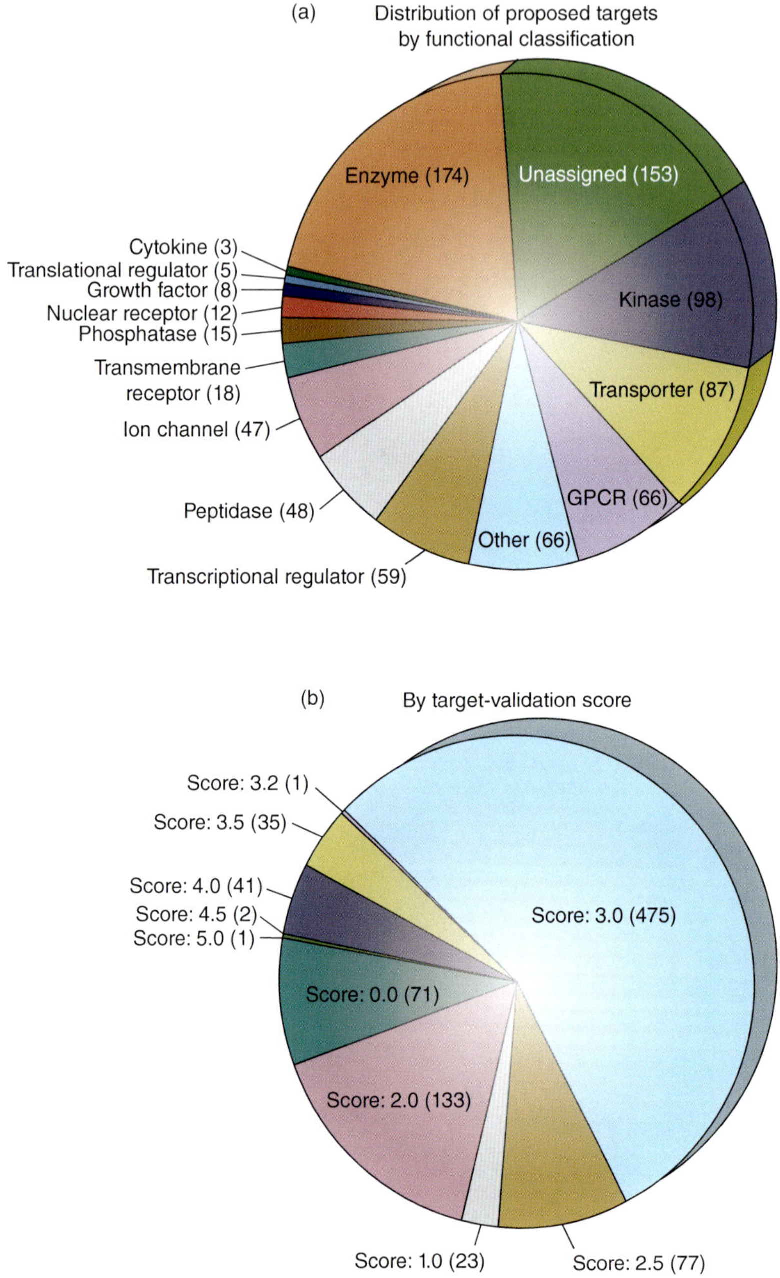

Figure 25.5. Distribution of proposed targets: (a) by functional classification, (b) by target-validation scores.

Assay	Cellular Function	HCS Analysis	HCS Library & Format
Bioparticle Uptake Inhib.	Phagocytosis	Particle Count	110k (chem) / 384-well
Beta-Catenin Activ.	Signaling	(Trans-) Location	200k (chem) / 384-well
Invadopodia Inhib.	Invasion, Cytoskeleton	Fluor. Aggregates/Structures	50k (chem) / 384-well
Cilia Modulators	Cytoskeleton, Developmental	Fluor. Aggregates/Structures	50k (RNAi) / 384-well
Insulin Promoter Modul.	Signaling, Gene/Protein Expr.	Fluor. Intensity / Sub-population	225k (chem) / 384-well
PML-Oncogenic Domain Activ.	Nuclear Foci	Fluor. Aggregates	380k (chem) / 384-well
Lipid Droplet Inhib.	Storage	Fluor. Intensity / Fluor. Aggregates	300k (chem) / 384-well
GPCRs (b-arrestin)	Signaling	Fluor. Aggregates	3 x 300k (chem) / 384-well 1 x 330k (chem) / 1536-well

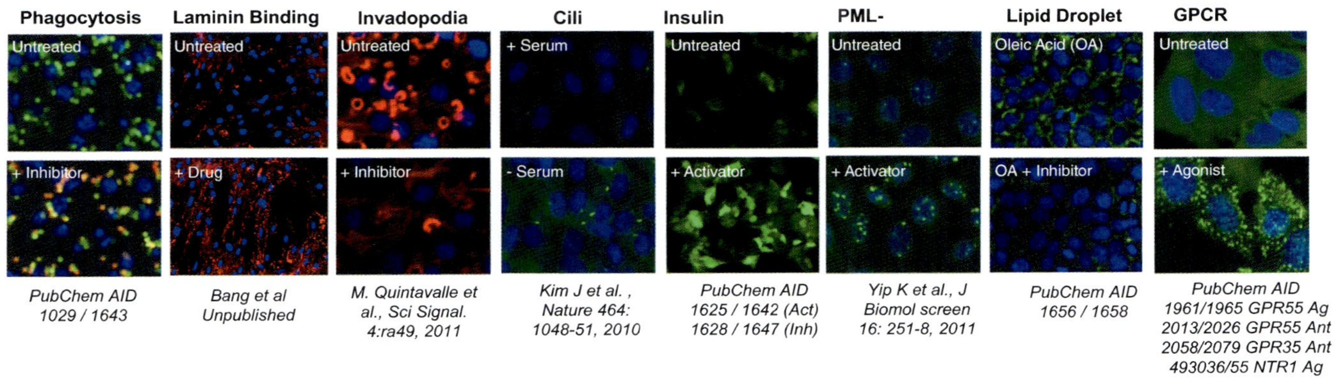

Figure 27.6. Examples of high-content screen (HCS) campaigns completed by the Prebys Center for NIH roadmap projects. The table details the HCS automated microscopic image-based assays that illustrate the breadth of phenotype or cellular function interrogated, the analysis conducted, the number of substances screened during each campaign, and the well density of the assay. The montage of microscopic images are false-colored vertically paired example images of untreated control versus their respective drug/inhibitor/activator/agonist treated samples. (Generally, the nuclei are stained blue with 4′,6-diamidino-2-phenylindole [DAPI], a fluorescent stain that binds strongly to DNA). The PubChem Assay Identifiers (AIDs) or literature references are given for each pair.

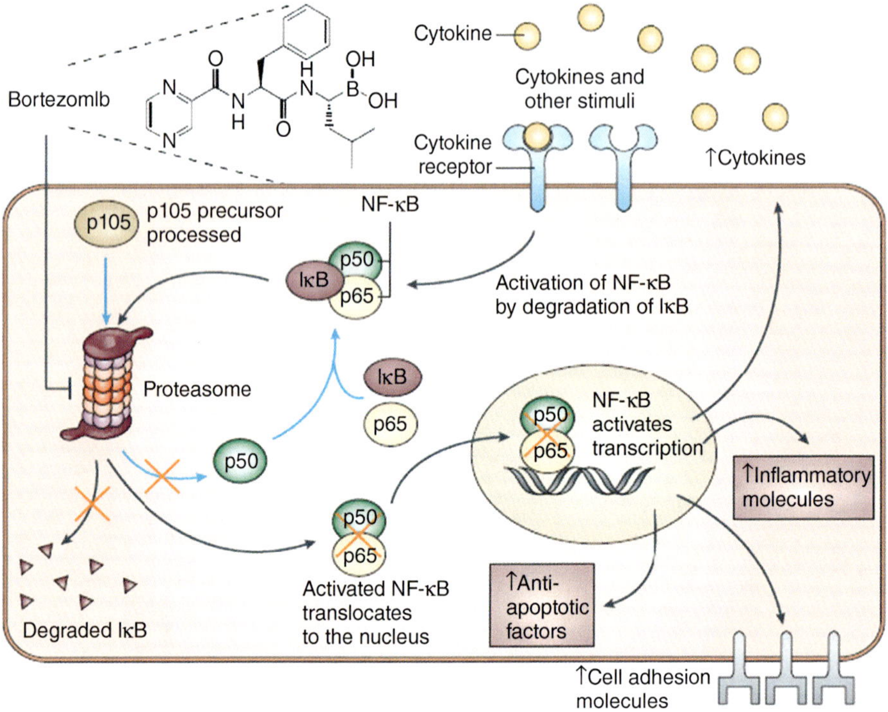

Figure 35.1. The Proteasome, NF-κB and bortezomib. The proteasome is a barrel-shaped multiprotein particle that destroys proteins marked for degradation by conjugation to ubiquitin. Binding of the transcription factor nuclear-κB (NF-κB) to the inhibitor protein IκB in the cytoplasm renders NF-κB inactive. Cellular stimuli, such as cytokines, antigens, oxidants, viruses, and other agents, trigger a cascade of signal transduction events that phosphorylate and ubiquitinate IκB, leading to its degradation by the proteasome, which in turn allows the liberation of NF-κB and its translocation into the nucleus. Once in the nucleus, NF-κB binds to promoter regions of genes that are involved in the activation of transcription, growth, angiogenesis and anti-apoptotic factors and cell-adhesion molecules. By inhibiting the proteasome, bortezomib inhibits the activation of NF-κB (orange crosses) and subsequent events that can promote tumor cell survival and proliferation. *Source*: Sánchez-Serrano, I. Success in translational research: lessons from the development of bortezomib. *Nat Rev Drug Discov*. 2006;5:107–114.

24

THE MYELIN REPAIR FOUNDATION ACCELERATED RESEARCH COLLABORATION™ MODEL: INNOVATIVE DISRUPTION IN BIOMEDICAL RESEARCH

Gali Hagel

MRF supporter and freelance writer, Burlingame, CA, USA

In the short time since it began funding research on multiple sclerosis (MS) in 2004, the Myelin Repair Foundation (MRF) has been widely recognized for its pioneering Accelerated Research Collaboration™ (ARC™) model and resulting achievements. Among its many honors has been *Fast Company* magazine's March 2011 ranking of MRF among the most innovative companies worldwide. The MRF—the only nonprofit to be included in the biotech category—earned the ninth spot in the Top Ten list of biotech companies, selected for its "unorthodox, highly collaborative approach to drug development" [1].

INTRODUCTION

The MRF, the world's largest nonprofit myelin repair research organization, is demonstrating that its ARC model significantly shortens the time to market for treatments and cures not just for MS but for all diseases. Its early successes have been driven by a

Collaborative Innovation in Drug Discovery: Strategies for Public and Private Partnerships, First Edition.
Edited by Rathnam Chaguturu.

strategic roadmap that expedites scientific discovery and validates it for commercial therapeutic development. Among the other innovations ARC brings to drug research, it also fundamentally challenges the traditional model of competition for publishing at the academic level by requiring its scientists to actively collaborate on a strategic, multidisciplinary plan to solve a larger, therapeutic problem. This organized form of open source collaboration leads to better-designed, more innovative experiments as well as constant sharing of discoveries, information, and ideas within the team. It also creates a framework for individual researchers to achieve greater, faster results than they could have focusing on their next experiment, working on their own in isolated laboratories.

Why was, and is, the ARC model needed? The facts confirm what patients and their families have known for years. The number of new medicines coming to market in the last 50 years has been dismally small. In 2010, for example, the FDA approved only 21 such medicines in spite of an annual expenditure of approximately $120 billion for all medical research. The cost in quality of life and real dollars of not delivering safe and effective drugs to chronically ill patients is enormous. Treatments and cures are making their way to patients at a glacial pace; in too many cases, this has cost patients their lives and it will continue to do so. This situation is entirely inconsistent with the massive investment by academia and industry of talent, time, and money in the discovery, development, and commercialization of candidate therapeutics. As MRF President and founder Scott Johnson summarizes the problem, "The system is failing us."

ARC is a work in progress, and part of the challenge the MRF has accepted is to advance its model beyond the basic science phase into the translational research phase, across the notorious valley of death. Various definitions of the terms "translational research" and "Valley of Death" can be found in the literature. From the perspective of the MRF and its ARC model, translational research is what happens between the discovery of a potential drug target in the academic lab and the acceptance of that target by big pharma or the biotechnology sector for commercial development. The translational research phase is critical because industry is requiring that drug targets developed in academic laboratories be validated before it agrees to advance the target through its own pipeline. In spite of the fact that myelin repair drug targets present unusually difficult validation issues, the MRF is addressing those issues with the goal of continuing acceleration past academia, past the translational research phase, and into pharmaceutical labs.

The term "Valley of Death" includes translational research and refers more broadly to the entire process: it is the gap between basic research in the academic lab and the advent of a safe, effective medicine on the market. This gap greatly limits capitalizing on the more than $40 billion currently being invested in academic medical research.

The MRF is demonstrating in practice that academia, nonprofit research organizations, and the pharmaceutical industry can collaborate to identify and achieve the fastest, most effective routes across the Valley of Death to get needed treatments to patients. How, specifically, is the MRF bridging this gap?

- By applying best business practices to academic medical research—setting and meeting objectives for its principal investigators, and putting a premium on accountability and performance
- By replacing competition with collaboration
- By adopting a targeted, interdisciplinary approach to research
- By developing innovative solutions to intellectual property ownership
- By engaging industry to facilitate commercialization
- By staying focused on the goal: effective and available patient medication
- By transforming the road from lab to patient into a win–win for all parties at all stages of the process.

ARC has generated serious interest among other nonprofits, academia, industry, and the government. Even NASA has explored how the model might help it solve pressing life science questions related to long-duration space flight. Media coverage has been extensive, with articles in publications ranging from *Fortune* and *Forbes* to *Business Week*, *The Wall Street Journal*, *The New York Times*, and more. The MRF's leadership in both the science of myelin repair and the development of translational research continues to be widely recognized.

THE BEGINNING

The early story of the MRF is in large part the story of Scott Johnson, who saw something terribly amiss in the process of getting drugs to patients and had the professional experience, personal motivation, and unique vision to effectively identify and disrupt the weaknesses entrenched in that process.

Johnson was diagnosed with MS in 1976, when he was 20 years old. At the time he was a civil engineering student at the University of California, Davis; with relatively mild symptoms, he went on to receive his BS in Civil Engineering from UC Davis and his MBA from the University of California, Berkeley Haas School of Business. From there, Johnson joined the Boston Consulting Group as a management consultant and later founded and headed up three successful start-ups.

On the day in 2001 that changed the trajectory of his life, Johnson happened across a short article in *Business Week* that described an experiment whose results indicated that myelin repair might be achievable. The possibility of reversing the damage to the myelin sheath that occurs in MS was especially exciting because the proposed treatment, unlike the available treatments at the time—all of which acted to broadly suppress the immune system and only slowed the progress of the disease—had the potential to restore lost function to patients. Given the side effects of the immune-suppressing drugs on the market, and the fact that medical science still didn't know what causes MS, repair seemed the far better bet.

True, the first four disease-modifying drugs had come on the market in the previous decade, and they did slow progression of the disease and reduce the number of relapses to some extent in about a third of patients. As far as they went, they were a huge boon to patients with MS (while at the same time representing a more than $10 billion annual market for pharma). But none of them halted the disease or reversed the damage it caused. All treatments lost efficacy after a period of time, and for patients with primary progressive MS, they didn't work at all.

The problem at its core was that the scientific community still didn't know what causes MS. While not accepted in all quarters, the predominant theory was (and continues to be) that MS is an autoimmune disease: that damage to the myelin sheath occurs when the blood–brain barrier is breached, and the immune system mistakes the myelin sheath for a foreign body and attacks it. Even assuming the autoimmune theory was correct, the precise mechanism by which this cascade of events occurs had not (and has not yet) been established.

This was consistent with Johnson's personal experience and that of virtually all MS patients. When he was first diagnosed, Johnson's doctor had told him it would be 25–30 years before a cure was found. Almost 30 years later, newly diagnosed patients were being told the same thing. No visible progress toward a treatment that would substantially arrest or reverse symptoms for all forms of MS, much less a cure, had been made.

Scott Johnson wanted to understand why this was the case. He recognized that undertaking to find a cure would likely be a far more long-term process than focusing on a way to reverse the damage caused by MS—in other words, discovering how myelin is created, how it is destroyed, and how it can be recreated. Johnson, being the problem-solving entrepreneur he was, accepted the challenge with a sense of urgency and a realistic appreciation of the difficulties that lay ahead.

With his strong entrepreneurial background, Johnson first formed a Business Advisory Board composed of diverse members with varied areas of expertise. The first meeting was held in January 2002; like all meetings of the Business Advisory Board it was informal. Johnson brought a pile of materials he had culled from the Internet, including information about the prestigious Gordon Research Conferences, the premier international conference on myelin repair. In the meantime, Johnson set out to learn what was and was not known about myelin biology. He needed to know who the experts were and how they did their work, and he needed to understand the process of medical research: the players, their relationships, and their respective incentives.

Johnson's research revealed a complex system in which academia and industry worked virtually isolated from each other, with different agendas, cultures, goals, and incentives. Academic researchers worked in silos, guarding rather than sharing valuable findings until they were published, sometimes years later. Just as detrimental, the same experiment might be performed multiple times by other researchers, wasting valuable time and money, and failed experiments were never published. The reward systems in academia—the recognition brought by publication, promotions, tenure, and continuation of funding—all contrasted with enhanced shareholder value in industry. Incredibly, academic research seemed to have no goal directly related to what should have been the obvious: patient treatment.

In 2002, Johnson attended the Gordon Conferences, meeting the world's experts and taking copious notes. His conclusion, after talking with many of the scientists involved, was consistent with the studies he had researched: while there was clear intellectual interest in myelin repair, there was no specific research focused on the practical goals of developing drug targets and moving them on through to development of successful, effective treatments for patients. In addition, conference attendees were interested in his idea of a coordinated, collaborative, multidisciplinary research effort. A footnote indicator of the nature of the abstract interest in the problem was that many of the researchers had never met a person with MS until they met Johnson. He asked some of the scientists whom they would select if they could assemble their dream team to conduct research on myelin repair, and why. To his surprise, many gave him detailed responses, indicating their hunger for the changes Johnson already suspected were needed: collaboration and support across disciplines and institutions. From there it was a matter of making it happen.

Johnson concluded that to identify viable drug targets within a reasonable period of time, a more efficient, results-oriented approach to myelin repair was desperately needed: one that would not simply satisfy the intellectual curiosity of researchers and ensure the continued flow of research funds, but that would be targeted specifically to finding effective patient treatments.

The solution would be to create a new research methodology: an outcome-driven, transparent construct that would shave years or even decades off the length of time between the initial phases of laboratory research and getting treatments to patients. With an entrepreneur's focus on results, he approached the problem as he would a start-up. He established the nonprofit Myelin Repair Foundation headquartered in Saratoga, California—appropriately, in Silicon Valley, the center of innovation in countless fields—as the vehicle for the dramatic change that was now his vision.

Challenges abounded in every direction.

INITIAL CHALLENGES

Those challenges were all the more daunting because they weren't just about setting up a medical research nonprofit: at their core they would require a major disruption of a long-standing and, in some quarters, closely guarded, status quo. The ARC model, which Johnson had already formulated, could not work as an abstract set of principles; it could succeed only if properly translated into practice. The way he and his team accomplished this provides a roadmap for practical application of the model.

Johnson would have to attract a small but strong initial staff that understood and believed in his idea and was willing to gamble that it would work. The researchers funded by the MRF would have to be world-class, from complementary disciplines, and willing to break out of the competitive way they and their colleagues had operated throughout their careers. They would also need to have buy-in from their academic institutions. Johnson would have to discover how to get around the conviction that excellence and recognition could belong only to those who worked competitively, concealing their discoveries until they would be lauded at publication at some point a year or two or three distant, and, at the same time, convince university technology transfer offices to share the intellectual property resulting from the joint efforts of their researchers.

One by one, Johnson tackled the list of challenges.

Management Team

Unlike many patients who have founded nonprofit research initiatives, Johnson was not wealthy and did not pour a personal fortune into this endeavor. Instead, he sought initial seed capital from friends and business colleagues who believed in his concept and the organization's mission. In fact, for 2 years he took no salary or compensation, and others involved also performed as volunteers. When an office was obtained in September 2003 (rent-free, courtesy of a landlord also frustrated by the slow pace of medical research!), Rusty Bromley, whom Johnson had recruited as his Chief Operating Officer, made the change from volunteering to being the first paid employee. Maintaining laser focus on the vision, mission, and goals of the MRF, Johnson and Bromley worked closely together to accomplish their objectives. As defined in one of the organization's founding documents, those objectives included (1) actively supporting the development and execution of the MRF strategic and annual research plans and process, (2) evaluating progress toward performance milestones, (3) raising funds, (4) ensuring all research expenditures were fully accounted for by member labs and universities, (5) making complete and accurate financial reports of all expenditures available to donors and appropriate government agencies, (6) providing regular scientific progress reports to donors and interested members of the media and the public, (7) identifying inventions appropriate for protection, (8) developing revenue streams to ultimately self-fund the MRF, and (9) establishing relationships with commercial entities with technology directly relevant to achievement of the MRF's mission to accelerate research and development (R&D) of therapeutic targets.

Principal Investigators

Johnson had already taken the initial steps toward identifying the potential members of a world-class team of researchers and labs. Principal investigators (PIs) were selected based on specific criteria, including their complementary knowledge and expertise, past contributions to understanding the key biological processes and interactions that control myelination, common interest in developing treatments to repair myelin, and willingness to collaborate with each other. Each PI and lab would be critical to the MRF's achievement of its first 5-year research objective: to attract commercial interest in myelin repair in general, and specifically in discoveries generated by the MRF team. An outstanding team came together:

- Dr. Ben Barres, Chair of Stanford University School of Medicine's Neurobiology Department and Professor of Neurobiology, Developmental Biology and Neurology

- Dr. Brian Popko, the Jack Miller Professor of Neurological Diseases, Director of the Jack Miller Center for Peripheral Neuropathy, and Associate Chair for Research in the Department of Neurology at the University of Chicago
- Dr. Stephen D. Miller, the Judy Guggenheim Research Professor of Biomedical Sciences, Professor of Microbiology-Immunology and Director of the Interdepartmental Immunobiology Center at Northwestern University
- Dr. Robert Miller, Vice Dean for Research at the School of Medicine, and the Allen C. Holmes Professor of Neurological Diseases in the Department of Neurosciences at Case Western Reserve University
- Dr. David Colman (now deceased), Director of the Montreal Neurologic Institute and Hospital, and the Wilder Penfield Professor of Neuroscience at McGill University.

The PIs would be eligible to receive MRF funding only if they agreed to collaborate with one another—scientists whom in some cases they might have previously viewed as competitors—and only if they met their scheduled objectives, which would have to further the specific goal of getting effective treatments to patients. Like those of the management team, the responsibilities of the PIs were clearly defined. PIs were expected to participate in the annual meetings, research reviews, and update meetings; participate in development and execution of the strategic and annual research plans; develop a budget necessary to successfully complete the plan; share MRF-funded research information; and open their lab to MRF research team members from other labs. The dynamics of this structure are reflected in Figure 24.1.

Boards and Internal Advisory Groups

Johnson assembled an outstanding Board of Directors with a broad range of business and legal expertise. Unlike that of most nonprofit boards, the function of the MRF Board was primarily business oversight and governance, not fundraising to support the organization. It was crucial to his plan to also have an active and involved Scientific Advisory Board (SAB) composed of eminent scientists who would add a broader, objective scientific perspective; advise on development of research plans; and help accelerate the research process. The SAB would evaluate the progress of the research team to keep them on track, and provide other appropriate scientific oversight. Having an outstanding SAB would ensure the highest-quality research results while helping to attract sophisticated donors who "got" this new research concept and would be willing to invest in it.

The MRF did not employ the traditional peer review process to select projects to fund. This was because peer review tends to tilt toward risk aversion, and the MRF was intent on taking bold, scientifically sound, risks. The MRF desired a more interactive team approach to designing and approving new experiments, which it dubbed "peer review 2.0." The members of the SAB and the MRF staff scientists, who individually and collectively had the intellectual horsepower to stringently evaluate the proposed research projects, would guide the MRF in its groundbreaking approach while

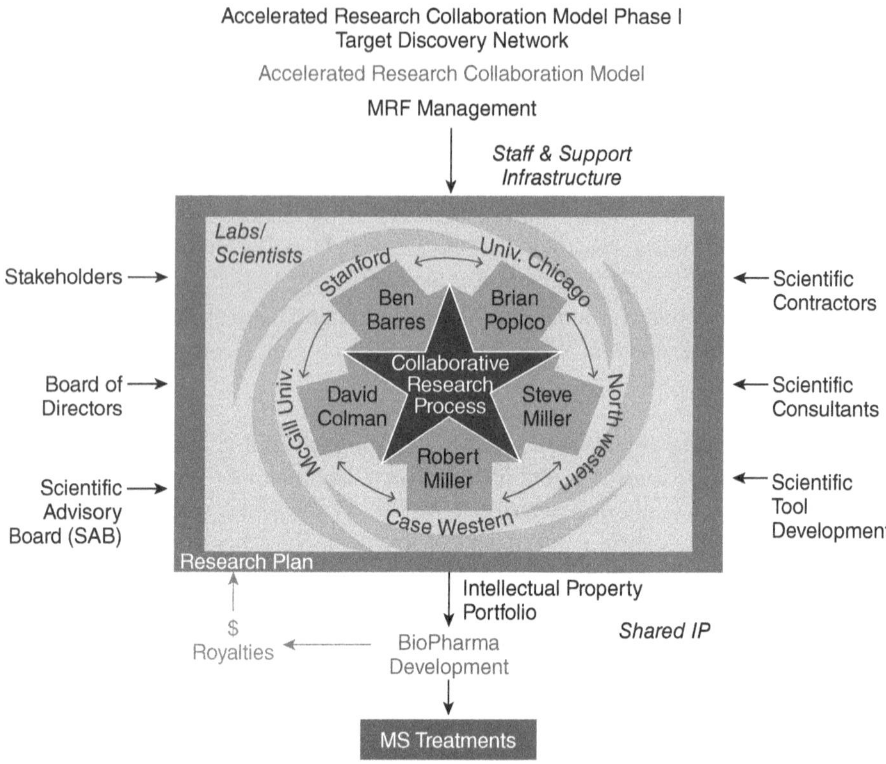

Figure 24.1. 2003 ARC model graphic. See color insert.

preserving, in a different form, the checks and balances that would otherwise be performed through the peer review process employed by other nonprofits and the National Institutes of Health (NIH).

Later came the formation of a Drug Discovery Advisory Board (DDAB), composed of individuals with diverse backgrounds and experience in biotech and pharma. The DDAB was created to provide guidance as the MRF segues from discovery to validation to licensing for development of drug targets. The same process used to select the PIs was used to select the members of the DDAB: the MRF first identified the skill sets necessary to best validate academic discoveries to industry standards, and then identified the preeminent individuals who possessed those skills and had proven track records in driving discoveries from bench to the clinic.

The MRF also has a Clinical Advisory Board (CAB) that provides guidance on the work necessary to gain approval for future human trials and helps oversee the MRF biomarker initiative. Because no one has attempted a myelin repair clinical trial, the CAB is also thinking through novel issues related to trial design and clinical end points (Table 24.1).

TABLE 24.1. MRF Core Capabilities and Competencies

A summary of the core capabilities and competencies that resulted from the way the MRF was organized and staffed from its inception include the following:

- Best-in-class team of PIs, research associates, postdoctoral investigators, graduate investigators, and technicians
- Broad range of complementary scientific skills and experience within the research team
- Capabilities and results optimized through professional collaboration
- IT infrastructure to support collaboration
- IP agreements with universities and industry
- Eventually sustainable funding from contributors, royalties, and licenses
- Scientific Advisory Board, Drug Discovery Advisory Board, and Clinical Advisory Board actively engaged in the research process

MULTIPLE SCLEROSIS AND MYELIN REPAIR

MS is a progressive and incurable neurological disease that affects approximately 2.5 million people. It most commonly hits young adults between the ages of 20 and 40, although pediatric MS and diagnosis after the age of 40 also occur. Its symptoms vary widely, sometimes dramatically, among patients. The most frequent are numbness, tremors, balance and coordination issues, bladder and bowel problems, blurred vision or blindness, nerve pain, swallowing problems, cognitive difficulties and, in the worst cases, paralysis. The degree of severity and speed of progression are wide-ranging and largely unpredictable. Complicating matters is that symptoms often appear, disappear, and then reappear again.

While symptoms differ, all MS patients share in common damage to their myelin, the fatty coating that surrounds and protects nerve fibers in the brain and spinal cord. Nerve signals travel along the myelin sheath, and when the sheath is damaged, the transmission of the signals is slowed or, worse, stopped altogether. This in turn triggers the disabilities that are the hallmark of MS.

No one knows what causes MS. Several possible risk factors have been suggested but not yet proven. Theories range from the Epstein–Barr virus and the human herpes virus-6 to a vitamin D3 deficiency, environmental factors, and others. There also appears to be a genetic predisposition to MS. Siblings and children of a person with MS run a 2.5% chance of contracting the disease, while the risk in the general population is approximately 0.5% in women and 0.3% in men. The genetic risk factor increases dramatically if more than one family member has MS. Nonetheless, MS is not considered a genetic disease because in addition to the genetic predisposition, the individual also needs to be exposed to the as-yet unknown triggering factor. The predominant theory is that MS is an autoimmune disorder in which the blood–brain barrier is breached and the body's immune system attacks the myelin sheath.

Whatever is not known about MS, and whatever causes the disease, what is certain is that it involves the destruction of the myelin sheath. To discover how myelin can be repaired, it is necessary to understand how it is created in the first place, and again, the mechanisms are not well understood. In healthy persons, remyelination occurs continuously, just as other cells in the body continuously replace themselves. It now appears that in the early stages of MS, spontaneous remyelination continues to occur, but that eventually the disease overwhelms the body's capacity for repair. The result is the formation of chronic lesions that can result in permanent disability.

The focus of the MRF's research is not a cure but something that can become a reality and help MS patients much more quickly: an effective treatment that will repair the damaged myelin sheath and reverse disability that has already occurred.

Importantly, many discoveries concerning myelin repair also benefit research in other diseases and disorders in which myelin destruction is also implicated, such as Alzheimer's, Parkinson's, and even schizophrenia. It is no wonder that myelin repair is so important, given that myelinated axons make up about 60% of the human brain.

ARC ACCOMPLISHMENTS

Management legend Peter Drucker understood the potential of the model as early as 2005, describing the Myelin Repair Foundation's approach as magnificent and urging observation of its progress. ARC was recognized in the September 2006 issue of *Fortune* magazine as a "lifesaving shakeup in the world of medical research" [2], an assessment that by all measures appears to be proving prescient.

In 2004, the MRF committed to raise $25 million within 5 years, and to ensure that at least 80% of its expenses would be dedicated to R&D. It more than met its promise. By 2009, the MRF had donations and pledges of more than $28 million, and R&D accounted for 81% of total expenditures. As of early 2012, the figures are equally impressive. Donations and pledges to date total $45 million, and R&D continues to account for 80% of total expenditures. Across nonprofits overall, achieving program expenditures of more than 70% is considered exceptional, and few of the largest disease-focused nonprofits spend more than 50% of total revenue on research. For patients, the depth and breadth of these discoveries provide multiple potential targets against which therapies may be developed and demonstrated.

Also as of early 2012, collaborations funded by MRF have resulted in the development of 24 new research tools—animal models and assays—with broad application to all neurological disease research; identification of more than 150 novel potential myelin repair targets; issuance of 5 U.S. patents; and publication of more than 90 scientific articles in peer-reviewed journals. In addition, the MRF has raised $45 million from individuals and organizations interested in the MRF's innovative model, and persons whose lives have been personally touched by MS.

According to the analysis by the Boston Consulting Group illustrated in Figure 24.2, within commercial R&D, as many as 360 potential therapeutic targets on average must enter the drug development pipeline to bring a single treatment to market. The MRF believes the quality of its targets could cut that number significantly. The MRF is working on two parallel paths to bring its discoveries closer to commercial development. It is (1) validating the most promising discoveries identified in its labs, and (2) investigating the potential of repurposing one or more discoveries that are currently in clinical development for other, non-MS disease indications.

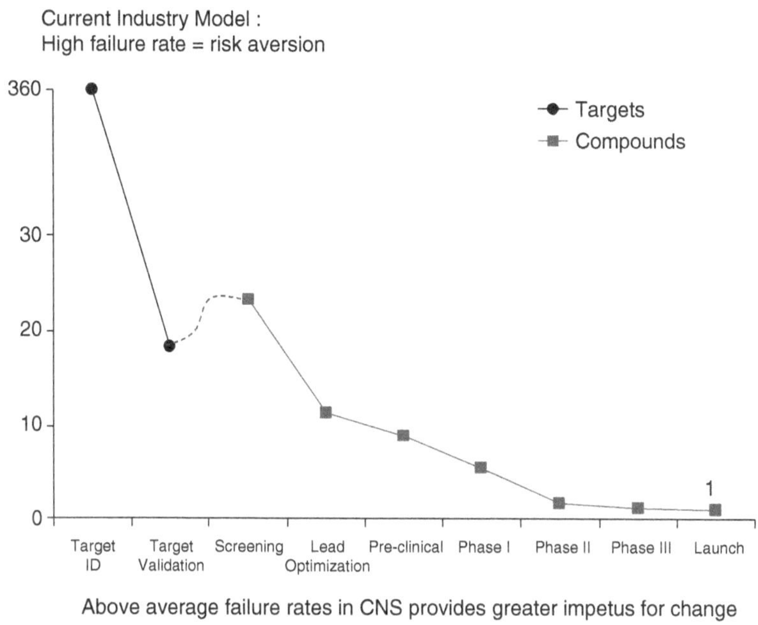

Figure 24.2. Current Industry Model.
Source: BCG client experience, BCG economic model study in Biomedical Industry Advisory Group (2005); CMR Internationql 2006/2007 Pharmaceutical R&D Fact Book.

THE ACCELERATED RESEARCH COLLABORATION MODEL

Before ARC, what happened in the lab, stayed in the lab. —Scott Johnson [3]

Beginning at the Beginning: The Traditional Research Model and Why It's Broken

Johnson and his colleagues set out to disrupt the conventional approach to medical research. Why did they believe disruption was needed? Because all available data confirmed what Johnson and others waiting for cures like him already knew from personal experience: the success rate of research to date—success being defined as bringing effective drugs to patients—was, in short, dismal. It wasn't for of lack of funding. The NIH is the country's leading funder of medical research, at its own labs and in academia and hospitals. In the 20 years between 1990 and 2010, the budget of the NIH quadrupled from $7.6 billion to $30.9 billion. Furthermore, basic scientific discoveries were in fact happening in university laboratories. In 2008, a typical year, more than 800,000 new discoveries in basic science were published in peer-reviewed scientific journals.

So why did only 0.6% of those discoveries actually result in patient treatments? A 2010 *Newsweek* cover article entitled "Desperately Seeking Cures" reported that "[f]rom 1996 to 1999, the U.S. Food and Drug Administration approved 157 new drugs. . . . From 2006 to 2009—the agency approved 74. Not among them were any cures, or even meaningfully effective treatments. . . ." [4]. In the last decade, the number of new drugs coming to market has remained flat.

Add to that the fact that for the first time, big pharma has actually reduced its R&D spending. On June 26, 2011, Reuters reported in an article by Ben Hirschler entitled "Drug R&D Spending Fell in 2010, and Heading Lower" that the $70 billion global expenditure on R&D of new medicines in 2008 and 2009 declined to $68 billion in 2010, a drop of almost 3% in just 1 year. It further reported that pharma is reducing its investment because of unacceptably low returns on that R&D.

The MRF team knew it was challenging powerfully institutionalized traditions and beliefs. The time was right, however: not only were patients frustrated, but so were many research scientists and donors. There had to be a better way, and the MRF's scientific "dream team" was receptive and eager to get started. There were institutional hurdles to jump, but everyone involved was convinced that the ARC model could effectively address each of them in turn.

The question the MRF asked was simple: What are the reasons for these unacceptable results, and what can be done about them?

The ARC Model: Key Features

The ARC model is a collaborative, goal-oriented, open source approach to research that creates more powerful, personally satisfying incentives and eliminates many frustrating limitations inherent in the traditional model. It is ironic that approaching the problem in a methodical and logical manner in a world so dominated by order, reason, and logic—that *managing* the problem—has been so tremendously disruptive. But it has been this very disruption that has accounted for the success of the MRF's work in this area.

To grasp just how revolutionary the ARC model is, it is important to understand the typical features of the traditional medical research model, and why promising scientific discoveries too often have collected dust in academic journals rather than benefiting patients. The sections that follow contrast the key features of the traditional model with those of the ARC model.

The Goal of Research *The traditional goal of research is scientific discovery, not patient treatment.* First and foremost, the goal of research pursued in academic labs has not been patient treatment but scientific discovery. This fundamental flaw reduces medical research to an intellectual exercise that caters not to patients but to the individual interests of researchers. Discoveries that may be translatable to effective treatments are made, inevitably, by accident in the sense that treatment was not the goal from the outset. Vitally important discoveries have been made this way, certainly, but while these successes have resulted in effective treatments, they have been all too rare.

This debate is not new; on the contrary, it has been going on for decades. The rationale for the conventional approach has been that researchers do better work and accomplish more when allowed to pursue subjects that interest them, and that discoveries are just as (or even more) likely to occur when scientists stumble on them while pursuing their areas of intellectual interest. This process by definition fails to focus on the patient, the cure, or the treatment, and is an indiscriminate methodology that only haphazardly identifies potential drug targets that may lead to actual treatments or cures.

The goal of research under the ARC model is patient treatment. ARC posits that the *sine qua non* of effective, efficient research is focus on a critical medical need. Meeting that need is the overarching goal. Achieving that goal requires capitalizing on the random discoveries that the current process produces while developing and executing an outcome-driven research plan focused on identifying therapeutic targets that will lead to patient treatments.

Accordingly, the PIs, along with management and the SAB, design and develop experiments whose ultimate goal is patient treatments—a goal achieved through research, but research oriented to druggability and informed by established timelines and other clearly defined benchmarks. Accountability is key. The PIs take ownership of their research plans, and the MRF holds them to those plans. Performance is a condition of continued funding from the MRF. Although the traditional model has not required research scientists to operate within such a clearly defined framework, the MRF PIs have reported that they are doing "better science" because, ironically, it provides for greater leeway and speed in building on each other's discoveries, and more energy and enthusiasm for the work itself.

Collaborative Research Trumps Competitive Research *The traditional model of research is competitive and restricted to silos.* The currency of scientific research is ideas, and academic scientists compete with each other to produce the most provocative and promising ideas. Getting credit for those ideas is the route to promotions and additional grant funding. This has created an embedded culture of competitiveness and secrecy in which individual scientists work in relative isolation, naturally focused on their specific field of expertise. Few scientists share results—or are permitted to share results—until they have completed their experiments and published the results of their work. The average length of time for a medical research paper to be published is 4–6 years from the time the scientific question to be answered is conceived. In the meantime, an experiment to answer the question must be formulated, funding obtained, experiments conducted, results written up, and the resulting paper or papers accepted for publication. It is only when a paper is published that other scientists struggling for answers in the same field can build on the reported findings. Pharmaceutical companies can identify promising drug targets only at this point, or when the relevant university files for patent rights in the discovery.

The result: research progress on complex diseases like MS, Parkinson's, and cancer, which require multiple experiments in multiple areas, is extraordinarily—and unnecessarily—slow.

TABLE 24.2. ARC Model: Critical Elements

1. The right team of scientists—PIs and SAB, DDAB, and CAB
   - Necessary individual and collective expertise
   - Committed and focused on MRF's mission, processes, and goals
   - Think "out of the box" to speed the time to discovery
   - Get things done according to plan
   - Collaborate with the team
   - Document and protect intellectual property
2. Jointly developed ARC process
   - Coordination of experiments
   - Communication with other MRF research teams and with MRF and advisory boards
   - Documentation of experiments
   - Timing and mechanisms for sharing findings and results
   - Evaluation of progress, measurement of results
   - Measurement of each lab's contribution to consortium
   - Consolidation of research and funding requirements
3. Jointly developed collaborative research plan
   - Defines what is and is not known about myelin-related biology and the interactions relating to how myelin can be repaired
   - Defines the appropriate set of prioritized experiments with defined targets
   - Sets milestones, targets, and metrics
   - Identifies responsibilities for projects and experiments by PIs and labs
   - Is updated as new results become available
   - Meets budgeted funding levels
4. Eventually sustainable funding of experiments and the collaborative process

The ARC model of research is collaborative and interdisciplinary. ARC brings together a first-class team of research scientists who work collaboratively as a condition of funding. Johnson understood that MS is a complex condition that involves neurological, immunological, and other disease processes. Accordingly, an interdisciplinary approach is deemed critical to success, and each PI selected to receive funding from the MRF is an expert in a different but complementary discipline (Table 24.2).

It cannot be emphasized enough that collaboration with researchers who had previously been competitors, and with researchers in different disciplines (such as immunologists and cell biologists) who otherwise never would have interacted or been aware of each other's discoveries, has represented a profound but welcome shift. The MRF provides an effective infrastructure to support the researchers in navigating this shift and maximizing the benefits they gain from it.

Active collaboration is facilitated through regular conference calls, during which each PI shares the successes and failures of his lab since the previous call, questions are asked, and ideas are shared and generated that can be applied immediately in the other labs to accelerate their research. Thus, discoveries are shared in real time without the substantial delays inherent in waiting for publication of scientific papers. In addition, the PIs and their research teams of graduate students and postdocs have three in-person meetings each year. These meetings include an annual conference at which the PIs and their team members present the results of the prior year's work to each other and to the MRF, the SAB, the DDAB, and the CAB. Along with invited distinguished researchers and guests in related fields, they engage in formal and informal discussions that inevitably help to propel their research forward.

As expected, these required interactions early on led to spontaneous, constant e-mail and telephone contact among team members. Some labs have even instituted regular cross-university lab meetings. Furthermore, students and postdocs visit other labs to learn from and collaborate with other team members. MRF staff scientists also visit each lab approximately every 6 weeks to ensure a thorough understanding of the status of all experiments and to identify any bottlenecks that the MRF can help to overcome.

The rewards of this interdisciplinary, collaborative approach are great: the PIs achieve results more quickly, and the cumulative effect of the exchange of ideas gives them a profound sense of satisfaction. Dr. Ben Barres has stated, "Within the [MRF] collaboration model, every time the team gets together, sparks fly. There's no question that better teamwork in science can significantly accelerate results" [5].

Dr. Stephen D. Miller has reflected on his work with fellow MRF PI Dr. Brian Popko, this way: "For a number of years, I worked just across town from Brian Popko at the University of Chicago. Because I am an immunologist and he is a neurobiologist, there was no reason for us to know each other or the nature of each other's work. Because truly understanding myelin repair in the context of multiple sclerosis can't happen without both areas of expertise, collaborating with MRF has put us on the same playing field. Collaborations with Brian and the other investigators have helped advance my work to a new level—a level that I believe could generate a real myelin repair treatment for patients" [6].

Adherence to Benchmarks *Many medical research nonprofits do not require adherence to stringent benchmarks as a condition of funding.* Many nonprofits are still hampered by a lack of business expertise, especially the absence of clear benchmarks—in other words, internal accountability (to and among management, researchers, and Advisory Boards) and external accountability (to donors and the public). Furthermore, most medical nonprofits have multiple, important, and competing—albeit related—missions, including education, patient support, research, and advocacy. For most large, brand-name disease nonprofits, research is only one of many organizational priorities. In fact, for many, research is less than 20% of total funds raised nationally. Most rely on the NIH peer review process to distribute funding and do not have scientific staff members with the pharma expertise required to best ensure the therapeutic relevance of the work they fund.

In the ARC model, continued funding requires adherence to benchmarks in the business and research plans. The ARC model recognizes that even a medical research nonprofit can be more efficiently and productively operated on the model of a for-profit start-up business. Thus, from the beginning the MRF's business and research plans have consistently reflected its singular focus on myelin repair and raising the funds necessary to accomplish that mission. A clear definition of its goals, plans, and benchmarks, all based on that mission, creates accountability among management and PIs.

Intellectual Property *Academic institutions have little incentive to invest in intellectual property protection for biological discoveries, and when they do invest, traditionally they are the exclusive owners of that intellectual property.* The academic institution employing the researcher who makes a biological discovery owns that discovery. Serious financial considerations are at play when deciding what to do with that asset: university technology transfer offices must carefully consider the cost–benefit ratio of each investment they make. Biological discoveries are inherently risky because they may or may not lead to licensable drug targets, and even if they do, the payoff to the university can be significantly delayed—usually by at least 15 years after patent expenses are incurred. In contrast, other technologies that universities patent and license (e.g., software) are easier to evaluate up front and offer faster payoffs. Furthermore, if multiple universities are involved in a patent, significant new hurdles are introduced, such as who will lead the patent effort, how expenses will be shared, and, most difficult, how any revenues will be shared.

Although the potential dollars to be earned from biological discoveries are not—indeed cannot be—known up front, the fact remains that drug targets can be some of the most valuable assets researchers produce for their academic employers, and effective protection of those assets is essential to all involved in each phase, from discovery to marketing of a treatment. Again, however, university technology transfer offices are often unable to make the sizable financial investment in patenting them in the first place because the risks are perceived as too high.

This bottleneck is a major contributor to the Valley of Death. When the time, talent, and effort of the university's best and brightest create a drug target whose potential the university does not promote, and pharma does not pick up (as a rule pharma does not consider conducting further drug development and clinical trials on unpatented discoveries), the results can be devastating.

Finally, the issue of protection of intellectual property comes down to a basic premise of traditional research: discoveries must be kept confidential until publication. It would be hard to overstate the adverse effects of the combination of the traditionally competitive nature of research and the view that sharing valuable patents and other intellectual property across cooperating institutions will defeat each individual institution's competitive edge.

Under the ARC model, universities continue to own the intellectual property generated by their scientists while reducing their financial risks. Under ARC, the relevant university or universities own the patent on each drug target to the extent they would be considered the inventors under U.S. patent law. Inventions made by more than one university's research team are shared by the relevant universities.

The MRF substantially reduces the risk for each university by applying for the patent on behalf of that university and paying all legal fees and costs associated with the application (or obtaining *pro bono* services from a law firm). In exchange, the university grants to the MRF exclusive worldwide licensing rights to the patent. Thus, it is the MRF, not the university, that identifies optimum potential development partners, and markets to and negotiates with those pharmaceutical companies to sublicense each drug target for commercial purposes. Revenues generated by the patent ultimately will be divided between the MRF and the participating university or universities.

Even though it gives the MRF the exclusive license to the patent, the university can still leverage its investment in the invention: it can use the invention and nonexclusively license it to other nonprofit research institutions for internal research and educational purposes (as opposed to diagnosis, treatment, or prevention of disease).

If the MRF decides not to apply for a patent, the university may decide to do so, as well as to control the licensing of the invention, all at the university's own cost. In that case, the university will receive all royalties from the patent, but it must offer each of the other member institutions (that were member institutions when the IP in question was invented) the opportunity to share in both the costs of and the royalties derived from the patent activity.

This aggressive approach is a key underpinning of the ARC model. It works so well because it is straightforward and spreads potential risks and benefits among the vested parties. Granting licensing rights in exchange for revenue-sharing provides potential financial advantages to universities that might otherwise elect not to participate at all and thereby forego all potential revenues. The creation of potential downstream revenues is also one of the most important benefits to the MRF. As discussed below, those revenues can be reinvested in research, potentially reducing the MRF's dependence on charitable donations.

Through this breakthrough mechanism, a win–win for the academic institutions and the MRF is achieved, pharma is given a greater incentive to pick up more discoveries, and ultimately, the patients will be the beneficiaries.

The Intersection of Funding and Structure: Eventual Self-Funding and Independence *Under the traditional model, the nonprofit continues throughout its existence to depend on donations for funding.* The traditional model assumes that charitable donations are the only feasible type of income for nonprofits. This exclusive reliance on donations undermines the stability of most nonprofits because it requires constant fundraising efforts, diverting staff attention and donor contributions from directly furthering the nonprofit cause. Financial planning and budgets must also assume levels of income that may or may not be realized. Most importantly, all these factors distract from and diffuse the ever-important focus on the mission: therapeutically relevant research that pharmaceutical companies (the only entities with the expertise and resources to turn academic discoveries into treatments for patients) will value enough to license and move forward to the clinic.

Under ARC, income from the intellectual property that results from research has the potential to allow the MRF eventually to become self-funding. Just as for-profit corporations are charged with maximizing shareholder wealth, the MRF identifies ways to leverage donations to create income-producing assets that in the long run will reduce reliance on donors, especially in questionable economies, thus "maximizing" donor satisfaction. The MRF's long-term plan—likely 20 years out or more—is to use income generated from patent licensing to fund future research, with the goal of ultimately becoming self-supporting as more discoveries are made and increasing numbers of patents are granted.

Understanding the Different Mind-sets of Academia and Pharma *Communication, collaboration, and formal ties between academic science and the* *pharmaceutical industry early in the research process have been less than optimal.* Traditionally, even as academic scientists have engaged in basic research in their own labs, big pharma has separately conducted its own R&D. The high cost of investing resources in basic science and disproportionately low success rate have exacerbated the situation, however, by incentivizing pharma to scale back its own R&D and require additional validation of discoveries before licensing drug targets emerging from academia.

The drivers of this disconnect flow quite naturally from the differing goals, cultures, and motivations of academia and industry. Basic academic research is driven entirely by individual interest in particular scientific problems and is dependent on the ability to obtain funding. Applied research, on the other hand, is driven by company goals, including the obligation to increase shareholder value, and therefore by what industry would argue are the practical needs of the marketplace.

In fact, academics have a fundamental lack of knowledge about what pharma does, its scientific challenges, and the scientific skills it possesses. Virtually all academics have always worked in an academic environment and have never had any interactions with business, let alone pharma. They often characterize scientists who leave academia as failed scientists. Their distrust of business is so deep that many feel that interacting with pharma would compromise their professional integrity.

Another factor contributing to the academia–industry disconnect is the previously noted culture of competitiveness and secrecy. Without an appropriate confidentiality agreement in place, it is difficult for academic and industry scientists to be completely open in communications that would require disclosure of competitive information. Even casual conversations at conferences can be stifled. One unfortunate result is that valuable information isn't shared, so problems often can't be fully explained. Another, of course, is that academic labs often generate data that do not meet the rigorous industry standards required by pharma.

This historical disconnect is especially counterproductive given the importance and necessity of the symbiotic relationship between university labs and the pharmaceutical industry. The latter is largely dependent on academic science for discoveries that can be developed into marketable drugs, and academic science is largely dependent on the pharmaceutical industry for that development. The ARC model is exploring ways to create partnerships that recognize the differences but work effectively within them.

The ARC model recognizes the need to inform academic research with the perspective of big pharma and biotech. The MRF has furthered its goal of partnering with industry to commercially develop a myelin repair treatment, in part by developing and conforming its own validation to industry standards. Creation of the DDAB and recruitment of key individuals have been essential to this process. In the same way that the SAB informs the basic research, the DDAB, staff members, and consultants with backgrounds and expertise in pharma and biotech are able to provide insights into the concerns, and the R&D standards, of industry. In a sense serving as stand-ins for actual industry presence, they have the knowledge to prioritize experiments and projects and advise on the kind of data needed. In this way the MRF has supplemented its extensive core basic research knowledge with the expertise needed to guide the basic research and validation process in ways that are likely to produce results of greater interest to pharma. It is hoped that by bridging this gap between academia and pharma, ARC can continue to facilitate greater conversation between academia and pharma.

All of the above-mentioned traditional features have contributed to the slow pace that so notoriously dominates the research and discovery landscape. In the process of pursuing patient treatments for MS, the MRF's model is not only challenging those features, but is also demonstrating its usefulness as a comprehensive model for medical research and drug development that significantly shortens the time to market for new medicines for all diseases.

WALKING THE TALK: HOW THE ARC PRINCIPLES WORK IN PRACTICE

The process by which ARC is implemented is critical to the success of the model, provides a roadmap as to how ARC works in practice, and is contractually agreed by the parties. The key elements of that process are enumerated in Table 24.3.

Annual Planning Process

The annual planning process must be thorough and meticulous, with the goal always of furthering the discovery of viable drug targets and facilitating their rapid movement into the pharmaceutical pipeline. The planning process therefore begins with an assessment of the achievements of the scientific team during the preceding year. The MRF scientific staff and the SAB and the DDAB review the relevance and value of each project, prioritize targets, identify actions to further validate the therapeutic relevance of each, and evaluate opportunities to improve the collaborative research process and its results.

Importantly, publicly available discoveries made outside the MRF, and their implications for the MRF's 5-year critical research path, are reviewed. If appropriate, the MRF updates its research path to eliminate activities that are no longer relevant and to exploit unanticipated new avenues of opportunity. The next step is to identify and prioritize the top questions that must be answered in the next 12 months in each

TABLE 24.3. Key Process Elements

- Values, roles, and responsibilities
- Strategic research goals
- Annual research review and research progress report
- Annual research plan and budget
- Four-month research reviews and plan update
- Monthly collaboration forums
- Continuous improvement

functional area of research. The research path inevitably will have many unanswered questions, so this prioritization and sequence of experimentation is critical. The final research plan for the upcoming year uses concurrent, parallel paths of investigation to maximize the scope and speed of discovery.

Each lab and researcher then selects individual questions, and the resulting projects and experiments, based on the list of priorities. The plan also sets forth milestones, a list of deliverables, budgets, and a timeline against which each project may be measured. In many cases, answers or discoveries from one lab are necessary to initiate the progress of another, and in fact many MRF experiments now require more than one lab to complete. Because of the interdependencies between labs, some are free to pursue lower priority questions first. Each lab posts results in a timely fashion so follow-on experiments by other MRF labs may commence promptly.

The MRF establishes a schedule by which funds will be transferred to each university, which in turn must provide the MRF with timely accounting for all MRF-funded expenditures. The procedure for this accounting is similar to that required by the NIH.

Each participating university is free to receive funding from the NIH or any other institution or governmental agency in addition to the funding provided by the MRF, and to engage in research generally outside the MRF group.

Implementing the Research Plan

Each PI is responsible for ensuring timely project execution, soliciting assistance with problems, and reporting progress to the MRF management team. The PI may request support from other MRF labs with the expertise and capacity to conduct experiments required to complete the project. This process may entail members of the research team from one lab working in another for an extended period.

Annual Research Review and Progress Report

MRF management and the SAB and DDAB conduct regular reviews and updates to the research plan and budget. Just as the MRF approves the plan to begin with, it must also approve any proposed modifications. Representatives from each lab participate in a progress review and planning meeting every 4 months or as necessary. SAB members are invited to participate. The format of these reviews is consistent with presentations made at scientific meetings. They may also be used as a forum to present and discuss results the MRF may wish to publish or present at external scientific meetings.

Monthly Collaboration Forums

In the ARC world, even 4 months can be too long to wait to compare notes. Accordingly, PIs and their teams also participate in a monthly conference call to discuss their progress, problems, and interim results. This monthly call not only fosters informal communications between the labs, but also serves the critical purpose of promoting real-time, collaborative problem-solving among the scientific teams. Some of the most productive discussions take place during these calls. Between calls, of course, all investigators are free to, and do, communicate frequently with each other to talk through challenges and successes as they occur.

Annual Conference

The PIs and their research teams; MRF management and staff; the members of the SAB, DDAB, and CAB; and invited guests from the scientific community all attend the annual MRF conference. The PIs and their teams present a summary of their research progress and achievements, and discuss how their progress addresses the high-priority questions identified in the annual research plan. Their presentations include information they have submitted, or plan to submit, for publication in leading journals. At this conference, the MRF recognizes outstanding scientific achievements, publications, and key contributions from members of the MRF research teams.

Each annual conference is spirited, creative, and energizing. As during the conference calls that have been taking place regularly all year, ideas and theories are openly shared, questions abound, and conversation is lively. The conference adds the opportunity for in-depth exploration of the year's successes and failures. It gives participants the chance to bounce theories off each other and take those theories to areas they would never otherwise have the chance to explore with colleagues in complementary disciplines. Impromptu exchanges occur in hallways, and dinner table discussions extend into the night. The excitement of sharing and leveraging information is palpable, and the resulting achievements speak for themselves.

Licensing of Intellectual Property to Pharma

As of the spring of 2012, the MRF has not yet reached the stage of actually entering into sublicensing arrangements with commercial enterprises for drug target development. It is hoped that when it does, both parties will seek terms that will accelerate that development. An additional goal will be to provide incentives to ensure that promising targets are not abandoned.

Publication

Consistent with the ARC model of collaboration and sharing, results from MRF-funded research are made publicly available as soon as both the research and management teams deem the results valid and ready for publication.

TRANSLATIONAL RESEARCH: A CREATIVE STRATEGY FOR MOVING DRUG TARGETS ACROSS THE CHASM

Getting from Bench across the Valley of Death to Bedside

The MRF can point to a robust portfolio of potential myelin repair drug targets—concrete evidence with productive outcomes—from its basic research program. This portfolio, however, is positioned at or near the beginning of the pharma pipeline. Perhaps one of the deepest divides that slows and often halts the development of new medicines is the Valley of Death—that rupture in the pipeline through which hundreds of thousands of basic scientific discoveries that lack the rigorous data sets required by industry simply disappear into the pages of countless scientific journals.

How is it possible that so much funding, talent, and productivity could be wasted in this way? What are the forces in academia and industry that have created such a "Catch-22"? In Scott Johnson's conversations and presentations on the need to reform the system, he makes clear his conviction that the players in the medical research field have the best intentions. This is an important point, and it is central to his philosophy and that of the MRF. Johnson believes that each scientist in academia and industry has a strong desire to produce effective medicines that can be delivered to the bedside. Because scientists are operating in silos—separate cultures with different goals—however, and because there is no unifying game plan or management infrastructure connecting players, their best intentions remain unfulfilled. For example, as cited earlier, academic scientists seeking answers to intriguing biological questions naturally assume their work will engage the interest of pharma, which will in turn translate those discoveries into medicines.

In a PBS NewsHour interview aired on September 23, 2010, Dr. Barres put it this way: "We work very hard for a few years, and we publish a paper in a good journal. And I always assumed that that discovery, if it were immediately relevant to a new treatment, would be picked up by one of the big drug companies [which would] then develop the next step. And, so, I thought I was doing my part" [7].

The Translational Research Center and the Translational Research Platform: How the ARC Model De-risks Discoveries and Delivers Validated Targets to Pharma

The ARC model is a work in progress driven by a strategy that continuously challenges the barriers to commercial development. Consistent with that strategy, in late 2011, the MRF opened the Translational Medicine Center headed by Dr. Jason Dugas, Senior Staff Scientist, under the guidance of Dr. Tassie Collins, the MRF Director of Translational Medicine. Focused on advancing myelin reparative therapeutics from basic research to successful commercialization, the Center is the logical follow-on to the early-stage ARC model. As its name suggests, it is the centerpiece of the MRF postacademia, industry-standard translational medicine platform.

With the advice and counsel of its Drug Discovery and Clinical Advisory Boards, the MRF has recruited a team of experts with more than a century of biopharma experience moving therapeutic compounds into clinical trials. They understand first-hand the financial and other challenges that demotivate risk-taking and innovation in the industry. They understand further that data produced in academic laboratories must be compelling and protected, and must meet the same rigorous standards that pharma applies to its own research programs. These standards are met through a detailed process that replicates findings many times in multiple animal models and assays, seeks concordance across those studies and, finally, demonstrates that what has been well documented in animal cell cultures and models can also be demonstrated in human cell cultures. After all, curing a rodent with neurological disease is not the end game.

Reproducibility of experiment results reported in medical journals is key. Substantial investments in setting up labs and dedicating financial, time, and human resources to duplicating assays based on discoveries reported in scientific journals often lead to the disappointing conclusion that the results are not in fact consistently reproducible. In the meantime, resources have been diverted from other promising targets. One example is that pharmaceutical giant Bayer's internal experiment results differ from results reported in the literature in an impressive 65% of its projects [8].

As a result, pharma modulates its risks by devoting resources largely to therapeutic approaches that have already been demonstrated to work. Until recently, myelin repair was considered impossible, so pharma naturally has instead been focused on immune-modulating treatments with a high rate of success. The MRF's challenge is to convince pharma to take a chance on novel myelinating therapeutics.

The MRF is meeting that challenge by reducing the risks for pharma. It is identifying and funding replication studies, as well as investing in the development of human tissue assays and biomarkers that do not yet exist for the purposes of studying myelin repair (Figure 24.3).

Working together with the nine academic investigators that the MRF sponsors, the MRF's Translational Medicine Center will provide complete packages of data—rodent *in vivo* and *in vitro* models and, where possible, *in vitro* assays using human cells—so they can be further reproduced by unbiased researchers and by industry. Put another way, the Center's business model centers around providing highly reproducible data for

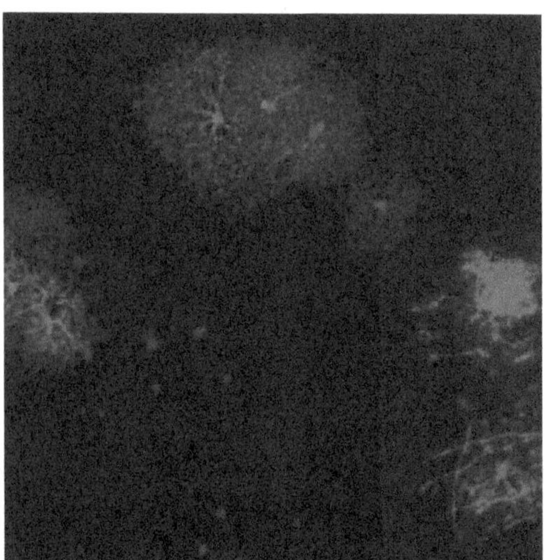

Figure 24.3. *In vitro* myelin image. See color insert.

the highest possible value. For all the obvious reasons, the MRF is as concerned with efficiency as are biotech and big pharma. The Center must be selective about the assays it chooses to optimize and therefore has focused on assays that are biologically complex and relevant to a large number of myelination targets. The assays at the Center will be easily accessed by industry for a nominal fee, providing greater incentive and ability to explore promising new therapeutic approaches with a minimum of time and cost investment.

Biotech can also benefit by accessing the Center and even potentially partnering with larger pharma companies. Furthermore, other academics have expressed interest in potential partnerships with MRF.

Even with a fee structure, all of this comes at a cost to the MRF because pharma will not have paid for the costs of developing and optimizing the assays and animal models, or even the costs involved in identifying targets. There is a potential for some cost recovery down the road, but in the meantime, the MRF is willing to take on the financial burden because motivating pharma and biotech to undertake R&D on myelin repair drug targets is essential to the MRF mission. (This, of course, proportionally intensifies the pressure on the MRF to attract the large donations on which it is dependent for all its activities.)

Importantly, in the future, the Center will be in a position to provide access to MRF-developed assays to, and evaluate discoveries made in, labs anywhere in the world. It is expected that this capability will accelerate the validation process for a myriad of myelin repair discoveries beyond even those achieved within MRF.

Ultimately, the MRF envisions bringing together a constellation of scientists whose work will have an impact on myelin repair and beyond. Providing access to the discoveries and brain trust at the MRF is an unprecedented and hugely significant service to industry that can exponentially accelerate the delivery of effective myelinating drugs to patients. At the same time, the MRF continues to invest in its own research labs to generate additional targets as well as set the stage for the design of myelin repair clinical trials for which, in this uncharted territory, new parameters must be established.

The establishment of the Center and the design of innovative relationships with pharma is further evidence that the MRF is flexible and nimble—and therefore in a unique position to respond quickly and effectively to challenges, and to help coordinate the entire ecosystem of medical research, whose players range from academics to other nonprofits, to the NIH, pharma, biotech, the FDA, and to physicians and patients.

As an outside entity with only one goal in mind, the MRF is not constrained by a need to establish academic careers or increase shareholder value. In fact, one could argue that this singular focus is perhaps what best characterizes the ARC model approach and what will lead to its ultimate success. Those who visit with Johnson in his Saratoga, California office are likely to see on his desk the words spoken by sixteenth-century poet Sir Philip Sidney: "Either we will find a way, or we will make one."

Johnson's personal commitment to Sidney's words cannot be underestimated as a, if not *the*, major contributor to the remarkable success of the MRF. Innovation is often traceable to the relentless perseverance of one individual, and ARC is no exception. The story of the MRF is largely the story of Johnson's conviction and persistent attention to every aspect of the pharma pipeline from basic science through target validation, drug discovery, clinical trials, and FDA approvals.

THE BANDWAGON: LEADING THE CHARGE

The MRF began its work with a small seed fund from a handful of entrepreneurs who recognized value in its model. Support for its efforts has grown to include a broad base of individuals, foundations, and corporations, including the Robert Wood Johnson Foundation, the Donaghue Foundation, the Ewing Marion Kauffman Foundation, the Penates Foundation, Omidyar Network, and Biogen Idec. Support continues to grow daily. Numerous supporters of the MRF do not have a special interest in MS or myelin repair. What they are interested in is the model, and in how the MRF is demonstrating the model by its successes in working toward its goal of bringing myelin repair to patients.

From *The Wall Street Journal* to *Business Week*, *Newsweek*, *Scientific American*, and other major publications, the work of the MRF is drawing widespread attention. In addition, to date, some 70 research and funding organizations have expressed interest in learning how the ARC model might help them leverage their own research dollars for a better outcome. Many observers believe it will soon become expected as donors and patients demand more accountability, and as the benefits become ever more apparent.

CONCLUSION

MRF's ARC model started out as a radical departure from traditional research. It is proving itself in the concrete results it is achieving every day. Along with a handful of other medical research organizations, the MRF is pioneering open source collaboration as a means of increasing the number of potential drug targets, de-risking them to make them more attractive for commercial development, and radically shortening the time it takes to get treatments to patients. By recognizing the incentives and limitations of academic scientists, commercial biopharma, and government regulators, ARC leverages the MRF's position as a nonpartisan, nonprofit organization with only one goal in mind: shortening the time to market for effective new medicines. The results have given all a reason to hope.

In this era of exponentially rapid biomedical and technological advances, patients can't afford to wait; patients are unwilling to wait; and patients should not have to wait.

A key strength of the MRF is to clearly define each roadblock and then assemble the best possible team to create an innovative solution. Scott Johnson has said, "We are problem solvers." This statement and the attitude it reflects aptly capture the reason for the MRF's success to date.

ACKNOWLEDGMENTS

The author wishes to thank the Myelin Repair Foundation for its support and cooperation, with special gratitude to Carol Menaker.

REFERENCES

1. *Fast Company*, March 2011, Issue 153, p. 126.
2. Schonfeld, E. One man's crusade for a cure, *Fortune*, September 21, 2006. (Originally published in *Business 2.0* magazine, May 2006, Vol. 7, Issue 4, p. 62.)
3. Myelin Repair Foundation, President's message, April 22, 2010. Available at http://myelinrepair.org/about/pres_message_archive/presidents_message_26.shtml.
4. Begley, S., Carmichael, M. Desperately seeking cures, *Newsweek*, May 31, 2010.
5. Myelin Repair Foundation, Statements from participants. Available at www.myelinrepair.org/documents/Prelim_Comments_Scientists.pdf.
6. Myelin Repair Foundation, Testimonials. Available at www.myelinrepair.org/donate/testimonials.shtml.
7. PBS NewsHour, For scientists, collaborative efforts could speed medical advances. Interview, September 23, 2010. Available at www.pbs.org/newshour/bb/health-july-dec10-cure_9-23/.
8. Mullard, A. Reliability of "new drug target" claims called into question. *Nat Rev Drug Discov.* 2011;10:643.

25

FROM CATALYSIS TO MASS ACTION: THE EVOLUTION OF CHDI FOUNDATION, A DRUG-DEVELOPMENT ORGANIZATION DEVOTED TO HUNTINGTON'S DISEASE

Allan J. Tobin

Formerly at CHDI Foundation, New York, NY, USA

Cure Huntington's Disease Initiative (CHDI) is a not-for-profit organization. The generosity and guidance of CHDI's donors (who prefer to remain anonymous) enable it to focus solely on drug development. It has become the largest single sponsor of Huntington's disease (HD) research, supporting more than 500 full-time scientists in contract research organizations (CROs), biotechnology companies, pharmaceutical companies, universities, and government laboratories. CHDI's scientific team now includes 30 scientists, with expertise ranging from cell biology and genetics to medicinal chemistry and pharmacology, working together with 14 project managers and administrative, business, and legal staff. More than 80% of CHDI's scientists have previously worked in biotechnology or pharmaceutical companies, further underscoring CHDI's therapeutic goals. Staff scientists work closely with collaborators to define goals, specify deliverables, develop experimental programs, monitor progress, and maintain a therapeutic focus.

From its beginnings, CHDI has devoted considerable resources to human studies, with the convergent goals of understanding the biology of disease progression and of preparing for clinical trials. Because clinical trials for a potentially neuroprotective drug

Collaborative Innovation in Drug Discovery: Strategies for Public and Private Partnerships, First Edition.
Edited by Rathnam Chaguturu.

will require sophisticated monitoring of early changes in brain structure and function, CHDI's human studies investigate HD-associated changes in cognitive and motor function, in biochemical readouts of cellular and brain activity, in patterns of gene expression, and in psychological assessments.

An essential part of CHDI's strategy is to be a "collaborative enabler" of HD research, particularly for projects that will bridge the gaps that separate basic discovery, drug development, and clinical trials. These activities include:

1. Funding of infrastructure (e.g., patient registries, repositories for biological samples and biochemical reagents, and animal models)
2. Sponsorship of clinical studies of HD progression
3. Development of new assessment tools and biomarkers, both for disease progression and for the engagement of drug targets
4. Creation and distribution of research tools (e.g., new reagents and new animal models)
5. Exploration of potential new targets and technologies
6. Convening of multidisciplinary workshops to discuss emerging areas of particular relevance to HD research and drug development
7. Support of new means of rapid and up-to-date communication among CHDI collaborators and, more widely, all HD researchers
8. Adoption of research contracts whose terms require the conduct of proposed experiments with regular reporting of results and sharing of information and reagents rather than traditional research grants.

CHDI has self-consciously experimented with different strategies for accomplishing its goals, always with an eye to its long-term mission. Early efforts emphasized individual research projects, roughly following the pattern employed by NIH and other foundations. As CHDI paid increased attention to anticipating the conversion of preclinical program to human trials, it has come to focus on fewer and more extended programs.

These extended programs may involve different kinds of individual collaborators and partnering organizations, including academic laboratories, CROs, and pharmaceutical and biotechnology companies. CHDI also works to lower the barrier to entry into HD research for individual research scientists and biotechnology companies and by underwriting academic research projects through the National Institute of Neurological Disorders and Stroke, the Huntington's Disease Society of America, and the European HD Network. It also supports the exchange of information and reagents within the HD research community.

The goal of CHDI remains constant—to develop therapies and to advance them to clinical trials—but its strategy and programs continually responds to advances in biomedical science and technology, to shifts in the biotechnology and pharmaceutical industries, and to organizational and financial changes in universities and government. Current challenges—both to CHDI and to other disease-directed foundations—include

1. How to handle vast amounts of data and information from the research community
2. How to use advances in systems biology to help the selection of therapeutic targets and to predict the effects of specific interventions
3. How to share information and knowledge among researchers, both in public and commercial settings
4. How to set up collaborative arrangements for "precompetitive" work—early stages of development during which competitors collaborate without compromising their future competitive positions, and
5. How to find contracting, licensing, overhead, and management arrangements that will best speed the path to effective therapies.

CHDI PROGRAMS DEPEND ON COLLABORATIONS WITH ACADEMIC, INDUSTRIAL, AND GOVERNMENT PARTNERS

Like other disease-directed foundations, CHDI has a single goal: to find treatments for a single disorder—HD—as quickly as possible. The fastest access to resources for specific studies is usually through providers who already have them, so CHDI endeavors to work with collaborators with direct experience in a needed technology. CHDI programs depend on collaborations with academic, industrial, and government partners.

CHDI is a "virtual" organization, meaning that it has no laboratories of its own: every scientific project is executed off-site, though with the direct participation of CHDI staff. The strategy is to employ already established capabilities in universities, CROs, and biotechnology companies—to "buy" rather than to "build."

CHDI supports the application of new technologies to HD-related problems, although not technology development *per se*. Nor does CHDI support work on general neuropathological mechanisms, unless they are directly involved in HD.

CHDI differs from most other foundations in that its own scientists directly contribute to supported research programs, working with both academic and industrial collaborators to develop goals, to choose experimental strategies, to specify laboratory tactics, and to write reports and papers. Similarly, CHDI project managers directly engage the scientists and managers of each collaborating organization, establishing and checking timelines, tracking materials, verifying deliverables, and monitoring personnel deployment. In many ways, then, CHDI resembles a virtual biotechnology company, though CHDI is directed toward a single indication and is not-for-profit.

The goal of all CHDI's efforts is to find treatments for HD. Once CHDI scientists decide to pursue a particular therapeutic target, they evaluate every feasible modality and every available technology. Recently explored approaches—in addition to small molecules—include antisense oligonucleotides, siRNAs, aptamers, peptides, single-chain antibodies, viral delivery, and cell-based therapies. Of course, as discussed later, competing priorities limit the number of concurrent programs.

CHDI's complex enterprise, with its many internal and external players, requires stable and user-friendly repositories for data and documents. Because CHDI is a multisite organization (with offices in New York, Los Angeles, and Princeton) with collaborators around the world, its success also depends on open and effective communication.

CHDI also cooperates closely with HD-directed governmental programs, including the National Institute of Neurological Disorders and Stroke (NINDS). For example, CHDI has provided supplemental funds to support an NIH program announcement, "Validation of Novel Therapeutic Targets for Huntington's Disease" (PAS-10-183). CHDI support allows NIH to fund up to four HD-related projects, whose applications are judged to be scientifically excellent but whose scores fall below those required for ordinary NIH funding. CHDI also collaborates with NINDS in funding an international consortium directed to developing cellular model HD with induced pluripotent stem cells ("iPS cells"). NIH funds the U.S. investigators of this consortium, while CHDI funds the non-U.S. investigators. Finally, CHDI and NINDS each supported the first long-term study of premanifest HD, "Neurobiological Predictors of Huntington's Disease" (PREDICT-HD). NINDS has recently issued a Funding Opportunity Announcement (FOA) to support studies based on the data and biological samples from this 10-year study. The net effect of CHDI's collaboration with NINDS has been to attract more investigators to HD research, to increase the opportunities for data mining and analysis, and to increase the total funding available for HD research in the U.S.

STUDIES OF PEOPLE WITH HD ARE A HIGH PRIORITY

HD is the clearest and possibly the oldest example of a dominant genetic disease: inheritance of one mutant version of the disease-causing variant—from an affected parent of either sex—is enough to cause the disease to appear. George Huntington presented the first description of HD in 1872, almost 30 years before the rediscovery of Mendel's revolutionary paper [1].

HD results from a single mutated gene (*HTT*) and the consequent changes in the functions and metabolism of a single large protein (called huntingtin, or HTT), which performs diverse cellular functions in essentially every cell of the body. HTT contains more than 3000 amino-acid residues, including a tract of glutamine residues (each encoded by a CAG sequence in the DNA), starting with residue #18. In people not affected with HD, the polyglutamine is from 15 to 25 residues long, whereas in people with HD it is longer, from 37 to 180 residues. Broadly speaking, the longer the polyglutamine tract, the earlier is the manifestation of disease. Because everyone has an *HTT* gene, it is incorrect to call it the "HD gene." Versions ("alleles") of the gene that have more than 36 CAG cause HD, and this chapter refers to any these pathogenic versions as an "HD allele." There are also a few reported cases of HD arising from people who carry "intermediate *HTT* alleles," with 27–35 CAG repeats. Increased longevity may increase the number of such cases, as suggested by M.R. Hayden.

At least 100,000 people in the United States carry the HD allele (with 36 or more CAG repeats), with approximately the same number in Europe. (The incidence of HD outside North America, Europe, and Australia is mostly unstudied. One goal of

Enroll-HD, described later, is to examine HD incidence globally.) Of these, about one-third have developed the characteristic HD movement abnormalities, which are the current basis of diagnosis.

Because of HD's dominant hereditary pattern, each affected person (and each unaffected or uncertainly affected sibling) has almost certainly watched a parent suffer the same fate, making the future all the more concrete and frightening. While HD afflicts individuals, it affects whole families, further amplifying the number of people whose lives are changed by HD. Although HD may have devastating effects on affected families, many people from these families have become HD activists, and many participate in human studies and clinical trials. The active engagement of the HD family community has been an important inspiration to many individual HD researchers, both among CHDI scientists and their collaborators and, more generally, within the entire HD research community.

The characteristic effects of HD on movement—dance-like gestures (chorea) and loss of balance—typically start around age 40 (earlier with larger numbers CAG repeats). Subtler signs of HD—irritability, apathy, and depression—may appear a decade earlier. Over the course of some 25 years, a person with an HD allele undergoes a devastating decline of physical, mental, and emotional capacities. As brain circuits fail, an affected person gradually loses the ability to control movement, to maintain motivation, to regulate mood, and to plan and execute life's everyday tasks.

HD provides an opportunity—almost unique among neurodegenerative diseases—to look prospectively for changes in people who have an HD allele but who have not yet developed the motor signs that are the basis for the current formal clinical diagnosis. Understanding that observations made in humans carrying the HD allele may give important hints for intervention, CHDI's clinical group has supported several studies of "premanifest" or "prodromal" HD. (Many HD clinical researchers use one of these two words ["prodromal" and "premanifest"] to designate the state of a person who carries an HD allele but who has not developed the specific motor signs currently necessary for a formal diagnosis. Prodromal HD includes alterations in the volume of specific brain regions, diminished cognitive capacity [particularly in "executive function"], and some characteristic emotional changes, including apathy, depression, and irritability.)

The development of an intervention will require detailed knowledge of how HD affects people who carry the HD allele. CHDI has therefore devoted much effort to programs that measure changes that accompany and underlie disease progression, particularly in early stages, where intervention is likely to be most efficacious. Human studies may also provide clues for target selection.

CHDI recognizes the importance of understanding the biology and the clinical progression in people who are carrying an HD allele. Several projects have sought to identify proteins, peptides, and small molecules—in plasma and in cerebrospinal fluid (CSF)—that changed with disease progression. In collaboration with the Human Proteome Organization, for example, CHDI sponsored a multicenter project to detail the CSF proteome (the collection of all proteins expressed in CSF) and to look for changes in premanifest and early HD. While biochemical biomarkers are unlikely to provide surrogate end points for clinical trials, they would certainly help investigators to follow effects of interventions and, ideally, to measure the engagement of selected drug targets.

CHDI has supported three comprehensive studies of disease progression in premanifest people—Prospective Huntington At Risk Observational Study (PHAROS), PREDICT-HD, and TRACK-HD. These studies have included brain imaging, motor changes, eye movements, cognitive testing, and psychiatric evaluation. Similar projects now include the selection of a cognitive battery, the development of a functional rating scale for early HD, and the formulation of a descriptive mathematical model for disease progression.

CHDI has also supported the infrastructure of the Huntington Study Group and the European HD Network, and it has funded two projects—COHORT (mostly in the US and Canada) and REGISTRY (mostly in Europe)—whose goals are to collect clinical information and biological specimens from people who carry an HD allele. A new project—Enroll HD—will unify COHORT and REGISTRY and extend this joint to Latin America with the aim of developing a global HD patient registry. Enroll-HD will build on the many successes of the North American and European efforts in standardizing assessments, collecting biological samples, and handling data. The goals of Enroll-HD are to refine current knowledge of HD's natural history, to help improve HD care throughout the world, and to provide a recruiting platform for future clinical trials.

CHDI HAS INCREASINGLY FOCUSED ITS EFFORTS INTO A FEW CORE PROGRAMS

CHDI originated in the "Cure HD Initiative" of the Hereditary Disease Foundation (HDF). This initiative began in 1997 as a way of providing support for projects that might contribute to understanding and treating HD. Rather than waiting for investigators to formulate relevant proposals, however, Ethan Signer (the initiative's founding Executive Director and now a Senior Scientific Advisor to CHDI) sought appropriate researchers to undertake projects that he and the initiative's advisory committee identified. During the time of the initiative, scientific work on HD expanded enormously, to more than 600 scientific publications in 2002 (compared with about 350 in 1997). The initiative provided support for each academic investigator as a contract with his or her university rather than as a traditional research grant or fellowship.

By 2003, when CHDI became independent of HDF, the initiative had provided support for more than 150 projects, almost all in academic laboratories. CHDI's donors soon supported three complementary organizations:

1. High Q Foundation, which supported basic research relevant to understanding HD pathogenesis and progression, including many of the project originally funded by the HDF initiative
2. CHDI Inc., which focused on drug discovery and drug development
3. HP Therapeutics Foundation, which—among its other activities—organized a project ("Systematic Evaluation of Therapies for HD [SET-HD]") to evaluate the feasibility of clinical trials for scores of suggested HD drugs.

In 2008, CHDI Inc. was renamed the "CHDI Foundation," which then supported the full gamut of HD research activities from basic discovery research through clinical development.

CHDI's philosophy and operations have dramatically changed from those of 1997 and 2003. Most importantly, CHDI leadership soon became aware of the need for new expertise, particularly in the areas of drug development and clinical trials. This realization led to the recruitment of a Chief Scientific Officer, a Chief Clinical Officer, and Vice-Presidents for Chemistry, Translational Biology, and Systems Biology, each with experience in drug development and each able to build teams with previously untapped expertise. As CHDI grew, it similarly recruited talented professionals with needed experience in law, operations, information technology, and human resources.

As its preclinical research portfolio grew, CHDI's leadership also came to realize the importance of focus. Each individual program might require the extended efforts of several CHDI science directors and 20 or more external biologists and chemists. With time, CHDI became increasingly concerned about its capacity: just how many projects could CHDI effectively pursue, given its established scale and resources?

In 2004, Robert Pacifici, CHDI's then-new Chief Scientific Officer, suggested narrowing CHDI's preclinical efforts to 20 potential drug targets and programs. With increased experience, the number of major preclinical programs diminished to about 15. These programs fall into five broad areas, listed in Figure 25.1. CHDI now devotes some 50% of its drug-development efforts to a program to reduce the concentration of mutant *HTT* RNA. Another 25% of CHDI's efforts address the posttranslational modification of HTT, seeking to alter its pathogenic action. The remaining projects fall into three broader programs, whose goals are to reverse or compensate for pathogenic changes in (1) synaptic function, (2) energy metabolism, and (3) autophagy (see Figure 25.1).

CHDI'S HIGHEST CURRENT PRECLINICAL PRIORITY

HD results (almost always) from an *HTT* allele with more than 36 CAG repeats. Most researchers think that HD results from the action of the abnormal HTT protein with a polyglutamine track of more than about 36 glutamine residues, though the possibility remains that abnormal RNA might itself be pathogenic, perhaps in addition to the abnormal protein. In either case, strategies that reduce mutant *HTT* RNA (with more than 36 CAG repeats) should reduce or prevent pathogenesis, whether HD results from abnormal RNA or from abnormal protein. Therefore, *CHDI's highest current preclinical priority is to reduce mutant HTT RNA and HTT protein.*

CHDI currently supports the efforts of Isis Pharmaceuticals (Carlsbad, CA) to use antisense oligonucleotides to reduce *HTT* RNA, as well as those of Alnylam Pharmaceuticals (Cambridge, MA) to use small interfering RNAs (siRNAs) to accomplish the same goal. In both cases potential therapeutic agents must be delivered directly to the brain. Following the approach taken in its trial for amyotrophic lateral sclerosis (ALS), Isis plans to inject antisense oligonucleotides into cerebrospinal fluid via intrathecal injection (into the fluid surrounding the spinal cord). Alnylam, in collaboration with

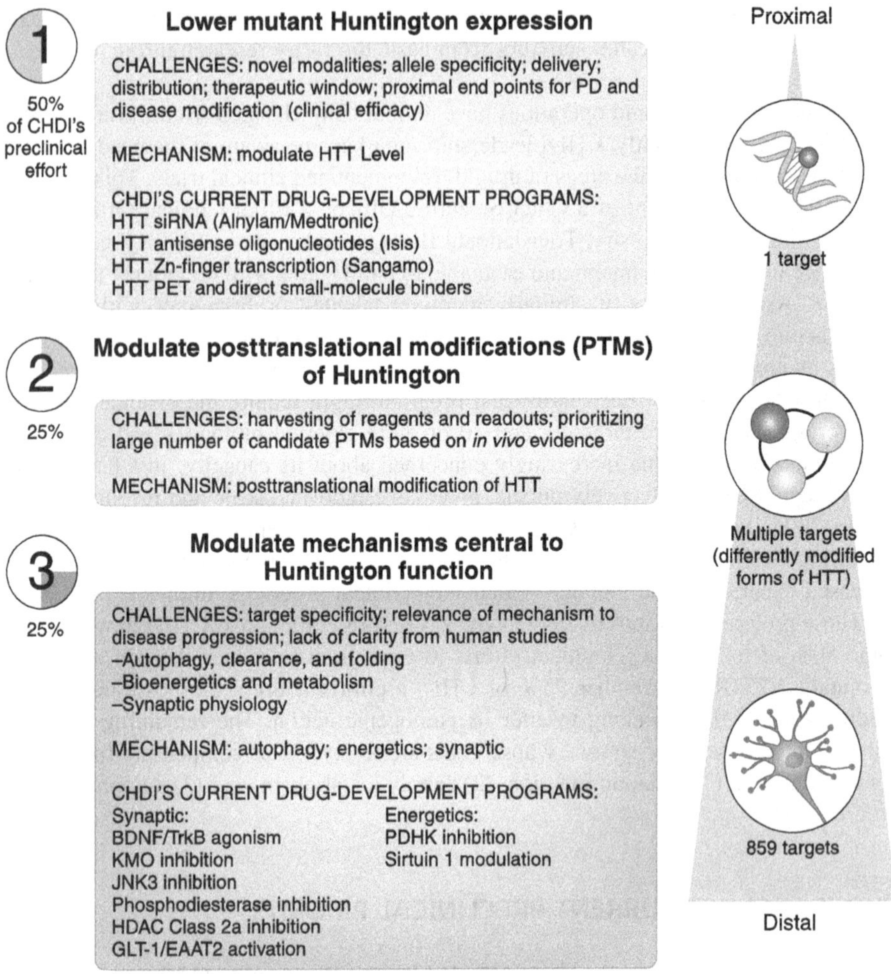

Figure 25.1. Drug development programs at CHDI.

Medtronic (Minneapolis, MN), plans an alternative approach: pumping siRNAs directly into the brain through a thin catheter. CHDI is also working with other investigators and companies to explore other ways of reducing *HTT* RNA (e.g., with the zinc-finger technology developed by Sangamo BioSciences, Richmond CA), of delivering DNA cargoes to brain (e.g., with viral and liposome-based methods of traversing the blood-brain barrier), and of measuring the effects of treatments on HTT levels (e.g., with HTT-binding ligands for positron-emission tomography [PET]). Some HTT-lowering strategies aim to distinguish mutant from wild-type *HTT*, to preserve normal HTT function while reducing the pathogenic actions of the abnormal protein and RNA (see Figure 25.1).

CHDI CONTRACTS PROTECT THE FREEDOM TO OPERATE IN THE ARENA OF HD RESEARCH

Working as a virtual organization, CHDI requires exchanges of materials and information among its varied collaborators, and hence seeks to ensure and *protect the freedom to operate in the arena of HD research.* The technology-transfer policies of both universities and companies, however, initially obstructed CHDI's ability to promote the exchange of materials among its collaborators, and, more broadly, among all HD researchers. Under the guidelines used by most funding organizations, both public and private, a funder does not have the right to share materials, or even information, among the researchers it supports. Since each CHDI project usually involves investigators from several institutions or companies, these limitations directly conflict with CHDI's fundamental strategy. To overcome these difficulties, CHDI has pioneered contracting practices that are unusual among disease-directed foundation.

One approach is to make use of CROs (companies that perform research specified by CHDI on a fee-for-service basis). These companies do not own the data (or the compounds) developed in CHDI projects, nor can they use data gathered in CHDI projects for other purposes. For these projects, then, a CRO works as if it were internal to CHDI.

CHDI also collaborates with pharmaceutical and biotechnology companies; however, and for them, sharing information and materials may be inimical to their goal of creating valuable compounds to sell. CHDI contracts, therefore, distinguish between "compounds," intended for therapy, and "tools," intended for research. CHDI contracts do not interfere with a company's ability to protect and sell its compounds, but CHDI contracts seek to avoid situations in which a collaborator can unilaterally decide to stop the development of a compound in which CHDI has an interest. With both commercial and noncommercial tools—such as recombinant DNAs, monoclonal antibodies, biochemical assays, or compounds not intended to be drugs—CHDI does insist on having a nonexclusive license for HD research along with permission to distribute to other CHDI collaborators for purpose of HD research.

For academic investigators, CHDI's concerns about freedom to operate have led to more complicated discussions. Since the passage of the Bayh–Dole Act in 1980, universities have controlled the intellectual property (IP) derived from government-supported academic research. In general, universities have made patented or patentable materials available to investigators in not-for-profit institutions with only limited restrictions. For example, many institutions adopted the Uniform Biology Material Transfer Agreement (UBMTA), which covers the use of materials or derivatives to noncommercial purposes. The UBMTA specifies that commercial use—of either the original material or its derivatives—requires a commercial license from the original provider and that the provider (i.e., the university) need not grant such a license.

To CHDI, this limitation is unacceptable: it would mean that materials developed with CHDI support would not automatically be available for the studies done in commercial entities, including CROs. Under the leadership of David Rankin (Chief Legal Officer), CHDI—unusually among disease-directed foundations—established a "bundle of rights" for each research contract. Each contract specifies that, in return for funding, CHDI has the right to transfer materials and information to other CHDI collaborators (after a specified time, to allow reasonable time for investigators to publish). Under these terms, now incorporated into CHDI contracts with more than 50 academic institutions, the university generally retains varying degrees of ownership of materials and data, but they grant a nonexclusive license to CHDI for the purpose of HD research. When potential therapeutics are involved, CHDI's research contracts specify criteria that will be used to evaluate opportunities for commercial development, for example, to bring therapeutics to patients as quickly as possible and to ensure that licensees will be adequately funded.

CHDI has also worked with Science Commons (http://www.sciencecommons.org) to develop standard and easily accessible forms for material transfer agreements (MTA). One of these MTAs, which CHDI has adopted, eliminates the UBMTA's restrictions on redistribution and on commercial use.

Universities and medical schools have increasingly embraced the goals of translational medicine—the conversion of basic discoveries in biomedical science into useful therapies. In doing so, however, they repeatedly face the yawning gulf—sometimes called the "translational gap" and, more dramatically, the "valley of death"—between the hatching of a promising therapeutic idea and the beginning of a clinical trial. This challenge has gradually led to warmer welcomes for private funders, including CHDI. In the last decade, most universities have realized that the best chance of making this valley fertile may depend on collaborations with disease-directed foundations and their research networks. Although a university forgoes higher indirect costs and immediate income from materials transfer, most universities have decided that foundation support is likely to improve the longer-term prospect for therapeutic development.

Even after securing the right to transfer materials from a collaborator, however, CHDI has still faced the possibility of infringing the IP of a third party. For example, a DNA construct for a genetically modified mouse may derive from plasmids or genes obtained under a license of which the researcher is unaware. CHDI scientists and project managers must therefore evaluate the intellectual property status not only of physical deliverables but also of the materials used in producing those deliverables. To the extent that CHDI operates as a virtual nonprofit biotechnology company, CHDI scientists and project managers, together with internal and external legal counsel, must attend more closely to IP matters than perhaps is usual among disease-directed foundations. These considerations (together with scientific merit and the effort required for their specific evaluation in each project) have contributed to CHDI's decision to limit the number of its supported programs.

CHDI'S EARLY DISCOVERY INITIATIVE SUPPORTS EXPLORATION OF NOVEL TARGETS AND MECHANISMS

Although CHDI's staff spends most of its time and effort with selected high-priority projects, CHDI also supports a few dozen early-stage projects geared toward the exploration of novel targets and mechanisms, as part of its Early Discovery Initiative (EDI).

The level of project management for EDI projects is much less than in later stage projects, and the decision time is also much shorter, usually within a month after receiving a completed proposal.

EDI supports exploratory projects with the goal of identifying novel therapeutic targets or pathogenic mechanisms. Funding for such projects is limited and usually extends for only 1 year. Led by Ethan Signer (Senior Scientific Advisor), EDI science directors work with applicants to define goals that may be particularly important to CHDI, including, for example, the evaluation of underexplored therapeutic targets, biomarkers of disease progression, and the exploitation of stem-cell biology.

SCIENTIFIC COORDINATION AND PROJECT MANAGEMENT INCREASINGLY RELY ON JOINT STEERING COMMITTEES

In every contract, CHDI scientists and project managers specify deliverables and anticipate potential IP issues. Because research often takes unexpected turns, advance specification of deliverables is difficult, and contracts have often required supplementation and amendment. For some collaborators, CHDI has developed a more flexible contractual arrangement with deliverables and experimental goals regularly updated by the joint steering committees (JSCs). A JSC typically comprises two science directors and one project manager from CHDI and the same number from the collaborating organization. Each JSC meets periodically (monthly or even weekly) by teleconference and quarterly face-to-face, with additional participants as needed for a particular discussion.

CHDI'S INTERNAL CULTURE SUPPORTS THE SPECIFIC NEEDS OF A DRUG-DEVELOPMENT ORGANIZATION WITHOUT ITS OWN LABORATORIES

Despite its apparent resemblance to a company, CHDI differs from a biotechnology or a pharmaceutical company in at least three respects:

1. Its motivation is speed to intervention rather than profit
2. Its goals solely concern HD
3. It has no competitors.

No matter who "invents" a successful treatment for HD, CHDI will have fulfilled its goal. Conversely, if CHDI's efforts should happen incidentally to produce a treatment for some condition other than HD, another organization, not CHDI, is likely to develop it.

Since 2004, Robi Blumenstein, CHDI's President, and Robert Pacifici, CHDI's Chief Scientific Officer, have led an ongoing internal discussion of how best to fit CHDI's operations and culture to its goals. One guiding principle, repeatedly emphasized, is that drug development is an iterative process, with dozens of repeated cycles of synthesis and evaluation. Success depends on the rapidity of each cycle and the wisdom of each sequential judgment. CHDI must therefore demand scientific, technical, and operational excellence, as well as effective communication both internally and externally.

CHDI has recruited the many kinds of professionals required to carry out drug-development campaigns in "virtual" mode. CHDI staff now includes some 60 people—science directors and project managers, as well as experts in informatics and information technology, operations, administration, human resources, financial management, business development, contracts, and IP (Figure 25.2). CHDI offers salary and other conditions of employment competitive with those received in industry, and it has succeeded in hiring excellent and experienced scientists and managers from large pharmaceutical companies and from smaller biotechnology companies.

As a not-for-profit organization, however, CHDI does not have the same tools as for-profit companies for recruiting, incentivizing, and retaining personnel. In particular, CHDI cannot offer stock or stock options to its employees, nor can the organization

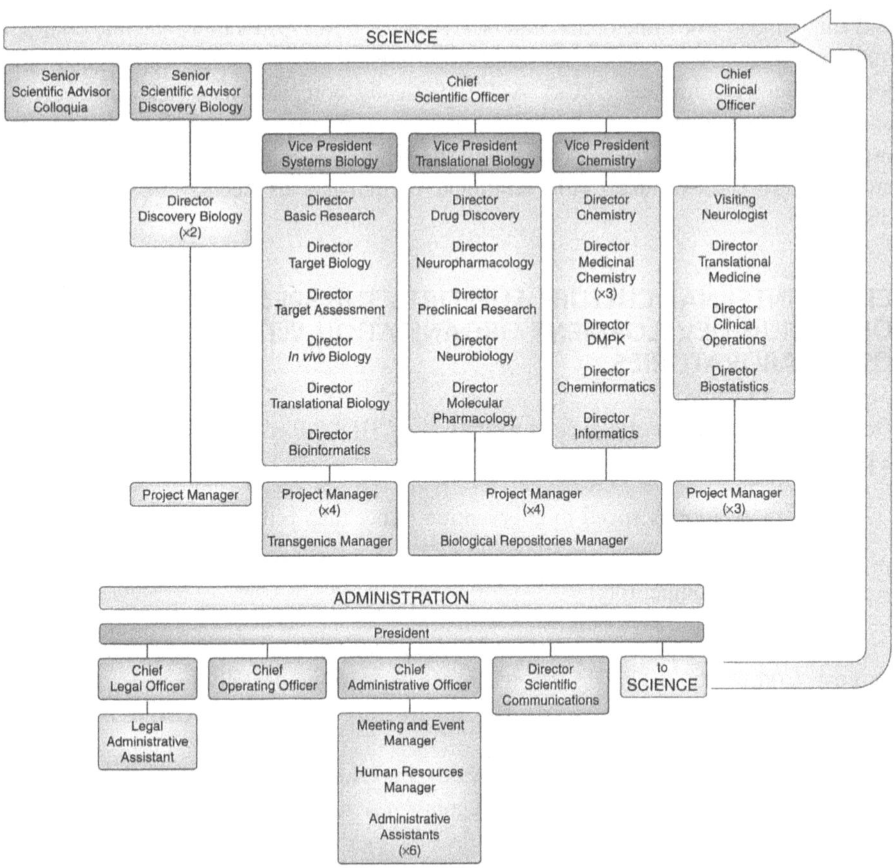

Figure 25.2. CHDI's organization.

use its sales or its stock price as a proxy for its achievements. Furthermore, CHDI is not a competitor with any company or individual: if *anyone* succeeds in developing an HD therapy, CHDI will have fulfilled its mission. So CHDI also seeks to provide incentives for cooperation, collaboration, and enablement with other organizations, both for-profit and not-for-profit.

CHDI's current bonus strategy therefore provides incentives to reinforce hard work and insightful thinking, not only for the individual programs in which an employee participates but also for other internal and external programs and, in addition, to recognize advances in HD therapeutics outside CHDI. For example, in 2008, the Food and Drug Administration (FDA) approved tetrabenazine (Xenazine®), in whose development CHDI did not directly participate, as a treatment for the treatment of chorea associated with HD—the first and only FDA-approved treatment for any symptom of HD. CHDI gave bonus contributions to employees's 401(k) plans. CHDI's message to its employees is that anything good for HD patients is also good for everyone at CHDI, even when CHDI's contributions have occurred only indirectly.

CHDI SCIENTISTS PARTICIPATE REGULARLY IN INTERNAL MEETINGS, ANNUAL RETREATS, AND AT THE ANNUAL HD THERAPEUTICS CONFERENCE

CHDI's science directors meet repeatedly with their supervisors to evaluate the status of each project. In addition to regular supervision, all CHDI scientists meet for 2 hours every week—in person and by teleconference among the three sites—to discuss one of the ongoing projects. A full review of all major projects occurs once a quarter. In addition to these regular meetings, all CHDI staff—scientists, project managers, and administrators—participate in an annual 2-day retreat to promote teamwork and to foster a better understanding of CHDI's goals, methods, and organization.

Finally, most of the scientists and project managers participate in CHDI's annual HD Therapeutics Conference, an unusual four-day scientific symposium that brings together a strong mix of academic, biotechnology, and pharmaceutical researchers. The conference serves to disseminate data, share ideas, initiate collaborations, and encourage cross talk. In the last few years, the conference's participants have increasingly come from industry, suggesting that more companies now see HD as a as a tractable problem with promising (and potentially profitable) therapeutics. For CHDI staff, the meeting also provides an extended opportunity for scientific exchanges with current and potential collaborators and with colleagues across CHDI itself.

CHDI OBTAINS EXTERNAL SCIENTIFIC ADVICE THROUGH ATTENDANCE AT SCIENTIFIC MEETINGS, CONSULTANTS, WORKING GROUPS, WORKSHOPS, AND A SCIENTIFIC ADVISORY COMMITTEE

With some 30 internal scientists working with more than 500 external collaborators, CHDI has enormous reach into scientific knowledge relevant to HD. Realizing the importance and relevance of research in other areas, CHDI scientists also attend widely diverse general scientific meetings. CHDI also engages non-HD scientists to participate in CHDI projects as consultants and in meetings of project-specific working groups.

CHDI also sponsors small multidisciplinary workshops on topics of particular interest in HD. Topics have ranged from new ways of imaging the brain to details of specific synaptic mechanisms. The workshops are unusual for scientific meetings in that they are small (fewer than 20 people, about 15 external and 5 from CHDI) with no slides, no speeches, and no fixed agenda. By allowing participants to talk *with* (rather than *at*) each other, these workshops promote new ideas and new approaches to problems relevant to HD. Workshop discussions often contribute to the establishment of new projects as well as to the modification of existing programs.

The model of small interactive workshops originated with the Hereditary Disease Foundation's founding president, Milton Wexler, as a way to attract new investigators to HD research. In contrast, CHDI workshops generally focus more narrowly, on topics of present or anticipated relevance to HD drug development. CHDI continues to experiment with workshop format and follow-up, striving to engage both CHDI and outside researchers in addressing CHDI's questions and programs. Recent workshop topics have ranged from RNA profiling to magnetic resonance spectroscopy to the role of exercise in neuroprotection.

CHDI regularly convenes a Scientific Advisory Committee (SAC), whose members come from the pharmaceutical and biotechnology industries. The SAC meets three times a year, to provide independent advice on specific preclinical programs to CHDI's president and donors. At each SAC meeting, CHDI scientists present and discuss two projects, so that over the course of about 2 years, the SAC can review essentially the whole preclinical portfolio and to make recommendations to guide each program either toward an Investigational New Drug (IND) application or toward termination.

CHDI'S CURRENT *IN VIVO* TESTING PROGRAMS FOCUS ON THE EVALUATION OF SPECIFIC TARGETS AND ON THE TESTING OF SPECIFIC HYPOTHESES

One special attraction of HD to researchers interested in neurodegenerative disorders is its simple genetics and the attendant prospect for creating a genetic model for drug testing. Using standard techniques for gene manipulation, researchers have transferred the normal ("wild-type") and the HD (mutant) versions of the *HTT* gene into different cells and into experimental animals, including roundworms (*Caenorhabditis elegans*), fruit flies (*Drosophila melanogaster*) and mammals (mice, rats, pigs, sheep, and monkeys). After such gene transfers, researchers can study differences between cells or animals that received wild-type *HTT* and those that received mutant *HTT*. To the extent that these differences mimic human HD, they also provided formal proof that the mutant *HTT* is the cause of the disease.

To follow, roughly, Robert Koch's principles for establishing the causal relationship between a microbe and an infectious disease, a geneticist would not only demonstrate that a mutant gene is associated with a disease but also that it is sufficient to cause the disease in an experimental model [3]. The lack of specific parallels in the phenotype of a transgenic animal and a human with HD, however, means that transgenic animals can never prove an *HTT* allele actually causes HD. More formally, the differences between humans and experimental animals imply that no animal can ever fully "model" human disease, though the similarities at the biochemical and cellular levels may be far more convincing than at the systems or behavioral levels.

Over 500 published papers have studied HD-like changes in presumed mouse models of HD and their modification by genetic and pharmacological manipulation. As of May 2013, PubMed searches found 374 citations for "R6/2 + 'Huntington disease.'" The totals for other mouse were 73 for R6/1 (another exon-1 model, but with slower development of neurological signs), 56 for N171 (another fragment model, containing exons 1–3), 40 for YAC models (full-length transgenic), 5 for BAC (another full-length transgenic), and 98 for knock-in mice). CHDI has encouraged and underwritten the creation and distribution of HD mouse models, most of which are now widely available.

Despite the proliferation of HD models, including genetically accurate knock-ins, no model can completely match the attributes of human HD. Yet, if the same pathogenic processes are at work in both a model and in the disease itself, the model may be useful in choosing drugs for human trials. A disease model is said to be "valid" if the same manipulations—genetic or pharmacological—ameliorate both the model and the human disease. As no such manipulation has been found, there is as yet *no* validated model for HD. Nonetheless, the parallels between mutant-*HTT*-dependent changes in a mouse and those in human disease are tantalizing, and many researchers point to those parallels to support their enthusiasms for a drug, a target, or a pathogenic pathway.

Even with the most robust readouts, however, evaluation of therapeutic effects has posed formidable challenges. For example, several groups reported a neuroprotective effect in mouse models of minocycline—an antibiotic that also (fortuitously) inhibits two enzymes involved in programmed cell death (see later discussion). But different laboratories obtained widely divergent results [4–7]. Despite this controversy, the positive mouse reports became a major argument in securing NIH funding for a minocycline trial in humans. (The trial showed no positive effect in people with HD [8].)

Discussion with individual investigators—mostly in universities—revealed a staggering lack of consistency in conditions of husbandry, tests employed, test equipment and protocols, genetic backgrounds, and even in the number of CAG repeats in the transgene. Attempts to standardize such studies produced little convergence and no resolution of divergent results.

CHDI therefore decided to take a different approach to *in vivo* testing starting in 2004. It seemed unlikely that academic investigators, even in consortium, would be able to converge on a common platform for drug evaluation. CHDI scientists therefore decided to engage PsychoGenics (Tarrytown, NY) as a CRO to evaluate alternative methods of husbandry and drug assessment, with the goal of standardization [9]. CHDI and PsychoGenics developed programs both for drug testing and for studies of the pathogenic process itself, under carefully controlled, standardized conditions of husbandry. CHDI-sponsored studies would not only guide choices of programs and drugs within CHDI but also would assess and standardize best practices for the entire HD research community.

To ensure that PsychoGenics make use of the substantial work already done with HD mouse models in the academic research community, CHDI assembled an advisory committee comprising several leaders of HD research, as well as experts in behavioral testing. The advisory group met several times a year with PsychoGenics scientists. The group discussed issues of experimental design, ranging from genetic background to husbandry to test design to statistics and group size, and then took on the interpretation of results and interpretation.

The drug-testing program evaluated some 100 compounds in R6/2 mice, and only three compounds showed any neuroprotective activity. In each case, however, other compounds that interact with the same target (e.g., the adenosine A2a receptor) did not show any neuroprotection, indicating that the positive effects resulted from an unknown mechanism.

From the point of view of drug development, these mouse studies were particularly disappointing. Not only did they not succeed in demonstrating the efficacy of a nominated drug or the relevance of its putative target, but they also failed to give information about why any particular drug may have failed. Many of the compounds tested were already used for other neurological and neuropsychiatric illnesses, so CHDI and PsychoGenics scientists assumed that the compounds reached their putative targets in the brain. The initial studies, however, provided no measurements of brain concentrations or of target engagement. Other tested compounds were not chemically stable, so researchers could not ascribe an effect or lack of effect to their putative targets.

CHDI's first 5 years' experience of *in vivo* assessment of drugs and targets in HD mouse models led to a better understanding of its strengths and limitations: (1) the credibility of any *in-vivo* result depends on its reproducibility, so that husbandry and testing must reflect systematic evaluation of best practices and meticulously described standard operating procedures; (2) for any drug or target, promising results from a single animal model in a single laboratory (or company) is not a compelling reason to undertake a clinical trial; (3) pharmacokinetics is an early priority: the effects of any compound depend not only on its ability to bind to its target but also on its ability to get to that target (presumably in the brain) at an effective concentration; (4) drug effects on general behavioral or neuropathological readouts are less useful than target-specific readouts; development of a drug requires knowing its mechanism of action: otherwise, there is no basis for measuring target engagement (pharmacodynamics) and no way of improving on the first-tried drug.

CONFLICTING RESULTS UNDERSCORE THE NEED FOR RAPID COMMUNICATION OF DATA AND METHODS

Some of the early divergent results from mouse studies quickly appeared in print, while others did not. The reticence of researchers to submit and of journals to publish replicative studies, whether confirming or nonconfirming, is well known [10, 11]. In view of the substantial pressure to choose drugs for clinical trials, the publication and public discussion of possibly conflicting results must be a priority for disease-directed foundations like CHDI. Critical evaluation is especially when those results (as in the case of minocycline) are relevant to decisions about clinical trials or even about preclinical drug development. To encourage rapid dissemination of data and public discussion of conflicting results and interpretations, CHDI's president Robi Blumenstein and its Communications Director Simon Noble have therefore worked closely with the Public Library of Science to create a new online journal, *PLoS Currents: Huntington Disease*. The goal of the new journal, which is supported by CHDI, is rapid and open sharing of new research data, analyses, and ideas, in a form that allows them to be permanently archived and citable.

In addition to the new journal, CHDI has undertaken several other ways of increasing communication of data, information, and knowledge concerning HD. CHDI science directors not only attend HD-related and general scientific meetings, but they also publish both research papers and reviews, usually in collaboration with an academic or commercial collaborator. CHDI has also established a website (HD Research Crossroads, http://www.hdresearchcrossroads.org), which provides continually updated information about HD-relevant targets and compounds, as well as news of recent papers and meetings relevant to HD researchers.

CHDI HAS MOVED AWAY FROM PHENOMENOLOGICAL SCREENS AND TOWARD A BETTER UNDERSTANDING OF MECHANISTIC TARGETS

The most striking characteristic of the brains of people who have died from HD is the massive loss of brain matter. By the time of death, the brain of a person with HD may show almost complete loss of the striata, an accompanying increase in the volume of the lateral ventricles, a wholesale loss of brain volume, and a general decrease in the thickness of the cortex. Imaging techniques show similar, though less striking, loss of brain matter even in early HD. From the very first discussions of neurodegenerative mechanisms, many basic scientists have inferred that HD results from accelerated cell death, particularly of the medium spiny neurons of the striatum.

In the late 1970s, cell biologists and geneticists became increasingly aware that the death of neurons in neurodegenerative disease had much in common with the death of neurons and of other cell types during development. The program for cell death, called apoptosis, involves a cascade of protein-digesting enzymes, called caspases, whose action ensure that the remains of dying cells do not trigger an inflammatory reaction in the surrounding tissue (Minocycline inhibits caspase-1 and caspase-3, two enzymes in the apoptotic pathway). A once-favored hypothesis for cell death in HD was that the mutant HTT inappropriately triggered the apoptosis program, leading to the premature death of specific classes of neurons.

As researchers learned about the molecular mechanisms of apoptosis, many researchers—hopeful that they would be able to stop or slow cell death in every neurodegenerative disease—began to look for specific ways to keep the process from starting and for molecules that would inhibit caspases and prevent them from destroying cells. Other researchers, more directly concerned with HD, worked specifically with cells that died upon the induction of mutant *HTT*. They developed cell-based screens for compounds that would reduce cell death, typically after the induction of mutant *HTT* in engineered cell lines.

CHDI supported many of these compound and genetic screens, both in academic laboratories and in companies. In the early days of these screens (before 2005), the HD research community, including CHDI, generally accepted the argument that a "phenomenological" (later called "phenotypic") screen, not directed against a particular mechanism, could identify molecules that reduced cell death broadly and possibly the death of neurons in neurodegenerative disease. Such screens monitored cell death or apoptosis-related events could lead to the rapid identification of small molecules that would rescue vulnerable neurons. This approach, however, provided little insight about how to move forward in drug development.

In 2001, NINDS assembled a collection of 1040 compounds, most of which had already been approved for human use by FDA or European Medicines Agency (EMA) [12, 13]. In partnership with several disease-directed organizations, including HDF's Cure HD Initiative, NINDS sponsored a consortium of 26 laboratories, which collectively performed 29 screens of these compounds against cellular and invertebrate "models" of neurodegenerative disorders, including HD, Parkinson's disease, ALS, familial dysautonomia, and spinal muscular atrophy. Almost 30% of the tested compounds had activity in at least one of the 29 assays, and 17 were active in three or more assays. None of the "hits," however, ultimately led either to mechanistic insights or to clinical trials. For example, several cannabinoids were effective in several independent assays, but they did not act through cannabinoid receptors [14].

CHDI's disappointing experience with phenomenological readouts—both in transgenic mice and in engineered cells—led to a determined avoidance of screens not related to specific mechanisms. While a few compounds appeared to influence the course of *HTT*-dependent changes, they did so by acting at unknown targets, for which there was little or no biochemical data. Most of the tested compounds failed to influence the course of *HTT*-dependent decline, but the reason for failure was not apparent. Without knowing what a potential drug is supposed to do, the chemists cannot seek to improve its action or its ability to arrive at its target, nor can the biologists look for other ways to manipulate the same biochemical or cellular system.

CHDI CONTINUALLY FACES CHOICES AMONG POTENTIAL THERAPEUTIC TARGETS

A particularly important goal for each CHDI program is to establish a "go/no-go" decision point, using a standard checklist for drugs and targets to go forward. Figure 25.3 shows an example of the application of the checklist to a compound under development as an inhibitor for kynurenine monooxygenase (KMO). Any target-directed program must establish, either genetically or pharmacologically, that blocking or enhancing the target molecule affects the course of pathogenesis in an animal model. Any drug-evaluation program must identify a compelling target and demonstrate target engagement by a potential drug or even a "validating ligand" (a compound that engages a target of interest, but that for other reasons could not serve as an actual therapeutic).

Profile of CHDI-00340246 lead

Structure and physicochemical properties

MW = molecular weight, should be <450	MW	290
TPSA = total polar surface area, should be <85 Å²	TPSA (Å²)	75
P = octanol-water partition coefficient: 2<logP<4.5	logP	3.13
CpK_a = calculated pKa, should be >7	cpK_a	3.5
Solubility should be >10 μg/ml	Thermodynamic solubility in 0.1 M PBS, pH 7 (mg/ml)	0.17

ADMET

Permeability/ER in MDCK-MRD1 cells	High/0.8	**Permeability/ER in MDCK-MRD1 cells** = ability of cells to traverse a monolayer of cells expressing PGP, P-glycoproptein, which transports compounds across the BBB from brain to blood; ER = ratio of efflux to influx should be <2
Human/Rat/Mouse Cl_{int} (μl/min/mg)	<29.5/<41.9/<101	**Cl_{int}** = intrinsic clearance rate determined in liver microsomes, should be <10 μg/min/mg
Cytotoxicity/Ames (μM)	>50/ND	**Cytotoxicity/Ames** = doses that cause cytotoxicity and mutagenesis by Ames test; should be >10 μM
Cerep Pharmacological Profile (μM)	>10	**Cerep pharmacological profile** = the IC_{50} of compound for other enzymes and receptors in the Cerep collection; should be >10 μM
hERG and Nav1.5 (μM)	>10	**HERG and Nav1.5** = IC_{50} for sodium and potassium channels, determined by patch clamp) whose inhibition can cause cardiac arhythmias; should be >10 μM

Biochemical and cellular assays

IC_{50} = concentration of compound, *in vitro*, at which KMO is 50% inhibited, separate values for human, rat, and mouse, should be <0.1 μM	Human KMO IC_{50} (μM)	0.0002
	Rat KMO IC_{50} (μM)	0.002
	Mouse KMO IC_{50} (μM)	0.002
Plasma shift = ratio of the IC_{50} in the presence and absence fo 3 mg/ml plasma proteins	Plasma shift human/rat/mouse	27/3/5
	IC_{50} human PBMC (μM)	0.045
	IC_{50} primary rat microglia (μM)	0.076
Cellular activity = IC_{50} of compound for KMO activity (from each species) expressed in human peripheral blood mononuclear cells (leukocytes), rat microglia, and in CHO cells	IC_{50} cellular human KMO activity (μM)	0.0083
	IC_{50} cellular rat KMO activity (μM)	0.29
	IC_{50} cellular mouse KMO activity (μM)	0.35
KYNU = IC_{50} of compound for kynureninase	IC_{50} KYNU (μM)	>30
KATII = IC_{50} of compound for kynurenine acetyltransferase, isoform II	IC_{50} KAT-II (μM)	>30

Pharmacokinetics in rats (10 mg/kg po)

Cl_p (L/hr/kg)	0.29 (5mg/kg iv)	**Cl_p** = *in vivo* plasma clearance of the compound
$T_{1/2}$ (hr)	2.34	$T_{1/2}$ = half-life of compound in the plasma
Plasma/brain/kidney/liver AUC (μM*hr/dose)	7.5/0.12/126/25	**Plasma/brain/kidney/liver AUC [area under curve]** = total amount of compound accumulating in each organ after a single dose
Kidney/plasma ratio	16.8	**Kidney/plasma ratio, etc** = ratio of concentration of the compound in each tissue to its concentration in plasma
Liver/plasma ratio	3.33	
Brain/plasma ratio	0.016	
T_{max} plasma (hr)	1.00	**T_{max} plasma (hr)** = time to reach maximum concentration in plasma
C_{max} plasma (μM/dose)	1.9	**C_{max} plasma** = maximum concentration reached in plasma after a single dose

In vivo toxicology (male Sprague Dawley rats)

Single-dose NOAEL (mg/kg)	500	**NOAEL** = No observable adverse effect level
Multi-dose NOAEL (mg/kg)	200	

Biochemical and cellular based on side-by-side measurements with CHDI-00240259, CHDI-00313209, and CHDI-00190327.
$n = 2$, September 22, 2011

Figure 25.3. Checklist for a potential therapeutic compound.

Furthermore, the dosing regimen must make the compound available to the target, presumably in the brain, at an effective concentration.

With over 1000 HD research papers now published each year and scores of CHDI projects providing additional unpublished data, CHDI scientists have had to choose among many competing targets and drugs. Despite its substantial funding, CHDI does not have the resources for rigorous confirmation and development of all the targets and drugs for which there are promising reports, either internally or externally.

An important challenge, then, is how and when to say "no." For each project, the lead science director, together with others involved in that project, evaluates available data and, for CHDI projects, designs follow-on studies. Besides continual input from the appropriate vice-president, each project is the subject of an organization-wide discussion at weekly 2-hour meetings that bring together all science directors and project managers, in person and by teleconference. Finally, as a project advances to the point where CHDI begins to consider an IND application, the project becomes the subject of discussion by a CHDI-convened workshop and, later, by the external Scientific Advisory Committee.

Saying "no" is often difficult. Scientists, both outside and inside CHDI, often see their own program more favorably than others may see it. And everyone wants a "cure," or at least something to do in the face of unrelenting decline. Nonetheless, CHDI has come to focus its resources on the targets and drugs that are most compelling and to attempt to evaluate those programs as quickly as possible. Not unexpectedly, CHDI's decisions on how to proceed—particularly the insistence on using best-practice husbandry and standardized testing protocols—have sometimes led to disagreements with external researchers.

CHDI'S TARGET-VALIDATION SCALE GUIDES THE PRIORITIZATION OF MORE THAN 800 POTENTIAL TARGETS

CHDI's initial effort at target identification and target validation started with a list of suggestions made in the literature or in the reports of CHDI-supported projects. Potential drug targets included molecules with one or more of the following characteristics:

1. A direct interaction with Htt itself
2. Altered expression in postmortem HD brain, in an animal model of, or in a cell expressing mutant Htt
3. A genetic or pharmacological manipulation that modified its expression would modify the pathogenesis in a cell or animal model of HD.

CHDI has assembled and evaluated thousands of published papers and unpublished progress reports concerning molecules with these characteristics. The harvest from this exercise was a set of more than 800 proteins that have some suspected involvement in HD. CHDI's leadership realized that it could fruitfully pursue only a few of these. How, then, to choose?

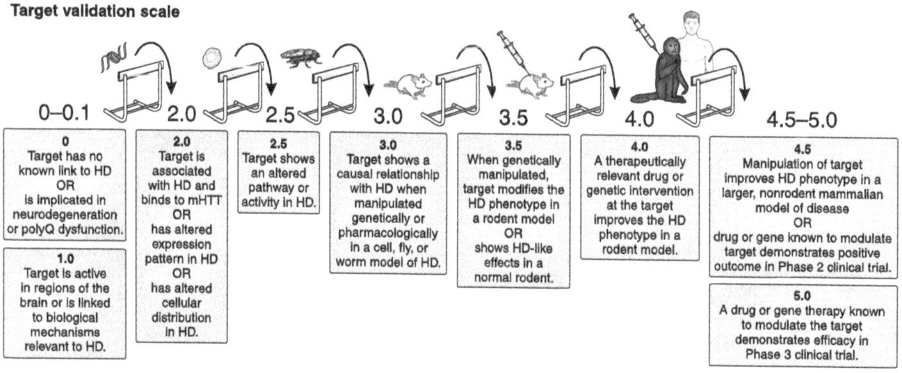

Figure 25.4. CHDI's target-validation scale.

An early step in choosing among targets was the establishment of a monotonic scale for target validation—the demonstration that a particular molecule (or set of related molecules) is involved in the disease process and that changing that target will change the course of the disease. CHDI's target-validation scale, shown in Figure 25.4, has helped CHDI scientists to choose programs to pursue by providing a way to compare available information among hundreds of potential targets. The scale ranks potential targets into bins according to the likelihood a molecule that binds to that target will affect the course of HD, for example, by inhibiting or enhancing enzymatic activity or by antagonizing or stimulating a receptor. The highest score—5—would go to a target known to modulate the disease in humans, but no such target is presently known. A few dozen targets (41) get a rating of 4, because pharmacological or genetic modulation alters the course of the neurodegeneration in a rodent HD model.

CHDI has assessed the target validation score of all 855 potential targets considered at the time of this writing (http://www.hdresearchcrossroads.org). The listing for each molecule contains the reason for its rating, with literature citations. More than half the nominated targets have a score of 3.0, meaning that genetic or pharmacological manipulation of the target mitigates the effects of mutant HTT in cell culture or in a yeast or invertebrate model, but have not yet shown such an effect in a rodent model—the necessary criterion for moving to a score of 4.0. As Figure 25.3 shows, the validation scale also has several intermediate levels. Figure 25.5 shows the distribution of the targets by target-validation score and by functional classification—membrane receptors, growth factors, transcriptional regulators, kinases, and so on.

ADVANCES IN SYSTEMS BIOLOGY SHOULD ALLOW A RATIONAL SELECTION OF THERAPEUTIC TARGETS AND PATHWAYS

The number of potential targets proliferated far beyond initial expectations, to the more than 855 molecules listed in HD Research Crossroads—far beyond the possibility of evaluating all in even a single animal model. CHDI has therefore grouped potential

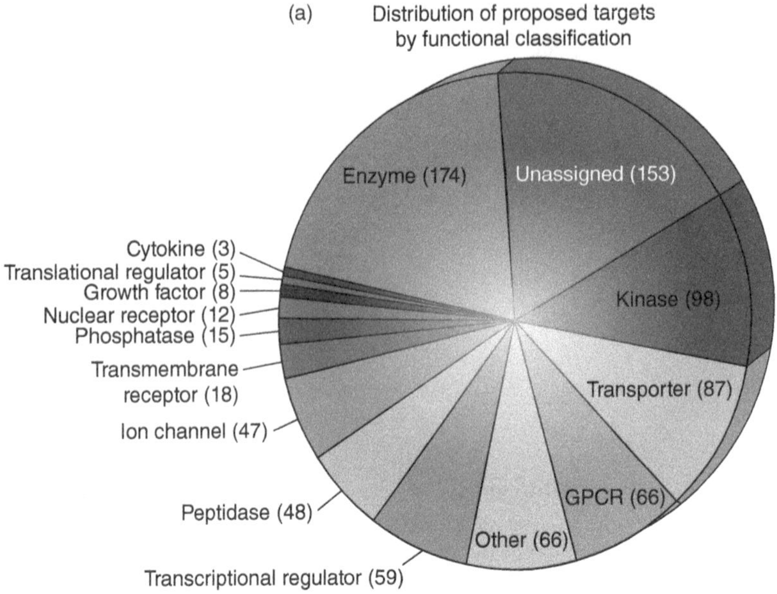

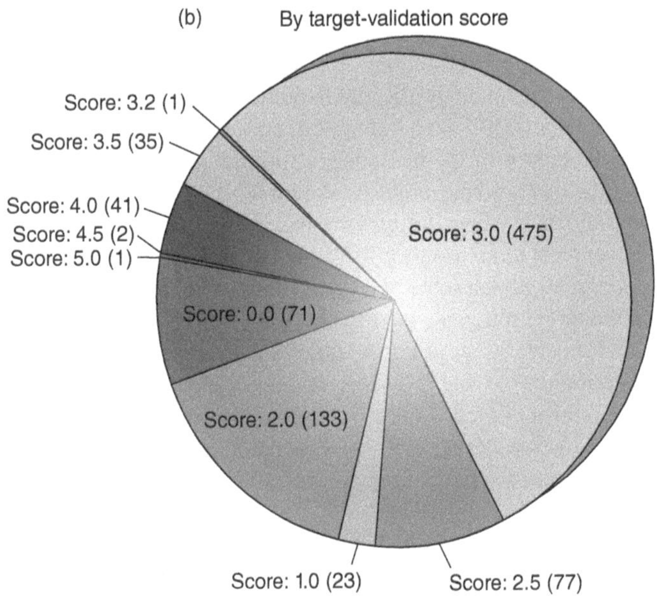

Figure 25.5. Distribution of proposed targets: (a) by functional classification, (b) by target-validation scores. See color insert.

targets into broad mechanistic groups, or "pathways," using categories, curated literature database, and other tools provided by Ingenuity Systems. Their analysis, called the HD Pathway Analysis Database, indicated that many potential targets fell into a relatively small number of biochemical or cellular pathways, as defined by several classification schemes ("ontologies"), including GO (Gene Ontology), IPA (Ingenuity Pathway Analysis), and KEGG (Kyoto Encyclopedia of Genes and Genomes). The pathways of greatest interest—and of greatest changes after perturbation with mutant Htt—included contributors to synaptic function and vesicular recycling, to intracellular transport, to autophagy, and to energetics.

Systems biology, which attempts to integrate the interactions among pathways and cellular mechanism and to connect seemingly disparate kinds of data into a single holistic model, provides a still more comprehensive basis for looking changes associated with disease and with drug treatment. The U.S. interstate highway network is an example of a system, while a particular coast-to-coast route is an example of a pathway (this illustration comes from Keith Elliston, CHDI's former vice president for systems biology). Even if a single pathway becomes blocked, the system provides for other routes. Just so, the systems of a cell or an organism can adapt to the perturbations that come from a disease. Systems biology, then, provides a framework for understanding just how cells and people maintain themselves for decades after a mutant *HTT* gene begins to make its evidently toxic products.

The goal of CHDI's systems-biology program is to define the networks and pathways perturbed by mutant *HTT* and to use this analysis to choose the therapeutic strategies that will restore or compensate for the changes that lead to disease. To accomplish this goal, CHDI is developing the infrastructure, resources, and capabilities necessary to enable systems-driven drug discovery. If these efforts are fruitful, it may be possible to finds ways of treating the entire "corrupted network" that results from an *HTT* allele and to repair the whole spectrum of HD pathology [15].

FUTURE CHALLENGES

CHDI's mission remains: to find therapies that will slow or stop HD, but its strategy and tactics continue to evolve. High priorities for the next few years are the following:

1. To take maximum advantage of systems biology, both to analyze existing data and to direct new experimental approaches
2. To understand more fully the early stages of HD progression, especially of early functional changes
3. To complete the experiments and the regulatory applications necessary for clinical trials of HTT-reducing therapeutics
4. To improve the sharing of data and knowledge not only among CHDI scientists and collaborators, but also among all HD researchers
5. To establish collaborations for precompetitive work that would benefit all types of HD drug development

6. To expand the repertoire of arrangements for contracting, licensing, and reagent distribution to speed the path to effective therapies
7. To continue the preparatory work required for future clinical trials, including the development of biomarkers, the analysis of trial design, the formulation of a new rating scale for early and prodromal HD, and the expansion of the family base by Enroll-HD.

ACKNOWLEDGMENTS

I am grateful to all my CHDI colleagues for many candid discussions, these last 10 years, about the many strategic and tactical challenges of drug development. I thank Katie Ris-Vicari for her help in preparing the figures and the following people for their comments during the preparation of this chapter: Maria Beconi, Robi Blumenstein, Celia Dominguez, Keith Elliston, Evelyn Galstian, David Howland, Ignacio Muñoz-Sanjuan, Eileen Neacy, Simon Noble, Robert Pacifici, Larry Park, David Rankin, Ethan Signer, Leticia Toledo-Sherman, and George Yohrling. I am particularly appreciative of the detailed critique and suggestions from Simon Noble.

REFERENCES

1. Huntington, G. On chorea. *Med Surg Rep.* 1872;26(15):317–321; reprinted in *J Neuropsychiatry Clin Neurosci.* 2003;15:109–112.
2. Fang, Q., Strand, A., Law, W., Faca, V.M., Fitzgibbon, M.P., Hamel, N., Houle, B., Liu, X., May, D.H., Poschmann, G., Roy, L., Stühler, K., Ying, W., Zhang, J., Zheng, Z., Bergeron, J.J., Hanash, S., He, F., Leavitt, B.R., Meyer, H.E., Qian, X., McIntosh, M.W. Brain-specific proteins decline in the cerebrospinal fluid of humans with Huntington disease. *Mol Cell Proteomics* 2009;8:451–466.
3. Koch, R. An address on bacteriological research. *Br Med J.* 1890;2:380–383.
4. Chen, M., Ona, V.O., Li, M., Ferrante, R.J., Fink, K.B., Zhu, S., Bian, J., Guo, L., Farrell, L.A., Hersch, S.M., Hobbs, W., Vonsattel, J.P., Cha, J.H., Friedlander, R.M. Minocycline inhibits caspase-1 and caspase-3 expression and delays mortality in a transgenic mouse model of Huntington disease. *Nat Med.* 2000;6:797–801.
5. Smith, D.L., Woodman, B., Mahal, A., Sathasivam, K., Ghazi-Noori, S., Lowden, P.A., Bates, G.P., Hockly, E. Minocycline and doxycycline are not beneficial in a model of Huntington's disease. *Ann Neurol.* 2003;54:186–196.
6. Mievis, S., Levivier, M., Communi, D., Vassart, G., Brotchi, J., Ledent, C., Blum, D. Lack of minocycline efficiency in genetic models of Huntington's disease. *Neuromolecular Med.* 2007;9:47–54.
7. Menalled, L.B., Patry, M., Ragland, N., Lowden, P.A., Goodman, J., Minnich, J., Zahasky, B., Park, L., Leeds, J., Howland, D., Signer, E., Tobin, A.J., Brunner, D. Comprehensive behavioral testing in the R6/2 mouse model of Huntington's disease shows no benefit from CoQ10 or minocycline. *PLoS ONE* 2010;5:e9793.
8. Huntington Study Group DOMINO Investigators. A futility study of minocycline in Huntington's disease. *Mov Disord.* 2010;25:2219–2224.

9. Menalled, L., El-Khodor, B.F., Patry, M., Suárez-Fariñas, M., Orenstein, S.J., Zahasky, B., Leahy, C., Wheeler, V., Yang, X.W., MacDonald, M., Morton, A.J., Bates, G., Leeds, J., Park, L., Howland, D., Signer, E., Tobin, A., Brunner, D. Systematic behavioral evaluation of Huntington's disease transgenic and knock-in mouse models. *Neurobiol Dis.* 2009;35: 319–336.
10. Young, N.S., Ioannidis, J.P., Al-Ubaydli, O. Why current publication practices may distort science. *PLoS Med.* 2008;5:e201.
11. Brunner, D., Balci, F., Ludvig, E.A. Comparative psychology and the grand challenge of drug discovery in psychiatry and neurodegeneration. *Behav Processes* 2012;89:187–195.
12. Heemskerk, J., Tobin, A.J., Bain, L.J. Teaching old drugs new tricks. *Trends Neurosci.* 2002;25:494–496.
13. Heemskerk, J., Tobin, A.J., Ravina, B. From chemical to drug: neurodegeneration drug screening and the ethics of clinical trials. *Nat Neurosci.* 2002;5(Suppl.):1027–1029.
14. Aiken, C.T., Tobin, A.J., Schweitzer, E.S. A cell-based screen for drugs to treat Huntington's disease. *Neurobiol Dis.* 2004;16:546–555.
15. Ainsworth, C. Networking for new drugs. *Nat Med.* 2011;17:1166–1168.

26

LESSONS FROM THE PAST AS A MEANS TO THE FUTURE: INSTITUT PASTEUR AS A MODEL STRATEGY

Spencer L. Shorte

Plateforme d'Imagerie Dynamique (PFID), Imagopole, Institut Pasteur, Paris, France

Science has no country, or rather its country encompasses all humanity. Knowledge belongs to all mankind. —Louis Pasteur c. 1891

Scientific discovery to benefit human health requires academic, industrial, and government to work together, but how?

The *Institut Pasteur* is a nonprofit private foundation named after Louis Pasteur, who made seminal breakthroughs contributing a founding basis to our modern understanding of biology, microbiology, immunology, and medicine (http://www.pasteur.fr; http://www.pasteur-international.org). From the beginning the mission of the institute has been: *excellence in fundamental biomedical research to the benefit of global public health*. For over a century the researchers of the Institut Pasteur have dedicated themselves to the study of host-pathogen biology, microorganisms, and vaccine development, maintaining the institute at the vanguard of the battle against infectious disease [1]. Consequently, Pasteurian researchers have been responsible for breakthrough discoveries, milestones enabling medical science to control public health threats, including virulent diseases such as rabies, diphtheria, tetanus, tuberculosis,

Collaborative Innovation in Drug Discovery: Strategies for Public and Private Partnerships, First Edition. Edited by Rathnam Chaguturu.

poliomyelitis, influenza, yellow fever, and the plague. Since 1908, 10 Pasteur Institute scientists have been awarded the Nobel Prize, the most recent in 2008 for medicine and physiology: the discovery of HIV. This proven success is concrete testimony to the remarkable nature of the Institute Pasteur that with the benefit of hindsight has become inspiration to many seeking to design optimized, reproducible, and robust models for successfully operating transnational scientific research frameworks and infrastructure [2–5]. Optimization of the sophisticated governance of such collaborative research frameworks is unarguably necessary to tip the balance in favor of useful scientific discovery, and it is in this sense that the Pasteur model is singularly compelling because in itself it presents us a somewhat remarkable example of how productive academic, industry, and government (AIG) partnerships can emerge at the grassroots level nourished by a dedication to the vision of fundamental research science. Indeed, the founding of the Institut Pasteur in Paris, and the synergistic emergence of the Institut Pasteur International Network (IPIN) has proven itself a robust and successful example of transnational collaboration during over 100 years, bringing it under scrutiny as a case study example of a transnational (transformational) scientific network [3, 4]. Understanding why the Pasteur model works so well as a transnational network might help to extract a recipe for improved AIG partnerships more likely to facilitate scientific discovery (including drug discovery) for the global benefit of human health and the environment. This chapter explains my views on this matter extrapolated from a small selection of learned works on the subject, to which I strongly urge the motivated reader to refer in looking for the details on this rich and fascinating history.

THE INSTITUT PASTEUR INTERNATIONAL NETWORK

The IPIN is a scientific community of regional centers and institutions (Table 26.1) spread out across 5 continents, and uniting 10,000 people, most of whom are recruited locally. It provides us one of the rare examples of a durable scientific network that operates successfully on an international scale, arguably because its founding mission has remained for 120 years a private institute dedicated to research, public health, and education. The Institut Pasteur's international mission dates back to its creation in 1887. The contributions that permitted the founding of the Institut Pasteur in Paris came from many countries across the world. Pasteur scientists in the institution's earliest days traveled around the world to study and combat epidemics, and this diaspora of Pasteurians led naturally to the founding of research and public health outposts that gave birth to the IPIN. Facilities were often located in, or near pandemic areas. The first Institut Pasteur outside France was created in 1891 in Saigon in Vietnam with the collusion of a Pasteurian cohort leadership: Eugene Etienne (an influential French colonialist party leader) and Albert Calmette (a military doctor) who aimed to vaccinate people against rabies and smallpox, by establishing locally the means for vaccine production [6]. In its success this model rapidly led to other institutes being founded in Tunisia and Algeria. That these expansions initially occurred in French colonial enclaves has brought many historians to question the imperialist subtext of the IPIN model [6]. While undeniably French colonialist interests were involved in these initiatives, it seems this

TABLE 26.1. The Chronology of the Institut Pasteur International Network

Institute Name	Country	Founded
Institut Pasteur Paris	France	1887
Institut Pasteur in Ho Chi Minh Ville	Vietnam	1891
Institut Pasteur in Tunis	Tunisia	1893
Institut Pasteur of Algeria, Algiers	Algeria	1894
Institut Pasteur in Lille	France	1894
Institut Pasteur of Nha Trang	Vietnam	1895
Institut Pasteur in Madagascar	Madagascar	1898
Institut Cantacuzene in Bucarest	Romania	1901
Institut Pasteur in Brussels	Belgium	1901
Institut Pasteur in Casablanca	Morocco	1911
Institut Pasteur Hellenic, Athens	Greece	1919
Institut Pasteur of Iran, Tehran	Iran	1920
Institut Pasteur of Dakar	Senegal	1923
Institute of Hygiene & Epidemiology, Hanoi	Vietnam	1923
Institut Pasteur in Saint-Petersburg	Russia	1923
Institut Pasteur of French Guyana, Cayenne	Guyane Francaise	1940
Stephan Angeloff Institute of Microbiology, Sofia	Bulgaria	1947
Institut Pasteur of Guadeloupe	France	1948
Institut Pasteur of New Caledonia, Noumea	New Caledonia	1954
Pasteur Centre of Cameroon, Yaoundé	Cameroon	1959
Institut Pasteur of Bangui	Central African Republic	1961
Cenci Bolognetti Foundation, Rome	Italy	1970
Institut Pasteur, Abidjan	Ivory Coast	1972
Institut Pasteur in Rome	Italy	1976
CERMES, Niamey	Niger	1978
Cantacuzino Institute, Bucharest	Romania	1991
Institut Pasteur of Cambodia	Phnom Penh	1995
Scientific Institute of Public Health, Brussels	Belgium	1996
Hong Kong University—Pasteur Research Centre	Hong Kong	1999
INRS—Institut Armand-Frappier, Quebec City	Canada	2004
Institut Pasteur Korea, Gyeonggi-do	South Korea	2004
Institut Pasteur of Shanghai	China	2004
Institut Pasteur in Montevideo	Uruguay	2006
Institut Pasteur of Laos, Ventiane	RDP Lao	2007

feature should not be in any way considered anything more than context; it has nothing to do with any prerequisite feature of the IPIN model itself. Supporting this view is the modern cartography of today's IPIN, which is distributed across five continents in a manner distinct from any perimeter that might reflect direction *per se* by French government, or diplomacy. Furthermore, the continuous growth of the IPIN during 120 years has been sustained at a remarkably constant rate of one new institute every 4 years (see figure and table). The average growth rate has progressed unabated by two world wars, three global economic depressions, and extreme patterns of change in the ebb and flow of French diplomatic policy on so-called colonial outposts. The IPIN growth trend does betray microcosm inertia due to socioeconomic factors (these are discussed in Reference 4), but overall, it is a linear trend. So, while undoubtedly French and other government interests did, do, and will play a part in the growth of IPIN, just as for any other AIG partnership founding transnational research infrastructure, it is unlikely that we will find therein the driving parameters, or impetus to this remarkable growth curve. Any role of influence at this level should be considered merely tactical when compared with the more fundamental constants to be found at the higher level of vision and strategy of the founding Pasteurians. The genesis of the IPIN as a model for transnational scientific research infrastructure can be considered from the perspective of a higher level by appreciating the history, vision, and context of the founding of the Institut Pasteur in Paris, where Louis Pasteur and his collaborators shared a common ideal in a mission aimed to enable the largest number of countries to benefit from scientific discoveries in the fight against infectious diseases. It is at this level that we find some understanding toward explaining the startling level of sustained IPIN growth.

CHANCE FAVORS THE PREPARED MIND

By the 1880s, somewhat toward the end of his career, Louis Pasteur had already enjoyed great success as a research scientist. His early breakthrough experiments using his knowledge as a chemist delineated the bases for industrial fermentation processes and led to his pioneering research in bacteriology. Pasteur's importance and contribution cannot be overestimated: he discovered the principle of sterilization, which came to be known as "pasteurization," and from among his other discoveries came the universal practice of surgical antisepsis, today the basis for modern hospital procedures and still the only useful practice to limit the risk and spread of nosocomial infection. However, the catalyst to the founding of the Institut Pasteur, considered by many as by far his most important discovery, was his invention of the rabies vaccine.

Rabies is a viral disease that causes acute encephalitis (inflammation of the brain) in warm-blooded animals. The disease is zoonotic, meaning it can be transmitted from one species to another, such as from dogs to humans, commonly by a bite from an infected animal. For a human, rabies is almost invariably fatal if postexposure prophylaxis is not administered prior to the onset of severe symptoms. The rabies virus infects the central nervous system, ultimately causing disease in the brain and death. Early-stage symptoms of rabies are malaise, headache, and fever, progressing to acute pain, violent movements, uncontrolled excitement, depression, and hydrophobia. The primary cause of death is usually respiratory insufficiency. In short, rabies is a miserable and painful death, and the dramatic details of how Pasteur came to produce a vaccine cure for humans set the stage for the founding of the institute that would eventually bear his name.

The practice of "vaccination" (inducing immunity from disease) has multiple origins that historically can easily be traced back to therapies developed in India, China, and the Turkish Ottoman Empire in the seventeenth to eighteenth centuries, but likely goes back much further to 1000 BC.* However, it was the Englishman Edward Jenner that brought the first scientific presentation of the method in its application to smallpox, presenting his results to the Royal Society in 1796. Nearly 100 years later, Louis Pasteur's contribution to understanding the mechanism by which vaccine inoculation could protect from infectious disease was emerging in the light of Germ Theory. In parallel to the studies of Robert Koch (who would soon publish his seminal "Koch's Postulates" in 1890) both scientists were working on the idea that specific pathogenic agents were responsible for infectious diseases and proposed that such agents could therefore be isolated and were likely players in induced immunity. In 1879, working on the disease called chicken cholera, Pasteur's contribution had exploited the chance observation that an accidental exposure of laboratory chickens to cholera culture isolated from diseased animals, cultured and then "attenuated," resulted in induced immunity. This linked Germ Theory to infectious agents in a way that would eventually help debunk the popular dogma of that era "spontaneous generation" of disease, thereby revolutionizing rational strategies for inoculation and vaccine production. However, that the cholera vaccine experiments had used only laboratory isolates and animals meant that there was no evidence to suggest this idea could be extended to humans. Pasteur was also working on developing an emerging rabies vaccine in a series of experiments in dogs inoculated with the infectious agent extracted from rabbit bone marrow. His results showed similarly that inoculation with attenuated preparations could protect healthy animals from becoming infected upon exposure to live (nonattenuated) preparations. However, as for chicken cholera, he remained far from any proof of efficacy in humans.

During the late summer of 1885 a 9-year-old Alsatian boy, Joseph Meister, had been badly bitten in an attack by rabid dog, which left him mortally injured. Despite being advised by a local physician that the young boy was certain to die, having no chance to escape the rapid onset of this devastating and uncontrollable infection, Joseph's distraught parents brought him to Paris demanding an audience with Louis Pasteur himself, having heard the popular rumors of his successful experiments curing dogs with the disease. However, Pasteur was a chemist, not a medical doctor, and had not begun any clinical experiments with humans. Nonetheless, he was moved by boy's case and the insistence of his distraught parents, perhaps relating to them through his experience of having lost three little daughters from tuberculosis and typhoid fever. In this dramatic context, and against the impassioned objection and contrary advice of one of his closest colleagues, the medical doctor Emile Roux, Pasteur agreed to try to help the boy. This decision was not taken lightly, and Pasteur knew full well that the treatment itself, an injection of rabbit bone marrow in which the virus was "attenuated," might kill the boy outright. It is not parenthetical to consider this dilemma from today's perspective where the regulatory authorities that dominate the necessary protocol controlling translational progression along the preclinical to clinical research path. In the absence of such authority in late nineteenth-century Europe, it was merely collegial ethics, the rule of common law, and one's own sense of personal responsibility that were to be taken into account on matters of regulatory affairs. Pasteur was risking the boy's life and his own stature as a scientist, researcher, and member of the *College de*

*http://en.wikipedia.org/wiki/Vaccination; http://en.wikipedia.org/wiki/Smallpox_vaccine.

France, where his maverick ways had already risked isolating him, despite his genius. So, after several weeks of daily treatment by excruciatingly painful injections performed in complete secrecy except for under the surveillance of two clinicians and the boy's mother, when Joseph survived, apparently cured, it was both a joyful triumph and considerable relief!

INSTITUT PASTEUR PARIS: A PRIVATE FOUNDATION WITH A PUBLIC HEALTH MISSION

The news of the extraordinary story of Joseph Meister spread rapidly throughout France and beyond, resulting in literally hundreds more patients arriving in Paris during the following months, seeking treatment. In the nineteenth century, rabies was prevalent across Western Europe, and much of the rest of the world too. It was a terrifying disease scourge, compelling the *zeitgeist* of that time (press and pundits) to deem Pasteur's discovery as more than remarkable and earning him the popular title "Benefactor of Humanity" [1]. The impact of these results was compelling, and on March 1, 1886, Pasteur relayed his results to the Academy of Sciences, revealing how he had succeeded in curing all except one of the nearly 400 rabies-infected patients that had come to him since hearing of the cure provided by his treatment. Pasteur seized this opportunity to request the Academy's support to create the institute of which he had so long dreamed: a research institute dedicated to the study of microbiology, delving into the experimental research analyses of infectious disease, and applying those results to better public health. These ideas would, of course, be illustrated by the new institute's commitment to specialization in the *development* and *production* of a vaccine against rabies. The Academy, astounded by the results, provided unanimous support, and even went so far as to launch an international fund-raising campaign toward establishing the Institut Pasteur, which was founded on June 4, 1887. A month later, laboratory construction began, while Jules Grevy, President of the French Republic, signed a government decree recognizing the Institut Pasteur as a "foundation directed to the public benefit" and approved its statutes [1, 7]. This compelling context gave impetus to the rapid progression of events, which made it for the first time possible for Pasteur to create infrastructure that adhered to his goal for a scientific institute that could operate in liberty, guided by a vision of fundamental biological research aimed to the benefit of applications in public health.

From the moment Grevy signed the statutes, the birth of the Institut Pasteur in Paris was assured financial benefit from the State. Particularly, the early construction work consumed all of the first public subscription, leaving nothing for operating costs (see Reference 4, and references therein, e.g., [8]). Furthermore, the initial lack of funds was later redressed upon successful development of an effective serum for diphtheria that assured a second public subscription (1890). Additional income was provided when the public purse was used to purchase the serum, destined to be used for public health and military field care. However, in the statute of the newly founded Institut Pasteur, from the very outset, was the insistence of autonomy and independence, particularly imbibed in the liberty to operate as a *private* institution with a *public* health mission recognized by the state as such, yet independent from it. This distinction that may seem trivial by comparison to the science that belied the context, was without precedent, and even today is rare if not still unique. The reason is simple: this positioning is intrinsically ambiguous, beckoning the question of how a private enterprise (albeit a nonprofit), generating its own lines of income, can at the same time consume public funding in a privileged manner, and yet still be independent in its governance. A legal quagmire of a sort, this fuzzy, mixed definition of charter and statute has served a founding principle, allowing the institute to maintain itself successfully during over 120 years, through multiple sources of funding, including private enterprise income; donations from private individuals, societies, and companies; and government funding. Consequently, among science historians and policy analysts the ambiguity of the legal statute of the Institut Pasteur has come to be considered a core premise of the Pasteurian model [3, 6]. The duplicity of the founding statute set precedent allowing the institute to access diverse (private and public) funds while still leveraging independence of governance from the state. This defining quality is argued to be the distinguishing characteristic that brought the model to propagate so directly and rapidly outwards into an extensive global scientific network, enduring and still growing for one hundred and twenty120 years.

PROPAGATION OF THE PASTEURIAN SPIRIT

The Institut Pasteur in Paris was inaugurated on November 14, 1888. From the very beginning, the institute was an exceptional instance of an international "biomedical research institute" formed before such a concept had ever been established. The institute heralded a sense of independent scientific autonomy, and imbued this innovative type of international biomedical institute with the mission to understand the new science of biology in medicine. The "Pasteurians" comprised a new breed of scientist dedicated to the study of microbiology, biology, and immunology (at that early stage: vaccine development). The phrase "Pastorian," first coined by Fleury (see footnote 9 in Reference 3), had captured the popular imagination within just a few years after the founding of the institute and came to be rapidly embedded in the enduring global lexicon. The expression testified the particular independence of the Institute Pasteur and its researchers, distinguishing them from the conventional, structured homogeneity of the universities and *Grandes Ecoles* dominating the French research and teaching system at that time. Indeed, to be a Pasteurian was to be identified as a part of that special breed, dedicated to charity for the greater good, driven by a faith that truth was to be found in a deep knowledge of science and experimental proof therein. Pasteur and his disciples expounded the application of scientific knowledge as providing the hope of continuously improving public health. Pasteur had suffered the loss of his own children to infectious disease, and his genius in science was made all the more profound by his tenacity that for many seemed to resemble more an ideology than a philosophy [3, 6]. In reference therein to Georges Canguilhem's assessment of Pasteurian microbiology, that he called "the end of medical theories" [9], Anne-Marie Moulin [6] writes:

> Pasteurian ideology, it is true, was difficult to grasp. The *Pastorians* displayed a remarkable intellectual mobility; they shifted their targets according to the context. They identified their practice with 'methods' versus theories, 'doctrines' versus systems, 'applications' versus pure science, non-profit research versus industry-linked research, according to circumstances. This elastic ideology animated the network of overseas Pasteur Institutes (itself, like Pasteur's work an empirical assay), which (again like Pasteur's work) acquired stable characteristics only secondarily.

So, from such analysis it may seem that Institut Pasteur has been built on a statute designed to be duplicitous, for scientists that shift according to circumstances! In this light the Pasteurian vision appears far from science. However, this could not be further from the truth inasmuch as both the institute and the great scientists that have hallowed its aisles are stalwart on certain matters, and these are to be found in the simple, but higher processes of the organization and its protagonists, for example, in its mission statement derived from exactly the same ideals and vision today as was the case 120 years ago. Throughout his research career, until his death, Louis Pasteur had followed his belief that experimental approaches provide scientific facts that together could yield truths, an ultimate, unassailable knowledge. Furthermore, he believed that all knowledge belonged to mankind as a whole, and that this fundamental premise would in itself be to the benefit of humanity. He believed strongly that such knowledge could only be achieved through independent analysis and scientific autonomy, and was committed to the propagation and dissemination of such knowledge by teaching and apprenticeship because he believed that knowledge belonged to humanity, and wherein its imperative was to benefit Public Health. It is these timeless values, encompassing the vision and clear mission upon which the institute was founded, that do not change; whereas, everything "below" remains open to question and revision: strategy, tactics, theory/dogma, and most everything else can be altered, except, of course, good scientific practice and method.

As Pasteur's health began to fail, the responsibility for the emerging institute was assumed by a handful of well-prepared researchers, more than capable of propagating the Pasteurian spirit. This vanguard of visionary Pasteurians was selected from the handful of those closest to Louis Pasteur himself, notably among them Emile Duclaux, Ilya Ilych Mechnikoff, Charles Chamberland, Jacques-Joseph Grancher, and Emile Roux. These Pasteurians set about the business of expounding the reputation of the institute with their groundbreaking science on infectious disease. From the original leadership and dogged spirit of these visionary scientists, their remarkable achievements so successfully brought into play to rapidly benefit public health, there emerged what came to be termed the Pasteurian ideology and model [4]. To be referred as Pasteurian not only was an accolade of distinction, but it also recognized their separateness from the French mainstream, and especially the academic system. A strong sense of religious faith dominated, where cleanliness (Pasteur invented the counsel to wash hands) was tantamount to godliness, in a world that Pasteur believed was pillared by four columns of humanity: faith, hope, charity, and *science*. These exact four words appear, one at each corner, of the ceiling to the crypt in Louis Pasteur's tomb housed in a subterranean vault of the *Bâtiment Musée* at the Pasteur Institute in Paris. They can be viewed freely by visiting the museum. Following the prolonged ill health of Pasteur, his death (1895) so soon after the founding of the institute was a great loss, but did not threaten the dynamic of the institute. Pasteurians were monastic in their commitment and drive toward the vision set out by Pasteur. As the first director, Duclaux forged this caste with his own brand of austerity and high expectations of his colleagues.

In his plans for founding the institute, Louis Pasteur had insisted upon the value of dissemination by rapid publication of findings, followed by embodiments of knowledge and methodology in teaching and training. So, following his death, it came down to the Pasteurians to continue this path under Duclaux, and eventually Roux, focusing on implementing forums to present and publish work independent from the mainstream. For example, the *Annals of the Institut Pasteur* is a journal launched by the institute under the guiding impetus of Emile Duclaux in 1887 that continues to this day. Teaching programs were also from the beginning integral to augmenting the influence of the institute [1, 4], where ranks of French and foreign researchers alike would come to Paris to learn the fundamental methodologies of general microbiology, vaccine production, and the rudiments of hygiene and public health. These students would carry with them their experiences to the far outreaches of the globe as scientists and clinicians, microbiologists studying diseases on mission in the field, and doctors attending the wounded in foreign wars or at the sites of disease outbreak among the colonial settlements scattered across the world since Napoleonic imperialism [6]. In the context of this diaspora, the IPIN emerged.

THE INSTITUTE PASTEUR INTERNATIONAL NETWORK TODAY

Today, a unique organization, the Institut Pasteur International Network, unites close to 10,000 staff from 32 institutions (including 8 World Health Organization [WHO] collaborating centers) spread across five continents (Figure 26.1). The Network is dedicated to the surveillance, prevention, and treatment of infectious diseases through biological research, education, and public health activities.

The Institut Pasteur in Paris is at the core of the International Network, but inasmuch as they are independent legal entities governed by their national authorities, they are not so much directed from Paris as guided in a multifarious relationship where hierarchical subordination to Paris is replaced by coordinate scientific and public health strategies for collaboration and surveillance. Implementation of such strategies is animated by the International Affairs Department of the Institut Pasteur, allowing a framework for the institutes of the International Network to associate in partnerships and cooperation agreements covering scientific research, training, and public health services, where they share common values and objectives, and do not necessarily involve Paris. Rather, united by the same values, the majority of IPIN institutes are independent from governance *per se* by the Institut Pasteur in Paris. Falling under the authority of the health officials of their respective countries, they belong voluntarily to this informal network, which has a unifying goal: scientific research in the service of human health.

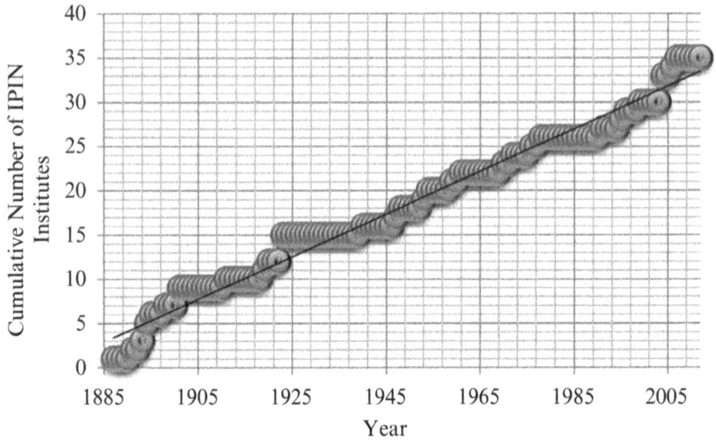

Figure 26.1. Growth of Institut Pasteur International Network over the years.

Toward these ends, an international Pasteur scientific community has gradually been formed without borders. All the institutes have signed a general declaration of scientific cooperation, based upon a network charter that reflects the fundamental principles adhered in the founding of the institute in Paris. For example, the training of foreign scientists, managers, and technicians is deemed a priority activity. Training and teaching flourishes in Paris in the dedicated Education Centre of the Institut Pasteur where more than 250 students every year come for short-term training courses from over 50 countries; furthermore, 1000 plus interns from over 70 countries receive high-level training through the IPIN, either in the International Network's institutes or through the courses and lectures given each year throughout the world by Pasteur scientists. In this context of mobility, communication and dissemination activities are assured between the IPIN's different institutions. Cross-sector programs support micro-networks of specialized research, in which several institutions join together to carry out cutting-edge research initiatives. While the IPIN aims locally to participate in improving the health of each respective country's population, combining their expertise through interinstitutional collaborations also aims to ensure that this activity contributes to developing knowledge within the international scientific community, and is exposed to critical external scientific review.

The network of institutes contributes together to the prevention and treatment of infectious diseases through

Research and technology development activities focusing on the main transmissible bacterial, parasitic, or viral diseases. Special relationships with the laboratories of other institutes belonging to the Network create a link between local concerns and the international scientific community. They also enable technology transfer and the implementation of leading-edge biological methods in environments that would otherwise be unable to have access to such developments.

Public health activities, including national reference centers and WHO collaborating centers, and participation in national programs to fight against infectious diseases. These activities also include field investigations, composed of both diagnostics and blood, tissue, and microbiological sample collections backed by the cooperation of the laboratories that then process numerous organic samples. These are essential for an accurate understanding of the environment and the epidemiological mechanisms of disease.

Service activities, including histopathology, clinical chemistry, microbiology of water and food, testing centers, centers for vaccinations, and, more recently, technology service centers providing chemical and genomic screening capacity.

Training activities, offered not only to personnel of the institutes, but also to external staff, technicians, and students, who may then use their experience in other regional, national, or international bodies.

These activities are currently exercised in the context of a focus on

- Major diseases with epidemic potential (e.g., HIV/AIDS, tuberculosis, malaria)
- Upstream research for new vaccines and therapies
- Emerging diseases (e.g., dengue fever, encephalitis, hemorrhagic fever, chikungunya)
- Security health activities (surveillance and outbreaks intervention)
- Surveillance and research on resistance to anti-infective therapies
- Neglected diseases (e.g., rabies, diarrhea, leishmaniasis).

In the context of the international network, the Institut Pasteur has established formal partnerships with public and private organizations on all five continents. The Institut Pasteur is thus able to multiply and stimulate funding for scientific programs, courses, forums, and support for young researchers in Paris and throughout the International Network. By doing so, the Institut Pasteur has become an essential and proactive player in the strategic issues involved in research and public health across the globe, and is one of the largest single bodies providing funding for research on neglected and emergent disease.

The Institut Pasteur policy for developing international activities focuses on strengthening public health and scientific research activities on a bilateral or multilateral basis. Although diligent, the institute is rather liberal, if not promiscuous, in its choice of partners and works with a broad range of third-party organizations, including public and private sector institutions, nonprofit foundations, international organizations (including the European Union and nongovernmental organizations, such as WHO). The International Network often follows the deep roots of the modern French diplomatic landscape, promoting the development of pan-continental regional clusters such as those grouped in South America, the Asia-Pacific, and North Africa. Indeed, in testimony to the success of this approach, there are more than 1100 collaborative biomedical and scientific projects currently in progress throughout the world.

CONCLUSION AND PERSPECTIVES

This chapter does not even hope to contribute to what others have strived to achieve on the subject of what has brought the Institut Pasteur in Paris to be the enduring success in biomedical research for public health that it is. Rather, it aims to bespeak a historic descriptive based upon the conclusions from a wealth of such analysis and to build an argument to explain the secret of the success of the Institut Pasteur that led so readily to the spawning of the IPIN.

First, the formula for this success is derived certainly from a combination of a simple but powerful and visionary mission statement applicable to all, without exception, and firmly founded in scientific philosophy, application, and method; while at the grassroots level maintaining a liberal flexibility for change and adaptation, specifically at the level of strategy and tactics (including management, financing, resource mobilization, policies, logistics, etc.). In short, do whatever needs to be done, with due diligence, to get closer to the mission goal, and make sure to have each decision and each action along this pathway reflect the imperative for good scientific practice.

Second, as Pasteur is famously quoted: "Chance favors the prepared mind." At the level of the grand research goals underlying multidisciplinary collaboration, in the sophisticated context of transnational networks bringing together academic, public, industrial, private, and government organizations, "Lady Luck" is squeezed out by systems approaches aiming to have "go/no-go" decisions as fast as possible. The Pasteur model is different because of its capacity to adapt flexibly. The Pasteur model creates a framework that embraces, enhances, and ultimately benefits from its disposition and capacity to support unpredictable processes along the knowledge discovery pipeline. Moreover, Pasteur has found a formula that allows this dynamic to scale from the Pasteurian research scientist at the bench up to an entire transnational framework of collaborative institutes, the IPIN. Indeed, there was no choice in this matter, since for the IPIN to realize its mission requires fundamentally that the framework can deal with unpredictable disease outbreak at a global geographical and epidemiological scale. My view is that all transnational collaborative AIG infrastructure strategies should do the same. This is especially true in collaborative drug discovery, where we need to be better able to leverage *hypothesis-driven* low-throughput, experimental approaches that generally are less favored by industry compared with *discovery-driven* high-throughput screening approaches. Of course, this is to say that frameworks and strategies imbibing a practical means to handle throughput so very necessary to industry need to be finessed sufficiently with respect to good scientific method, so as to be valuable in *scientific quality*, as much as quantity. However, more important still, discovery pipelines must accommodate and derive scientific value from experimental paradigms that fail unpredictably, producing aleatory findings. We must remember that, in the scientific process, questioning a hypothesis sometimes requires multiple attempts at experimental design because at first no single design seems to answer all the positive and negative controls necessary to allow a confident conclusion. This can be an evolutionary process, requiring time and multiple failures before the scientist arrives at the right experimental design to yield a clean answer. Somewhere along this route every research scientist knows that you will observe unpredictable things. Sometimes these are extremely valuable to the discovery process. The Pasteur model teaches us that to preserve good science requires scalar strategies that do not suffocate this very fragile part of the scientific process: not at the bench, and especially not at the transnational network or infrastructure levels.

What does this all mean for collaborative drug discovery in the future? Certainly, Institut Pasteur Paris and the IPIN show us that AIG partnerships can be very productive and offer a powerful means to scientific and public health success. New improved strategies in drug discovery may benefit to be less structured by the needs of any single interest lobby (e.g., industrial pharmaceutical companies that have until now had a tendency to set, if not, dominate the agenda to transnational collaborative road maps). Road maps and management must be more flexible, and management strategies more multilateral. The ability to be less systematic and more flexible in strategy and tactical management will allow science to show itself. We need less black and white structured organization in AIG partnerships and improved flexibility to manage collaborative research in the hazy gray area where real scientific breakthrough is most abundant. This assembled volume of chapters provides us both a means and impetus to go in this direction.

REFERENCES

1. Schwartz, M. The Pasteur Institute: 120 years of research in microbiology. *Res Microbiol.* 2008;159(1):5–14.
2. Charle, C., Schriewer, J., Wagner, P. *Transnational Intellectual Networks: Forms of Academic Knowledge and the Search for Cultural Identities.* Frankfurt: Campus Verlag; 2004.
3. Moulin, A.-M. The Pasteur Institute's International Network: scientific innovations and French tropisms. In: Charle, C., Schriewer, J., Wagner, P., editors. *Transnational Intellectual Networks: Forms of Academic Knowledge and the Search for Cultural Identities.* Frankfurt: Campus Verlag; 2004.
4. Hage, J., Mote, J. Transformational organizations and institutional change: the case of the Institut Pasteur and French science. *Socioecon Rev.* 2008;6(2):313–336.
5. Mote, J., Whitestone, Y., Jordan, G., Hage, J. Innovation, networks, and the research environment: examining the linkages. *IJFIP* 2008;4(3):246–264.
6. Moulin, A.-M. Patriarchal science: the network of the overseas Pasteur Institutes. In: Petitjean, P., et al., editors. *Science and Empires.* Dordrecht: Kluwer; 1992, pp. 307–322.
7. Institut Pasteur, Statutes. Available at http://www.pasteur.fr/ip/easysite/pasteur/en/institut-pasteur/statutes.
8. Seidell, A. Chemical research at the Pasteur Institute. *J Chem Educ.* 1926;3:1217–1239.
9. Canguilhem, G. L'effet de la bactériologie dans la fin des théories médicales. In: Dozon, J.-P., Fassin, D., editors. *Idéologie et rationalité dans l'histoire des sciences de la vie.* Paris: Vrin; 1977, pp. 55–77.

27

SEEDING OPEN INNOVATION DRUG DISCOVERY AND TRANSLATIONAL COLLABORATIONS TO LEVERAGE GOVERNMENT FUNDING: A CASE STUDY OF STRATEGIC PARTNERSHIP BETWEEN SANFORD-BURNHAM AND MAYO CLINIC

Thomas D.Y. Chung,[1] Sundeep Khosla,[2] Andrew D. Badley,[2] and Michael R. Jackson[1]

[1]*Sanford-Burnham Medical Research Institute, La Jolla, CA, USA*
[2]*Mayo Clinic, Rochester, MN, USA*

PHARMACEUTICAL INDUSTRY STAGNATION AND THE CLARION CALL FOR NEW BUSINESS MODELS

Over the last decade and especially in the past few years, there has been increasing scrutiny of the stagnant productivity of the pharmaceutical industry. In a recent provocative opinion on "crowdsourcing" of pharma drug development, Mintz [1] suggests that thinning drug pipelines, burgeoning development costs, and a lack of innovation condemn the "traditional" pharmaceutical drug development process as too costly and

Collaborative Innovation in Drug Discovery: Strategies for Public and Private Partnerships, First Edition. Edited by Rathnam Chaguturu.

inefficient to survive in its current form. According to Munos [2], the costs of pharmaceutical research and development (R&D) to generate new drugs has dramatically increased to ~$50 billion per year, while the number of new drugs have remained invariant at 25–30 per year.

Pharmaceutical executives have blamed this poor productivity on increased regulatory scrutiny, chronic indications leading to longer R&D cycles, escalating labor costs, and patent expirations leading to increased generics competition (estimated at 17% of all sales by 2014 compared with 10% in 2008), which have led to massive job cuts (>200,000 in past 4 years), corporate restructuring, and mergers and acquisitions. While these will control costs and provide products in the short term, none improves long-term productivity or fosters the spark of innovation needed for future advances. Dire forecasts estimate loss of sales from patent expirations between 2010 and 2014 of $204 billion [3] and between 2011 and 2015 to exceed $250 billion [4] (cutting revenues by as much as 41% [1]). However, Munos [2] and Kaitin [5] have suggested that big pharma's current woes are rather the result of fundamentally flawed business practices established in the 1960s that have guided drug development and that have remained largely unchanged for the past half a century.

The current estimates for the R&D cycle time (10–15 years [3–6]) and costs for taking a drug from discovery to market now range widely, depending upon indication complexity and amortization of failures: estimates range from $1.32 billion [5] to $1.3–2.35 billion [3] to $3.7–12 billion [4], though most of the costs (75% [7]) are due to high failure rates. Indeed, Abbasi et al. [4] suggest the high R&D cost structure (>18% of sales), high failure rate [3, 8], long cycle time, and high R&D spend per drug make it a dismally performing, high-risk low-reward industry. When compared with orthogonal high-technology businesses such as fabrication of semiconductor devices, where the average complexity has grown in 20 years from each device containing 25,000 to more than 2 million transistors while development time and costs have been dramatically reduced with greater than 98% success of devices post fabrication, pharma has shown little progress. They cite a fundamental reason for pharma's failure is a lack of extensive qualification and lack in validation of designs by running the massive simulations and "what-if" scenarios that are the norm in industries with high success rates.

Clearly, industry, disease advocacy groups, government, and investment communities have sounded clarion calls to find new innovative business models for drug discovery and development [6]. These include private–public partnerships such as the formation of pharma-sponsored incubators and centers of innovation consortia [9], deeper and more translational collaborations [10, 11], and strategic alliances [5] with academic [12] and hospital centers. Drug repurposing (or repositioning, reprofiling) [13–17] has also been proposed as solution to the low productivity, and this angle has been pursued by several biotechnology companies as well as by multiple academic groups that have arranged selective access to sets of "shelved development compounds" from pharma companies [18, 19]. More recently, open source drug discovery [20–22], open access [23], and crowdsourcing, as well as a number of nonprofit pharmaceutical development corporations [24, 25] have made their mark in the less profit-driven pursuit of therapies for rare diseases and neglected diseases of the developing countries. The productivity of the Office of Orphan Products Development, in which 41 FDA-approved therapies have advanced with only a budget of $14 million, is noteworthy.

NIH ROADMAP MOLECULAR LIBRARIES INITIATIVE SEEDS OPEN INNOVATION FOR THE DISCOVERY OF CHEMICAL PROBES

Sanford-Burnham's Experience and Lessons Learned

It was in the context of a perceived lack of innovation and unsustainability in the pharmaceutical industry, as well as a poor track record of basic research discoveries yielding meaningful translation into new therapies for human diseases, that Elias A. Zerhouni, MD, Director of the National Institutes of Health (NIH), announced the Molecular Libraries Initiative (MLI) creation on June 15, 2005. With $88.9 million dollars of funding, a Molecular Libraries Screening Centers Network (MLSCN) comprising nine institutions from seven states was launched. This was a component of the NIH Roadmap's "New Pathways to Discovery" initiative to advance the understanding of biological systems and to build a better "toolbox" for twenty-first-century medical researchers [26, 27]. This "open innovation" initiative was controversial: it mandated that the primary and confirmatory activity data generated from high-throughput screens conducted on a publicly disclosed Molecular Libraries Small Molecule Repository (MLSMR) of ~100,000 compounds by the MLSCN would be deposited and publically accessible in a PubChem database, managed by the National Library of Medicine at the NIH [28]. Assays for the program would be sourced from academic investigators through an abbreviated grant application and peer review process. Successful proposals were assigned to an MLSCN center that subsequently collaborated with the investigator to complete assay development, resulting in robust, scalable assays that could be used to screen the entire MSLMR.

The Burnham Institute for Medical Research was selected as an MSLCN center based on the singular vision of the Chief Executive Officer, Dr. John C. Reed, who had invested and established an automated chemical screening core lab, the San Diego Center for Chemical Genomics, around several integrated Beckman FX workstations, as well as strategically recruiting dedicated staff with the functional skill set and specific experience in high-throughput screening (HTS) and drug discovery. The MLI was a heuristically valuable 3-year pilot whereby the individual center learned to function as a network, the NIH Roadmap managers tested and developed operational management processes and performance metrics and clarified deliverables and definitions of success, and the NIH learned what needed further improvement. The key collective lessons learned during this "pilot" phase were (1) the need for the network to provide significant assay development mentoring and educational outreach to improve the quality and technical feasibility of assays, as well as supportive preliminary results for quality grant applications, (2) development of clear work plans and go/no go decision gates during progression of hits toward selection of a chemical probe, and (3) the need for additional synthetic chemistry resources and access to new screening modalities. Through participation in the MLSCN, the Burnham Chemical genomics team was able to optimize processes, solidify broad expertise in all signal generation and detection technologies, complete hiring and training of staff, and most importantly learn to manage a portfolio of collaborations with principal investigators (PIs) of varied temperaments and styles of engagement and communication. During this pilot period core competencies in managing projects were developed, specifically dealing with the significant diversity among the PIs with respect to their knowledge of or interest in assay development and screening and their overall willingness to deliver to a timeline. These accumulated core competencies and operational experiences were key factors in our ability to attract future collaborations as well as enable effective execution and delivery on future milestone-driven collaborations.

In the next sections, we describe where we envision the renamed Conrad Prebys Center for Chemical Genomics (Prebys Center) at the (also renamed) Sanford-Burnham Medical Research Institute (SBMRI) would most contribute in new models of academic drug discovery and development. We will describe the capabilities and infrastructure we have developed and invested in that align with our ambitions and validate our ability to contribute meaningfully to new processes by which the early stages of drug discovery are executed. We will additionally describe some cutting edge approaches the Prebys Center has invested in with regard to creating "disease in a dish" models that may well provide future translational technologies and physiologically relevant yet renewable and propagatable model systems of disease. We will also provide an example where we have advanced the translation of an early MLPCN (Molecular Libraries Probe Production Centers Network) probe into a potential proof-of-concept preclinical lead candidate. We will articulate how the Prebys Center's accumulated experiences and core competencies developed and refined with the NIH Roadmap's Molecular Libraries Initiative and Programs poise us to leverage NIH funds toward advancing probe and drug discovery initially in a precompetitive environment through strategic collaborations.

Finally, we will describe Mayo Clinic's experience, whose initial focus has been at the other end of the NIH Roadmap initiative. They were an early awardee of the Clinical and Translational Science Awards (CTSAs), and we will discuss how this award and the announcement of the National Center for the Advancement of Translational Sciences (NCATS) provided the context and motivation for the Mayo-Sanford-Burnham strategic collaboration. We will share how we selected each other as partners, the key cultural alignments and complementarities in core competencies that yield a mutually beneficial alliance whose impact is greater than the sum of the individual parts. And finally we will share some of the operational and governance models that have worked well in our translational research collaboration.

The Prebys Center as the Engine of New Drug Development Beyond Basic Research

Traditionally pharmaceutical companies and academia have sought collaborations that generally provided research funds in the form of sponsored reach agreements that typically focused on "Target Identification or Basic Research" as shown in the lowermost bracket "Academia, Scientific Institution, Not-for-profit" to the right of the leftmost stack of activity boxes entitled "The Recent Past" in Figure 27.1. In the future model of drug discovery and development, academic or not-for-profit organizations will have to extend their contribution further along the process, advancing compounds through more costly investment phases to larger, preclinical animal studies where proof-of-concept in an organism can be attained. This is shown in the lowermost right bracket, which extends upwards now close to the Preclinical development box, and into the region of the vertical gray-shaded oval, which indicates where a project must now advance ("The Future") for a venture capitalist and/or a pharma company to take on a project.

An approximate timeline and attrition rate for compounds (sideward funnel) as they transition through the drug discovery and development phases or pipeline [29] is also shown in Figure 27.1, and the left box indicates where drug discovery needs to proceed to a proof-of-concept in an animal model to become valuable enough for a pharma company, venture capitalist or clinical partner to invest to bring the compound into a proof of clinical concept, or first in man stage (second box). The Prebys Center has the competencies to advance compounds into at least small animal studies; however, we lack clinical trial capabilities and the institutional funding to advance projects this far.

Prebys Center's Key Strengths, Expertise, and Experience

Through its participation in the NIH Roadmap, the Prebys Center has almost 8 years of experience in conducting and completing chemical probe and drug discovery projects from target/assay concept, through assay development, HTS, and hit validation, "hit-to-lead," and "lead optimization" to yield optimized drug leads or preclinical development candidates.

- More than 50 professional staff and management comprising pharmaceutical and biotechnology veterans with 450+ person-years of industrial experience in HTS and high-content screening (HCS), as well as all associated activities of drug discovery and early development.
- Performed >350 assay development projects (1° and 2° assays) and completed 125 HTS campaigns on large chemical libraries (>350,000 compounds) for both NIH- and NCI-sponsored as well as pharma/biotech collaborators. These assays represent
  - A large diversity of target classes and assay formats (Figure 27.2a)
  - With 39% cell-based versus "simple" biochemical formats (including primary cells and organelles)
  - With 11% phenotypic HCS automated imaging formats (including induced pluripotent stem cell (iPSC) and *Caenorhabditis elegans*) (Figure 27.2b)
  - All signal generation and detection modalities (except radiometric).
- Designated as a "comprehensive" center (capable of both screening and lead generation chemistry) for NIH MLPCN, delivering 52 new chemical probes and depositing more than 33.6 million test data to PubChem and the NCI Chemical Biological Consortium (CBC).

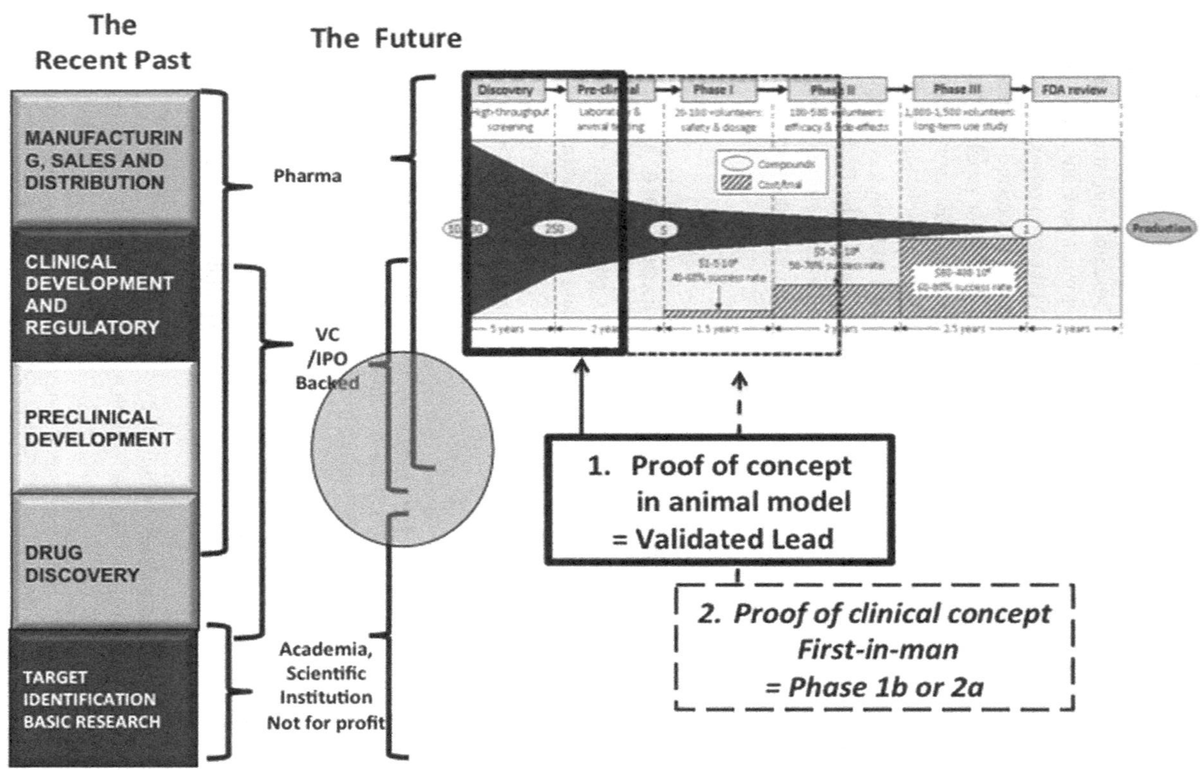

Figure 27.1. Engines of new drug development. Stacked blocks (on left) representing stages of traditional ("the Recent Past") drug discovery are shown with brackets illustrating where academia (not-for-profit), venture capital (VC/IPO-backed), and big "Pharma" currently participate. The brackets under "The Future" show the extension of academia's role further into drug discovery toward preclinical development, while both Pharma and VC have retracted toward later stage of participation. The drug discovery and development process graphic in the upper right is from Reference 29 and illustrates the approximate durations and attrition for compounds (sideways funnel) as they progress through the phases/stages from early discovery to FDA review. Box 1 (solid border)—POC in animals: high-value "transactable" outcome achievable by and appropriate to academia. Box 2 (dotted border)—POC in man: higher-value "transactable" outcome achievable by and appropriate to clinical partners/pharmaceutical/biotechnology companies.

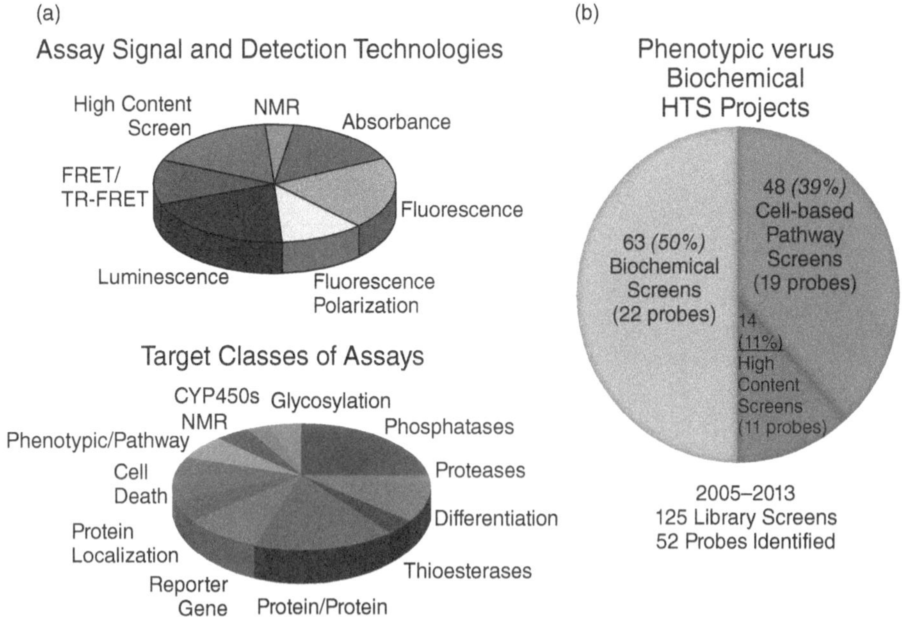

Figure 27.2. Assay signal and detection technologies, target classes, and formats. Pie charts for the distribution of (a) signal generation and detection technologies used for assays developed and classes of targets represented by assays developed, and (b) distribution of the 125 HTS campaigns completed by the Prebys Center during the MLSCN/MLPCN by assay format: biochemical and phenotypic (pathway and high-content automated microscopy image-based), with productivity with respect to chemical MLP probes derived from these formats.

- Proven track record of successfully collaborating with >130 PIs, representing >42 academic, nonprofit medical research, and commercial institutions in >31 geographic locations and 3 foreign countries (Figure 27.3), including several of the Mayo Clinic centers (Scottsdale, AZ; Rochester, MN; and Jacksonville, FL).
- Proven capability and infrastructure to manage 50 concurrent projects, to meet/exceed timelines, milestones, metrics, and deliverables while providing timely communication, data loads to PubChem, project reviews/meetings, and reports to stakeholders.
- One of most advanced state-of-the-art laboratory automation platforms and bio-coastal infrastructure in a not-for-profit institution to support drug discovery and innovation that represent a combined $44 million investment in the Prebys Center (Figure 27.4).
  - Workstation and modular "plug & play" fully integrated 1-arm (La Jolla, CA) and 3-arm (Orlando, FL) robotic HTS platforms

- Metabolex and Vala
- Touro University, USF, USC, Stanford, UN, Caltech
- UCSF, UCSB, UCLA, UCI, UCSD
- City of Hope, LIAI, Salk, Gladstone, Vollum, RTI
- UNM, KU, UTSW, UTA, UTMDA,
- U Minn, U Mich, UI, UVA, Vanderbilt, SRI, Duke, UNC, UVA, VCU, ***Mayo Clinic***
- U Penn, U Pitt, Temple, UVM, Umass, Mt. Sinai, JHCC, Einstein, Sloan-Kettering
- McGill, Justus-Leibeg, U. Padua

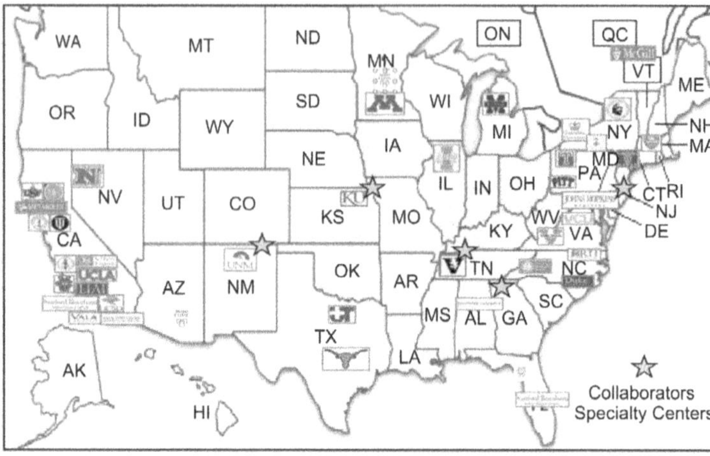

Figure 27.3. Prebys Center collaborators from the NIH Roadmap. The geographical locations in the the United States and Canada (Italy, Germany, and China not shown) of the more than 42 academic, nonprofits, and commercial institutions that the Prebys Center has collaborated with during the NIH Roadmap, also including the three Mayo Clinic centers collaborating with the Prebys Center currently.

- ○ The largest installation among academic institutions of five fully integrated and two offline noncontact nanoliter acoustic drop ejection devices and capabilities (Figure 27.4 and Figure 27.5)
- ○ A greater than 900,000 small molecule chemical and natural product extract collection
- ○ Medicinal chemistry, cheminformatics, and exploratory pharmacology
- ○ Tissue culture facilities and advanced image algorithm development for high-content iPSC assays.

Overview of the Screening Resources at the San Diego Campus. In San Diego, the drug discovery facilities are located in approximately 20,000 sq. ft. of lab space. A major portion of the space is dedicated to central assay development and HTS lab space, liquid handler workstations, and various plate readers (as well as an image-based high-content screening system). The centerpiece of the screening facility is a HighRes Biosolutions (HighRes) 1-POD system (Figure 27.4), as well as two Beckman Coulter core systems. The ultrahigh-throughput screening (uHTS) system in San Diego has a theoretical maximum throughput of ~320,000 wells per day. When combined with the two Beckman Coulter core systems, an overall theoretical maximum throughput of approximately 475,000 wells/day is achievable for plate reader-based assays at the San Diego site.

Overview of the Screening Resources in the Orlando Campus. Sanford-Burnham has established an uHTS screening system in Orlando, Florida,

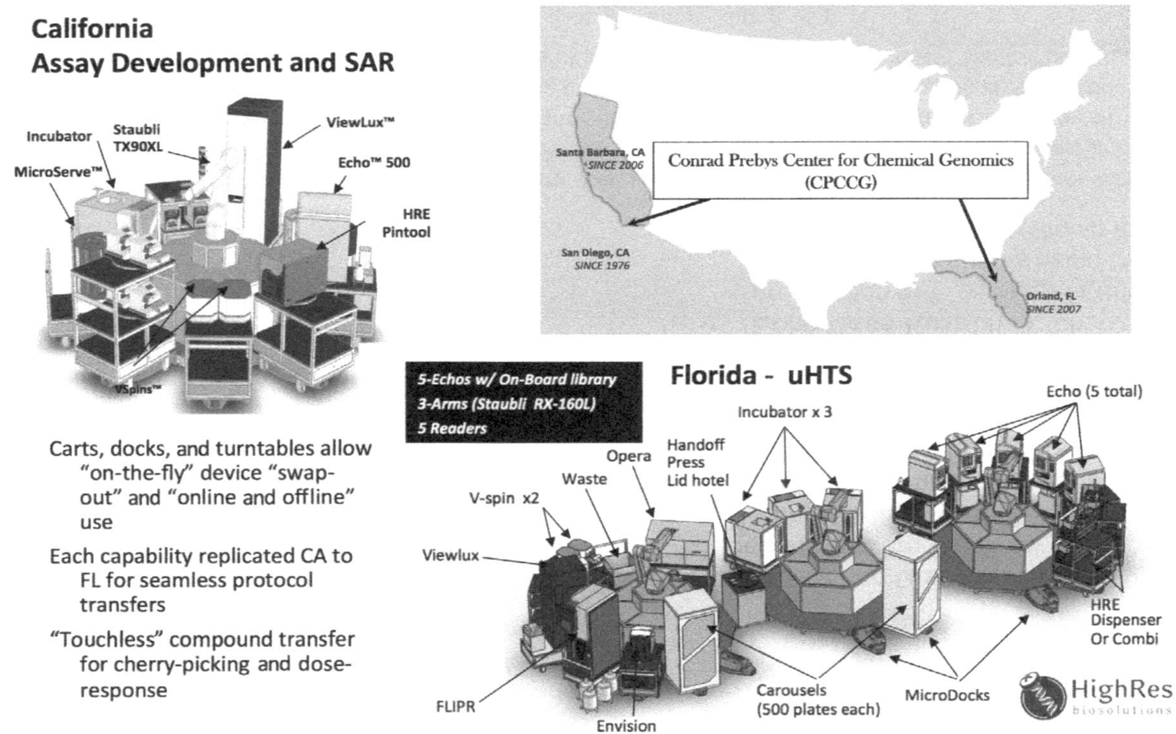

Figure 27.4. Functional and geographic redundancy and symmetry in assay development (CA) and ultrahigh-throughput screening (uHTS—FL) systems. The locations and computer-aided design (CAD) representations of the key functional components and layout of the "Assay Development and Structure–Activity Relationships (SAR)" support the 1-POD (single Stäubli robotic arm) in San Diego, CA campus of the Prebys Center and the uHTS frontline production screening 3-POD (three larger Stäubli robotic arms) in our Orlando, FL campus. The system allows for maximal reconfigurability through rollable MicroCarts™, which mate to MicroDocks™ that supply all services, IP address, and system awareness to the master controller. The two systems are designed to work seamlessly together with assay development on the 1-POD, being translatable to the 3-POD system for uHTS execution and SAR support, then returning to the 1-POD system. However, the systems at each location are fully capable of supporting the full complement of screening modalities and capacities (i.e., functionally redundant) in the event of a catastrophic natural disaster on either coast—earthquake or wildfire in California or hurricane in Florida.

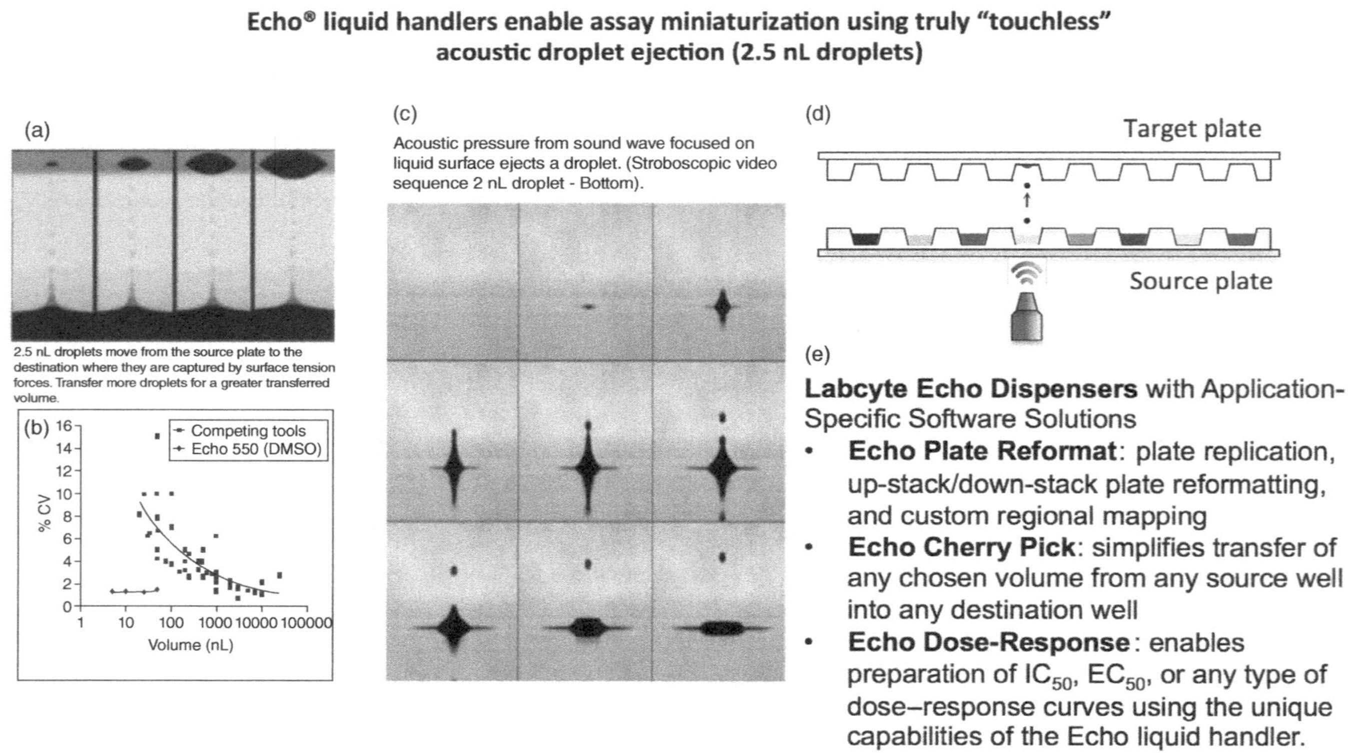

Figure 27.5. Noncontact liquid transfer with Echo® 555 acoustic drop ejection technology (ADE) from Labcyte (Sunnyvale, CA). Composite figure illustrating from left to right, (a) ability to transfer a wide range of volumes with ADE, (b) the accuracy and precision of ADE versus conventional liquid transfer, (c) stroboscopic sequence of a droplet ejection of 2 nL, (d) configuration of the "source" microtiter plate (384- or 1536-well) containing master stocks (e.g., 100% DMSO stock of tested compound) of solutions and empty receiving "target" microtiter plate into which 2.5 nL–2.5 μL can be transferred, and (e) Echo software applications useful for liquid transfer operations.

occupying a 175,000-sq ft building since April of 2009. This uHTS system has a theoretical maximum throughput of ~2.2 million wells/day, with the capability to perform almost any assay, whether cellular or biochemical plate reader, or cellular image-based assay. This system currently exceeds the capability of most HTS systems in industry or academia, and provides a unique resource. Approximately 6500 sq ft has been provided for the uHTS facility in Orlando, including biosafety level 2 (BSL-2) tissue culture labs and compound management space. The facility is adjacent to the Exploratory Pharmacology and Medicinal Chemistry Cores. The HighRes Biosolutions 3-POD-based system at Sanford-Burnham's Orlando campus consists of three linked robot PODs. At the center of each POD is a Stäubli RX-160L robot with a CS8 controller.

Hit Confirmation and SAR Support Capabilities. The Echo555 (Labcyte) acoustic dispensers (Figure 27.5) available at both the San Diego and Orlando sites are invaluable for hit picking during reconfirmation, as well as for direct serial dilutions (dose–response) in structure–activity relationship (SAR) studies. Echo instruments perform contact-free transfer of 2.5-nL droplets directly from a source to a destination plate using focused energy of acoustic waveforms. Solution from any well within a source plate can be transferred into any well of the destination plate. The majority of our assays are performed in 1536-well plates with the assay volumes between 3 and 8 μL that require withdrawals of low-nanoliter volumes and which consume only a negligible portion of the 3- to 5-μL aliquots of compounds in the source plates. This helps to overcome the uneven depletion of wells associated with conventional liquid handling cherry-picking instruments. As a working copy of source plates is kept online for less than 6 months, they are usually replaced with a fresh copy prior to a noticeable depletion of compound wells. The ability to dispense small-volume aliquots into the assay plates enables cherry-picking of hits directly from compound source plates containing 2- or 10-mM stock solutions of library compounds without an intermediate dilution into aqueous solution. This shortcut helps save on costs associated with intermediate plates and reduces potential compound solubility issues. A majority of compounds with limited solubility in aqueous solution are known to come out of solution during the intermediate-dilution stage required with conventional liquid-handling instrumentation. Dispensing nanoliter volumes directly into assay plates circumvents this step, maintains high dimethyl sulfoxide (DMSO) concentration, and helps prevent precipitation.

Furthermore, the wide range of transfer volumes possible with acoustic dispensing allows dose-dependent (over 3 logs) confirmation of hits directly from source compound plates as part of our workflow. This approach is especially valuable for culling out nonselective or assay-interfering compounds through side-by-side tests in primary and counterscreen assays. The relative ease of use and precision of dispensing inherent to Echo® instruments led us to establish acoustic compound dispensing as the main workflow for compound handling within SAR studies. This approach allows us to dramatically reduce compound consumption extending the use of stock solutions through multiple secondary assays. The Labcyte Access workstation significantly simplifies the logistics of handling serial dilution of large number of compounds. Another benefit of Echo dispensers, their ability to dispense aqueous solutions, is instrumental for performing mechanism-of-action studies in which both compound and a known ligand for the target are cross-titrated in the assays plates to determine their interaction, while minimizing depletion of these often critically limited reagents.

High-Content Screening Capabilities, Facilities, and Image Management Infrastructure. We provide herein a rather extensive exposition of our capabilities in this area, as this is a particular strength and differentiator of the Prebys Center. As Figure 27.2b exemplifies, the 14 HCS campaigns yielded a disproportionately high number of 11 novel probes. In particular, novel nonpeptide small molecule probe agonists for the orphan GPR55 and GPR35, as well NTR1 agonists were significant developments. This finding was in agreement with a recent claim by Swinney and Anthony [30] that more first-in-class drugs have arisen from phenotypic and image-based assays. The Prebys Center regards intimate association of HCS with patient-derived induced pluripotent cells to a strategic focus toward assays that reflect relevant translational biology. Among the current academic/nonprofit screening centers, we are the only center that has completed multiple complex HCS campaigns in both 384-well and 1536-well formats of large compound libraries (Figure 27.6) and deposited those data into PubChem.

Assay development for HCS is a multidisciplinary endeavor combining cell-based assays, multistep fixation and staining protocols, multicolor imaging, and custom algorithm development for image segmentation, feature extraction, and correlation of the multiparametric data to meaningful biology. The expert HCS staff, located at the San Diego site while supporting both the San Diego and Orlando locations, executes and/or supports assay development, screening, and data analysis/mining for image-based high-content primary and secondary assays, where the readout is typically based on images from high-throughput microscopy systems. Due to the variety of image-based assay projects encountered at CPCCG, the HCS facility's systems provide capabilities covering fluorescence confocal, fluorescence wide-field, and trans-illumination imaging at magnifications ranging from 3.5× to 60× air and water immersion objectives (0.15 to 1.2 NA). Specifically, the HCS equipment in San Diego includes a spinning-disk confocal Opera QEHS system (PerkinElmer) with on-the-fly image analysis and an InCell1000 imaging system (GE). Examples of a wide range of full MLSMR phenotypic HCS campaigns and representative images are shown in Figure 27.6.

A variety of image processing, data analysis, and visualization tools are available for multiparametric imaging data analysis and mining. With these software packages, the HCS team can rapidly apply available HCS solutions and algorithms, as well as develop new HCS solutions, algorithms, and tools for the wide variety of HCS projects to generate image montages as previously mentioned. HCS assays are suitable to support SARs by quantitative determination of compound potency from phenotypic dose–response curves that the software can aggregate for presentation. An example of an IC_{50} curve for a phenotypic G-coupled protein receptor assay based on the redistribution of GFP-tagged β-arrestin (Figure 27.7) also illustrates the unexpected finding that at the highest concentrations of agonist, the β-arrestin while remaining condensed, redistribute into smaller punctae compared with larger granules at the initial saturating doses.

Assay	Cellular Function	HCS Analysis	HCS Library & Format
Bioparticle Uptake Inhib.	Phagocytosis	Particle Count	110k (chem) / 384-well
Beta-Catenin Activ.	Signaling	(Trans-) Location	200k (chem) / 384-well
Invadopodia Inhib.	Invasion, Cytoskeleton	Fluor. Aggregates/Structures	50k (chem) / 384-well
Cilia Modulators	Cytoskeleton, Developmental	Fluor. Aggregates/Structures	50k (RNAi) / 384-well
Insulin Promoter Modul.	Signaling, Gene/Protein Expr.	Fluor. Intensity / Sub-population	225k (chem) / 384-well
PML-Oncogenic Domain Activ.	Nuclear Foci	Fluor. Aggregates	380k (chem) / 384-well
Lipid Droplet Inhib.	Storage	Fluor. Intensity / Fluor. Aggregates	300k (chem) / 384-well
GPCRs (b-arrestin)	Signaling	Fluor. Aggregates	3 x 300k (chem) / 384-well 1 x 330k (chem) / 1536-well

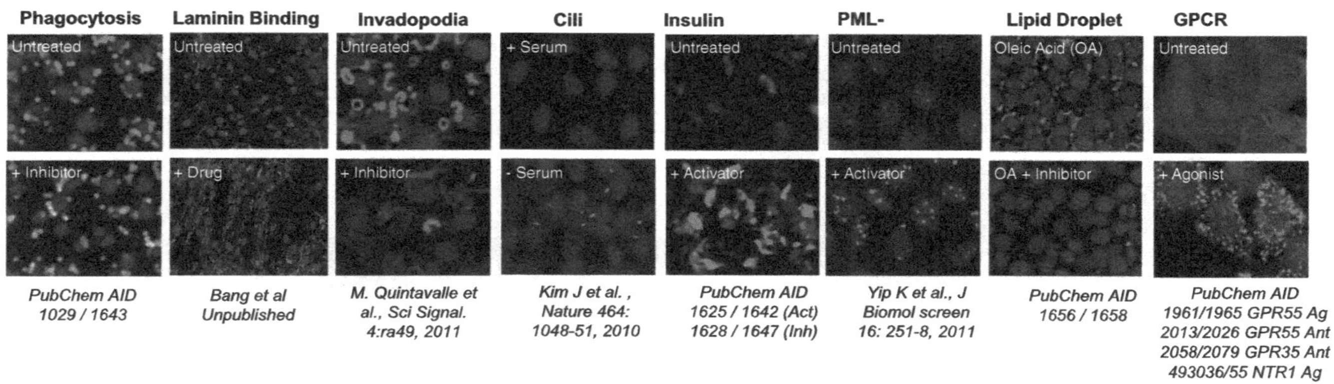

Figure 27.6. Examples of high-content screen (HCS) campaigns completed by the Prebys Center for NIH roadmap projects. The table details the HCS automated microscopic image-based assays that illustrate the breadth of phenotype or cellular function interrogated, the analysis conducted, the number of substances screened during each campaign, and the well density of the assay. The montage of microscopic images are false-colored vertically paired example images of untreated control versus their respective drug/inhibitor/activator/agonist treated samples. (Generally, the nuclei are stained blue with 4′,6-diamidino-2-phenylindole [DAPI], a fluorescent stain that binds strongly to DNA). The PubChem Assay Identifiers (AIDs) or literature references are given for each pair. See color insert.

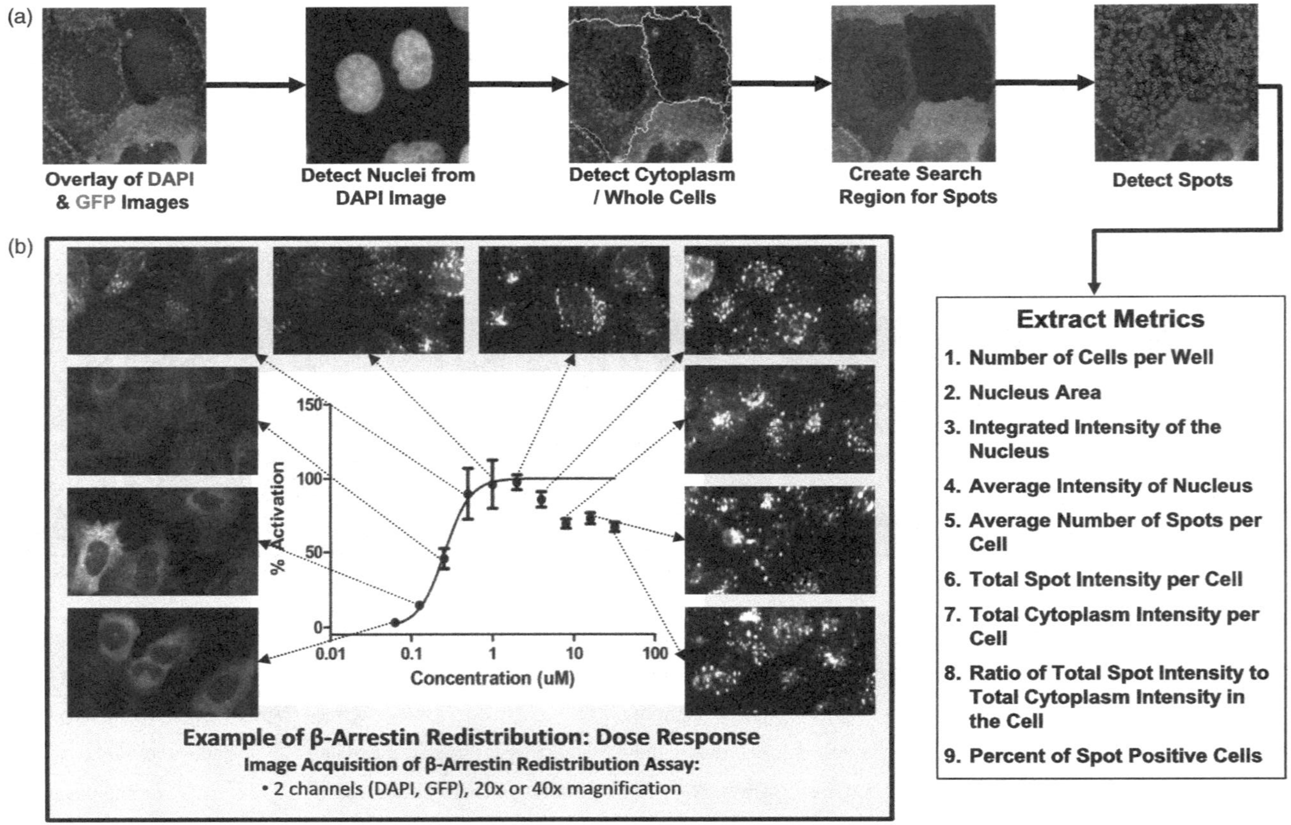

Figure 27.7. Image analysis for β-arrestin redistribution HCS assay for G-protein-coupled receptors. Composite montage of images that graphically illustrate (a) results of the image processing algorithms (object location, image segmentation, masking, feature extraction, and quantitation) to the multiple types of metrics and parameters that can be calculated on an object-by-object and stochastic basis, and (b) phenotypic response corresponding to a specific concentrations of test compound along a dose–response curve.

To streamline HCS processing, visualization, and data management, the San Diego HCS team implemented a fast access local storage system (LSS) dedicated to HCS based on the Columbus image-data management system (PerkinElmer). The Columbus system currently provides ~10 TB (scalable to >50 TB) of fast access storage at the San Diego site and handles image and data file management for currently ongoing HCS projects. A similar LSS dedicated to image-based HCS has been established at the Orlando site providing currently ~12 TB of fast access storage. The LSSs at both sites are connected via Sanford-Burnham intranet and provide Sanford-Burnham's scientists with web-based access to image visualization, analysis, and data visualization tools. In addition to the Columbus-based LSSs, a temporary HCS LSS (Dell PowerEdge 2950 connected to MD1000 external RAID storage systems) with a current usable storage capacity of 15 TB (scalable to >40 TB as needed) is available for temporary storage of images and data files as well as execution of image analysis outside of Columbus. The LSSs at both sites is connected through the Sanford-Burnham intranet and fully accessible from both locations via remote desktop and secured drive sharing. In addition to our LSSs, we currently have access to 20 TB of offsite storage at the San Diego Supercomputing Center (SDSC), which currently has 1 PB of online disk and 25 PB of tape storage. These large-scale redundant facilities are in close bandwidth proximity to supercomputing nodes that can be employed to process the image data. The HCS images are transferred via a 1-giga-bit (*Internet2*) line to imaging repository at the SDSC during or at completion of probe/drug discovery projects. Access to the supercomputer network for the Orlando site has been established via a 1-giga-bit line through the University of Central Florida. This infrastructure is diagrammed in Figure 27.8.

Compound Management Facility. The Prebys Center has access to the MLSMR compound collection,* founded by NIH in 2004 specifically for the Molecular Libraries Program via contract to Galapagos/Discovery Partners International (now part of Evotec). This collection was assembled and is being continuously updated and expanded (currently ~370,000 compounds). The business rules† used to select compounds and storage protocols are (1) compounds have acceptable physicochemical properties, including DMSO and water solubility; (2) compounds are devoid of highly reactive groups; (3) compound contribute to the diversity of the library, yet, at the same time, allow building clusters of four to five close analogs for instant SAR data for screening hits; (4) compounds are >90% pure as judged by liquid chromatography/mass spectrometry; and (5) >10 mg of compounds are available from the primary supplier. The collection also includes compound from commercial vendors and research labs around the world, for example, diversity-oriented compound sets produced within Molecular Library Initiative's Centers of Excellence in Chemical Methodologies and Library Development.‡ Full-library copies and quarterly library updates are distributed to all MLPCN centers. Hit sets identified in primary screening performed by the centers

*See http://mli.nih.gov/mli/compound-repository/.

†See http://mli.nih.gov/mli/compound-repository/mlsmr-compounds/.

‡See http://www.nigms.nih.gov/Research/SpecificAreas/CMLD/Pages/default.aspx.

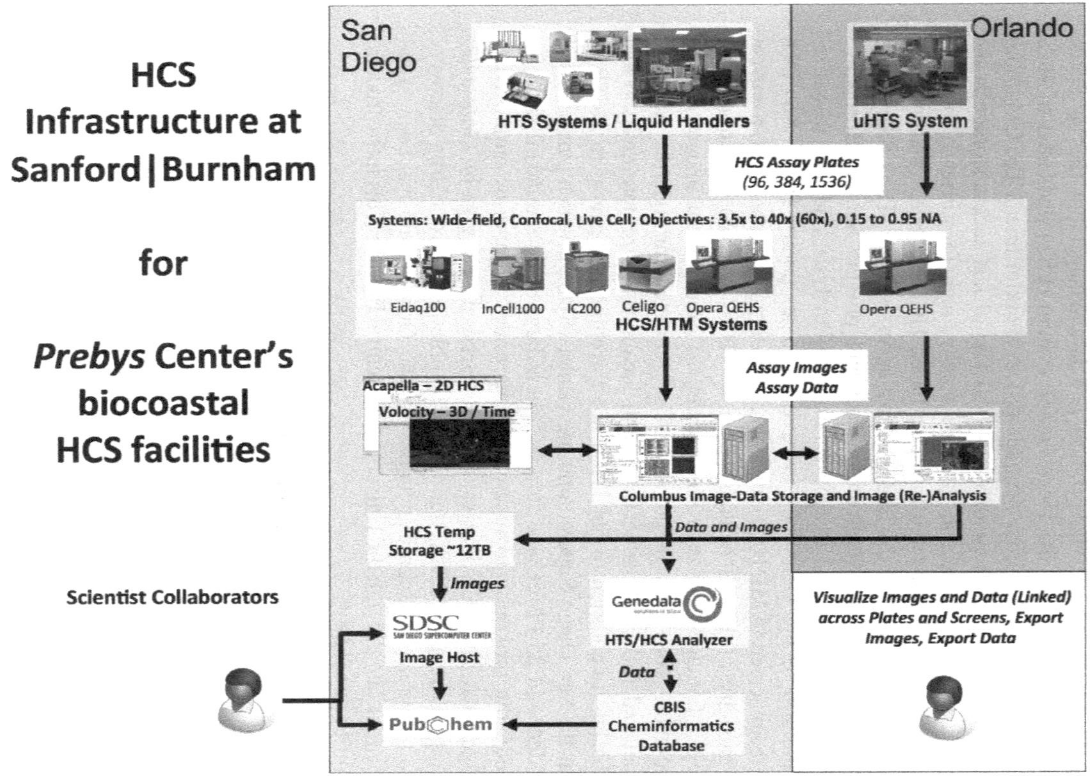

Figure 27.8. HCS infrastructure at Sanford-Burnham for the Prebys Center's biocoastal automated image-based facilities. Schematic of the relationships, data transfer processes, and direction of image data flow among the HTS, uHTS systems containing HCS modules, as well as offline HCS and image-based devices, to the image processing servers for image analysis, local storage, and transfer of reduced data to informatics and cheminformatics databases both local and public (PubChem) and final hosting of images and data to end users.

are cherry-picked and resupplied in 10-mM 10-μL aliquots upon HTS center's requests, allowing cost-effective and expedient hit confirmation and other follow-up studies. MLSMR deposits the structures of all compounds available in the repository into PubChem, NCBI's database of small organic molecules and their biological activity. MLPCN screening centers link to these substances when depositing their HTS data into PubChem, creating unparalleled constantly updated selectivity data sets for MLSMR compounds that assist in triaging out generally promiscuous compounds. As a continuing member of the MLPCN, the Prebys Center will continue to have access to MLSMR cherry-pick services until the formal end of the MLPCN on May 31, 2014, an important consideration in prosecution of HTS campaigns of the MLSMR under the new Funding Opportunity Announcements (FOA) PAR-12-058 and PAR-12-059 which are serving as the successor to the expired R03 mechanisms to access the MLPCN.

Sanford-Burnham's Internal Compound Collections include a large library of 321,259 compounds selected from a pool of over 3 million compounds from 5 chemical vendors. Selection strategy was based on cheminformatics tools to ensure novelty and diversity balanced with good representation per scaffold, as well as minimal compound overlap (<5% identical compounds) with the MLSMR collection. Appropriate filters were also applied to exclude undesired functional groups and ensure good general physicochemical properties of compounds. Additionally, the Prebys Center has the ChemBridge DIVERSet Library of 50,000 compounds a number of smaller focused compound collections, and these libraries total over 900,000 samples available for screening.

The Prebys Center has an extensive experience ordering fresh powder resupplies from the original vendors for validated hits from the MLSMR collection as well as performing the structure-based searches and matching these results to available vendors. A good proportion of these analog compounds originate from former Eastern-bloc companies, and CPCCG has established good vendor relationships, batching ordering procedures, to streamline and coordinate discounts, customs-clearing, and shipping procedures and costs. CPCCG also has in place standardized QC standard-operating procedures for compound identity and purity by ^{1}H-NMR, ^{13}C-NMR, and LC-MS, purification by LC and MS-triggered fractionation, and experienced medicinal and synthetic chemists who can resynthesize commercially sourced compounds if the original vendors are depleted.

Medicinal Chemistry Facility and Capabilities. Sanford-Burnham's state-of-the-art Medicinal Chemistry facilities operate on both San Diego and Orlando campuses, and are organized as a transparent, single institute function.

The San Diego Medicinal Chemistry Facility. It consists of~2000 sq. ft. of wet-lab space. The laboratories are well equipped for synthetic chemistry functions, with 12 chemical fume hoods and a wide array of equipment for conducting various chemical synthetic procedures including medicinal chemistry to synthesize and modify organic molecules for both initial lead generation and subsequent SAR studies. The laboratories are also equipped to perform parallel synthesis of small molecule libraries and the synthesis of natural product analogues. The laboratory has state-of-the-art instrumentation, equipment, and glassware to perform organic reactions capable of producing milligram to multigram quantities. In particular, the Syrris AFRICA microfluidic synthesis system enables entirely automated continuous flow synthesis, analysis, aqueous workup, reagent injection, and product collection. Thus, hundreds of compounds can be synthesized, worked-up, and analyzed daily, including those requiring multiple steps. The lab has dedicated mid-scale and larger scale preparative HPLCs and analytical LC-MS with autosamplers, capable of purifying hundreds of compounds daily. Purity analysis and quantification of final products performed on either of two HPLC-MS systems with photodiode array detectors are available for the routine analysis of chemical reactions and final compound purity. We also have a triple-quadrupole LC-MS/MS system for the precise quantification and identification of compounds for pharmacokinetic analyses and other applications requiring high sensitivity and precision. A centralized NMR core equipped with high-field NMR spectrometers, as well as a highly automated 400 MHz instrument for small molecule analysis, is available near the facility.

The Orlando Medicinal Chemistry Facility. It includes four 16 ft. fume hoods and ~3100 sq. ft. of synthetic chemistry laboratory space specifically designed and equipped for hit-to-probe optimization and medicinal chemistry programs. In addition, there is a separate flexible lab space specifically designed for automated synthesis instrumentation and analytical instruments. A separate 250-sq ft suite houses the 400 and 500 MHz NMR systems. Dedicated synthesis equipment includes an automated microwave system, automated reaction blocks, a UV-guided, mass-triggered autopurification system, walk-up LC-MS analysis capacity, and a Syrris AFRICA microfluidic synthesis system. There is sufficient bench space and hood space for up to 12–14 chemists, with the capability to expand up to 24 hoods.

The Pharmacology Facility. The Exploratory Pharmacology Shared Resource at Sanford-Burnham's Orlando site is immediately adjacent to and contiguous with the Medicinal Chemistry lab. The pharmacology space includes bench and equipment space, tissue culture resources, and a "hot" room for radioligand work. This isotope lab is available for the development and execution of hERG and other assays. Analysis of drug levels and metabolite identification is made using high-performance hybrid triple quadrupole/linear ion trap system with excellent dynamic range and sensitivity. A flexible ADME/T assay high-throughput/high-capacity system for automated liquid handling facilitates efficient automation of compound management, sample processing, and assay execution with an integrated multimode plate reader. There is a separate tissue culture facility equipped with two 4-ft biosafety cabinets, microscopes, cell culture incubators, and other minor lab equipment necessary for the efficient growth and maintenance of cell lines. The Exploratory Pharmacology lab also performs routine basic drug metabolism and pharmacokinetic studies in mice and rats, as well as acute and subchronic toxicology studies including gross necropsy and detailed histologic analysis of compound effects in conjunction with the Molecular Pathology core at Sanford-Burnham.

Cheminformatics/**In Silico** ***Screening/Computational Modeling Facility.*** This San Diego facility applies cheminformatics tools and computer-aided molecular design approaches to support various aspects of molecular probe discovery and development, which include SAR analysis, similarity search, compound selection (for hit follow-up and target-based library design), library diversity analysis and design, pharmacophore modeling, QSAR, scaffold hopping, virtual docking and screening, conformation analysis, molecular dynamic (MD) simulations, protein homology modeling, structure-based design and ADME/T prediction. To support these activities, Sanford-Burnham has assembled a broad spectrum of industry-standard software packages: Accelrys's Discovery Studio 2.1 (DS), Accelrys's Pipeline Pilot and Catalyst, CambrigeSoft's ChemDraw, Schrödinger's modeling/computing package, Chemical Computing's MOE, Leadscope, SilmlationsPlus's ADMET Predictor, and Eidogen-Sertanty's TIP and EVE.

The Cheminformatics core also supports compound registration, inventory, HTS data collection and analysis, using the Cheminnovation CBIS software. The core's Vendor Compound Database (VCD) along with ChemNavigator, facilitates the analog selection and ordering process accessible through CBIS interface. This constantly updated database from reputable commercial compound vendors currently contains over 12 million compounds. The chemical information on each compound in the database is annotated with key physical properties, such as AlogP, MW and t-PSA calculated, and properties can be filtered with user-built settings, for example, "Lipinski's rule of 5." Intensive computing of virtual screen is carried out using clusters. FlexX is run on PBS cluster with 16 nodes, each with 2 CPUs. Schrödinger software is hosted on Linux-based Dell precision Workstation equipped with 3-D stereo monitor for enhanced visualization to aid in structure-based design.

Project and Portfolio Management. All drug discovery and chemical genomics projects at Sanford-Burnham are managed by experienced project managers who apply industry best practices and utilize state-of-the-art project management tools. CPCCG has implemented @Task™, a web-based project management software system to improve the efficiency of our workflows and operations. @Task enables project pipeline visualization, capacity planning, and portfolio forecasting that enhances the ability of the center to efficiently manage project-related tasks and avoid bottlenecks in personnel and resources utilization. The software also serves to centralize information and project-related documents. Proprietary projects, collaborations, and thematic portfolios are managed professionally and robustly through password-protected login with restricted hierarchy of read/write/edit access depending upon functional roles and access permissions. Project-specific view-only access can be provided on a need-to-know basis to ensure maintenance of proprietary information on a multiproject, multi-portfolio level or individual project/portfolio level (Figure 27.9).

FUTURE SUSTAINABLE, RENEWABLE DISEASE IN A DISH MODEL FROM IPSC TECHNOLOGY

A key unmet need in drug discovery and development are better predictive *in vitro* cellular models of disease, which translate into *in vivo* animal models then ultimately

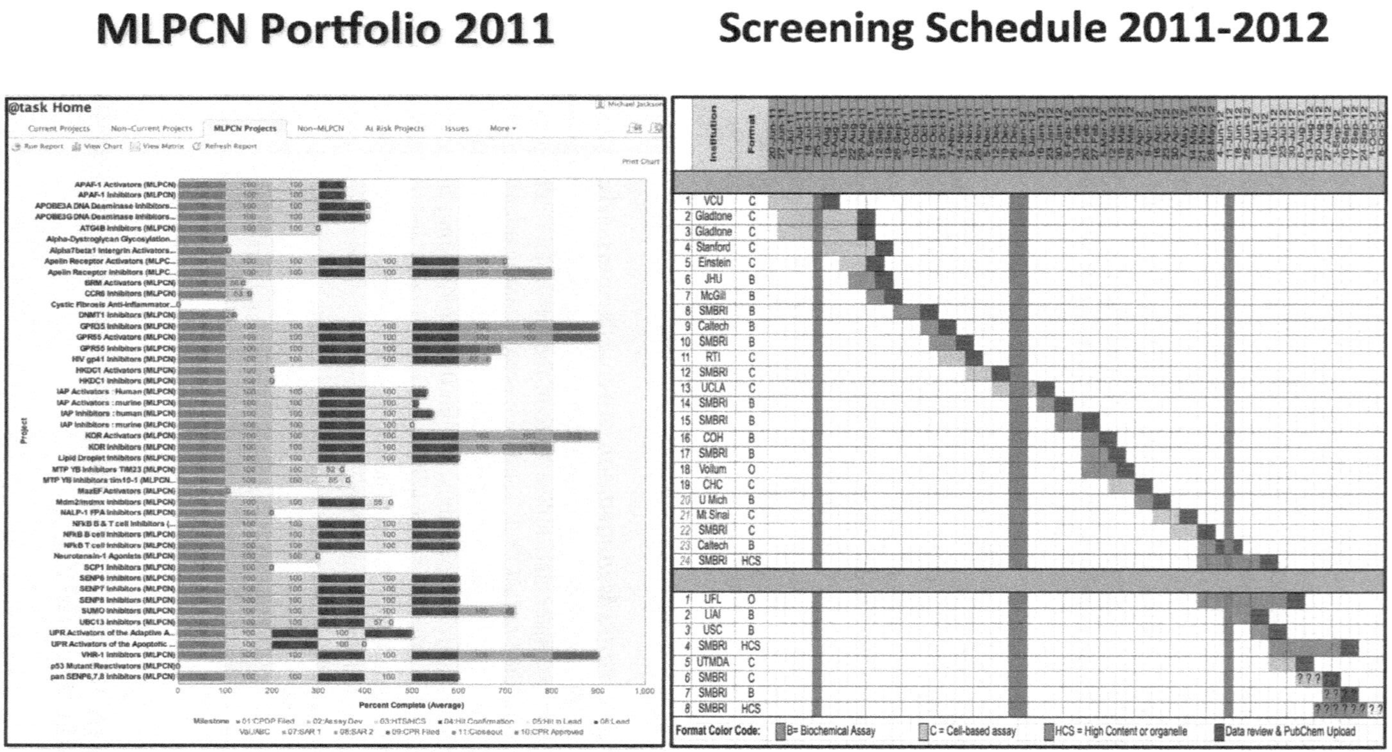

Figure 27.9. Prebys Center's web-based project and portfolio management system @task (http://www.attask.com). Example "Pipeline" view and dynamics report with respect to stage and percentage completion at each stage for multiple projects (left) and a Gantt-like view of scheduled HTS campaigns (right). Collaborator institutions are listed at left of chart with assay format type (C, cell-based; B, biochemical; HCS, high-content screens; and O, organisms or organelle, with respective grayscale shading—dark gray boxes represent required data upload to PubChem, and "?" indicate uncertainty in the schedule due to indeterminate reagent status).

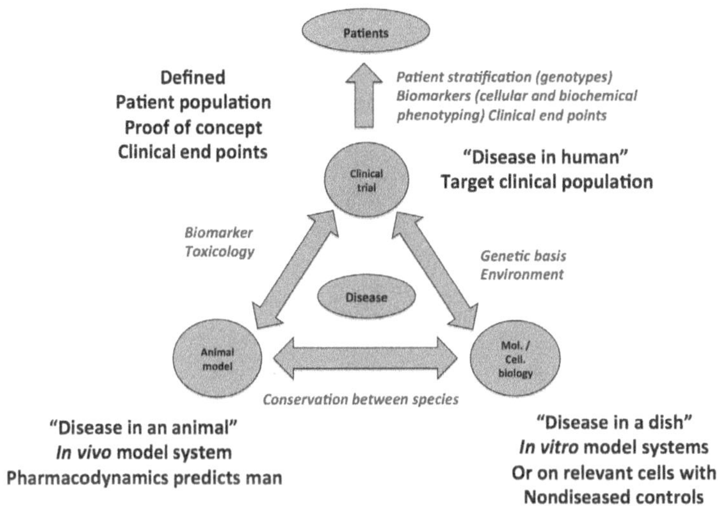

Figure 27.10. The "translational triangle." The triangle of models representative of a disease and of the binary connections between *in vitro* cell-based models ("disease in a dish"), *in vivo* animal models ("disease in an animal"), and *in situ* human models ("disease in a human") of disease in which lead molecules are tested, ultimately to benefit patients. The italicized light-gray text highlights the main relevant factors interrogated at each connection.

into man. The cell is a fundamental unit that represents the full spectrum of intracellular biology and all pathways therein. The components of this idealized "Translational Triangle" which connects a molecular defect to a dysregulated cellular pathway and ultimately to a diseased organism is illustrated in Figure 27.10, where some of the key factors and considerations between the nodes are articulated in italicized gray text.

However, immortalized cell lines have undergone a number of mutations and adaptations that render them far from representative of the original cell from which they were derived. Recent advances in the generation of stem cell technology in which patient-derived somatic cells (from disease-carrying patients and nonaffected relatives) are reprogrammed using transcription factors to generate stem cells, so-called iPSCs can provide an almost unlimited supply of an individual's cells. Furthermore, these iPSCs can be further differentiated into almost any cell type of the human body through defined protocols involving growth factors and chemical inhibitors [31]. In this way, so-called disease-in-a dish models can be generated in which the patient-derived cells can be generated that display a dysfunction reflecting their genetics. This is especially relevant to many rare childhood orphan diseases, which are monogenic and hypomorphic and present only in the homozygous state. In these cases, iPSCs can be derived from matched diseased and unaffected (heterozygous or defect free) siblings or parents; thereby producing ideal genetics-matched controls. iPSC lines provide a renewable source of cells of specific disease cellular phenotype suitable for use in cell-based HCS phenotypic assays, and for further follow-up in the hit validation and late-stage

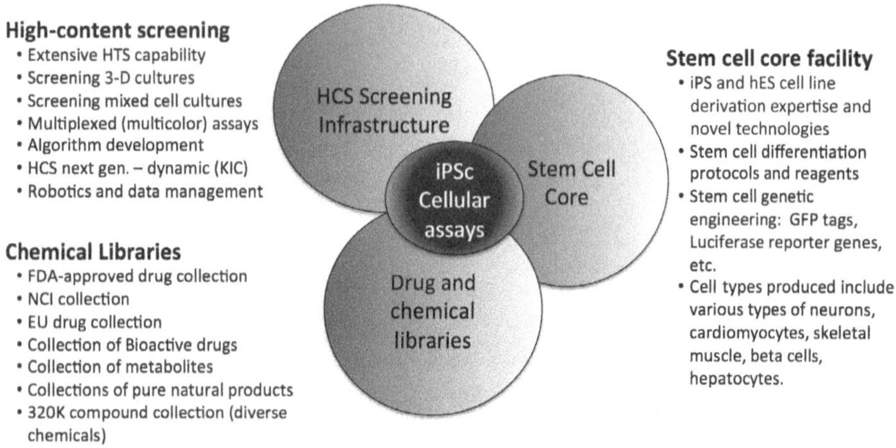

Figure 27.11. Sanford-Burnham Prebys Center's "disease-in-a-dish" capabilities and unit components. The Venn diagram schematically illustrates the individual core functionalities and components resident in the Prebys Center necessary to conduct innovative development and execution of cell-based phenotypic assays based on iPS cellular technologies.

compound SAR and optimization studies. The Prebys Center has the relevant component of infrastructure, expertise, and suitable compound collections that could be useful toward rapid drug repurposing or describing new biology and completely novel chemical scaffolds, based on discovery of novel, "unobvious" biology that could only be enumerated and reduced to practice with the combination of sophisticated know-how and reagents, used in conjunction with automated image-based phenotypic assays that the Prebys Center excels in. These key components are diagrammed in Figure 27.11.

ADVANCING PROJECTS TOWARD PRECLINICAL CANDIDACY

Thus far we have established and demonstrated that the Prebys Center has solid core competencies and infrastructure to complete chemical probe generation. As we have noted, the Prebys Center has a cadre of medicinal chemists who have had responsibility for major drug discovery and development programs in both biotech and pharma, where they have taken compounds through all steps of preclinical study into early Phase 1 trials. We present here a seminal example of how the Prebys Center has taken an early probe ML028, an inhibitor of tissue nonspecific alkaline phosphatase, from the MLSCN and performed lead optimization to dramatically improve its relevant drug-like pharmacokinetic properties and compound exposure.

The original probe, ML028, had modest potency in a model of cellular mineralization in vascular smooth muscle cells (VSMCs), very high clearance and poor exposure (area under the curve—AUC). Thus, it was unsuitable for use in animal efficacy studies (Figure 27.12).

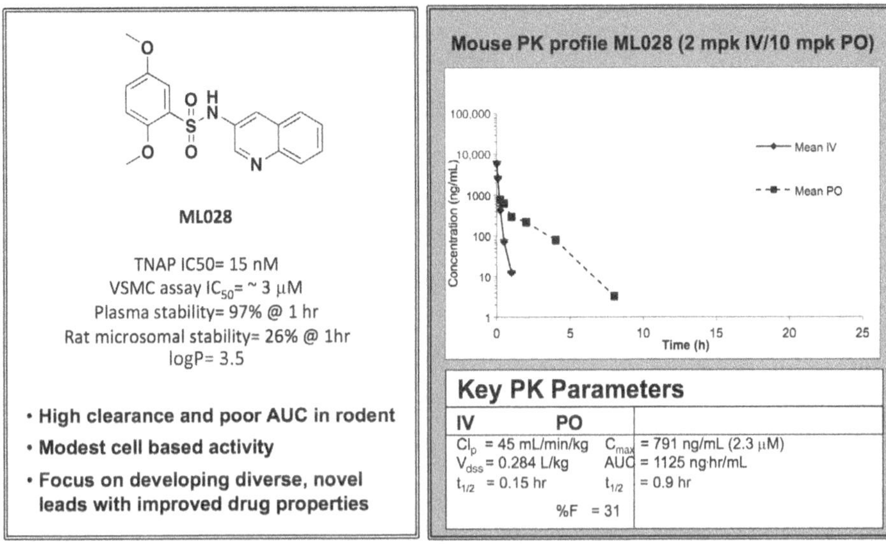

Figure 27.12. Summary of the characteristics of the tissue nonspecific alkaline phosphatase (TNAP) chemical probe ML028 from the MLSCN. The chemical structure and key potency and pharmacologic parameters of this starting early chemical probe for TNAP are presented in the left hand panel. A pharmacokinetic profile and derived parameters illustrate the rapid clearance when ML028 was dosed either orally (PO) or intravenously (IV) in rodents, making this compound unsuitable for animal POC studies.

The Prebys Center medicinal chemists performed several rounds of lead optimization in close collaboration with the exploratory pharmacology group, who evaluated improvements in drug-like parameters, and the biology team who performed the initial cell-free biochemical enzyme assay to monitor potency, as well as developed a potency assay in whole blood (a biomarker), so that the actual potency of the compound could be determined in this relevant physiological compartment. As shown in Figure 27.13, the final optimized lead compound SBI-425 exceeded every parameter of the "ideal profile" compound, and every suboptimal drug-like parameter in ML028 (in light gray text) was significantly improved (in underlined text), especially efficacy in whole blood (50- to 100-fold improvement), microsomal stability (3-fold improvement), and compound exposure (greater than 5-fold improvement). This compound is now undergoing extensive proof of concept (POC) studies in animals with confidence that the ADME/T and PK properties are sufficient for meaningful results.

Next Steps and Lessons Learned

The ML028 probe was deposited in PubChem in 2007, but it was not until 2011 when one of the coauthors (Dr. Jackson), upon reviewing the portfolio of Prebys Center projects, felt that this probe met criteria for potential translational success meeting the following criteria: (1) developability of the "hits" or leads with regard to

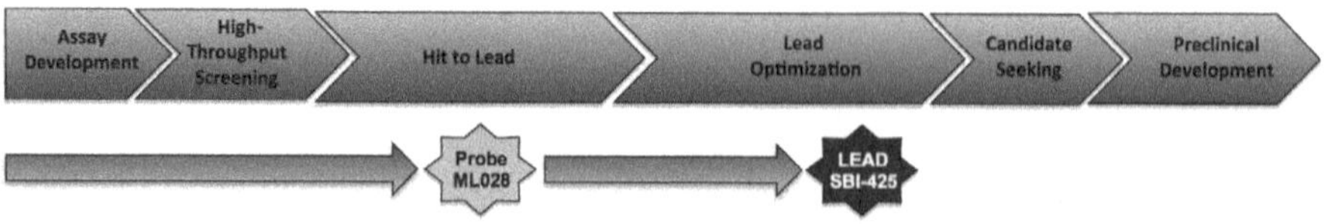

Parameter	Ideal Profile	ML028	SBI-425
IC_{50} in TNAP enzymatic assay – nM (human))	<100	***15***	***14***
IC_{50} whole blood assay (pH7.5) μM (mouse/human)	<1	6.8	***0.128/0.06***
IC_{50} in mineralization of cultured VSMCs - μM	5	3	ND
Microsomal stability (HLM, RLM) % remaining @1 hr	>75%	26	***85%***
$t_{1/2}$ (rodent) - hr	>1	0.9	***2.3***
PO bioavailability (rodent)	> 50%	31%	***58%***
Clearance (rodent) - mL/min/kg	<30	45	***5***
Plasma Protein binding	< 95%	>98%	<95%
Exposure (AUC, rodent) - ng·mL/kg	3,000	1125	***84,818***
Solubility - μg/ml	>10	<10	>10
HERG binding – "clean" at dose (μM)	>50	ND	>10
IC_{50} P450 inhibition 5 major isozymes (μM)	>10	<10	>10
Selectivity over other targets (IAP,PLAP) - fold	>10	***>100***	***>100***

Values in bold light gray were suboptimal; Values in ***bold underlined italics*** have been improved through HTL & LO

Figure 27.13. Lead optimization of the TNAP probe ML028 into POC lead SBI-425. The light gray chevron bars summarize the process stages and the block arrows show the progress through these stages that ML028 and SBI-425 reached. The figure summarizes the Parameters and values of the "Ideal" profile, where ML028 started, and what was achieved after several iterative rounds of lead optimization (LO) by the medicinal chemistry and biology SAR support teams at the Prebys Center. All parameters which had suboptimal values for ML028 are noted in light gray text, while the superior parameter values after LO are underlined in the table.

drug like properties, clear SAR, creation and retention of intellectual property (IP), (2) probability of securing a development partner (pharma foundation, etc.) presenting an exit path or a licensing event or codevelopment agreement, and (3) strength of the disease-specific "translation triangle" (see Figure 27.10), that is a path to clinic, relevance of target to the disease process, and availability of quality animal models. The next steps beyond lead optimization into candidate seeking and preclinical development are well known to the Prebys Center team (Figure 27.14) and much of it can be outsourced with close management by our chemists and project management team. However, the key challenge for not-for-profit laboratory-based research institutes is how to fund these nontraditional research activities. We continue to partner with the NCI CBC and NeXT mechanisms for cancer-related indications and we look to partner with disease-specific foundations or biotechnology companies with a focus on target to lead activities.

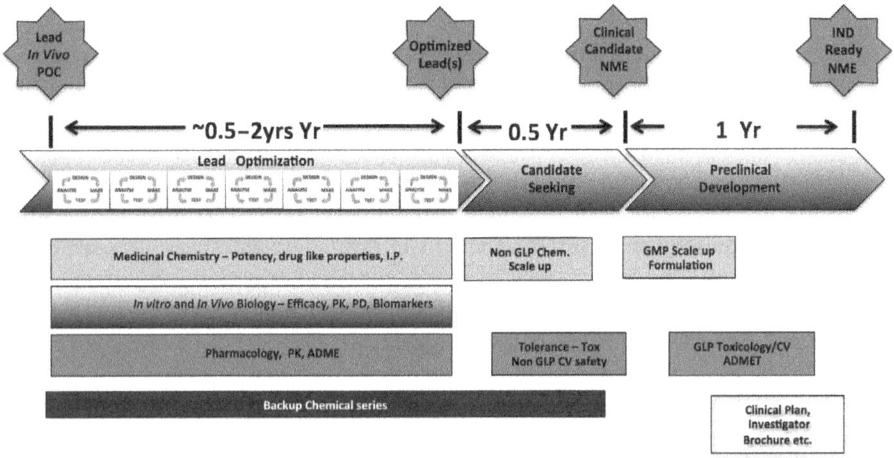

Figure 27.14. Lead optimization to IND drug discovery processes, timeline, and milestones. The chevron bars at the top summarize the three main stages from lead optimization to preclinical development, with loops of "Design, make, test, analyze" representing the iterative SAR rounds that seek to improve potency, selectivity, and pharmacological properties of compounds until a suitable preclinical lead is achieved. The "typical" duration in each stage are noted above each bar. The shaded bars below each stage summarize the additional types and duration of parallel activities that must be conducted for each stage.

OPPORTUNITIES FOR THE PREBYS CENTER (S-B) AND ITS POTENTIAL PARTNERS

The translational capabilities of the Prebys Center that are key differentiators among academic and nonprofit screening centers are

1. A pharma/biotech seasoned staff highly experienced with engagements with both academic and commercial partners
2. State-of-the-art uHTS facilities including recently assembled "fresh" diverse chemical library
3. A track record of successfully screening "tough" assays
4. Extensive assay development expertise in all HTS-relevant formats, including HCS/phenotypic and thermal melt assays, developing an average of two to four secondary assay for every primary HTS assay executed
5. A proven track record of generating drug leads
6. A proven track record of attracting drug discovery and *in vivo* probe development grants, and conducting active outreach to ensure solid preliminary data and working HTS assays
7. A not-for-profit cost and capital structure

8. Existing firewalls for data and compound handling
9. Operational expertise with professional, pharma quality, and disciplined project and portfolio management (of up to 50 projects/year) to timelines and milestones.

Furthermore, conducting drug discovery projects is part of the overall mission of the Sanford-Burnham Medical Research Institute, as evidenced by SIB investment of >$40 million to establish the Prebys Center's infrastructure and state-of-the-art uHTS and HCS systems. Our proven track record of attracting drug discovery grants helps leverage precious drug development funds from nongovernment sources. And while a translational focus can generate milestone payments allowing value creation (Figure 27.15), the more compelling reason to be conducting such studies at the institute is that they are commensurate with our mission as a not-for-profit biomedical research institute and our institute's motto: "From Research the Power to Cure." The Institute's nonprofit cost structure and success in attracting drug discovery grants does allow it the freedom to explore "riskier" less well-validated innovative projects. Unlike many pharmaceutical

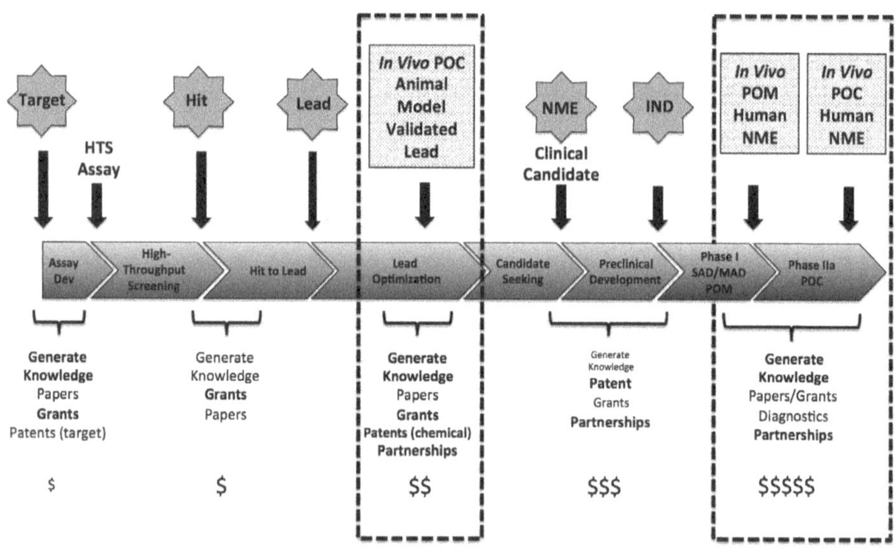

Figure 27.15. Drug discovery and early development key stages and transaction points. The shaded chevron bars list the major stages of the drug discovery (expanded compared to Figure 27.13 and Figure 27.14). The major milestones or deliverables are noted in gray-shaded stars for each major stage transition point indicated by black solid downward arrows. The brackets below the stages illustrate what are generated in terms of knowledge generation, grants, patents, and potential partnerships. The two major transactable points are enclosed in bold dotted line boxes for *in vivo* animal POC or *in vivo* human proof-of-mechanism (POM) and POC with the final new molecular entity (NME) or drug candidate. The relative number and size of "$" signs are indicative of the potential value of licenses and payments.

or biotechnology companies, our portfolio does *not* need to be overly focused on specific indications of high commercial value *nor* mandate that targets be fully validated prior to embarking on a project. Indeed this presents a potential partner the opportunity to broaden its target base, by engaging the Prebys Center to work on exciting new targets not yet validated, but amenable to HTS and use of the resultant chemical biology probes to explore the nascent biology in complementary fashion to genetic knockouts. This allows de-risking of incompletely validated targets. Ironically, the original mandate and goal of the NIH MLPCN was exactly along these lines. The Prebys Center has demonstrated that via a chemical genomic approach it can bring value through de-risking early drug discovery activities and specifically chemically validating innovative targets *in vivo*, providing a launching pad for a *de novo* drug discovery project in a commercial setting.

STRATEGIC PARTNERSHIP WITH MAYO CLINIC

Even with an institute the size of Sanford-Burnham, with 90 PIs and over 500 PhD scientists, the Prebys Center's industrialized screening and drug discovery capacity and bandwidth to execute 35–50 HTS campaigns per year exceed the number of quality drug discovery project proposed and funded by the institute's faculty. Therefore, we have sought strategic partners who bring additional drug discovery projects based on advances in their research. The ideal partner for Sanford-Burnham would bring not only new targets, reagents, and animal models but most importantly a clinical perspective of how the drug discovery projects in the portfolio would be tested in the clinic to generate human POC. Hence we sought a partner that has both cultural as well as organizational complementarity that could complement our strength in early drug discovery. The scientific leadership of Sanford-Burnham was aware of the outstanding experimental clinical capabilities and track record of the Mayo Clinics. The Mayo clinic is renowned for conducting and managing innovative clinical trials; they have had notable success in receiving a Clinical and Translational Science Award, as well as a scholarly and world-class basic and clinical research program. The deep well of physician-scientists who are actively engaged in patient care was clearly aligned with Sanford-Burnham's bench-to-beside mission. We reasoned that while Mayo Clinic excelled on these fronts, the Prebys Center excelled in driving early-stage drug discovery activities and attracting grant funding (through a disciplined process of Outreach and early engagement with partners) (Figure 27.16). In this way the Prebys Center strives to secure funding for high priority early translational activities, specifically de-risking the technical barriers toward an HTS program.

In seeking a partner Sanford-Burnham looked to address some of the challenges encountered in executing the MLPCN portfolio. In particular Sanford-Burnham was seeking to establish a strong, long-term, and direct trusting relationship with a clinical organization, and to avoid a "throw it over the wall" mentality that we had experienced from some of the assay providers in the MLPCN program in which issues of irreproducibility, scalability, and experimental design rapidly fell on the shoulders of the Prebys Center to solve. Indeed in a retrospective analysis of the first 4 years of

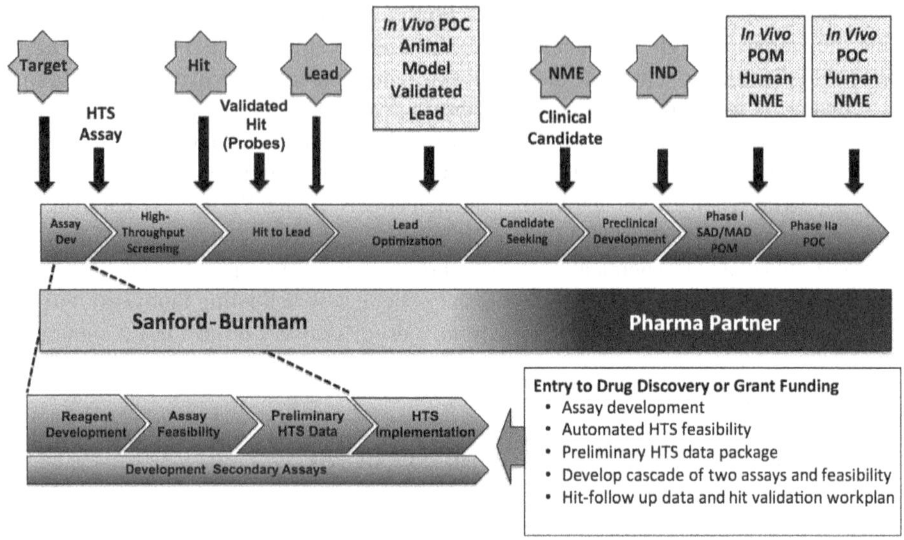

Figure 27.16. Role and core competency of the Prebys Center in assay development, HTS feasibility, and generation of preliminary data to seed early drug discovery. The early role and relationship of Sanford-Burnham Prebys Center (and strategic partner) to the potential pharmaceutical partner. The "Assay Development" chevron has been expanded into further substeps to detail their sequence, and the Berger dots in the "Entry to Drug Development or Grant Funding" box note the five deliverables that the Prebys Center delivers to its strategic partners.

executing MLPCN programs, we identified that the single most critical factor that delayed execution of projects to pre-agreed timelines was a lack of reagent readiness from the PIs, in terms of both robustness and quality control. The need to scale reagents to sufficient quantities all too frequently became a bottleneck as the amount of reagent required to complete a primary HTS campaign, subsequent replicate hit confirmation, and potency profiling needs reagents for 750,000 to 1 million data points. This was often further confounding for cell-based assay when target engineered cell lines were not validated for long-term expression or functional stability, were difficult or slow growing, or were configured for 5- to 10-day assay durations. The second most prevalent factor that delayed projects was the complete lack of appropriate secondary (counterscreen, orthogonal, or selectivity) assays or insufficient robustness and capacity of these downstream assays to support the rapid triage of typically of a few thousand or hundreds of hits in full dose response, and even a few tens of compounds during elucidation of SAR and lead optimization.

By offering a suite of competencies and end points ("Entry to Drug Discovery or Grant Funding" box in Figure 27.16) to the Mayo Clinic, we looked to de-risk many of the technical barriers to success.

Translational Science Activities at Mayo Clinic

Mayo Clinic has had a long tradition of clinical and translational research, having had an NIH-supported General Clinical Research Center (GCRC) that was funded over 40 years ago. The GCRCs, at Mayo and at other academic medical centers, provided much of the infrastructure needed for clinical research, including outpatient and inpatient units, skilled nursing and technical support, facilities for specialized research procedures, and subsidized core laboratories for a range of assays. During Dr. Elias Zerhouni's tenure as Director of NIH, support for the GCRCs was subsumed into Clinical and Translational Science Awards (CTSAs) that provided funding not only for the GCRCs (now termed clinical research units [CRUs]) but also for educational programs for the training of clinical investigators. In addition, these awards also included resources to enable institutions to engage the community in providing input into research directions and protocols and to set up programs that provided regulatory and informatics support to investigators.

In December 2011, under Dr. Francis Collins' leadership as NIH Director, the administration of the CTSA program moved to a newly created institute, the National Center for Advancing Translational Sciences (NCATS). As summarized by Dr. Collins, NCATS had a bold, new vision: "To catalyze the generation of innovative methods and technologies that will enhance the development, testing, and implementation of diagnostics and therapeutics across a wide range of diseases and conditions" [32]. The underlying driving force for this vision of NCATS, which now highlights drug discovery and development as a central initiative, was the growing realization that despite dramatic advances in our fundamental understanding of various diseases, the path from these basic discoveries to actual drugs for our patients was far too long (~13 years), too prone to failure (~95% of the time), and too expensive (~$1 billion per successful drug) [32].

The NCATS vision is certainly consistent with the goals of most academic medical centers, including Mayo, which emphasize bringing new treatments to patients. Traditionally, however, these academic medical centers have focused largely on the basic "discovery" phase and much less on the drug discovery aspects, partnering instead with pharmaceutical companies for the latter. As noted by Dr. Collins, however, relying solely on industry to bring new drugs to market may be less than ideal, particularly in an era of declining pharmaceutical research budgets and patent expirations [32].

Historically, translational research in the GCRC and subsequently in the CRUs at Mayo focused on better understanding human physiology and mechanisms of disease. Thus, over the years, Mayo investigators have made major contributions to the literature defining the pathogenesis of diverse disease processes such as diabetes, osteoporosis, sleep disorders, HIV-related immune dysfunction, determinants of vaccine responsiveness, and others. These traditional forms of clinical investigation are clearly critical for a better understanding of disease mechanisms in humans that continues to drive new therapeutic approaches. However, in order for these activities to become better aligned with one of the central missions of NCATS—specifically, helping bring new drugs to market—there was a critical deficit within Mayo that had to be addressed: Mayo had virtually no in-house effort to develop and test compounds modulating molecules or pathways that its investigators had identified, through basic laboratory or CRU studies, as potential therapeutic targets.

Interestingly, despite this deficit, Mayo Clinic's history in the development of clinically relevant drugs in fact dates back to the first half of the latter century when Mayo researchers played an integral role in the clinical development of drugs, including streptomycin [33, 34] and cortisol [35]. In recognizing that the landscape of biomedical research is changing, as outlined previously, leadership at Mayo took the opportunity to reevaluate capabilities for drug development. It rapidly became apparent that investigators have a long-standing track record in biomedical discovery leading to novel therapeutic targets, and that investigators have expertise in early phase and licensing clinical trials, but that they were limited on site resources with which to develop lead compounds and optimize lead compounds for the development of future drugs.

With this as a background, stakeholder interviews were held to learn of individual experiences with external agencies that offered such services. These sessions identified several concerns with prior interactions, and identified areas that would be essential for successful drug screening activities: assistance was required in screening assay development, and investigators required access to large chemical libraries. Moreover, once hits were identified in primary screens, assistance was needed in developing and implementing secondary and tertiary screens, and advancing hits to leads was often difficult due to a lack of medicinal chemistry or *in vitro* toxicology. Therefore, in considering ongoing partnerships for drug development, Mayo CTSA Research Resource leadership sought the following resources from potential partners: (1) availability of multiple screening platforms, (2) expertise in the development and testing of primary and secondary screening modalities, (3) access to large chemical libraries of diverse sources of compounds, (4) access to medicinal chemistry expertise and synthesis capabilities for translating hit to lead compounds and performing structural function analyses, and (5) capability for early phase *in vitro* and *in vivo* toxicology.

Case Study of the Sanford-Burnham's Prebys Center–Mayo Clinic Strategic Partnership

The Mayo Clinic investigators remain free to explore any and all potential collaborations for drug development. Mayo Clinic's research leadership has endorsed the following model for drug discovery, and this model is in early-stage testing in pilot collaborations (Figure 27.16). Lead therapeutic targets are identified through Mayo investigators through traditional investigator-initiated funding mechanisms. An institutionally supported partnership has been developed wherein investigators interact with Stanford|Burnham Drug Discovery personnel to review the appropriateness of the target for therapeutic targeting, the intellectual property space, and design the optimal assay format. These proposals are reviewed in a competitive request for applications (RFA) format. This funding provides support for assay development, pilot screens, and determination of the robustness of the screening platform. The background data concerning the basic biology of the target protein, along with the validated screens, are then used to generate applications for federal funding for drug screening. We have already achieved the initial aggressive milestone for two "fast-tracked" proposals that yielded two joint grant applications for the October 5, 2103 PAR-12-058 FOA deadline, just 3 months from the initial engagement and the inaugural Joint-Steering Committee meeting on July 20, 2013. In one of these cases, the Joint Steering Committee (JSC) had agreed to conduct a pilot screen "at-risk," that is, prefunding and interesting validated hits have emerged.

Projects that successfully attract an R01 PAR-058 grant result in a collaborative initiative between Sanford-Burnham and the Mayo Clinic in which drug screening and hit validation occurs. A large chemical library is screened (the ~370,000 compounds of the MLMSR) to identify hits, which are verified through secondary and tertiary screens. Hits identified are then used by the Mayo investigator to determine their effects on the biology being studied. If successful, future collaborative applications for medicinal chemistry and toxicology will be created with the goal of advancing hits to lead compound status.

Perceived advantages of this model of drug discovery in comparison to the traditional pharma model include the following. First, the opportunity for rapid translation from lead target identification to initiation of drug development activities. This contrasts with the traditional pharma model where lead targets become the focus of drug developments only after multiple years of basic biology have occurred. Second, while traditional pharma focuses on targets with large perceived markets, this collaborative model allows for niche targets to be developed. Third, while traditional pharma pays for the full development cost of every stage of drug development, costs associated with drug development, in this partnership model, this cost is mitigated at least in part by the availability and receipt of NIH research grants. Finally, within the partnership model, relatively little institutional financial investment is required; therefore, the validity of the science alone can be used to drive metric of success or failure. In the pharma model, large corporate investments, and the need to generate returns on investments, may limit the more innovative high-risk programs entering the portfolio and the willingness to terminate projects that underperform.

Commentary on Cultural Factors and Governance that Characterize This Succeeding Partnership. In researching this case study, one of the authors (Dr. Chung) retrieved a fascinating case study on the Mayo Clinic by the *Commonwealth Fund* [36] in 2008 entitled "Mayo Clinic: Multidisciplinary Teamwork, Physician-Led Governance . . ." which highlights the Mayo Clinic's exemplary status among the 15 case studies of the Commonwealth Fund Commission's study on "High Performance Health Systems" as having all of the six key attributes of an "ideal" health-care delivery system. Indeed, they cite that the Mayo has served as a model for other institutions such as the Cleveland Clinic in Ohio and the Lahey Clinic in Massachusetts. While most of these attributes are focused on primary health-care delivery delivery to patients, two attributes, peer review and teamwork for continuous innovation, resonate soundly with regard to Mayo's basic and translational research collaboration with the Sanford-Burnham Medical Research Institute's Prebys Center. The article also emphasized that the Mayo Clinic is the world's oldest and largest integrated multispecialty group medical practice, combining clinical practice, education, and research at the regional, national, and international levels to the benefit ultimately of patients. The article also cites that Mayo's leadership believes in the critical importance of creating and maintaining a learning organization and thereby fosters a scholarly environment for research and education, which includes translational research and technological innovation. By these measures, our respective organizations were an ideal strategic fit with organizational and cultural alignment with Sanford-Burnham's "DNA of team-based and highly collaborative science" as evidenced by our success in competing for large collaborative multicenter/multidisciplinary grants. This also matches well with the Prebys Center's history of collaborating with multiple PIs from many different institutes both in the United States and abroad (Canada, Italy, Germany, and China). Finally, they cite Mayo's culture of clear accountability, initiative, and centrally held resources that prevent turf and infighting, and a tradition of consensus decision making so that resources and operations are deployed to serve the mission and cohesive functioning. The transparency, timeliness, collegiality, and informality of the discussions during the JSC meetings and teleconferences and subsequent e-mails engender consensus yet very timely decision making in an overall governance structure of adjudication rather than micromanaging individual projects that preserve local autonomy while encouraging a collaborative team science approach [37, 38]. The composition of the JSC core represents the multidisciplinary nature of our respective teams, and the personal relationships solidified by site visits and joint teleconferences, we believe are and continue to be critical to effectiveness. The current remaining pilot projects continue to yield success, and we have recently jointly expanded our collaboration with additional projects. This will provide enough projects in the grant submission pipeline to have a reasonable expectation of being awarded several grants (based on the 8–12% current success rates of all NIH grants). We hope to move these awarded projects into screening and early drug discovery with the long-term goal of mutually building a pipeline of drug discovery programs, some of which will reach the preclinical then clinical stages in the several years.

CONCLUDING REMARKS

The strategic partnership we have described here with the Mayo Clinic is just one example of several that the Prebys Center has engaged in with partners from academia, foundations, health-care partners, governmental funding agencies, and pharma and biotech companies. One common feature of these partnerships is that our Prebys Center serves as Sanford-Burnham's Drug Discovery Center (or "engine") that executes on a portfolio of projects (central box in Figure 27.17), some of which will yield lead compounds. We emphasize that the best partners have sufficient depth of science to present a portfolio of several impactful and technically feasible projects, since the burden of managing additional projects is incremental, but more importantly, a multiproject portfolio distributes risk and allows continued progress of the overall pipeline despite the stalling or termination of individual projects.

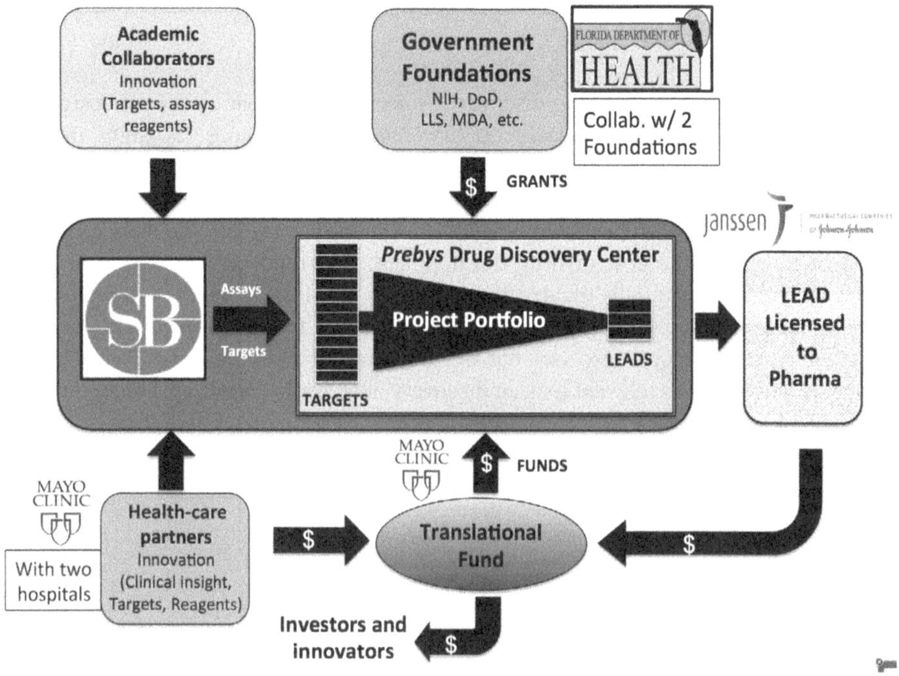

Figure 27.17. Preferred business model for the Sanford-Burnham and the Prebys Center with strategic partners. The central feature is the strategic partnership to be governed by a collaboration or partnership agreement with the academic, health care, government, foundations, and the Sanford-Burnham (indicated by solid black arrows pointed inwards toward the central salmon colored rounded rectangle from the various stakeholders), from which in collaboration with (usually under a joint steering committee [JSC] arrangement between S|B and the partner) the partner reviews and approves assay/target proposals to develop a portfolio of projects for which the Prebys Center will launch (with leveraged funding—"translational fund") early drug discovery projects, some fraction of which will ultimately yield high-quality transactable "leads" that could be licensed to pharma. The model accesses the broad scientific rationale and robustness of the innovation of the stakeholders, while the reduction to practice of HTS to screen large or targeted small molecular collections will be provided by the validated core competency and excellence of the Prebys Center. The other innovative feature is a translational fund which allows flexible management of operational funds originally from seed investments of strategic partners (e.g., Mayo Clinic), but replenished and accrued to through successful grant awards from NIH, direct investments from disease-specific venture philantrophy, and ultimately returns on investments represented by licensing events to pharma. The strength of this translational fund is that it allows flexibility of the allocation of these funds to maintain a steady-state flow of project through the pipeline (along with smoothing out the operational "burn rate"), while allowing strategic redeployment toward new technology investments to improve the drug discovery operations, process, and science through funding internal project or external "innovators." Additionally, funds could be deployed to fund "riskier" yet highly innovative projects with less validated targets. Finally, as this "virtuous" cycle of priming the drug discovery engine yield further and further growth and return on investment, these could be shared with original seed investors, further increasing the attractiveness of this strategic partnership model.

The scientifically based drug discovery projects entering our center are designated with solid black arrows, whereas solid black arrows containing "$" signs indicate potential sources of funding or use of funds. Central to this funding flow is an accruing "translational fund" that is used to fund the Prebys Center to execute on approved projects. Government and charitable foundation monies represent the source of fund leverage to reduce the cost of drug development. The funding arrow from pharma/biotech represent potentially sizable licensing fees if a lead is transacted, which will expand the translational fund, a portion of which could be deployed to investors that have funded research or to also seek out and fund innovators or innovative development of new technologies as relevant to drug discovery and development. This model has the elements of sustainability by distributing the risk of funding and risk of drug discovery among a portfolio of projects rather than individual single projects. We are using this model with the eight stakeholders shown in Figure 27.17.

REFERENCES

1. Mintz, C. Crowdsourcing pharma drug development. *Life Science Leader* 2013.
2. Munos, B. Lessons from 60 years of pharmaceutical innovation. *Nat Rev Drug Discov.* 2009;8:959–968.
3. Paul, S.M., Mytelka, D.S., Dunwiddie, C.T., Persinger, C.C., Munos, B.H., Lindborg, S.R., Schacht, A.L. How to improve R&D productivity: the pharmaceutical industry's grand challenge. *Nat Rev Drug Discov.* 2010;9(3):203–214.
4. Abbasi, T., Fernandes, P., Vali, S., Singh, G. Innovate while derisking drug development–yes we can! *Drug Discov World* 2012;Summer:9–18.
5. Kaitin, K.I. Deconstructing the drug development process: the new face of innovation. *Clin Pharmacol Ther.* 2010;87(3):356–361.
6. Wizemann, T., Robinson, S., Giffin, R. *Breakthrough Business Models: Drug Development for Rare and Neglected Diseases and Individualized Therapies: Workshop Summary.* Washington, DC: The National Academies Press; 2008.
7. Lessl, M., Douglas, F. From technology transfer to know-how interchange. *Wissenschafts-management* 2010;2:34–41.
8. Lendrem, D.W., Lendrem, B.C. Torching the haystack: modelling fast-fail strategies in drug development. *Drug Discov Today* 2013;18(7–8):331–336.
9. Cressey, D. Traditional drug-discovery model ripe for reform. *Nature* 2011;471(7336): 17–18.
10. Wadman, M. US Translational-Science Centre gets underway. *Nature* 2012;481:128.
11. Wadman, M. NIH encourages translational collaboration with industry. *Nat Rev Drug Discov.* 2010;9(4):255–256.
12. Reitz, A.B., Czupich, K.M. The challenge of drug discovery in the 21st century. *Open Conf Proc J.* 2010;1:49–53.
13. Grau, D., Serbedzija, G. Innovative strategies for drug repurposing. *Drug Discov Devel.* 2005;8(5):56–61.
14. Collins, F.S. Mining for therapeutic gold. *Nat Rev Drug Discov.* 2011;10(6):397–397.

15. Bradley, D. Why big pharma needs to learn the three "R"s. *Nat Rev Drug Discov*. 2005; 4(6):446–446.
16. Huang, R., Southall, N., Wang, Y., Yasgar, A., Shinn, P., Jadhav, A., Nguyen, D.-T., Austin, C.P. The NCGC pharmaceutical collection: a comprehensive resource of clinically approved drugs enabling repurposing and chemical genomics. *Sci Transl Med*. 2011;3(80):80ps16.
17. Mucke, H.A. Drug repositioning: extracting added value from prior R&D investments. *Insight Pharma Reports* 2010.
18. Mullard, A. Could pharma open its drug freezers? *Nat Rev Drug Discov*. 2011;10(6): 399–400.
19. Mullard, A. Drug repurposing programmes get lift off. *Nat Rev Drug Discov*. 2012;11: 1–2.
20. Judd, D.B. Open Innovation in Drug Discovery research comes of age. *Drug Discov Today* 2013;18(7–8):315–317.
21. Årdal, C., Røttingen, J.-A. Open source drug discovery in practice: a case study. *PLoS Negl Trop Dis*. 2012;6(9):e1827.
22. Zdrazil, B., Blomberg, N., Ecker, G.F. Taking open innovation to the molecular level—strengths and limitations. *Mol Inform*. 2012;31(8):528–535.
23. News in brief. Introducing the NIH's drug database. *Nat Rev Drug Discov*. 2011;10(6):403–403.
24. Juliano, R.L. Pharmaceutical innovation and public policy: the case for a new strategy for drug discovery and development. *Sci Public Policy* 2013;40(3):393–405.
25. Hale, V.G., Woo, K., Lipton, H.L. Oxymoron no more: the potential of nonprofit drug companies to deliver on the promise of medicines for the developing world. *Health Aff*. 2005;24(4):1057–1063.
26. Zerhouni, E.A. Translational and clinical science—time for a new vision. *N Engl J Med*. 2005;353(15):1621–1623.
27. Zerhouni, E. The NIH roadmap. *Science* 2003;302(5642):63–72.
28. Check, E. NIH open-access plans draw fire from both sides. *Nature* 2005;433(7026): 561–561.
29. Colvin, M., Maravelias, C.T. Modeling methods and a branch and cut algorithm for pharmaceutical clinical trial planning using stochastic programming. *Eur J Oper Res*. 2010; 203(1):205–215.
30. Swinney, D.C., Anthony, J. How were new medicines discovered? *Nat Rev Drug Discov*. 2011;10(7):507–519.
31. Marchetto, M.C.N., Winner, B., Gage, F.H. Pluripotent stem cells in neurodegenerative and neurodevelopmental diseases. *Hum Mol Genet*. 2010;19(R1):R71–R76.
32. Collins, F. Reengineering translational science: the time is right. *Sci Transl Med*. 2011;3:1–6.
33. Feldman, W.H., Hinshaw, H.C., Moses, H.E. The treatment of experimental tuberculosis with Promin (sodium salt of p,p′-diamino-diphenyl-sulfone-N,N′-didextrose sulfonate): a preliminary report. *Proc Staff Meet Mayo Clin*. 1941;16:187–190.
34. Hinshaw, H.C., Feldman, W.H., Pfuetze, K.H. Treatment of tuberculosis with streptomycin: a summary of observations on one hundred cases. *JAMA*. 1946;132:778–782.
35. Hench, P. The reversibility of certain rheumatic and non-rheumatic conditions by the use of cortisone or of the pituitary adrenocorticotrophic hormone. In: *Nobel Lectures, Physiology or Medicine; 1942–1962*. Singapore: World Scientific Publishing Company; 1999, pp. 311–341.

36. McCarthy, D., Mueller, K., Wrenn, J. Mayo Clinic: multidisciplinary teamwork, physician-led governance, and patient-centered culture drive world-class health care. *Commonw Fund* 2009;27:1308.

37. Arar, N.H., Nandamudi, D. Advancing translational research by enabling collaborative teamwork: the TRACT approach. *J Res Interprof Pract Educ*. 2012;2(3):303–319.

38. Falk-Krzesinski, H.J., Börner, K., Contractor, N., Fiore, S.M., Hall, K.L., Keyton, J., Spring, B., Stokols, D., Trochim, W., Uzzi, B. Advancing the science of team science. *Clin Transl Sci*. 2010;3(5):263–266.

Part V

ACADEMIC SCREENING CENTERS COME OF AGE

Open access high-throughput drug discovery in the public domain is a Mount Everest in the making. —*Rathnam Chaguturu*

Collaborative Innovation in Drug Discovery: Strategies for Public and Private Partnerships, First Edition.
Edited by Rathnam Chaguturu.

28

FINDING THE MIDDLE GROUND: DRUG DISCOVERY TECHNOLOGY IN THE ERA OF ACADEMIC SCREENING CENTERS

Nathan S. Blow

BioTechniques, New York, NY, USA

Since the early 1990s, using high-throughput (HT) chemical screens for the purpose of lead compound discovery has been the realm of the pharmaceutical industry. As small molecule libraries expanded in size and end point assays increased in complexity, HT lead discovery screens required the use of an increasingly sophisticated array of large-scale robotics systems, small well volume plates (either 384- or 1536-wells), rapid liquid transfer and plate washing devices, and high-content imaging instrumentation capable of reading cellular phenotypes. However, the overall results of this brute force, HT approach are questionable. The number of lead compounds in late-stage clinical trials has not significantly risen in the past ten years, leading to smaller pharmaceutical drug pipelines even at a time when older drugs are coming to the end of their patent protection life [1]. However, a paradigm shift in the manner by which chemical screens are performed and by whom they are initiated has occurred in recent years [2]. This shift is due in large measure to the emergence and continuing expansion of academic chemical screening facilities that possess expertise and technological sophistication on par with major pharmaceutical centers. The number of academic screening centers around the world has continued to rise for the past five years, and the trend does not appear to be temporary. In 2009, the Society for Biomolecular Sciences Reference Library listed 64 screening centers in their Academic Screening Facility Directory.

Collaborative Innovation in Drug Discovery: Strategies for Public and Private Partnerships, First Edition. Edited by Rathnam Chaguturu.

However, by late July 2011 that number had risen to 84 facilities according to the Society for Laboratory Automation and Screening directory.* Such a steady growth rate over a short period of time is most certainly due in part to the increasing availability of advanced screening technologies, instrumentation, and chemical libraries that can be applied at varying levels of throughput depending on the specific needs of an investigator or center—key elements for academic screening facilities. In this chapter the recent evolution of academic screening centers will be examined, with a specific focus on the effects technology and methods developments are playing in shifting the priorities of researchers when it comes to their chemical screening efforts. In addition, the possible roles that academic screening facilities might play in the development of drugs in the coming decades will also be explored.

LARGE-SCALE SCREENING CENTERS AND LEAD COMPOUND DISCOVERY

Increasing availability, not to mention flexibility, of HT robotic systems and screening technologies has enabled several academic centers to implement chemical screening platforms with throughput and capacity on par with their pharmaceutical counterparts [2]. However, there are differences: due to the needs of academic screening centers, the ability to perform a wide range of screening applications (both in terms of compound library size and assay design) on a particular platform is critical as different investigators using the facility will have differing needs (size of screen, end point assay, etc.) when performing their particular HT screen.

Since the implementation of their chemical biology platform, the Broad Institute located in Cambridge, Massachusetts has been involved in a wide array of chemical screening projects. For example, in 2009 Gupta et al. explored the possibility of identifying inhibitors of epithelial cancer stem cells using a HT screening approach [3]. Following the development of an assay demonstrating that breast cancer cell populations induced into epithelial–mesenchymal transition were indistinguishable from cancer stem cell populations, the authors screened a small molecule library of 16,000 compounds containing both commercially available small molecules and natural extracts; they identified 32 compounds that exhibited toxicity toward the induced breast cancer cells [3]. Although only a subset of the Broad Institute's complete chemical collection of nearly 500,000 compounds [4], studies such as this provide an example of how proper assay design with a smaller collection of molecules can provide new biological insights and possible lead compounds for clinical testing—another key factor for academic centers, where it is crucial to keep both screening time and costs as low as possible.

The varied nature of screens performed at academic centers can be seen in a somewhat larger screen that was also conducted at the Broad Institute chemical biology platform searching for novel antimicrobial agents. Here Moy et al. [5] screened the nematode *Caenorhabditis elegans* for survival when infected with the human pathogen

*http://www.slas.org/screeningFacilities/facilityList.cfm.

Enterococcus faecalis. The authors designed an assay wherein infected worms were placed into individual wells of 384-well plates using a large particle sorter. Live animals were screened for increased survival with 37,200 small molecules and natural product extracts using the dye SYTOX Orange, which only enters cells with damaged membranes [6]. The screen resulted in the identification of 28 compounds not previously reported to have antimicrobial effects, and gave further insight into the cellular mechanisms of action of these antimicrobial agents.

As more academic facilities enter into the world of chemical screening, the need to understand and minimize the sources of error from HT screening (i.e., false-negative and false-positive results [7]) has never been greater. This concern has led some researchers to develop HT screening applications wherein the numbers of false positives or false negatives could potentially be decreased significantly. In 2006, Inglese et al. reported a quantitative HT screening approach where greater than 60,000 compounds were titrated in series in 1536-well microplates [8]. Using low-volume dispensing, high-sensitivity detectors, and robotic plate handling, the authors were able to generate concentration-response curves to determine structure-function relationships in a screen of pyruvate kinase modulators. The ability to generate such dose-response curves for such a large number of samples and small molecules, thereby minimizing false negatives within an HT screen, is a direct result of the development of automation technologies that enable the more widespread use of 384- or 1536-well plates.

HT screens undertaken now at centers such as the Broad Institute are being accomplished using technologies that, although similar in design, differ from what has been previously applied in pharmaceutical settings. Several academic screening facilities have invested in and utilize "modular designs" for their screening instrumentation to enable a diverse array of screening assays and applications to be performed. For example, instrument development companies such as HighRes BioSolutions have constructed screening platforms where key instrumentation is placed on carts that can be moved or connected ("docked") for specific assays. Such approaches increase the potential configurations of the screening platform (although some elements in these configurations, including the larger robotic arms, are fixed), therefore enabling a greater number of possible screening scenarios. Larger instrumentation, including imaging systems such as the FLIPR Tetra from Molecular Devices, can also be added when called for in a specific assay.

While pharmaceutical companies and larger academic screening centers might perform 60 or more HT screening campaigns in a given year [2], these are not the only players now when it comes to chemical screening; smaller centers are also on the rise around the globe. For example, with a compound library collection of 55,000 small molecules and natural products, the University of California at Santa Cruz chemical screening center is capable of screening up to 30,000 compounds per day, relying on technologies including PerkinElmer Janus systems for liquid handling, a BioTek plate washer, a Matrix Wellmate stacker, and various imaging systems, including Molecular Devices ImageExpress for automated image analysis to accomplish their goals.*

*http://chemscreen.ucsc.edu/equipment-list.

CHEMICAL GENOMICS

The availability of commercial systems to perform HT screens, as well as interaction among scientist in different disciplines, is fueling the continued expansion of another form of chemical screening, the chemical genomic, or chemogenomic, screen. Chemogenomic screens seek to determine a drug's effect by exploring the treatment of that drug on a set of different strains or isolates [9].

Here yeast has proven a fertile ground in which to apply chemogenomic screens owing to the generation and availability of large yeast deletion collections [10]. Such collections can be used to screen compounds and generate fitness profiles that can then be compared and used as a starting point to identify gene relationships [11]. In their review, Hoon et al. describe the various approaches, including forward screens such as haploinsufficiency profiling, homozygous profiling, and multi-copy and overexpression-suppression profiling, as well as reverse approaches such as toxicity-suppression profiling, that have been developed to either define the mechanism of action of a drug on a yeast biological process or identify a particular drug target. As noted in the previous section, cost is always an issue and chemical genomics screens using large numbers of strains and/or small molecules can be cost prohibitive and labor intensive, not to mention that in many cases there is only a limited amount of compound available to test. To overcome these issues, Pierce et al. developed a chemogenomic approach for yeast using the so-called TAG4 barcoding array that resolves fitness profiles for individual yeast strains that have been pooled by extracting changes in the abundance of the individual strains measured using the barcodes. Thus, the approach reduces the physical screening numbers as well as the volume of compound used in each screening assay [12].

Chemogenomic screens do have the ability to reveal novel insights into the mechanism of action of a particular compound. However, the ability to derive meaningful insights from such a chemical genomics screen is critically dependent upon developing the proper assay. A 2010 study by Kapitzky et al. demonstrates that crossing species boundaries when performing fitness profiling can provide such novel insights [9]. Here the authors created a screening platform using libraries of *Saccharomyces cerevisiae* and *Schizosaccharomyces pombe* haploid deletion mutants. By screening 727 *S. cerevisiae* and 438 *S. pombe* strains against 21 compounds, a DNA damaging agent that had a conserved mode of action with human cells was identified.

Chemogenomic screens are not exclusively used with organisms (such as yeast) where large collections of deletions mutations have been generated. A more recent example of the value that can be derived from performing chemical genomic profiling in other organisms can be see in the work of Yuan et al. and their profiling of antimalarial therapies and molecular targets [13]. Here, Yuan and colleagues from the National Institutes of Health Chemical Genomics Center and Columbia University attempted to both (1) identify new drugs with activity against *Plasmodium falciparum*, the malaria parasite, and (2) understand the mechanism of drug activity and resistance at the molecular level. The rationale for performing such a chemical genomics screen in this case is that resistance to antimalarial drugs has been demonstrated to occur (i.e., chloroquine resistance) and a more complete understanding of the underlying molecular mechanisms of drug activity could be beneficial in predicting or even eliminating such resistance in the future.

In their chemical screen, Yuan et al. surveyed 61 parasite lines against 2816 compounds that were registered for human or animal use at eight five-fold dilutions in a growth assay resulting in over 170,000 total assays and more than one million data points. From this, 32 active compounds were found. What separated this study from previous efforts to identify new antimalarial compounds using HT screening [14, 15] was the genetic analysis performed by the authors to understand the molecular bases of these 32 active compounds. Initially, authors performed pairwise comparisons to determine correlated responses and potential signature groups, and then followed these analyses with genome-wide association studies using candidate single nucleotide polymorphisms and microsatellites to explore parasite response. The effort of tying together the chemical screening data and the genome-wide association study data resulted in a clearer understanding of the possible roles the 32 active compounds and provided new opportunities to target different *P. falciparum* cellular pathways than did previous efforts.

The potential health value of efforts such as those of Yuan et al. to bring forward lead compounds targeting the malaria parasite is obvious. A challenge prior to the rise of academic screening centers was the sheer number of diseases that needed to be targeted and the need to "team" with experts in these particular fields—a so-called "crowdsourcing" effort that in some ways lies at the heart of many academic screening efforts and centers at the moment (see "Crowdsourcing Efforts," below).

CHEMICAL LIBRARIES

For technology developers, the move into robotic instrumentation, rapid plate washers and readers, and HT assay analysis instruments has actually been a logical outgrowth of the emergence of larger chemical compound libraries [16]. Since the early 1990s, the number and size of chemical screening libraries has expanded on a yearly basis. This expansion has been facilitated by (1) an increasing availability of commercial chemical collections, (2) the implementation of an NIH-funded effort to generate small molecule collections, and (3) new combinatorial chemistry approaches for the creation of small molecules.

Small molecule collections are commercially available for researchers and screening centers to purchase. For example, the company ChemBridge offers more than 800,000 drug-like and lead-like compounds, with subsets of their libraries available for specific applications as well as highly diverse libraries aimed at primary HT screening applications. Other commercial suppliers offer additional compound libraries based, in some cases, on differing structures and synthesis protocols.

The development of new compounds to supplement and expand current small molecule library collections has been aided in recent years through publicly funded efforts from the National Institutes of Health Small Molecule Repository and the Molecular Libraries Initiative. This initiative, launched in 2004, comprises three main efforts: (1) the Molecular Libraries Probe Production Centers Network (MLPCN), (2) the development of a comprehensive database of chemical structures and related genetic information called PubChem, and (3) a technology development effort focused on increasing chemical and assay diversity and making instrumentation advances. At the moment, there are seven MLPCN centers around the United States that have contributed to the more than 350,000 compounds that currently constitute the network's chemical collection (updates on recent probe entries into the MLPCN collection can be found at http://www.ncbi.nlm.nih.gov/books/NBK47352/). The development of this resource has accelerated and enhanced HT screening efforts at both larger and smaller academic screening centers. In fact, a recent study by Zhang et al. noted that the Molecular Libraries Initiative screening collection was more chemically diverse and had greater biogenic bias than comparable commercially available small molecule collections; however, the authors also suggested that the collection has not yet been fully optimized for biochemical diversity [17].

While small molecule libraries continue to grow, the question of how much chemical space is being covered by these libraries remains open. Chemical space is estimated to be in excess of 10^{60} even when examining only smaller (sub 500 dalton) carbon-based compounds [18]. Given that most small molecule collections at academic, or even large pharmaceutical, centers rarely exceed 2 million, there would appear to be a large amount of chemical space missed in current libraries. This exclusion of a large percentage of chemical space might be the result of two factors: (1) the costs of screening chemical libraries remains fairly high and actually increases with library size (see previous section), and (2) the rate of synthesis of new compounds is slow, as is the ability to store and sort larger collections of molecules. The refinement of combinatorial chemistry approaches to small molecule design has been a major factor in the generation of chemical libraries of increasing size and complexity.

Several combinatorial chemistry approaches for library generation are aimed at testing entire collections of small molecules against a specific target in a single reaction vessel. These approaches, so called selection-based approaches, rely on the ability to capture the identity of (deconvolute) a specific compound acting on a particular target. One such selection-based approach, first described by David Liu and colleagues in 2001, takes advantage of DNA strands that act as molecular barcodes when attached to the small molecules composing a library [19]. The key to the use of DNA as a barcode for such synthetic libraries actually lies in the recent advances that have been made in DNA sequencing technology. Next-generation sequencing instrumentation now has the capacity to generate hundreds of gigabases of data in a single instrument run, from single molecules in some cases, sequencing the equivalent of several human genomes, to high coverage [20–22]. This depth of sequencing when applied to a DNA-barcoded synthetic small molecule library not only allows investigators to identify active compounds, but it also enables researchers to assess small abundance changes in library members upon exposure to a target [23], thus providing an additional level of information during a chemical screen.

The major benefit when it comes to generating selection-based libraries is that the size of such screening libraries can be greatly increased relative to their traditional counterparts, since without a way to identify library members all the small molecules must be individually tested in single wells. As an example of the potential size of selection-based libraries, in 2009 researchers from GlaxcoSmithKline reported the design of a DNA-encoded small molecule library containing more than 800 million small molecules tethered to double stranded DNA barcodes [24]. As a proof of concept experiment to demonstrate the possibility of using affinity selection to identify active compounds, the authors screened for inhibitors of the enzymes Aurora A kinase and p38 MAP kinase. In both cases, positive controls as well as unique, novel inhibitors were identified from the small molecule collections [24].

Virtual screening of an expanded chemical space using chemoinformatic approaches is also possible due to the expansion of databases and resources such as PubChem and ChemBank [25, 26]. Here researchers can take advantage of the over 450,000 bioassay records in PubChem targeting more than 7000 genes that link to over 1.8 million substances [25]. While computer-aided drug discovery is a valuable tool, previously described issues with false-positive and false-negative numbers in HT screens (see section above) have to be taken into account when performing virtual screens.

CROWDSOURCING EFFORTS

Academia, lacking the downstream financial concerns often associated with pharmaceutical drug discovery, provides an opportunity for researchers from a number of specialties to come together to work on a specific problem. In fact, the possibility of fostering further interactive dialogues on the topic of drug discovery was at the heart of the National Institutes of Health Molecular Libraries Initiative (see previous section). "The NIH aimed not only to provide a suitable framework for the development of such 'chemical probes' but also to encourage a cultural shift toward open, cross-disciplinary communication within multiple scientific and technological disciplines to facilitate probe discovery" was the wording associated with one of the major goals of the initiative, as framed in a 2009 commentary on evaluating NIH chemical probes published in *Nature Chemical Biology* [27]. During the first five years of the Molecular Libraries Initiative, 64 chemical probes (small molecules capable of probing the biochemical properties of proteins) were identified. Following this initial effort, the 64 chemical probes were to be evaluated by 11 outside experts in a crowdsourcing analysis. By crowdsourcing the analysis of the chemical probes, it was easier to assess the quality of the output from these initial academic screening efforts and get insights into the probes from experts in preclinical drug discovery [27]. In the end, the experts had either medium or high confidence levels for 48 out of the 64 probes. The crowdsourcing probe validation program initiated in this commentary may prove to be only the starting point for new crowdsourcing efforts in the world of chemical screening and drug discovery.

Crowdsourcing is not the exclusive domain of the academic world, in fact the latest trend in HT screening involves greater collaboration among all parties, and open sourcing between pharmaceutical companies and academic screening facilities.

A 2010 analysis showed that the cost to bring a drug to market, including all failures leading up to marketing, currently exceeds \$1.8 billion [2]. Additional analyses have shown that key patents, which are set to expire between now and 2014, are estimated to put more than $209 billion in annual revenue for pharmaceutical companies at risk [2]. According to Paul et al., without a significant investment in research and development leading to new drug products, this loss in revenue will not be able to be replaced, thus placing our current model for drug discovery and development in jeopardy [1].

In light of these commercial challenges faced in the drug discovery field, particular attention is now being paid to lead discovery and methods to enhance lead discovery productivity. Here open-source efforts have started to pop up in the drug discovery world. A number of collaborations between pharmaceutical companies and academic centers as well as other nonprofit institutions have sprung up in the past two years. Willingness to engage in nonproprietary collaborations can be seen in moves made in recent years by several large pharmaceutical companies. In 2010 for example, GlaxoSmithKline and Novaritis of Basel deposited 328,100 compounds directed against the malaria parasite into the European Bioinformatics Institute ChEMBL Negelected Tropical Disease archive.* Such precompetitive data sharing initiatives are growing when it comes to drug discovery, but the challenge of how best to organize and develop such projects remains an open debate.

In February 2011, a group of researchers and policy makers came together to discuss the idea of how best to form public-private partnerships for precompetitive drug development from screening to validation (efforts are summarized in [28]). The discussions revolved around the best way to create an open-source environment whereby both the public and private sectors could work together on specific diseases and screening approaches to speed drug discovery [28].

Several open source efforts have been established in recent years. The Alzheimer's Disease Neuroimaging Intiative is a public-private partnership testing the ability of various imaging modalities (MRI, PET) to measure the progression of the disease, while GlaxoSmithKline invested $8 million to set up an open lab in Spain to study malaria. Such collaborative efforts represent the first steps toward what could be a more global crowdsourcing approach to early drug discovery efforts in the future.

Technology innovation and methods development, along with the need to refine the current model of drug discovery, have led to a new era in HT chemical screening. Today, researchers from many different disciplines of life science can take advantage of the power of small molecule screens to both inform our understanding of basic biological processes (i.e., chemogenomics screens) and identify compounds with the potential to play a role in global health care. The changes described in this chapter most likely represent only the early steps of a field that will continue to evolve and change in the years to come.

REFERENCES

1. Paul, S.M., Mytelka, D.S., Dunwiddie, C.T., Persinger, C.C., Munos, B.H., Lindborg, S.R., Schacht, A.L. How to improve R&D productivity: the pharmaceutical industry's grand challenge. *Nat Rev Drug Discov*. 2010;9(3):203–214.

*http://www.nature.com/nbt/journal/v28/n7/full/nbt0710-631.html.

2. Blow, N.S. High-throughput screening: designer screens. *Nat Methods* 2009;6(1):105–108.
3. Gupta, P.B., Onder, T.T., Jiang, G., Tao, K., Kuperwasser, C., Weinberg, R., Lander, E.S. Identification of selective inhibitors of cancer stem cells by high-throughput screening. *Cell* 2009;138:645–659.
4. Blow, N.S. Back to school. *Biotechniques* 2011;50(1):19–21.
5. Moy, T.I., Conery, A.L., Larkins-Ford, J., Wu, G., Mazitschek, R., Casadei, G., Lewis, K., Carpenter, A.E., Ausubel, F.M. High-throughput screen for novel antimicrobials using a whole animal infection model. *ACS Chem Biol.* 2009;4(7):527–533.
6. Gill, M.S., Olsen, A., Sampayo, J.N., Lithgow, G.J. An automated high-throughput assay for survival of the nematode *Caenorhabditis elegans*. *Free Radical Biol Med.* 2003;35: 558–565.
7. Thomas, C.J., Auld, D.S., Huang, R., Huang, W., Jadhav, A., Johnson, R.L., Leister, W., Maloney, D.J., Marugan, J.J., Michael, S., Simeonov, A., Southall, N., Xia, M., Zheng, W., Inglese, J., Austin, C.P. The pilot phase of the NIH Chemical Genomics Center. *Curr Top Med Chem.* 2009;9(13):1181–1193.
8. Inglese, J.M., Auld, D.S., Jadhav, A., Johnson, R.L., Simeonov, A., Yasgar, A., Zheng, W., Austin, C.A. Quantitative high-throughput screening: a titration-based approach that efficiently identifies biological activities in large chemical libraries. *Proc Natl Acad Sci U S A.* 2006;103(31):11473–11478.
9. Kapitzky, L., Beltrao, P., Berens, T.J., Gassner, N., Zhou, C., Wuster, A., Wu, J., Babu, M.M., Elledge, S.J., Toczyski, D., Lokey, R.S., Krogan, N.J. Cross-species chemogenomic profiling reveals evolutionarily conserved drug mode of action. *Mol Syst Biol.* 2010;6:451.
10. Giaever, G., Chu, A.M., Ni, L., Connelly, C., Riles, L., Véronneau, S., Dow, S., Lucau-Danila, A., Anderson, K., André, B., Arkin, A.P., Astromoff, A., El-Bakkoury, M., Bangham, R., Benito, R., Brachat, S., Campanaro, S., Curtiss, M., Davis, K., Deutschbauer, A., Entian, K.D., Flaherty, P., Foury, F., Garfinkel, D.J., Gerstein, M., Gotte, D., Güldener, U., Hegemann, J.H., Hempel, S., Herman, Z., Jaramillo, D.F., Kelly, D.E., Kelly, S.L., Kötter, P., LaBonte, D., Lamb, D.C., Lan, N., Liang, H., Liao, H., Liu, L., Luo, C., Lussier, M., Mao, R., Menard, P., Ooi, S.L., Revuelta, J.L., Roberts, C.J., Rose, M., Ross-Macdonald, P., Scherens, B., Schimmack, G., Shafer, B., Shoemaker, D.D., Sookhai-Mahadeo, S., Storms, R.K., Strathern, J.N., Valle, G., Voet, M., Volckaert, G., Wang, C.Y., Ward, T.R., Wilhelmy, J., Winzeler, E.A., Yang, Y., Yen, G., Youngman, E., Yu, K., Bussey, H., Boeke, J.D., Snyder, M., Philippsen, P., Davis, R.W., Johnston, M. Functional profiling of the Saccharomyces cerevisiae genome. *Nature* 2002;418:387–391.
11. Hoon, S., St. Onge, R.P., Giaever, G., Nislow, C. Yeast chemical genomics and drug discovery: an update. *Trends Pharmacol Sci.* 2008;29(10):499–504.
12. Pierce, S.E., Fung, E.L., Jaramillo, D.F., Chu, A.M., Davis, R.W., Nislow, C., Giaever, G. A unique and universal molecular barcode array. *Nat Methods* 2006;3(8):601–603.
13. Yuan, J., Cheng, K.C., Johnson, R.L., Huang, R., Pattaradilokrat, S., Liu, A., Guha, R., Fidock, D.A., Inglese, J., Wellems, T.E., Austin, C.P., Su, X.Z. Chemical genomic profiling for antimicrobial therapies, response signatures, and molecular targets. *Science* 2011;333: 724–729.
14. Gamo, F.J., Sanz, L.M., Vidal, J., de Cozar, C., Alvarez, E., Lavandera, J.L., Vanderwall, D.E., Green, D.V., Kumar, V., Hasan, S., Brown, J.R., Peishoff, C.E., Cardon, L.R., Garcia-Bustos, J.F. Thousands of chemical starting points for antimalarial lead identification. *Nature* 2010;465:305–310.

15. Guiguemde, W.A., Shelat, A.A., Bouck, D., Duffy, S., Crowther, G.J., Davis, P.H., Smithson, D.C., Connelly, M., Clark, J., Zhu, F., Jiménez-Díaz, M.B., Martinez, M.S., Wilson, E.B., Tripathi, A.K., Gut, J., Sharlow, E.R., Bathurst, I., El Mazouni, F., Fowble, J.W., Forquer, I., McGinley, P.L., Castro, S., Angulo-Barturen, I., Ferrer, S., Rosenthal, P.J., Derisi, J.L., Sullivan, D.J., Lazo, J.S., Roos, D.S., Riscoe, M.K., Phillips, M.A., Rathod, P.K., Van Voorhis, W.C., Avery, V.M., Guy, R.K. Chemical genetics of Plasmodium falciparum. *Nature* 2010;465:311–315.

16. Schnur, D.M., Beno, B.R., Tebben, A.J., Cavallaro, C. Methods for combinatorial and parallel library design. *Methods Mol Biol.* 2011;672:387–434.

17. Zhang, J., Lushington, G., Huan, J. Characterizing the diversity and biological relevance of the MLPCN assay manifold and screening set. *J Chem Inf Model.* 2011;51(6):1205–1215.

18. Bohacek, R.S., McMartin, C., Guida, W.C. The art and practice of structure-based drug design: a molecular modeling perspective. *Med Res Rev.* 1996;16(1):3–50.

19. Gartner, Z.J., Liu, D.R. The generality of DNA-templated synthesis as a basis for evolving non-natural small molecules. *J Am Chem Soc.* 2001;123(28):6961–6963.

20. Zhao, J., Grant, S.F. Advances in whole genome sequencing technology. *Curr Pharm Biotechnol.* 2011;12(2):293–305.

21. Ng, S.B., Nickerson, D.A., Bamshad, M.J., Shendure, J. Massively parallel sequencing and rare disease. *Hum Mol Genet.* 2010;19(R2):R119–R124.

22. Thompson, J.F., Milos, P.M. The properties and applications of single-molecule DNA sequencing. *Genome Biol.* 2011;12(2):217.

23. Kleiner, R.E., Dumelin, C.E., Liu, D.R. Small-molecule discovery from DNA-encoded chemical libraries. *Chem Soc Rev.* 2011;40:5707–5717.

24. Clark, M.A., Acharya, R.A., Arico-Muendel, C.C., Belyanskaya, S.L., Benjamin, D.R., Carlson, N.R., Centrella, P.A., Chiu, C.H., Creaser, S.P., Cuozzo, J.W., Davie, C.P., Ding, Y., Franklin, G.J., Franzen, K.D., Gefter, M.L., Hale, S.P., Hansen, N.J., Israel, D.I., Jiang, J., Kavarana, M.J., Kelley, M.S., Kollmann, C.S., Li, F., Lind, K., Mataruse, S., Medeiros, P.F., Messer, J.A., Myers, P., O'Keefe, H., Oliff, M.C., Rise, C.E., Satz, A.L., Skinner, S.R., Svendsen, J.L., Tang, L., van Vloten, K., Wagner, R.W., Yao, G., Zhao, B., Morgan, B.A. Design, synthesis, and selection of DNA-encoded small molecule libraries. *Nat Chem Biol.* 2009;5(9):647–654.

25. Xie, X.Q. Exploiting PubChem for virtual screening. *Expert Opin Drug Discov.* 2010; 5(12):1205–1220.

26. Seiler, K.P., George, G.A., Happ, M.P., Bodycombe, N.E., Carrinski, H.A., Norton, S., Brudz, S., Sullivan, J.P., Muhlich, J., Serrano, M., Ferraiolo, P., Tolliday, N.J., Schreiber, S.L., Clemons, P.A. ChemBank: a small molecule screening and chemoinformatics resource database. *Nucleic Acids Res.* 2008;36:D351–D359.

27. Oprea, T.I., Bologa, C.G., Boyer, S., Curpan, R.F., Glen, R.C., Hopkins, A.L., Lipinski, C.A., Marshall, G.R., Martin, Y.C., Ostopovici-Halip, L., Rishton, G., Ursu, O., Vaz, R.J., Waller, C., Waldmann, H., Sklar, L.A. A crowdsourcing evaluation of the NIH chemical probes. *Nat Chem Biol.* 2009;5(7):441–447.

28. Norman, T.C., Bountra, C., Edwards, A.M., Yamamoto, K.R., Friend, S.H. Leveraging crowdsourcing to facilitate the discovery of new medicines. *Sci Transl Med.* 2011;3(88):88mr1.

29

OPEN INNOVATION-BASED DRUG DISCOVERY IN EUROPE: SOME EXAMPLES OF NATIONAL AND TRANSNATIONAL EUROPEAN INITIATIVES INTEGRATING CHEMISTRY, BIOLOGY, AND TECHNOLOGY PLATFORMS

Philip Gribbon

European ScreeningPort GMBH, Hamburg, Germany

Industrial drug discovery and development activities play an important role in the economy of Europe. The pharmaceutical sector is responsible for the direct employment of some 640,000 people, of which some 115,000 are directly involved in research and development (R&D). It is estimated that for each direct job in an R&D function, some 3 to 4 additional jobs are supported in services and associated roles [1]. While major pharmaceutical companies typically operate on a global basis, with research activities occurring within Asia, Europe, and North America, European-based companies still made up 6 of the top 15 positions in terms of revenues in 2010, with Novartis (2), Sanofi (4), GlaxoSmithKline (5), AstraZeneca (6), Bayer Schering (13) and Boehringer Ingelheim (14) represented. The trade body representing industrial pharmaceutical research is the European Federation of Pharmaceutical Industries and Associations (EFPIA) drawn from across the 30 members of the European Free Trade Area.

Collaborative Innovation in Drug Discovery: Strategies for Public and Private Partnerships, First Edition.
Edited by Rathnam Chaguturu.

Collectively, members of EFPIA invested some €27 billion into R&D in 2010 compared with €37 billion in the United States. These R&D investments are broadly in line with the size of the respective markets, with the European market representing some 29.2% of the world total, compared with 42% within the United States. However, the market for new products (launched between 2005 and 2009) in Europe falls to 22% of worldwide sales, compared with 61% in the United States, indicating a greater barrier within Europe to the acceptance of the products of innovation, namely new medicines [1].

Regardless of the huge investment in R&D, a recurring theme in recent presentations from members of the scientific, medical, and financial communities has been disappointment over the lack of the new treatment options capable of addressing patients' unmet needs [2]. This situation has arisen despite the year-on-year increases in expenditure on pharmaceutical R&D and the opportunities presented by the increased application of genomics and other "omics-based" sciences within the drug discovery paradigm. Nevertheless, the rate of successful submissions of new molecular entities to the U.S. Food and Drug Administration (FDA) and the European Medical Agency has remained essentially stagnant [3]. To address this situation, new strategies are being developed in the pharmaceutical industry with the objective of overcoming the limitations of the current R&D process, which historically was based on an integrated operating model [4]. In the integrated model, core competencies were developed and kept in-house to prevent the distribution of knowledge and intellectual property (IP) and thus assure a competitive advantage. Gradually, the pharmaceutical industry has seen increased outsourcing of what are considered as the more routine tasks, such as liability profiling of compounds and conduct of clinical trials [5]. Although outsourcing can lead to efficiency gains, its impact on the rate of delivery of successful new molecular entities appears to be limited. Improved outsourcing strategies, although necessary, do not address issues associated with, for example, selection of the optimal disease-relevant therapeutic targets for "difficult" indications such as Alzheimer's disease.

EUROPEAN PHARMACEUTICAL INDUSTRY RESPONSE

The pharmaceutical industry has reached the conclusion that it needs to transform its R&D operation and find more efficient ways to innovate. At the same time academic biomedical research institutions are recognizing the benefits of forming deeper partnerships with the pharmaceutical industry: there are tools, technologies, experience, and financial resources within pharmaceutical companies that can help academic researchers to translate their ideas and research into drugs [6]. This understanding is leading to an opening of organizational structures leading to the formation of strong cross-links between academia and pharmaceutical industry. The umbrella term for this new R&D model is "open innovation." Academic institutions and bridging organizations are looking to identify new ways to partner in a more collaborative manner [7]. Almost by definition, the strategies pursued by the various organizations to achieve this goal are diverse, as are their expectations and concerns. One significant outcome has been the initiation of academic small molecule drug discovery groups based around university core facilities, cross-institutional bridging organizations, and public–private partnerships [8]. This new breed includes structural biology-based drug discovery organizations, molecular screening facilities, and medicinal chemistry-focused research centers [9]. All had the remit of developing academic disease biology projects beyond the traditional end-point stage of target validation, in order to create high quality preclinical assets that would be of interest to the pharmaceutical industry. The activities of these academic-based organizations are still dependent on the interest and in some cases, financial support of the pharmaceutical industry, not only for the clinical stages of drug development but also as active partners in, for example, the counter-financing of earlier stage hit finding projects [10].

RESEARCH CHARITIES AND DRUG DISCOVERY: EXAMPLES FROM THE UNITED KINGDOM

In the United Kingdom, research charities, both disease focused (e.g., Cancer Research U.K., British Heart Foundation) and more general in outlook (e.g., Wellcome Trust) are important sources of support for academic based drug discovery [11, 12]. In 2010, research charities made up a significant proportion of the total U.K. R&D funding contribution at higher education institutions (15%). Cancer Research U.K. has in excess of 70 projects in its drug discovery portfolio, of which over 50 involve small molecule approaches. Several Cancer Research U.K. projects are partnered with pharma, most notably a recent three-year alliance set up with AstraZeneca to develop targets related to cancer metabolism [13]. The Wellcome Trust has a number of programs that involve the development of therapeutics, including the Translation Awards, for development of new biomedical products that are close to market. The Trust's Strategic Translation Awards fund the commercialization of novel therapeutics, vaccines, diagnostics, as well as platform or regenerative medicine technologies and associated topics that align with its scientific strategy [14]. The most relevant Wellcome funding scheme in terms of small molecule therapeutics is the Seeding Drug Discovery Initiative. The project-based program is accessed via a competitive, biannual peer review process. Academics with novel target biology insights or preexisting chemical starting points are provided with funds to enable them to progress through screening, medicinal chemistry, and pharmacological profiling. The aim is to develop projects to the lead candidate point, after which third-party funding is secured and the project advanced through preclinical and clinical stages [15]. Compared with the full costs of an internal pharmaceutical industrial project, the Seeding Drug Discovery funding rates are modest (1–3 million GBP per project over 2 to 3 years). However, outsourcing of activities such as screening, medical chemistry, and liability profiling is encouraged, allowing grant holders to use low-cost suppliers of these services. Project teams are expected to use expert professional advisors, typically with a background in the pharmaceutical industry, to guide the projects and ensure optimal alignment of outcomes with the requirements of eventual pharmaceutical industry development partners. As would be anticipated with the Wellcome Trust, diseases of the developing world are well represented within the existing portfolio, with projects addressing dengue fever, malaria and visceral leishmaniasis. However, funding of other more "mainstream" indications, which nevertheless represent clear unmet needs, is within scope. Examples include projects addressing Alzheimer's, glioblastoma, and hepatitis C infection. In these cases, the Wellcome Trust is able fund research into unusual or novel mechanisms that may not be under active consideration by the pharmaceutical industry. The recipients of funding are predominantly drawn from the U.K. academic sector, with Universities of London, Manchester, and Cardiff well represented. However, biotechnology and pharmaceutical industry–funded projects include work at GlaxoSmithKline on hospital-acquired infections, and the Genomic Institute of the Novartis Foundation on diseases of the developing world.

EXISTING EUROPEAN-WIDE INFRASTRUCTURES AND PUBLIC–PRIVATE PARTNERSHIPS

There are relatively few operational European-wide academic infrastructures in the area of chemical biology and screening, although several initiatives are in the pipeline to remedy this deficit. The most notable existing infrastructure is that organized through the ChemBioNet network of European academic institutions, which provide an open-access distributed chemical biology platform to academic and industrial organizations [16]. The group is based in Germany (Leibnitz-Institut für Molekulare Pharmakologie, Max-Delbrück-Center for Molecular Medicine, Helmholtz-Centre for Infection Research, University of Konstanz) and Norway (University of Oslo) with associate partners drawn from across Europe. The aim of the organization is to provide biologists and chemists access to tools, including compound libraries and a range of high-throughput screening technologies. The organization has access to a compound library of >20,000 small molecules, which is distributed across the members institutions [17]. In one sense, the ChemBioNet represents a pilot project for larger scale projects planned within Europe, such as EU-OPENSCREEN, which will be discussed later.

Europe's centerpiece public–private program focused on drug discovery is the Innovative Medicines Initiative (IMI) which brings together the pharmaceutical industry (through EFPIA), European academia, small-to-medium enterprises (SMEs), patient groups, regulatory agencies, and hospitals, with the aim of increasing the competitiveness of the European Union (EU) pharma sector through the creation of new precompetitive research tools [18]. Originally envisioned in 2004, the IMI concept passed through an extended period of development before the first call for proposals in 2008. The budget for the final program was in excess of €2 billion, dwarfing other ongoing projects. Half of funding is provided by the European Commission under the Framework 7 Health umbrella, with the balance provided by members of the EFPIA group of pharmaceutical companies, mainly in the form of in-kind contributions. The rationale behind setting up IMI aligns with the requirement to raise European R&D productivity, and the strategic approach is to initiate very large-scale projects to help increase understanding of disease and improve the selection of targets in the discovery phase and patients during the clinical phases [19]. All IMI projects cover precompetitive topics, which by definition exclude activities that will lead to the direct approval of a medicine or vaccines; instead, the projects look to improve the overall efficiency of the drug discovery process. The original IMI areas of interest in 2008 dealt with efficacy, safety, knowledge management, and education and training. These were then broken down into themes such as "predictive pharmacology," and "validation of biomarkers," around which specific call topics for research proposals were then published. In contrast to regular EU Framework programs, following peer review of the submitted grant proposal only a single consortium would be selected per topic. Consortia of academics would be paired with a collection of between 4 and 15 pharmaceutical companies to create integrated academia–industry teams. In 2011, there were 23 running IMI projects representing some €190 million of funding, with around 15% of the participants coming from the SME category and the rest a mixture of academia and patient groups. Although at the time of writing, the projects have been running for a relatively short time, some early success stories have been highlighted. For example, in schizophrenia, the identified bottlenecks are the lack of relevant animal models, insufficient tools to determine treatment efficacy, and widespread use of outdated clinical trial methodology. To address these deficiencies, the IMI NEWMEDS consortium has amalgamated data from 67 trials representing 11 compounds and in excess of 23,000 patients from 25 countries to create the world's largest single repository of (anonymized) patient data in schizophrenia [20]. One of their approaches will be to develop improved clinical imaging approaches using functional MRI and PET in combination with machine learning methods in order to better define and quantify patient responses to drug treatment and allow more efficient design of clinical trials. The critical link between in-vivo models and the clinic is tackled through the development of new animal models involving quantify responses using brain recording and improved standardized behavioral tests. In the area of drug safety, a second consortium "Safe-T" has been formed to address the problem of drug-induced toxicities in the kidney, liver, and vascular tissues [21]. The Safe-T consortium is aiming very high with the expectation that their identified safety biomarkers will be suitable for use in medical product development and will gain associated European Medicine Agency and FDA approvals. An important component of IMI has been the emphasis on training and knowledge management, with four projects in this area receiving funding to date. Education and training projects cover the entire lifecycle of drug discovery and development, from basic science to pharmacovigilance and health economics. In the PharmaTrain consortium, the 15 pharmaceutical industry partners work with a group of 22 academic partners and 14 learned societies, as well as 3 regulatory authorities [22]. A PharmaTrain devised syllabus for postgraduate studies in drug discovery and medicine development has now been implemented at 15 universities in Europe. Direct teaching is supported by an e-campus facility and standardized syllabus. A completely new master's degree in medicines development has been created as part of the proposal.

The IMI research priorities were updated in 2011 in response to the significant changes in pharmaceutical industry priorities in the four years since the first call [23]. The new priorities are very extensive and include pharmacogenetics, rare disease, systems approaches, active pharmaceutical ingredient development, formulations, stem cells, and imaging techniques. Of particular interest to early stage discovery is the new priority, "Beyond high throughput screening—pharmacological interactions at the molecular level." Although the content of an associated call has yet to be fully elucidated, there is some early indication that this topic could lead to the creation of a centralized European compound collection and associated screening center. Proposals also include a chemical library drawn from a group of pharma companies (e.g., six companies each contributing 50 k compounds). This would then be augmented by a similarly sized collection (200–300 k) of newly synthesized compounds, and screens would be run for targets derived from both academic and pharmaceutical companies. At first glance, initiating compound screening–related projects would appear to go against the spirit of the original IMI mission to work exclusively on precompetitive projects. However, it is likely that such a center would address the issues that current screening technologies are finding difficult to tackle, such as defining appropriate screening platforms to identify small molecule modulators of protein : protein interactions or identifying novel targets related to protein ubiquitantion and assessing their intrinsic "druggability."

The IMI initiative has not been without criticism from certain sectors of the academic community, particularly in relation to a perceived imbalance in the favor of the industrial partners in the eventual ownership of intellectual property derived during IMI projects [24]. As the program has developed though, it is clear from the degree of involvement from academics across Europe that most of the initial misgivings have been allayed and open questions around the IP accounted for. The current position is summarized in the statement "In the course of an IMI project, partners who generate results are the owners of corresponding information and IP rights" while any IP that is bought into a project by a partner remains under their exclusive ownership. This represents a position on IP that is similar to the bulk of European-funded public–private cooperations. Dealing with IP issues from the outset is critical to the success of IMI, as with any other public private partnership [25].

FUTURE EUROPEAN DRUG DISCOVERY RESEARCH INFRASTRUCTURES

Private sector drug discovery requires extensive infrastructures, such as compound libraries, screening facilities, protein and chemical analytical instrumentation, biobanks, *in-vivo* facilities, and imaging facilities. Although similar capabilities are required in academia and the public sector, very few publically funded organizations possess the deep pockets required either to initiate such facilities or cover ongoing running costs. The European Strategy Forum on Research Infrastructures (ESFRI) plays an important coordinating role in the development of large-scale research infrastructures within Europe [26]. The mission of ESFRI is primarily to support policy making and facilitate international cooperation with respect to existing infrastructure such as synchrotrons, biobanks, libraries, and telescopes. The Forum also provides preparatory funding for the development of new infrastructures to address Europe's future scientific and economic needs. Within the Biological and Medical Sciences field, there are a number of planned Research Infrastructures that may impact upon European academic drug discovery either directly or indirectly toward the end of this decade [27].

EUROPEAN ADVANCED TRANSLATIONAL RESEARCH INFRASTRUCTURE IN MEDICINE

The most advanced at the point of writing is the European Advanced Translational Research Infrastructure in Medicine (EATRIS). The aim of EATRIS will be to provide academic and industrial partners access to high-technology infrastructure, services, and training to progress projects from the discovery through the clinical stages [28]. In common with similar initiatives planned in the United States (e.g., National Center for Advancing Translational Sciences), projects entering EATRIS will be selected on the basis of their innovativeness, clinical relevance, and capacity to impact medical needs. The anticipated EATRIS Centers will address specific indication areas, including cancer, infectious diseases, neurological diseases, cardiovascular diseases, and metabolic diseases. Multiple infrastructural elements are envisioned within EATRIS to support all the main therapeutic mechanisms, including cell and gene therapies, small molecule drug discovery, vaccine development, molecular imaging (SPECT and PET), and biomarker discovery. In the area of small molecules the intention is for EATRIS to establish lead discovery and optimization facilities to identify small molecules for specific targets and then progress through a classical drug discovery process, including medicinal chemistry optimization, biophysical characterization, and structural determination. The timescale for EATRIS to become operational is strongly dependent on obtaining governmental support from member states to finance the build and operational phases of the project. Current estimates show initial operations commencing in 2012 with full implementation in 2013 onwards.

EU-OPENSCREEN

A second European-wide infrastructure currently being developed under the auspices of ESFRI is EU-OPENSCREEN. This broadly constituted group includes academic institutions in 13 countries as full members and institutions in 8 further countries as associate members [29]. The EU-OPENSCREEN concept is in the preparatory phase at the time of writing, and it is intended to focus on creating the scientific and business case to secure funding for the build and operational phases. Although the proposed model has still to be finalized, the intention is to create a network centered around existing academic screening facilities that would be supported by their parent institutions. Capacity would be added, in the form of upgrades to the existing centers or the construction of new screening facilities. One central part of the project is the development of common standards for the creation, deposition, and exchange of screening data generated within diverse screening centers. Consistency in data standards for compound biological data has been something of an ongoing issue for the pharmaceutical industry, and regularly rears its head in the aftermath of company mergers when efforts are put in place to integrate the various legacy compound activity databases [30].

A second area of EU-OPENSCREEN activity is related to the creation of a compound screening library, which would be made up of a combination of commercial compounds as well as proprietary compounds drawn from chemistry laboratories across

European academia. Traditionally, academically generated compounds would only have received limited exposure to bioassays at the host institution and normally only for a limited number of indications. The advantage of exposing academic compounds to a broad panel of targets across multiple indications is that previously unidentified activities would become elucidated [31]. Considerable efforts need to be made to create a sustainable IP framework to deal with results from such screens so that the interests of involved parties (originating chemist, disease biologist, and screener) are all accommodated. A significant part of EU-OPENSCREEN's ongoing work is to create the legal framework within which such results could be generated and then successfully exploited, with the minimum of administrative or legal overhead.

EUROPEAN PUBLIC DATA SOURCES FOR ACADEMIC DRUG DISCOVERY

The Chemical Group at European Molecular Biology Laboratory (EMBL) has been addressing the requirement to create databases of both publicly and privately originated bioactivity data. This ChEMBL project builds and makes publically available databases of drug discovery–related data sets covering the range from target discovery to clinical studies [32]. At the time of writing, the ChEMBL databases contain in excess of 5.6 million compound activities related to over 1 million distinct compounds and 8700 targets. Effective stewardship of compound activity databases requires rigorous quality control procedures that in practice are more difficult to implement if generation of primary data occurs within third-party organizations. The associated quality metric devised by ChEMBL is a Confidence Score, which assigns a value of 0 for un-curated data entries, and 9, where the compound : target interaction has been defined with a high degree of confidence. Within this continuum, primary screen–level data generated in biochemical format assays against defined targets receives a score of 4, which from this author's perspective appears to be a pragmatic assessment of the intrinsic value of data of this type. In addition to experimentally derived results, 2-D structures and calculated property information (cLogP, Lipinski parameters, etc.) are also available in the database. The core of the data collection is based upon manually extracted data sets from peer-reviewed journal sources, including the Journal of Medicinal Chemistry, and Journal of Natural Products. A powerful aspect of the ChEMBL database approach is that compound activity data originally drawn from an industrial organization (Inpharmatica/Biofocus) were also made publically available as part of the original database construction, following sponsorship from the Wellcome Trust [33]. A key differentiating factor between ChEMBL and other well-used public data sources such as PubChem or ChemBank is that the latter are mainly composed of primary screening results, as opposed to IC50 and EC50 values. In the future, this should allow the ChEMBL database to be used as part of virtual profiling of compounds where academics are able to mine the public data sources to identify possible selectivity or confirmatory bioactivities for compound hits emerging from their in-house screening projects. This concept is well established within private industry; for example, the Cerep Bio-Print® resource contains expertly curated *in-vitro* and *in-vivo* data covering some 2500 marketed drugs and standard compounds [34]. The BioPrint® *in-vitro* database contains results from 137 assays covering many of the classical targeted gene families of interest to pharma, including kinases, nuclear hormone receptors, and so on. There are several advantages to using such data sources as part of the decision making process; for example, profile data from novel hit compounds can be compared with the database in order to infer potential clinical effects that may be associated with a particular activity sub-profile; or if the target is available within the *in-vitro* panel, then compound activities can be used to identify possible pharmacaphores that could serve as starting points for the creation of new compounds with improved properties. It could be argued that commercial databases may suffer from two practical drawbacks with respect to their use by academic or public–private organizations. Firstly, the financial costs associated with gaining access to the database can be high and may well be beyond the typical "shoestring" budgets associated with academic screening projects. Secondly, the classes of targets and mechanisms that are of interest to many academic investigators, for example those involving modulating protein : protein interactions, epigenetic status, or transcription factor activity, are not as well represented by the "classical" targets contained in databases such as BioPrint®. This second point is also recognized by pharma companies who are expanding their screening activities into these less well defined target classes, for example ubiquitination and sumoylation related targets currently represent a current "hot topic" in pharma, much as HDACs (histone deacetylases) and kinases were in previous decades [35]. Public databases such as the ChEMBL resource will continue to develop in scope, as more academic screening centers look to deposit their results into these repositories. As the "open innovation" trend continues, it seems likely that more commercial as well as public organizations will increasingly contribute to (and exploit) publicly available compound bioactivity data. The IMI public–private Open PHACTS consortium is also working in this topic [36]. In addition to storing data on efficacy, liability, absorption, distribution, metabolism, excretion, and toxicological properties of compounds, the Open PHACTS consortium also aims to create new software tools to interrogate the data sets, including a "Target Dossier" that will allow visualization of complex relationships, a "Chem-Bio Navigator" tool for analyzing medicinal chemistry concepts across multiple targets, and a "Polypharmacology" browser that will allow users to explore complex chemical biology concepts and the selectivity properties of drug-like molecules. This new focus on sharing key biological data between the public and private sector is extremely welcome, as it offers a chance for organizations to learn from the successes and failures of others.

THE BIOPHARMA INITIATIVE IN GERMANY: EXPLORING NEW MODELS FOR INDUSTRY–ACADEMIA PARTNERSHIPS

In 2008, the German government initiated the Biopharma Initiative as a concerted response to the perception that Germany was no longer in a preeminent position as the "pharmacy of the world," a role that its pharmaceutical industry had arguably occupied for the majority of the previous three decades [37]. The strategic aim of the initiative was to create new funding and organizational structures to promote interactions between academia and industry in a way that would allow the best ideas from academia to be combined with the strength of pharmaceutical industry partners. After a review process involving an international panel, three separate consortia were selected to share €60 million of government funding over three years, with an additional €40 million earmarked for a further two years, depending on the progress of the individual consortia. This expectation was that government funding would be matched by private funding from industry and the investment community. The three successful consortia were: the Max Planck Lead Discovery and Development Centre; Neuroallianz, based around the University of Bonn; and New Drugs for Neurological Diseases (Neu^2). In addition to monitoring the delivery of individual project portfolios, each consortium has a different stated operational model, which should give the German government an opportunity to identify how organizational structures correlate with successful outcomes. The model for Neuroallianz, which is focused on neurodegenerative diseases, maintains the equality of academic and industrial partners in all stages of development. This group has a strong component of training and skills development, with academics learning drug discovery techniques from their industry counterparts. The Max Planck Lead Discovery Centre does not have a particular indication focus, but instead has created an infrastructure in which riskier early-stage academic sourced projects can be partnered with private or industry funding to progress them rapidly to the point where they are marketed to pharma partners. The Neu^2 consortium, with which the European ScreeningPort (ESP) is associated, operates as competence consortium with academic, SME, and industrial partners having the capability and capacity to address all stages of the value chain from target identification to marketing. It has an acute focus on multiple sclerosis as an indication and is looking to identify treatments that supersede the efficacy and safety profile of marketed treatments, rather than creating me-too versions of, for example, existing anti-inflammatory drugs. A typical project in this competence network would have a target selected on the basis of the work of a clinician or academic; assay development and screening performed by the ESP; lead optimization, structural studies, and medicinal chemistry support by consortium members such as Evotec; preclinical studies at a pharmaceutical industry partner; and clinical studies at a university medical center; followed by commercialization and marketing by the pharmaceutical partner. All the competencies to deliver these activities are held within the consortium. The Neu^2 consortium is open, so that new members can join and existing leave, a situation which creates a dynamic environment. Selection of new partners and projects for funding occurs with the aim of creating a balanced portfolio of early-, mid-, and late-stage opportunities. After less than 3 years of full operation, it remains too early to properly gage the success of any of the BioPharma Initiative consortia, however, it is certain that one key criterion will be the rate of progression of projects into the clinic and new treatments into the hands of doctors and patients.

EUROPEAN SCREENINGPORT

The European ScreeningPort is an example of an active public–private partnership that provides academic organizations with access to drug discovery services. As such it can provide useful examples of the challenges involved in both setting up such an infrastructure and operating within a network of academic collaborators.

The ESP was initiated in 2008 with funding from Germany's Science and Education Ministry (BMBF), the City of Hamburg, and Evotec GmbH, a mid-sized biopharmaceutical company with sites in Germany, U.K., India and the United States. The ESP's intention is to progress academic projects through to so-called "value inflection points." For example an academic who has been investigating the biological function of a novel target for several years may wish to establish the amenability of using a small molecule approach to modulate the function of the target (druggability determination) in order to then secure public or private funding for large-scale high-throughput screening. In another case, an academic may have run an in-house screen that identified several attractive hit series, which require selectivity and liability profiling to create a data set to precipitate venture capital funding of a spin-out company. Projects representing these and similar scenarios appear on a weekly basis at the door of the European ScreeningPort. The question of which organizational operating model would be best suited to answering these diverse needs was examined exhaustively during the ESP's initiation, a process that would occur at the outset of any similar enterprise. An initially attractive proposal was to operate the ESP as a virtual company, with a skeleton staff of expert project managers who would create project plans in consultation with academic customers and employ a network of third-party suppliers of the "physical" services in order to deliver project content. Analogous models are used successfully in a number of public, private, and non-profit organizations that may be focused on a single area of unmet medical need (e.g., the Medicines for Malaria Venture [38] and Drugs for Neglected Diseases Initiative [39]), or those with an acute disease focus (such as CHDI on Huntington's disease [40]). However, in the case of the European ScreeningPort, after discussion with potential academic clients, it became clear that being able to provide physical access to dedicated screening infrastructure was a critical requirement. Three main arguments were identified. Firstly, full quality control of the screening outputs is only possible when there is direct operational oversight of the experimental process. Secondly, providing constant justification to the academic client of the precise value produced at each point in the project management "virtual" process is a time-consuming and essentially non-productive activity. Thirdly, and probably most importantly, "virtual" project management models do not allow for very small projects to be easily initiated, as the suppliers of the physical contract research services typically do not wish to get involved in risk-sharing exercises and prefer a direct transactional model. This third point represents a barrier that would limit activities such as the creation of data sets to support grant proposals on what is essentially a "pro-bono" basis. The consequence would be that only "funded" projects could be easily progressed. As anybody who operates in the academic screening environment is well aware, it is far more challenging to argue that large-scale funding for an academic drug discovery project is justifiable if the applicant lacks any type of initial screening data confirming the potential druggability of their target. The role of the ESP in assisting in the procurement of funding for its customers through the generation of supporting data sets from pilot screens is critical to the success of its business model as a public–private partnership. This service is performed under the banner of the "Visiting Scientist Scheme," whereby academics can embed a pre- or postdoctoral researcher within the ESP for 2–3 weeks, during which they can develop assays under the guidance of ESP scientists and screen small collections of compounds against their target before integrating the data set into the next round of grant proposals. Typically reagents are supplied by the visiting scientists, with ESP providing access to chemical libraries and screening hardware. It is sometimes tempting, following the pilot screen of some 5000 compounds, for academic investigators to over-interpret the results of a screen and look to progress the putative hit series into lead seeking. However, a strong steer is always given that the best course of action is to screen a larger library in order to identify more hit series to increase the chances of success in the long run.

Assays entering the European ScreeningPort are normally partially developed at the academic laboratory before transfer to the ESP. In situations where the academic labs are using a broad range of possible assay and consumable formats, there is a risk that the transfer process is extended while instrument alignments and reader settings are re-optimized. The solution at the ESP is to drive appropriate standardization of the academic assays being developed by partners. This is achieved by a combination of education (e.g., ESP scientists give presentations at events such as EMBL Master Courses in Chemical Biology and Screening) and providing "free" samples of plates to academic customers at the outset of a project so that when an assay is transferred to the ESP automated platform, the amount of re-teaching of robot handling and re-optimization of reader settings is minimized.

The design of the physical screening infrastructure at the European ScreeningPort was optimized in order to maximize the efficient use of equipment. The system is centered on four standalone Cell Explorer modules (PerkinElmer), which are linked via a conveyer belt to allow transfer of plates. In terms of the screening formats, the system can handle 96-, 385-, and 1536-well plates. In common with many other medium-sized screening organizations there is a preference for low-volume 384-well assays as a default, as the effort involved in adapting assays to 1536-well format is disproportionate to the benefit, when the maximum screening library size is typically 250,000 compounds.

The ESP compound reformatting module is built around an Echo 550 acoustic dispensing system (Labcyte) and includes two temperature-controlled plate stores (1000 position and 500 positions, Liconic), a plate centrifuge (Agilent), and a Multidrop Micro liquid dispensing system (Thermo Fisher). Compound reformatting operations for primary screening, cherry picking, and dose response creation can take place within the module. Applications requiring larger volumes (e.g., preparation of daughter plates) use conventional tip-based systems (PlateMate+ - Matrix) run offline in batch mode. For low volumes ($<0.5\ \mu L$) the automated compound handling system has a high degree of performance in terms of compound transfer speed, efficiency, and throughput. However, all of these capabilities and capacities have required a relatively high capital investment by the ESP, and the reality is that few if any academic groups have access to funding that would allow similar arrangements in their own laboratory. The unintended negative consequence, therefore, of having expensive infrastructure at the ESP with no counterpart in academia is to complicate the process of transferring protocols both to the ESP at the outset of a project and back to the academic lab at project conclusion. Addressing such technical issues comes to the heart of the challenges involved in developing a centralized screening facility that can efficiently serve the needs of a dispersed community of academic groups. Although it is not practical to provide all academic partners with "loan" instruments so that they can develop lower volume assays in house, a number of Multidrop Micros have been purchased by ESP and provided to academics on a temporary basis to accelerate the process of assay development. In our experience, the most significant bottleneck in screening projects is not the time taken to perform the physical screen itself, but the effort involved in transferring assays from academic sites to the ESP and then adapting the assay to an automated format. Any advantage that can be gained by exposing the academic assay development team to the technical formalities and underlying constraints of high-throughput screening experiments is an advantage. Of course, this process should not be taken too far so that biological relevance of assays is compromised by a simple focus on performance metrics such as throughput or Z'.

The liquid handing and reader module on the ESP screening system contains the following components: Envision plate reader (Perkin Elmer), Biotek plate washer (Biotek), and a Flex-drop dispensing system. Unlike with the Echo system, in our experience many of the potential academic customers have access to fluorescent microplate readers that are capable of several common optical readouts (fluorescence intensity, fluorescence polarization, and luminescence). The ESP instrument also has time-resolved fluorescence resonance energy transfer (trFRET) and AlphaScreen (Perkin Elmer) readout modes, which are less commonly found in academic labs due the specialized nature of the detection system and costs involved for both purchasing hardware and covering ongoing reagent costs. This can be a disadvantage for projects, as these proximity-based readout modes represent powerful approaches for monitoring protein : protein interactions, which are increasingly emerging as important drug target mechanisms. The consequence is that the bulk of assay development for such projects must typically occur within the ESP laboratories, and the opportunity to perform follow-up studies at the academic site can also be restricted. A similar situation has arisen for other specialized readout technologies at the ESP, for example the MesoScaleDiscovery Sector 6000, an electro-chemi-luminescent reader used by the Biomarker group of the ESP. The MesoScaleDiscovery instrument has a relatively low penetration into the academic market, possibly due to the high costs involved and the preference for common alternatives like ELISA. Classical approaches such as ELISA are understandably more familiar to the user, but are also intrinsically less sensitive, unsuitable for multiplexing, and cannot be easily applied to assays involving complex biological matrices such as serum or cerebrospinal fluid. It seems likely that this incompatibility between the technologies used at centralized screening sites and the equipment in use at academic providers of assays is not just an issue for the European ScreeningPort, but also for other similar organizations. The reverse situation has also been observed for assays based on tracking fluorescent proteins. Many academic laboratories have developed sensitive live cell imaging assays or biochemical FRET-based assays founded on cell lines, plasmids, and similar reagents that are regarded (however mistakenly) in a sense as "common property" among the academic community. What is not appreciated within some academic groups is that when a compound screening project is initiated with the explicit intention of identifying starting points for a drug discovery project (as opposed to a chemical biology basic research study), then a licensing agreement must be set up with the owners of the IP. In our experience, the license costs associated with fluorescent proteins can be considerable and in many cases are not supportable under the budget. The consequence is that other readout approaches must be recommended to academic partners even when they already possess a viable assay format built around fluorescent proteins. Suitable alternatives include luciferase-based reporters or immunocytochemistry (ICC) coupled to fluorescence imaging using an automated microscope. A disadvantage is that both of these alternatives preclude the use of live cell imaging and therefore the ability to track potential relevant temporal changes in assay signal with time. However luciferase readouts are highly sensitive, can be less prone to optical interference by compounds, and involve short read times. Similarly, the signals from ICC treated cells remain stable for several weeks, which enables endpoint data to be gathered separately from the primary assay. Automated cell washing and fixation is essential when performing large-scale ICC assays on multiple plates. At the ESP, this has been enabled using an adapted Biotek (Biotek) plate washer capable of switching between four reagent reservoirs, to allow fixation, washing, and additions to occur within a single operational sequence. Additional ventilation has also been installed for this instrument to comply with safety requirements when using formaldehyde and other hazardous materials during fixation.

The 3rd Cell Explorer module on the ESP screening system caters for incubation of cell plates under various temperature and humidity settings. The module contains three cell incubators (Liconic), with a combined capacity of 300 plates. An additional Multidrop micro (Thermo-Fisher) is normally used for plating of cells onto compound spotted plates. In general, cell-based assays constitute 50% of the total portfolio at the ESP, which mirrors the current norms in industry. Many of the cell-based assays are phenotypic in nature, relying on reporter gene readouts such as luciferase.

The 4th module on the ESP screening system is designed to support hit profiling studies and high content screening (HCS) applications. The module is built around a fully equipped Opera (Perkin Elmer) automated confocal microscope system and includes the Janus 4-channel Varispan and a 96/385/1536 compatible Janus MDT liquid handler, both from PerkinElmer. Within academia there is increasing interest in developing HCS assays, albeit with the caveats outlined above covering the use of fluorescent proteins and license issues. The typical design of a project critical path at the ESP would use a bulk optical readout for the primary screen in a biochemical, (e.g., 250,000 compounds) followed by hit confirmation in the same assay format for up to 1% of the compounds. Secondary HCS assays during hit profiling would then be employed on the top 500 compounds in 11-point dose response in triplicate corresponding to some 80×384 plates (including controls and counter assays). Even with a very rapid instrument such as the Opera, plate reading can take up to 1 hour, depending on the number of image fields required per plate to reach the appropriate assay statistical quality threshold. In our opinion, this precludes the use of HCS for primary screening, except in the case of small library collections (<10 k).

In conclusion, we believe that the modular design of the ESP screening system satisfies our requirements when working with an extended network of academic customers. The design allows several activities involving multiple projects in parallel, such as primary screening, compound reformatting, and hit profiling. One possible drawback from a broader compatibility, rather than logistics, perspective is that the more advanced instrumentation on the system (e.g., acoustic dispenser or alphascreen detection) cannot be found in typical academic labs which serves to create extra workloads at the ESP site as assays require additional input during development and adaptation.

CONCLUSION

Is this chapter, the role of partnerships between academia and industry in the area of drug discovery and development has been explored. The mechanism by which partnerships are enabled are diverse, ranging from the large-scale IMI consortia with budgets of tens of millions of Euro and close involvement a large sections of the pharma industry, down to small scale druggability trials where an academic with little or no funding turns up at the door of a screening center looking to create a data set that will drive forward a grant application to support a large-scale screen. Looking to the future, when considering some of the planned activities in Europe, for example EU-OPENSCREEN or EATRIS, it appears that Europe is re-capitulating some of the large scale initiatives in the United States such as the NIH Molecular Libraries Initiative network or the U.S. Translational Research Centers networks. However, it is clear that the proponents of these new planned European infrastructures are using the experience of others to formulate new working models that should build and improve on the useful aspects while avoiding most of the major pitfalls. The IMI initiative as a model for public–private cooperation in life sciences is especially interesting, as it has no obvious counterpart in the U.S. or Asia and includes a very strong industry component. Again it is too early to determine how successful IMI will be, but early indications based on the number of applicants from academia support the attractiveness of the model as well as the desire of academics to engage with industry to reach solutions to their common challenges.

REFERENCES

1. EFPIA. The Pharmaceutical Industry in Figures (2011) EFPIA publication. 2013. Available at http://www.efpia.eu/facts-figures. Accessed January 27, 2014.
2. Pammolli, F., Magazzini, L., Riccaboni, M. The productivity crisis in pharmaceutical R&D. *Nat Rev Drug Discov*. 2011;10:428–438.
3. Munos, B. Lessons from 60 years of pharmaceutical innovation. *Nat Rev Drug Discov*. 2009;8:959–968.
4. Paul, S.M., Mytelka, D.S., Dunwiddie, C.T., Charles, C., Persinger, C.C., Munos, B.H., Stacy, R., Lindborg, S.R., Schacht, A.L. How to improve R&D productivity: the pharmaceutical industry's grand challenge. *Nat Rev Drug Discov*. 2010;9:203–214.
5. Clark, D.E., Newton, C.G. Outsourcing lead optimisation—the quiet revolution. *Drug Discov Today* 2004;9(11):492–500.

6. Tralau-Stewart, C.J., Wyatt, C.A., Kleyn, D.E., Ayad, A. Drug discovery: new models for industry–academic partnerships. *Drug Discov Today* 2009;14:95–101.
7. Hunter, J., Stephens, S. Is open innovation the way forward for big pharma? *Nat Rev Drug Discov*. 2010;9:87–88.
8. Inglese, J., Johnson, R.L., Simeonov, A., Xia, M., Zheng, W., Austin, C.P., Auld, D.S. High-throughput screening assays for the identification of chemical probes. *Nat Chem Biol.* 2007;3:466–479.
9. Lazo, J.S., Brady, L.S., Dingledine, R. Building a pharmacological lexicon: small molecule discovery in academia. *Mol Pharmacol*. 2007;72:1–7.
10. Frearson, J.A., Collie, I.T. HTS and hit finding in academia—from chemical genomics to drug discovery. *Drug Discov Today* 2009;14:1150–1158.
11. Moran, N. Public sector seeks to bridge "valley of death". *Nat Biotechnol*. 2007;25: 266.
12. Hughes, B.A. Wellcome experiment in seeding drug discovery. *Nat Rev Drug Discov*. 2010;9(3):178–180.
13. CRUK. Cancer Research UK Research Strategy Report 2011. 2012. Available at http://www.cancerresearchuk.org/prod_consump/groups/cr_common/@fre/@gen/documents/generalcontent/res-strategy-progress-2011-12.pdf. Accessed January 27, 2014.
14. Clark, D.E. Outsourcing lead optimization: the eye of the storm. *Drug Discov Today* 2010;16:147–157.
15. Wellcome Trust. Wellcome Trust Seeding Drug Discovery program information. 2013. Available at http://www.wellcome.ac.uk/Funding/Technology-transfer/Awards/Seeding-Drug-Discovery/index.htm. Accessed January 27, 2014.
16. Nelson, S.L. Academic HTS: diverse portraits. *Drug Discov Today Technol*. 2008;5: e29–e33.
17. Lisurek, M., Rupp, B., Wichard, J., Neuenschwander, M., von Kries, J.P., Frank, R., Rademann, J., Kühne, R. Design of chemical libraries with potentially bioactive molecules applying a maximum common substructure concept. *Mol Divers*. 2010;14:401–408.
18. Kamel, N., Compton, C., Middelveld, R., Higenbottam, T., Dahlén, S.E. The Innovative Medicines Initiative (IMI): a new opportunity for scientific collaboration between academia and industry at the European level. *Eur Respir J*. 2008;31:924–926.
19. Goldman, M. Reflections on the innovative medicines initiative. *Nat Rev Drug Discov*. 2011;10:321–322.
20. IMI NewMeds Consortium. IMI NewMeds Consortium program information. 2014. Available at http://www.newmeds-europe.com/en/news.php?hs=id3&display=none. Accessed January 27, 2014.
21. Matheis, K., Laurie, D., Andriamandroso, C., Arber, N., Badimon, L., Benain, X., Bendjama, K., Clavier, I., Colman, P., Firat, H., Goepfert, J., Hall, S., Joos, T., Kraus, S., Kretschmer, A., Merz, M., Padro, T., Planatscher, H., Rossi, A., Schneiderhan-Marra, N., Schuppe-Koistinen, I., Thomann, P., Vidal, J.-M., Molac, B. A generic operational strategy to qualify translational safety biomarkers. *Drug Discov Today* 2011;6:600–608.
22. Bjerrum, O.J. Excellence in education and training advances competitiveness of the pharmaceutical industry in Europe. *Eur J Pharm Sci*. 2011;14:174–175.
23. IMI. IMI revised strategic agenda document. 2011. Available at http://www.imi.europa.eu/sites/default/files/uploads/documents/SRArevised2011.pdf. Accessed January 27, 2014.
24. Sinha, G. Spat over IMI funding and intellectual property. *Nat Biotechnol*. 2011;29:473.

25. De Rijck, K., Goldman, M. Clarifying knowledge ownership in Europe's medicines initiative. *Nature* 2010;466:1040–1041.
26. Pero, H. Research infrastructures of pan-European interest: the EU and global issues. *Nucl Instrum Methods Phys Res A* 2011;626:S69–S71.
27. van Dyck, L. A new global strategy for the life sciences in Europe. *EMBO Rep.* 2005;6(3):194–197.
28. Becker, R., van Dongen, G.A.M.S. EATRIS, a vision for translational research in Europe. *J Cardiovasc Transl Res.* 2011;4:231–237.
29. EU-OPENSCREEN. EU-OPENSCREEN flyer. 2011. Available at http://www.eu-openscreen.de/fileadmin/user_upload/PowerPoint/EU-OPENSCREEN_flyer_for_website.pdf. Accessed January 27, 2014.
30. Gaulton, A., Overington, J.P. Role of open chemical data in aiding drug discovery and design. *Future Med Chem.* 2010;2(6):903–907.
31. Hibert, M.F. French/European academic compound library initiative. *Drug Discov Today* 2009;14:723–725.
32. Gaulton, A., Bellis, L., Chambers, J., Davies, M., Hersey, A., Light, Y., McGlinchey, S., Akhtar, R., Atkinson, F., Bento, A.P., Al-Lazikani, B., Michalovich, D., Overington, J.P. ChEMBL: a large-scale bioactivity database for chemical biology and drug discovery. *Nucleic Acids Res.* 2012;40:D1100–D1107.
33. Houlton, S. Wellcome boost for open-access chemistry. *Nat Rev Drug Discov.* 2008;7: 789–790.
34. Price, D.A., Blagg, J., Jones, L., Greene, N., Wager, T. Physicochemical drug properties associated with in vivo toxicological outcomes: a review. *Expert Opin Drug Metab Toxicol.* 2009;5(8):921–931.
35. Grzegorz, N., Rolfe, M., Harper, J.W. Drug discovery in the ubiquitin—proteasome system. *Nat Rev Drug Discov.* 2006;5:596–613.
36. OpenPHACTS Consortium. Information on the Open PHACTS consortium. 2014. Available at http://www.openphacts.org/about-open-phacts/about-open-phacts. Accessed January 27, 2014.
37. BMBF. Information on the BioPharma initiative. 2014. Available at http://www.bmbf.de/en/10540.php. Accessed January 27, 2014.
38. Nwaka, S. Drug discovery and beyond: the role of public-private partnerships in improving access to new malaria medicines. *Trans R Soc Trop Med Hyg.* 2005;99:20–29.
39. Ioset, J.R., Chang, S. Drugs for Neglected Diseases initiative model of drug development for neglected diseases: current status and future challenges. *Future Med Chem.* 2011;3(11): 1361–1371.
40. CHDI Foundation. Information on the CHDI initiative. 2014. Available at http://chdifoundation.org/. Accessed January 27, 2014.

30

IN SICKNESS AND IN HEALTH: A SHOTGUN MARRIAGE THAT IS FLOURISHING

Horst Flotow and Alex Matter

Experimental Therapeutics Centre, Singapore

In the last decades, the overall cost of bringing a new drug to the market has ballooned to an estimated $800 million to $1.5 billion. Clearly, even for large pharmaceutical industry companies ("big pharma") with deep pockets, this is a high-stakes gamble, particularly when one considers that, even if everything goes according to plan, there is a lead time of 10–12 years for new drugs to get to the market. Combine the long lead time with a high rate of attrition (on average only one in ten drug candidates entering clinical trials even makes it to the market) and add in that only one in three marketed drugs recovers this expenditure, the picture emerging is not particularly encouraging. At the same time, the productivity of big pharma has stayed flat, with few medicines with novel mechanisms of action being registered, among a flurry of "me-too" drugs (i.e., new drugs with mechanisms of action or stated indications that are substantially the same as existing medications). Overall then, costs have been soaring to unsustainable levels, productivity has stayed flat and a serious innovation and productivity gap has opened up [1, 2]. While big pharma has been counting more and more on biotechnology companies ("biotech") to create innovation and fill this void, that approach has not been completely successful either, for a number of reasons. Although biotech may be more innovative and nimble to exploit approaches big pharma cannot, biotech has a shorter time in which to do so and is (usually) strapped for funds.

Collaborative Innovation in Drug Discovery: Strategies for Public and Private Partnerships, First Edition.
Edited by Rathnam Chaguturu.

One alternative approach that has emerged more recently is the formation of public–private collaborative partnerships for strategically advancing drug discovery. Traditionally, public-sector research has concentrated on basic research that elucidates signaling pathways and mechanisms of disease and culminates in the identification of potential drug targets. These potentially druggable targets and pathways were then picked up by big pharma and further developed in in-house drug discovery programs (applied research), culminating in a marketed drug (in those rare successful projects). It is becoming clear that this is an overly simplistic drug discovery model and that the contributions of the academic and industrial research extend well beyond their traditional realms [3–7]. In a recent analysis of the contributions of public sector research to drug discovery, Stevens et al. have shown that public sector research has made substantial contributions to drug discovery [8]. We will discuss how drug discovery as a whole can benefit from expanding and streamlining public–private partnerships (PPPs) as an avenue to improve productivity and the innovation flow, and restock the pipelines with novel medicines.

PUBLIC–PRIVATE PARTNERSHIPS

The Early Days: Insulin

PPPs are nothing new. Perhaps the earliest example of this is the discovery of insulin as a treatment for diabetes mellitus at the University of Toronto in the 1920s, which is also an early example of translational research. The discoverers subsequently collaborated with Eli Lilly and Company, which refined the isolation and production of insulin to a commercial product. Several incremental improvements, mainly within industry, were made to improve isolation procedures, stability, and so on, and insulin became one of the first wonder drugs [9, 10]. The early 1970s witnessed the advent of genetic engineering and recombinant DNA technology. It wasn't long before this technology made the leap from the academic laboratories where it was invented and was applied in industry. One of its first commercial applications was in the manufacture of human insulin (Humulin), which thus became the first recombinant drug on the market [11–13]. The move from isolating insulin from an animal source to producing it recombinantly in bacteria is clearly a paradigm shift in drug manufacturing rooted in academic research and subsequently applied in industry.

Fundamental Characteristics

PPPs have not always provided a smooth and untroubled transition from basic discovery to industrial application. There have been numerous legal (and scientific) battles between academia and industry, one of the more well known being the legal wrangles over the rights to human growth hormone between Genentech and Stanford University (see [14] for a summary). In fact, academia and industry have some fundamental characteristics that lead to rather different priorities and expectations of the outcome of academic-industrial collaboration (Table 30.1).

TABLE 30.1. Fundamental Characteristics and Strengths of Academia and Industry

Public (Academia)	Private (Industry)
"Upstream basic research"	"Downstream applied research"
Driven by thirst for knowledge, not so much milestones and timelines	Driven by economic imperatives (customers and shareholders) and meeting milestones and timelines
Competition for funding, recognition, etc	Competition for customers/sales
In-depth knowledge in biology, chemistry, etc.; Mostly specialized and more interested in a journey of discovery, for example, know more and more about less and less	Focus on the end result in the broader context of profitability, not on the journey
Publish or perish	Maintain competitive advantage, intellectual property portfolio, and secrecy
Funded by the public	Funded by investors and shareholders, fueled by high margin products
Preclinical and clinical basic research	Preclinical applied research, clinical development expertise

The fundamental difference revolves around money—traditionally the principal source in most partnership conflicts, and PPPs for drug discovery are no different. The major unmet medical needs of many developed Western and Asian countries that drive industrial drug discovery research and development are prioritized based on commercial incentives. In the public sector, on the other hand, basic drug discovery research is prioritized in many ways, depending on the interests of the investigators, funding sources, and so on, and generally less on whether there is any commercial interest in the disease [15, 16].

Nevertheless, while at first glance these differences appear to be two sides of the same coin, it can also be argued that opposites attract. A (well-managed) PPP where the two partners respect and complement each other is at the core of many successful drug discovery projects. We will discuss a number of examples of academic-industrial collaborations and the lessons learned from them.

Example 1: CGP57148/STI571/Imatinib/Glivec/Gleevec for the Treatment of Chronic Myelogenous Leukemia (CML)

CML is a myeloproliferative disorder with a well-characterized molecular etiology. The disease progresses through three distinct phases: chronic, accelerated, and blast phase. CML accounts for around 15% of all adult leukemias and is characterized by the presence of the so-called Philadelphia chromosome. The actual molecular defect is a reciprocal translocation and fusion of the break-point cluster region (BCR) gene on chromosome 22 and the Abelson (ABL) tyrosine kinase on chromosome 9. The resultant BCR-ABL fusion protein mediates the development and maintenance of CML

[17–21]. So, CML is a disease with a well-characterized molecular etiology that was still being treated with conventional chemotherapy.

It was also recognized in the mid-1980s that mutated or deregulated protein tyrosine kinases were oncogenes and therefore potential targets for therapeutic intervention in oncology. As a result, Ciba-Geigy had initiated medicinal chemistry programs to make protein kinase inhibitors in a number of therapeutic areas, including cancer. Started in the mid-1980s, this is the program that resulted in the synthesis of CGP57148, now known as Glivec (or Gleevec), in 1992 [22]. In parallel, clinical researchers in academia had started to develop BCR-ABL–driven model systems for the study of the disease biology of CML. The critical aspect of these models was that they enabled the testing of compounds in a way that was ultimately predictive of clinical outcomes [23]. Fortuitously, the academic (clinical) and industrial scientists (Druker and Lydon) contacted each other to begin a collaboration that resulted in the iterative cycles of innovation between academic clinical and industry scientists. This resulted ultimately (hundreds of people involved!) in the registration of Glivec for the treatment of CML in 2001, and the era of targeted therapies was born. Further innovative clinical research with Glivec, including the discovery of novel imaging biomarkers for CML and the stepwise expansion of indication range, followed and is still ongoing. In 2008, Glivec was registered for numerous additional indications, including c-Kit positive gastrointestinal stromal tumor, acute myeloid leukemias, and Philadelphia chromosome–positive acute lymphocytic leukemia. In addition, Glivec was found to be effective in inhibiting the ABL kinase-driven pathogenesis of pox viruses; platelet-derived growth factor receptor–driven hypereosinophilic syndrome and chronic myelomonocytic leukemia; Langerhans cell histiocytosis, chordoma and dermatofibrosarcoma protuberans; as well as some novel non-cancer indications such as stroke, asthma, and fibrosis [24–48].

Glivec: Challenges and Lessons Learned

Some of the fundamental challenges (common to all drug discovery projects) that were encountered and overcome included

- The development of an *in vitro* assay that is representative of the *in vivo* conditions and allows the determination of the extent of modulation or inhibition of the selected target by the drug that is sufficient to elicit the desired effect (e.g., cell cycle arrest, cell death)
- A good understanding of the role of the target in the pathophysiology of the disease
- The molecular epidemiology of the target in representative sample sizes and the potential for testing hypotheses in clinical studies; for example, by marker analysis.

As a first-in-class, revolutionary treatment, Glivec was viewed as more risky than a me-too drug going down a well-trodden path (of least resistance). The scientific discoveries revealed a number of important lessons for drug discovery and how to carry out successful academic-industrial collaborations. Although Glivec was a revolutionary concept designed to show proof-of-concept for targeted anti-cancer therapy, it became a breakthrough (and eventually a mega-blockbuster) drug. Although a welcome surprise, this was all entirely unforeseen by scientists, clinicians, and marketing professionals alike. The key was that Glivec changed the paradigm of the disease by turning CML from a fatal disease into a chronic disease requiring life-long treatment (thankfully, the side effects experienced by most patients allowed this). Clearly, the fact that Glivec provides important (life-saving) patient benefit is extremely valuable—driving down regulatory review times and enhancing access to clinicians, payors, and patients. As pointed out by Druker and by Lydon, the collaboration would not have gotten off the ground if there had been protracted negotiations over the research collaboration agreements (e.g., over royalties); instead the scientists were allowed to get on with generating the data [49, 50].

Biologically, it was a surprise to see that BCR-ABL inhibition led to selective death of the leukemic cells. This was probably the first example illustrating the concept of pathway addiction, where one particular pathway is responsible for the survival of the cells and its inhibition will result in cell death via apoptosis. Ironically, the emergence of resistance is the best validation for targeted anticancer therapy, spurring the discovery of new compounds able to overcome this.

The conventional wisdom that kinase inhibitors cannot be selective was shown to be partially true, but also worked to the researchers' advantage, allowing many indications to be captured. The molecular mechanism of action of Glivec provided another surprise. Kuryon at the Rockefeller University was able to show that Glivec freezes the inactive state of the enzyme and therefore is not directly competing with the ATP-binding to the active site as most protein kinase inhibitors do. The discovery of functional biomarkers (white blood cell, cytogenetic response, etc.) in academic labs and the clinics played a decisive role. Molecular epidemiology became the decisive foundation of modern clinical research in oncology and virology.

Glivec was the result of possibly one of the first examples of a truly synergistic (and on-going) partnership between academics, clinicians, patients, regulators, and industrial partners.

Example 2: CGP 73547/Atazanavir/Reyataz® for the Treatment of AIDS

Acquired Immune Deficiency Syndrome (AIDS) appeared on the public health radar in the early 1980s. AIDS was first noticed predominantly in gay and intravenous drug user communities and initially received lower priority as a "gay disease." However, by the middle of the eighties it became clear that this was a disease transmitted through the exchange of bodily fluids, including blood transfusions, and was in no way restricted to gay men and drug users [51, 52]. This finally prompted authorities to declare a national emergency and give AIDS the attention it needed, providing, particularly in the United States, the impetus to get the resources needed to investigate AIDS. An emergency program under the leadership of the National Cancer Institute (NCI) and the National Institute of Allergy and Infectious Diseases was launched, with a mobilization of funds and intellectual resources to allow free and immediate access to any tools that industry needed. As a first result of this massive effort, the FDA licensed molecular diagnostics for the detection of HIV in 1985, allowing blood banks to secure blood transfusions and laying the foundation of molecular epidemiology. Industry started working on drugs to treat AIDS, and in 1987, zidovudine, a reverse transcriptase inhibitor, became the first registered AIDS drug. The first protease inhibitor, Roche's saquinavir, was approved in 1995. As with most drugs against infectious diseases, resistance quickly developed, and a number of additional protease inhibitors with improved potency and spectrum of activity were developed. Combination treatments were developed, based on viral load measurements in the blood, that decisively improved the quality of life and chances of survival of seropositive patients [53–55].

Much of the speed with which the scientific community developed the diagnostics and treatments for AIDS is due to the partnership between academic and industrial partners, and this can be illustrated by the development of atazanavir today, a first line protease inhibitor for the treatment of AIDS. In the late eighties, Ciba-Geigy started work on HIV protease inhibitors, producing CGP 73547 in 1995. This was then licensed to Bristol-Myers Squibb in 1997 and registered as atazanavir in 2003. Although Ciba-Geigy was late out of the starting gate, research in many academic clinical centers showed atazanavir to be an effective once daily treatment equivalent in potency to efavirenz, the gold standard treatment at the time. Additional studies showed that resistance to atazanavir is due to an I50L mutation, reducing the likelihood of cross-resistance with any other protease inhibitors used in the clinic. Furthermore, atazanavir is the first HIV antiviral with lesser effects on lipid parameters, thus making it particularly useful as part of a combination treatment [56, 57].

Atazanavir: Challenges and Lessons Learned

Similar to CML, the discovery of effective combinations treatments transformed AIDS from an acutely fatal disease into a chronic disease, changing the fundamentals of the market and, again, this was entirely unforeseen at the time by most experts. Today, the median survival rate after diagnosis of AIDS has gone from an estimated 6 to 19 months (without treatment for HIV or any of the secondary infections), to today's life expectancy for a newly diagnosed HIV-infected person of about 20 years. This has had a clear effect on the bottom line of industry, in fact bringing some industrial players back into working on infectious diseases. Clearly, atazanavir would never have emerged in such a short time without the partnership between NCI, academics, clinicians, patients, regulators, and industrial partners. The vocal involvement of patient advocacy groups demanding treatments for a disease added an incredible impetus and kept progress in the public eye.

The national emergency of AIDS pushed academia and industry closer together, and reengaged industry in working on infectious diseases for the benefits of patients. It has also shown the dangers of complacency and neglecting diseases affecting those who are perceived as marginal in society. In the age of globalization, any infectious disease can and will travel around the globe in a relatively short time. The worldwide effort to diagnose and treat HIV infections and AIDS has brought this disease under control in highly industrialized countries; the situation is, however, far from satisfactory in developing countries, in particular sub-Saharan Africa [58].

Example 3: Coartem for the Treatment of Malaria

Malaria is a mosquito-borne infectious disease, endemic in much of the tropics and the southern hemisphere. Although 40% of the total world population is at risk, malaria is primarily an African tragedy. Of the 300–500 million clinical cases per year, 80% of them occur in Africa, along with more than 90% of the at least 1 million deaths per year (mostly in children). The economic burden to African countries is immense and has had a serious impact on the ability of the affected countries to improve their economies. Also, given the parasite's ability to acquire resistance to the currently used therapies (and the mosquito vectors acquiring insecticide resistance), there is a (seemingly) never-ending battle to find novel and effective treatments [59–63].

Artemisinin-based combination therapies are the only effective treatments in drug-resistant areas and are recommended by the World Health Organization (WHO) as first line therapy. Novartis' artemisinin-based combination therapy, Coartem, came out of a ground-breaking (in more ways than one) academic-industrial collaboration and represents the first fixed-dose artemisinin combination prequalified by the WHO and included on the WHO Model List of Essential Medicines in 2002 [64–69]. The active ingredient in Coartem, artemisinin, is a plant-derived product from sweet wormwood (Artemisia annua), which was used in an ancient Chinese medicine. Artemisinin was discovered in 1972 by Chinese scientists and used to treat malaria-stricken soldiers during the Vietnam War. In the 1980s and 1990s, semisynthetic derivatives of artemisinin were tested in combination with lumefantrine by the Academy of Military Medical Sciences in Beijing. A collaboration with Ciba-Geigy (later Novartis) brought this to market in 1994 and was the first collaboration of its kind in Chinese history. Following significant technology transfer from Ciba-Geigy to their Chinese partners, Coartem is now manufactured by Novartis as well as Chinese manufacturers according to Good Manufacturing Practices, with 70 million doses produced and distributed worldwide in 2008.

Coartem: Challenges and Lessons Learned

There was a true synergy between the Chinese inventors and the experts of CIBA-Geigy/Novartis. The Chinese inventors contributed their knowledge on novel active ingredients, while the Swiss scientists brought their know-how in drug development, manufacturing, formulation and clinical trial expertise to the table [67]. Importantly, the WHO, at many different levels, very actively supported this partnership. Furthermore, championship at the international level was found through Medicines for Malaria Venture (MMV) in Geneva, a non-governmental organization (NGO), with their global connections in developing countries.

MMV has contributed to a major extent to providing a pipeline of new treatment modalities for this disease, largely neglected because the commercial prospects of any medical intervention (diagnostics, therapy, prevention) will be modest, at best. PPPs can get around this major obstacle by securing funding that is independent of commercial prospects (the Bill and Melinda Gates Foundation supports MMV, funding from the United Nations flows to WHO, and major contributions from the Global Fund to fight AIDS, tuberculosis, and malaria underpin the complex manufacturing and logistics of Coartem).

Effective treatments for malaria not only contribute to relieving risk-exposed populations from suffering and death, but also improve the economic outlook for these countries. Such medicines are then one of the pillars on which developing countries can base their growth.

PUBLIC–PRIVATE PARTNERSHIPS IN SINGAPORE

Singapore has over the last 25 years built up impressive research capabilities in the engineering and biomedical sector, which in turn has attracted many big pharmaceutical companies to set up manufacturing facilities as well. The completion of Biopolis, a biomedical research hub, has encouraged several pharmaceutical companies, including GSK, Merck, and Novartis to set up research centers there as well.

The Novartis Institute for Tropical Diseases (NITD) is the first of these as well as an example of a PPP. Located in the Biopolis complex in Singapore, it was established in 2002, with the 200 million Swiss Franc total cost jointly covered by Novartis and the Singapore Economic Development Board (a government entity) in a 10-year PPP. NITD has managed to attract additional funding from major private foundations (Wellcome Trust, MMV, the Bill and Melinda Gates Foundation) and established a global network with international tropical disease institutes and academia. Initially carrying out drug discovery centered on therapeutics for tuberculosis and dengue, malaria was added recently and has yielded some promising results in a fairly short time [70]. The antimalarial project is funded primarily by the Wellcome Trust, plus the Singapore Economic Development Board, MMV, and in-kind funding from Novartis. The malaria project started with two main goals: a one-dose cure for Plasmodium falciparum and a curative modality for Plasmodium vivax.

A portfolio of ten initial projects was started in 2007 with a consortium of institutions including the Swiss Tropical Institute (STI), the Biomedical Primate Research Center in Rijswijk, NL, the Genomics Institute of the Novartis Foundation, NITD, and several other academic research groups, with oversight provided through MMV.

In 2009 a new potential antimalarial development candidate, NITD 609, emerged from this collaborative venture. NITD 609 resulted from a cellular screen of a large library of natural products provided by the industrial partner, Novartis. The compound was lethal to cultured parasites, and *in vivo* tests after chemical optimization have confirmed its potency. Three doses rapidly and completely eliminated malaria parasites in infected mice—far exceeding the "gold standard" for efficacy at this stage of development [71–76]. NITD 609 is on course to enter the MMV drug development pipeline with Phase 1 clinical trials currently on-going (fall of 2011).

In addition to NITD, Biopolis hosts the biomedical research institutes funded by the Singapore government's Agency for Science and Technology Research. Of these research institutes, the Experimental Therapeutics Centre (ETC) was established in 2007 as a translational biomedical research institute. ETC has established capabilities in the area of bioanalytics and early stages of drug discovery (protein production, assay development, high throughput screening, and medicinal chemistry). ETC has core funding from Step 2011–2015 over 5 years with additional funding through competitive grants. ETC's mission is to carry out drug discovery through translational research and development with all projects done in collaborative partnerships based on strong external partners. ETC has also formed a strong network of local and international partners in industry and academia, which has seen several projects progress to preclinical development and the out-licensing of a number of infectious disease diagnostics. Singapore is now trying to move into the next phase of drug discovery using public–private collaborations and building state-of-the-art infrastructure.

FRUITFUL AREAS TO CONSIDER FOR PUBLIC–PRIVATE PARTNERSHIPS

New Approaches

The pharmaceutical industry has become rather risk averse as a result of very high attrition rates of even late-stage projects. Novel approaches that may or may not work generally carry a much higher risk than do me-too drugs. This represents an area where PPPs can involve industry at the leading edge and provide academics with the possibility of carrying out applied research leading to novel therapeutic approaches. Glivec, as discussed above, is an example of a successful PPP. This approach may also help the drug discovery and development process to manage the inherent risks while improving productivity and adding innovation and creativity.

Public private partnerships will become increasingly important, especially if personalized medicine takes off. As the "low hanging fruits" have mostly been picked, smaller (in terms of epidemiology) therapeutic indications that are well defined in genetic terms are being pursued by industry. These indications will benefit from academic expertise in their genetics and pathophysiology to make headway in finding medicines with significant patient benefit.

Drug Repurposing

Publicly funded research organizations generally are very good at basic research activities like target identification and validation. Such new targets can be tested using old drugs and can lead to novel approaches to drug discovery using high throughput screening as well as novel therapeutic indications for existing registered drugs. To be most productive, however, this requires an interactive PPP.

MAJOR INDICATIONS WITHOUT ANY REAL OR PERCEIVED COMMERCIAL BENEFIT

Indications or diseases lacking sufficient commercial potential are neglected (or have been abandoned) by the established pharmaceutical industry and are therefore often referred to as neglected diseases [77]. Publicly funded research in developed countries has not been able to pick up the slack, as funding bodies often prioritize grant funding along the same lines as industry. It is an unfortunate fact that all major tropical diseases are neglected diseases, with the three major ones being

- HIV/AIDS (32 million infected, 2.0 million dying each year)
- Malaria (more than 500 million infected and more than 1 million dying per year)
- Tuberculosis (2 billion latently infected, and 9 million new cases and 2 million dying each year.

As mentioned above, drug discovery research and development requires the mobilization of significant resources. Driven by the urgency and needs infected populations, new organizations and programs (based mainly around PPPs) are emerging to tackle neglected diseases, including:

- NGOs devoted to one or several diseases, such as MMV, Pediatric Dengue Vaccine Initiative, Drugs for Neglected Diseases, and many others
- Private companies pursuing corporate citizenship programs
- Private foundations stepping in to fund projects, staff, and whole institutions (Wellcome Trust, Bill and Melinda Gates Foundation)
- Government-funded institutions (U.S. Agency for International Development)
- International organizations (WHO TDR, UNAIDS, EU Framework programs)
- Academic consortia driving open-source, web-based research and development.

It is immediately obvious that highly complex organizations have emerged, beyond the traditional simple collaborative arrangements between two partners. To reach critical mass scientifically and be able to mobilize the resources, PPPs or consortia have formed. They include participants from the government sector, private foundations, NGOs, and companies, as well as consortia of researchers made up of academic institutions, government agencies, hospitals, and industrial partners that are driving research and development. Essentially, these are extensive public–private collaborations for drug discovery and there are now numerous examples (see the example of Coartem above) of alternative, successful modes of financing drug discovery and development that may contain important lessons for the pharmaceutical industry.

CONCLUSIONS AND "TAKE-HOME" MESSAGES

It is becoming abundantly clear that the old paradigm of a pharmaceutical company performing drug discovery research and development in splendid isolation is obsolete.

Indeed, this may be one of the factors in the decline in productivity of the pharmaceutical industry in general. As t he possibilities for coming up with meaningful innovation are still (almost) boundless, a new approach may be needed.

Successful drug discovery and development occurs in dynamic layers of networks, and well-planned PPPs offer many advantages to both parties. While the attrition rate from start to finish may not be any better than a more traditional strategy, this approach will cast a wider net at the cheaper upstream discovery end of the process. With a better understanding of the biology coming from the experts in the public domain, this approach can hopefully refine the targets and yield better drug candidates more quickly in the future.

Management of expectation is critical to success. Thus academics should not view industry simply as a source of funding and resources, and industry should not view academia simply as a source of ideas, samples, and tools. In general, there is a need to have a sound and joint understanding of the genetics and pathophysiology of a disease and the means to clinically test robust hypotheses (rigorous science driving patient stratification and analysis of the drug pharmacodynamic candidate) and this is best achieved by a true PPP.

To start off, both parties need to focus on the desired outcomes from such collaborations. It is also best to start off with as simple an arrangement as possible, perhaps in the form of a research collaboration agreement, that doesn't get bogged down in details and allows the first steps to be taken without too much worrying about outcomes and economic benefits. It should be noted that overly bureaucratic barriers to technology transfer (put up by either side) will encourage a risk-averse culture. By the time all the i's are dotted and t's crossed, the proposed project may be obsolete or no longer of interest to one or both parties. The old saying of "if you think you have a gold mine, show me some gold" applies to public–private drug discovery partnerships as well. Until a novel target or technology platform is applied or marketed, its value is difficult to assess, and royalty "negotiations" are premature. There has to be a certain amount of trust and good faith between the potential partners that they want to work together to address an unmet medical need with a novel approach so that the collaboration can get off the ground.

REFERENCES

1. Bunnage, M.E. Getting pharmaceutical R&D back on target. *Nat Chem Biol*. 2011;7: 335–339.
2. Malakoff, D. Can treatment costs be tamed? *Science* 2011;331:1545–1547.
3. Buck, E., Mulvihill, M., Iwata, K.K. Pursuit of personalized anticancer therapy: leveraging collaboration between academia and the biotech/pharmaceutical industry. *Mt Sinai J Med*. 2010;77:358–365.
4. Chin-Dusting, J., Mizrahi, J., Jennings, G., Fitzgerald, D. Outlook: finding improved medicines: the role of academic-industrial collaboration. *Nat Rev Drug Discov*. 2005;4: 891–897.

5. Johnston, S.C., Hauser, S.L., Desmond-Hellmann, S. Enhancing ties between academia and industry to improve health. *Nat Med.* 2011;17:434–436.
6. Lander, B., Atkinson-Grosjean, J. Translational science and the hidden research system in universities and academic hospitals: a case study. *Soc Sci Med.* 2011;72:537–544.
7. Macarron, R., Banks, M.N., Bojanic, D., Burns, D.J., Cirovic, D.A., Garyantes, T., Green, D.V.S., Hertzberg, R.P., Janzen, W.P., Paslay, J.W., Schopfer, U., Sittampalam, G.S. Impact of high-throughput screening in biomedical research. *Nat Rev Drug Discov.* 2011;10: 188–195.
8. Stevens, A.J., Jensen, J.J., Wyller, K., Kilgore, P.C., Chatterjee, S., Rohrbaugh, M.L. The role of public-sector research in the discovery of drugs and vaccines. *N Engl J Med.* 2011;364:535–541.
9. Heller, S., Kozlovski, P., Kurtzhals, P. Insulin's 85th anniversary—an enduring medical miracle. *Diabetes Res Clin Pract.* 2007;78:149–158.
10. Heinemann, L. New ways of insulin delivery. *Int J Clin Pract.* 2010;64:29–40.
11. Black, W. Drug products of recombinant DNA technology. *Am J Health Syst Pharm.* 1989;46:1834–1844.
12. Bristow, A.F. Recombinant-DNA-derived insulin analogues as potentially useful therapeutic agents. *Trends Biotechnol.* 1993;11:301–305.
13. Dingermann, T. Recombinant therapeutic proteins: production platforms and challenges. *J Biotechnol.* 2008;3:90–97.
14. Rimmer, M. Genentech and the stolen gene: patent law and pioneer inventions. *Bio-Sci Law Rev.* 2003;6:198–211.
15. Patino, R.M. Moving research to patient applications through commercialization: understanding and evaluating the role of intellectual property. *J Am Assoc Lab Anim Sci.* 2010;49:147–154.
16. Thangaraj, H., van Dolleweerd, C.J., McGowan, E.G., Ma, J.K.-C. Dynamics of global disclosure through patent and journal publications for biopharmaceutical products. *Nat Biotechnol.* 2009;27:614–618.
17. Faderl, S., Talpaz, M., Estrov, Z., Obrien, S., Kurzrock, R., Kantarjian, A.H.M. The biology of chronic myeloid leukemia. *N Engl J Med.* 1999;341:164–172.
18. Kurzrock, R. The biology of chronic myeloid leukemia. *N Engl J Med.* 1999;341: 164–172.
19. Kurzrock, R., Kantarjian, H.M., Druker, B.J., Talpaz, M. Philadelphia chromosome-positive leukemias: from basic mechanisms to molecular therapeutics. *Ann Intern Med.* 2003; 138:819–830.
20. Pierotti, M.A., Negri, T., Tamborini, E., Perrone, F., Pricl, S., Pilotti, S. Targeted therapies: the rare cancer paradigm. *Mol Oncol.* 2010;4:19–37.
21. Sawyers, C. Chronic myeloid leukemia. *N Engl J Med.* 1999;340:1330–1340.
22. Pearson, M.A., Fabbro, D. Targeting protein kinases in cancer therapy: a success? *Expert Rev Anticancer Ther.* 2004;4:1113–1124.
23. Druker, B.J., Sawyers, C.L., Kantarjian, H., Resta, D.J., Reese, S.F., Ford, J.M., Capdeville, R., Talpaz, M. Activity of a specific inhibitor of the BCR-ABL tyrosine kinase in the blast crisis of chronic myeloid leukemia and acute lymphoblastic leukemia with the Philadelphia chromosome. *N Engl J Med.* 2001;344:1038–1042.

24. An, X., Tiwari, A.K., Sun, Y., Ding, P.-R., Ashby, C.R., Jr., Chen, Z.-S. BCR-ABL tyrosine kinase inhibitors in the treatment of Philadelphia chromosome positive chronic myeloid leukemia: a review. *Leuk Res.* 2010;34:1255–1268.

25. Baker, S.J., Reddy, E.P. Targeted inhibition of kinases in cancer therapy. *Mt Sinai J Med.* 2010;77:573–586.

26. Broxterman, H.J., Georgopapadakou, N.H. New cancer therapeutics: target-specific in, cytotoxics out? *Drug Resist Updat.* 2004;7:79–87.

27. Bumbea, H., Vladareanu, A., Cisleanu, D. Chronic myeloid leukemia therapy in the era of tyrosine kinase inhibitors. The first molecular targeted treatment. *J Med.* 2010;3:162–166.

28. Burkard, M.E., Jallepalli, P.V. Validating cancer drug targets through chemical genetics. *Biochim Biophys Acta* 2010;1806:251–257.

29. Collins, I., Workman, P. New approaches to molecular cancer therapeutics. *Nat Chem Biol.* 2006;2:689–700.

30. Cullinane, C., Dorow, D.S., Kansara, M., Conus, N., Binns, D., Hicks, R.J., Ashman, L.K., McArthur, G.A., Thomas, D.M. An in vivo tumor model exploiting metabolic response as a biomarker for targeted drug development. *Cancer Res.* 2005;65:9633–9636.

31. Cullinane, C., Natoli, A., Hui, Y., Conus, N., Jackson, S., Bruggen, J., Manley, P.W., McArthur, G.A. Preclinical evaluation of nilotinib efficacy in an imatinib-resistant KIT-driven tumor model. *Mol Cancer Ther.* 2010;9:1461–1468.

32. DePrimo, S.E., Huang, X., Blackstein, M.E., Garrett, C.R., Harmon, C.S., Schoffski, P., Shah, M.H., Verweij, J., Baum, C.M., Demetri, G.D. Circulating levels of soluble KIT serve as a biomarker for clinical outcome in gastrointestinal stromal tumor patients receiving sunitinib following imatinib Failure. *Clin Cancer Res.* 2009;15:5869–5877.

33. Flamant, S., Ritchie, W., Guilhot, J., Holst, J., Bonnet, M.L., Chomel, J.C., Guilhot, F., Turhan, A.G., Rasko, J.E.J. Micro-RNA response to imatinib mesylate in patients with chronic myeloid leukemia. *Haematologica* 2010;95:1325–1333.

34. George, S., Blay, J.Y., Casali, P.G., Le Cesne, A., Stephenson, P., DePrimo, S.E., Harmon, C.S., Law, C.N.J., Morgan, J.A., Ray-Coquard, I., Tassell, V., Cohen, D.P., Demetri, G.D. Clinical evaluation of continuous daily dosing of sunitinib malate in patients with advanced gastrointestinal stromal tumour after imatinib failure. *Eur J Cancer* 2009;45: 1959–1968.

35. Gorre, M.E. BCR-ABL point mutants isolated from patients with imatinib mesylate-resistant chronic myeloid leukemia remain sensitive to inhibitors of the BCR-ABL chaperone heat shock protein 90. *Blood* 2002;100:3041–3044.

36. Hochhaus, A., Druker, B., Sawyers, C., Guilhot, F., Schiffer, C.A., Cortes, J., Niederwieser, D.W., Gambacorti-Passerini, C., Stone, R.M., Goldman, J., Fischer, T., O'Brien, S.G., Reiffers, J.J., Mone, M., Krahnke, T., Talpaz, M., Kantarjian, H.M. Favorable long-term follow-up results over 6 years for response, survival, and safety with imatinib mesylate therapy in chronic-phase chronic myeloid leukemia after failure of interferon treatment. *Blood* 2007;111:1039–1043.

37. Kaiser, J. Combining targeted drugs to stop resistant tumors. *Science* 2011;331:1542–1545.

38. Paterson, S., Smith, K., Holyoake, T. Is there a cloud in the silver lining for imatinib? *Br J Cancer* 2003;88:983–987.

39. Peng, B. Pharmacokinetics and pharmacodynamics of imatinib in a phase I trial with chronic myeloid leukemia patients. *J Clin Oncol.* 2004;22:935–942.

40. Press, R.D., Galderisi, C., Yang, R., Rempfer, C., Willis, S.G., Mauro, M.J., Druker, B.J., Deininger, M.W.N. A half-log increase in BCR-ABL RNA predicts a higher risk of relapse in patients with chronic myeloid leukemia with an imatinib-induced complete cytogenetic response. *Clin Cancer Res.* 2007;13:6136–6143.
41. Radich, J.P., Dai, H., Mao, M., Oehler, V., Schelter, J., Druker, B., Sawyers, C., Shah, N., Stock, W., Willman, C.L., Friend, S., Linsley, P.S. Gene expression changes associated with progression and response in chronic myeloid leukemia. *Proc Natl Acad Sci U S A.* 2006; 103:2794–2799.
42. Roumiantsev, S., Shah, N.P., Gorre, M.E., Nicoll, J., Brasher, B.B., Sawyers, C.L., Van Etten, R.A. Clinical resistance to the kinase inhibitor STI-571 in chronic myeloid leukemia by mutation of Tyr-253 in the Abl kinase domain P-loop. *Proc Natl Acad Sci U S A.* 2002;99: 10700–10705.
43. Sawyers, C.L. Imatinib induces hematologic and cytogenetic responses in patients with chronic myelogenous leukemia in myeloid blast crisis: results of a phase II study. *Blood* 2002;99:3530–3539.
44. Sawyers, C.L. Even better kinase inhibitors for chronic myeloid leukemia. *N Engl J Med.* 2010;362:2314–2315.
45. Shah, N.P., Nicoll, J.M., Nagar, B., Gorre, M.E., Paquette, R.L., Kuriyan, J., Sawyers, C.L. Multiple BCR-ABL kinase domain mutations confer polyclonal resistance to the tyrosine kinase inhibitor imatinib (STI571) in chronic phase and blast crisis chronic myeloid leukemia. *Cancer Cell.* 2002;2:117–125.
46. Spiera, R.F., Gordon, J.K., Mersten, J.N., Magro, C.M., Mehta, M., Wildman, H.F., Kloiber, S., Kirou, K.A., Lyman, S., Crow, M.K. Imatinib mesylate (Gleevec) in the treatment of diffuse cutaneous systemic sclerosis: results of a 1-year, phase IIa, single-arm, open-label clinical trial. *Ann Rheum Dis.* 2011. doi: 10.1136/ard.2010.143974.
47. Stegmeier, F., Warmuth, M., Sellers, W.R., Dorsch, M. Targeted cancer therapies in the twenty-first century: lessons from imatinib. *Clin Pharmacol Ther.* 2009;87:543–552.
48. Talpaz, M. Imatinib induces durable hematologic and cytogenetic responses in patients with accelerated phase chronic myeloid leukemia: results of a phase 2 study. *Blood* 2002;99: 1928–1937.
49. Druker, B.J. Perspectives on the development of imatinib and the future of cancer research. *Nat Med.* 2009;15:xv–xviii.
50. Lydon, N. Attacking cancer at its foundation. *Nat Med.* 2009;15:xix–xxiii.
51. Greene, W.C. A history of AIDS: looking back to see ahead. *Eur J Immunol.* 2007;37: S94–S102.
52. Merson, M.H., O'Malley, J., Serwadda, D., Apisuk, C. The history and challenge of HIV prevention. *Lancet* 2008;372:475–488.
53. Bailey, A.C., Fisher, M. Current use of antiretroviral treatment. *Br Med Bull.* 2008; 87:175–192.
54. De Clercq, E. The history of antiretrovirals: key discoveries over the past 25 years. *Rev Med Virol.* 2009;19:287–299.
55. Tozzi, V. Pharmacogenetics of antiretrovirals. *Antiviral Res.* 2010;85:190–200.
56. Gianotti, N., Soria, A., Lazzarin, A. Antiviral activity and clinical efficacy of atazanavir in HIV-1-infected patients: a review. *New Microbiol.* 2007;30:79–88.
57. Havlir, D.V., O'Marro, S.D. Atazanavir: new option for treatment of HIV infection. *Clin Infect Dis.* 2004;38:1599–1604.

58. Moreno, S., Lopez Aldeguer, J., Arribas, J.R., Domingo, P., Iribarren, J.A., Ribera, E., Rivero, A., Pulido, F., On behalf of The HIV 2020 Project. The future of antiretroviral therapy: challenges and needs. *J Antimicrob Chemother.* 2010;65:827–835.
59. Djimdé, A., Lefèvre, G. Understanding the pharmacokinetics of Coartem®. *Malar J.* 2009;8:S4.
60. Gamo, F.-J., Sanz, L.M., Vidal, J., de Cozar, C., Alvarez, E., Lavandera, J.-L., Vanderwall, D.E., Green, D.V.S., Kumar, V., Hasan, S., Brown, J.R., Peishoff, C.E., Cardon, L.R., Garcia-Bustos, J.F. Thousands of chemical starting points for antimalarial lead identification. *Nature* 2010;465:305–310.
61. Guiguemde, W.A., Shelat, A.A., Bouck, D., Duffy, S., Crowther, G.J., Davis, P.H., Smithson, D.C., Connelly, M., Clark, J., Zhu, F., Jiménez-Díaz, M.B., Martinez, M.S., Wilson, E.B., Tripathi, A.K., Gut, J., Sharlow, E.R., Bathurst, I., El Mazouni, F., Fowble, J.W., Forquer, I., et al. Chemical genetics of Plasmodium falciparum. *Nature* 2010;465:311–315.
62. Kokwaro, G. Ongoing challenges in the management of malaria. *Malar J.* 2009;8:S2.
63. Ogutu, B.R. Malaria management—progress made and challenges still to face. *Malar J.* 2009;8:S1.
64. Abdulla, S., Sagara, I. Dispersible formulation of artemether/lumefantrine: specifically developed for infants and young children. *Malar J.* 2009;8:S7.
65. Barnes, K.I., Chanda, P., Ab Barnabas, G. Impact of the large-scale deployment of artemether/lumefantrine on the malaria disease burden in Africa: case studies of South Africa, Zambia and Ethiopia. *Malar J.* 2009;8:S8.
66. Falade, C., Manyando, C. Safety profile of Coartem®: the evidence base. *Malar J.* 2009;8:S6.
67. Lefèvre, G., Marrast, A.C., Grueninger, H. Novartis Malaria Initiative: best practice example of pharmaceutical industry's engagement in the fight against malaria. *Ann N Y Acad Sci.* 2011;1222:19–29.
68. Makanga, M., Krudsood, S. The clinical efficacy of artemether/lumefantrine (Coartem®). *Malar J.* 2009;8:S5.
69. Premji, Z.G. Coartem®: the journey to the clinic. *Malar J.* 2009;8:S3.
70. Palmer, H., Diagana, T. Re-starting malaria research at the Novartis Institute for Tropical Diseases (NITD). *J Biotechnol.* 2007;2:1337–1338.
71. Guantai, E., Chibale, K. How can natural products serve as a viable source of lead compounds for the development of new/novel anti-malarials? *Malar J.* 2011;10:S2.
72. Keller, T.H., Shi, P.-Y., Wang, Q.-Y. Anti-infectives: can cellular screening deliver? *Curr Opin Chem Biol.* 2011;15:529–533.
73. Rottmann, M., McNamara, C., Yeung, B.K.S., Lee, M.C.S., Zou, B., Russell, B., Seitz, P., Plouffe, D.M., Dharia, N.V., Tan, J., Cohen, S.B., Spencer, K.R., Gonzalez-Paez, G.E., Lakshminarayana, S.B., Goh, A., Suwanarusk, R., Jegla, T., Schmitt, E.K., Beck, H.P., Brun, R., Nosten, F., Renia, L., Dartois, V., Keller, T.H., Fidock, D.A., Winzeler, E.A., Diagana, T.T. Spiroindolones, a potent compound class for the treatment of malaria. *Science* 2010;329:1175–1180.
74. Wells, T.N.C. Is the tide turning for new malaria medicines? *Science* 2010;329: 1153–1154.
75. Wells, T.N. Natural products as starting points for future anti-malarial therapies: going back to our roots? *Malar J.* 2011;10:S3.
76. Yeung, B.K.S., Zou, B., Rottmann, M., Lakshminarayana, S.B., Ang, S.H., Leong, S.Y., Tan, J., Wong, J., Keller-Maerki, S., Fischli, C., Goh, A., Schmitt, E.K., Krastel, P., Francotte, E.,

Kuhen, K., Plouffe, D., Henson, K., Wagner, T., Winzeler, E.A., Petersen, F., Brun, R., Dartois, V., Diagana, T.T., Keller, T.H. Spirotetrahydro β-carbolines (spiroindolones): a new class of potent and orally efficacious compounds for the treatment of malaria. *J Med Chem.* 2010;53:5155–5164.

77. Nathan, C., Goldberg, F.M. Outlook: the profit problem in antibiotic R&D. *Nat Rev Drug Discov.* 2005;4:887–891.

31

A FLEXIBLE MODEL FOR COMPOUND MANAGEMENT FACILITIES TO STIMULATE COLLABORATIONS IN THE LIFE SCIENCES

David Camp

Griffith University, Nathan, Queensland, Australia

GLOBAL CONTEXT

An enormous cache of human biology remains to be explored following completion of the Human Genome Project (HGP). The expectation is that selective modulators for new targets will be discovered that, in some cases, will be translated into novel therapeutics. Indeed, the identification of new structural classes is one of the many drivers *en route* to understanding biological systems and developing innovative, safer therapeutics with novel modes of action. The colossal number of molecules, estimated to be in excess of an estimated 10^{62} compounds [1] composed of the atoms that make up current molecular therapies, effectively ensures a vast reserve of structural diversity can be mined to achieve this goal. However, despite this amazing diversity, it would nevertheless be impossible to synthesize even one molecule of each member from this set considering there is estimated to be "only" 10^{50} atoms on Earth [2]. The real challenge lies in the selection of small molecules that can actually probe or otherwise modulate human biology in a therapeutically useful manner. In this respect, biologically valid small molecules (organic compounds with a molecular weight <500 composed mainly

Collaborative Innovation in Drug Discovery: Strategies for Public and Private Partnerships, First Edition.
Edited by Rathnam Chaguturu.

of carbon, hydrogen, and oxygen atoms, but typically also nitrogen, sulfur, and fluorine, and less often chlorine, bromine, and phosphorus) that interact with biology space (e.g., the surface of a protein) have become critical tools for understanding important cellular events and biological pathways involved in health and disease.

Until the completion of the HGP, however, only a few hundred human proteins had been studied in detail via small molecules [3]. Because many of these compounds were found/discovered serendipitously in animal models or cell lines, more modern approaches had to be developed to keep abreast of the massive amount of data emanating from the HGP and millions of compounds that could be produced by combinatorial chemistry. The resulting paradigm shift was driven by advances in molecular biology and robust automation so that it is now possible to undertake high-throughput screening (HTS) of 100,000 compounds per day against an isolated biomolecular target (usually a recombinant protein) or cell line to facilitate the efficient discovery of many more useful compounds. The screening of large and structurally diverse small molecule libraries that impede the function of specific target proteins is now commonplace in industrial drug discovery programs [4]. The same tools (compounds) and techniques (automated screening) are also being used by academia in the emerging interdisciplinary field of chemical biology to identify new proteins and map biochemical pathways in the context of functional assays [5–7].

Current wisdom suggests that the quest to identify new molecules that can ultimately be developed into drugs will be facilitated via a combination of chemical biology and phenotypic screening, typically pursued in a basic research environment, and a concerted drug discovery program which, for the most part, is only possible with the financial backing of industry. Chemical biology employs small molecules to interrogate biological systems and pathways. In contrast to classical genetics where manipulation occurs at the DNA level, small molecules generally modulate protein function, and not the gene locus or mRNA, by inducing conformational changes or competing for naturally occurring protein–ligand or protein–protein interaction sites resulting in altered activity. This allows temporal study of signaling pathways and the ability to wash out probes to study reversible inhibition.

Phenotypic screening enables the discovery of compounds which modulate biological systems in a target- and mechanism-agnostic manner, and is complementary to biochemical HTS where compound libraries are evaluated against an isolated protein target. Phenotypic drug discovery directly interrogates complex biological systems composed of multiple or unknown biochemical components and pathways via imaging of cells using high-content screening (HCS) techniques. The phenotypic approach is reminiscent of drug discovery strategies that predate the advent of molecular biology and HTS. HCS, like HTS, has benefited from advances in imaging technologies, assay miniaturization, robust automation, bioinformatics, and the ready availability of large compound libraries. Additionally, phenotypic screening profits from facile access to physiologically relevant biological systems (model organisms together with primary, differentiated, and stem cells).

Significant time and effort can be dedicated in academia to identifying small molecule modulators for proteins, receptors, DNA, and RNA that further explore the underlying biology of a specific system—a luxury not readily amenable to commercial reality. However, while the activities of academia are broader than those of industry, both sectors are inextricably linked as the industrial pipeline of candidate drugs comes to rely increasingly more upon knowledge generated by academic researchers. This trend is expected to continue given the many and complex issues currently facing the pharmaceutical industry. Hence, academic research that aims to comprehend the underlying biology of a disease state, validate drug targets, or otherwise add value to basic research is becoming progressively more important.

Early-phase drug discovery is also pursued by many academic screening centers in addition to curiosity-driven research as the underlying infrastructure, skill sets, and procedures required to unearth a new modulator for chemical biology or phenotype screening programs is identical to the identification of a lead molecule for potential development into a pharmaceutical. The possible development a marketable product in the form of a therapeutic following screening of a target that was identified through a university's own basic research makes the acquisition of infrastructure for screening molecular libraries extremely attractive. In this scenario, a hit is usually taken as far down the value-added chain as possible before partnering with industry.

The inclusion of drug discovery screening programs in academia has no doubt been aided by the HGP, which has provided many potential new drug targets outside the "low-hanging fruit" for which the vast majority of current therapeutics have been developed, and the recent ruling in the case of the University of Rochester versus Pfizer [8]. In the latter, discovery of COX-2 protein by the university was insufficient to file a patent claim on the basis that the target could be inhibited using a compound yet never actually disclosing any compound. The real invention was deemed to be the compound (Celebrix) developed by Searle (later Pfizer), which brought about a reduction in acute pain and inflammation via specific inhibition of COX-2. This ruling effectively means that potentially druggable targets found through basic research emanating from the "-omics" sciences are not patentable *per se* but require a compound to demonstrate usefulness. HTS and HCS therefore afford a mechanism to validate the discovery of a new target through disclosure of a compound that can inhibit, or otherwise modulate, its action. The benefits to academic and other nonindustry organizations are potentially enormous. Payment of milestones or substantial licensing revenues is possible depending, *inter alia*, on the size of the intended market for the drug, the novelty of the compound, how far along the value-adding pipeline the compound has progressed, and the manner in which the intellectual property (IP) was negotiated with an industry or product development partner.

It is interesting to note that while many academic organizations have developed a genuine capability for HTS, far less have invested in compound management. This is no doubt due, in part, to the fact that the combined cost to acquire a sizable library of more than 100,000 compounds and the supporting infrastructure can be prohibitive. Put simply, the value proposition for establishing a dedicated compound management capability to complement HTS where the library is less than 100,000 compounds does not necessarily justify the outlay, especially for the academic sector. To obviate this limitation, several governments have used public funds to establish centralized repositories to augment already established screening centers or newly created networks. This allows the screening sites to access small molecule libraries at a far lower cost than if they were purchased from commercial vendors and also obviates the need to initially purchase and then maintain additional infrastructure.

Two of the more high-profile publicly funded schemes include the U.S. National Institutes of Health (NIH) National Center for Advancing Translational Sciences (NCATS) [9], which had its genesis in the NIH Roadmap Molecular Libraries Program (MLP) [10], and the pan-European effort, EU-OPENSCREEN [11]. Both the U.S. and European initiatives aim to extend knowledge of human biology via investing in the identification and validation of new targets implicated in various disease states, and perturbing these targets with small molecule libraries. Ultimately, the objective of this chemical biology-based approach is to catalyze the wealth of information arising out of the HGP into tangible health benefits.

The NIH Roadmap was conceived in 2002 and, following funding to several programs in 2004, became the world's first truly national chemical biology initiative [12]. The NIH invested heavily in equipment and infrastructure to support basic research and development in order to achieve its goals. This initially included, *inter alia*, establishment of a Small Molecule Repository (SMR) that collated and supplied compounds to a network of 10 screening centers known collectively as the Molecular Libraries Screening Centers Network (MLSCN) [13]. The MLSCN was launched in mid-2005 as a pilot phase to the MLP. Targets for screening were funded by grants to predominately academic researchers to identify and develop assays for novel targets. The synthesis of chemical libraries that expanded the chemical space occupied by commercial libraries was also supported in the pilot phase. The immediate goals of the MLSCN were to expand the availability and use of chemical probes contained in the SMR to explore the function of genes, cells, and pathways in health and disease. Importantly, annotated information on the biological profiles of compounds was made available through PubChem, the NIH's publicly accessible database. The Molecular Libraries Probe Production Centers Network (MLPCN) superseded the MLSCN pilot phase in mid-2008. In addition to providing capacity to screen over 300,000 small molecules that had been collated over the previous 3 years using a variety of assay formats, the MLPCN significantly expanded its core capabilities to include medicinal chemistry and informatics related to the development of promising probes and library production. The MLPCN's combined resource is presently distributed among nine sites, four of which hold "comprehensive" status in that they prosecute both HTS and probe-related chemistry, three that undertake screening only and two that are specialist chemistry sites [10].

The MLP has positioned itself in the precompetitive arena and does not have the remit to embark on a fully-fledged drug discovery program *per se*. Rather, it aims to understand complex human biology through the use of small molecule modulators and place all data in the public domain. However, the NIH may still carry out early phase drug discovery should a promising probe be identified with limited commercial potential. In this respect, the success of the MLP was underscored with an announcement in early 2011 that the program had attained its first landmark achievement: FDA approval to conduct a Phase 1 clinical safety study for a sphingosine-1-phosphate receptor 1 (S1P1) agonist. The original probe report was submitted to PubChem in 2007 by the Scripps Research Institute and has since undergone further development outside of the

MLP by Scripps and Receptos Inc. The clinical trial has been initiated as a potential treatment for multiple sclerosis.

EU-OPENSCREEN is a pan-European network that currently involves partners form 12 countries (Austria, Czech Republic, Denmark, Finland, France, Germany, Italy, the Netherlands, Norway, Poland, Spain, and Sweden). The EU-OPENSCREEN initiative effectively leverages existing national networks like Germany's ChemBioNet [14] and other specialist technology platforms such as the FIMM High Throughput Center in Finland [15] into a seamless well-equipped entity that permits researchers in academia and SMEs to access resources for the development of bioactive small molecules. Planning is already well under way with the goal that EU-OPENSCREEN will be fully operational from 2015.

Like the MLP, the activities of EU-OPENSCREEN precede commercial development. EU-OPENSCREEN will be open to all European organizations involved in chemical biology and is committed to open access. It is envisaged that a legal entity will be created in an effort to facilitate collaboration among members and interactions with stakeholders and external users. In a departure from the MLP IP model, EU-OPENSCREEN is currently developing a flexible IP framework that attempts to balance rapid knowledge sharing and exploitation activities. Thus, researchers are able to protect knowledge gained following use of EU-OPENSCREENs resources before publishing in the database.

There are many other national initiatives in addition to the NCATS. Some like the Chimiothèque Nationalé (France) [16] are part of EU-OPENSCREEN and have been established to allow academic institutions to prosecute HTS and phenotypic screening of small molecule libraries. They are typically less expansive than the NIH effort and may occur as self-organized consortia that provide a range of chemical, screening, and informatics resources such as the Centre for Drug Research and Development (Canada) [17], CGC (Germany) [18], and the UK Drug Discovery Consortium [19]. Alternatively, they may take the form of an institute that makes resources and infrastructure available to the broader scientific community like the RIKEN Program for Drug Discovery and Medical Technology Platform in Japan [20], the Institut Pasteur Korea [21], Tel Aviv University (Israel) [22], the Dundee Drug Discovery Unit (Scotland) [23], Drug Discovery Centre at Imperial College (England) [24], the Broad Institute of MIT and Harvard (USA) [25], the Kansas University HTS Laboratory (USA) [26], or the Memorial Sloan-Kettering Cancer Center HTS Core Facility (USA) [27] to name a few from a large and expanding list. Additional academic screening centers can be found on the website for the Society for Laboratory Automation and Screening [28].

Finally, although not a publicly funded initiative, and somewhat outside the scope of this chapter, the Eli Lilly Phenotypic Drug Discovery (PD^2) Initiative [29] is worth mentioning as it provides another mechanism whereby academic compound collections may be screened. Lilly introduced the PD^2 initiative because of the slew of issues facing the pharmaceutical industry today. Lilly believes that the solution to future innovation lies, in part, through collaborative research with universities, institutes, and biotechs using a "neoclassical" approach to drug discovery. It is Lilly's belief that the phenotype drug discovery paradigm can accelerate the search for high-quality drug candidates and, to this end, they have established a panel of assay "modules" which represent therapeutic areas of long-term strategic interest to the company in oncology, neurological disorders, and metabolic diseases. Each module is composed of a phenotypic lead generation assay which is followed by relevant biochemical and cellular assays designed to quickly delineate a compound's activity profile and potential for future development.

External investigators can upload and submit structures of their compounds via a secure site where they are evaluated for novelty, lead- and drug-likeness. If a molecule passes this decision gate, investigators are invited to submit a physical sample to Lilly for testing in the PD^2 panel. In exchange, the investigator receives a full report of the biological data. Once testing is completed, Lilly evaluates the data and determines whether to initiate discussions leading to a possible collaboration. Importantly, the investigator (or their organization) retains all IP rights in the compound and is free to publish the data if mutual consent is not reached or the molecule does not show any promise.

AUSTRALIAN CONTEXT

Australia has a small population by world standards (23M), which can limit the availability of public funds to support the establishment of systemic infrastructure, functional capability (e.g., staff costs and equipment maintenance), and project-related costs (reagents, labware, etc.) across multiple sites like the erstwhile MLSCN or MLPCN.

Highly competitive funding schemes from the federal government have helped cultivate a quality basic medical research environment [30]. This has contributed, in part, to an outstanding scientific output over the past 50 years and allows Australia to continually punch above its weight in science and technology, contributing 3% of global medical research publications each year and establishing 10 times the number of World Health Organization (WHO) "collaborating centers" than would be expected for a population that accounted for 0.3% of the world's total [31]. Moreover, Australia currently ranks seventh in the world for the number of science research publications per capita [32]. The largest number of research publications originates from the medical and health sciences (which account for approximately one-third of papers published) and the biological sciences (approximately one-fifth of publications).

This international standing in medical research is evident among Australia's six Nobel laureates for physiology or medicine. Five were born, educated, and performed the work for which they were awarded the Nobel Prize in Australia: Sir Frank Macfarlane Burnett (1960), John Eccles (1963), Peter Doherty (1996), Barry Marshall, and Robin Warren (2005). The sixth, Howard Florey (1945), graduated from the University of Adelaide with an MB, BS in 1921 before being awarded a Rhodes Scholarship to continue his studies in England.

Other notable Australian efforts include the WHO smallpox eradication program led by Frank Fenner from the Australian National University that freed the world from this devastating disease. The program removed smallpox that threatened 60% of the world's population, killed every fourth victim, and scarred or blinded most survivors. Over 3.5 million patients have been treated with the colony-stimulating factors discovered by Don Metcalf and his colleagues at the Walter and Eliza Hall Institute of Medical Research. These naturally occurring proteins accelerate white blood cell regeneration in patients treated for cancer. More recently, the development of an innovative neuraminadase inhibitor for influenza infections by Peter Colman, Graham Laver, and Mark von Itzstein has become part of the defensive armory against the threat of avian influenza, while a team led by Ian Frazer at the University of Queensland developed a vaccine for cervical cancer that could protect millions of women. Yet, despite these and other success stories, preclinical development is not as strong as its basic research counterpart, and the immediate challenge is to translate this high world standing into internationally competitive industrial niches.

Australia's 3% contribution to global medical science has not translated into a local industry that accounts for 3% of global drug development [33], in part because of the relatively small number of drug candidates entering preclinical trials every year [34]. Indeed, only four FDA-approved small molecule pharmaceuticals have been largely developed in Australia, that is, *Eryc, Doryx, Kapanol*, and *Relenza*, with the first three being formulations rather than *de novo* developed drugs. In effect, investment in basic research does not realize its full potential, and there is a clear opportunity cost if Australia becomes a country that supplies disease knowledge but fails to supply therapeutics.

As evidenced by *Relenza* [35], Australia has the necessary skill set to prosecute small molecule drug development but, with the advent of the HTS and HCS, began to suffer from a lack of systemic infrastructure which prevented modernization and the integration of new skills into a truly competitive capability. The region's chemical and life science communities stand to profit from entry into the HTS and phenotypic screening paradigms. HTS and HCS of appropriately constructed compound libraries can facilitate the discovery of small molecule modulators against validated targets. In turn, these lead or probe molecules may act as the starting points for value adding through existing centers of expertise in structural biology, modeling, synthesis, ADME, toxicology, scale-up, animal models, and clinical trials. Demonstration that a compound has promising pharmacological activity, favorable ADME characteristics, and acceptable toxicity facilitates its progression along the pharmaceutical value chain. The existing capabilities dovetail with HTS and phenotypic screening methodologies and allow a more mature IP position to be obtained before partnering.

Over the past 10–15 years, a few farsighted Australian research organizations and State governments have invested to create the infrastructure necessary for pharmaceutical discovery; however, the public institutions possessing this infrastructure are under enormous financial pressure and, without the support of funding agencies, are unable to provide the recurrent funding on a scale necessary to build a sustainable Australian preclinical capacity. Consequently, translation of basic research fails to occur or drug discovery IP is sold to overseas interests too early and for too little value.

A paucity of systemic infrastructure to underpin the early discovery phase in the HTS and phenotypic screening paradigms potentially means that Australia would not be in as strong a position to exploit the "-omics" revolution as it relates to small molecule drug discovery and the understanding of biological interactions through new interdisciplinary sciences like chemical biology. The region would benefit enormously if the screening of small molecules was facilitated to unearth probes that allow robust interrogation of biological systems and concomitantly generate leads for drug discovery that act as the starting point for value adding via existing centers of expertise so that a more mature IP position could be obtained before partnering.

Preexisting Conditions in Australia Facilitating Establishment of the QCL

HTS was introduced into Australia during the early 1990s via a collaborative drug discovery program between the pharmaceutical company, AstraZeneca, and Griffith University. Griffith University screened natural product extracts against predominately biochemical assays supplied by AstraZeneca in an effort to provide innovative leads for further development. Although well funded, the exclusive nature of this arrangement did not permit screening of compounds from academia or SMEs against targets that may have impinged on AstraZeneca's business interests.

By 2001, approximately 45,000 biota samples had been collected from megadiverse countries in accordance with the United Nations Convention on Biological Diversity (UN CBD). Two extracts were taken from 35,000 samples. The ensuing 70,000 extracts along with 2000 natural products and 1000 synthetics were subsequently stored as stock solutions in microtubes equipped with a 2-D barcode. While the introduction of this labware facilitated cradle-to-grave tracking, increased flexibility, and efficiency, it nevertheless suffered from the fact that microtube racks (or occasionally specific tubes) had to be retrieved manually.

It was not until 2003, some 10 years after commencement of HTS at Griffith University, that an open access platform became available through a consortium based in the State of Victoria. This platform was located at the Walter & Eliza Hall Institute (WEHI) of Medical Research in Melbourne and, in addition to automation, incorporated a library of 100,000 compounds in 384-well plate format obtained from commercial vendors. Automated screening began in South Australia a few years later when Bionomics integrated small molecule drug discovery into previously established technology platforms in the genomics of epilepsy, angiogenesis, and breast cancer.

By 2007, the exclusive agreement between AstraZeneca and Griffith University had ceased allowing the contribution of HTS resources, together with the WEHI and Bionomics, into a federal government-sponsored program: the Cooperative Research Centre for Cancer Therapeutics (CRC CTx). The CRC CTx fills a frustrating gap for Australian cancer researchers through provision of funds that permits a promising small molecule lead to progress to the regulatory preclinical stage. Small molecule leads for new targets are typically identified following HTS of an expansive compound library. The strong undercurrent of basic research also prompted other groups with exemplary track records in child-related cancer at the Children's Cancer Institute of Australia in New South Wales and G-protein-coupled receptors (GPCRs) at Monash University to seek funding via different mechanisms to establish HTS and phenotypic screening capabilities.

The screening of large diverse chemical libraries to identify small molecules capable of modifying the behavior of a target protein or a cellular phenotype effectively means that biology and chemistry are really two sides of the same coin in the HTS and phenotypic screening paradigms. It is as pointless to acquire massive chemical libraries that cannot be evaluated in biochemical, cellular, or organismic assays, as it is to have highly automated and efficient methods to screen a biological system without small molecules. Given the traction among several groups to invest in screening platforms, it therefore became necessary to consider the creation of a compound management and logistics facility from a national perspective. A centralized repository that met the national need would clearly benefit from economies of scale that smaller facilities scattered throughout the country could never achieve.

Griffith University took the lead in acquiring this infrastructure in collaboration with other interested stakeholders, partly because there was already a clear need to automate the 70,000 natural product extracts and partly because there was a considerable knowledge base available through AstraZeneca's Compound Management Network, which was implementing its global, automated compound management facilities. It was fortuitous that planning for the compound management facility coincided with a major national effort in Australia to improve acquisition and access to systemic infrastructure. The scheme, known as the National Collaborative Research Infrastructure Strategy (NCRIS), commenced in 2004 with a broad outline of its scope. Importantly, while the NCRIS scheme aimed to identify and address critical gaps in infrastructure, it also had the remit to increase collaboration among researchers in the same space and reduce the duplication and suboptimal use of resources arising from a lack of coordination. Essentially, the NCRIS program would not provide funds to individual organizations or institutions for the same facilities to be established at multiple sites.

The research community could participate in the process throughout 2005 prior to the release of an Exposure Draft of the Strategic Roadmap. During 2005–2006, grassroots support was received from 14× universities, 6× medical research institutes, 6× publicly funded research agencies [36], and 2× major national resource facilities resulting in a proposal for an integrated compound management and HTS to be included in the priority area of "Translating health discovery into clinical applications" (THD). Although this concept was included in the 2006 Strategic Roadmap [37], the NCRIS Committee noted that responses to the Exposure Draft of the Roadmap indicated not only a strong support for the THD capability area from a large cross-section of the research community but also a diversity of opinions as to appropriate investments. The NCRIS Committee suggested stakeholders continue to work toward clarification of the issues and needs and that this capability be reviewed for possible implementation in 2007.

Concomitantly, the Griffith University group, who was based in Queensland, applied through their State Government's Smart State Research Facility Fund (SSRFF) in 2005 for the infrastructure. The proposal was successful and ultimately led to the establishment Queensland Compound Library (QCL). This was again fortuitous as the scheme that followed on from NCRIS, the Super Science Initiative (SSI), was delayed until 2008 with in-principle funding earmarked for mid-2012. The postponement was no doubt due, in part, to the global financial crises that afflicted many economies prompting governments to offer stimulus packages and curtail other investments.

Importantly, it should be noted that the SSRFF is not a grant but rather a loan made by the State government to Griffith University. The loan is very generous with repayments commencing after the tenth year and staggered over the next 20 years. Moreover, one-third of the loan is forgiven if the facility reaches agreed Key Performance Indicators (KPIs) so that two-thirds need only be paid in the final 20 years. As part of the loan agreement, the State also owns a percentage of any profits generated by the QCL. This percentage is worked out using a formula that takes into account the up-front investment made by the State and other stakeholders to establish the facility. The SSRFF and Griffith University helped provide resources at a critical time for Australian life science research. Without this scheme, it is doubtful whether a national compound management and logistics facility with an equipment base like that of the QCL would have been established.

It should also be noted that the exchange rate of the AUD to the USD and GBP in 2007, when equipment was being ordered and successively paid, was not particularly strong and did not allow the most comprehensive package to be obtained. The compromises made early on, especially with the sample processing robot, came back to haunt the facility. It was fortuitous then, that the SSI, which had been delayed by 2 years, was reactivated at this critical juncture and the QCL was able to access additional funds. Indeed, the exchange rate from late 2008 onwards improved dramatically and, at the time of writing this chapter, had reached parity, with the USD facilitating the acquisition of an expanded platform.

Stakeholder Involvement

Despite funding from the Queensland government, a road show was nevertheless organized by the Griffith University team during the latter part of 2006 that included stopovers at Melbourne, Sydney, Brisbane, Adelaide, and Canberra in an effort to provide the NCRIS Committee with additional information to clarify outstanding issues and needs from THD stakeholders. The road show provided invaluable feedback from a national perspective. Although much of the feedback was not directly related to establishment of an integrated compound management and HTS capability *per se*, it did capture the national zeitgeist of what any facility that was established or supported by NCRIS should adhere to. These included

- Benchmarking of a facility's performance against KPIs ensure they operated at world's best practice.
- Adherence to an IP model could redress the lack of incentives to progress innovative discoveries to the market. Stakeholders believed that inventors should be given complete ownership of any IP they may create. No claim to IP arising out of any discovery made through use of an NCRIS-funded facility should be made. It was argued that this would provide a protected environment for progression of promising commercial ventures in a timely fashion without reach-through clauses.
- All organizations, whether academic, public/privately funded research institutes or industry should be able to access facilities to prosecute basic or applied research.

- Projects that are funded in part or in full by NCRIS facilities must first be submitted to an independent panel of experts to assess the quality of science in an effort to maintain a high standard of research funded from the public purse.
- Specific project costs (reagents, consumables, etc.) would typically be borne by the user, while other overheads would be assessed on a case-by-case basis. In this model, industry may be charged at full cost recovery while academic groups could bear project-related costs only.

Specific feedback from potential users of the QCL at existing and planned HTS sites as well as academic chemistry departments echoed the same general sentiments. Quarantining IP was considered essential for all users; that is, the facility should not invoke reach-through clauses that laid claim to IP generated by users. Failure to do so would potentially negate patronage.

Storage that operated at world's best practice was also high on the agenda. Chemists did not want to submit precious samples that, in some circumstances may have represented the world's supply, only to have it degrade through inadequate curation or handling. Likewise, screeners needed to be sure that any compounds being accessed had not decomposed over time for the same reasons. In addition, the HTS sites highly recommended the inclusion of kit capable of dispensing nanoliter (nL) volumes given a general move toward cell-based assays and biochemical assays in low volume 384-well and 1536-well microtiter plate formats.

Development of IP Model for the Queensland Compound Library

The clarion call from stakeholders regarding IP was that ownership should rest with users that made commercially viable discoveries and not in any way be diluted by the QCL. This was not unexpected as Australian academia and industry were sensitive to the fact that IP had, in the past, not always been in a mature position before partnering. While this may be more of an issue with value adding, the fact that several facilities could lay claim to IP as value was added would ultimately erode the inventor's IP base, make successive/downstream contracts unwieldy, and stymie innovation. In this sense, users unanimously agreed that the remit of facilities like the QCL was in the provision of critical infrastructure and services that enabled research.

Handling of IP Associated with Chemical Structures

Stakeholder feedback indicated one potential chemistry depositor had taken great pains to preserve the IP of their corporate library over the preceding 20 years. Their library had been screened by various agrochemical companies resulting in several collaborative projects and patents for animal health and crop protection. While this organization was supportive of making their compound library accessible for human biomedical research, it was absolutely imperative from their perspective that chemical structures would not be divulged to prospective collaborators unless some prior agreement had been reached. The onus on IP protection effectively meant that the facility would not receive structures to obviate any inadvertent disclosure.

Some academic chemists who feared divulging structures would weaken their position to enter into a collaborative partnership with a biologist also took up this position. It should be noted, however, that the vast majority of academic chemists were comfortable with submitting structures with the facility as great pains were taken to communicate that such a facility would only work if there was a spirit of cooperation among all users.

However, while some organizations or individual chemists were not in a position to divulge structures, their comfort level was much higher for submitting fingerprints and physicochemical profiles (log P, molecular weight, number of H–bond donors and acceptors, etc.). Fingerprints are an abstract form of the structure that allow substructure and similarity searching, but not exact matching. All fingerprints are held by the facility and searching for similar structures or common substructures is undertaken by facility staff. Physicochemical profiles, on the other hand, are provided to users that have an interest in selecting lead- or drug-like molecules [38–40].

To assist fingerprints and physicochemical profiles to be produced in a secure environment, the QCL developed a software package that could be copied to the hard drive of a desk or laptop computer. This software is provided gratis to chemistry depositors and permits structures to be read within an organization's own network. This avoids commercially sensitive data being transmitted over the Internet and residing, even momentarily, on the QCL's server. The output can be produced in various file formats that allows editing before submission.

Handling of IP Associated with Screening

Because the QCL is dedicated to the delivery of a compound management and logistics service, it does not undertake any screening and all biological data resides with the screening group. The right to access compounds is only possible after an agreement is signed that prohibits any attempts at structural elucidation of compounds or passing the compounds onto a third party.

Given structures of compounds are not always divulged, there is a strong possibility that screening all, or part of, the selection will be done blind. In this scenario, the screening group will hold biological data associated with a compound (even if just an identification and/or batch number) and would initiate collaborations with chemistry groups that deposited molecules, which most effectively modulated the biological system.

The facility also prepares usage reports for chemistry depositors upon request. This allows a chemist to contact groups that may have screened their compounds but, for various reasons, did not pursue collaboration. This affords a mechanism for both groups to potentially copublish data related to the compounds deposited by the chemist and screened by the biologist.

The UN CBD and Handling of IP Associated with Traditional Knowledge

Some stakeholders raised the issue of compliance with the United Nations Convention on Biological Diversity (UN CBD), as well as access rights, traditional knowledge, and

IP of indigenous peoples. These considerations are generally associated with natural product drug discovery programs. In the former case, the facility places the onus on the user depositing natural product extracts to guarantee, via specific clauses in a standard agreement, that the original biota has been collected in accordance with the UN CBD.

This is particularly important as Article 15 of the UN CBD recognizes the sovereign rights of states (countries) over their natural resources and that the authority to determine access to genetic resources rests with national governments and is subject to national legislation. Article 15 also specifies that access shall be on mutually agreed terms (Article 15. 4), subject to the prior informed consent of the contracting party (Article 15.5) and the benefits arising from the commercial and other utilization of genetic resources shared on mutually agreed terms (Article 15.7).

The Commonwealth of Australia is a signatory to the UN CBD. As such, the facility is duty bound to comply with legislation enacted by the States that give effect to Article 15. Therefore, the facility

- Will accept extracts, molecules, or other genetic material derived directly from the biosphere where the depositor warrants that access and benefit sharing agreements have been entered into with the applicable authorities
- Will suspend access to any samples in doubt until *bona fide* documentation is sighted
- Will not accept extracts, molecules, and other genetic material derived directly from the biosphere where the depositor cannot warrant that access and benefit sharing agreements have been entered into with the applicable authorities.

In the latter case, the facility recognizes traditional knowledge and the IP rights of indigenous peoples where collecting specimens was undertaken with this knowledge. As such, it will suspend access to any samples derived from the biosphere where a dispute arises following collecting activities related to access rights by indigenous peoples, traditional knowledge, or indigenous IP. The suspension will remain in effect until a satisfactory settlement is reached between the parties concerned. Confirmation in writing attesting to the settlement must be signed by all parties involved and sighted by the facility.

Summary of IP Model

A unique IP model was developed in response to feedback from a range of stakeholders to meet the current Australian need. Preexisting IP of users and IP generated through use of the facility would remain intact and not be diluted. Because the facility was established as an inclusive national resource that aimed to synergize interactions, not only between chemists and biologists, but also a variety of organizational types, the model had to strike a balance between access to compounds deposited by chemists and how biologists could use them. The model also had to be sensitive to the needs of all users from academia to industry. In this respect, a flexible framework that let the user decide the level of comfort associated with submission or posting of data was employed.

Thus, chemists from industry would be permitted to submit molecules with a limited amount of detail (typically only a compound identification number), while those from academia were free to divulge structures.

Biologists are at liberty to choose the chemistry groups that they would prefer to partner with on downstream collaborations. This is particularly relevant for academic groups where partnering with research leaders with strong track records enhances proposals to the two leading funding schemes in Australia.

It is important to note that the facility does not become involved in any negotiations regarding further collaboration, publication, or IP between users, and it becomes the remit of the biology and chemistry owners to progress a project. This model allows synergies to develop and mature into projects that are prosecuted in a way best suited to the partners. As a result, molecules submitted by chemists may be tested further to interrogate biological function as part of a chemical biology project or form the basis of a drug discovery program.

IP considerations also take into account global conventions like the UN CBD and IP rights of indigenous peoples. Policies for both circumstances are posted on the QCL's website for additional clarity, while specific mention is made in all agreements with chemistry depositors regarding their obligations under the UN CBD should they submit samples derived from the biosphere.

Development of an Operating Model

Viewed simply, the QCL encourages organizations to deposit and/or access samples. Deposits may be made via one of three categories: open, restricted or closed. Chemists depositing molecules in the open category have not encumbered the sample in any way and it is free to be used in all assays. Restricted compounds are essentially the same except the chemistry owner may have ongoing collaborations with, for example, an agrochemical company. In this instance, the owner can still make the compounds available for screening but restrict usage to nonagrochemical targets/assays. The closed mechanism, as the name suggests, means that the chemistry owner does not permit access directly via the QCL so that the samples are essentially invisible. Closed samples may still be accessed following written permission being received from the owner.

To allay concerns related to IP, there is no requirement to submit structural details. This would ideally suit chemists that deposited samples into the closed and possibly restricted categories. Chemists that deposit closed sample sets would then pursue collaborations with biologists. This proactive mechanism is available to all chemistry depositors to facilitate screening of their compounds by life scientists outside of the QCL's member base.

Biologists, for their part, may access all open and restricted category compounds (provided their screen does not impinge on any restrictions in the latter case). The benefit is that potentially novel compounds may be found following screening as the samples have been sourced from academia for the most part. The novelty stems from the fact that academic chemists work in niche areas that give their group a point of difference and therefore, in the best-case scenario, produce novel structures or, more generally, molecules that are not normally available through commercial vendors. Importantly, biologists are not able to loan or give the samples they accessed from the QCL to any other party. Neither are they permitted to reverse engineer a sample's structure. They do, however, have the final say in deciding what chemistry group they may wish to partner with.

Because the facility has to cater for a range of organizational types with different business models, a few basic options were developed that allowed value for money based on circumstances or need. In all cases, an organization pays a "member contribution" that allows access to the facility and any "fee-for-service" work.

The simplest is *Ruby* whereby an organization can choose to be a member on a year-to-year basis. The member organization generally pays a low membership contribution to deposit or access compounds. The QCL covers all consumable (microtubes, septum seals, DMSO) and staff costs required to prepare the sample. This option is best suited to universities or institutes with staff that wish to deposit or access relatively small numbers of samples.

Sapphire is geared for members that may wish to access or deposit much larger sample sets. Here the membership contribution is higher per annum to cover the ongoing effort of facility staff, but is balanced by lower fee-for-service costs. Several industry organizations have chosen this option, as it is highly flexible and can be tailored to meet their specific needs.

The final option, *Emerald*, is an all-inclusive alternative that rolls deposit and access into a simple "set and forget" formula. The member contribution allows all staff from the organization to deposit a large number of samples and access any of the "open" and "restricted" compounds in the collection. Larger universities with independent research groups prefer this option as it allows staff to access the facility at very affordable rates.

Establishing the Infrastructure and Developing the Work Stream

Excellent HTS facilities have been set up throughout Australia that, collectively, are equipped with a wide array of screening and imaging technologies to prosecute drug discovery and chemical biology programs. The advent of a dedicated compound management facility to augment this investment benefits from economies of scale and adoption of a common process for sample preparation and management. The consolidation of small molecules into a central repository also achieves a greater coverage of chemistry space compared to isolated collections, facilitates entry into the HTS and phenotype screening paradigms as the collection is sizeable and can be reformatted directly *en masse* into labware used by biological research teams, provides a mechanism to add value to molecules that may otherwise have lain neglected by way of publications or patents and generally synergizes collaborations between chemists and biologists.

Four basic storage concepts were commercially available at the time to address storage of samples: large aisle-based sample stores that could occupy entire floors; paternoster (vertical) systems; compact (fixed-position) stores that can be expanded to meet future demand and; highly compact, scalable, more mobile stores.

Design

Feedback provided through fora like NCRIS and the existing HTS sites indicated that the facility had to operate at world's best practice. In this sense, the end product could not be cobbled together from low cost components that could not address the preservation of sample integrity for extended periods (5+ years) and delivering sample requests in a timely manner.

Additionally, a scoping exercise was undertaken to gage initial capacity and capability requirements from likely users of the QCL. This exercise indicated that chemists were comfortable with submitting sufficient quantities of sample that would allow preparation of 1 mL of a 5 mM dimethyl sulfoxide (DMSO) solution. The actual concentration was based on literature reports that balanced storage at ambient temperature against issues associated with freeze-thaw cycles when samples are maintained at low (−20°C) temperatures [41–45].

The 1 mL volume and storage at c. 19–20°C for stock solutions helped refine the design and logistics for the QCL. Thus, a system configuration consisting of three main building blocks (microtube storage, sample processing unit and microtiter plate storage) was ultimately chosen. Importantly, each of the main elements had to be scalable so that additional modules could be added to cater for increased capacity and throughput with minimal disruption to on-going operations. In addition, environmentally controlled tube and plate storage had to be integrated with sample processing to ensure compound integrity without compromising flexibility. Other instrumentation, like an acoustic droplet ejector (a capability voiced through user feedback), was then added to this base provided compatibility with the modular design concept was maintained. The final configuration was groundbreaking in that it was the first system to fully integrate modular, scalable components into a fully functional facility.

Because the scoping exercise indicated the initial influx of samples would be a modest 50,000–100,000 compounds, the QCL design principle was governed by modularity and scalability. This would allow the facility to meet the immediate need and also address future proofing. Ultimately, the decision was made to proceed with the more compact, mobile systems because of budget and space constraints. The choice of sample processing unit was also determined using the same principles of modularity and scalability. In addition the sample processing unit had to be highly adaptable to allow the interchange of components over time in an effort to keep abreast of current trends and meet user demand.

Modular components that allowed a stepwise increase in capacity or throughput, if required, were selected. The initial configuration comprised 2 × TTP Labtech comPOUND® microtube stores, each capable of storing 100,000 microtubes, a combined Hamilton Storage Technologies (HST) asmStore and asmPlateServerST that could store and retrieve of 2160 standard microtiter plates fitted with SealTite™ lids, and an Agilent BioCel® sample processing unit (Figure 31.1). The BioCel® acted as the "master" unit to the "slave" storage modules. This arrangement utilized VWorks to interact with the software on both the tube and plate stores. In effect, orders could be placed through VWorks to store and retrieve samples held in microtubes or microtiter plates via the comPANION™ and PlateServer respectively.

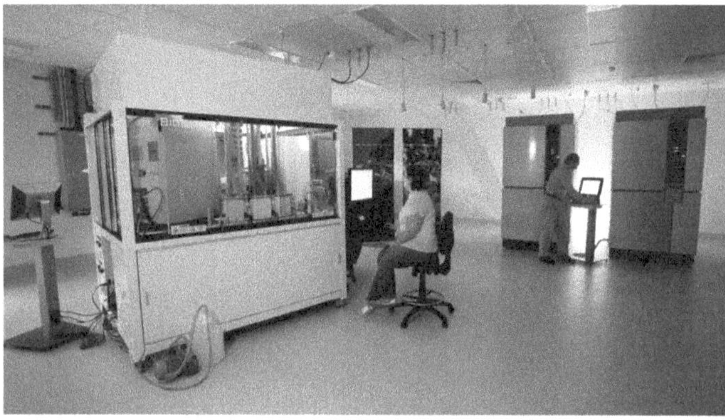

Figure 31.1. Modest beginnings: the QCL's capacity in early 2008 included storage of samples in 200,000 microtubes and 2000 microtiter plates. A sample processing robot (foreground) could utilize both types of labware. Courtesy of Chris Stacey, Griffith University.

Microtube Storage

Samples stored as microtubes are individually cherry picked from the stores, obviating unnecessary freeze thaw cycles intrinsic to plate-based systems. Moreover, racks of retrieved tubes can be arrayed in user-specified formats to simplify subsequent plate creation. Each tube carries an individual 2D barcode (Figure 31.2), which is read on the way into and out of the store, making it virtually impossible to obtain the wrong sample. Every microtube used in the QCL has a unique 14 × 14 alphanumeric 2D barcode. This matrix affords 3,600,000,000,000,000 individual barcodes, that is, 3.6 quadrillion unique permutations. Thermo-ABgene, suppliers of microtube labware used in the QCL, warrant no two will have the same code.

Compounds submitted by chemists are initially transferred to a microtube and then dissolved in dry DMSO. The microtubes are then deposited into the store for curation under a dry nitrogen atmosphere for up to 5–6 years. The format free environment facilitates cherry picking of individual microtubes for reformatting into microtiter plates. Microtube subsets for retest or counter screens are accessed as easily as the entire set is for a primary screening campaign.

Microtubes are retrieved and collated by the comPANION™ at the rate of approximately 96 tubes every 8 minutes. This was considered more than adequate for an initial compound collection of 200,000 samples. Higher rates are obviously necessary in an industrial setting if over 1,000,000 samples are screened to progress a project at the same rate as a smaller collection. Here, dedicated buildings containing several high-speed pick-and-place robots are employed. This was clearly outside the scope of the QCL's budget and anticipated influx of compounds following the scoping exercise. Moreover, because the QCL was designed around modular, scalable kit, the capacity could be expanded to accommodate a further 200,000–400,000 microtubes if and when the need arose.

Figure 31.2. Microtubes showing the 2-D barcode used to track sample usage and location. Courtesy of Trine Barfod-Jensen, Griffith University.

It is imperative to maintain a low humidity environment because of the potential for precipitation and degradation of samples due to ingress of adventitious water. At best, the presence of water can precipitate compounds but, in the worst case scenario, promote degradation, especially if a strong acid, such as trifluoroacetic acid, was used to cleave a molecule from a solid support and/or purify the compound prior to curation [46]. Precipitated compounds may be rescued by forcing them back into solution, typically through ultrasonic techniques [47]. Degraded compounds, on the other hand are discarded.

A direct consequence of precipitation is that chemistry may not be screened at the expected concentration resulting in a higher rate of false negatives. Compound management groups have recognized this issue over the past 10 years and responded by ensuring all curation and liquid handling manipulations are carried out under an inert atmosphere (nitrogen) or at a low relative humidity (10%). Analytical equipment such as an Echo™ 550 acoustic auditor resident within the BioCel® can determine the extent of water uptake in microtiter plates while LC-UV-ELSD- (CLND)-MS- (NMR) techniques scan for potential degradation products. The QCL employs the nonbracketed analytical techniques. Ideally, every microtube sample would be analysed prior to storage in conjunction with on-going random selection of between 5% and 10% of the entire library per annum. Complete analysis of the whole library is cost prohibitive for a relatively small facility like the QCL.

Sample Processing

Agilent BioCel® units undertake all reformatting and replication tasks. *Reformatting* is where an aliquot is taken from a particular storage format, such as microtubes, and transferred to another format, like a 384-well microtiter plate. Microtitre plates can also be reformatted in a process known as quadranting. Here, 96- or 384-well plates may be reformatted into higher density 384- (4×96) or 1536- (4×384) well plates respectively. *Replication*, as the name suggests, is a direct transfer of an aliquot held in one format to a destination in the same format, for example, the preparation of daughter plates from working plates where both would be the same plate density and layout.

Agilent and other third party devices are positioned around a radial robot that allows rapid processing of racks of microtubes and plates retrieved from the storage units. Two × VPrep® precision liquid handlers allow for simultaneous processing of both septum-sealed tubes and microtiter plates. A Labcyte Echo® 550 acoustic dispenser provides noncontact, low volume (nL) dispensing capability. Because liquid is transferred through an acoustic mechanism in the latter case, consumables are not required for operation, making the Echo series and related products an extremely attractive alternative to syringe-based technologies.

Nanoliter dispensing is becoming increasingly more popular as assays are miniaturized ever further into low volume 384-well plates and 1536-well plates. Some cell based assays like those run using FLIPR® technology also benefit from acoustic droplet ejection, by reducing the volume of DMSO present, a causative agent for cell death. The Echo® 550 can deliver miniscule volumes of concentrated solution (up to 10 mM) rather than larger volumes of more dilute solutions typically used to probe molecular targets with other technologies.

To complement the environmental control in the storage units the BioCel® robots undertaking plate-based operations are equipped with an inert box in which "plates in waiting" reside, and a PlateLoc® thermal sealer with integrated inert gas purging that enables plates to be sealed under dry nitrogen.

The initial configuration balanced microtiter plate and microtube storage with sample processing of both types of labware on the same BioCel. However, as alluded to earlier when establishing the facility (section 5.0), the initial set-up was not a complete solution due to the poor exchange rate of the AUD with most foreign currencies. When storage capacity was increased to accommodate an influx of additional samples beyond that originally catered for (section 6.1), the ability of the sample-processing robot to maintain throughput was compromised and turn-around times for the preparation of assay ready plates and processing of microtubes suffered. To ameliorate the situation, two additional BioCel® robots were purchased in 2012 from SSI funds. This allowed collation of microtubes (Figure 31.3) and microtiter plates (Figure 31.4) to be spread across two dedicated robots while preparation of assay ready plates could continue on the first BioCel®.

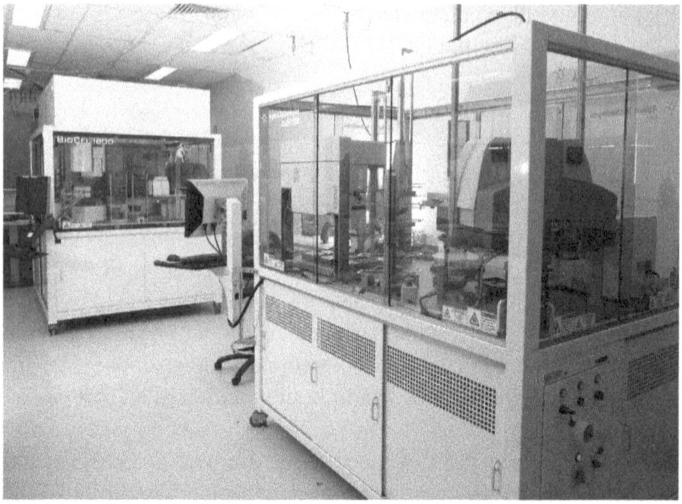

Figure 31.3. Additional sample processing unit dedicated for microtubes (foreground). The original sample processing robot is in the background. Courtesy of Chris Stacey, Griffith University.

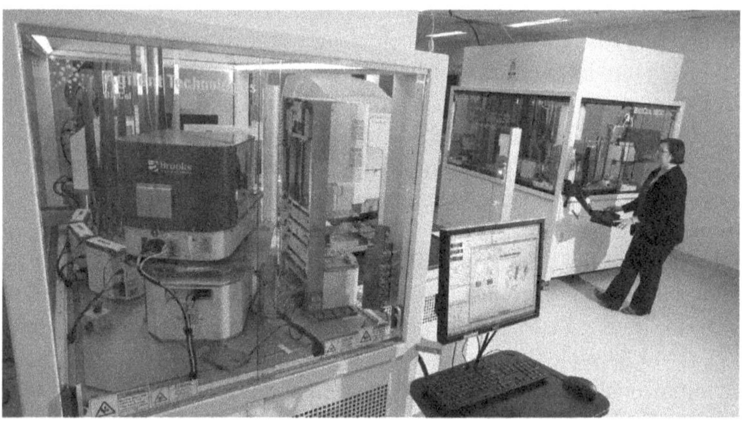

Figure 31.4. Additional sample processing unit dedicated for microtiter plates (left). The Echo ADE has been moved from the original sample processing robot (background) to a docking table at the back right hand corner. Courtesy of Chris Stacey, Griffith University.

The highly flexible nature of the BioCel® lends itself to the above proposal; "gross modularity" facilitates scalability (i.e., additional BioCel® robots can be incorporated into the facility) while the "fine modularity" allows components to be interchanged between the two decks to meet the current need/demand. In the QCL's case, the planned introduction of a dedicated BioCel® for microtubes allowed greater throughput to be achieved via acquisition of several plate hubs. Servicing of the Echo has also been facilitated by removal from the existing platform and placement on a docking table attached to the BioCel® dedicated to plate collation/processing. An off-line Bravo can also be positioned on the plate-centric robot as shown or kept off-line as a back-up unit for periods of peak demand.

The 3 × BioCel® units have a high degree of redundancy. Processing of assay ready plates, for example, can now be undertaken on all three platforms.

Microtiter Plate Storage

Microtubes are typically reformatted into a deck of "working" plates to facilitate downstream production of assay ready plates using multichannel 96- or 384-well pipetting heads. Assay ready plates are plates that are compatible with the screening technology and to which compound has been added. The target protein (or cells) and other reagents are added to the chemistry. Once made, working plates can be reused numerous times to afford multiple decks of assay ready plates. Because working plates may be accessed several times, they are maintained in the same dry nitrogen atmosphere as the microtubes and typically discarded after 6–12 months, after which a new set of working plates is made.

The QCL employs the HST active sample manager (asm) for microtiter plate storage and retrieval. The asm is a compact, modular, localized storage system that provides storage and retrieval for up to 2200 microtiter plates. The asm PlateserverST can unseal and deliver plates via a handoff arm directly onto the BioCel's deck. SealTite® lids facilitate fully automated and repeated sealing and unsealing of microtiter plates, with no heat and adhesives to contaminate compounds in storage. SealTite lids are manufactured from stainless steel with a 100% DMSO-compatible inner liner that conforms to the shape of the microtiter plate wells.

The asm provides a controlled environment between ambient (room) temperature conditions and −20°C. An inert gas environment is being employed by the QCL to maintain a low humidity within the storage environment. These environmental options afford optimum conditions for compound storage and help preserve sample integrity.

Work Stream

The typical QCL work stream (Figure 31.5) commences with submitted samples that are prepared as 5 mM stock solutions in microtubes and then reformatted into working plates that are ultimately replicated as assay ready plates. Each microtube is identified by a unique 2D barcode, which is read on the way into and out of the stores to assist cradle-to-grave tracking of all samples. A 200 μL disposable tip, 96 channel pipetting head is used to initially add dry DMSO directly to the microtubes to initiate the solubilization process.

Solubilization is facilitated with a custom plate "rotisserie" designed by Agilent (Figure 31.6). This device has proven extremely efficient in solubilizing samples and can accommodate up to 10 × microtube racks (960 samples) per run. It has provided the QCL with a low cost alternative to kit based on sonic pulses.

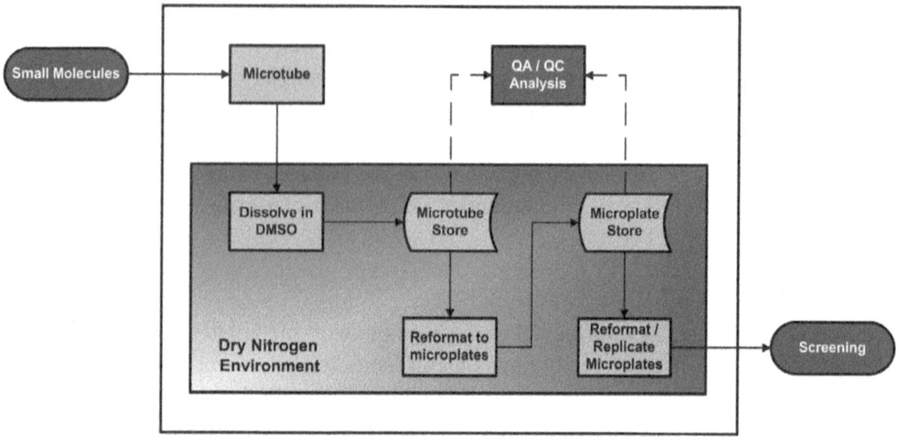

Figure 31.5. Workstream from submission of small molecules to preparation of assay-ready plates. Reprinted with permission from *Struct Chem.* 2010;21(5):1117–1129, © 2010 Springer.

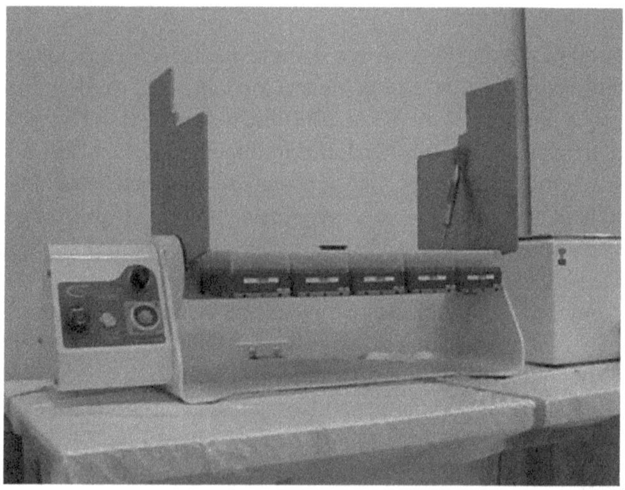

Figure 31.6. Solubilization of samples in DMSO is facilitated by a bespoke "rotisserie" design piece.

Microtubes are stored at 19–20°C in a low humidity, nitrogen rich environment and can be individually picked from the tube store. Microtubes allow individual samples to be accessed unlike plate-based operations in which additional samples are unnecessarily removed from the environmental chamber. Storage at 19–20°C, which is just above the freezing point of DMSO, obviates unnecessary freeze-thaw cycles. This ameliorates potential precipitation of compound due to the nonideal behavior of DMSO-water mixtures should ingress of water occur [48]. Racks of retrieved tubes can be arrayed in user-specified formats to simplify subsequent plate creation.

Microtubes are typically reformatted into "working" plates, which are maintained for 12 months in the HST PlateStore. A 50 μL fixed tip 96 channel pipetting head is employed to prepare working plates in either 96- or 384-well plate formats. The fixed tips enable piercing of the pre-split septa caps thus obviating the need to constantly remove and replace caps under an inert atmosphere and thus save on consumable costs. Liquid handling devices located on the deck of the BioCel® transfer microliter and nanoliter volumes of samples from the working plates into the assay ready plates. A 384-pipetting head allows plates to be replicated in an efficient manner while an acoustic dispenser permits volumes as low as 2.5 nL to be transferred with coefficients of variance $\leq$ 5% for cell based screens that may be sensitive to DMSO in the assay [49].

Every rack of microtubes that is assembled, every working plate, every assay ready plate and every cherry picking plate is labeled with a linear barcode to address cradle to grave sample tracking via a bespoke data base that was developed in-house known as the QCL Manager.

Intended Outcomes

The Queensland Compound Library was established to synergize interactions between Australian chemists, biomedical researchers, and their international colleagues. By promoting collaborations, value can be added to the already excellent basic medical research, synthetic organic chemistry and natural product expertise in the region. Chemists can deposit compounds into a central repository thereby allowing biologists access to a unique suite of molecules in screen-ready microtiter plates.

The consolidation of Australian chemistry into a central repository results in a greater coverage of chemistry space than any single collection in the country, public or private, could achieve. Biologists can requisition all, or a subset of, the open-access compounds from the small molecule repository to evaluate against their targets. Because the QCL provides format free storage of microtubes, it can deliver any number of molecules in a wide range of plate types and formats to screening groups.

Extensive Use of a National Resource

The QCL has grown dramatically since commencing operations in mid-2008. Within 1 month of the equipment being commissioned, 150,000 compounds were submitted for storage necessitating the immediate acquisition of another tube and plate store. An additional 115,000 compounds were deposited from a variety of organizations 12 months later. A fourth tube store was introduced in early 2010 to accommodate the growing demand, and in 2012 additional capacity was provided by means of double density store that could house 200,000 samples, bringing the total number of microtubes to 600,000. It was gratifying to observe that the additional storage modules were readily integrated within 2–3 days, thereby minimizing disruptions to ongoing operations.

In addition to the growing number of samples being deposited for processing into microtubes, plate making commenced almost immediately and has continued unabated ever since. Almost 20,000 microtiter plates have been produced from microtubes or replicated from working plates in the 5 years since the QCL commenced operations.

We have observed a strong demand from users for assay-ready plates to be prepared using the acoustic dispenser. Plate formats for this type of activity have typically been in low-volume 384-well plates or 1536-well formats. Volumes requested range between 20 and 100 nL. The Echo has proven to be extremely reliable in meeting demand. VWorks has allowed cherry picking of samples from 384-well plates and reformatting into 1536-well formats.

Since commencing operations, some users realized that their facilities were inadequate for long-term storage and have submitted plate-based legacy collections. This format necessitated acquisition of kit that permitted cherry picking of microtiter plate wells. The acquisition of an Agilent Bravo addressed the need for well-to-well cherry picking. Because samples held in plate-based libraries are minor compared with those in microtubes, a trade-off between efficient cherry picking and throughput was made in favor of the latter. After 5 years in operation, all samples are submitted for microtube storage, even those from commercial suppliers. The QCL works with both the chemistry owner and vendor to enable samples to be either placed directly, or easily transferred, into microtubes.

FACILITATE COLLABORATIVE RESEARCH

Biological screening is one of the strongest mechanisms to engage the biology and chemistry research communities. Screening the combined collection has enormous potential to negate missed opportunities for both chemists producing novel molecules and biologists in quest of chemical starting points for validated targets. Lead compounds identified following HTS or phenotype screening may be progressed as an academic investigation or prosecuted further to deliver a therapeutic, agrochemical, or other marketable product. It becomes the remit of the biology and chemistry groups to progress a project in the QCL model, and potential collaborations are negotiated directly between the two groups.

Several groups have utilized the services of the QCL to prosecute international collaborations. The Eskitis Institute at Griffith University, for example, has developed a natural product screening set (Nature Bank) in which extracts and fractions are enhanced with lead- and drug-like components. The screening set comprises over 600 × 384-well plates. Copies have been sent to various industry partners, nonprofit organizations, and universities for screening against cancer and anti-infective targets. Similarly, the Commonwealth Scientific and Industrial Research Organization (CSIRO) has opened its unique library for biomedical and agrochemical research to industry and nonindustry partners alike.

Enable Value Adding and Prosecution of a Commercial Outcome

Australia is strong in terms of the quality of its basic science funded by the Australian Research Council and National Health and Medical Research Council. Its major weaknesses are the lack of mechanisms and strong incentives for translating the innovative discoveries through to commercial outcomes, and the paucity of a culture of entrepreneurial and risk-taking activity which can promote developmental research. To achieve this, inventors are given 100% ownership of the IP they create. No claim to IP arising out of any discovery by the facility is made in this model. This provides a protected IP environment to facilitate progression of promising commercial ventures in a timely fashion.

This model is ideal for organizations that are most active in the early discovery phases of the drug development pipeline, and represents the space where the majority of Australian industry and academia currently reside. Some chemists have taken advantage of the proactive storage mechanism to pursue collaborations of their chemistry with large pharmaceutical companies and biotechs. The QCL is not part of any negotiations, and contracts are a matter between the user and a potential collaborator. The fact that the QCL does not impinge any IP rights can bring additional surety to negotiations.

Benefits to Industry

Industry can utilize the facility in a number of ways. First, an organization can deposit compounds for screening against biological targets. This strategy exposes the corporate collection to selected targets that can potentially lead to fruitful new avenues of research with a strong IP position and, ultimately, a marketable product. Several industry organizations use the QCL in precisely this manner although the actual mechanism to achieve this goal can vary.

Some organizations have acquired commercial libraries that are screened against in-house targets. As part of their business model, these organizations will sell a copy of their library to selected partners in exchange for the first right of refusal if a promising hit is discovered. Other organizations have invested heavily in medicinal chemistry and produced novel molecular libraries. In both examples, all compound management activities are effectively outsourced to the QCL in both business models. This allows the organizations to benefit from automated storage and retrieval at optimal conditions without having to invest in expensive infrastructure worth several millions of dollars, employment of dedicated staff, training, and payment of preventative maintenance contracts or repair costs. Moreover, in each case, the corporate compound collections are deposited as closed sets, and no structural details are provided. Being closed samples effectively ensures that the molecules are invisible to all other users of the facility and places the onus of screening onto the organization.

Second, the QCL's collection of open-access compounds can be evaluated against a propriety target that has been identified in-house to further probe biological function or identify chemical starting points for a drug development program. Access to unique molecules produced by academia has the potential to identify an innovative lead molecule for downstream optimization. The Lilley PD^2 Initiative works along a similar model in that chemistry is selected after passing appropriate filters for a series of assays in any given module.

Benefits to Academia and Other Nonindustry Organizations

The metrics that drive academia are vastly different from those that drive industry. Success is measured, *inter alia*, by the number and quality of peer-reviewed publications, typically in journals, the ensuing number of citations contributing to the "H-index" [50], and funding awarded through competitive grants.

Successful academic screening programs should be a joint venture between two highly motivated groups of scientists: a biology team working on their targets and a medicinal chemistry group working on chemical optimization of hits in their area of expertise. As is the case for industry, the screening of a unique set of compounds sourced from academic laboratories opens up the possibility for serendipitous discovery of new modulators for biological targets and pathways. Novel structures (and structural classes) with new modes of action can be published in high-impact journals that, in turn, can lead to increased citations with concomitant boosting of the researcher's H-index and grant income.

In instances where the target is indeed druggable, such as the COX-2 protein mentioned earlier, then the potential rewards in disclosing a modulator have enormous implications in terms of future income streams for both institutions. A spin-off can be formed to dedicate time and resources to chemical optimization, ADMET, and proof of concept in animal models so that a solid IP position is obtained before partnering prior to clinical trials.

Benefits to Chemists and Biologists

There is no value in storing molecules in a cupboard or freezer after they have been synthesized or isolated. In rare circumstances, limited kudos may be gained from loaning a sample to another researcher for, say, checking the physical characterization of a natural product synthesized in the laboratory against the originally isolated material. The deposition of compounds into a central repository, however, increases the potential value of a chemist's compounds via exposure to a range of biological targets that would not otherwise be possible.

It is difficult to justify screening 10 s to 100 s of compounds on infrastructure geared for several orders of magnitude more compounds. A screen becomes increasingly wasteful if small numbers of compounds are screened in an *ad hoc* fashion using labware that is geared for up to 1536 simultaneous assays. The empty wells are still filled with reagents, a protein or cells, all of which may have a significant cost component attached to them. Consumables like tips are also used unnecessarily as not all automated pipette heads can pick up tips at selected positions that match the samples in the microtiter plate. By depositing compounds into a central repository, chemists can have their molecules incorporated into a larger set that may number 10,000–25,000 samples. This is also a more attractive proposition for researchers in the life sciences that can choose to screen the whole library or a subset. A central repository therefore allows chemists to create critical mass and make the resulting suite of small molecules more attractive for screening. Ultimately, the collation of small disparate collections into a larger entity facilitates construction of a structurally diverse library that can provide valuable insights to biologists.

Storage can be via one of two modes for chemists in the QCL model, that is, passive or proactive. Passive storage occurs when the chemist submits samples for *potential* access by biologists. Here the biologists drive the selection of compounds. This may be based on physicochemical profiles or structures known to be active in specific target classes. Proactive storage, on the other hand, is where the chemist pursues third-party collaborations with, for example, an industrial partner. In proactive mode, the QCL is able to readily reformat the chemist's samples into microtiter plates for biological evaluation by the collaborator. This mechanism dramatically increases the likelihood of an individual collection becoming the starting point of downstream projects.

Finally, for academics (and particularly chemists), there are an increasing number of journals dedicated to chemical biology and automated screening. Some like *Nature Chemical Biology* have impact factors beyond that of the most esteemed chemistry journals. Others like *ACS Chemical Biology* that have been in circulation for less than 10 years have already achieved impact factors on par with, or exceeding, the most cited (and, some would argue, most prestigious) journals in organic and medicinal chemistry. By way of example, *The Journal of Organic Chemistry* and *Journal of Medicinal Chemistry* from the ACS stable are both highly regarded journals with premier citation rates for their disciplines, and have been published for 79 years and 57 years, respectively. The 2011 impact factor for *The Journal of Organic Chemistry* was 4.564 while that for the *Journal of Medicinal Chemistry* was 5.614. By contrast, *ACS Chemical Biology* achieved an impact factor of 6.446 after 7 years. It is interesting to speculate whether *ACS Chemical Biology* may someday overtake the flagship publication of the American Chemical Society, the *Journal of the American Chemical Society*, now in its 136th year and with a 2012 impact factor of 10.677.

CONCLUDING REMARKS

The establishment of the Queensland Compound Library will be a valuable tool in stimulating basic biomedical research and drug discovery in Australia. The combination of a unique IP model, preexisting, successful HTS infrastructure and a fully integrated compound handling system will create an environment that encourages collaborations between different groups of scientists.

The integration of environmentally controlled plate and tube storage with sample processing ensures that compound integrity is maintained without compromising flexibility. The modular, scalable approach has already allowed capacity building to be realized with minimal disruption to ongoing operations.

Perhaps the single largest challenge that lies ahead for the QCL is to convert the groundswell of support into affordable access for potential users. Ideally, there would be a low barrier for access to users of the facility, but this would typically only be possible if support similar to that provided by the NIH in the MLP was received. Several options are available through different levels of government in Australia to provision operating expenses for limited periods. Savings in this area can be passed on to users of the facility to improve accessibility. In the interim, a competitive cost structure that distinguishes between industry and nonindustry organizations has been devised that balances access for users with continued viability of the facility.

REFERENCES

1. Bohacek, R.S., McMartin, C., Guida, W.C. The art and practice of structure-based drug design: a molecular modeling perspective. *Med Res Rev*. 1996;16(1):3–50.
2. Spring, D.R. Diversity-oriented synthesis; a challenge for synthetic chemists. *Org Biomol Chem*. 2003;1(22):3867–3870.
3. Hopkins, A.L., Groom, C.R. The druggable genome. *Nat Rev Drug Discov*. 2002;1(9): 727–730.
4. Janzen, W.P., Popa-Burke, I.G. Review: advances in improving the quality and flexibility of compound management. *J Biomol Screen*. 2009;14:444–451.
5. Lipinski, C.A., Hopkins, A. Navigating chemical space for biology and medicine. *Nature* 2004;432:855–861.
6. Stockwell, B.R. Exploring biology with small organic molecules. *Nature* 2004;432: 846–854.
7. Dobson, C.M. Chemical space and biology. *Nature* 2004;432(7019):824–828.
8. Malakoff, D. Judge turns Rochester's golden patent into lead. *Science* 2003;299: 1638–1639.
9. NIH National Center for Advancing Translational Sciences. Available at http://www.ncats.nih.gov/about/org/organization.html. Accessed January 25, 2014.
10. NIH Molecular Libraries Program. Available at http://mli.nih.gov/mli/secondary-menu/news-events/. Accessed January 25, 2014.
11. EU-OPENSCREEN. Available at http://www.eu-openscreen.de/. Accessed January 25, 2014.
12. Zerhouni, E. The NIH roadmap. *Science* 2004;302:63–72.
13. Austin, C.P., et al. NIH Molecular Libraries Initiative. *Science* 2004;306(5699): 1138–1139.
14. ChemBioNet. Available at http://www.chembionet.info/. Accessed January 25, 2014.
15. FIMM. Available at http://www.fimm.fi/en/. Accessed January 25, 2014.
16. Hibert, M.F. French/European academic compound library initiative. *Drug Discov Today* 2009;14:723–725.
17. Center for Drug Research and Development. Available at http://www.cdrd.ca/. Accessed January 25, 2014.
18. Max Planck Society Chemical Genomics Centre. Available at http://www.mpg.de/512146/pressRelease20050124. Accessed January 25, 2014.
19. UK Drug Discovery Consortium. Available at http://www.ukddc.org/. Accessed January 25, 2014.
20. RIKEN Program for Drug Discovery and Medical Technology Platform. Available at http://www.riken.jp/en/research/labs/dmp/. Accessed January 25, 2014.
21. Institut Pasteur Korea. Available at http://www.ip-korea.org/. Accessed January 25, 2014.
22. Tel Aviv University. Available at http://www.tau.ac.il/∼nchts/. Accessed January 25, 2014.
23. Dundee Drug Discovery Unit. Available at http://www.drugdiscovery.dundee.ac.uk/. Accessed January 25, 2014.

24. Drug Discovery Centre at Imperial College. Available at http://www3.imperial.ac.uk/biomedeng/networks/htt/drugs. Accessed January 25, 2014.
25. Broad Institute of MIT and Harvard. Available at http://www.broadinstitute.org/. Accessed January 25, 2014.
26. Kansas University HTS Laboratory. Available at http://www.hts.ku.edu/index.php. Accessed January 25, 2014.
27. Memorial Sloan-Kettering Cancer Center High-Throughput Screening Core Facility. Available at http://www.mskcc.org/mskcc/html/52147.cfm. Accessed January 25, 2014.
28. Society for Laboratory Automation and Screening. Available at http://www.slas.org/screeningFacilities/facilityList.cfm. Accessed January 25, 2014.
29. Eli Lilly Phenotypic Drug Discovery Initiative. Available at https://openinnovation.lilly.com/dd/. Accessed January 25, 2014.
30. Department of Education Science and Training. Mapping Australian science and innovation. Canberra: Australian Government; 2003.
31. Dwyer, T. Treat medical science as one of our major icons. The Australian Society for Medical Research; 2003.
32. How Canada performs. Available at http://www.conferenceboard.ca/hcp/details/innovation/scientific-articles.aspx. Accessed January 25, 2014.
33. Grant, J. Sustaining the virtuous cycle. Canberra: Australian Government; 2004.
34. Department of Industry Tourism and Resources. Pharmaceutical industry action agenda: local priority—global partner. Canberra: Australian Government; 2002.
35. von Itzstein, M. The war against influenza: discovery and development of sialidase inhibitors. *Nat Rev Drug Discov*. 2007;6:967–974.
36. Simpson, G. Drug discovery: a case for national coordination. *Chem Aust*. 2005;72(7): 22.
37. Department of Education Science and Training. National collaborative research infrastructure strategy strategic roadmap. Canberra: Australian Government; 2006.
38. Hann, M.M., Oprea, T.I. Pursuing the leadlikeness concept in pharmaceutical research. *Curr Opin Chem Biol*. 2004;8(3):255–263.
39. Lipinski, C.A., Lombardo, F., Dominy, B.W., Feeney, P.J. Experimental and computational approaches to estimate solubility and permeability in drug discovery and development settings. *Adv Drug Deliv Rev*. 1997;23(1–3):3–25.
40. Oprea, T.I., Allu, T.K., Dara, D.C., Rad, R.F., Ostopovici, L., Bologa, C.G. Lead-like, drug-like or "pub-like": how different are they? *J Comput Aided Mol Des*. 2007;21: 113–119.
41. Kozikowski, B.A., Burt, T.M., Tirey, D.A., Williams, L.E., Kuzmak, B.R., Stanton, D.T., Morand, K.L., Nelson, S.L. The effect of room-temperature storage on the stability of compounds in DMSO. *J Biomol Screen*. 2003;8(2):205–209.
42. Kozikowski, B.A., Burt, T.M., Tirey, D.A., Williams, L.E., Kuzmak, B.R., Stanton, D.T., Morand, K.L., Nelson, S.L. The effect of freeze/thaw cycles on the stability of compounds in DMSO. *J Biomol Screen*. 2003;8(2):210–215.
43. Matson, S.L., Chatterjee, M., Stock, D.A., Leet, J.E., Dumas, E.A., Ferrante, C.D., Monahan, W.E., Cook, L.S., Watson, J., Cloutier, N.J., Ferrante, M.A., Houston, J.G., Banks, M.N. Best practices in compound management for preserving compound integrity and accurately providing samples for assays. *J Biomol Screen*. 2009;14:476–484.

44. Zitha-Bovens, E., Maas, P., Wife, D., Tijhuis, J., Hu, Q.N., Kleinoder, T., Gasteiger, J. COM-DECOM: predicting the lifetime of screening compounds in DMSO solution. *J Biomol Screen*. 2009;14(5):557–565.
45. Bowes, S., Sun, D., Kaffashan, A., Zeng, C., Chuaqui, C., Hronowski, X., Buko, A., Zhang, X., Josiah, S. Quality assessment and analysis of Biogen Idec compound library. *J Biomol Screen*. 2006;11:828–835.
46. Hochlowski, J., Cheng, X., Sauer, D., Djuric, S. Studies of the relative stability of TFA adducts vs non-TFA analogues for combinatorial chemistry library members in DMSO in a repository compound collection. *J Comb Chem*. 2003;5:345–349.
47. Nixon, E., Holland-Crimmin, S., Lupotsky, B., Chan, J., Curtis, J., Dobbs, K., Blaxill, Z. Applications of adaptive focused acoustics to compound management. *J Biomol Screen*. 2009;14:460–467.
48. Catalan, J., Diaz, C., Garcia-Blanco, F. Characterization of binary solvent mixtures of DMSO with water and other cosolvents. *J Org Chem*. 2001;66(17):5846–5852.
49. Grant, R.J., Roberts, K., Pointon, C., Hodgson, C., Womersley, L., Jones, D.C., Tang, E. Achieving accurate compound concentration in cell-based screening: validation of acoustic droplet ejection technology. *J Biomol Screen*. 2009;14:452–459.
50. Hirsch, J.E. An index to quantify an individual's scientific research output. *Proc Natl Acad Sci U S A*. 2005;102:16569–16572.

Part VI

INTELLECTUAL PROPERTY AND TECHNOLOGY TRANSFER

Everything that can be invented has been invented. —Charles Duell

Collaborative Innovation in Drug Discovery: Strategies for Public and Private Partnerships, First Edition.
Edited by Rathnam Chaguturu.

32

SUCCESSFUL TECHNOLOGY TRANSFER: LESSONS FROM THE TRANSLATIONAL MEDICINE RESEARCH COLLABORATION

Assem S. el Baghdady

AlphaBeta Pharma Group, Leatherhead, Surrey, UK

The Translational Medicine Research Collaboration (TMRC) case study discussed here describes a successful venture in technology transfer in medical research and drug development. It illustrates many of the issues and concepts discussed in the chapter entitled "Unlocking the Market Potential of Academic Research." It also gives insight to some of the complications, unexpected developments, as well as success factors that occur in a real-life setting. It will examine the first 5 years of activity at TMRC, and its executive arm, Translational Medicine Research Initiative Ltd. (TMRI). This case study is relevant because of the unprecedented scale of investment, the model to share risk and reward, and involvement of pharma, several universities, National Health Services, and a regional economic enterprise agency.

Translational medicine is an emerging field that focuses on using biomarkers, new "measurable proteins," to predict, prevent, diagnose, and treat diseases. This relatively young area of biomedicine uses a patient-driven approach to drug development. TMRC was an ambitious, unique, and innovative collaboration between the Scottish Enterprise (SE), four universities in Scotland, UK (Dundee, Aberdeen, Edinburgh, and Glasgow), four National Health Service Trusts in Scotland (Dundee, Grampian, Greater Glasgow, and Lothian), and the pharmaceutical giant Wyeth.* They developed a model for

*Pfizer acquired Wyeth Pharmaceuticals in October 2009.

Collaborative Innovation in Drug Discovery: Strategies for Public and Private Partnerships, First Edition. Edited by Rathnam Chaguturu.

working together to create a fully integrated network to deliver world-leading clinical research.

BUSINESS PREMISE FOR CREATING TRANSLATIONAL MEDICINE RESEARCH COLLABORATION

The TMRC was set up with the intention of transferring academic research more easily and more rapidly to new commercial companies. There was a common scientific aim, but some differing objectives with respect to the results that needed to be managed. Academia searches for answers to fundamental research questions, which may or may not have direct commercial value. Industry in turn expects to realize tangible results from academic research that can be directly applied to their business needs. Industry's increased pressure to produce shareholder value forces them to concentrate their research investment on commercially promising projects. Academic research is essential in the process of arriving at commercially viable knowledge but takes a very long time, and, being research, success is far from certain.

In addition to different expectations, there is frequently a mismatch in management styles between the pharmaceutical industry and academic research groups. Large corporate management, which is in general cautious and product oriented, may find that the informal, exploratory, even somewhat "geeky" lab culture within academia incompatible, and *vice versa*.

Although both parties want to bridge these two worlds and collaborate, the reality is usually that there are too many separate groups, pursuing different objectives. TRMC and TMRI were set up to address this problem. The objective was to create strength and efficiencies through collaboration and coordination.

There were several existing conditions that made the foundation of TMRC and TMRI a logical step, and which were fundamental to their success:

Medical Schools at Renowned Universities

Scotland's four medical schools—Edinburgh, Dundee, Glasgow, and Aberdeen—have existed for centuries, and each enjoys a well-deserved prestigious reputation. However, they also had some of the cultural insularity and rivalry that this implies. Each institution has great talents, but due to size constraints, the individual university medical schools cannot compete on the world stage in ways to match their history or reputation. It became clear that they would accomplish far more through research collaboration. By joining forces through the TMRC they would be more competitive and better able to match efforts of larger research organizations.

National Health Service Trusts

The four National Health Service (NHS) Trusts in Scotland have coordinated their clinical patient work for decades, but work in a centralized manner yields a vital advantage: data management. Each of the approximately 5 million patients under these NHS trusts has a unique identification code, referred to as a Common Health Index (CHI) number, which links an individual to his or her complete medical history and to family members' information, and is stored in a central database. This system enables hospitals and doctors to follow all patients, despite address changes, name changes due to marriage, immigration, and so on. The NHS Trust charters include a mandate to carry out clinical research, which can range widely from reducing incidence of bedsores and tackling infection in wards to developing protocols for effective treatment. The chief scientist of the NHS administers this.

With this powerful database, several broad research initiatives could be undertaken. One project that predates TMRC was spearheaded by Prof. Andrew Morris, who later became chairman of the TMRC, studied a disease cohort of 300,000 patients with diabetes to study genetic makeup and incidence of the disease among family members. Another project from 2004, 2 years before the TMRC initiative, "Generation Scotland," was initiated by David Porteous, Professor of Genetics at Edinburgh University, to track diseases within families, to see if the manifestation of the disease in one individual patient remained an isolated incident or whether it could indicate causes for genetic or behavioral trends that affected family members.*

The Originators of the Idea

The Scottish Government Health Directorate (SGHD) was fortunate to have a brilliant and dynamic chief scientist, Professor John Savill (now Sir John Savill). Alongside that role, he was also the head of the Medical School at the University of Edinburgh. These two organizations have different research missions. The SGHD drives public health strategy for Scotland, and is responsible for the Scottish NHS, whose focus is on clinical research, intended to benefit individual patients. Academia's research generally focuses on precompetitive research.[†] It was John Savill who looked to the Medicines Research Council,[‡] the Wellcome Foundation[§] and the separate medical schools for funding, in an attempt to create a critical mass to undertake the size and type of research projects that would better serve the market, and attract industry investment. Savill convinced the different parties to combine their resources—scientific talent, patient data, and clinical experience, funding, and physical facilities—to a more synergistic result.

*This experience greatly improved the processes of patient preconsent and patient education regarding participation in studies and in managing privacy issues.

[†]Philip Cohen had pioneered a model of precompetitive research lab with involvement at a distance from Roche.

[‡]The Medicines Research Council is a publicly funded government agency responsible for co-coordinating and funding high-impact medical research in the United Kingdom. Research funded by the MRC has produced 29 Nobel Prize winners to date.

[§]Established in 1936, The Wellcome Trust is a global charitable foundation, independent of political and commercial interest, dedicated to improving health. It employs more than 500 staff and has an annual budget of approximately £600m ($952.5m). It supports researchers, new product development, and patient-oriented research advances, among other activities.

Scottish Enterprise

SE is a regional development agency. Its mission is to create a platform for industrial and scientific growth and economic development for Scotland, and to better utilize the existing talent, infrastructure, facilities, and resources, including engaging the local small- and medium-sized company base and creating spinout companies from successful projects. SE's aims were aligned with those of what was to become the TRMC. Headed by Chief Executive Jack Perry, SE offered crucial matching funds to develop the infrastructure needed to conduct the kind of world-class research that would attract investment from industry. In addition, they showed strong continued support as TMRC projects progressed: for instance, an initial £9.5 m ($14m) SE loan turned to a grant, with £2.7m ($4m) nonrepayable and £6.8m ($10m) repayable.

Wyeth Pharmaceuticals

With the arrival of Frank Walsh, head of Global Research at Wyeth (who is Scottish by birth and American by residence but never lost his accent!), essential input and funding materialized in support of projects for translational medicine. He had been looking for a strategy for streamlined collaboration with academic institutions and hospitals. Wyeth was interested in participation in the TMRC initiative because it was different: it effectively combined resources and cooperation of the medical schools, the NHS Trusts, and the SE. Therefore, Walsh succeeded in committing Wyeth to invest substantial funding (Figure 32.1).

So, all the stars were aligned: the timing was right, willing partners were ready to combine their efforts, and far-sighted champions had emerged to drive the initiative, ". . . and there was light." The four stakeholders forming the TMRC were composed of the four Scottish medical schools, represented by their deans, four NHS Trusts;* SE, represented by its CEO Jack Perry, and Wyeth.

HOW DID THEY COLLABORATE?

The TMRC had a relatively large governance structure, of about 25, chaired by Savill, and later by Morris. It consisted of a Scientific Review Board and a Steering Committee. Senior scientists of the institutions, including the NHS, led the Scientific Review Board. Its role was to consider the application of the funds and the academic merits of the projects. The Steering Committee was composed of the heads of R&D from the four NHS Trusts and senior management from Wyeth. Its role was to secure funding and continued participation of all parties.

TRMI Ltd. was established in 2006 as the dedicated delivery and exploitation vehicle for the TMRC [1]. TMRI Ltd. consisted of 10 people, with Janet Brown serving

*The four NHS participants had access to approximately 85% of Scotland's patient population data and their medical histories.

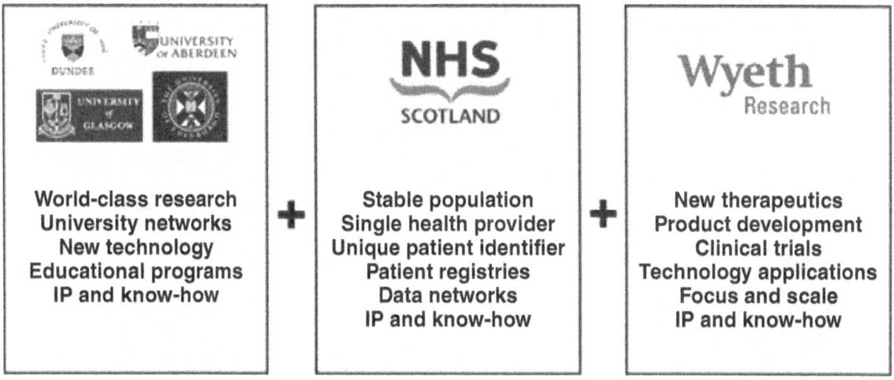

Figure 32.1. Stakeholders in the Translational Medicine Research Collaboration (TMRC).

as its first chairperson. Its primary responsibility was management of TMRC's finances, grants, governance, and IP, and it assumed joint responsibility with partners for management of projects. TMRI in partnership with the Wyeth team ensured that projects progressed according to timelines and budget. Also, it played an important role as mediator and central project manager for the different parties (Figure 32.2).

All parties contributed intellectual property (IP) to projects. IP that predates any collaboration remained with its contributor. Wyeth and TMRI owned IP resulting from TMRC collaborations jointly unless it is relevant to Wyeth drug candidates. Wyeth owned new IP in the therapeutic field, and TMRI owned any new IP in the diagnostic or other fields. If IP that predates partnership is required to exploit new IP, TMRI and Wyeth provided cross-licenses and had arrangements to divide royalties through TMRI.

HOW WERE PROJECTS SELECTED?

Initially, projects were proposed from the universities and NHS, with the anticipation that they might benefit Wyeth. The large number of projects helped to spread the risk if one would fail, but that also meant that resources were not harnessed together optimally. At this stage, IP revenue was not a selection factor. TMRI served as a matchmaker, using the criteria that the projects benefited all parties: publications and further research for the universities; benefit to patients for the NHS; and interesting discovery for preclinical research and development for Wyeth. The Scientific Review Board, consisting of the heads of the university medical schools, decided which projects to proceed with. The Steering Committees reviewed the projects for approval and funding, and monitored them periodically.

By and large, the TMRC, Wyeth, and TMRI participants were kindred spirits, who believed that their structure and method was the optimal way to move relevant research

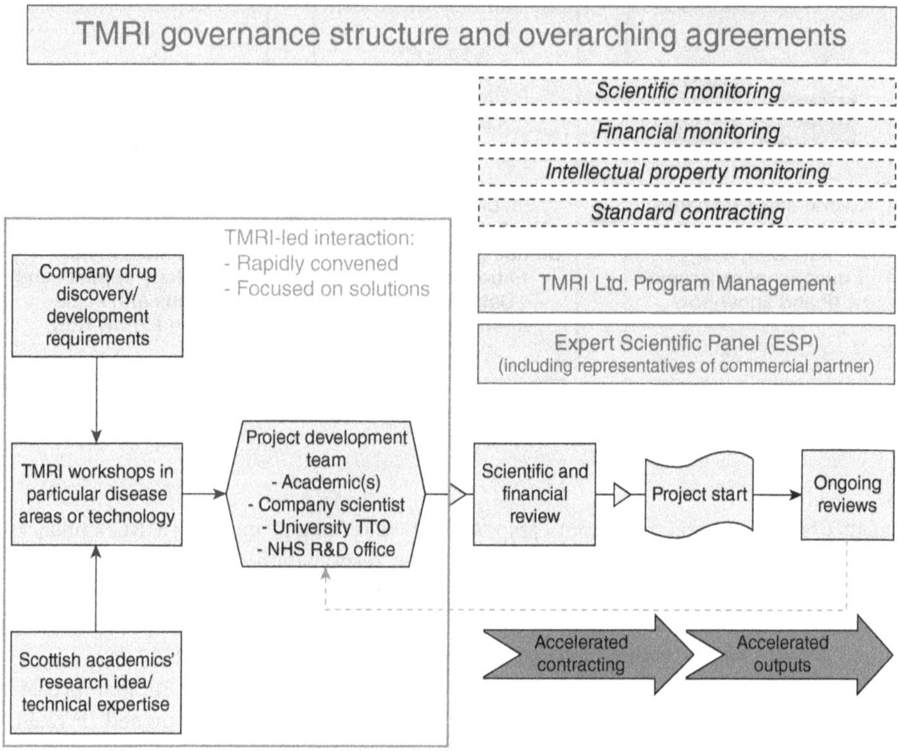

Figure 32.2. Governance structure of the Translational Medicine Research Initiative Ltd.

forward quickly. In an environment of trust and shared objectives, they had open, frank discussions that focused on finding the optimal solutions to problems.

INITIAL RELATIONSHIP OF TMRI AND WYETH AND THE TECHNOLOGY TRANSFER OFFICES

Early on in the relationship, there were efforts to generalize the cooperative model so that other industrial players could join. Each medical school still had its own Technology Transfer Office (TTO) that was focused on IP transfer. Although this was emphatically not the mission of the TMRI, it was viewed somewhat as a competitor by the TTOs! By and large, when potential funding by commercial pharmaceutical companies appeared, each university TTO, without the benefit of a centralized command and control structure, pursued it. One could use the analogy of a cycling team in the Tour de France: the team consists of players who also want personal glory beside the team recognition, and once the finish line is in sight, cooperation ceases as each individual dashes to the prize.

SE's significant matching funding for 5 years was an important signal of commitment and professionalism. It enabled the continued and unified efforts of the same talents, research questions, and conditions for a program of research.

WHAT DID THE TMRC ACCOMPLISH?

The TMRC has produced tangible results. The first TMRC went to completion with close to £26.6m ($37m) of TMRC-related funding invested in Scotland. This project delivered successful results; it was renewed as TMRC2. Despite the acquisition of Wyeth by Pfizer, with the inevitable change of individuals in key positions and reduction of diseases covered, well-planned collaborative projects delivered the promised results.

One of the TMRC's important accomplishments was to make significant progress on consistent Terms and Conditions for the projects. Before this, each party developed complex new agreements for each project, and the involvement of large numbers of people from legal departments and company management led to long delays and loss of momentum. The standard Terms and Conditions were reflected in all agreements between Wyeth and the TRMC. As the administrative and executive body, with only 10 members, TMRI acted on behalf of the Scottish partners to mediate agreements between Wyeth and the other stakeholders. It could secure senior level buy-in, making negotiations smoother and faster.

Companies liked a single point of contact and relationship management to gain access to academics, resources, and patients.

Over the course of 5 years, Wyeth/Pfizer invested around £22m ($49m) into various projects and provided £4.6 m ($8.5m) for TMRC start-up and operating costs. A second-stage Grand Challenge has projects planned under a new TRMC2 agreement. SE provided £13.3m ($20m) in grants, and converted an initial loan of nearly £10m ($15m) to a grant. Overall, 74 translational projects were funded, including five Stage I Grand Challenge projects to support studies on 25 diseases. Around 75% involved patients or their samples, and around 50% involved the core biomarker lab. Around 50% had an imaging element. These projects were efficient and fast: 39 second-stage projects were contracted within 4 months. Another significant accomplishment was that there was a minimum of time overrun: 3 months (20%).

This efficiency and shorter overrun was due largely to TMRI's coordinated project leadership approach, with a compact executive team whose purpose was to get the projects completed. When needed, they involved third parties with niche skills. Also, they understood the requirements of Pfizer's R&D planning, accounting cycles, and other internal processes, so that the efforts of the Scottish research partners would be planned optimally.

The strong genetic component and excellent patient data put them in a strong position to carry out the kind of precompetitive research that simply cannot be done in another way.

The collaboration yielded several completed projects, which delivered clear impact to Wyeth/Pfizer. They facilitated clear decisions of whether to proceed on drug programs and provided various biomarkers for trials, animal models, and knowledge to fulfill FDA regulations. TMRC's output to date has yielded more than 100 publications, 14 technical disclosures, and some early IP output, including licenses on an animal model and four patent applications.

These successes enabled it to leverage significant third-party funding. The Core Laboratory is a facility that has been upgraded to industrial rigor to conduct research in four areas: bioinformatics, proteomics, assay development, and genomics. Staffed with about 25 people, it provided support for TMRC.

FINANCIAL ASPECTS: EVERYBODY IS A WINNER!

TMRI was the first private company jointly owned by Scottish Academic Health Sciences Collaboration (SAHSC) partners. Its partners did not conceptualize it that way at first. Primarily it was conceived as a vehicle to access funding. Although not all partners welcomed it, many top-levels and operational representatives dedicated huge amounts of time, effort, and goodwill to the company and its executive team. All of the £41m ($60m) investment into TMRC flowed through TMRI, which had an important role in ensuring that this money was deployed responsibly according to the Collaboration Agreement. TMRI had a 5-year operating cost of £2.6m ($4m), representing 6% of the total TMRC expenditure. It proved to be a sound financial manager. Through various activities and initiatives, the TMRI team directly saved the collaboration in excess of £3m ($4.5m) (primarily through reducing SE investments), and contributed to £4m ($6m) of additional savings to Pfizer for reinvestment in projects.

Although IP income was below the initial estimates, TMRI was responsible for securing the £5.3m ($8m) European Regional Development Fund grant. TMRI developed business plans and led negotiations to enable the conversion of the £9.5m ($14m) SE loan into a grant (£2.7m [$4m] of which is nonrepayable, and £6.8m [$10m] which was repayable).

VALUE FOR THE NATIONAL HEALTH SERVICE TRUST

Some important process improvements came about through the key role of the TMRI team in various projects for efficiencies in contracting and delivery. For instance, TMRI supported a "lean" review of approvals processes in NHS Lothian, resulting in a marked increase in the proportion of studies receiving approval within 30 days (from 38% to 96%).

POTENTIAL LESSONS FOR THE FUTURE

Change Is the Only Constant

The TMRC is a unique model. Looking back, some lessons can be learned to help in the future. One factor is that private sector companies are subject to market forces that can change their business priority and organization model. Pfizer acquired Wyeth, and that did affect the collaboration. SE was strongly dependent on Wyeth as a single customer, which had allowed Wyeth to influence terms of agreements, and so when conditions changed at Wyeth/Pfizer, the collaboration and feasibility of the projects suffered to the detriment of the Scottish team, despite their responsiveness and commercial attitude.

A Pinch of Self-Criticism Is Always Healthy

The TMRC was at times unduly bureaucratic. The different governance groups did not maximize top-level representatives. For instance, the Scientific Review Boards had too much operational focus. The Steering Committee was too focused on finance and SE funding.

Another lesson is that the TMRC had unrealistic expectations of the amount and timing of IP income. It was anticipated that there would be £7 m ($10 m) income within the initial 5 years. The reality is that TMRC studies were mainly early-stage research studies that were supported based on their scientific merits and perceived benefit to Wyeth translational medicine, not on projected IP revenue. For the year 2009–2010, the reported IP income across all disciplines for all the four universities in TMRC was £4.86 m ($7 M) (against total income of around £192 m [$340 m] for that period, and in light of the billions of prior R&D investment). Delivering validated IP, negotiations deals, and the generation of license income requires significant investment and time.

Reflections on TMRI as a Vehicle for Scotland

The Translational Medicine Research Initiative Ltd. was set up to be an initiator and implementer of change. It has had mixed outcomes. It did have several derivative successes. One of these is that the formation of TMRI led to the development of the business plan for a new approach to clinical research in Scotland, and it supported top Scottish representatives in their presentations to Ministers for a Scottish Network of Academic Medical Centers, which became the SAHSC. It has formed a marketing and business development arm called Health Science Scotland to work with pharmaceutical companies in setting up very early-stage projects, and match needs with academic research facilities.

While it is difficult to translate collaborations with global pharma companies without a major national presence into sustainable jobs, TMRI and TMRC's coordination to fund and launch research projects supported many competitively remunerated positions for scientists and skilled staff in Scotland.

CONCLUSIONS

Collaborations of academia, NHS, government, and the industry are not only possible but also could be very successful to all if there is a true will. Having a single point as the negotiations lead was effective for some agreements, but not for all, in particular for external negotiations. TMRI delivered several legal negotiations, including the standard project agreement. These are very important technology transfer components, and many past attempts at collaboration between industry and academia have fallen far short of their potential because they could not resolve these issues.

The benefit of a centralized management and control structure is to utilize the scale of large funds and to make maximum use of the different players' resources. A critical mass of resources and of funding is necessary to realize any useful progress and to reduce overheads in the project's management. This is true for both the public sector as well as commercial pharmaceutical R&D programs, which can sometimes also lack consistent approaches on R&D across therapeutic areas in the same company or even across different compounds in the therapeutic area.

The lack of understanding of each party's role, responsibility, and boundary could bring the whole collaboration project to a complete halt or at least keep the stakeholders from achieving their maximum potential simply by delaying the whole process. While the feedback from the pharmaceutical industry is that they have a strong preference for a single point of contact and solid relationship management to gain access to academics, resources, and patients, the university TTOs felt that their role had been taken over, and their reaction seriously threatened the whole project.

The human factor is as important as the funding commitments or the relevance of the science. This project would not have seen the light of the day without the visionaries like John Savill, Frank Walsh, Jack Perry, and others.

It would be a missed opportunity to set aside the lessons learned from the TMRC and TMRI. Again, it is vital that universities are aware of the requirements that pharma has on study design and methodology, and align their work with this as much as possible.

Now that the universities, NHS Trusts, and industry have experienced the power and benefit of collaboration, it is recommended that they continue to use this model, at least as a starting point, rather than return to smaller-scale, fragmented research programs. In effect, the TMRI has served its purpose, and now its approach is adopted by the parties themselves, and by some TTOs. Even we in AlphaBeta Pharma have adopted most of this model and its spirit in the private sector with very promising results. Future collaborations will require more structured and strategic models of SAHSC-led engagement based on key strengths and resources available. The TMRI model is being adopted for England, with a similar independent company established.

Participation in TMRC has undoubtedly heightened commercial awareness in the university research groups. It resulted in some completed projects, and some tangible results from projects that had to be terminated. A combined TTO approach, similar in scale and sophistication to TMRI, seems to be the most promising model, whether structured around a geographic area, scientific specialty, or technique. However, it remains to be seen whether the combination of highly motivated, capable program champions and visionaries can be replicated so that the TMRI model will continue, or whether individual academic centers will, for various reasons, prefer to retain their autonomy, and work along the traditional lines of their separate TTOs.

ACKNOWLEDGMENTS

The author would like to thank the following for their insights and contributions to this case study: Dr. John M. Hall (Lead Partners Ltd., former CEO of TMRI, UK), Dr. Ross McLennan (Brains Unlimited, the Netherlands), Dr. Julia Brown (Scottish Enterprise, Glasgow, UK), Dr. Elizabeth Rattray (University of Aberdeen, UK), and Dr. Nigel Paul (University of Edinburgh, UK).

REFERENCE

1. Translational Medicine Research Initiative (TMRI) Ltd. TRMC 5 year overview. Executive summary presentation June 2011.

33

CHALLENGES AND OPPORTUNITIES IN COMMERCIALIZING ACADEMIC DRUG DISCOVERIES

Christopher Paschall

University of Virginia Licensing and Ventures Group, University of Virginia, Charlottesville, VA, USA

The push to free intellectual property (IP) from the management hands of the federal government actually began in the era of World War II. The great industrialist, inventor, MIT professor, and founder of Raytheon, Vannevar Bush, published a report for President Truman called "Science—The Endless Frontier." It was apparent to government leaders and the nation at the time that academic researchers were critically important to the success of the Manhattan Project and defense of the country. The prescient Bush also recognized the possibilities for universities bolstering the economy by transferring its understanding of basic science to industry for application. Bush's report was instrumental in the formation of the National Institutes of Health (NIH) and the National Science Foundation (NSF), and provided substantial increases in funding by the federal government for basic research at universities [1]. As a result of the increased funding for basic scientific research, universities began to emphasize cutting-edge research programs and recruit faculty members capable of securing federal research grants. By the 1960s and 1970s, it was apparent research and inventions developed under federal grants were not being commercialized as efficiently as possible. Without a unified federal policy and with the reluctance of the federal government to cede ownership of inventions to the funded organizations, flow of government-funded inventions to the

Collaborative Innovation in Drug Discovery: Strategies for Public and Private Partnerships, First Edition. Edited by Rathnam Chaguturu.

private sector was very limited. Furthermore, government would only grant nonexclusive licenses to inventions created under its funding. In such an environment, companies were hesitant to invest in and develop new products if competitors could also obtain licenses and sell the same products, especially in the pharmaceutical sector. While taxpayers supported the federal research enterprise, they were not benefiting from products developed with federal research dollars or the economic development that could have occurred with the manufacture and sale of those products.

BAYH–DOLE ACT

The topic of commercializing academic drug discoveries cannot be properly treated without at least a minimal understanding of the seminal piece of legislation that spawned the university technology transfer industry, the Bayh–Dole Act. Codified on December 12, 1980, the Bayh–Dole Act (also known as P.L. 96-517 and The Patent and Trademark Law Amendments Act) provides the legal framework for which federally funded university technologies are commercialized. Prior to implementation of the Bayh–Dole Act, the federal government retained title and ownership of any inventions that were developed with federal grants. The federal government did not grant exclusive licenses to patents that were filed on such inventions, only nonexclusive licenses. One can imagine the pharmaceutical industry's reluctance to license patents from the federal government under the pre-Bayh–Dole conditions, and the data supports the industry's hesitancy. In a 1968 study, it was found that not a single pharmaceutical patent owned by the federal government had ever been licensed, and by 1980, only 5% of the government's 28,000 patents had ever been licensed [1, 2].

By 1980, the time was ripe for a policy that permitted universities and small businesses to elect ownership of inventions made under federal funding and to become directly involved in the commercialization process. Importantly, the new policy would also permit exclusive licensing of federally funded IP. Thus, the Bayh–Dole Act was born.

Quantitatively, the results of the Bayh–Dole Act have been astonishing [3]. In 1980, U.S. universities were granted approximately 200 patents and executed a negligible number of licenses. According to the 2010 AUTM Licensing Survey, in 2009, universities filed over 18,000 patent applications, executed over 5000 license and option agreements, witnessed the formation of 651 new companies, saw the introduction of 657 new products on the market, and realized $2 billion in income from licensing activity [1–3]. Some of the more significant university inventions commercialized over the past 30 years include Emtriva, which was developed by Emory University researchers and is taken by 94% of HIV patients, Lyrica, which was developed by researchers at Northwestern University for fibromyalgia and is Pfizer's second leading drug, and Remicade ®, which was developed by researchers at New York University and is a leading therapy for patients with rheumatoid arthritis.

To achieve the same level of success as some of the products listed above, it is important for universities and their respective licensees and collaborators such as pharmaceutical and biotechnology companies to be cognizant of some of the important regulations of the Bayh–Dole Act. First, the provisions of the act apply to all inventions conceived or first reduced to practice in the performance of a federal grant or contract [4]. This is true even if the Federal government is not the sole source of funding for either the conception or the reduction to practice. Therefore, if an invention is developed with both federal and corporate funding, the presence of the corporate funding does not alleviate or attenuate the obligations and regulations under Bayh–Dole. Second, any company that obtains an exclusive license to a patent that involves sales of a product in the United States must substantially manufacture the product in the U.S [4]. Waivers of this rule may be available, but the company and university must demonstrate that reasonable but unsuccessful efforts have been made to find a company that will manufacture the product in the US, or that manufacture in the US will not be economically feasible. Third, universities must give licensing preference to small businesses (fewer than 500 employees) [4]. However, if a large company has provided research support for the development of the invention, that company may be granted the license. Fourth, universities may not assign their ownership of inventions to third parties, with few exceptions [4]. Finally, under certain circumstances, the government can require the university to grant a license to a third party or the government may take title and grant licenses itself (called "march-in rights") [4]. In the history of Bayh–Dole, no federal agency has ever exercised its march-in rights.

TECHNOLOGY TRANSFER OFFICE PHILOSOPHY

Ask ten different universities for a philosophy on technology transfer and one is likely to receive ten different answers. In fact, if ten members of the administration, faculty, and staff within one university are asked the same question, one is probably destined to receive ten different responses. A university's technology transfer office philosophy is often shaped by the composition and vision of the administration, the research strengths of the faculty, the local and regional economic and entrepreneurial environment, the leadership of the technology transfer office, and historical successes and failures.

While the emphases vary between and within universities, most university technology transfer offices share a similar core mission—to disseminate knowledge through competent transfer of technology into the marketplace for the benefit of society and to support the university in its research and educational endeavors by providing an unrestricted source of income through licensing and commercialization of the university's intellectual assets [5–9].

Practically speaking, the technology transfer office accomplishes its mission by engaging in the following activities:

1. *Serve the Faculty as Customers.* Take a poll of the major research university technology transfer offices and many of them will express that outstanding customer service to the faculty is a primary goal of the office. Great customer service often yields more and higher quality invention disclosures as well as fosters the entrepreneurial spirit of the research community.

2. *Negotiate and Execute License Agreements and Other Contracts.* A core function of the technology transfer office is the negotiation, drafting, revising, and execution of IP license agreements, confidentiality agreements, material transfer agreements, inter-institutional agreements, and occasionally, research collaboration agreements. The licensing professional typically negotiates both the financial and legal terms to be incorporated in the agreement.
3. *Expedite Local or Regional Economic Development.* Many universities, especially state and public universities, are charged with spurring and improving local and regional economic development. Federal, state, and local governments often translate economic development as company formation and job creation. In fact, the Bayh–Dole Act requires universities to give IP licensing preference to small businesses or start-up companies. Furthermore, the Bayh–Dole Act obligates licensees of federally funded inventions to manufacture substantially such inventions in the United States.
4. *Facilitate Start-Up Company Formation.* The technology transfer office plays an essential role in the creation of new companies. The office can help in the actual company formation (e.g., drafting and filing the articles of incorporation), advise on the creation of a business plan, identify entrepreneurs, consultants, and board members that can help initiate and grow a fledgling company, and locate funding sources suitable for the company's business. Given the early-stage nature of university inventions, and oftentimes, the unknown or ill-defined market opportunity, a start-up company can be an attractive, and perhaps, only avenue for incubating and cultivating a new, unproven technology.
5. *Protect University Intellectual Assets.* The technology transfer office is responsible for ensuring that new inventions are protected appropriately, be it through patents, copyrights, or material control. The office usually manages prosecution of patents from initial filing to expiration, registers copyrights when necessary, and organizes transfer of tangible materials to other entities through material transfer agreements or nonexclusive licenses.
6. *Generate Revenue.* While university administrators often stress the importance of revenue generation, the topic can be quite challenging and complex for the technology transfer professional. University technology transfer offices agree that revenue generation is important, but they also acknowledge that revenue is a lagging indicator of an office's performance and that the office has the least control over the amount of revenue that is generated in a given year. Large revenue-generating deals are frequently completed years before any actual revenue is realized, and many times by a professional who no longer works for the office.
7. *Disseminate of Knowledge through New Product Creation.* Universities are centers of knowledge creation and discovery. While the traditional route of knowledge dissemination at a university is through peer-reviewed publication, the technology transfer office can disseminate knowledge in a more practical way—by creating and developing opportunities for new products that

incorporate university research. The taxpayer realizes a return on his or her tax dollars when federal research money is used in the development of new products that benefit the public. The university technology transfer office improves the probability that such products make it to market by requiring licensees to commercialize diligently inventions developed with federal research dollars.

8. *Explore Business Development and Marketing Opportunities.* Unlike companies that can create opportunities around markets, universities must find or create markets for their opportunities. Technology transfer offices spend a significant amount of time marketing their technologies, trying to find the best opportunity to develop and commercialize the invention. The marketing process includes direct contact to companies and indirect advertisement on websites such as the technology transfer office's webpage or third party hosts such as the Kauffman Foundation's iBridge^{SM} Network or yet2.com.
9. *Support Translational Funding Infrastructure.* Many major research universities now have translational funding sources to fund projects that have commercial potential or that may not be suited for traditional NIH or other basic science funding. The translational funds can be supported by the schools, private foundations, or local or state governments. Similar to a study section that reviews grant proposals for the NIH, these translational funds have boards or executive committees that review submissions for technical merit, IP position, and potential for significant market penetration. The technology transfer office generally participates on these boards and provides important due diligence and market analysis to ensure that the funds go toward the most deserving and promising projects.
10. *Solicit Invention Disclosures.* A pipeline of invention disclosures is vital for the health of a technology transfer office. Faculty researchers can sometimes be confused as to what merits an invention. The licensing professional not only stays in close contact with faculty and remains up-to-date on inventive activity at the university, he or she can also advise researchers on the standards for what comprises an invention. He or she can also encourage them to add value to the invention by performing specific experiments, for example.
11. *Make strategic Investments in the Patent Portfolio.* What do IBM, GE, and university technology transfer offices have in common? They all want more money to spend on patents! Unfortunately for all, the patent budget is finite, and at universities, it is usually severely limited. Universities make extensive use of the provisional patent filing system to save on costs upfront, but at some point, decisions have to be made as to which opportunities will receive patent funds. The technology transfer office does its best to keep as many opportunities alive as long as possible. However, the harsh reality is that some opportunities will be dropped. The office is responsible for making these decisions, and they are primarily based on the strength of the IP covering the opportunity, the market potential, and prospects for further development.
12. *Provide IP and Compliance Support to University Administration.* As the primary expert on all things IP at a university, from patents and copyrights to

infringement and export controls, the technology transfer office aids university administrators when IP issues arise. Offices at the university that may receive advice include the office of sponsored programs and research compliance, the office of general counsel, and the conflict of interest office. Issues that appear commonly range from IP disposition in a research contract sponsored by a corporation to resolution of a conflict in a faculty start-up company.

The above is certainly not an exhaustive list, but it should give the reader a flavor for what the university technology transfer office does and what it emphasizes as important for its mission.

THE UNIVERSITY'S IP POLICY

The university's IP policy governs the ownership and disposition of IP created by faculty, staff, and students. While there is great diversity among the details in university IP policies across the country, most policies address at least the following three issues: treatment of different categories of IP and ownership of each type, disclosure obligations, and distribution of royalties.

Because of the Bayh–Dole statute, universities treat patentable inventions developed with federal funds very similarly—the university typically asserts ownership in such inventions and elects title from the federal government. Likewise, universities usually will not take ownership in IP that is commonly referred to as scholarly works. Examples of scholarly works include textbooks, plays, music, and scientific articles.

Unfortunately, the bright lines of university ownership begin to blur when one examines ownership of nonscholarly copyrightable works, software, IP developed under a consulting agreement, or IP created outside the area of expertise of the university researcher. In large part, universities assert ownership in nonscholarly copyrightable works and software assuming that university funds, facilities or resources are used. The university treats such works as "works made for hire" under the U.S. Copyright Act.

IP created under the scope of a consulting agreement can be difficult to manage and philosophies on the disposition and ownership of such IP vary significantly among universities. Generally, the university is not a party to the consulting agreement—it is entered into between a researcher and a company. Some universities review consulting agreements and others do not. Likewise, some schools will assert ownership in consulting IP if substantial university resources are used, and many others will allow ownership to vest in the individual inventor (who may assign ownership to a company through a consulting agreement). Since each school is different, it is advisable to check the school's respective IP policy [5–9].

Along the same lines, universities differ in their view of inventions developed by researchers that fall outside their area of expertise or are created with external funds. Often, it falls to a senior administrator or committee, such as the vice president for research or a university-wide IP committee, to make ownership determinations of inventions that fall within these gray areas.

Another important aspect of the university IP policy is an express description of the disclosure obligations of faculty and employees. Because much of the research at universities is funded by the federal government or corporate entities, and because the federal government and corporate partners require prompt disclosure of any inventions developed under their respective funding, universities obligate researchers to disclose any and all inventions to the university technology transfer office. The university IP policy also may direct where and how such invention disclosures are to be made.

Section 202(c)(7)(B) of the Bayh–Dole Act requires universities to share any royalties received from federally funded inventions with the inventor. Like most aspects of university IP policies, exact distribution of royalties varies from school to school. However, a cursory survey of university IP policies reveals that most universities give 25–40% of royalties received from commercialization of IP to inventors. The remainder is split among the university, and the inventor's respective lab, department, and school, with the university receiving the lion's share of the remainder. Some universities implement a sliding scale where the percentage distribution changes as the royalties received increases [5–9].

OWNERSHIP OF IP (*STANFORD v. ROCHE*)

Major research universities assert ownership of IP developed by faculty and staff in two ways. First, like corporations, universities take ownership by virtue of the employment agreement between the faculty or staff member and the university. Second, most universities and research institutions are funded by grants from the federal government. The Bayh–Dole Act allows universities and other nonprofit entities to elect title to inventions developed under the sponsorship of a federal grant, contract or cooperative agreement.

There are several issues for universities to consider with respect to the ownership avenues listed above. Unfortunately, many universities do not have "airtight" employment agreements with their faculty and staff. Creating new IP is considered a passive activity at universities and therefore, securing ownership of IP is not of primary concern. The shortcoming of these "soft" ownership arrangements between universities and their faculty and staff was highlighted in the case of *Stanford v. Roche*, which was decided by the Supreme Court in 2011.

As a brief summary of the case, Mark Holodniy was a professor at Stanford University in 1988, and he invented an HIV-detection kit with federal grant money. During his time at Stanford, he also served as a researcher at a company called Cetus (now Roche). Holodniy signed a confidentiality agreement with Cetus, and the confidentiality agreement contained assignment language for any IP developed during his time at the company. By signing the confidentiality agreement, he granted Cetus rights to anything he invented. Cetus/Roche commercialized the invention and the HIV test kits invented are now used in hospitals and clinics worldwide. In 2005, Stanford sued Roche claiming that, under Bayh–Dole, Stanford owned all of the IP invented by Holodniy. Stanford asserted that the Bayh–Dole Act should trump any other contract Holodniy signed.

Unfortunately for Stanford, the Supreme Court disagreed and said that the assignment agreement Holodniy signed with Cetus/Roche took precedent. Stanford's employment agreement turned out to be a major issue for them. At the time, Stanford's employment agreement said that inventors "agree to assign" (future tense) inventions to Stanford, whereas Cetus's assignment agreement said that the inventors "do hereby assign" (present tense). University ownership of faculty inventions does not vest automatically by way of the Bayh–Dole Act—universities must either elect title to federally funded inventions or take ownership by way of an employment agreement that explicitly assigns ownership of inventions to the university. Since the decision, universities have been modifying their IP and employment policies to assert ownership of IP. Many universities now have employment and IP policies that look like corporate policies [10, 11].

Other issues that arise for universities in asserting ownership of IP include ownership of inventions made by undergraduates, inventions developed under nonfederally funded research grants, and standards by which a university decides when to take ownership of IP. Typically, universities do not take ownership of any IP developed by faculty and staff that is considered a scholarly work. Textbooks are a primary example of scholarly works. Universities also do not ordinarily assert ownership in inventions developed by undergraduates. The rationale for such a position is that undergraduates pay to attend universities and therefore, it would be unfair for the university to take ownership of IP on top of tuition. The one possible exception is where an undergraduate is participating in a federally funded research project and the Bayh–Dole provisions apply. However, some universities still will not assert ownership in such circumstances. Each university has its own standard by which ownership in nonfederally funded inventions is determined. Often, these decisions are made by examining the funding source of the project or the facilities and resources used by the inventor. A university executive or committee is usually charged with making the final decision [5–9].

STAFFING FOR SUCCESS

University technology transfer offices perform a wide variety of highly specialized functions related to the patenting and licensing of inventions. In addition, these offices also perform a vital responsibility at their institutions related to the formation of research partnerships with industry, and in negotiating the exchange of research materials and research tools (Figure 33.1).

A variety of skill sets are required of the technology transfer professional. He or she must be an expert in or have a working knowledge of all forms of IP (patents, copyrights, etc.), negotiating, contract drafting, government compliance, business development, marketing, valuation, and of course, science and technology. Because of the assortment of technologies managed by the university technology transfer office, the range of skills required for the job, and the "cradle-to-grave" business model that most offices employ, a very diverse composition of professionals make up the technology transfer industry. The vast majority of professionals have advanced degrees in

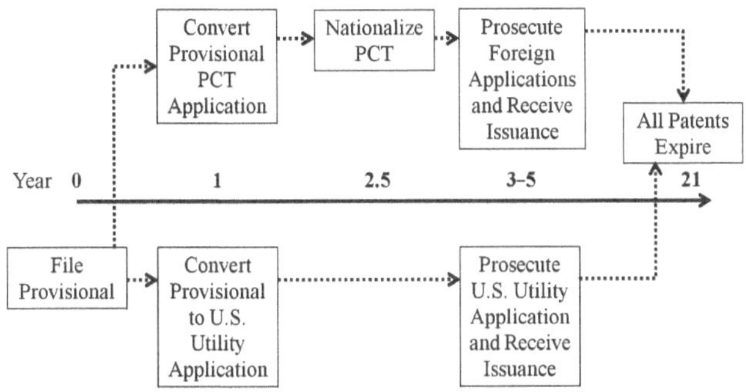

Figure 33.1. Overview of life of a typical university patent application.

science, law, or business, and many have specialized certifications such as registered patent attorneys or agents, Certified Licensing Professionals (CLP™), or Registered Technology Transfer Professionals (RTTP).

PROTECTING DRUG DISCOVERIES: INTELLECTUAL PROPERTY

Patenting Basics

The patent system in the Unites States is derived from Article 1, Section 8 of the Constitution, which reads:

> The Congress shall have power . . . To promote the progress of science and useful arts, by securing for limited times to authors and inventors the exclusive right to their respective writings and discoveries.

The seminal legislation defining patentable subject matter was passed in 1793, the second iteration of the U.S. Patent Act. The Act defined patentable subject matter as follows:

> [A]ny new and useful art, machine, manufacture or composition of matter and any new and useful improvement on any art, machine, manufacture or composition of matter.

It is important to understand that U.S. patent law confers an exclusionary right on the inventor, not a positive right. In other words, a patent owner can exclude others from making, using, selling, offering for sale, or exporting a product that is covered by the patent. Patents are treated like a piece of real estate, in that the patent owner can sell, assign, license or transfer the patent to another entity [12–16].

U.S. patent laws afford inventors and patent owners several routes with which to secure patent rights. Many U.S. research universities take advantage of the provisional patent application. A provisional application is a low-cost way to preserve a priority filing date at the USPTO. The provisional application only requires a specification or written description—it does not require formal claims. Provisionals are attractive options for universities because the university tech transfer office can file peer-reviewed manuscripts, conference abstracts or presentations, or grant applications as a provisional patent application for approximately $125 (as of 2012). Provisional patent applications are good for one year, at which point the owner or tech transfer office must decide whether to convert the provisional to a nonprovisional application in the form of a utility U.S. application, a PCT application, or some other application filed directly in a foreign country.

For university technology transfer offices, the most likely next step for patentable technologies to be pursued is the filing of a U.S. utility application. The U.S. utility application must contain a description of the technology that is detailed enough for one of ordinary skill in the art to, for example, make or use the technology. The application must also contain claims, an abstract, and an appropriate list of inventors. The U.S. utility application is prosecuted at the USPTO, and it can take up to five years to receive an issued patent. The term of the patent, under current U.S. law, is twenty years from the earliest filing date. It is important to note, particularly for pharmaceutical patent applications, that the term can be extended through Patent Term Adjustment or Patent Term Extension as a result of delays at the USPTO (not delays at the FDA or any other regulatory body).

The technology transfer office may want to delay the decision regarding in which countries to pursue patent protection as it waits to locate a suitable licensee or commercialization partner. In such a case, the technology transfer team may decide to convert a provisional application to a PCT. A PCT application, which stands for Patent Cooperation Treaty, acts as a placeholder because it preserves the right of the patent owner to file in any number of PCT member countries through a single mechanism. Like the U.S. utility application, the PCT must contain a detailed description of the invention, an abstract, claims, and a list of inventors. The PCT is good for thirty months (or eighteen months from the provisional conversion date). At the end of the PCT term, the applicant must decide in which countries it wants to nationalize. Nationalization means the PCT application is filed directly in a member country (e.g., U.S., Japan, Europe, Canada, etc.). The nationalized application is then prosecuted in each country as a normal, utility application. Alternatively, the applicant can forego the PCT process altogether and file patent applications directly in foreign countries.

Finally, how does an inventor obtain a patent after the application has been filed? While the details of patent prosecution are too complex to address here, a patentable invention is one that is deemed by the USPTO to be useful, novel, and nonobvious. The specific laws regarding patentability can be found in sections 101, 102, and 103 of section 35 of the United States Code. Foreign patent authorities have slight distinctions on each of the three requirements for patentability, but generally speaking, the requirements are very similar across the globe. The usefulness or utility of an invention is the lowest hurdle to overcome with respect to patentability. An invention is useful if it can actually be used (no perpetual motion machines) and provides some sort of benefit to the public. To be patentable, an invention must also be novel or new. That is, the invention cannot have been invented previously or by a different inventor. Lastly, an invention must be nonobvious to receive patent protection. The nonobviousness bar is often the most subjective and difficult to overcome. However, in the United States, patent examiners and courts use the so-called Graham factors to determine obviousness. The Graham factors, which are listed below, provide a methodical framework for patent examiners and courts to analyze the prior art and decide whether a new invention is the result of a nonobvious step taken by the inventor:

1. Scope and content of the prior art
2. Level of ordinary skill in the art
3. Differences between the claimed invention and the prior art
4. Objective evidence of nonobviousness.

The courts have also outlined examples of factors that show objective evidence of nonobviousness, or secondary indicia of nonobviousness, which can include commercial success, long-felt but unmet need in the art, and the failure of other inventors working on the same problem [12–16].

Biological or Drug Targets

Arguably the first step in the drug development process is the identification of a target that in some way contributes to the pathology of a disease [17]. The target can be a protein, enzyme, or receptor with etiological significance. Universities are particularly adept at identifying biological targets of diseases for two reasons. First, university researchers are funded primarily for basic sciences through NIH, and identifying targets is mostly a basic science activity. Second, universities typically have strong research programs in basic biology, pathophysiology, and life sciences in general. Since the university is not trying to develop products, but to move the boundaries of knowledge space outward, it can focus intensely on target identification and disease etiology without concern for the risks of such research endeavors (i.e., "fishing expeditions").

What does the university technology transfer office do with a newly identified disease target? This question has been the source of great controversy and debate over the past twenty years. A seminal case on the subject was concluded in 2004 that guides the decision making process for most universities. Researchers at the University of Rochester identified the COX-2 enzyme and its role in inflammation. Pfizer, and its acquiree Searle, developed the first COX-2 inhibitor, Celebrex®, for the treatment of osteoarthritis and other painful conditions associated with excessive inflammation. Rochester had filed a patent on the COX-2 target and mechanism and sued Roche asserting that its patent precluded Pfizer or any other company from developing and selling drugs that modulated the activity of COX-2. In the end, the courts concluded that Rochester's "reach-through" claim to its drug target was invalid because it failed to identify and describe any drugs that could inhibit the COX-2 enzyme in its patent [18].

Because of the Rochester case and other business considerations, universities are very reluctant to patent drug targets, with a few possible exceptions. Companies and the market see the targets as having very little value—the primary value is realized in the compounds that modify, inhibit, or promote the activity of the target, depending on the pathological mechanism [17–19]. Targets themselves have a very short life of usefulness, which is typically the length of time needed to identify a compound effective at modulating the target. Given such a short period, companies are hesitant to license drug targets and universities have a very difficult time commercializing them. A company may be willing to license a target if it is validated, plays a role in a major unmet medical need, or affords the company a jumpstart on its competition in the drug discovery process. While universities routinely identify targets with major unmet need implications, the other two hurdles are more difficult to traverse. Universities generally do not have the expertise or resources to validate targets to the extent required by pharmaceutical companies. In addition, it is the rare and aberrant case that a university would be able to license a drug target well ahead of publication to give a company a head start on the competition in drug development.

Drug Screening Hits

Once a valid drug target is identified, the next step in the development process is the search for chemical compounds or biological molecules that can affect the target and treat the disease [17]. University researchers have access to numerous compound libraries through the NIH or pharmaceutical companies and perform high throughput screening (HTS) to discover drug hits. More often than not, hits are not drugs; they are simply compounds that act on the given biological target. A hit must go through rigorous experimentation and chemistry modifications to become a drug. This so-called hit-to-lead process is not done frequently in the university setting. However, as large pharmaceutical companies become more risk averse and outsource more early-stage drug development to universities and small companies, such hit-to-lead optimization is likely to become normal fare in university research labs [20–24].

Most technology transfer offices, even knowing that hits are not likely to be the final drug product, will often file patent applications on them. There are several reasons for this approach. First, a hit is likely to yield the parent structure of the final drug product, so the university can claim in its patent a compound and a method of treatment, wherein the compound is a generic structure of the hit(s), and a clever patent attorney can propose derivatives and analogs that may cover the final drug substance. Second, university researchers are unlikely to progress past the hit stage. Posthit drug development requires extensive medicinal chemistry, pharmacology, and toxicology, and many universities simply do not have this expertise. Furthermore, it is difficult to secure government funding for such work, so universities cannot afford to pursue it. Third, once a compound is actually identified, the technology transfer office reasons that it has a product to license (as opposed to just a target), and they will market the opportunity aggressively. Many universities will license hits to a spinout company and the company can secure the appropriate funding to perform the necessary chemical modifications, pharmacology and toxicology testing, selection of the final drug candidate, and initiation of preclinical and clinical testing [17, 20–24].

Biomarkers

Universities are uniquely positioned to discover new biomarkers that can initiate new drug development research or augment current drug development efforts. In broad terms, a biomarker can be a protein, a hormone, a naturally occurring molecule, a nucleotide or gene sequence, a pattern of genetic changes, or a combination of the above, and it is an indicator of a normal or aberrant biological process, a disease state, or a response to a therapeutic regime [17]. Importantly, a biomarker itself can even be a drug target. While a company may carry out some biomarker discovery work as part of its overall basic research activity, universities have access to large pools of detailed clinical samples, a wide array of basic science researchers, and grant funding to support biomarker discovery research. In other words, biomarker discovery is a strength of the university in the overall scheme of drug development relative to the pharmaceutical and biotechnology corporations.

Biomarker research and discovery is becoming increasingly important in a time when the boundary between therapeutics and diagnostics is progressively blurred. The drive toward personalized medicine inherently stresses the importance of the identification of biomarkers that can help physicians accurately compartmentalize patient populations that may be suitable for a specific course of treatment. As drugs and biologics are commercialized more frequently with companion diagnostics, and as the therapies themselves target subsets of patients with a given disease, the value of biomarkers increases dramatically.

So how do universities capture the value of biomarkers and exploit their inherent advantage in biomarker discovery? Can the university protect IP around biomarkers and if so, are corporate partners or licensees willing to pay for what biomarkers potentially bring to the drug development process? Given the broad definition of biomarkers and the many areas in which they play a role, there are no easy answers to the questions posed. However, there are well-established spaces where universities can extract value for their biomarkers, and other defined areas known to deliver minimal return on investment in biomarkers.

Examples of biomarkers that are patentable or otherwise protectable and that can help the university realize value for its research include a serum based IVD that obviates the need for invasive or exploratory measures, an epigenetic fingerprint that identifies a genetic disorder quantitatively, thus minimizing the need for subjective tests administered by a clinician (e.g., autism), or a marker that stratifies a subpopulation of patients primed for a specific course of therapy (e.g., Genentech's HER-2 biomarker and Herceptin®) [17]. On the other hand, a well-known type of biomarker that is notoriously difficult to commercialize is the biomarker that acts as a drug target. The classic example of such a biomarker is the COX-2 enzyme, which was discussed in the previous section. In addition, biomarkers that act as surrogate endpoints in clinical trials are also difficult for universities to leverage. The safe harbor provisions of the Hatch-Waxman Act preclude universities from patenting such biomarkers because companies can use the biomarkers in clinical trials without the risk of patent infringement.

Drug Repurposing

As a result of the biological and basic life science expertise in academia, university researchers are especially proficient at finding new indications for existing drugs. A sample discovery pathway might look like this—an academic researcher uncovers a new mechanism for the pathology of a disease, the researcher peruses the literature to find known molecules or drugs that act on the mechanism, the researcher identifies and obtains the drug that he or she hypothesizes may deliver the proposed therapeutic effect, and the researcher tests the drug in cell culture or animal models to test the hypothesis. If the hypothesis is proven correct through experimentation, the researcher will often disclose the invention to the technology transfer office and the licensing team will file a patent application and market the technology accordingly.

The above scenario is very common in universities and many technology transfer offices have executed very successful deals around such inventions. New uses for known drugs can have several valuable advantages for potential licensees, but there are also disadvantages that need to be considered in the overall business strategy. A major advantage of exploiting a new use for an existing drug is the known safety profile of the drug. Commercializing a known drug for a known indication removes multiple years of preclinical testing and may obviate the need for a Phase 1 clinical trial. Often, start-up companies are the initial licensees for such opportunities, and they are immediately attractive to investors or potential acquirers because they are holding a clinical asset. Admittedly, there are disadvantages to known drugs, but most of the disadvantages can be overcome or minimized. One difficulty with commercializing a known drug in a new indication is securing patent claims outside the United States. In particular, it is difficult to obtain patent coverage in Europe on a new indication for a known drug. The university technology transfer office must work with foreign counsel to craft claims carefully to protect the asset. The process can be expensive, but usually, valuable claims can be obtained in foreign countries with some effort and compelling data. Even in the U.S., novel methods of using known drugs can be subject to obviousness rejections at the U.S. Patent Office. One way to overcome such rejections and preserve value is to invent new IP around the drug for the new indication. For example, the academic researcher or licensee may need to develop new dosing strategies, routes of administration, or formulations to patent and commercialize the opportunity. Obviously, such work will add to the cost of commercializing the opportunity, but the cost will likely pale in comparison to the preclinical work that would be needed otherwise. Finally, the technology transfer office and licensee will need to be conscientious of the potential for off-label use of a known drug. Physicians can prescribe most any drug for any condition, so the licensee will need to work with the FDA to ensure that the label on the drug for the new indication adequately protects the asset [25–28].

PATENT REFORM

The America Invents Act (AIA; H.R. 1249; Public Law 112-29) was enacted September 16, 2011 and will take full effect on March 16, 2013. AIA is the most significant change to United States patent law in nearly 60 years. The reforms have been the subject of much debate among university technology transfer offices, but it is unlikely that the full impact of the change in law will be realized for several years. Below is a list of the prominent changes implemented through AIA.

First Inventor to File

The United States has moved from an absolute first inventor system to a first inventor to file system. Each application is given an effective filing date and prior art will be considered before the effectively filing date. Importantly for universities, the one-year grace period remains in effect, but only for an inventor's own disclosures (e.g., scientific publications or presentations). Unfortunately, inventors will no longer be able to swear behind prior art. In the past, university researchers have been able to swear behind prior art by presenting notes from a lab notebook, which provided evidence that the invention had been conceived before the publication of the prior art. Swearing behind is no longer an option under the new law.

Prosecution Fees

The Director of the USPTO will now have the authority to adjust fees to cover administrative costs of the office. Under this section, a new micro-entity is also established. Micro-entities will be able to pay fees that are reduced by 75%. The establishment of the micro-entity is significant for universities because universities will be considered micro-entities under the new law.

Best Mode

Disclosure of the "best mode" will still be a requirement of patentability. The USPTO will still have a duty to issue patents that disclose the best mode of operation or use of the patented subject matter. However, failure to disclose best mode will not be a viable argument for invalidating an issued patent.

Prior User Rights

The new law expands the prior user rights defense to all patent types, not solely business method patents. The prior user rights defense says that an entity using an invention more than one year before a second inventor files for a patent on the same invention can continue using the invention after the second inventor is granted a patent as long as the first entity did not derive the invention from the second inventor. The benefit to universities of this rule change is that patents derived from inventions developed at universities are exempt from the prior user defense. In practice, it remains to be seen whether this will be a true benefit to the universities given the fact that the prior user defense has been around since 1999 for business method patents and no defense has ever successfully asserted prior user rights, nor has the court ever ruled on it.

Derivations

Derivation proceedings will replace interferences. In a derivation proceeding, the plaintiff will have to prove that a claim was derived from another inventor.

Damages

In many aspects, the new law limits damages that can be awarded in an infringement suit. The court will have to outline a rationale for the calculation of damages, limiting damages to the specific contribution over the prior art, and the court will be barred from levying enhanced damages in a close case on liability.

Third Party Challenges

The new law broadens the ability of third parties to challenge the validity of a patent. First, any third party will be able to submit any publication with a description of relevance to the patent office to be considered in the patent examination process during pendency. Second, any third party will be able to challenge the validity of at least one claim in an issued patent as long as the request is filed within nine months of issuance. Third, after the nine month, postissuance window, a third party may file an "*inter partes* review" in which the third party can bring prior art patents and printed publications to challenge only the novelty or nonobviousness of the issued patent. *Ex parte* re-exam proceedings do not change under the new law.

MANAGING THE PATENT BUDGET: SAYING NO TO FACULTY

While a very few of the large, major research universities have a patent budget sufficient to file patent applications on most invention disclosures submitted to the university technology transfer office, the overwhelming majority of academic centers must consider carefully the inventions that are pursued because of a limited patent budget. Unfortunately, the situation requires the university and its technology transfer office to be selective and turn down numerous products of academic research. There are two primary ways that the technology transfer office can maximize returns on its patent investments: it can drive down the cost of filing patent applications and prosecute the applications that have the potential for the most success.

Prosecuting a patent requires both fixed and variable costs. Fixed costs, such as filing fees and other administrative fees imposed by the patent offices, obviously cannot be altered, but must be considered in the context of quantity of applications the university files each year. The most obvious opportunity for cost savings comes in the variable costs, which includes attorney fees and time. Universities have taken several approaches to minimize variable costs. For one, universities may form relationships with preferred firms, where the firm gives a discount to the university for guaranteed business or a high volume of work. Some law firms will work for a flat fee, as opposed to the billable hour, so that the university can more easily budget and forecast its patent expenses. The flat fee model converts the variable costs to fixed costs, and can save the university money when a large number of applications are filed. Another strategy that is becoming more common in the university is the addition of in-house patent counsel. Significant cost savings can result from using in-house attorneys for patent prosecution because the law firm overhead and partner profit are eliminated from expenses immediately. Opponents of in-house counsel have always argued that patent quality is sacrificed because a small number of attorneys prosecute a wide range of patent applications. However, the universities using in-house counsel have reported no drop off in patent quality (e.g., Emory University and the University of Virginia).

With respect to quantity of applications the university files each year, there are two extreme positions that the university can take to minimize patent expenses. On one end, the university can file a large number of provisional patent applications, heavily market the inventions covered by the provisionals, and only move forward in prosecution with the licensed cases, with perhaps a few unlicensed cases that are particularly attractive. On the other extreme, the technology transfer office can file a very small number of applications on inventions that it thinks have significant commercial potential and prosecute those applications all the way to the end. In reality, most technology transfer offices fall somewhere in the middle of the extremes. Regardless of the quantity of filings, the technology transfer office will at some point be required to make a decision on whether to keep a case alive.

Again, there are numerous ways to determine which cases will be pursued. Some universities, mostly smaller universities, use patent review committees to determine which technologies advance in prosecution. Often, the patent review committees are composed of faculty members and possibly entrepreneurs and business people. Universities that market technologies assiduously use market feedback received from potential licensees to select the inventions that will be kept alive and pursued. Technology transfer offices at larger universities can generate formal commercial assessments that inform the ultimate decision makers on whether to pursue a technology. Such commercial assessments may include a scoring system, and are likely to assess the technology based on the strength of the IP, the market and market potential, the innovation or technology itself, the regulatory climate in the technology space, and the likelihood of receiving funding to further develop the technology.

Regardless of how the decision is made, the technology transfer office must be careful to include the faculty inventor in the decision-making process and inform the inventor of the rationale if his or her case is not pursued. Technology transfer offices can no longer make decisions in isolation. Some faculty may want to pursue technologies on their own if the university decides not to pursue it, in which case they can request that the assignment to the technology be released to them. Most universities will release technologies to faculty inventors under a well-defined set of criteria and obligations, and some entrepreneurial faculty will be successful on their own.

THE IMPORTANCE OF TRANSLATIONAL FUNDING

With federal government granting agencies' shrinking budgets and paylines for NIH grants at historic lows (some as low as 7%), it is critical for entrepreneurial faculty to

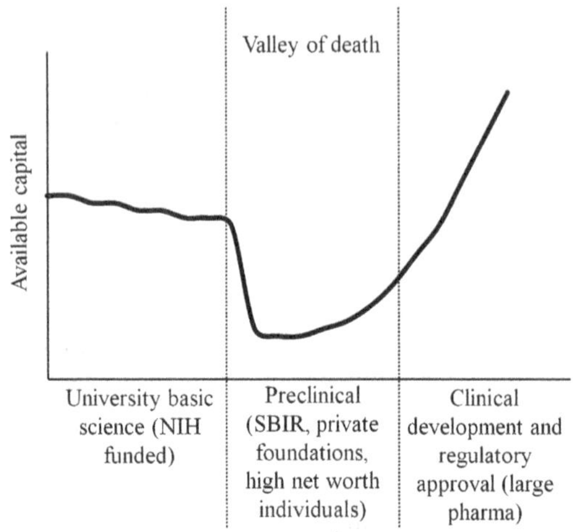

Figure 33.2. "Valley of death" in university technology transfer. Universities can fund basic science research with traditional NIH grants, and large pharmaceutical companies are willing to fund later stage clinical development. Early-stage clinical development, which may include animal studies, toxicology, lead optimization, and medicinal chemistry, can often be quite difficult to fund.

seek and obtain funding from nontraditional sources, especially for research that has potential commercial applications. Translational funding is an ideal source to fill in the gaps between traditional, basic science funding, and, in the pharmaceutical world, clinical development. Translational funds are designed to build upon knowledge gained from basic research, add value to a technology, and help move it forward in the commercial pathway, enabling a technology to traverse the so-called "Valley of Death" as depicted in Figure 33.2.

Perhaps the most prominent sources of translational funding for universities are the federal government's Small Business Innovation Research grant (SBIR) and the Small Business Technology Transfer program (STTR) [29]. As a brief overview, both programs are congressionally mandated and require many federal agencies with extramural research budgets to set aside a percentage (0.3–3.2%) of their total extramural research budget for contracts or grants to small businesses. There are two major differences between the SBIR and STTR. First, under the SBIR, the primary investigator (PI) must be an employee of the small business, whereas, under the STTR, the PI may be an employee of the small business or the collaborating nonprofit research institute or university. Second, the STTR program requires that the small business collaborate with a nonprofit research institution or university, while the SBIR only encourages such partnerships. For STTRs, the small business must perform at least 40% of the work and the nonprofit partner must perform at least 30% of the work, with the remaining 30% done by the small business, the nonprofit partner, or a third party. With SBIR grants, 67% of the work on a Phase I grant must be performed by the small business, and 50% of the work must be completed by the small business in Phase II. Both programs grant

TABLE 33.1. Brief Comparison of the SBIR and STTR Programs

	SBIR	STTR
Congressionally mandated budget set aside	2.5–3.2%	0.3%
Award amounts	Phase 1: up to $150K for 6 months Phase 2: up to $1M for 2 years	Phase 1: up to $100K for 6 months Phase 2: up to $750K for 2 years
Subcontracts	Phase 1: <33% Phase 2: <50%	Phase 1: <60% Phase 2: <60%
Academic partner	Recommended	Required
PI Employment	>50% at company	No requirement
Applicant/awardee	Company	Company
Participating agencies	DOD, DOE, DHHS (NIH), NASA, NSF, DOA, NIST, NOAA, DOEd, DHS, DOT, EPA	DOD, DOE, DHHS (NIH), NASA, NSF

funds in two phases. For SBIR, Phase I awards last for six to twelve months and can total up to $150,000, and Phase II awards can last two years and total up to $1,000,000. For STTR, Phase I awards last for six months and can total up to $100,000, and Phase II awards can last two years and total up to $750,000. A comparison of the SBIR and STTR programs is outlined in Table 33.1.

Admittedly, funding from both programs to universities is indirect as the actual awardees are small businesses. However, many university early-stage technologies can benefit tremendously from these programs. Technologies that are too applied, risky, or advanced for traditional funding mechanisms (e.g., NIH R01 grants) can be spun into a start-up company and developed technically with SBIR or STTR funds until they are market ready. It should be noted that, unlike basic science review committees that award funding based on the strength of the hypothesis and experimental protocol, review committees for SBIR/STTR grants and contracts award funding based on the likelihood of commercial success, the proposed pathway for product development, and the skill and experience of the team leading the efforts [29].

Other significant translational funds come from the innovation oriented private foundations. Some of these foundations include the Muscular Dystrophy Association (MDA), the Michael J. Fox Foundation, the Juvenile Diabetes Research Foundation (JDRF), the Bill & Melinda Gates Foundation, and the Coulter Foundation [9, 29–31]. Obviously, the above list is not complete, but these nonprofit entities represent a growing body of organizations aggressively seeking disease-specific cures and treatments or commercialization of beneficial pharmaceuticals and diagnostics through translational funding. The private foundations fund a broad spectrum of projects, from more basic science efforts to translational research, product development, commercialization, and even clinical trials. The translational funds are invaluable for university researchers advancing technology through the commercialization pathway. Unlike the SBIR/STTR program, the university itself receives the award from the foundation, so no corporate partner is required. Interestingly, some foundations contractually obligate the university to share any royalties derived from products that are developed with foundation funding with the granting foundation. Such an arrangement is not unexpected given the perpetually difficult fundraising climate and the demands of donors for a return on their donations.

A small, but rapidly expanding translational funding source can be found within the confines of the university. It has been estimated that the top twenty-five research universities are sitting on nearly $300 billion in their endowments in 2012, with Harvard's endowment alone exceeding $30 billion. With venture capital drying up in the biotechnology sector, universities are turning to their hefty endowments to fund translational research and start-up ventures. Two trailblazers in this area of endowment-funded translational research and venture investing have been the University of Michigan (UM) and the University of Medicine and Dentistry of New Jersey (UMDNJ). In 2011, UM decided to tap its endowment to fund start-up companies that were spun-out from the school. UM limited its investments to follow on funding from more traditional early-stage investors, who performed the normal due diligence and analysis of any start-up opportunity. Around 2005, UMDNJ created a $200 million venture fund from its endowment. UMDNJ funds start-up companies with seed investments up to $500,000.

Finally, many state governments have formed or commissioned translational funds as part of economic development efforts. The Georgia Research Alliance (GRA), Virginia's Center for Innovative Technology (CIT), the Michigan University Commercialization Initiative (MUCI), and Washington's Life Science Development Fund (LSDF) are just a few examples of state-based translational funding sources. The details of the obligations to the university or start-up company vary tremendously, but most of the state funds incentivize a new venture to remain in the state and hire employees from within the state as well.

SCIENTIFIC VERSUS COMMERCIALLY RELEVANT MILESTONES AND DEVELOPMENT PLANS

The primary difference between traditional, basic science funding and translational funding is the content and purpose of the specific aims or milestones. Whereas researchers score well in most traditional funding mechanisms when the specific aims guide hypothesis-driven research and answer fundamental questions of basic science, the milestones in translational funds are crafted to propel a technology to commercially viable and valuable inflection points. One way to look at the difference between the two funding models is that traditional, basic science funding is "push-based" while translational funding is "pull-based." In a push-based model, the explorative aims are designed to push the boundaries of science and knowledge, elucidating previously unknown mechanisms or uncovering the physiological importance of a newly discovered protein. The push-based model starts at the beginning of the scientific method with a question or observation, and moves through the traditional means of scientific inquiry by developing a testable hypothesis, experimentation, analysis, re-experimentation if necessary, and ultimately, a report in a peer-reviewed journal. Reviewers of grants submitted to scientific study sections tend to review the merits of the science, the strength of the hypothesis, the logic of the experimental design, and even the academic pedigree of an investigator. Review committees on translational funds, on the other hand, are inclined to approach applications more like a venture capitalist. In other words, these reviewers start at the back end with the commercial endpoints, and work in reverse to assess the merits of an application. They are less likely to emphasize scientific hypotheses and academic pedigree, but are likely to be concerned about market potential, IP position, exit strategy, and the probability of reimbursement by a third party payer.

It should be noted that both funding mechanisms serve a unique purpose and are necessary to assimilate knowledge, explore new frontiers, and develop new products that benefit humanity. One model is not necessarily better than the other, but it can be argued that both live in a symbiotic relationship with each other.

MARKETING AND LICENSING NEW TECHNOLOGY

Marketing

The prevailing wisdom in university technology transfer is that the best marketing strategy for university-based technologies is to exploit the contact list of the faculty inventor. In other words, the faculty inventor knows the technology space quite well, is knowledgeable about the products that can be derived from any IP, and may have a high quality list of potential partners for commercialization. While the above scenario may be true for the majority of university technologies, what is the tech transfer professional to do with the remaining portfolio, where a faculty member is a basic researcher but has serendipitously discovered a technology with tremendous applicability and potential? The tech transfer office must convert itself into a sales force and be the principal marketer of the newfound invention.

How does the technology transfer office market its technologies? First, the tech transfer office will work with the faculty inventor to draft a nonconfidential summary of the technology. Often, the invention disclosure submitted to the tech transfer office is highly technical, with minimal information on the potential market or product that can be derived. The tech transfer office must determine whether a product or service can be created from the invention, whether the product or service confers an advantage or benefit to a potential licensee or customer, and what markets can be exploited and penetrated. The tech transfer professional will incorporate the technical background, any relevant and compelling data, benefits and advantages, and potential market into a nonconfidential summary that will be used as the primary marketing piece for the technology. Second, the tech transfer office will identify a target list of companies and contacts and send a marketing package to them. The marketing package will definitely include the nonconfidential summary, but may also include a publication or patent application if available. It should be noted also that the technology transfer office typically engages in so-called passive marketing. That is, they post nonconfidential summaries of available technologies on their website and third party databases such as Flintbox®, iBridge℠ Network and the AUTM Global Technology Portal. Third, the tech transfer office will follow up with any marketing "hits" that result from the campaign, and either put the interested party in contact with the inventor for a technical discussion, or even begin negotiations for an option or license.

The marketing strategy outlined above is a generic overview of how the technology transfer offices market most of its technologies. However, there are nuances in strategy, depending on the technology, that must be appreciated. For instance, life science technologies such as pharmaceuticals, diagnostics, and biologicals, may require a much different strategy from physical science inventions such as devices, semiconductors, materials, or software. If a faculty inventor has developed a novel compound and discovered a disease on which the compound seems to be efficacious, the tech transfer office will search for pharmaceutical companies looking for new compounds in the disease area. It is true that new conditions or diseases may be identified in the course of clinical testing in which the new compound works, but at the outset, the marketing job is fairly straightforward. Marketing physical science inventions is generally not as easy. For example, let us say that an inventor has developed a new polymer with very attractive properties like biocompatibility, super hydrophobicity, and unique time-dependent deformation characteristics. How does a tech transfer office market a material with such a wide range of appealing features? The tech transfer professional may need to consult with the inventor, advisors, and industry experts to brainstorm different uses for the material. It may not be unusual to uncover dozens of uses and potential applications, so the tech transfer professional is then tasked with identifying companies and contacts for each use and tailoring the marketing campaign accordingly.

Financial Considerations

After marketing is completed and a licensee has been found, the next phase in the commercialization process is negotiation of the financial terms of the agreement. The details of the financial terms will be discussed in a later section, but it is important to understand the two primary extremes regarding financial considerations in the license agreement. A university's emphasis in financial considerations depends on the risk tolerance of the university and the financial support that it gives the technology transfer office. On one end of the spectrum, if the university does not financially support the technology transfer office and the technology transfer office has a small appetite for risk, deals are likely to be weighted more toward the front end. That is, the tech transfer office will likely seek higher upfront fees and perhaps even milestone payments in exchange for a smaller than market rate royalty. These universities tend to "eat what they kill." University tech transfer offices that must support their own budget on their own royalty income tend to fall into this category. The other end of the spectrum includes universities that financially support the technology transfer office, or where the technology transfer office has an endowment that supports its operations, and the university has

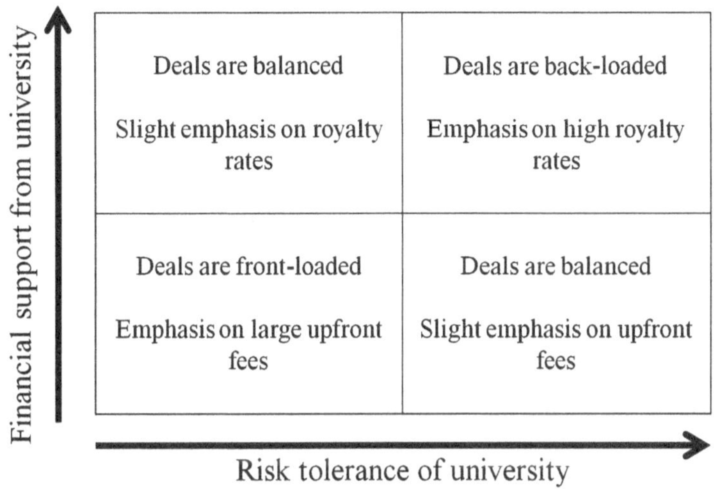

Figure 33.3. Risk tolerance and financial support matrix for university tech transfer offices.

the capacity to withstand high-risk ventures. Such universities can weight deals toward the back end, with a chance for a great upside with a higher royalty rate. In other words, these universities "swing for the fence" when it comes to deal making. The large, well established research institutions in the U.S. likely fall into this category. The spectrum of financial support and risk is outlined graphically in Figure 33.3.

Of course, many universities abide in the middle ground, and not all can be categorized neatly by financial support and risk tolerance.

Socially Responsible License Terms

As of May 2012, nearly 200 universities, research institutions, and nonprofit organizations had endorsed the "Nine Points to Consider in Licensing University Technology." Crafted in March 2007 by twelve leading research universities in the U.S., including Stanford, Harvard, Yale, CalTech, and WARF, the Nine Points is an attempt to memorialize best university licensing practices and to challenge universities to include socially responsible terms in its license agreements. The Nine Points are as follows:

1. Universities should reserve the right to practice licensed inventions and to allow other nonprofit and governmental organizations to do so.
2. Exclusive licenses should be structured in a manner that encourages technology development and use.
3. Universities should strive to minimize the licensing of "future improvements."
4. Universities should anticipate and help to manage technology transfer-related conflicts of interest.
5. Universities should ensure broad access to research tools.

6. Enforcement action should be carefully considered.
7. Universities should be mindful of export regulations.
8. Universities should be mindful of the implications of working with patent aggregators (i.e., "patent trolls").
9. Universities should consider including provisions that address unmet needs, such as those of neglected patient populations or geographic areas, giving particular attention to improved therapeutics, diagnostics, and agricultural technologies for the developing world.

Many universities have committed to socially responsible licensing through the endorsement or implementation of the Nine Points. The final point of the Nine Points is very important for pharmaceutical, diagnostic, and biotechnology inventions. Universities that adhere to the Nine Points typically forego the filing of patent applications in developing countries. They may even waive royalties or reduce royalties for sales of pharmaceuticals or biotechnologies in developing countries in which such technologies could be useful to the population. As universities continue to grow into global centers of education, they will be expected to utilize new knowledge and understanding for the public good, whether it be locally or globally [26, 32, 33].

NEGOTIATING FOR SUCCESS

Financial Considerations

As with most licensing agreements for pharmaceutical or biotechnology inventions, the licensor (i.e., the university) expects to share in the revenue derived from the IP licensed to the licensee. The details of the revenue structure vary by agreement and technology, as do valuations. Ideally, the license is structured such that the licensee is incentivized to commercially exploit the IP and the university shares in the licensee's success proportionally [34]. A pharmaceutical or biotechnology licensing agreement is likely to contain at least the following financial provisions: an upfront fee, running royalties, milestone fees, and consideration for sublicensing if the licensee is granted the right to sublicense. Some licenses may contain other financial terms such as minimum annual royalties or maintenance fees. Most university license agreements also require that patent expenses be reimbursed by the licensor.

Given the early-stage, high-risk nature of most university pharmaceutical or biotechnology intellectual assets, often the financial structure of a license is back loaded. In other words, upfront fees are typically small compared with the entire pool of pharmaceutical deals. Royalty rates serve as the bulk of the value realized by the university, and they vary dramatically, from 12–20% if an opportunity is composed of a novel compound, preclinical animal efficacy data, and favorable toxicology and pharmacokinetic results, to 1–5% if a target has been identified or a new method of use on an existing drug has been discovered [17]. Milestone payments are generally linked to clinical stages of development (e.g., submission of IND, initiation of Phase 1 clinical trial), and they serve to mitigate financial risk for both parties.

One other financial term that is considered in licenses to start-up companies is equity. Ordinarily, start-up companies are strapped for cash at the outset, so universities often take equity in consideration for an IP license in lieu of an upfront cash payment. Both parties must be careful in structuring a deal that grants upside to the university and makes the start-up company attractive to potential investors and acquirers. University philosophy varies with respect to equity dilution. Some universities insert anti-dilution provisions up to a certain amount of equity or debt financing secured by the company. Other universities forego anti-dilution in exchange for a larger equity portion upon execution of the license.

Diligence

Next to the financial provisions of a licensing agreement, the diligence terms are probably the most important and critical for a university. There are two primary reasons for the inclusion and necessity of diligence terms in a license to commercialize university technologies. First, the Bayh–Dole Act requires that licensors of federally funded inventions take effective steps to achieve practical application of the technology and to ensure that it becomes available for public use to the greatest extent possible [4]. Second, the university itself does not typically have the capacity to abide by these provisions, so it must obligate the licensee to use commercially reasonable efforts to bring products based on the technology to market.

In pharmaceutical and biotechnology licenses, diligence requirements vary based on the stage of development at which the asset is licensed. For an early-stage opportunity, diligence requirements are likely to include a business or development plan that outlines the pathway for commercialization of the technology, dates for the initiation of preclinical animal testing sufficient to support an IND filing, first dosing in a patient in Phase 1, 2, and 3 clinical trials, the receipt of regulatory approval in the U.S. or a foreign country, the first sale or disposition of a product incorporating the subject IP, and a sales threshold. Universities may require start-ups to achieve additional milestones pertaining to fundraising. For example, it is not uncommon for a university to require a start-up company in which it holds equity to raise at least one or two rounds of capital in debt or equity financing.

While licensees generally push back, universities require that the majority of the diligence milestones listed above be achieved by a certain date. Granted, selecting dates for the completion of milestones in an early-stage, high-risk environment with a development timeline on the order of ten years is impossible to do with much accuracy. Universities recognize this fact, and strive hard to negotiate reasonable timelines that simultaneously suit the licensee's development plan and move the technology to the market place as quickly as possible.

TERMINATION AND RETURN OF CLINICAL DATA

Most universities allow a licensee to terminate a license agreement at will for no cause. As of the writing of this chapter, AUTM's official position on the subject is identical.

On occasion, universities charge a termination fee to offset some of the cost associated with re-licensing a case or offsetting expected revenues. The university also wants the right to terminate the license agreement, but its termination most often is the result of a breach of the terms and conditions of the contract by the licensee, which may include licensee's failure to make a payment, failure to achieve a diligence milestone, or failure to continue to use best commercial efforts to bring a product to market.

Most importantly, in a pharmaceutical or biotechnology license agreement, the university wants rights to any clinical data collected during the course of development of a product that is covered by the university's IP. Clinical data may include results of *in vitro* studies, toxicology experiments, as well as pharmacokinetic and pharmacodynamic profiles and efficacy reports. Once the license is terminated and the rights to license to the patent are returned to the university, the university assesses the value of its asset based on such clinical data. Furthermore, the university wants the right to grant a third party the use of the clinical data in the case that a third party wants to license and commercialize the asset. The right to use the clinical data can be granted in numerous ways. For example, the original licensee can assign or grant a license to the university to the clinical data, the original licensee can agree to grant a license to the clinical data to a third party with a *bona fide* interest in developing the technology, or the original licensee can agree to allow a third party to reference its IND for the asset.

LICENSING TO START-UPS VERSUS ESTABLISHED COMPANIES

Decision Process

There is no hard-and-fast rule when a university decides whether to license a technology to an established company or a start-up company. The decision does not necessarily depend on the stage of development of the technology. Large pharmaceutical companies may be willing to license an early-stage biomarker if they believe it represents an opportunity for a novel mechanism of action, and a start-up may be the most suitable partner for a drug that is already in clinical trials.

The university tech transfer office works with the inventor and perhaps a group of advisors to determine the best way to commercialize the technology. Often, the tech transfer office markets the technology to existing companies to gauge interest and receive feedback from industry experts to help guide the decision making process. Given the early-stage, high-risk, cutting-edge nature of many university technologies, a market may not yet be available for the opportunity. If an established company is not willing to help build the market or forage into new territory, the opportunity may be licensed to a start-up company for additional incubation and development.

What criteria are necessary to license a technology to a start-up company? Obviously, a good technology is a great place to start. Platform technologies, novel small molecules, and therapeutic antibodies with a companion diagnostic are just some of the technologies that can be successfully commercialized through a start-up. In addition to a great technology, the company needs an experienced management team and advisory board. In areas like Boston, San Diego, Silicon Valley, or the Research Triangle Park, experienced entrepreneurs and managers may not be difficult to find. Universities outside the areas listed above may have more of a challenge finding management talent, but the task is not impossible. A cooperative and enthusiastic faculty inventor also improves the odds that a start-up will be successful. The role of faculty members will be discussed in more detail below, but their input and guidance, especially in the early years of the company, on the technical side of development is absolutely critical. The faculty member also plays a vital role in writing SBIR and STTR grants to fund the technical development of an opportunity with nondilutive capital.

Faculty as Founders or Consultants: Conflict of Interest

Faculty inventors choose to what extent they want to be involved in a new start-up venture. Some faculty go as far as to leave their academic position to join the company full time, while others remain as far from the start-up company as possible. A middle ground occupied by many faculty has them filling the role of consultant or scientific advisor if the university allows it. While most universities permit their faculty to allocate 20% of their time to consulting, conflicts can arise for the faculty member when he or she maintains academic duty and consults for his or her start-up company. In fact, many universities do not allow academic founders to consult for the start-up company because of the conflict (e.g., the University of North Carolina).

What conflicts can faculty inventors and entrepreneurs face in the creation and pursuit of start-ups? First, as has been alluded to, faculty members serving as consultants for their start-ups can create a major issue for the university, especially in terms of ownership of IP. Most often, a technology licensed to a faculty start-up company falls into the research domain of the faculty member. As a result, it's nearly impossible to separate the faculty member's academic research activity from the company's research endeavors. One, then, can imagine a scenario where a faculty member is using his or her consulting time to perform research for the company in a technology space that is identical to his or her academic research. In this case, if new IP is developed, what entity owns the IP? Is it the company or the university? What constitutes consulting time as opposed to academic time or institutional responsibilities? Complicating matters is the fact that consulting agreements ordinarily do not name the university as a party—the agreements are executed between the consultant and individual inventor (some universities review a faculty consulting agreement as part of the conflict management process). Furthermore, it is a practice of academic physicians and surgeons, at least those without research programs, to consult for companies and then assign any IP created under the consulting agreement to the company in exchange for a flat fee or small royalty.

How does the university reconcile all of these seemingly disparate and complex relationships? The primary strategy for asserting and securing ownership of university inventions is through the employment agreement between the institution and the faculty member as well as the university IP policy. Most universities assert ownership of inventions developed in the faculty member's area of expertise that were developed using significant university funds, facilities, or resources. Therefore, if a faculty member is consulting for his or her start-up company, but using university lab space or research grants, the university typically asserts ownership to any IP developed. The university typically agrees to license any such IP to the start-up company for fair consideration. If, however, a research faculty member develops an invention during consulting time using company resources at a company facility and the invention falls outside the researcher's bailiwick, the university may not assert ownership. Similarly, if a surgeon develops a new medical device under the scope of a consulting agreement and does not use significant university funds, facilities or resources, the institution will not assert ownership. As is apparent from the scenarios above, conflicts can arise under consulting arrangements, but most universities manage the conflicts quite well.

A second major area of conflict that can arise for faculty entrepreneurs is in the area of compensation paid in the form of equity in the start-up. The faculty inventor can receive equity in two ways: as a founding member of the company or through the licensing agreement between the start-up and the university (assuming equity is conferred as consideration for the license). The conflict arises when the faculty member tries to receive both. Most universities preclude such "double-dipping" by having the faculty founders waive any equity shares due under the license agreement.

A third area of conflict emerges when a faculty member owns equity in a start-up company and performs research for the company in his or her laboratory. The major ethical concern in this situation is that it appears that the faculty member is enriching himself or herself using university facilities and resources. The tax problem, particularly for state universities, is that such research may constitute "for-profit" activity, and the university has a limit as a nonprofit entity as to how much "for-profit" research can be conducted in its facilities. Another issue in this scenario concerns the graduate students studying under the faculty member. The faculty member may not be able to require the graduate student to perform company research since, for example, the student is likely to be supported primarily by grants from the federal government.

While all of the above conflicts can be difficult and complex for the university and faculty, most of the major research universities now have highly competent conflict of interest committees or offices devoted solely to managing such conflicts. Conflicts are not necessarily detrimental for university faculty members. In fact, some tech transfer offices admit that conflicts are positive and desirable. Nevertheless, the conflicts must be managed and are now being managed much better by university administrators than they were in the past.

The University's Role in Start-Ups

The university not only acts as a licensor to start-up companies, but also serves as an IP advisor, a scientific or technical consultant, a technology marketer, an enthusiastic pitch-maker, an alternative sources fundraiser, an entrepreneur and management headhunter, and always a champion. Occasionally, the university acts as an investor in the company, or provides space for the start-up in its incubator or accelerator facility (which makes it a real estate agent and broker). The relationship between a start-up and the university does not begin and end at the execution of the initial license agreement. Since most universities are equity participants in start-ups, they have a vested interest in the success of the company. Not only does the university want to push products to market based on the technology licensed to the company, it also wants to spur economic development in its locale or region. In fact, many universities provide discounts on payments required under the license agreement to start-up companies that remain local, hire locally, and manufacture products locally. Such practices are especially common at large, public, state institutions.

THE IMPORTANCE OF A POSTLICENSING PROGRAM

"Staying around the Table"

Every licensing professional knows that a deal is not complete when the contract is signed and the ink is dry. In the ideal scenario, the execution of the license agreement is the beginning of what could be a twenty-year relationship or more (i.e., throughout the life of a patent). The university maintains its relationship with the licensee primarily through a formal or informal postlicensing program.

With the university's emphasis on diligent commercialization of its technologies, the tech transfer office tries to keep abreast of licensee developments through progress and financial reporting, consistent communication with the company counterpart, and through shared management of the IP portfolio. University licenses commonly obligate the licensee to submit quarterly, semi-annual, or annual progress reports to ensure that the technology continues to move to market. If a product is on the market, the university requires that the licensee shares financial reports to justify royalties paid under the license agreement. With respect to patent prosecution, the university and licensee generally share responsibility. A reasonable way to think of the relationship is as follows—the university maintains legal control of the patents being prosecuted while the licensee has functional control of the prosecution strategy. While licensees often request total control over patent prosecution, the university is reluctant to do so as owner of the intellectual asset. The university is charged with exploiting the IP to the fullest extent possible, so it may want to reserve the right to keep open opportunities for other licenses in other fields of use on a given piece of IP. Nevertheless, in most instances, the university tech transfer office defers to the licensee to make decisions regarding prosecution strategy.

License Modifications and Resolution of Conflicts

Even the best-laid plans to commercialize a technology can go awry, and in the world of pharmaceuticals and biotechnology, adjustments and modifications to the original business or development plan are inevitable. Accordingly, the diligence milestones or their respective dates are often renegotiated during the postlicensing period. Universities are generally amenable to revising the diligence requirements, provided that the licensee has used its best commercial efforts to achieve the original milestones. Universities recognize that the original milestones were, in essence, a "best guess" at the time of execution. The parties may reconvene, agree upon a new reasonable set of milestones, and amend the agreement to reflect the changes.

Perhaps the area that stokes the most contention in the postlicensing period is the financial obligations of the licensee. If the technology transfer office suspects that a licensee is not reporting sales of a licensed product correctly, and as a result, not submitting appropriate royalty payments, it may initiate an audit of the books and records of the licensee. The details of the audit provisions vary slightly among universities. Most allow for the university to conduct at least one audit per year by a reputable firm. If the auditors find that the licensee has underpaid the university licensor by 5% or more, the licensee will pay for the cost of the audit in addition to all back royalties owed.

REFERENCES

1. Council on Government Relations. The Bayh-Dole Act: a guide to the law and implementing regulations. 1999. Available at http://www.ucop.edu/ott/faculty/bayh.html. Accessed February 10, 2014.
2. Collins, A.P. Tracking innovation: an examination of university research and its relationship to early stage venture capital. M.S. Thesis, MIT, June 2003.
3. Odza, M. From the ivory tower to the marketplace: the Bayh-Dole Law and the myth of better mousetraps. Available at http://www.columbia.edu/cu/21stC/issue-3.1/odza.html. Accessed February 10, 2013.
4. 35 USC § 200. Available at http://www.law.cornell.edu/uscode/text/35/200. Accessed December 13, 2013.
5. Massachusetts Institute of Technology. Policies and procedures: a guide for faculty and staff members, 13.0 Information Policies. Available at http://web.mit.edu/policies/13/13.1.html. Accessed February 10, 2013.
6. University of Florida Intellectual Property Policy. 2013. Available at http://www.research.ufl.edu/otl/pdf/ipp.pdf. Accessed February 10, 2013.
7. Harvard University, Office of Technology Development. Intellectual Property Policy. Available at http://otd.harvard.edu/resources/policies/IP/. Accessed February 10, 2013.
8. Emory University Intellectual Property Policy. Available at http://policies.emory.edu/7.6. Accessed February 10, 2013.
9. http://www.innovation.virginia.edu. Accessed February 10, 2013.
10. A summary of *Stanford v. Roche*. 2011. Available at http://otm.illinois.edu/features/2011-07-12/summary-stanford-v-roche. Accessed February 10, 2013.
11. McEwen, J.G., O'Conner, S.M., McCarthy, J.E., Jr., Ebner, S.W. The impact of Stanford v. Roche on technology licensing under Bayh-Dole. *Procurement Lawyer* Spring 2012;47(3):5–16.
12. Mueller, J.M. *Patent Law*, Third edn. New York: Wolters Kluwer Law & Business; 2009.
13. Flower, A.C. *Intellectual Property Technology Transfer*. New York: BNA Books; 2006.
14. Port, K.L., McManis, C.R., McElwee, T.P., Hammersley, F.M. *Licensing Intellectual Property in the Digital Age*. Durham: Carolina Academic Press; 1999.
15. Pressman, D. *Patent It Yourself: Your Step-by-Step Guide to Filing at the U.S. Patent Office*. Berkeley: NOLO; 2011.

16. IP Handbook of Best Practices. Available at http://www.iphandbook.org/index.html. Accessed February 10, 2013.
17. Mehta, S.S. *Commercializing Successful Biomedical Technologies: Basic Principles of the Development of Drugs, Diagnostics and Devices*. New York: Cambridge University Press; 2011.
18. Bohrer, R.A. Reach-through claims for drug target patents: Rx for pharmaceutical policy. *Nat Biotechnol*. 2008;26(1):55–56.
19. Kirschbaum, J.B. Does it make sense to patent university drug discovery targets? June 8, 2008. Available at http://mysdscience.com/group/scienceentrepreneurs/forum/topics/2110706:Topic:2035. Accessed December 13, 2013.
20. Kneller, R. The importance of new companies for drug discovery: origins of a decade of new drugs. *Nat Rev Drug Discov*. 2010;9(11):867–882.
21. Freeman, M.W., Dervan, A.P. The path from bench to bedside: considerations before starting the journey. *J Investig Med*. 2011;59(5):746–751.
22. Stevens, A.J., Jensen, J.J., Wyller, K., Kilgore, P.C., Chatterjee, S., Rohrbaugh, M.L. The role of public-sector research in the discovery of drugs and vaccines. *N Engl J Med*. 2011;364(6):535–541.
23. Miller, J.R., Thanabal, V., Melnick, M.M., Lall, M., Donovan, C., Sarver, R.W., Lee, D.Y., Ohren, J., Emerson, D. The use of biochemical and biophysical tools for triage of high-throughput screening hits—a case study with Escherichia coli phosphopantetheine adenylyltransferase. *Chem Biol Drug Des*. 2010;75(5):444–454.
24. Blumenstyk, G. Big pharma finds a home on campus. *Chron High Educ*. 2011;57(42):A1–A5.
25. Newman, D.J., Cragg, G.M. Natural products as sources of new drugs over the 30 years from 1981 to 2010. *J Nat Prod*. 2012;75(3):311–335.
26. Lowell, J.E., Earl, C.D. Leveraging biotech's drug discovery expertise for neglected diseases. *Nat Biotechnol*. 2009;27(4):323–329.
27. Ouellette, L.L. How many patents does it take to make a drug? Follow-on pharmaceutical patents and university licensing. *Mich Telecomm Technol Law Rev*. 2010;17(1):299–336.
28. Sampat, B.N. Research locally, diffuse globally? American universities, patents and global public health. *J Int Aff*. 2010;64(1):69–82.
29. Small Business Innovation Research (SBIR) and Small Business Technology Transfer (STTR) Programs at the National Institutes of Health. Available at http://grants.nih.gov/grants/funding/sbir.htm. Accessed February 10, 2013.
30. Pienta, K.J. Successfully accelerating translational research at an academic medical center: The University of Michigan-Coulter Translational Research Partnership Program. *Clin Transl Sci*. 2010;3(6):316–318.
31. Begley, S., Carmichael, M. Desperately seeking cures. *Newsweek* 2010;155(21/22):38–42.
32. Kumar, M.N. Ethical conflicts in commercialization of university research in the post-Bayh-Dole era. *Ethics Behav*. 2010;20(5):324–351.
33. Caulfield, T., Harmon, S.H.E., Joly, Y. Open science versus commercialization: a modern research conflict? *Genome Med*. 2012;4(2):1–11.
34. Stim, R. *Profit from Your Idea: How to Make Smart Licensing Deals*. Berkeley: NOLO; 2010.

34

THE PIVOTAL ROLE OF THE ACADEMIC ENTREPRENEUR AND THE ENTREPRENEURIAL UNIVERSITY IN GLOBAL LIFE SCIENCES

Donna Marie De Carolis

Charles D. Close School of Entrepreneurship, Drexel University, Philadelphia, PA, USA

Our global economy and the capabilities that underpin technological and societal advancement are being transformed by seismic trends that impact the way that new technologies and products are discovered and developed. Traditional approaches to research and development (R&D), in every industry, are being uprooted. Most notably, the discovery, development, and commercialization of new technologies are now rarely accomplished by one individual, one company, one country, or one university department. We are in the throes of a paradigm shift in the discovery and commercialization of innovations—a paradigm shift that affects all players. This is the era of global open innovation, of global R&D.

In its original conceptualization, open innovation is described as the deliberate inflow and outflow of knowledge to facilitate innovation; it constitutes a strategy of collaborating with competitors, government, customers, and suppliers to create, develop, and commercialize knowledge. The underlying assumption is that companies can no longer rely solely on internally generated knowledge, but must seek and incorporate external sources of knowledge to innovate [1].

Open innovation is now bolstered by the digitization of information and the dynamic advances in computing and telecommunications technologies. The

Collaborative Innovation in Drug Discovery: Strategies for Public and Private Partnerships, First Edition. Edited by Rathnam Chaguturu.

combination of open innovation and information technology in drug discovery is exemplified by Novartis' bold move in 2007 to share its genomics research in type 2 diabetes with the world by posting their knowledge on the Web for other life sciences and academic centers to utilize. After investing several years and millions of dollars into this research, the company noted that the only way to invent new medicines in this area would be a global effort.

In the twenty-first-century competitive landscape, open innovation no longer provides a competitive advantage; it provides competitive parity. Companies, universities, governments, countries—all must share knowledge to innovate and advance, particularly in the high technology sectors such as aerospace, pharmaceuticals, computing, communications equipment, and scientific instruments.

As bastions of pioneering ideas, universities continue to play a critical role in the knowledge economy as evidenced by the continual funding, collaboration, and reliance on university inventions by both industry and government. Universities, however, are not immune to the paradigm shift in technology commercialization. They are a vital part of the innovation ecosystem—equally affected by the competitive, technological, and global trends transforming industries, business models, and paths to innovation advancement.

The paradigm shift in technology commercialization for universities cannot be fully understood out of context. Universities, and in particular, their technology transfer offices and procedures, do not exist in a vacuum. The larger, encompassing global trends in the knowledge economy are directly related to how universities will continue to generate and disseminate knowledge.

THE GENERATION OF KNOWLEDGE REPRESENTS THE *RAISON D'ÊTRE* FOR UNIVERSITIES

Yet the paradigm shift in scientific and technological discoveries and innovations mandates an extension of the basic mission of universities. Research and teaching have traditionally constituted the mission of universities. Flowing from these core values, universities have developed strong policies, structures, and cultures to protect and nurture research and teaching. Concurrently, the archetype of the academician mimics the university culture: the researcher and professor who, as experts in their fields, advance knowledge and disseminate that knowledge to the larger academic community and to the classroom. Indeed, even the study of medicine and related cures was, in universities, an academic pursuit devoid of any commercial connection, until the beginning of the twentieth century with the birth of the pharmaceutical industry.

The second half of the twentieth century, however, brought about the third mission of universities: economic development [2]. While economic development captures a broad array of activities, in the academic scientist arena it largely focuses on the commercialization of knowledge to improve regional or national economic performance. Such commercialization activities simultaneously benefit the university in many ways as well, from finances to increased research revenues to reputation.

The notion of academic entrepreneurship materialized concurrently with universities' commitment to economic development. Academic entrepreneurship may be broadly defined at two levels. First, academic entrepreneurs, in a university setting, are individuals who are intrinsically motivated to bring their knowledge to society and/or the marketplace. Their passion for creativity and innovation moves beyond discovery and invention into implementation and application. Second, at the university level, academic entrepreneurship may be defined as an organizational culture that supports innovative faculty who desire to disseminate their knowledge. It is the involvement of academic scientists and universities in commercially relevant activities such as industry-university collaborations, university-based incubators, start-ups by academics, and university-based venture funds.

If embraced by universities, the path of a culture of academic entrepreneurship is not an easy one, given resistance from internal constituencies. Still, evidence is mounting that university-based innovations create jobs directly and indirectly through company creation [3]. Although federally funded research constitutes the largest part of university research budgets, the global competitive landscape points to the increasing role of corporate-sponsored research for universities. Particularly in life sciences, "global" is the operative word in drug discovery.

THE TRADITIONAL AND CHANGING ROLE OF TECHNOLOGY TRANSFER OFFICES IN THE CONTEXT OF ACADEMIC ENTREPRENEURSHIP

Several events over the past few decades have contributed to the emphasis on economic development as part of the university mission and to the proliferation of technology transfer offices. The "first" technology transfer office, WARF (Wisconsin Alumni Research Foundation), was founded at the University of Wisconsin in 1924. WARF was designed as an entity that could patent technologies invented by Wisconsin faculty, license them, and then return the license royalties to the university. While WARF remains a separate entity, its structure gave rise to the university technology transfer office as we know it today.

Subsequently, and slowly, a few other universities established technology transfer offices for the purpose of disseminating innovations for the good of society. However, a major obstacle to the transfer of university-generated innovations was the federal government, who held the intellectual property rights to all federally funded research. The Bayh–Dole Act of 1980 ignited the proliferation of university technology transfer offices by allowing universities to retain the right to intellectual property ownership of inventions generated through federally funded research. The Cohen-Boyer biotechnology patents in 1978 provided further impetus for a changing academic landscape in terms of commitment to economic relevance. Those patents revolutionized drug discovery and development as they not only disrupted the knowledge base of the pharmaceutical industry, but they also stimulated new attitudes in academic laboratories—attitudes that included a propensity to bring knowledge to market and, for some researchers, an entrepreneurial focus on commercializing their inventions. Thus, since Bayh–Dole, the number of university technology transfer offices has increased and their role and importance has expanded. The Association of University Technology Managers (AUTM) is a nonprofit group that was created as the number of technology transfer offices in universities grew. AUTM is a nonprofit association with a membership of about 3000 individuals from about 300 universities, research institutions, and teaching hospitals, as well as businesses and government organizations. AUTM provide numerous services to its membership, including education, conferences, benchmarking data, and various statistics.

Still, the majority of university inventions are not ready for prime time. Most are early-stage innovations, especially in the life sciences. University inventions require significant resources to develop them from prototype or proof of concept to actual product—in other words, a commercialization process. The "valley of death" is a phrase used to describe the uncertain and difficult path for early stage innovations as they attempt to garner support and resources for commercialization. Technology transfer professionals manage the transfer of knowledge from the laboratory to the marketplace—their role is to find the mechanism that will navigate the innovation successfully through the valley of death, whether it be to license that innovation or to form a start-up. It is also the office in which inventions are evaluated for patent and copyright protection and for their ability to be commercialized. The technology transfer office executes licensing agreements to companies and individuals interested in moving early-stage university innovations through the commercialization process.

There are many measures used to assess the effectiveness of the technology transfer office, including but not limited to the number of disclosures to the technology transfer office from researchers, number of patents filed, number of license agreements, number of companies formed, royalties from license fees, royalties and cash from equity arrangements, and the number of products successfully brought to market. There are also qualitative measures of the performance of a technology transfer office, such as faculty satisfaction, its role in attracting and retaining excellent professors and graduate students, and its ability to attract corporate-sponsored research [4]. More recently, the activities of the university technology transfer office are associated with job creation, start-up formation, and societal welfare, as they facilitate the transfer of valuable university innovations to the marketplace for the public good.

Over the past 20 years or so, there has been a proliferation of interest in academic entrepreneurship and the role of technology transfer offices from management scholars. This literature is too vast to analyze and summarize here and the reader is directed to a comprehensive review article that organizes and synthesizes this research [5]. In this article, Rothermel et al. summarize and synthesize scholarly investigations that address research questions pertaining to academic entrepreneurship and technology transfer offices. Several studies investigate the exogenous factors, such as access to venture capital and regional economic indicators that may impact the university and the technology transfer office. Other studies look at factors endogenous to the university, such as tenure policies, research dollars, university policies, and incentives that impact the productivity or success of a technology transfer office. Finally, there are a set of studies that investigate the characteristics of the technology transfer office itself as antecedents of performance, such as number of personnel and age of the office. These studies are not limited to U.S. universities, as many of them investigate the university knowledge diffusion process in European and Asian settings [5].

The expanse and growth of interest in academic entrepreneurship and university technology transfer underscores and validates the significance of the critical role that universities play in innovation diffusion and technology commercialization as well as in the generation and dissemination of knowledge.

Yet, the role of the technology transfer office in the efficient transfer and commercialization of university knowledge is also questioned. Critics of technology transfer offices cite the lengthy time between disclosure and licensing or patent application. While AUTM provides valuable statistics on licensing income, revenues, and patent data from university technology transfer offices, there is a lack of information on the time intervals from disclosure to patent or commercialization. This brings up questions of the adequacy of staffing technology transfer offices and the types of research, particularly market research, that are expended on innovations. For example, it has been suggested that there might be a difference in time to license/patent between those innovations whose inventors have a preexisting tie to a company versus those who do not. The implication is that those inventions with existing corporate interest are those with the necessary market research to hasten licensing or start-up [6].

Probably the most controversial response to the questions of efficiency and performance of university technology transfer offices is the recently proposed "Start-Up Act." Under this piece of legislation, which generally endorses several policies intended to enhance new venture formation in the United States, one particular clause would allow university inventors to basically seek, on their own, corporate partners to license or to form a start-up. The Start-Up Act seems to be in direct conflict with the existence of university technology transfer offices.

In conclusion, the role of academic entrepreneurship and university technology transfer offices, while considered to be vital, still needs to be enhanced to improve the flow of inventions to the marketplace.

CONTRIBUTIONS AND FUNDING OF LIFE SCIENCES UNIVERSITY RESEARCH

University drug discovery contributes to economic and societal welfare through new drug approvals. A recent empirical study published in the New England Journal of Medicine found that between 1990 and 2007, 19% of all priority reviewed drugs and 6% of standard review drugs were originally created in public-sector research universities. The study also illustrated that over the past 40 years 153 FDA-approved drugs were discovered at least in part by public sector research. Of note was the fact that mostly all the innovative vaccines that have been commercialized over the past 25 years were created by public sector research [7].

The federal government continues to support the majority of funding for academic R&D. While it provides the majority of funding, the percentage of funding for academic

R&D by the federal government has not been increasing on an inflation-adjusted basis. The federal government provided $32.6 billion or 59% of the $54.9 billion of academic spending on science and engineering R&D in 2009. This $32.6 billion of federal funding in 2009 was 4.2% above the level of the previous year. The rates of growth in 2008 and 2007 were 2.8% and 1.0%, respectively. Over the previous 10 years, growth in federal funding for academic R&D has been steady, averaging 3.3% annually for the 5-year period of 2004–09 and 7.3% annually for the 10-year period of FY 1999–2009.

On the other hand, adjusting for inflation, the annual growth in nonfederal funding for academic R&D has averaged 4.8% over the last 5 years; the corresponding inflation-adjusted growth rate for federal funding of academic R&D was 0.8% over the same period.

Nonfederal funding of academic R&D includes the following sources of funds: university and college institutional funds (since 1991, accounting for 20% of sources of academic R&D), state and local government funds (about 7% of source of funding for academic R&D in recent years), and industry (the smallest share of academic R&D funding, at about 6%). As mentioned earlier, industry funding of academic R&D experienced a 3-year decline between 2001 and 2004, but has been steadily increasing, and reached $3.2 billion in 2009 [8]. Over the past 20 years, funding of life sciences research has seen a significant (6%) increase in share of total academic R&D funding, while the physical sciences share has dropped 3% over that same period.

In sum, these funding trends suggest a steady increase of industry funding and a slowing of government funding for university R&D, with an increased interest in the life sciences as compared to the physical sciences. For most universities, pharmaceutical, medical device, and life sciences account for about 50% of licensing revenues and therefore are a critical element of commercialization efforts.

GLOBALIZATION AND OPEN INNOVATION IN THE LIFE SCIENCES INDUSTRY

Currently the fields of pharmaceuticals, biotechnology, medical devices, and diagnostics form the underpinnings of what is referred to as the Life Science Industry Complex (LSIC). This industry sector is estimated to be worth about a trillion dollars in global sales. The Organization for Economic Cooperation and Development (OECD) predicts that the incorporation of biotechnology into a vast array of products and processes will result in commercialized products representing 35% of chemical output, 80% of pharmaceutical and diagnostic output, and 50% of agricultural output [9].

The traditional underlying knowledge base of the pharmaceutical industry was chemistry; biotechnology is largely based on molecular biology and other areas. What we have witnessed over the past several years is the convergence of pharmaceutical and biotechnology drug discovery and development. Over the past 30 years, pharmaceutical companies have increasingly collaborated with, taken equity in, or acquired biotechnology companies to gain access to innovative technologies. Additionally, the LSIC is enabled and driven by technological advances in other areas, such as large scale computing and computational biology, and nanotechnology.

Technological advances in drug discovery have transformed the search for new molecular entities, through painstaking hand-design of molecules and robotic ultra high throughput screening systems that can screen more than a hundred million reactions (e.g., how a molecule will attach to a protein) in ten hours, at a millionth of the cost of traditional ways of screening potential drug compounds.

The pressures facing the pharmaceutical industry are unrelenting: the high cost and lengthy process of drug discovery and development, the lack of promising pipelines, major blockbuster drugs losing patent protection and the subsequent competition from generics, the threat of competition from emerging markets, and pressures to lower the cost of prescription drugs.

Consolidation within the LSIC has not resulted in stronger pipelines and blockbuster drugs. The trend now is collaboration. It is interaction, networking, and sharing of knowledge that will hasten the drug discovery process. Communication and information technology support the transparency and sharing of critical information in the drug discovery process. Not so long ago it was unheard of to have two pharmaceutical companies collaborating in drug discovery. There was too much concern over intellectual property, trade secrets, and competition. This has changed. Pharmaceutical companies are partnering with biotechnology companies and academia to drive innovation.

Offshoring of manufacturing, information technology, marketing, and other back-office functions has increased across industries. Recently, global innovation offshoring has also begun to increase. Industries such as software, information technologies, and electronics have been establishing R&D centers and collaborations in emerging economies such as India and China.

As the governments of emerging countries (e.g., China, India, Taiwan, Singapore, South Korea) continue to strategically develop policies and investments in the life sciences sector, the opportunity for collaboration, offshoring, and competition will increase significantly. In particular, as education, funding, human capital, and strategic government policy investments are amplified in emerging economies, there will be more opportunities for drug discovery to occur outside of the United States. We have already seen outsourcing of clinical trials and manufacturing in the drug development stage [9].

A recent survey of 282 life science executives and in-depth interviews with selected companies corroborates the importance of open innovation for the future of this industry [10]. One of the major findings of this investigation is that life science companies who are leaders in innovation create the right culture, are more engaged in open innovation, and make better use of data. Basically, these leading companies are working hard to create a culture that supports entrepreneurial thinking by recognizing achievements and allowing failures. They have a more flexible perspective on intellectual property and are open to a wide range of concepts and new ways of operating. The surveyed companies felt that this new approach to innovation improves research, development, and the use of existing intellectual property.

THE INNOVATION IMPERATIVE: ROLE OF THE UNIVERSITY AND TECHNOLOGY TRANSFER IN INDUSTRY COLLABORATIONS

The traditional organization design and culture of the twentieth century university may no longer suffice in the twenty-first-century competitive landscape. The external forces igniting economic relevance for universities are accelerating. Rapid and dynamic scientific and technological advances are global in scope. Industry and governments seek the pioneering basic knowledge generated at universities to generate technologies, products, and services. Many universities have already addressed this new innovation imperative and most others have, at least verbally, expressed their commitment to economic relevance. In many scientific and technology sectors, universities are indispensable partners in light of the long-term nature of commercializable products. In the life sciences area, there are numerous university-industry partnerships, as illustrated in Table 34.1.

TABLE 34.1. Major Publicly Disclosed Collaborations Since 2008

Industry Partner	Academic Institution	Sector/Strategy
Agilent	Berkeley	Synthetic biology
AstraZeneca	University College London	Stem cells
AstraZeneca	University of Manchester	Inflammatory disease
AstraZeneca	University of Pennsylvania	Alzheimer's
Bayer	UCSF	Master R&D agreement
Genentech	UCSF	Neurodegenerative diseases
Gilead	Yale	Oncology
GSK	MaRS Innovation	University consortium
GSK	University of Manchester	Inflammatory disease
GSK	Harvard	Stem cells
GSK	Undisclosed (Global)	10 "Academic Superstars"
J&J	MIT	Oncology
J&J	Sanford-Burnham	Alzheimer's
Novartis	Harvard	Stem cells
Pfizer	CTI: SF, NYC, Boston	University consortium
Pfizer	UCSF	20 PI-sponsored projects
Pfizer	Washington University in St. Louis	Immunology
Sanofi	Harvard	Cancer, diabetes, inflammation
Sanofi	Columbia University	Diabetes
Sanofi	Salk Institute	Gene therapy
Sanofi	Stanford	Stem cells
Sanofi	UCSF	Master R&D agreement
Takeda	Sanford-Burnham	Obesity
UCB	Harvard	Neurology, immunology
Veridex	Mass General Hospital	Circulating tumor cells
Zambon	UCSF	Drug delivery

Source: http://www.blogoup.com/blog/2011/6/13/a-look-at-pharma-academic-collaborations-from-the-sidelines.html.

The global forces show no signs of abating; they are gaining strength. As industries, companies and economies respond and contribute to the heightened innovation imperative, so too must universities [11].

THE QUESTION IS, "HOW?"

Two developments in the corporate sector affect how universities and university technology transfer activities adjust to and participate in a changing world economy and social order. First, corporations have reinvigorated spending, particularly in science-based technologies. Second, companies recognize the need to reach out to external partners, including but not limited to universities, to re-invigorate their traditional approaches and knowledge bases: hence, the acceleration of "open innovation."

The competition for "suppliers" of knowledge is global and plentiful and is certainly not limited to universities.

Despite examples of successful university-industry collaborations, there are two basic challenges to these relationships. The first is the culture difference. Corporations deal in a global competitive environment. While their missions may incorporate an intent to serve society in some way (e.g., life sciences companies) they must be profit driven to do so. They are performance driven; there is no tenure. Universities and the professors that inhabit them are trained and rewarded on generating new knowledge and insights. Universities value public dissemination of knowledge. Long-term research projects are the currency of academia.

The second challenge in university-industry relationships lies in the governance structure of the collaboration. These collaborations are not legitimate joint ventures that create a new entity; they are not market contracts or hierarchical relationships. For the most part, university-industry relationships operate like strategic alliances, and as such all contingencies and behaviors cannot be documented, detailed, or measured.

A recent study investigated the factors that determine success in university-industry collaborations [12]. Defining success in terms of learning and technological outcomes, they found that four variables are associated with successful university-industry partnerships.

The first is flexibility in university intellectual property policies. University-industry collaborations are intended for technology commercialization, and technology ownership rights are critical to commercialization. For both parties, patent ownership and licensing are vehicles for revenue generation as well as for competitiveness and enhanced reputation, albeit in different ways. Universities are reluctant to grant intellectual property rights to industry partners due to the loss of both revenue and the freedom to disseminate the knowledge. Companies view such university approaches as rigid and not accommodating of their needs. Rigid university policies toward intellectual property discourage university-industry partnerships.

Second, the presence of individual corporate and university champions for each project seems to play a positive and influential role in these collaborations. These champions on both sides serve as ambassadors and sustain good relationships. They monitor and communicate activities and progress between the firm and the university, and provide guidance and strategic direction.

Finally, both communication effectiveness and trust play a significant role in mutual learning and successful outcomes. This is particularly true of the nature of the communication from the university to the corporate partner. Longer-term relationships tend to generate more trust.

As mentioned, alliances are more informal in nature as all contingencies and expectations may not easily be spelled out, particularly in the case of science-based collaborations. Companies typically yield control over research dollars and resources to universities. A firm's trust in its research partner is a critical ingredient in success.

UNIVERSITY-INDUSTRY PARTNERSHIPS AND UNIVERSITY TECHNOLOGY TRANSFER

Universities and their technology transfer offices must adapt to changing global competitive conditions. Universities have much to offer in terms of their specialized knowledge, physical infrastructure, supply of students, and their unconstrained approach to research (academic freedom). Companies benefit from university knowledge and bring funding and opportunities to commercialize academic research.

So, how can universities adapt to the changing paradigm of scientific discovery, described earlier? First, universities need to rethink their approach to technology transfer and become more "customer friendly"—probably a term not frequently used in academia. Embedding a culture of academic entrepreneurship can be accomplished by updating reward systems to researchers to encourage industry collaboration. Building a university-industry ecosystem that encourages relationships between companies and professors, departments, and students will also contribute to increased research funding and technology commercialization. It gives students the opportunities to be engaged with and understand the importance and challenges of commercializing new technologies. Similarly, universities can support early-stage research with seed funding for prototypes and proof of concept.

In their relationships with companies, and based on the findings of Betts, 2011, the appointment of individuals as champions or liaisons will further engage industry and academia. A central point of contact for each collaboration facilitates communication and trust [12].

Companies too need to recognize that universities value long-term relationships, while companies frequently view their industry engagements as transactional [13]. When companies include universities as strategic partners, when they engage senior level managers to work with universities on specific projects, and when they incorporate universities as part of their R&D strategies, then positive outcomes for both partners become more of a reality.

Universities continue to play a critical role in the advancement of scientific and technological innovation, and companies will continue to leverage university knowledge. The relationships between universities and companies in global life sciences will continue to evolve into more collaborative, long-term, and flexible types of arrangements in the twenty-first-century global economy.

REFERENCES

1. Chesbrough, H. The era of open innovation. *Sloan Manage Rev*. 2003;44(3):35–41.
2. Etzkowitz, H. *The Second Academic Revolution: MIT and the Rise of Entrepreneurial Science*. London: Gordon and Breach; 2000.
3. Sparking Economic Growth. Available at http://www.sciencecoalition.org/successstories/.
4. Frequently asked questions. Association of University Technology Managers. Available at http://www.autm.net/FAQs. Accessed August 17, 2012.
5. Rothermel, F., Agung, S., Jiang, L. University entrepreneurship: a taxonomy of the literature. *Ind Corp Change*. 2007;16(4):691–791.
6. Swamidass, P.M. A proposal to improve our understanding of innovation using the Association of University Technology Managers Licensing Activity Survey. Kaufmann Symposium on Entrepreneurship and Innovation; 2008. Available at http://www.kauffman.org/uploadedFiles/ResearchAndPolicy/EntrepreneurshipData/2008data/association-of-university-technology-managers.pdf. Accessed December 26, 2013.
7. Stevens, A.J., Jensen, J.J., Wyller, K., Kilgore, P.C., Chatterjee, S., Rohrbaugh, M.L. The role of public sector research in the discovery of drugs and vaccines. *N Engl J Med*. 2011;364(6):535–541.
8. National Science Foundation. Science and Engineering Indicators, 2012. Available at http://www.nsf.gov/statistics. Accessed July 20, 2012.
9. Mroczkowski, T. *The New Players in Life Science Innovation: Best Practices in R&D from around the World*. Upper Saddle River, NJ: FT Press; 2011.
10. Kilestra, P. Reinventing biopharma: strategies for an evolving marketplace. London: Economist Intelligence Unit Report; 2011.
11. Bagley, R.O. Strengthening the link between industry and university research. 2012. Available at http://www.forbes.com/sites/rebeccabagley/2012/08/01/strengthening-the-link-between-industry-and-university-research-part-2/. Accessed December 26, 2013.
12. Betts, S.C., Santoro, M.D. Somewhere between markets and hierarchies: controlling industry university relationships for success. *Acad Strateg Manage J*. 2011;10(1):19–43.
13. Wright, R. How to get the most from university relationships. *MIT Sloan Manage Rev*. 2008;49(3):75–80.

Part VII

THE FINAL FRONTIER

No one can whistle a symphony; it takes a whole orchestra to play it. —*Halford E. Luccock*

Collaborative Innovation in Drug Discovery: Strategies for Public and Private Partnerships, First Edition.
Edited by Rathnam Chaguturu.

35

THE CORE MODEL: DRUG DISCOVERY AND DEVELOPMENT VIA EFFECTIVE TRANSLATIONAL SCIENCE AND PUBLIC–PRIVATE COLLABORATION

Ibis Sánchez-Serrano

The Core Model Corporation, S.A., Panama City, Panama

Drug discovery and development via collaboration is as old as the pharmaceutical industry itself [1]. Since its beginnings back in the middle to late 1800s, the pharmaceutical industry has established extremely close and productive research relationships with universities and research institutes in France, England, Germany, Japan, and the United States, among other nations, which provided industry with an immense wealth of knowledge, ideas, and innovation that were translated by industry into important and revolutionary medical products [2]. Ever since, the impact of academic discoveries has influenced every aspect of drug development, from the initial identification of targets to the understanding of biochemical and genetic pathways within the cell; from the elucidation of mechanistic paradigms involved in specific diseases to the creation of diagnostic tools and biomedically oriented devices. Academia has pioneered scientific knowledge, complemented, and validated the more focused, sophisticated, and refined work that is performed in industry; while industry, on the other hand, has always taken basic scientific research to the next level, to applied science or technology, therefore creating great health care and social and economic benefits for society.

Collaborative Innovation in Drug Discovery: Strategies for Public and Private Partnerships, First Edition. Edited by Rathnam Chaguturu.

During the last few decades, the influence of academic research has become vital to the biopharmaceutical industry, as exemplified by the creation and growth of the biotechnology industry as well as by the way in which long-established pharmaceutical companies are developing new pharmaceuticals, capitalizing on cutting-edge technologies and discoveries that saw daylight in academic settings (e.g., genomics, postgenomics, proteomics tools, and nanotechnology). The biotechnology tools that have been created in academia and further developed in industry have facilitated a better and unprecedented knowledge of how the cell works at the atomic and molecular levels, which has had a tremendous impact on our understanding of diseases and the creation of novel drugs. A remarkable example is the creation of monoclonal antibodies, a work that started out in academia and that later revolutionized medicine when optimized and applied by industry. Though the pioneering work on monoclonal antibodies in, say, cancer drugs, such as alemtuzumab, imatinib mesylate, cetuximab, trastuzumab, and more recently, bevacizumab (see Table 35.1), was performed in academia, none would have become therapeutic applications if industry had not taken the financial risk to develop them and contribute their know-how and marketing capabilities. Furthermore, these drugs would not have come to the market if it had not been for a very symbiotic relationship, a constant flow of scientists, ideas, experimental results, and materials, between the academic laboratories where these antibodies were originally invented and engineered and the respective companies that optimized and developed them further.

In the last few years, the pharmaceutical industry has been facing many remarkable challenges, such as a significant decrease in productivity in spite of the ever increasing investment in research and development (R&D), loss of marketing exclusivity of some important and highly profitable drugs due to massive patent expirations, major safety setbacks with drugs in development or already in the market, higher regulatory hurdles for drug approval, poor public image, and so on. Accordingly, the pharmaceutical industry has responded to this crisis by adopting a series of strategies, such as increase of price of medicines, patent extension, reformulation, severe cuts in R&D and labor, mergers and acquisitions—none of which can sustain growth in the medium and long-terms, and which, unfortunately, have severe direct and indirect effects in the pockets of patients worldwide. In fact, the last wave of mergers in the recent years has compounded, instead of solving, the pharmaceutical industry's problems; because, these mergers create not only more managerial layers but also great disruptions in the cultures and programs of these large firms. As a result, it will be impossible to manage them directly from the top, as used to be done, or in a "closed innovation" fashion, in which ideas in these companies were generated internally and taken from concept to commercialization using vertically integrated resources. Instead, an "open innovation" model—one in which new product innovation originates from both internal and external ideas—has become a major component in how drug discovery is being executed by the pharmaceutical industry in recent years.

In addition to the academia-industry relationship, other three components, namely government agencies, public and private funders/investors, and advocacy/philanthropic organizations have played a very important role in drug discovery and development. These elements deserve some discussion and some credit, as the drug creation machinery would not work without them.

TABLE 35.1. Examples of Monoclonal Antibodies Approved by the Food and Drug Administration

Antibody	Brand Name	Company	Approval Nate	Type	Target	Indication (Targeted Disease)
Abciximab	ReoPro	Eli Lilly	1994	Chimeric	Inhibition of glycoprotein IIb/IIIa	Cardiovascular disease
Adalimumab	Humira	Abbott Laboratories	2002	Human	Inhibition of TNF-α signaling	Several auto-immune disorders
Alemtuzumab	Campath	Genzyme	2001	Humanized	CD52	Chronic lymphocytic leukemia
Basiliximab	Simulect	Novartis	1998	Chimeric	IL-2Rα receptor (CD25)	Transplant rejection
Belimumab	Benlysta	GlaxoSmithKline	2011	Human	Inhibition of B- cell activating factor	Systemic lupus erythematosus
Bevacizumab	Avastin	Genentech/Roche	2004	Humanized	Vascular endothelial growth factor (VEGF)	Colorectal cancer, age related macular degeneration (off-label)
Brentuximab vedotin	Adcetris		2011	Chimeric	CD30	Anaplastic large cell lymphoma (ALCL) and Hodgkin lymphoma
Canakinumab	Ilaris	Novartis	2009	Human	IL-1β	Cryopyrin-associated periodic syndrome (CAPS)
Cetuximab	Erbitux	Bristol-Myers Squibb/Eli Lilly/ Merck KGaA	2004	Chimeric	Epidermal growth factor receptor	Colorectal cancer, head and neck cancer
Certolizumab pegol[19]	Cimzia	UCB	2008	Humanized	Inhibition of TNF-α signaling	Crohn's disease
Daclizumab	Zenapax	Genentech/Roche	1997	Humanized	IL-2Rα receptor (CD25)	Transplant rejection
Denosumab	Prolia, Xgeva	Amgen	2010	Human	RANK ligand inhibitor	Postmenopausal osteoporosis, Solid tumor's bony metasteses
Eculizumab	Soliris	Alexion Pharmaceuticals	2007	Humanized	Complement systemprotein C5	Paroxysmal nocturnal hemoglobinuria
Efalizumab	Raptiva	Genentech/Merck Serono	2002	Humanized	CD11a	Psoriasis
Gemtuzumab	Mylotarg	Wyeth	2000	Humanized	CD33	Acute myelogenous leukemia (with calicheamicin)

(Continued)

TABLE 35.1. (*Continued*)

Antibody	Brand Name	Company	Approval Nate	Type	Target	Indication (Targeted Disease)
Golimumab	Simponi	Johnson & Johnson/Merck & Co.	2009	Human	TNF-α inihibitor	Rheumatoid arthritis, Psoriatic arthritis, and Ankylosing spondylitis
Ibritumomab tiuxetan	Zevalin	Spectrum Pharmaceuticals	2002	Murine	CD20	Non-Hodgkin lymphoma (with yttrium-90 or indium-111)
Infliximab	Remicade	Janssen Biotech, Inc./Merck & Co.	1998	Chimeric	Inhibition of TNF-α signaling	Several autoimmune disorders
Ipilimumab (MDX-101)	Yervoy		2011	Human	Blocks CTLA-4	Melanoma
Muromonab-CD3	Orthoclone OKT3	Janssen-Cilag	1986	Murine	T cell CD3 receptor	Transplant rejection
Natalizumab	Tysabri	Biogen Idec/Élan	2006	Humanized	α4 integrin	Multiple sclerosis and Crohn's disease
Ofatumumab	Arzerra		2009	Human	CD20	Chronic lymphocytic leukemia
Omalizumab	Xolair	Genentech/ Novartis	2004	Humanized	Immunoglobulin E (IgE)	Mainly allergy-related asthma
Palivizumab	Synagis	MedImmune	1998	Humanized	An epitope of the RSV F protein	Respiratory syncytial virus
Panitumumab	Vectibix	Amgen	2006	Human	Epidermal growth factor receptor	Colorectal cancer
Ranibizumab	Lucentis	Genentech/ Novartis	2006	Humanized	Vascular endothelial growth factor A (VEGF-A)	Macular degeneration
Rituximab	Rituxan, Mabthera	Biogen Idec/ Genentech	1997	Chimeric	CD20	Non-Hodgkin lymphoma
Tocilizumab (or Atlizumab)	Actemra and RoActemra		2010	Humanized	Anti-IL-6R	Rheumatoid arthritis
Tositumomab	Bexxar	GlaxoSmithKline	2003	Murine	CD20	Non-Hodgkin lymphoma
Trastuzumab	Herceptin	Genentech	1998	Humanized	ErbB2	Breast cancer

Source: http://en.wikipedia.org/wiki/Monoclonal_antibody_therapy; accessed January 26, 2014.

To illustrate the great potential that academia–industry/public–private collaboration has in bringing an important drug to the marketplace, this chapter discusses the development of anti-cancer drug bortezomib (Velcade; Millennium Pharmaceuticals) by a biotechnology company—Myogenics/ProScript—started by academics from Harvard University, and the effective use of translational research in this process. I dissect the key academia-industry/public sector-private sector interactions that made the development of this drug a success despite many barriers and offer a model to explain how and why bortezomib was approved in record time. Areas for public–policy initiatives to improve translational research in general are highlighted. The chapter concludes with other examples of public–private sector interactions in drug research that have had varying degrees of success, as some lessons can be learned from them too.

SUCCESS IN TRANSLATIONAL RESEARCH: LESSONS FROM THE DEVELOPMENT OF BORTEZOMIB

Success in bringing novel drugs to the market is not only a function of the quality of the science, but also of the collaboration between academia and industry, the organization and management of R&D, the public policies that regulate scientific research, and the personal connections among key players. With the aim of understanding the key factors underlying success in drug discovery and development, here we focus on the effective use of translational research and the key academia–industry/public–private interactions in the development of bortezomib (Velcade) [3], an innovative anticancer drug for the treatment of multiple myeloma created by a company started by academics from Harvard University, Myogenics—later renamed as ProScript and then acquired by LeukoSite and Millennium Pharmaceuticals, respectively. A model to explain how and why bortezomib was approved in record time is presented, and areas for public-policy initiatives to improve drug discovery and translational research in general are highlighted.

Cancer is the second-leading cause of death in the United States after heart disease, and the amount of funding that goes into cancer research—public and private—is higher than for any other disease [4–6]. But despite private and public efforts, the process of bringing effective cancer drugs into the market still faces many challenges. Thus, anticancer research and drug development provides an excellent illustration of the relationships and interactions between: basic, applied and translational research (see Box 35.1 for definitions); academia and industry; and public and private funding. The analysis here focuses on the route leading to the proof-of-principle of the therapeutic potential of bortezomib in humans to probe the key interactions between early-stage academic biotech spin-offs, academia, industry, government agencies, public and private investors and advocacy groups.

Translational Research

Translational research is particularly interesting, because it gains from and provides valuable information to both basic and applied research (Box 35.1). In addition, translational research has the potential to promote the creation of new potential therapeutics

Box 35.1. Basic, Translational, and Applied Research

The definitions used to describe the biological research that is performed in the domains of universities and companies vary widely. Here, basic research is defined as the activity concerned primarily with the elucidation of the biological mechanisms and physicochemical processes of living organisms, irrespective of whether the findings of this research may eventually have potential therapeutic applications. The term translational research is used to refer to the translation of findings from the "bench to the bedside". In other words, translational research takes basic and preclinical findings and moves them into humans. Applied research is used to illustrate the development of commercial therapeutic applications, as related to health-care interventions, with the objective of creating health, financial and social benefits.

that are riper for commercialization, thereby decreasing economic risks (due to possible failure) and development costs.

At present, there is a funding gap, at least in academia, for translational research, and especially for clinical trials. So there could be a significant number of projects and potential drugs that have shown promise in preclinical studies that are languishing in academic settings because of the lack of adequate funding to take them into the clinic. (For a more thorough listing of project areas that might inform about a gap in funding for translational research, please see the Rapid Access to Intervention Development (RAID) website in Further information, and for discussion of the Developmental Therapeutics Program at the NCI, see [7]). In many cases, these drugs may not taken up by pharmaceutical or biotech companies because they are not economically or strategically viable, or simply because they neither target large markets nor promise high returns. However, initial approval for smaller markets, while providing benefits to patients, could open the doors for later approvals in larger, and therefore more profitable, markets and provide the necessary validation for "first-in-class" drugs that can be used in several other medical applications. Moreover, trials funded by smaller biotechnology companies could potentially be carried out more effectively if additional funding for translational studies were available. Realistically, small biotechnology companies can barely afford to develop one or two products at the same time, and additional basic or translational research cannot be carried out without the risk of running out of funds before subsequent rounds of funding (see Box 35.2, note 1). It is, therefore, important to understand the impediments that translational research encounters today, and whether there are adequate incentives to invest or participate in it.

The Bayh–Dole Act of 1980 entitles U.S. universities to the intellectual property rights of discoveries made using federal funding. The rationale behind this act was to increase patenting of discoveries and the acceleration of economic growth through the creation of "high-tech" firms that license these technologies from the university. However, licensing these technologies does not necessarily mean that more funding has been channeled into their translation into concrete commercial products (see [8] for a description of the impact of the Bayh–Dole Act). In addition, although the involvement of academic scientists in translational research and entrepreneurial activities may be

Box 35.2. Translational Research: Bortezomib

Note 1: The example of bortezomib clearly indicates the value of a company's "translational" investment in assays relevant to clinical development. The scientists working for Peter Elliott and Julian Adams validated an assay for peripheral blood mononuclear cell proteasome activity and its modulation by bortezomib that critically influenced the selection of the clinical schedule, and allowed a tight dose escalation scheme leading to a very efficient Phase I program. This should highlighted as evidence favouring investment by drug companies in this type of assay, particularly for "first-in-class" agents such as bortezomib.

Note 2: The history of boronates as pharmaceutical agents is interesting; they were developed as serine-protease inhibitors by Dupont/Merck. However, the compound failed in Phase II clinical trials as a treatment for emphysema, and so boronates acquired a bad reputation among medicinal chemists. Adams's team linked a boronate group to Myogenics lead compounds, which showed considerably increased efficacy against the proteasome, and so created a novel chemotype.

Note 3: Multiple myeloma is for the most part a bone marrow disease and the second most common cancer of the blood [representing 1% of all cancers and 2% of all cancer deaths]. It is estimated that in the United States alone there are 40,000–50,000 people with multiple myeloma, and 14,000 patients develop the disease each year. The average life expectancy of these patients is ~4.3 years [see the Multiple Myeloma Research Foundation in further information].

Note 4: Agreement reached for up to $38 million from Hoechst Marion Roussel, plus royalties paid to ProScript on sales of products deriving from this partnership: http://www.archive.hoechst.com/deutsch/news/95/019_95.html.

Note 5: US$20 million in equity investment from Roche Group to ProScript, plus royalties on sales of products resulting from this collaboration; http://www.prwire.com?cgi-bin/stories.pl?ACCT=105&STORY=/www/story/38582.

Note 6: From a business point of view it must be stressed that although the conclusion to focus on cancer was influenced by these results, bortezomib was first developed as an anti-inflammatory agent and was licensed to Hoechst Marion Roussel for that purpose [arthritis]. The decision to pursue cancer as a business model was a fall-back position because Hoechst Marion Roussel [HMR] chose to focus upstream of the proteasome in inflammation. Ultimately, HMR dropped proteasome inhibitors for inflammation and cancer and returned its license rights on the drug to ProScript.

more evident, even fostered, in areas such as biotech and high-tech "clusters," these clusters are exceptions; their formation is a complex, expensive and lengthy process and their impact on economic acceleration and growth is unknown or, at least, not easily measurable. In many respects, in the United States and in Europe, the participation of academic scientists in commercial activities may still be seen as a distraction or deviation from their academic duties, and many academic scientists prefer not to leave university circles to pursue opportunities as entrepreneurs, because of the risks involved, economic and otherwise.

Intellectual property rights, although doubtless necessary, can also be an impediment in the process of translating basic science into commercial products. Of course, most companies and investors will not finance translational science and develop drugs without a strong intellectual property position to protect their investment. But after collaboration conversations have started, the terms and conditions of the licensing and technology transfer agreements can often reach absurd levels for either or both parties, and interesting and highly viable projects can reach an impasse, slowing down the translational process. Indeed, there is a perception among people working in academic and biotechnology spin-off companies that a significant amount of the obstacles that they encounter come from the universities themselves, especially when dealing with issues related to ownership and economic dividends.

So, how did bortezomib, a first-in-class drug that originated in an academic biotechnology spin-off, make it to market so rapidly? This is a particularly interesting question considering the fragile funding base of its originator, the considerable obstacles (internal and external) encountered, and the risks involved in pursuing a new molecular target—the proteasome—with a new and ill-famed class of inhibitor (boronates; see Box 35.2, note 2).

Properties of Bortezomib

Bortezomib (also known as MG-341/PS-341/LDP-341/ML-341) is a first-in-class proteasome inhibitor for the treatment of multiple myeloma, an incurable cancer of the blood (see Box 35.2, note 3). The proteasome is a hollow and cylindrical enzymatic complex present in both the cytoplasm and the nucleus of all eukaryotic cells and is necessary for the degradation of over 80% of the cell's proteins. This is an important step in the regulation of other cellular functions, including signal transduction pathways that regulate cell growth and proliferation [9–11].

In blocking the proteasome's catalytic active site, bortezomib inhibits an important cellular mechanism that regulates the cell cycle via the activation of nuclear factor-κB (NF-κB) [11–13] (Figure 35.1). Preventing NF-κB activation leads to apoptosis and renders malignant cells more vulnerable to chemotherapy and radiation [11]. Although not a cure for multiple myeloma, bortezomib has had significant effects in prolonging the life of patients who received at least two prior therapies and demonstrated disease progression on the last therapy [14, 15] (see also Multiple Myeloma Research Foundation in Further Information). The following is a short and simplified account of how this drug came to the market.

Discovery and Development of Bortezomib

In 1992, Alfred Goldberg decided to use the growing basic knowledge on the proteasome to create a biotechnology company focused on one goal: using inhibitors to block the proteasome, with the aims of investigating the physiological roles of the proteasome and translating basic proteasome research into a therapeutic application (the history of

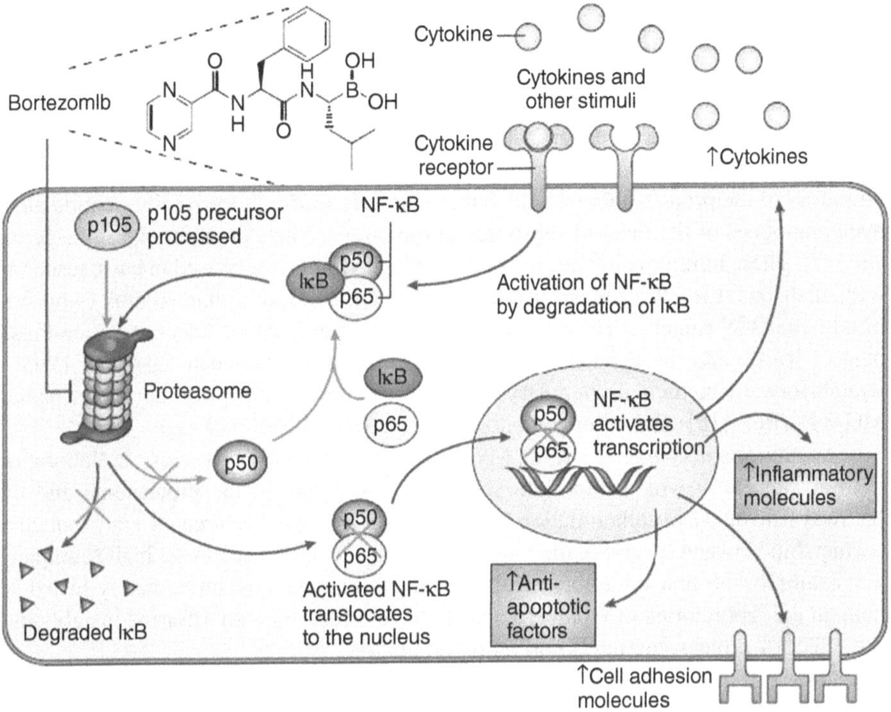

Figure 35.1. The Proteasome, NF-κB and bortezomib. The proteasome is a barrel-shaped multiprotein particle that destroys proteins marked for degradation by conjugation to ubiquitin. Binding of the transcription factor nuclear-κB (NF-κB) to the inhibitor protein IκB in the cytoplasm renders NF-κB inactive. Cellular stimuli, such as cytokines, antigens, oxidants, viruses, and other agents, trigger a cascade of signal transduction events that phosphorylate and ubiquitinate IκB, leading to its degradation by the proteasome, which in turn allows the liberation of NF-κB and its translocation into the nucleus. Once in the nucleus, NF-κB binds to promoter regions of genes that are involved in the activation of transcription, growth, angiogenesis and anti-apoptotic factors and cell-adhesion molecules. By inhibiting the proteasome, bortezomib inhibits the activation of NF-κB (orange crosses) and subsequent events that can promote tumor cell survival and proliferation. *Source*: Sánchez-Serrano, I. Success in translational research: lessons from the development of bortezomib. *Nat Rev Drug Discov.* 2006;5:107–114. See color insert.

proteasome research is far too extensive for this article; see reference [16] for a description of the history and key players). The company, Myogenics, was founded in 1993, and its objective was to target the ubiquitin-proteasome pathway to slow down the process of muscle wasting (cachexia) associated with fast protein degradation.

To create the company, Goldberg formed partnerships with two scientists at Harvard: Kenneth Rock, an immunologist and pathologist who had collaborated with Goldberg on studies of proteasome and antigen presentation; and Michael Rosenblatt, who brought a wealth of experience in drug development. Tom Maniatis, co-discoverer of the NF-κB pathway, who collaborated with Goldberg on gene transcription linked to the proteasome [12], subsequently also joined.

In 1993, the company hired its first CEO, Frans Stassen, who came from Ciba-Geigy and had significant experience in drug development. A team of enzymologists (led by Ross Stein, who came from Merck) was also employed, and created the first inhibitors of the proteasome, which would eventually lead to bortezomib: peptide aldehyde analogues of the favored substrates of the proteasome's chymotrypsin-like active site [17]. These inhibitors including MG132, which is still widely used in basic research, were distributed freely to many academic researchers. Next, Julian Adams (who was hired from Boehringer as Head Chemist, and who later became Executive Vice President of R&D) and the team of chemists that he led used, between 1994 and 1995, a straightforward medicinal chemistry approach to create a dipeptide boronate, named MG-341 (Ref. [18]) and other inhibitors such as MG-519/PS-519.

An important characteristic of Myogenics was its extremely close collaboration with academia, in which important scientific knowledge on the proteasome and the pathways involved in inflammation and apoptosis was quickly created via productive partnership between several of the founders at Harvard and the scientists in the company. For example, the first inhibitors created by the company were immediately tested on cells in the laboratories of Goldberg and Rock, and this research afforded insights into the effect of proteasome inhibition on inflammation.

One of the initial key findings from the laboratories of Rock and Goldberg was that blocking the proteasome *in vivo* did not immediately alter the normal life of the cell immediately [16]. Ensuing studies carried out in collaboration between Goldberg and Maniatis's laboratory at Harvard, such as the discovery by Vito Palombella and colleagues that the proteasome is very important in the activation of NF-κB [12], which is in turn involved in the inflammatory response [19], led to a change of focus in the company: from muscle wasting to inflammation, and a corresponding change in company name, from Myogenics to ProScript. In August 1994, Avram Hershko suggested to Adams that the company investigate cancer as a potential disease target (J. Adams, personal communication), and investigations carried out by the company scientists showed that the proteasome inhibitors blocked the proliferation of cancer cells *in vitro*, bearing in mind the role of NF-κB in gene regulation. This strongly increased the interest in cancer, although the company continued to be focused on inflammation.

At this time, there was considerable tension between the founders, the company scientists and the Scientific Advisory Board (SAB) owing to differences in opinion on the direction and strategies that the company should pursue: muscle wasting, inflammation or cancer. A step forward in the cancer direction was the establishment of a collaboration between ProScript and Beverly Teicher from the Dana Farber Cancer Institute. Teicher was introduced to the company by Bruce Zetter (Harvard), who had been, in turn, introduced by Goldberg to the company as a SAB member. The group at ProScript then initiated their first proof-of-concept study in cancer in collaboration with Teicher in 1995. At this time, Teicher was studying angiogenesis and the role of toxic agents such as alkylating agents in cancer cells and provided ProScript with the first tumor mice models. By 1997, the group had shown that bortezomib, known as PS-341 at the time, inhibits tumor growth and metastasis in a mouse model of lung cancer; the results were published in 1999 [20]. However, there was considerable skepticism about the drug based on what was considered its potential toxicity in humans.

In addition, Adams, through Maniatis, met David Livingston, a leading figure in the mechanistic field of oncology, and asked him to join ProScript's SAB. Livingston's role was crucial in steering the company to thought leaders in the cancer field, such as Kenneth Anderson from Dana Farber. At the end of 1995, ProScript established a collaboration agreement with Hoechst Marion Roussel (HMR) to develop orally active anti-inflammatory and anticancer agents based on ProScript's ubiquitin-proteasome inhibition technology (see Box 35.2, note 4). A year later the company signed a drug discovery/development collaboration with Hoffmann-La Roche (Nippon Roche) on compounds to treat cachexia, muscle wasting associated with cancer (see Box 35.2, note 5). These important collaborations were secured by ProScript's second CEO, Richard Bagley, who was crucial not only in establishing these collaborations but also in further negotiations between ProScript and these companies, including the recovery of rights related to proteasome inhibitors from HMR.

Between 1996 and 1997, ProScript approached the NCI for collaboration. During this period, the NCI was interested in looking for new chemotherapeutic agents and had a large collection of cell lines in which PS-341 could be tested. This was a direct collaboration with Edward Sausville, Head of the Developmental Therapeutics Program (DTP) at NCI and Chair of the NCI Decision Network (the body that makes decisions on the commitment of NCI funds to new drug development initiatives arising either from NCI or from outside) and his team. After this collaboration and the initial data that resulted from it, the company started to focus increasingly on cancer, although it continued to pursue inflammation as well (see Box 35.2, note 6).

Through 1997, ProScript continued to collect animal data and established collaborations with several academic researchers [21], among them, Christopher Logothetis, Chairman of Genitourinary Oncology at the MD Anderson Cancer Center (MDACC). Logothetis was introduced to ProScript by David McConkey (MD Anderson) and was invaluable early on for the progression of ProScript's trials (P. Elliott, personal communication). Through Logothetis, ProScript met Howard Soule, Chief of Science Officer at CaP Cure, now the Prostate Cancer Foundation.

After encouraging results from the NCI and animal studies carried out by the pharmacology group led by Elliott at ProScript [22], the company won unanimous approval and funding from the NCI, CaP Cure and two academic groups, the Memorial Sloan Kettering Cancer Center (MSKCC) and the University of North Carolina (UNC) to conduct Phase 1 clinical trials. After successful initiation of the Phase 1 trial at MDACC, two additional trials were started. The first was at Memorial Sloan Kettering with David Spriggs and the second was with Robert Orlowski at UNC, who was introduced to the company by James Cusack and Albert Baldwin (UNC). Both trials were funded by the individual institutions, which shows that it is possible to find external funding for trials of promising drugs and thereby avoid the rapid depletion of the limited resources of a small biotechnology company. The trial at UNC focused on hematological malignancies and it was this trial that demonstrated (in 2000) that bortezomib was active in multiple myeloma [23] (see Box 35.3, note 1).

Although bortezomib worked exceptionally well in animal models of inflammation, especially rheumatoid arthritis, ProScript realized that the therapeutic index was not large enough for chronic administration. Moreover, despite promising progress in cancer, such as the initiation of Phase 1 trials, ProScript's funds had almost depleted

Box 35.3. Translational Research: Factors Leading up to the Development of Core Model

Note 1: The actual timetable of events is as follows: National Cancer Institute [NCI] clinical candidate [Decision Network stage III: DNIII]. NCI approves funding for clinical trials on 8 June 1998. MD Anderson Cancer Center starts first trial on 7 October 1998. Memorial Sloan Kettering Cancer Center starts a second trial on 15 February 1999. New York University starts a third trial [using NCI funding] on 26 July 1999. The University of North Carolina starts fourth trial on 8 November 1999 [Peter Elliott, personal communication].

Note 2: Julian Adams in an interview in Myeloma Today [Autumn 2002, UK; http://www.myeloma.org/myeloma/newsletter.jsp?type=detail&id=1029]. Besides Dana-Farber Cancer Institute there were collaborations with other hospitals and academic institutions at the same time.

Note 3: Millennium was granted "Fast Track Status" by the FDA on June 2002. Millennium filed an NDA on January 21, 2003, under the provisions of Subpart H Accelerated Approval of New Drugs for Serious or Life-Threatening Illnesses. On 10 March 2003, FDA accepted application and granted Priority Review Status. On 13 May 2003, bortezomib was approved under Fast-Track Status [http://investor.millennium.com/phoenix.zhtml?c=80159&p=irol-newsArticle&ID=411937&highlight=]. Bortezomib generated revenues to Millennium of US$ 59.6 million in 2003, Millennium Pharmaceuticals Annual Report 2003, Form 10-K, p.1 [http://media.corporate-ir.net/media_files/irol/80/80159/reports/ar03.pdf]. In 2004, bortezomib had net product sales of US $143 million, Millennium Pharmaceuticals Annual Report 2004, Form 10-K, p.1 [http://library.corporate-ir.net/library/80/801/80159/items/144438/MLNM_80159/_AR_033105.pdf].

Note 4: Key people interacted before and after the LeukoSite/Millennium acquisitions are included because the "core" never disintegrated, even though legally ProScript had become part of LeukoSite and then Millennium and several of its original members were no longer present. By the time ProScript was sold out, the company had already gained so much weight and momentum with its science and the establishment of a tight network interaction with the "bridge" and the "periphery", that even within Millennium it continued to exercise powerful "centripetal" influence. For example, the key Phase I clinical trial at University of North Carolina, which showed safety and efficacy of bortezomib in humans, was planned and coordinated before the acquisitions took place. In addition, Phase II key players, such as Kenneth Anderson—who brought along the Multiple Myeloma Research Foundation and the International Myeloma Fund—had already been contacted years before by Adams and colleagues, as suggested by David Livingston.

Note 5: When collaborating with people in a nonexclusive way, those parties may have a deep knowledge of what is happening in a specific field, but this is not intended to imply that there will be "leakage of information". By contrast, these collaborations could be very fruitful, because they could help avoid unproductive paths and suggest new perspectives.

Note 6: The company distributed bortezomib from early on, and the group continued to do so, because first, because the company did not have the resources to carry out all the studies, and high-profile academic groups gave the drug more credibility. [Peter Elliott, personal communication].

Note 7: From a broader perspective, the "Core Model" also has important implications for economic growth. It has been proposed that it is the sate of technology that drives economic development [34]. However, the present analysis on biomedical research can be extrapolated to any major economic problem and demonstrate that it is trade [in this case of assets-knowledge and technology] and not the state of technology itself that drives economic growth. The "Core Model" explains how and why, and it can be generalized and used in other fields.

by June 1999. The company had received its initial funding from HealthCare Investment Corporation and Dillon Read Venture Capital. ProScript was unable to secure subsequent funding for several reasons, including the pioneering nature of their technology, and that targeting the proteasomal apparatus with a drug that was considered to be too toxic was viewed as too risky, especially when taking into account the cost of the ensuing clinical trials. In addition, from a venture capital point of view, there were no suitable comparable drugs whose success could be used to provide support for taking further risks. The fact that HMR, during restructuring to form Aventis, dropped proteasome inhibitors exacerbated the negative feeling about further investing into the company.

So, ProScript, like many other small biotechnology firms, fell victim to the financial market psychology of the moment and, faced by a funding shortage, reduced its staff and SAB. Adams, who had become bortezomib's champion, and the ProScript team made many efforts to promote the drug and establish collaborations with more than 50 companies, all of which declined the drug. Eventually, HealthCare Ventures decided to incorporate ProScript into another of their portfolio firms, Cambridge-based LeukoSite. The company team, together with some private investors, tried to buy the drug from HealthCare Ventures for US$2.4 million in cash (Julian Adams, personal communication). But in July 1999, ProScript was sold to LeukoSite for US$2.7 million (approximately 187,000 newly issued shares of LeukoSite's common stock valued at $2.3 million and $430,000 in cash) [24].

Three months later, Millennium bought LeukoSite for US$635 million, because it was interested in LeukoSite's pipeline, especially CamPath (which also made it to the market), but had no interest in bortezomib [25]. Unsurprisingly, passing from company to company created major disruption to the bortezomib project. However, the ProScript team did not give up. Adams had regular meetings with Mark Levin, CEO at

Millennium, to persuade him to keep the project alive and provide the necessary funding. In August 2000, the UNC clinical trial demonstrated that bortezomib erased all signs of cancer from a 47-year-old woman who months before was in the advanced stages of multiple myeloma [23]. Given this data, Millennium decided to make bortezomib Millennium's most funded drug (ML-341), and Julian Adams and others from ProScript quickly assumed leadership roles at Millennium.

Adams' group decided to team up with Anderson, a multiple myeloma expert, and his group (specifically Paul Richardson and Teru Hideshima) at the DFCI to conduct Phase 2 clinical trials at the institute. Consequently, there was an exchange of results and ideas back and forth between Millennium and Anderson and colleagues to discovery more about the molecular mechanisms that made multiple myeloma more susceptible to bortezomib. At Millennium, David Schenkein, Dixie Esseltine, Barry Greene, Michael Kauffman, and others, played an important role. This translational process was accomplished in the following way: basic research → clinical settings → patients' feedback → return to basic research to gain further understanding of the molecular mechanisms involved (see Box 35.3, note 2). In addition, through Anderson, Adams established contact with the Multiple Myeloma Research Foundation and the International Myeloma Fund, two advocacy groups that provide information and support to multiple myeloma patients and their relatives, that strongly supported the cause of bortezomib.

After the conclusion of Phase 2 clinical trials [14], bortezomib was approved in record time in May 13, 2003 by FDA under Fast-Track Application (see Box 35.3, note 3) as an injectable small molecule for the treatment of multiple myeloma. Millennium continued Phase 3 clinical trials and carried out the marketing of the drug.

LESSONS LEARNED: THE "CORE MODEL"

The way in which Myogenics/ProScript established several important collaborations with outside academics and agency/advocacy groups to move bortezomib forward was unusual. Although some companies form such kinds of collaborations in their programs, MyoGenics/ProScript did this exceptionally well and systematically, and so particularly benefited from these collaborations at critical points, both when the company needed scientific knowledge to move forward and when lacking the necessary economic resources. The example of bortezomib emphasizes the potential power of maximizing such collaborative approaches and is useful in providing insights to policy makers, scientists, investors and the public on how the process of drug development can be optimized.

Bortezomib, despite being a first-in-class drug that could have been shelved many times, managed not only to reach the market, but also to do so very quickly, in contrast to many well-funded efforts in industry. Here, I propose that this can be explained using a "Core Model" (Figure 35.2), which defines and structures the roles and bi-directional interactions of the parties involved in the process of drug development.

In the development of bortezomib, this model could be described as follows (see Box 35.3, note 4). The "core" is represented by internal people and resources such as Goldberg, Maniatis, Rock, Rosenblatt, Adams, Elliott, Palombella, managers, other internal scientists, and private investors. The "bridge" is represented by immediate collaborators, such as Teicher, Sausville, Logothetis, Soule, Spriggs, Orlowski, Anderson, Scientific Advisory Boards members, Millennium, and so on; and other private companies (via the simultaneous nonexclusive collaboration with external scientists or via nonexclusive SAB members that work simultaneously with these kinds of companies; see Box 35.3, note 5). The "periphery" is represented by CaP Cure, the Multiple Myeloma Research Foundation, the International Myeloma Fund, Memorial Sloan Kettering, Dana Farber Cancer Institute, NCI, NIH, MD Anderson, University of North Carolina, FDA, and so on (see Table 35.2).

Communications with the periphery were established through individual people involved. These players traded assets (i.e., materials, animal models, knowledge, connections) and in so doing, they advanced each other's research. Examination of their publications before, during and after using the first generation inhibitors, such as MG132 or lactacystin and bortezomib, reveals that the study of these drugs led to a better understanding of their mechanisms of action. The fact that the company decided to distribute bortezomib to outside researchers (see Box 35.3, note 6), especially when it reached a high level of economic distress, was not risk free. For example, the company took the risk that outside collaborators would have patented discoveries that, even if not valid or dominated by company patents, may have hindered the company and distracted resources. Although sharing the drug with outside scientists may have represented a tremendous loss of potential revenues for the company, it opened crucial doors that eventually led to bortezomib reaching the market. Figure 35.2b shows how these players interacted within the "Core Model" framework.

Although the science behind the development of bortezomib "worked," many people including well-established companies did not believe in bortezomib's commercial potential and rejected it, considering the drug to be too toxic, or the market too small. However, bortezomib is marketed today for several reasons. First, the Myogenics/ProScript "core" had an outstanding, although small, scientific staff that made a number of highly influential basic discoveries, and distributed the initial inhibitors for free to academic investigators, which led to rapid acquisition of knowledge in this area. Second, the collaborations established through the "bridge" accelerated the generation of knowledge necessary for speedy approval. Third, once at Millennium, ProScript's team rapidly secured valuable resources from Millennium for additional clinical trials and marketing. Fourth, ProScript's core used public resources within the "periphery" very efficiently, for example, NCI, cancer advocacy groups, the hospitals carrying out the clinical trials, and so on. ProScript developed bortezomib through a "Core" *modus operandi* using *knowledge transfer* (collaboration established with external people to exchange assets), *knowledge integration* (incorporation and assimilation of external assets), and *knowledge translation* (the conversion of all, internal and external, assets into a commercial therapeutic product).

Bortezomib would not have been successful in a large pharmaceutical company, as ProScript used a strategy that was very different from that taken by most of big pharma companies, which for the most part are not *centripetal* from outside of their organizations, that is, they generally keep assets within the company. ProScript critically engaged and used the biology at every turn and made a convincing case for the

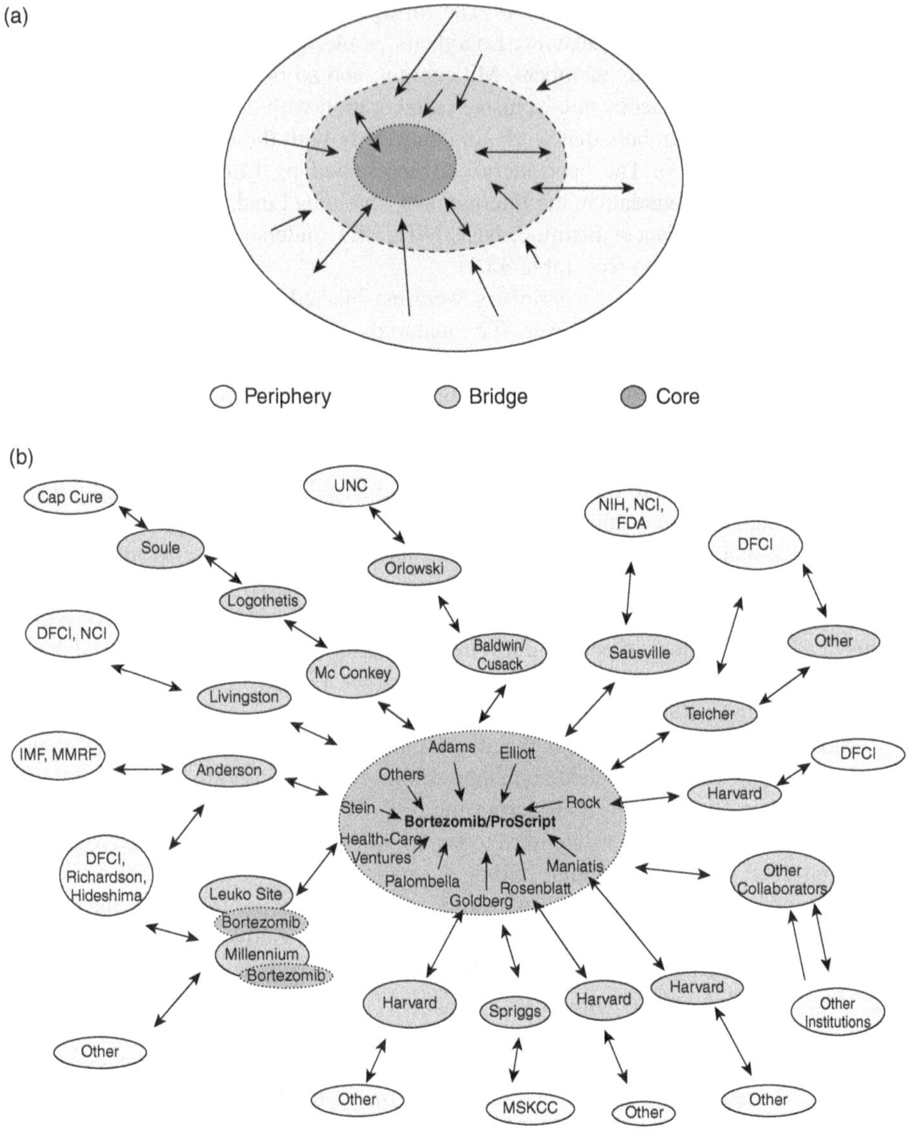
(a)
Periphery
Bridge
Core
(b)
Cap Cure
Soule
Logothetis
UNC
Orlowski
NIH, NCI, FDA
DFCI
DFCI, NCI
Mc Conkey
Baldwin/ Cusack
Sausville
Other
Livingston
Teicher
IMF, MMRF
Anderson
DFCI
Harvard
Adams
Elliott
Others
Stein
Bortezomib/ProScript
Rock
Health-Care Ventures
Maniatis
Palombella
Rosenblatt
Goldberg
DFCI, Richardson, Hideshima
Leuko Site
Bortezomib
Millennium
Bortezomib
Other Collaborators
Harvard
Spriggs
Harvard
Harvard
Other Institutions
Other
Other
MSKCC
Other
Other

Figure 35.2. The "Core Model." (a) This model has three major elements: the core, the bridge and the periphery. In a biotech start-up, the core represents the company's internal people and resources, who are hired because they have assets that are directly related to the core's objective: making drugs. The core needs a strong leader, capable of keeping the enterprise focused and able to secure collaboration with external people. The ideas of the core are protected by patenting and secrecy. The bridge represents the immediate collaborators of the core and the private institutions to which the core has indirect access through the external collaborators. The bridge contains external scientists interested in similar problems or whose research would be enriched as a result of the collaboration. It also includes consultants and Scientific Advisory Board (SAB) members (non-Founders) working in exclusive and nonexclusive ways. The periphery contains the institutions/agencies interested in what the core has to offer for the benefit of society as well as the funding and regulatory structures that support the core and the bridge. The periphery is an open, public and cooperative system. The goal of the core is to absorb efficiently and legally as much relevant knowledge and information as possible from its surroundings in three ways. via the leverage of the assets, professional backgrounds and connections of the people within the core; via the assets, connections, and expertise of the external collaborators within the bridge; via the support, relevant public knowledge and know-how within the periphery. (b) Selected people with roles in the development of bortezomib, with their roles illustrated using the core model.

TABLE 35.2. A Summary of the Key Players in the Development of Bortezomib, Their Roles, and Contributions[a]

Individual	Institution	Role	Assets
Alfred Goldberg	Harvard University, Dept. of Physiology	Founder, SAB	Muscle wasting research, discovery of the proteasome
Tom Maniatis	Harvard University, Dept. of Biology	Co-Founder, SAB	Research on IκB, NF-κB. Experience at the Genetics Institute, SABs member, Editor of prestigious journals, introduced Vito Palombella
Kenneth Rock	Harvard University, DFCI	Co-Founder, SAB	Cancer immunology and research on NFkB/Major Histocompatibility Complex Class I (MHC)

(Continued)

TABLE 35.2. (*Continued*)

Individual	Institution	Role	Assets
Michael Rosenblatt	Harvard University, BIMC	Co-Founder, SAB	Blood cancer research and business experience at Merck leading drug development and bringing a drug to the market for the treatment of osteoporosis. Consultant to venture capital firms, SABs member
Richard Bagley	Myogenics/ ProScript	CEO, SAB	Experience at BMS, SmithKline, CEO of ImmuLogic
Julian Adams	Myogenics/ ProScript	Head of Chemistry, SAB	Experience at Merck and Boehringer Ingelheim in drug discovery
Ross Stein	Myogenics/ ProScript	Head of Biochemistry	Enzymology, experience from Merck
Vito Palombella	Myogenics/ ProScript	Head of Cell and Molecular Biology	Research on Tumor Necrosis Factors (TNF), cancer immunology, interferon, NFkB. Maniatis' lab
Peter Elliott	Myogenics/ ProScript	Head of Pharmacology Project Leader	Pharmacology and experience in drug development from Alkermes, Inc. and the Glaxo Group Ltd
David Livingston	Harvard University, DFCI	Collaborator, SAB	Cancer expert: elucidation of pRB protein, work on BRCA1, BRCA2, BACH1, and BARD1. SAB Chairman at NCI, other SABs member, EPA
Beverly Teicher	Harvard University, DFCI	Collaborator	Toxicology, angiogenesis, animal (mice) models, cell cultures, simultaneous collaboration with Immunogen, Genzyme and Lilly
Edward Sausville	NCI	Collaborator	Head of the CTEP, NCI. NCI's scientific expertise and resources (in the form of cell lines and animal models), NIH/FDA
Christopher Logothetis	MD Anderson Cancer Center, TX	Collaborator	Head of Clinical Trials, MD Anderson TX. Founder of CaP Cure. Leading expert in urogenital cancer
Howard Soule	CaP Cure	Collaborator	Director of CaP Cure, influence and weight on advocacy group, coordination of clinical research and trials, multiple connections with regulators, SABs member

TABLE 35.2. (*Continued*)

Individual	Institution	Role	Assets
Robert Orlowski	University of North Carolina-Chapel Hill	Collaborator	Interested in cancer and proteasome research. His father, Marian Orlowski, is co-discoverer of the proteasome. Connections with ASH
David Spriggs	Memorial Sloan Kettering Cancer Center	Collaborator	MSKCC's Chief of Developmental Chemotherapy. Internationally recognized oncologist. Interested in solid tumors (gynecology) and translational research.
Kenneth Anderson	Harvard University, DFCI	Collaborator	Expertise in multiple myeloma, connections with the multiple myeloma Research Foundation and International Myeloma (BOD member)
Paul Richardson	Harvard University, DFCI	Collaborator	Expertise in multiple myeloma, clinical trials aspect
Teru Hideshima	Harvard University, DFCI	Collaborator	Expertise in multiple myeloma, clinical research
Kathy Giusti	Multiple Myeoloma Research Foundation	Collaborator	President of MMRF. Multiple myeloma patient. Sales executive at Merck and GD Searle (Pfizer). Strong advocate to the multiple myeloma cause
Susie Novis	International Myeloma Fund	Collaborator	Head of IMF. Husband died to multiple myeloma

[a]ASH, American Society of Hematology; BIMC, Beth Israel Medical Center; BMS, Bristol Myers Squibb; BOD, Board of Directors; CTEP, Cancer Therapy Evaluation Program; CEO, Chief Executive Officer; CSO, Chief Scientific Officer; DFCI, Dana Farber Cancer Institute; EPA, Environmental Protection Agency; FDA, Food and Drug Administration; IMF, International Myeloma Fund; MSKCC, Memorial Sloan Kettering Cancer Center; MMRF, Multiple Myeloma Research Foundation; NCI, National Cancer Institute; NIH, National Institutes of Health; SAB, Scientific Advisory Board.

drug, and so obtained support from the public sector at a critical juncture. The interest in bortezomib from the NCI was unusual: the NCI was going to pursue the molecule even if private funding was not forthcoming because of the drug's unique biological effects. Unlike many biotechnology start-ups or middle- or late-stage companies, ProScript was not only focused on economic profits—it initially concentrated on a small and narrowly focused indication, rather than "holding out" for a more lucrative but scientifically less supportable indication. The company prominently engaged the patient advocacy community (CaP Cure) when starting the first Phase 1 clinical trials, and when the time came to recruit patients for definitive clinical trials, ProScript's "core" was key in involving the Multiple Myeloma Research Foundation and the International Myeloma Fund.

NEW DIRECTIONS FOR THE ACADEMIA–INDUSTRY PERSPECTIVES

The relationship between academia and industry is generally understood in a unidirectional way, with basic science being translated into applied science. There is also the perception that in the academia–industry relationship, academia is exploited without receiving adequate benefits. However, upon close examination, it becomes clear that many scientists see translational research as a legitimate academic activity and that the academia–industry relationship is bidirectional. For example, academic research is stimulated by the questions that industry generates, which usually fall outside the scope, capabilities and economic interests of the companies. Indeed, one may be suggested that the process of academia producing ideas that are translated into commercial products is cyclical; academia provides answers to the questions created by new commercial products that, in turn, could lead to more commercial products. Although there is secrecy and proprietary knowledge in the process of developing a drug, once the drug is marketed, the mechanisms involved in targeting the disease become public knowledge, resulting in more questions that could be investigated by academia.

Considerations related to conflict of interest and the pharmaceutical industry benefiting from public investment in research have generated a debate about the nature of the interactions between industry and academic/public institutions and researchers and the propriety of such interactions. One of the important points emphasized by the "Core Model" is that these interactions are bidirectional in ways that benefit all parties and science in general, as well as providing society with new beneficial medicines. Such benefits are often missing from the debate.

In the case of bortezomib, scientists worked back and forth between academia and industry. The problems in the clinic gave rise to more research in the private sector and vice-versa. This kind of interaction allowed the science to be integrally linked to the clinical studies, and to in turn allow the clinical studies to drive the science. It was a synergistic interaction that has the potential, as illustrated here, to autocorrect directions that will not be productive. The transfer of knowledge between academia and industry allowed a better understanding of multiple myeloma, of bortezomib's mode of action, and the mechanism in which the proteasome is related to other key pathways that regulate the cell cycle. Indeed, there has been an explosive interest in the proteasome and its role not only in cancer but in other diseases as well. Many investigators are seeing bortezomib's antitumor effects in other types of cancer and other agents created by Myogenics/ProScript are currently in clinical trials. Second-generation drugs have been developed in different laboratories and are awaiting funding to enter clinical trials.

In summary, academia has one characteristic that is very important to industry: it creates "full-stories" in terms of how living organisms work. Academia is a dynamic and open system that allows for rapid interchange of information among people from all over the world. This constant flow of people and ideas enriches scientific research and promotes progress. In other words, economic and social progress is achieved via a trade of assets and knowledge (Box 35.3, note 7).

PUBLIC–PRIVATE COLLABORATION

In spite of the obviousness of the great productivity of the academia–industry interaction [1], the relationship between academia and industry in recent years has been not only uneasy but also under attack [26]. But there is a reason for this.

Since the creation of the Bayh–Dole Act of 1980, mentioned earlier, which encourages universities to license discoveries made with federal funds to private industry and benefit financially from this, the university-industry relationship has become very stormy and plagued by multiple scandals. In general, industry has a bad reputation among academic scientists, as academic scientists see industrial research as money oriented, scientifically less rigorous, and compromised by high-stake financial interests: in other words, they see it as the "dark side." For many academics, "industry-sponsored research presents several significant problems for universities. For example, it is not the cash cow that many suppose. In general, companies pay for research that benefits them and their shareholders, not for the undirected curiosity-driven research that is at the heart of the academic enterprise" [27].

On the opposite side, industry considers that some academic scientists have shown a "certain business appetite . . . and (an) aggressiveness regarding intellectual property issues (that) are about as loud as those from the (industrial) direction" [28] that academics so criticize. And not a few industrial researchers consider academic research as unfocused, lacking in practical sense, and a slave to the personal and political games that sometimes characterize the academic settings, not to mention the great lack of regard that many industrial scientists feel toward their academic colleagues.

As a result, today, "officially," the collaborations that seem to work best for the pharmaceutical industry are not the ones performed directly with academic groups, but rather with the small, entrepreneurial biotech companies that have derived from universities, through which a number of academics who believe passionately in their products will gradually disengage from the university and spin-off companies and commute back and forth from the academic and industrial bases. But even in these types of collaborations, intellectual property concerns become a huge issue and without doubt, judging by events, very fertile soil for all sorts of misappropriation of information and subsequent scandalous and disgraceful legal battles.

Among the most sensitive territories for interaction between academia and industry are teaching hospitals in the field of clinical research as many academic scientists feel that after industry-sponsored clinical trials have become successful, industry does not quite give them the credit they feel they deserve—which, not surprisingly, has created the perception that industry is prone to exploitation.

But aside from these issues, from a practical and scientific point of view, the reciprocal negative attitude on both sides of the academia–industry equation is nonsensical given the innumerable benefits—targets, genes, and drugs such as insulin; human growth hormone; contraceptives; chemotherapeutic agents paclitaxel, imatinib mesylate, and bortezomib; monoclonal antibodies; statins; and so forth—that have been the crowning achievement of close collaborative efforts between academia and industry.

There are other areas of utmost importance that are currently shaping drug research and which have important beneficial consequences for society, which have been possible only thanks to a very close collaboration between academia and industry. A prime example is the Human Genome Project, an initiative that was sponsored by the public and private sectors and which mapping work took place at universities, government-sponsored agencies, and pharmaceutical companies all over the world. This successful mega project has made possible the identification and understanding of diseased genes and genetic markers, with obvious importance in health care. In fact, the vast array of genomic and proteomic tools, such as the ever sophisticated DNA and protein sequencing technologies and DNA-chip technology, to mention just a couple examples, would never have realized their full scientific and commercial potential had it not been for an uninterrupted osmosis between academia and industry.

Another relevant example is the Single Nucleotide Polymorphisms Consortium (SNP's Consortium),* an open and cooperative project created in 1999 by ten large British pharmaceutical companies and the Wellcome Trust Philanthropy. The SNP's Consortium later also included the Human Genome Project to generate a widely accepted, high-quality, extensive, publicly available map using SNPs as markers evenly distributed throughout the human genome, and which has been quite useful to scientists in the public and private domains all over the world in understanding better the genetic causes of important cellular processes and maladies.

In the United States, the National Cancer Institute's somatic cell line panel, through which scientists can test the effect of experimental drug agents on 60 cell lines at the same time, has become an invaluable resource for academic scientists, early-stage biotechnology firms, and pharmaceutical companies, especially when determining the initial potential applications to clinical trials based on a specific drug candidate. One further example of an important academic-industry collaborative project is the establishment of the C-Path Institute (Critical Path Institute) at the University of Arizona, which, in collaboration with the U.S. Food and Drug Administration (FDA), SRI International (an independent, nonprofit research institute conducting client-sponsored research and development for government agencies, commercial businesses, foundations, and other organizations), the National Cancer Institute, and industry, has become a major step in implementing the ideas behind the FDA Critical Path Opportunity List

*Single nucleotide polymorphisms, or SNPs (pronounced "snips"), are DNA sequence variations that occur when a single nucleotide (A, T, C, or G) in the genome sequence is altered. For example, a SNP might change the DNA sequence AAGGCTAA to ATGGCTAA. For a variation to be considered a SNP, it must occur in at least 1% of the population. SNPs, which make up about 90% of all human genetic variation, occur every one hundred to three hundred bases along the three-billion-base human genome. Two of every three SNPs involve the replacement of cytosine (C) with thymine (T). SNPs can occur in coding (gene) and noncoding regions of the genome. Many SNPs have no effect on cell function, but scientists believe others could predispose people to disease or influence their response to a drug. Available at http://www.ornl.gov/sci/techresources/Human_Genome/faq/snps.shtml.

initiative and in helping industry to accelerate the process of drug development via the sharing of research information, the development of drug safety and efficacy tests, and the identification and development of biomarkers [29]. And the list of examples can continue ad infinitum.

Perhaps a lack of understanding about each player's roles, goals, and interests, has created a barrier between the academic and the industrial sectors, and this is something that needs to be addressed with urgency in our search for better and more efficient ways in which to discover and develop drugs.

It is true that a new target or chemical structure identified by academia is usually hundreds of millions of dollars and years away from ever making it to the market as usually, the original idea or compound is not the one that succeeds in making it through the iterative process of drug development, the process whereby companies test, fail, try again, and occasionally succeed in optimizing a potential drug candidate.* However, it would be difficult for industry to come up with such knowledge in the first place because its role is to create drugs, not to pursue basic or open-sky research—this is not only the role of academia, but it is also the area in which seminal and important knowledge for further applications arises.

An important example of how the academia–industry collaboration can affect the fate of a pharmaceutical company is the following. In the late 1970s, when the earliest biotechnology companies were being created in the United States, a number of large and long-standing leading European pharmaceutical companies, especially German ones such as Hoechst and Bayer, failed to take full advantage of this revolution for two reasons: first, German universities had long fallen behind American and British universities; second, despite that these companies established some relationship with the leading American universities, their relationships, as has been indicated elsewhere [30], were difficult. For instance, in the early 1980s, Hoechst established a collaborative arrangement with a Harvard University affiliate, Massachusetts General Hospital (MGH), through which the company made a ten-year, $70 million investment in a genetic laboratory (the Molecular Biology Department) in return for exclusive licensing rights to MGH's inventions. Though this agreement came under fire in the United States because it was considered a foreign firm "skimming the cream off of a publicly built and maintained resource" to which American taxpayers were contributing about $25 million annually [31], the deal had important consequences for all parties involved. For MGH this was a beneficial deal because of the creation of a biotech lab supported by the largest and most prestigious pharmaceutical company in the world . . . even if it gave rise to controversy about the appropriateness of this type of collaboration between an academic institution and a pharmaceutical company. And for Hoechst, it represented a great opportunity to access the most innovative biotechnology research of the time (we need to remember that because of environmental protests in Germany and elsewhere in Europe, biotechnology research at that time was stalled).

*Interview with Jackie Hunter, GlaxoSmithKline, June 2007. In fact, it is believed that the average number of small molecules that make it from the chemist to the clinic is one in five hundred or one in one thousand in terms of synthesis. Although this is already an extremely low number, some scientists believe that for diseases such as central nervous system disorders, the numbers are even lower.

But a few years later, Hoechst dropped its collaboration with this teaching hospital, which was a key figure in the further development of biotechnology, and missed out on great opportunity: when it tried to catch up on biotech, it was already too late. By the late 1980s, this company had fallen behind and eventually, through a succession of mergers, disappeared into Sanofi-Aventis.* In contrast, companies such as Wellcome, Ciba, and Roche and American ones such as Merck and Lilly pursued the emergent biotechnology [30] and created a platform for biotech start-ups and learned from them. This strategy immediately placed them at the forefront of research and development and paid off years later.

In recent years, the way in which some companies have approached their collaborations with academia has had mixed results. Some companies, such as Monsanto and Novartis, invested heavily in a direct collaboration with Washington University in St. Louis and the University of California at Berkeley, respectively, with the high expectation of getting some important patents out of their relationships [32]. At the end of the day, both companies ended up disappointed because only a small number of patents, and by no means remarkable ones, came from the collaborative partnerships. Other companies have made smaller investments in specific departments or researchers in different universities, and the collaborations seemed to work, as was the case with Amgen and Lilly funding particular labs at the Massachusetts Institute of Technology (MIT). In another approach, companies such as Novartis and Merck have established research institutes close to universities like Harvard; MIT; Tufts; the University of California, San Diego; the University of North Carolina, Chapel Hill; and the University of California, Berkeley; and also with Cambridge and Oxford universities in the United Kingdom, in the hope of being closer to academia and to keep an eye on what is taking place there.

So now pharmaceutical and large biotechnology companies are importing innovation from academia indirectly, essentially via university technology transfer offices, which act as a liaison in the academia–industry interface, and more so via early-stage biotech enterprises that have derived from academic settings, although recently, in 2008, a novel collaboration between Janssen Pharmaceuticals and Vanderbilt University for the discovery of new drugs for the treatment of schizophrenia took place. Janssen has agreed to pay Vanderbilt University $10 million as an upfront fee for a licensing and research agreement. Janssen will gain access to existing compounds that act on metabotropic glutamate receptors (mGluRs) as well as any additional compounds that Vanderbilt discovers over the next three years.

This collaboration, together with several others that took place in 2008, such as Pfizer's Biotherapeutics and Bioinnovation Center at the University of California, San Francisco, has led to the realization on the industry's part that after all, academia–industry partnering is paramount in drug discovery and development [33], and the reasons for this are simple: without the input of academia's novel ideas and pathways and without this kind of open-minded search for how things function, the development of drugs at the pharmaceutical industry level would too often find itself at a dead end,

*Bayer, which imitated Hoechst in this respect, did so too late and was not able to succeed. Nowadays Bayer does not even figure among the top ten pharmaceutical companies.

where companies cannot actually make sense of the data that they observe when testing promising compounds because they are looking too narrowly at the use of a drug they are developing—and a rule of thumb in science is that one usually finds things where one is not looking. It is not surprising that oftentimes, companies are unable to convert some chemical agents into a useful, patentable product or pass them through clinical trials.

Though at times research in academia can be fiercely competitive, the pressure to show provable performance is much higher in industry than in academia, and this pressure has a negative impact on productivity. To illustrate, if one is pressured to follow a line to the production of a product—for instance, to find an inhibitor for a particular enzyme such as a kinase*—one's line of investigation will be so straight and narrow that one may not realize that the inhibitor or its pathway is doing something completely different from what one was seeking. This happens frequently, and it takes someone working with a completely different train of mind to see the new direction of the inhibitor and say, "Oh! We just did that kinase in flies and worms, and we found that it does something completely different!" Without this kind of information, companies would spend years and millions of dollars finding the right inhibitor. However, despite this, industry curtails expenses by giving deadlines and telling its researchers, "If you don't find a solution by the deadline, throw it out." This is very wasteful, and the history of drug discovery, from aspirin to imatinib mesyalte and, more recently, bortezomib, all of which were initially put aside by major pharmaceutical companies, demonstrates that this is not a productive way to foster drug development.

The relationship between industrial and academic research needs to be seen as a symbiotic one. Pharmaceutical or applied research is a second stage in science that has as a goal the creation of health care and social and economic benefits for society and that goes along in parallel or together with the more open-ended exploratory research. On the other hand, the exploratory research has to fit in with the more directed applied research to find the proper path. And in my opinion, nowhere is this more evident than in translational research.

To improve efficiency in the academia–industry interface, basic research students at universities—those who are interested in medical problems—should be trained to think about human diseases and human disease phenotypes;† in other words, in the course of their studies of mutants of flies or worms or behavioral read-outs in mice, they should be trained to ask themselves whether their findings may have some implication for human disease. This is important because in their communication with applied researchers in companies, they could apply the relationship between a target and proteins in pathways to a specific disease. This necessarily implies that scientists would have to read literature that they may not generally read, such as articles related to

*A kinase is a type of enzyme that transfers phosphate groups from high-energy donor molecules, such as ATP, to specific substrates.

†According to *Dorland's Illustrated Medical Dictionary*, 31st ed. (New York: Elsevier, 2007), a phenotype is "the observable morphological, biochemical, and physiological characteristics of an individual, either in whole or with respect to a single or a few traits, as determined by a combination of the genotype and the environment."

mutations and targets that are involved in human diseases, and encourage participation in seminars given by medically oriented scientists who are working on clinical research. The kind of literature on which scientists within this gray area between biology and medicine would be focused includes three sources of information having a medical impact: first, the genetic mapping of heritable human disease; second, the use of model organisms, where a favorite gene is knocked out in a mouse and then characterized and a particular phenotype is observed, which would allow them to make a link to a particular disease; and third, the actual examination of gene families, then finding new members with new phenotypes and trying to correlate them with a particular disease. If a pharmaceutical company is developing such and such kinds of inhibitors, such as a tyrosine kinase or histone deacylases inhibitor, the people at the academic center can look at the phenotypes in animal models either by using those inhibitors or by using gene knockouts that create a particular disease state or something that would mimic a disease state in animal models and thus allow a better understanding of the biology behind a particular disease, which could have a great impact on the development of drugs to treat or cure them.

This is very important because there is a great need for the understanding of the molecular basis of diseases. Indeed, many drug candidates go to the clinic without companies having a full understanding of the disease, and it is not surprising, then, that when these agents are taken to the clinic, they fail and are subsequently abandoned. The problem is that if something has failed in a clinical trial, it is very hard to rescue, unless there is a really strong person in the company who insists that it is possible to do so. So if scientists were able to better understand the molecular causes of disease, then they would be more successful at creating better patient stratification or basic patient selection and also have a better idea of which route they are going to follow and so forth. There is certainly a major need for development in this area. And to move in that direction, it is necessary to better understand, for instance, molecular changes in cancer cells and molecular symptoms of diseases; it would be necessary to have a better read-out of what genes are and are not being expressed genome-wide, more accurate analyses of changes in tumors, better screening for genomic fingerprinting of amplifications and deletions, and so on. This can be done mostly in academia, and if this were done at least for cancer, it would dramatically change the money spent on unsuccessful clinical trials not only in cancer but also in other diseases.

To make faster and better progress in developing new pharmaceuticals, therefore, we have to understand diseases better, and we need to increase our understanding of the basic mechanisms that control their behavior. We need simple models for disease and simple physiological read-outs before treating patients; this is the reason why scientists go for inhibitors or their equivalents and try them in cell lines: in doing so, they have a molecular asset. This approach is very important in developing strategies to treat subsets of particular diseases, which is an ongoing process, at least at the pharmaceutical industry level. The pharmaceutical industry has never really grappled with basic problems, and they go all the way from the most primitive screening into what they call discovery research, but obviously, that is a long way from basic research. It is therefore important that people in basic research be better prepared to communicate any insight they have on this front to people who are doing the clinical trials. Companies should, then, try to characterize, in collaboration with academia, disease at the molecular level as much as possible and be more open and better prepared to accept input from academic scientists.

IMPROVING TRANSLATIONAL RESEARCH

At the center of a discussion on increasing funding for translational research is the crucial question of who will provide such funds. The amount of funding that is allocated every year to it by the government is not enough for a sustainable progress in translational research, let alone to create the appropriate incentives to attract students, technicians, postdocs, and professors. On the other hand, at times, the private sector sees no incentives to invest in research that is not necessarily related to their drug portfolio, although in the last few years this attitude has been changing. For this reason it is important that public policy initiatives focus promptly on increasing the federal and industrial funding that goes into translational research in academia, including clinical trials. Indeed, the serial creation of academia-based translational research centers would contribute greatly to bring drugs to the market more efficiently. Funding for taking projects to proof-of-principle is much needed, because at times some potential drug products are at too early a stage to capture venture capital or industrial interest and funding. Furthermore, policies should encourage better collaboration between academia and pharmaceutical/biotechnology companies and better relationships between biotech start-ups and NIH/NCI/FDA (to help them design better clinical trials, hence increasing chances of success). Less heavy regulations and paperwork at the NIH/NCI/FDA and acceleration in the implementation of new policies at these agencies will also have a positive impact in the overall efficiency of the system. Finally, public policies should promote more involvement in the drug development process by the public, advocacy groups and private foundations through educational programs, and encourage the provision of more reliable information and transparency by the biotech and pharmaceutical industries.

REFERENCES

1. Sánchez-Serrano, I. *The World's Health Care Crisis: From the Laboratory Bench to the Patient's Bedside*. Amsterdam: Elsevier; 2011.
2. Weatherall, M. *In Search of a Cure: A History of Pharmaceutical Discovery*. New York: Oxford University Press; 1990.
3. Sánchez-Serrano, I. Success in translational research: lessons from the development of bortezomib. *Nat Rev Drug Discov*. 2006;5:107–114.
4. von Eschenbach, A.C. A vision for the National Cancer Program in the United States. *Nat Rev Cancer*. 2004;4:820–828.
5. AAAS Report XXX. Research & Development FY 2006, Intersociety Working Group. American Association for the Advancement of Science. Part II, Chapter 8, TABLE II-9. 2005.

6. Brower, V. The squeaky wheel gets the grease. *EMBO Rep*. 2005;6(11):1014–1017.
7. Monga, M., Sausville, E.A. Developmental therapeutics program at the NCI: molecular target and drug discovery process. *Leukemia*. 2002;16(4):520–526.
8. Mowery, D.C., Nelson, R.R., Sampat, B.N., Ziedonis, A.A. *Ivory Tower and Industrial Innovation: University-Industry Technology Transfer before and after the Bayh-Dole Act*. Stanford, CA: Stanford University Press; 2004.
9. Ciechanover, A. Proteolysis: from the lysosome to ubiquitin and the proteasome. *Nat Rev Mol Cell Biol*. 2005;6(1):79–87.
10. Hershko, A., Ciechanover, A. The ubiquitin system. *Annu Rev Biochem*. 1998;67:425–479.
11. Adams, J. The proteasome: a suitable antineoplastic target. *Nat Rev Cancer* 2004;4(5):349–360.
12. Palombella, V.J., Rando, O.J., Goldberg, A.L., Maniatis, T. The ubiquitin-proteasome pathway is required for processing the NF-κB 1 precursor protein and the activation of NF-κB. *Cell*. 1994;78(5):773–785.
13. Adams, J. Development of the proteasome inhibitor PS-341. *Oncologist*. 2002;7(1):9–16.
14. Richardson, P.G., et al. A phase 2 study of bortezomib in relapsed, refractory myeloma. *N Engl J Med*. 2003;348:2609–2617.
15. Richardson, P.G., et al. Bortezomib or high-dose dexamethasone for relapsed multiple myeloma. *N Engl J Med*. 2005;348:2487–2498.
16. Goldberg, A. Introduction to the proteasome and its inhibitors: biochemistry and cell biology. In *Proteasome Inhibitors in Cancer Therapy*, J. Adams, editor. Totowa, NJ: Humana Press Inc; 2004, pp. 17–38.
17. Stein, R.L., Ma, Y.T., Brand, S. Inhibitors of the 26s proteolytic complex and the 20s proteasome contained therein. US Patent 5,693,617. 1995.
18. Adams, J., et al. Boronic ester and acid compounds. US Patent 5,780,554. 1995.
19. Barnes, P.J., Karin, M. Nuclear factor-κB: a pivotal transcription factor in chronic inflammatory diseases. *N Engl J Med*. 1997;336:1066–1071.
20. Teicher, B.A., Aran, G., Herbest, R., Palombella, V.J., Adams, J. The proteasome inhibitor PS-341 in cancer therapy. *Clin Cancer Res*. 1999;5:2638–2645.
21. Palombella, V.J., et al. Role of the proteasome and NF-kappaB in streptococcal cell wall-induced polyarthritis. *Proc Natl Acad Sci U S A*. 1998;95:15671–15676.
22. Adams, J., et al. Proteasome inhibitors: a novel class of potent and effective anti-tumor agents. *Cancer Res*. 1999;59:2615–2622.
23. Orlowski, R., et al. Phase I trial of proteasome inhibitor PS-341 in patients with refractory hematologic malignancies. *J Clin Oncol*. 2002;20:4420–4427.
24. Bioline. 2005. Available at http://www.massbio.org/pubs/Biolone_v12_n3.pdf. Accessed December 26, 2013.
25. Langreth, R., Moukheiber, Z. Medical merlins. *Forbes*. 23 May 2003.
26. Wilson, D. Harvard Medical School in ethics quandary. *New York Times*. March 2, 2009. Available at http://www.nytimes.com/2009/03/03/business/03medschool.html?pagewanted=2. Accessed December 26, 2013.
27. Hall, Z.W., Scott, C. University–industry partnership. *Science*. 2001;291:26.
28. Kennedy, D. Industry and academia in transition. *Science*. 2003;302:1293.

29. An audience with . . . Ray Woosley. *Nat Rev Drug Discov*. 2008;7:884.

30. Chandler, A.D. *Shaping the Industrial Century: The Remarkable Story of the Evolution of Modern Chemical and Pharmaceutical Industries*. Cambridge, MA: Harvard University Press; 2005.

31. Mattera, P. Hoechst: the toxic brewmasters. *Multinational Monitor* 5. 1992. Available at http://www.multinationalmonitor.org/hyper/issues/1992/05/mm0592_11.html. Accessed December 26, 2013.

32. Lawler, A. Last of the big-time spenders? *Science*. 2003;299:330.

33. Hughes, B. Pharma pursues novel models for academic collaboration. *Nat Rev Drug Discov*. 2008;7:631–632; see also Novel pharma-academia collaborations continue. *Nat Rev Drug Discov*. 2009;8:97.

34. Solow, R.M. A contribution to the theory of economic growth. *Q J Econ*. 1956;70:65–94.

36

USING MARKET-DRIVEN COLLABORATION TO ACCELERATE INNOVATION IN BIOMEDICINE

Elizabeth Iorns

Science Exchange Inc., Palo Alto, CA, USA

The research landscape is currently undergoing an evolution driven by an "innovation crisis" that has left the pharmaceutical industry facing empty pipelines, patent cliffs and unsustainable development costs [1, 2]. Looking toward other industries for inspiration provides an opportunity to identify new collaborative models that may facilitate an innovation revolution that will enable faster, cheaper and better drug discovery in biomedicine.

A single scientist can no longer master all the techniques required to make major discoveries (many biomedical papers now have more than 10 authors [3]). Therefore, driving more collaboration between scientists is both necessary and highly beneficial. First, it increases the efficiency of research spending by promoting more complete utilization of resources and expertise, and enabling economies of scale. Second, it increases the speed of research because it can be conducted in parallel, and there is no downtime required for nonexperts to learn new techniques. Third, it increases the quality of research by enabling experts to conduct specialized, complicated techniques, which often have a steep learning curve. Improving quality is more important than ever—the quality of published research has recently been questioned, with more than

Collaborative Innovation in Drug Discovery: Strategies for Public and Private Partnerships, First Edition.
Edited by Rathnam Chaguturu.

70% of published results estimated to be irreproducible [4–6]. Using experts not only improves the quality of research produced, but it removes single investigator bias, which may help to improve reproducibility of the results.

The collaborative economy is one model that shows great promise to facilitate increased scientific collaboration. Fueled by social networks and payment systems, the collaborative economy empowers individuals to buy goods and services from each other, rather than large corporate entities. The collaborative economy is highly efficient, allowing better use of existing resources and expertise, and generates higher quality products by enabling the use of specialized equipment and highly trained individuals. Interestingly, dramatic increases in collaboration have been observed in numerous industries where the barriers to adoption of a collaborative economy are far greater than in scientific research (e.g., sharing spare rooms via Airbnb or personal cars via Getaround). This widespread adoption has been enabled by aggregating demand at centralized online marketplaces, establishing trust via integration with social networks, and promoting transactional ease with payment platforms.

EXAMPLES FROM OTHER INDUSTRIES

In the last five years, numerous companies have emerged to enable the collaborative economy to transform several industries (Table 36.1).

The most well-studied is probably the transportation industry, which has seen companies like Lyft, RelayRides and Getaround enable car-sharing on a widespread scale. A University of California study found that every car-sharing vehicle replaces 9–13 vehicles [7], reducing the idleness of cars and car ownership overall. This obviously has tremendous benefits for the environment but also challenges the transportation industry's existing players, many of which have adapted with innovative collaborative solutions of their own. For example, multiple automobile manufacturers have launched their own car sharing networks, including BMW's "DriveNow," Volkswagen's "Quicar," and Daimler's "Car2Go," creating rental markets for their previously purchase only automobiles. Zipcar was one of the early movers in this space, and was acquired by Avis for $500 million demonstrating the enormous value of the collaborative economy in the transportation industry.

Following closely behind the transportation industry—in terms of impact from the collaborative economy—is the accommodation industry. In this space homeowners are disrupting hotels, renting out their spare rooms or entire homes via online marketplaces like Airbnb, HomeAway and VRBO. Although only founded five years ago Airbnb now has more rooms available for rent than the entire Hilton group [8]. This market is estimated to be worth more than $26 billion [9].

Other examples of industries undergoing disruption from the collaborative economy include the labor market, where virtual workers can be hired on demand via oDesk and Elance, tasks and errands can be run using TaskRabbit's, and businesses can manage contractors and freelancers with Workmarket; and the banking industry, where consumers can lend to each other at lower rates via Prosper and Lending Club.

TABLE 36.1. Examples of Industries Transformed by Companies Creating Collaborative Economies

Industry	Example Company
Transportation	Lyft: A peer-to-peer ride-sharing network with a community of background-checked drivers who offer rides in their own vehicles. Raised over $80M from Andreessen Horowitz.
	RelayRides: A peer-to-peer car-sharing marketplace where people who need a car can "rent" one from an individual vehicle owner. Raised over $30M from Google Ventures, and General Motors Ventures.
	Getaround: Similar to RelayRides, a peer-to-peer car-sharing marketplace that enables car owners to rent their cars. Raised $19M from Menlo Ventures.
	DriveNow: Zipcar-like car-sharing membership program started by BMW and Mini. Similar initiatives have been started by Volkswagen (Quicar) and Daimler (Car2Go).
	Uber: A mobile application that provides on-demand access to black car and limo operators. Raised $307M from Google Ventures, Goldman Sachs and Jeff Bezos.
	BlackJet: Membership based program from the creators of Uber that provides "private jet convenience and reliability, at a fraction of the cost".
Accommodation/Space	Airbnb: Peer-to-peer marketplace for renting spare rooms and entire homes. Raised $326M and recently valued at over $2B.
	HomeAway: Marketplace for vacation homes. Listed on NASDAQ (AWAY).
	ShareDesk.net: a peer-to-peer marketplace matching unused workstations, offices and desks with free-floating workers.
	ParkingPanda: Online marketplace for parking enabling parking owners to drive extra traffic when yield is low and improve pricing when the lot is full. Raised ~$5M.
Labor	oDesk: An online workplace that enables businesses to find, hire, manage, and pay talented independent professionals via the Internet. Raised $44M from Benchmark Capital.
	Elance: Online marketplace to access 2 million freelance workers. Raised $95M from Kleiner Perkins.
	TaskRabbit: Online and mobile marketplace that helps people outsource their errands and tasks. Raised $38M from Founders Fund.
	Work Market: An enterprise-class platform and marketplace for managing contractors and freelance labor. Raised $15M from Spark Capital.
Banking	Prosper: Peer-to-peer lending marketplace with 2M customers. Raised over $120M from Sequoia Capital.
	Lending Club: Peer-to-peer lending platform that offers an alternative to banks with great rates for both borrowers and investors. Has over $2B in personal loans.

WHAT HAS LED TO THE RISE OF THE COLLABORATIVE ECONOMY?

The major driver of the collaborative economy has been advancement in technology and its widespread adoption. Social networking, often integrated into collaborative economies, facilitates peer-to-peer transactions by matching supply and demand, and promotes trust by providing transparency around the profiles of the interacting individuals. The creation of online payment systems has been essential for enabling transactions to be easily completed between buyers and sellers.

A second important driver has been the ongoing economic pressure that individuals and businesses face. The ability to monetize excess idle inventory and expertise, and increase financial flexibility by only accessing what is required rather than over-investing in expensive capital equipment and full time hires, has driven adoption of the collaborative economy in other sectors.

These drivers are also applicable to scientific research. Companies in the collaborative economy have developed platforms that help people create wealth from excess and underused assets such as cars, houses and their own spare time. Can the same sort of model be applied to accelerate innovation in biomedicine?

THE EVOLUTION OF SCIENTIFIC COLLABORATION

Scientific collaboration is currently undergoing an evolution as the research landscape changes from an individual pursuit to the era of team science. The different types of collaboration used in scientific research are summarized in Table 36.2 with the key features of each approach identified.

Traditionally, basic research has been conducted by a single scientist or a small team of scientists within a single laboratory, mostly within academia. The scientist(s) would conduct the majority of required experiments themselves, even if they did not initially have the necessary expertise or equipment. If they could not conduct an experiment themselves, they would attempt to find a collaborator in another academic lab to help them using a barter system. This barter system essentially involves one scientist asking for a favor from another scientist, with the potential upside being co-authorship on any publications that are produced by the work. This type of collaborative arrangement depends heavily on personal networks developed by scientists.

As research has become increasingly specialized and complex, this DIY/barter system has ceased to function effectively. It is no longer possible for a single scientist to master all the techniques or purchase all the equipment required to conduct their experiments. Collaboration allows the many technical components of a research project to be conducted by multiple experts, which improves the efficiency and quality of the research conducted. However the current bartering model of collaboration is not optimally efficient because the platforms to facilitate it and the incentives to promote it are not fully optimized. Middle authorship on a possible future publication may not incentivize the best experts to conduct experiments and it is difficult for scientists to maintain sufficient personal networks to enable effective bartering for all of the complex experiments required. Frequently, even an entire university cannot provide all the expertise

TABLE 36.2. Sharing Practices in Biomedical Research

Type	Example	Features
Nonmarket collaboration (aka Bartering)	Exchange of experiment for co-authorship on future publication.	Free for "requester" May be suboptimal "provider" due to poor incentives for them to contribute. Professional experience not likely (no fixed deliverables or turnaround time etc negotiated up front). Control over IP not retained. Costly for "provider" in terms of expenditure on reagents and time. Largely limited to personal network of "requester".
Outsourcing partnership	Exchange of experiment (usually large project based) for monetary compensation and sometimes co-owned intellectual property (IP).	Professional experience (clear milestones and deliverables negotiated up front). Possibility of shared risk and upsides through IP sharing agreements. Legal fees invested in complex agreements can be costly and time consuming. Experts and resources available are limited to partners, reducing flexibility and efficiency.
Market-driven collaboration	Exchange of experiment for monetary compensation.	Experts and resources available are limited only by the participants in the marketplace. Identifying and engaging best experts for specific experiments is quick and easy. Professional experience (clear milestones and deliverables negotiated up front). Control over IP is retained.

necessary for a research project, as over 60% of papers are now co-authored by investigators from multiple institutions [10].

The amount of collaboration required in research will continue to increase. This will be driven by many factors including the need for more complex and large scale instrumentation to delve deeper into biological and physical processes; the need to pool knowledge with others to make original significant advances, and an increasing desire for research that crosses multiple disciplines.

So with large teams of scientists, often based at remote institutions, increasingly needing to work together to solve complex problems, there is a demand for new tools to help facilitate collaboration. Specifically, there is an increasing need for tools that allow researchers to easily find and access other scientists with the expertise required to advance their research projects. To operate most efficiently, these tools also need new methods to reward researchers for participating in these collaborations. A market-driven collaborative economy seems to be an obvious solution.

MARKET-DRIVEN COLLABORATION FOR RESEARCH

The digital marketplace model provides a mechanism for researchers to list their expertise so that other researchers can easily find them and request collaborations. Several sites that allow for scientific networking exist including Mendeley.com, ResearchGate.net, Academia.edu and BiomedExperts.com, but these are not specifically designed to facilitate formal transactions between scientists. Science Exchange (http://www.scienceexchange.com) offers a transparent fee-for-service marketplace, delineating a clear incentive for experts to participate in collaborations that they may not have previously been interested in. If we look at the factors that have facilitated the collaborative economy in other sectors, the social network and payments system provided by Science Exchange may allow market-driven collaboration in science to flourish. One other factor—the ongoing economic pressure that is currently facing individual scientists and funders—may also serve to promote adoption of a market-driven collaboration system in science.

THE EMERGENCE OF VIRTUAL RESEARCH INSTITUTES

In fact, the basis for a market-driven collaboration system has already been established and is flourishing in the private sector. This has been driven by the availability of scientific expertise on a fee-for-service basis from contract research organizations (CROs). The rapid rise in CROs has enabled research to be conducted by organizations, which would previously have had to invest in wet lab space and the associated research personnel. This "virtual" method of conducting research is highly efficient and has been widely embraced by the biotech sector [11], where biotech start-ups outsource experimental work rather than build in-house capabilities. This is very cost effective as it allows these companies to avoid the purchase of expensive equipment and the hiring/training of staff for highly specialized one off techniques. This model is also increasingly used by pharmaceutical companies, which have cut in-house research staff and have started relying on more efficient mechanisms of accessing expertise via outsourcing. One example is AstraZeneca, which has outsourced its entire clinical pharmacology research to the CRO Quintiles [12].

This model has been slower to be adopted by academia, which predominantly still relies on bartering to facilitate collaboration. Given the significant funding restrictions that are now in place, researchers will need to adapt to become more efficient and ensure they are monetizing their assets effectively to fund their research programs. One academic sector that has moved to a market-driven system is core facilities at academic research institutes which house specialized equipment and expertise for fee-for-service use by investigators.

Many core facilities and CROs have begun creating the basis for an effective scientific collaborative economy by listing at Science Exchange. By aggregating demand, establishing trust via a reputation system and integration with publication databases, and promoting transactional ease with a payment platform, Science Exchange is providing the central components that have enabled the collaborative economy to thrive in other sectors.

An important point to note is that while the rise of CROs has enabled major efficiencies to be gained, the model as depicted earlier, is not a true collaborative economy of peer-to-peer transactions. This requires the individuals who are sharing their expertise and resources to be available for engagement on an individual level via a marketplace. This type of individual engagement has enabled other industries to be vastly more collaborative. Examples of this include home-share rentals on mediocre websites, which have been common for more than a decade. However, it took Airbnb increasing engagement by encouraging individuals to transact with each other via their platform, to grow this into a $26 billion industry. Another example is Etsy, which has dramatically grown the market for handmade goods by encouraging individuals to create storefronts on their platform and sell directly to their peers. This innovation is ultimately where Science Exchange offers the most promise to accelerate innovation in biomedicine. By providing a platform for individual scientists to offer their expertise, a collaborative economy for scientific research can thrive.

BENEFITS OF MARKET-DRIVEN COLLABORATION IN BIOMEDICINE

The major benefit of market-driven collaboration for scientific research is the increase in the amount of collaboration that will occur due to the alignment of incentives. Using a monetary transaction system motivates the best experts to share their expertise because they are fairly compensated to do so. Likewise, researchers seeking expertise are provided with a professional service and experience that comes from a payment model. This type of system can dramatically increase the amount of collaboration occurring in a specific industry. This has been clearly demonstrated by companies like Airbnb and Lyft, which have dramatically grown home- and car-sharing, respectively, from activities limited to personally connected individuals, to billion dollar industries.

Driving more collaboration between scientists through a collaborative economy will accelerate innovation by increasing the efficiency of research, allowing more research to be conducted per dollar spent, and increasing the speed at which discoveries are made. One of the biggest advantages of increased collaboration is improvement in the quality of research. Irreproducibility, particularly in preclinical research, has emerged as a major challenge to innovation in biomedical research.

Most scientific advancement occurs not through blockbuster papers, which definitively solve a problem, but through careful and incremental progress, building on prior work. However, recent work by the pharmaceutical companies Bayer Healthcare [5] and Amgen [6] indicate that 70–80% of academic preclinical biomedical studies are not reproducible. This highlights a significant problem that must be addressed to enable scientific progress. To try to solve this problem, various organizations are now funding replication studies to identify reproducible research [13]. The most prominent example is the Cancer Biology Reproducibility Project [14], which is replicating 50 recent high-impact cancer biology studies using the Science Exchange network. Notably, the market-driven collaboration model of the Science Exchange network has been critical for delivering and publishing these replication studies. Using monetary compensation as an incentive for participation has enabled rapid delivery of high-quality replication experiments. Quality control by independently replicating key experimental results will be an important component to enable successful innovation in biomedicine.

BARRIERS

Collaboration between academia and industry is essential for translating the basic discoveries made in academic labs into applications that benefit the public. A new model that enables the connection of individual researchers to the right expertise at the right time is required to enable this effectively. However, there are significant hurdles to overcome in the form of institutional bureaucracy, inertia, and incentive systems. By examining the success of other industries that have adopted collaborative models, we learn that the platform must enable individualized decision-making, provide trust, and ensure transactional ease. It is possible that efforts that promote high level partnerships between two large organizations fail to deliver on their promise simply because they don't allow individuals to operate freely with the right incentives. A phrase known as "consortium fatigue" [15] has already emerged to describe the disappointments of many high level public-private partnerships.

The collaborative economy works in sectors where it "builds on real customer need and creates experiences that are human and personal" [16]. This individualized personal component, which is trust-enabling, is essential for the success of this approach. Trust requires knowledge of the individuals collaborating via the platform. It previously seemed unthinkable that ordinary people would let strangers stay in their homes or get in their car, but the power of social networks that have enabled deep knowledge of an individual without a prior in-person relationship, has provided a "reputation" that enables the trust required for these transactions to occur. Monetary compensation provides the incentive. In a similar way, a platform for market-driven scientific collaboration must also provide this individualized approach—researchers must trust the scientists they will be working with.

The final barrier that must be eliminated is the institutional bureaucracy that is inherent to large organizations such as universities and pharmaceutical companies. The most valuable assets at any research institute are clearly the expertise and resources that are housed there. Yet these assets are currently incredibly difficult to access. For market-driven scientific collaboration to succeed, research organizations will have to empower their individual scientists to operate with freedom in the research-on-demand marketplace.

SUMMARY

The collaborative economy has changed the behavior of individuals in ways that could not have been imagined a decade ago. By providing online platforms that aggregate demand, promote trust, and ease payment, even the most risky industries have created enormous peer-to-peer markets. Biomedical research can take advantage of this model too. By moving to a collaborative economy model, we can enable efficient collaboration between the best scientific experts, which will ultimately drive the innovation, productivity and quality control needed to accelerate biomedical research.

REFERENCES

1. Paul, S.M., Mytelka, D.S., Dunwiddie, C.T., Persinger, C.C., Munos, B.H., Lindborg, S.R., Schacht, A.L. How to improve R&D productivity: the pharmaceutical industry's grand challenge. *Nat Rev Drug Discov*. 2010;9:203–214.
2. Herper, M. The truly staggering cost of inventing new drugs. *Forbes*. (2/10/2012).
3. King, C. Multiauthor papers: onward and upward. *Thomson Reuters*. 2013.
4. Zimmer, C.A. Sharp rise in retractions prompts calls for reform. *The New York Times*. (4/16/2012).
5. Prinz, F., Schlange, T., Asadullah, K. Believe it or not: how much can we rely on published data on potential drug targets? *Nat Rev Drug Discov*. 2011;10:712.
6. Begley, C.G., Ellis, L.M. Drug development: raise standards for preclinical cancer research. *Nature*. 2012;483:531–533.
7. Martin, E., Shaheen, S. The impact of carsharing on household vehicle ownership. *ACCESS Magazine*. 2011;38:22–27.
8. Empson, R. CEO Brian Chesky says Airbnb will be filling more room nights than all Hilton Hotels by the end of 2012. *TechCrunch*. (12/12/2012).
9. Botsman, R., Rogers, R. *What's Mine is Yours: The Rise of Collaborative Consumption*. New York: Harper Business; 2010.
10. OECD. International co-operation in science. OECD Science, Technology and Industry Scoreboard 2009, 4.3.
11. Carroll, J. Virtual is beautiful in biotech. *Fierce Biotech*. (3/20/2012).
12. Taylor, C., Connors, G. Quintiles announces clinical pharmacology alliance with AstraZeneca. Quintiles Library. (2/11/2009).
13. Wadman, M. NIH mulls rules for validating key results. *Nature News*. (7/31/2013).
14. Iorns, E. Reproducibility Initiative receives $1.3M grant to validate 50 landmark cancer studies. Science Exchange blog. (10/16/2013).
15. Papadaki, M., Hirsch, G. Curing consortium fatigue. *Sci Transl Med*. 2013;5:200fs35.
16. Gardner, J. What is the new sharing economy? *Forbes*. (7/30/2013).

37

THE COST OF TAKING EYES OFF THE TRUE END USER: FOCUS ON PATIENT NEEDS AND OUTCOMES

Deborah E. Collyar

Patient Advocates in Research, Danville, CA, USA

(IM)PATIENT SIDE OF DRUG DEVELOPMENT

The public thinks that medical research is funded to make people's lives better by producing patient outcomes that improve health in a seamless process of harmony and data sharing, from discovery and development through delivery of care. *If only it worked this way*. It frequently seems like the "system" focuses on everything *but* better patient outcomes.

Medical research has made progress in many health conditions and diseases and patients appreciate that. Each new improvement, however, comes with costs to the patient community and to individual patients. There are more survivors in some types of cancers now than in previous decades, but the new "chronic disease" focus from cancer research is not what they expected. Cancer patients are faced with long-term consequences, just as patients with diabetes, Parkinson's and many other illnesses have discovered [1].

The public believes in the traditional "goal" to "cure" diseases and medical conditions, but when they become patients, they learn that the "goal" has virtually nothing to do with the way the health-care or research systems actually work (at least in the

Collaborative Innovation in Drug Discovery: Strategies for Public and Private Partnerships, First Edition.
Edited by Rathnam Chaguturu.

United States). Fear mixed with hope is too often replaced by confusion and mistrust. Whether or not a patient is presented with positive or negative odds (statistics) for his/her medical condition, there is little help in applying those averages to a given patient's personal situation. Being a patient is not for the weak of spirit—it often turns into a lifelong endurance race.

There is no time to mince words about the problems in medical research. Far too many people believe in conspiracy theories regarding drug development, which are regularly reinforced through online outlets [2]. These perennial doubts are fueled by the unfortunate fact that most patients face at some point in their medical experience—they are merely the last consideration in how research is done, and how health care is actually organized [3].

It is clear in 20+ years of working with government, academia, industry, and the medical research community that the drug development process is far too fragmented and complex for anyone to actually create conspiracies that keep effective treatments from patients. It seems far more likely that those in the development process are focused on misguided priorities and incentives that end up putting patients last in line. There also is no "evil" person who works in the milieu, intentionally misdirecting promising leads, but researchers in academia and industry work within systems that clearly put self-interest ahead of solving problems for human beings. It is past time to fix these fundamental problems and build a cohesive and comprehensive medical research system based on patient needs, and that will produce partnerships that profit from providing significant value to patients' lives.

WHAT'S WRONG WITH THIS PICTURE?

Current drug and device development rarely create the kind of patient outcomes that most people are looking for as they search for relief and cures. The public continues to believe in medical research and supports continued government funding as well as payments for treatments in the hope that people will be healthier and can avoid life-threatening illnesses.

Recently, cracks in the ivory towers have publicly exposed the underbelly of drug development with issues like reproducibility [4] and scientific misconduct [5]. Drug development seems driven by fame and money (instead of patients) for investigators, institutions, and companies. Patients and the public expect this mind-set from companies, but are becoming increasingly repulsed by its reality in academia and from their own clinicians [6]. While people are rightfully outraged, these problems are by-products of a broken research system.

The thing we call a medical research "system" really isn't, starting with the traditional chasm between discovery research and health delivery. This is easily illustrated by the fact that most patients don't know what clinical trials are, or how research has improved overall health or the condition of their greatest concern [7].

The current research system is made up of a conglomeration of disparate pieces, many of which perform with competing goals (both explicit and implicit), and with virtually no purposeful connections. Main players include those in academia and industry members, along with government servants and clinicians who perform clinical trials in both academia and local communities.

Where are patients in the development process? They are paternally held up as the ideal, like mom and apple pie, but when the development process itself is examined, there has traditionally been no place for them. Patients who enroll in clinical trials are thought of as a means to an end, and their needs are trumped every time by narrow scientific questions, the quest for "significant" *p*-values that add publications to curriculum vitae and market share.

This perspective, developed over a 20-year period of working within the research system, may sound harsh to some, but until the research system bases drug development on science advances and patient needs, no amount of systemic tweaking will move the fragments onto the right track. It will take true partnership with a surprising stakeholder list and patient-focused expertise to resolve development barriers that keep us from producing relevant and timely solutions for people. This becomes critical with the implementation of the Affordable Care Act (ACA), which stresses a "values-based" payment system and Accountable Care Organizations (ACOs) [8].

CURRENT DRUG DEVELOPMENT ROLES

There are differences between "industry" (a community of biotech, device, and pharmaceutical companies) and "companies" which patients usually see as drug developers. "Academia" applies to the field of education, while "universities" or "institutions" are organizations where individual "academic" investigators and clinicians are housed, as opposed to "community" clinicians in private practice (who also play a pivotal role in drug development). Likewise, collective "patients" are different from each individual experience, and "patient advocates" in this chapter refer to those who represent the collective patient viewpoint.

The role of academia often appears early and late in drug development. Academic scientists produce leads. Miscommunication often starts when they claim to have found the "answer" in their publication—discovery is only the beginning of the quest and must be validated to be useful. Institutions quickly use the potential promise to negotiate intellectual property (IP) agreements when companies follow up on the lead. This role developed as a result of the Bayh–Dole Act in 1980, which has now outlived its usefulness [9]. While some academics are interested in preclinical development, they rarely possess the skills, or work in an institution that has the capacity to perform this step. Academic clinicians sometimes get involved in early-phase clinical trials. More complexity surfaces when completely different institutions and investigators are involved in this developmental stage. Negotiations between institutions and companies start over with each trial, and create headaches and time warps for all [10].

Clinicians from the community are also involved in drug development, primarily by opening up clinical trials in their private practices and collectively enrolling the majority of patients who participate in those trials [11]. Very little credit is given to them, but they contend with daily pressures of conflicting messages, multiple regulations, unmet medical needs, staff for clinical trials and data, cost containment, and doing no harm to their patients. Their voice must be included in improvement efforts. Even their efforts could be more focused by enrolling most of their patients, instead of a small minority.

Industry develops new products in (hopefully) ethical and safe ways so that they can market drugs and devices after achieving Food and Drug Administration (FDA) approval on what the product label can say. While discovery used to take place in companies, most now rely on academia for drug leads. These leads come to their attention through publications and personal relationships, and are then pursued with investigators and institutions. Once contracts are signed, experimental results must be replicated before preclinical development starts. Most leads end here due to failure to reproduce, which creates little value and much disappointment for all, including patients [12]. For the minority of leads that show promise, companies then prioritize the agents they will follow, based on market share and other feasibilities. Preclinical experiments with cells and animals focus on toxicology, chemistry, and production. FDA applications are produced before clinical trials are planned. Project managers implement these steps and identify clinicians to conduct the first trials while progressing through the approval process.

Government agencies provide supportive research infrastructure, funding, and protect patient safety and efficacy. Key entities include the FDA for drug and device development, formal guidance, and label approval for new products before they can be marketed. The Office of Human Research Protections (OHRP) regulates how those who receive federal funds conduct clinical trials through approvals of informed consent forms that must be given to prospective research participants. The Office of Civil Rights (OCR) plays an indirect role in research through its Health Insurance Portability and Accountability Act (HIPAA) regulations on patient privacy. The National Institutes of Health (NIH) not only fund much of the discovery work done in institutions, they also review proposals in federally funded institutions for preclinical and clinical research, guide investigators on regulatory requirements, and contract with some companies to test new agents and sometimes hold Investigational New Drug applications (INDs). Some NIH institutes approve protocols, produce informed consent templates, and provide Central Institutional Review Board (CIRB) approvals before trials can be opened. The loop is closed with NIH consensus panels on specific diseases, dissemination strategies produced with funds from the Agency for Healthcare Research and Quality (AHRQ), and a public health focus from the Centers for Disease Control and Prevention (CDC). See Table 37.1 for more information.

Patients play a critical role in the development process that is rarely acknowledged in a meaningful way. Without actual patients, no clinical trials would complete their objectives, and yet the system treats them as "human subjects," with the similar regulatory language that governs lab animal safety [13]. Today's patients consider themselves "research participants" and no longer accept antiquated mind-sets. The latest buzz words in medical research call for more "patient-centeredness" or attention to "patient outcomes." It is now time to turn these concepts into more than standard rhetoric or press release material.

TABLE 37.1. Federal Agencies Involved in Medical Research (Mission from Each Website)

Federal Agency	Acronym	Website	Purpose/Mission
Agency for Healthcare Research and Quality	AHRQ	http://www.ahrq.gov/	To improve the quality, safety, efficiency, and effectiveness of health care for all Americans
Centers for Disease Control and Prevention	CDC	http://www.cdc.gov/	To protect America from health, safety, and security threats, both foreign and in the United States
Food and Drug Administration	FDA	http://www.fda.gov/default.htm	FDA is responsible for protecting the public health by assuring the safety, efficacy, and security of human and veterinary drugs, biological products, medical devices, our nation's food supply, cosmetics, and products that emit radiation. FDA is also responsible for advancing the public health by helping to speed innovations that make medicines more effective, safer, and more affordable and by helping the public get the accurate, science-based information they need to use medicines and foods to maintain and improve their health. FDA also has responsibility for regulating the manufacturing, marketing, and distribution of tobacco products to protect the public health and to reduce tobacco use by minors. Finally, FDA plays a significant role in the nation's counterterrorism capability. FDA fulfills this responsibility by ensuring the security of the food supply and by fostering development of medical products to respond to deliberate and naturally emerging public health threats.
National Institutes of Health	NIH	http://www.nih.gov	To seek fundamental knowledge about the nature and behavior of living systems and the application of that knowledge to enhance health, lengthen life, and reduce illness and disability
Office of Civil Rights	OCR	http://www.hhs.gov/ocr/office/index.html	To protect people from discrimination in certain health-care and social service programs
Office of Human Research Protections	OHRP	http://www.hhs.gov/ohrp/index.html	Provides leadership in the protection of the rights, welfare, and well-being of subjects involved in research conducted or supported by the U.S. Department of Health and Human Services (HHS)

INDUSTRY AND COMPANY MISTAKES

Companies changed their focus many years ago from "creating solutions for patients" to maximizing profit through mergers, acquisitions, and blockbuster drugs (not effective therapies). Many turned their backs on past industry founders and the ways they reaped their rich rewards for risk-taking approaches. Since then, risk-averse behavior is the norm, and potential markets rule, often independent of demand or promising scientific leads [14]. The results—recent polls show that industry executives are now regarded as greedy tycoons who focus on short-term self-interest, not far removed from those who caused the global banking crisis [15].

The blockbuster model produced maximum short-term rewards, but it appears to be over [16]. Companies mistook successful blockbusters as a marker of their own business acumen rather than low-hanging fruit, and started picking markets instead of scientific leads. Unfortunately, focus on short-term gains allowed a financial viewpoint to take hold, reducing expenditures and following market parameters instead of biology. Many promising biotech agents were also bought and shelved to eliminate even the hint of competition that might affect their "cash cows" [17].

For some reason, the patent clock was initially ignored, reminiscent of attitudes in the 1920s. Companies hired lawyers to protect the status quo and many new products became "me-too" drugs [18]. Powerful lobbies pressured FDA to not only reduce bureaucracy, but to lower safety and efficacy standards [19]. Meanwhile, marketing departments created demand out of vapor with "direct-to-consumer" advertising, offering great promise with very little value attached [20].

Refreshing exceptions exist, but headlined breakthroughs such as Herceptin® and Plavix® lose their luster after being touted as innovations, decades after their market release. For over 15 years, emerging science has focused on genomics, but cells also live in dynamic environments that affect their functions as well [21]. Myopic focus on genes, spurred by a desire for simplicity by companies and lucrative grant funding for academics, has produced disappointing progress for patients. At best, even drugs like Gleevec® that target a single gene will eventually stop working in the human body [22] (Box 37.1).

All of these tactics are concerning, and public and patient trust has been shaken, even without much knowledge about (1) dogged resistance to publishing *all* clinical trial results and sharing data [23]; (2) ghost writers and scientific misconduct

Box 37.1.
Additional trust issues patients and the public have regarding industry:

1. Dogged resistance to publishing *all* clinical trial results and sharing data [23]
2. Ghost writers and scientific misconduct accusations [24]
3. Trials closed and agents eliminated without warning instead of appropriate go/no go decisions before enrolling trial participants [25]
4. Outdated study designs and lack of pharmacodynamic (PD) analysis [26].

TABLE 37.2. Sample of FDA-Approved Drugs and Their Status

Drug	Approval	Rescinded	Disease Indication	Manufacturer
Temafloxacin	1992	1992	Lower respiratory tract infections, genital and urinary infections	Abbott Laboratories
Sparfloxacin	1996	2000	Bacterial infections	Bayer AG
Alatrofloxacin	1997	2000	Broad-spectrum antibiotic	Pfizer
Trovafloxacin	1997	2000	(14) Types of infections	Pfizer
Grepafloxacin	1997	1999	Bacterial infections	Glaxo Wellcome
Gatifloxacin	1999	2008	Respiratory tract infections	Bristol-Myers Squibb

accusations [24]; (3) trials closed and agents eliminated without warning instead of appropriate go/no go decisions before enrolling trial participants [25]; and (4) outdated study designs and lack of pharmacodynamic (PD) analysis [26].

Another concern is the alarming number of rescinded drug approvals due to trial methods that do not produce better drugs. According to Outterson et al., "Among the 61 New Molecular Entity (NME) approved antibiotics, 26 were withdrawn (43%), a rate far higher than was observed among non-antibiotics (13%)" [27]. A sample of these drugs is included in Table 37.2. This acute situation should be more alarming than the impending health crisis we may face, as evidenced by this excerpt from the National Physicians Alliance and Public Citizen's Health Research Group, "The consequences of inappropriate trial designs in the setting of life threatening diseases are even more severe—death from ineffective treatments when effective therapies exist. FDA has released warnings regarding [antibacterial] drugs [like Tygacil®] that have increased deaths in serious and life-threatening diseases when approved through non-inferiority trials using outcomes other than mortality" [28].

Successful campaigns that scare everyone about super-resistant bugs have worked well, but old research methods that focus on bugs in a test tube and healthy patients will not lead to better patient outcomes, even if FDA eliminated all requirements [29]. This is similar to the reliance cancer research has toward cell lines, and both disease fields rely on mouse models that often cannot truly emulate the human body. Additionally, manufactured markets accelerate mistrust and prove to patients that their well-being is not a priority. Was it really a surprise that studies show potential health concerns for men taking drugs to eliminate "low T"? [30]. As Yogi Berra opined, "It's déjà vu all over again" when you consider historic health risks for women who take hormone replacement therapy [31].

Even marketing finesse cannot justify the questionable "value-based pricing" tactics of some companies [32]. Their justification of the "value to society" for drugs like Avastin® has created an avalanche of exorbitant drug prices in the United States (but not in other countries) and extreme ill will with patient communities and the public. Even after skillful lobbying, the FDA rescinded approval for metastatic breast cancer, citing lack of efficacy and concerns about toxicity [33]. Though the value proved low in this case, the pricing model remains, bankrupting some families who pay a hefty penalty for hope [34]. Many feel totally comfortable equating this highly aggressive economic model with the "Robber Barons" of yesteryear [35].

ACADEMIC AND INSTITUTIONAL MISTAKES

Unfortunately, the feudal structure still used by most academic institutions has morphed into a culture of ego-driven departments and labs that often reward those who excel at gaming the system instead of fostering teams who build on scientific findings [36] (Box 37.2).

Virtually every institutional, investigator, and individual incentive in academic research is misaligned with goals of the public that support medical research funding. Even though many still purport to rise above the tawdry affairs of industry, academics seem to have embraced the worst aspects of competition in many cases.

The ease of counting publications for tenure and potential IP seem to outweigh the quest for qualified science. This fever for highest impact factor publications has led to many shortcuts in the scientific process, which at best lead to reproducibility issues, and at worst, endorse fraudulent actions [37]. The resulting environment seeks to avoid risk and limits opportunities for young investigators who prefer to publish all of their experiments, not just the best one.

Senior principal investigators (PIs) chase the "latest greatest" discoveries to win the next grant or Nobel Prize title, instead of truly learning how biological factors interact within an integrated human body. Postdocs who actually perform experiments that lead to novel discoveries are expected to field accusations if publications are questioned. PIs are also expected to lead expanding labs with no ongoing management skills training. Fewer PIs are able to honorably succeed in a system that rewards egos in a "winner-take-all" approach. Their successes, against the odds, should be highlighted in best practice exercises and publications.

Open data sharing is also resisted in many academic circles. Biology is fueled by fears of being "scooped" [38] even though other fields, such as chemistry and physics, have long since discovered how to give people their due without withholding information that could accelerate progress so desperately needed [39]. Additional issues that develop on regulatory requirements, clinical trials operations, and hospital profits muddy the role of research even further.

All these concerns are clearly prioritized ahead of research participants and future patients. In fact, very little ever reaches patients, and academia's reputation has been tarnished to the brink of permanence [40]. It is critical to transform the culture toward a more inclusive and reproducible approach to the discovery, development, and delivery of patient therapies before the public stops supporting research funding.

Box 37.2.
Virtually every institutional, investigator, and individual incentive in academic research is misaligned with goals of the public that support medical research funding.

Box 37.3.
Too many government employees act as regulators, even when their job description does not include that function.

GOVERNMENT MISTAKES

Government agencies and programs are usually seen as obstacles, sometimes for good reason, but they support research infrastructure, fund ideas, and provide needed platforms that consider patient safety and efficacy. While good intentions created the mechanisms that protect and serve patients, the resulting rules are often overrun by a bureaucratic system that takes on a life of its own, creating an environment of checklists, procedures, paperwork, and penalties that can stymie progress (Box 37.3).

Too many government employees act as regulators, even when their job description does not include that function. In these cases, a facilitation model would bear more fruit. Ambiguous guidance that receive varying interpretations and conflicting messages from different employees also have to be avoided.

There also seems to be a dysfunctional relationship between some federal programs and external investigators and institutions that receive funding and monitoring from the same government staff. Mixed messages on who controls the process, whether internally or with external partners, is a common complaint as well [41]. Reliance on outdated policies and grant structures also continue to reward isolation and competition, rather than fostering teamwork and collaborations. Due to the vast spectrum of opinions and special interests, it is also standard for federal agencies to be criticized for most of their actions or inactions. A good example includes the FDA's recent decision to treat generic manufacturers like the rest of the industry [42].

RE-FOCUS ON PATIENT NEEDS

Translational research is a term now used in many ways, but it should always mean to truly transform scientific discoveries into clinical tools that solve real patient problems. This must include development processes that focus on human biology and function, not just on genes or test tubes, cell lines, and animal models that rarely mimic the complexity within a human body.

Current buzzwords such as "patient-centered" give the impression that companies, institutions, and government entities are eager to improve their images with patients, their advocates, and the public. After years of false hopes and promises, it will take specific and transparent steps to mitigate experiences that have caused many groups and families to harbor lasting fears and suspicions about the entire medical research enterprise. As stated in Chapter 13 of this book, "Producing mediocre drugs and offering them at unaffordable prices may invite such backlash."

Too often, advocacy organizations and patients have been used to front specific agendas that camouflage industry and academic self-interests by raising the "do what's best for patients" banner. These surface-level campaigns can dissolve into specious situations once probing questions are posed. Many patient community members are no longer willing to play along with companies, institutions, or opportunity-seeking individuals. Resources like HealthNewsReview.org [43] and AllTrials.net [44] also help warn people of questionable motives and keep them updated on relevant issues.

It is also time to stop treating patients as if they are their illness and cannot understand anything. Patients are whole people whose life skills stay with them through their medical experience. Some become patient advocates who are highly educated about research, want to help improve outcomes, and can act as peers throughout the research process [45].

Patients and the public expect the medical system to honor and respect their needs, and to be able to participate in research and health-care delivery. Coordinated actions that re-focus intentions, approaches, policies, and cultures while including patient input at every point will help reestablish the trust that is necessary to sustain public support and market share.

ROADS TO SOLUTIONS

Placating people won't work anymore, and disappointment in lackluster results will soon turn into formal inquiries or worse. The good news is that all of these problems can be transformed into a coordinated productive system that produces equitably priced, innovative products, yielding positive patient outcomes and acceptable profits [Box 37.4].

The ACA set the stage for transforming U.S. "fee for service" health care into reimbursement for actions that actually help a person who has a medical condition or illness through "value-based" payments and ACOs. This first step offers much-needed relief for patients, and can also benefit payers, companies, institutions, and clinicians who match more effective services and products to patient outcomes [46]. Researchers who include these concepts into their development plans will reap appropriate rewards, while those who ignore them may receive no reimbursement for new therapies that don't help patients.

The Patient-Centered Outcomes Research Institute (PCORI), funded through ACA, has already begun to change some research mind-sets by offering grant funding and a minimum standard of methods to those who are ready to bridge the traditional gaps between research data and health-care delivery. Groups such as the Core Outcome Measures in Effectiveness Trials (COMET) Initiative [47], the NIH Health Care Systems Research Collaboratory [48], and the Green Park Collaborative [49] are providing critical information on Core Outcome Sets (COS), pilot projects, and Effective Guidance Documents (EGDs) that incorporate these elements into "how-to" approaches for specific medical conditions. These third-party initiatives provide patient representatives, clinicians, researchers, payers, institutions, and companies opportunities to understand unique perspectives and to merge common purposes in coherent ways.

Other new initiatives and resources also exist to help forward-thinking researchers adapt rapidly to more effective research methods. Health and Human Services (HHS)

Box 37.4.
Possible solutions include

- Embrace "value-based" payments and ACOs and incorporate into development plan
- Include Patient-Reported Outcomes (PROs) and other outcome measures in trial design
- Include inclusive data sharing plans throughout development
- Establish and use data collection standards that allow comparisons
- Define terms for data sharing and minimize limitations
- Incorporate recommendations from third-party reviews
- Collaborate with government initiatives for translational research
- Create registries that collect real patient data and outcomes (can also be used as control arms if broad enough)
- Develop Next-Generation Sequencing standards for molecular profiling
- Fund, require, and reward reproducibility efforts with third parties
- Test all cell lines before using
- Invest in PI and investigator management training
- Establish ongoing board certification for PhDs and MDs/PhDs
- Change career criteria (tenure) to reflect quality, collaboration, and multiple first authors
- Publish best practices of those who truly exhibit quality behaviors
- Incorporate experienced patient advocates/representatives into all aspects of drug development
- Include patient feedback in early drug development for feasibility
- Expand advocacy offices to include research and development, not just marketing or policy
- Partner with patient groups who have biobanks, registries, and so on
- Develop combination therapy designs and methods for development and approval
- Create models for smaller markets
- Set Intellectual Property (IP) minimum standards

created the Office of the National Coordinator for Health Information Technology (ONC HIT) that promotes innovative projects and policies that add value, and can include new research concepts as well. The Institute of Medicine (IOM) has also considered many of these topics, as evidenced through their reports [50]. NIH has created a National Center for Advancing Translational Sciences (NCATS) which runs the Clinical and Translational Science Awards (CTSA) program, and provides therapeutic applications through its BRIDGs Program [51]. This is similar to the cancer-specific NCI Experimental Therapeutics (NExT) program [52].

Infrastructure improvements might also come from government or professional organizations. Professional associations, such as the American Society for Clinical Oncology (ASCO), are creating collaborative initiatives like the CancerLinQ "rapid learning system [that] will harness technological advances to connect oncology practices, measure quality and performance, and provide physicians with decision support in real time" [53]. Funding proposals from government sources and private foundations should require steps to (1) develop reproducibility efforts (with appropriate resources for independent lab testing); (2) test all cell lines before use; and (3) use a small percentage of each grant to build PI management skills or additional funds are returned to the granting agency. These examples illustrate changes that can help support longer-term changes that reorient the medical research system to patient needs and outcomes. Exploration into field-relevant certification or "boards" for established PhDs and MD/PhDs to keep their skills up to date, just as MDs must do to be able to practice their trade, is also worth pursuing.

Institutional changes are critical and will cover too many volumes to publish a full list here. Clearly, a shift toward quality publications (rather than impact factor) has to occur, as does an atmosphere that accepts multiple first authors, rewards collaboration and reproducibility, discourages empire-building, promotes input from patients and their advocates, and creates metrics that measure improvements in patient outcomes. Resources and pilots already exist for most of these categories, such as the Reproducibility Project [54], The Reproducibility Initiative [55], which can help transform the culture to redirect efforts to activities that lead to better patient results.

Academics can expand nascent steps to fully incorporate patient advocates and representatives as peers from the germ of a research idea through completion instead of using these representatives as window dressing to check off a box with grants that require patient participation, such as those issued by the PCORI [56]. Patient advocates who understand research and its structures offer cogent advice, and are ready to partner even more with institutions, companies, and government. Patient advocates with research expertise increasingly guide clinical research through feedback on early trial design, protocol development, informed consent improvements, accrual strategies and plain-language study result summaries. Some also influence discovery and early product development within academia and companies [57–59].

Industry seems to be a late adopter to open data, but has now created their "Principles for Responsible Clinical Trial Data Sharing" [60]. To ensure that this is not another public relations exercise, the ethical philosophy of Dr. Seuss, "I meant what I said and I said what I meant," [61] should be applied. For example, replace ambiguous language with specific definitions that explain what "proprietary data" are, and minimize exclusion categories. Companies must show that they are not only willing to change their words, but also put their money where their mouths are, so to speak. For instance, move advocacy relations offices out of marketing and into research and development to offer more substantive partnership opportunities with patient advocates who understand research. More partnerships with rare disease advocacy groups can also help find clues and complete faster clinical trials, achieving answers that can reach patients more quickly [62].

More sophisticated data modeling that turns voluminous data into usable knowledge and adaptive study designs that incorporate patient input early in development are also needed. Combination therapies should also be embraced, along with rapid adaption to breakthrough models in modest markets, such as those used for Gleevec® [63]. Since the majority of today's research projects fall into this range, real-world economic models that create more nimble R&D investment strategies and cost-effective clinical research can gain faster FDA approval and market acceptance. Success in today's marketplace means more than IP contracts, FDA approval, and reimbursement coding (both of which can be better aligned with computer technology and banking standards) [64]. Patient acceptance due to better outcomes is also a critical success factor, and can be won with broader patient and social media strategies.

Patient goals do not include chronic conditions and illnesses, even for those facing life-threatening diseases like cancers. They want to get well and hopefully be cured without long-term consequences from disease or its treatments. By working together, these goals can become realities.

REFERENCES

1. Turner, J., Kelly, B. Emotional dimensions of chronic disease. *West J Med.* 2000;172(2):124–128.
2. List of conspiracy theories. Available at http://en.wikipedia.org/wiki/List_of_conspiracy_theories#Medicine. Accessed December 13, 2013.
3. Anderson, J.M. Empowering patients: issues and strategies. *Soc Sci Med.* 1996;43(5): 697–705.
4. Lehrer, E. The truth wears off: is there something wrong with the scientific method? *The New Yorker* December 13, 2010. Available at http://www.neurofly.com/NeuroSeminar_files/2011-2.pdf. Accessed December 13, 2013.
5. Groeger, L. Big pharma's big fines. March 11, 2013. Available at http://projects.propublica.org/graphics/bigpharma. Accessed December 13, 2013.
6. Bhattacharjee, Y. The mind of a con man. 2013. Available at http://www.nytimes.com/2013/04/28/magazine/diederik-stapels-audacious-academic-fraud.html?pagewanted=all&_r=1&. Accessed December 13, 2013.
7. The need for awareness of clinical research. 2011. Available at http://www.nih.gov/health/clinicaltrials/providers/awareness.htm. Accessed December 13, 2013.
8. About the law. 2013. Available at http://www.hhs.gov/healthcare/rights/. Accessed August 13, 2013.
9. Markel, H. Patents, profits, and the American people—the Bayh–Dole Act of 1980. *N Engl J Med.* 2013;369:794–796.
10. Gambrill, S. Efficient contracting processes for clinical trials . . . from the legal perspective. 2012. Available at http://www.lifescienceleader.com/doc/efficient-contracting-processes-for-clinical-0001. Accessed December 13, 2013.
11. Yates, J.W. Clinicians in community practice are major contributors to clinical trials in cancer prevention, early detection, and treatment. *CA Cancer J Clin.* 2003;53:69–72.

12. Rehman, J. Cancer research in crisis: are the drugs we count on based on bad science? 2013. Available at http://www.salon.com/2013/09/01/is_cancer_research_facing_a_crisis/. Accessed December 13, 2013.
13. Federal Policy for the Protection of Human Subjects ("Common Rule"). Available at http://www.hhs.gov/ohrp/humansubjects/commonrule/. Accessed December 13, 2013.
14. The pharmaceutical golden era: 1930–60. Available at http://pubs.acs.org/cen/coverstory/83/8325/8325golden.html. Accessed December 13, 2013.
15. Hirsch, M.L. Side effects of corporate greed: pharmaceutical companies need a dose of corporate social responsibility: authentic happiness, self-knowledge and legal policy. *Minn J Law Sci Technol.* 2008;9(2):607–636.
16. Hunt, V., Manson, N., Morgan, P. A wake-up call for big pharma. 2011. Available at http://www.mckinsey.com/insights/health_systems_and_services/a_wake-up_call_for_big_pharma. Accessed December 13, 2013.
17. III. Pharmaceutical mergers. Available at http://www.ftc.gov/bc/hcindex/pharmmergers.shtm. Accessed December 13, 2013.
18. Oudard, S., Tombal, B. Prostate cancer: enough "me too" drugs. 2013. Available at http://www.medscape.com/viewarticle/812243. Accessed December 13, 2013.
19. Shlaes, D. The Infectious Diseases Society sounds off on the FDA and skin infections. 2010. Available at http://antibiotics-theperfectstorm.blogspot.com/2010/11/infectious-diseases-society-sounds-off.html. Accessed Deccember 13, 2013.
20. Wolinsky, H. Disease mongering and drug marketing. *EMBO Rep.* 2005;6(7):612–614.
21. Esteller, M. Epigenetic changes in cancer. 2011. Available at http://www.the-scientist.com/?articles.view/articleNo/30739/title/Epigenetic-Changes-in-Cancer/. Accessed December 13, 2013.
22. Pan, J. Overcoming Gleevec-resistance by blocking oncogene addiction in malignant hematologic cells. *J Hematol Oncol.* 2012;5(Suppl. 1):A6.
23. LaMattina, J. An open letter to the PhRMA CEO on pharma's image. 2013. Available at http://www.forbes.com/sites/johnlamattina/2013/04/20/an-open-letter-to-the-phrma-ceo-on-pharmas-image/. Accessed December 13, 2013.
24. Eban, K. Dirty medicine. 2013. Available at http://features.blogs.fortune.cnn.com/2013/05/15/ranbaxy-fraud-lipitor/. Accessed December 13, 2013.
25. Chauhan, C., Smith, M.L. Closing an on-going clinical trial: when is it betrayal of participants? *Breast Cancer Res Treat.* 2013;142:461–462.
26. Sanfilippo, F., Santonocito, C., Foëx, P. The last truth for beta-blockers in non-cardiac surgery is still missing! *Minerva Anestesiol.* 2013. [Epub ahead of print] PMID: 24193178.
27. Outterson, K., Powers, J.H., Seoane-Vazquez, E., Rodriguez-Monguio, R., Kesselheim, A.S. Approval and withdrawal of new antibiotics and other anti-infectives in the U.S., 1980–2009. *J Law Med Ethics.* 2013;41(3):688–696.
28. Bettigole, C., Carome, M.A. Letter from the National Physicians Alliance and Public Citizen's Health Research Group. 2013. Available at http://npalliance.org/wp-content/uploads/Response_to_FDA_Federal_Register_Notice_8-20-13-FINAL.pdf. Accessed December 13, 2013.
29. Ledford, H. FDA under pressure to relax drug rules. 2012. Available at http://www.nature.com/news/fda-under-pressure-to-relax-drug-rules-1.11936. Accessed December 13, 2013.
30. O'Riordan, M. Testosterone therapy linked with adverse CVD events. 2013. Available at http://www.medscape.com/viewarticle/813833. Accessed December 13, 2013.

31. Gardner, A. Study shows HRT even riskier than thought. 2010. Available at http://health.usnews.com/health-news/family-health/cancer/articles/2010/10/19/study-shows-hrt-even-riskier-than-thought. Accessed December 13, 2013.
32. Baker, R.J. Pricing on purpose: how to implement value pricing in your firm. 2009. Available at http://www.journalofaccountancy.com/Issues/2009/Jun/20091530.htm. Accessed December 13, 2013.
33. FDA Commissioner announces Avastin decision. 2011. Available at http://www.fda.gov/newsevents/newsroom/pressannouncements/ucm280536.htm. Accessed December 13, 2013.
34. Berenson, A. Avastin: a cancer drug shows promise, at a price that many can't pay. 2006. Available at http://www.nytimes.com/2006/02/15/business/15drug.html?_r=2&oref=slogin&. Accessed December 13, 2013.
35. Brummer, A. Like bankers/robber barons: special report: the deadly sins of greedy pharmaceutical companies. 2012. Available at http://www.thisismoney.co.uk/money/markets/article-2174537/CITY-FOCUS-The-deadly-sins-greedy-pharmaceutical-companies.html. Accessed December 13, 2013.
36. Swazey, J., Anderson, M., Louis, K. Ethical problems in academic research. 1993. Available at http://www.americanscientist.org/issues/issue.aspx?id=934&y=0&no=&content=true&page=1&css=print. Accessed Deccember 13, 2013.
37. Hiltzik, M. Reproducibility: science has lost its way, at a big cost to humanity. 2013. Available at http://www.latimes.com/business/la-fi-hiltzik-20131027,0,1228881.column?goback=%2Egde_4536998_member_5800996943907794948#%21. Accessed December 13, 2013.
38. Pearson, H. Competition in biology: it's a scoop! 2003. Available at http://www.nature.com/news/2003/031127/full/news031124-9.html. Accessed December 13, 2013.
39. arXiv. Available at http://en.wikipedia.org/wiki/ArXiv. Accessed December 13, 2013.
40. Shaw, C. Postdoc dilemma: to leave or not to leave academia? Live chat. 2013. Available at http://www.theguardian.com/higher-education-network/2013/jun/18/postdoc-dilemma-leave-stay-academia. Accessed December 13, 2013.
41. Criticism of the Food and Drug Administration. 2014. Available at http://en.wikipedia.org/wiki/Criticism_of_the_Food_and_Drug_Administration. Accessed December 13, 2013.
42. Serebov, M. FDA: generic industry mature enough for full responsibility. BioWorld. November 12, 2013.
43. HealthNewsReview.org.
44. AllTrials.net. Available at http://www.alltrials.net/. Accessed December 13, 2013.
45. Collyar, D. How have patient advocates in the United States benefited cancer research? *Nat Rev Cancer*. 2005;5(1):73–78.
46. Japsen, B. Obamacare will bring drug industry $35 billion in profits. 2013. Available at http://www.forbes.com/sites/brucejapsen/2013/05/25/obamacare-will-bring-drug-industry-35-billion-in-profits/. Accessed December 13, 2013.
47. Core Outcome Measures in Effectiveness Trials Initiative. 2013. Available at http://www.comet-initiative.org/.
48. NIH Health Care Systems Research Collaboratory. Demonstration projects for pragmatic clinical trials focusing on multiple chronic conditions (UH2/UH3). 2013. Available at http://grants.nih.gov/grants/guide/rfa-files/RFA-RM-13-012.html. Accessed December 13, 2013.
49. Green Park Collaborative—USA. Available at http://www.cmtpnet.org/publicprivate-partnerships/greenparkcollaborative/. Accessed December 13, 2013.

50. Institute of Medicine reports. Available at http://www.iom.edu/Reports.aspx. Accessed December 13, 2013.
51. National Center for Advancing Translational Science (NCATS). Transforming translational research. 2013. Available at http://www.ncats.nih.gov/files/factsheet.pdf. Accessed January 27, 2014.
52. National Cancer Institute (NCI) Experimental Therapeutics (NExT) Program. A unique partnership with the NCI to facilitate oncology drug discovery and development. Available at http://next.cancer.gov/. Accessed January 27, 2014.
53. CancerLinQ. Available at http://www.asco.org/quality-guidelines/cancerlinq. Accessed December 13, 2013.
54. Spies, J.R., et al. The Reproducibility Project. 2012. Available at http://centerforopenscience.org/. Accessed December 13, 2013.
55. The Reproducibility Initiative. Available at https://www.scienceexchange.com/reproducibility. Accessed December 13, 2013.
56. Patient-Centered Outcomes Research Institute (PCORI). Funding opportunities. Available at http://www.pcori.org/.
57. Katz, M.L., Archer, L.E., Peppercorn, J.M., Kereakoglow, S., Collyar, D.E., Burstein, H.J., Schilsky, R.L., Partridge, A.H. Patient advocates' role in clinical trials: perspectives from Cancer and Leukemia Group B investigators and advocates. *Cancer*. 2012;118(19): 4801–4805.
58. Roth, D.A. Third seat at the table: an insider's perspective on patient representatives. *Hastings Cent Rep*. 2011;41(1):29–31.
59. Landy, D.C., et al. How disease advocacy organizations participate in clinical research: a survey of genetic organizations. *Genet Med*. 2011;14(2):223–228.
60. The biopharmaceutical industry is firmly committed to enhancing public health. 2013. Available at http://www.phrma.org/phrmapedia/responsible-clinical-trial-data-sharing#sthash.SO0O4INk.dpuf. Accessed December 13, 2013.
61. Dr. Seuss quotes. Available at http://www.goodreads.com/author/quotes/61105.Dr_Seuss. Accessed December 13, 2013.
62. Adenoid Cystic Carcinoma Research Foundation. Strategy. 2014. Available at http://www.accrf.org/.
63. Pray, L.A. Gleevec: the breakthrough in cancer treatment. 2008. Available at http://www.nature.com/scitable/topicpage/gleevec-the-breakthrough-in-cancer-treatment-565. Accessed December 13, 2013.
64. Restoring value to biopharmaceutical R&D. 2012. Available at http://www.mckinsey.com/insights/health_systems_and_services/restoring_value_to_biopharmaceutical_r_and_38d. Accessed December 13 2013.

38

TO LEASH OR UNLEASH THE POWER OF PUBLIC–PRIVATE COLLABORATION: IN HEALTH AND DISEASE

Hakim Djaballah

Memorial Sloan-Kettering Cancer Center, New York, NY, USA

In this chapter I will provide my own personal perspectives on how and why we must learn to be facilitators toward common goals of collaboration, and orient ourselves toward comprehending the real unmet needs to combat disease: to literally become our village. I want to ask whether the view of learning as "becoming our village" can really be accommodated by the present-day institutions, such as the financiers and the large pharma companies, or whether we should begin creating more diverse options to support not just discovery efforts, but getting an education in development and commercialization of drugs.

IT TAKES A LIFETIME TO BECOME A VILLAGE: ACADEMIA IS A VILLAGE

I have always been fascinated by the concept of interpreting which came first, the chicken or the egg? From an experienced analyst's point of view, it is a matter of probability resulting in a 50:50 chance of either one coming first; a tough position to be in as you are not really sure whether to buy the chicken or the egg. If you then provide the same analyst with information derived from a high-profile publication claiming that

Collaborative Innovation in Drug Discovery: Strategies for Public and Private Partnerships, First Edition. Edited by Rathnam Chaguturu.

in fact the egg came first as a result of an archeological dig somewhere in the world discovering some eggshell remains, which analysis confirmed as very old using carbon 14 dating, the analyst, in turn, would now claim that there is a 99.9% chance that the egg came first because he altered his algorithms to take into account the new information; the decision to buy the egg is now made for you and would facilitate your transaction. However, a few months later, another high-profile publication refutes the earlier finding by providing evidence that the eggshell found in that particular dig was in fact of dinosaur origin and not of chicken. They used the most sensitive DNA sequencing methods to make the determination. The same analyst, once more, changes his algorithms by incorporating the new piece of information and would now predict that the chicken came first because it could not have been the egg; your choice now is to sell the egg and buy the chicken but at a substantial loss as the value of the egg dropped by 99%. In essence, you would have been better off splitting your assets between the chicken and the egg; in which case you minimize your losses and potentially maximize some profits. Imagine now a different scenario, where you seek the information from experts in the fields of genetics, evolution, and archeology. Your rely on their expert opinion and when combined they tell you that indeed the egg came first as a part of an evolutionary chain of events within the bird's village over millions of years resulting in today's chicken; therefore, the value proposition is in the chicken and not the egg. The egg was merely a temporary tool for the bird's village as it evolved and became one. The bird village became one over millions of years and during the process it allowed diversification and creativity to maintain both the health and growth of its communities. We have a beautiful diversity of birds today (unfortunately many of them are still fighting for their survival and avoiding extinction), but no signs of the big ones such as dinosaurs.

The analogy I would employ here, for the public–private collaboration also referred to as "academia–pharma," is that of a vertically fried salami sausage with a horizontally and perfectly fried egg. Academia is a network of independent contractors, representing the sausage, driven by curiosity and the sheer quest of learning and explaining various manifestations in a multitude of disciplines within the vast biological kingdom. They have acquired the deepest basic understanding of biology but within a very narrow scope, and their findings are not and will never be immune to artifacts, refutes, and controversies. Their source of investment is primarily the federal government through peer-reviewed grants (awarded through the National Institutes of Health in the United States), sometimes supplemented by donations or philanthropy dollars, and on occasion financed directly by the universities themselves or by the private sector for specific projects (referred to as sponsored research). An academic researcher can receive hundreds of millions of dollars of research funds during a professional career; not a bad incentive for the host university or institution, which, incidentally, would receive up to the same amount in indirect costs. What I find fascinating is that academia is indeed a very well-established village and has evolved to work hard and maintain its homeostasis as projected by many universities or research institutions across the world. Collaboration is the heart and soul of academics, bringing together folks with different expertise and disciplines toward a simple goal of obtaining funding, performing the work, and publishing the findings in the hope of receiving more funding, securing tenure, and of course becoming famous, with an eye on the Noble Prize among others. Academia's other critical roles are to keep teaching and educating the next generation of scientists and provide breakthrough discoveries, where some have already changed our ways and standards and others failed, as one would expect. I would sum it up as a simple daily life of an academic village.

CAN PHARMA BECOME A VILLAGE?

Pharma companies and the rest of the private sector, on the other hand, are a sharp contrast and would fit beautifully into the fried egg concept. They all have a nucleus corporation controlled primarily by a board of directors and the investors, who are at the mercy of the financiers through their analysts' predictions. This nucleus is the egg's yolk; it controls new ventures, company growth, acquisitions, mergers, and unfortunately, downsizing. The sole driving force of the yolk is survival at all costs with huge profit margins, even if it means destroying the community and subsequently the village. The egg white represents the vast applied knowledge within the companies, which lacks the deep understanding of basic biology, as have the academics. Within it you would typically observe the "pigeonhole" concept where pharma companies would segregate their scientists into well-defined job descriptions to perform predefined specific tasks within their allocated silos, in turn creating the pyramid progression of information and decision-making chain of command. The more valuable a project become, fewer eyes would be privy to it.

This model worked well for these corporations, but it did not make a village. In the mid to late 1990s, I observed an interesting modification to the model, in which medicinal chemistry and all associated chemists became powerful within these organizations as they became they key determinants of a project success, and in some cases demanding the respect they deserve from their biologist counterparts. Organizations had to act fast and introduce the concept of "matrix teams," where you have two coleaders now instead of one, and these roles are always represented by biology and chemistry. This is not a move toward becoming a village; rather. matrix management was adopted by these organizations for their survival's sake. Perhaps it was meant to introduce a sense of community, leading to improved collaboration and cooperation across the different silos. I view it more as an institutional adaptive response to maintain core competencies and competitive edge, with the caveat of inadvertently creating a new conflict of loyalty between management and project leaders over project survival, resource allocation, and potentially the mere death of the silos.

ACADEMIA AS A DISCOVERY PARTNER: PREY, A DECOY, OR BOTH?

Not long after that, we witnessed yet again another attempt by corporate pharma to push for becoming a village to survive; this time they introduced centers of excellence for discovery dedicated solely to one specific therapeutic area, self-contained, and with remits of producing several new chemical entities over a predefined period of time; a bold move from the centralized pyramid model, which they spent years building and refining to a decentralized new concept of an "octopus" model, with one head (corporate) controlling up to eight arms, some of them being the newly created communities with shared interest and convergent goals. Though the octopus is the most intelligent invertebrate, its brain is formed by 500 million large neurons (it is estimated that our brains are made of ~100 billion smaller neurons); it is a solitary and territorial hunter. It can detach an arm to distract its attacker and escape; the limb will continue to wriggle, gaining the attention of the predator and keeping it occupied, and the severed tentacle will grow back in no time. It is hardly a notion of a true sense of a community evolving into a village, rather metamorphosing into a sleek and agile corporation able to eliminate any of its arms or centers of excellence without compromising its overall survival.

I started the previous sections by describing pharma as the perfect fried egg and concluded by saying it is actually more of an octopus; I hope that I managed to convince you how fast it has evolved over the past two decades to survive. I salute the companies who fell during this compelling war of the fittest; among them were giants like Wyeth (1860–2009), Warner Lambert (1856–2000), and Schering Plough (1851–2009) to name few. Their souls may rest in peace. These unfortunate events unleashed a mass exodus of highly specialized scientists from the private sector and into academia and nonprofit institutions and constituted a true game changer for academia as it arms itself with appropriate and needed expertise at little or no up-front investment. The question now is whether this specialized workforce can successfully operate in academia, bearing in mind the clear distinction between "basic" and "applied" research combined with the inherent control of the principal investigators. The reverse is also true in my experience. Over the past 10 years at Sloan-Kettering, I recruited 11 highly skilled PhD-level scientists, at different times, to train them in my lab to develop and execute drug discovery screens; with the exception of one scientist who came to me from big pharma, the rest had an academic pedigree of training across multiple disciplines. Though all of them could execute detailed experimental procedures, very few were trained to develop concepts or be critical and vigilant about the work they were doing, with the associated naivety problem of believing and not questioning published work (including the one scientist trained in pharma). This experience validates the notion that academia ought to seek and build its own expertise in the discovery process, bearing in mind that some of these are not your typical academic disciplines, such as medicinal chemistry and chemical screening.

Pharma corporations are the octopuses of the twenty-first century, solitary in nature but the most compelling hunting adversary you can imagine, with mimicry abilities overcoming those of chameleons. It would not hurt at all if pharma designates one of its octopus arms to become the newly created center of excellence for external discovery (CEED) or in other words a new business development effort targeting academics, nonprofit institutions, and in some cases, start-up companies. This is the perfect hunting tool for valuable assets: under the framework of outreach and collaboration, they lure their prey with promises of lucrative deals, in some cases worth millions of dollars as we have read from press releases, and the much-needed help and expertise to develop drugs; in turn, the companies would either enhance their drug discovery and development efforts, resell the assets at a profit, or merely shelve them and thus eliminate any competitive threats—short or long term.

THE REAL PHARMA STRATEGY FOR ACADEMIC COLLABORATION

Throughout history, evidence of the medicine men or village healers or the mighty powerful sorcerers, sought by many to perform miracle treatments, can be found in every village in every corner of the Earth. It cannot be a coincidence or an accident that these healers existed back then and still exist today. Some attribute this simply to becoming a true community within the village to protect their own; others kindly remind us of the sheer animosities between villagers: hence the healing powers can easily become sickening ones to the rival villages. What is amazing is the way healing has evolved over the centuries, independent discoveries made through trial and error, recorded, and passed on from one generation to the next, nearly always using natural products native to the area.

With this in mind, I am yet to be convinced that the backbone fabric of today's public–private collaboration is real or possesses solid foundations; it seems to be more of a temporary marriage of convenience achieving short-term goals to pharma and making academia their guest of leisure by showering academics with gifts and promises. The pharma industry has already begun dismantling its own research infrastructure as the octopus looses one arm to escape predator threat, only to grow it back with a completely different focus and strategy in mind well in line with what their financiers expect them to do. Pharma is becoming synonymous to the antique pickers; they travel the world in search of items to restore and resell at a higher cost, knowing there is a market of collectors for them, or simply clean them up, polish them, and sell them in their retail stores. Academics, as the hoarders of such items, can themselves sell them directly to the collectors or, as is typically the case, at an auction and for the highest bid with or without reserve.

Moreover, I am concerned as to the dire consequences to academia; to better manage the collaboration and the inevitable outcome, we are in serious need of highly skilled and devoted facilitators to act as counselors and, more importantly, provide the most hated "reality check" to the academic investigator by communicating the honest merits of their discoveries and/or research to have an impact in an applied environment leading to a drug. Academics are often blind to quality assurance and quality control criteria, to other researcher's findings, and most times deaf to constructive criticism; besides, who are we to critique these investigators with such impressive track record of funding and an admirable list of publications? It is, unfortunately, one of the toughest jobs as it makes one become the most hated individual at the respective institutions.

THE UNSUNG HEROES OF COLLABORATION: THE ULTIMATE GAME CHANGER

By definition, a collaborative effort means synergy of sort and not duplication of efforts to successfully achieve common goals. As far back as the late 1980s and during my PhD years at the University of Leicester in England, I do recall collaborations between the university and several companies, including ICI (now Astra Zeneca). My studies were actually funded through a scholarship provided by the Celltech Group PLC (1980–2004). I have also been privileged to collaborate with the Carlsberg Research Center in Denmark, a brewery company; they provided me with a peptide library arrayed in several 96-well microtiter plates and enabled me to perform my first medium-throughput screen against the proteasome as a target, in a period where there were no fluorescence-based plate readers. It was the most time-consuming and complicated screen I have ever performed to date. So there is no novelty about these public–private collaborations, as they were routinely initiated, performed, and completed, enabling the private sector to seek and obtain specific answers to given problems. The question is what has changed since then?

The game changer in my opinion came about when the NIH roadmap was introduced, championed by Drs. Elias Zerhouni and Francis Collins, former and current NIH directors, respectively, against all the odds and the outcry of pharma complaining that this will create a new breed of competitors subsidized by the federal government. Almost a decade since the NIH-funded screening center became operational, hundreds of screens were performed against a wide range of targets, some of which pharma would have never considered in their discovery portfolios, and the data from which are publically available for all to browse and learn from. It is the most critical initiative the NIH and ultimately the public sector has ever done to help lower the discovery barrier and begin the elimination of the "stigma" associated with it. We have all heard the saying "what do you bring to the table?"—Sometimes we are happy to hear it and other times we are not sure how to respond. Well, these days we are happy to hear it, especially when it is coming from the pharma CEED representatives. We can do most of the things you are expert in and more, up to Phase 1 and 2 clinical trials.

BEYOND SCHOOLING, LET'S NOT FORGET THE UNHEARD VOICES IN THE VILLAGE

Stardom and richness are the only catalysts I am aware of that can make an awareness campaign on behalf of the unheard voices in the village; sometimes politicians can provide similar attributes, but not the heroes who can lend their voices to fighting diseases for which we have no treatment nor funding enabling basic and applied research to be carried out. These voices are those of health charities, which aim to cure disease, treat and support our sick and disabled, seek improvements and enhancements in treatments, and most importantly, promote public understanding and awareness of particular health risks, diseases, and disabilities. It is amazing how the village comes around to help them raise much-needed funds through charitable activities and events. These health charities more often than not represent disease areas that pharma finds to be economically unattractive and would never recoup their investments, let alone make huge profits; academia wants to research them but is faced with limited funding. In my opinion, this is the perfect scenario for academia–health charities partnerships, as both share the same vision and are part of the same village. It is dictated by a sense of urgency and very well-defined goals of achieving milestone-enabling phases toward conquering and defeating disease.

Health charities, like the Alliance for Lupus Research, Doctors Without Borders, and the Michael J Fox Foundation, to name a few, have provided many nano- to microscale funding opportunities to academia and start-up companies within a well-defined network pushing for collaboration and partnership among the groups to achieve the set goals. The sheer existence of such a "microcosm" for research with small purses and independent of pharma is a major accomplishment, shows that partnerships are the way to progress common goals, and once again brings to bear the power of the village to look after their own. Large and well-funded disease charities, such as the Bill and Melinda Gates Foundation, should be playing a wider role in funding both basic and applied research across the board. I caution that they may well begin to show the telltale signs of pharma behavior, putting the cart before the horse in terms of deliverables and expectations and defeating the whole purpose for their existence to help fight global disease. Throwing money at a disease does not necessarily mean that you will conquer it; spending money strategically and in a "microcosm" fashion, on the other hand, would build networks, cooperation, and strong partnerships.

FUTURE PERSPECTIVES

I started this chapter by introducing the concept of a village, and it does take one to discover and develop a drug. This has been demonstrated many times over throughout history but has yet to be demonstrated in our modern times. Pharma companies are not villages and will never be; their financiers and shareholders would not allow them to become one. On the other hand, academia is a village and has evolved over many years to modernize their concepts and adjust as time goes by. Synergy and common goals are the basis of any successful collaboration; I just cannot see this happening in a true sense between pharma and academia. Pharma is under constant pressure to deliver and make huge profits, and it is in a desperate mind-set, whereas academia is currently short of cash due to the current phase of economical uncertainties and would do anything, I assume within reason, to bring the hard cash into its coffers. The sausage is attracted by the wavy octopus arm not knowing it could easily be a decoy once the octopus figures out the value of the sausage as either food or just a toy to play with, and ultimately leave it hanging on a wobbling arm.

On the bright side, academia can do better. As Zerhouni and Collins helped demonstrate by reducing the discovery "stigma" it once had and leveling the field of play with pharma, academia can execute its own discovery projects, progress them to clinical stages, and then perhaps begin to think about forming cooperatives with other academics, enabling further development and ultimately commercialization. This concept has already begun, as we have seen academia-based drug discovery institutes being erected everywhere across the United States and the world. The success of these initiatives would depend 100% on the leadership of these institutes and most importantly on their in-house facilitators who would vet the incoming projects as to their true applied-research potential (even if it has been manifested by several high-profile publications and prizes in fear of upsetting the principal investigator as has been the trend thus far). Principal investigators will not, however, become merely managers of such multire-source learning processes for discovery and development. They will be concerned with the long term, with whom each pharma company is becoming involved, with how all that learning adds up to becoming in the process. They will maintain contact with pharma over many years. They must take responsibility for posing the difficult questions: about their research and how impactful it will become. This must take place again and again, over the very long time it takes to engage seriously with serious matters to join the race to fight global disease. A drug discovery and development village is not built in a day.

In summary, I am optimistic that academia will do the right thing to help with the global fight against disease, in partnership with the disease-focused health charities, and be sensitive to the geo-economical value of a single U.S. dollar. I am confident that pharma will exit the discovery research part of their business, leaving it wide open for academic-based discovery to take over; but I would caution against celebrating premature success and academic greed (the two devils which can seriously harm progress, and in turn destroy the village).

INDEX

Collaborative Innovation in Drug Discovery: Strategies for Public and Private Partnerships, First Edition. Edited by Rathnam Chaguturu.

Wiley Series on Technologies for the Pharmaceutical Industry

Sean Ekins, Series Editor

Computational Toxicology: Risk Assessment for Pharmaceutical and Environmental Chemicals
Edited by Sean Ekins

Pharmaceutical Applications of Raman Spectroscopy
Edited by Slobodan Šašić

Pathway Analysis for Drug Discovery: Computational Infrastructure and Applications
Edited by Anton Yuryev

Drug Efficacy, Safety, and Biologics Discovery: Emerging Technologies and Tools
Edited by Sean Ekins and Jinghai J. Xu

The Engines of Hippocrates: From the Dawn of Medicine to Medical and Pharmaceutical Informatics
Barry Robson and O.K. Baek

Pharmaceutical Data Mining: Approaches and Applications for Drug Discovery
Edited by Konstantin V. Balakin

The Agile Approach to Adaptive Research: Optimizing Efficiency in Clinical Development
Michael J. Rosenberg

Pharmaceutical and Biomedical Project Management in a Changing Global Environment
Edited by Scott D. Babler

Systems Biology in Drug Discovery and Development
Edited by Daniel L. Young and Seth Michelson

Collaborative Computational Technologies for Biomedical Research
Edited by Sean Ekins, Maggie A.Z. Hupcey, and Antony J. Williams

Predictive Approaches in Drug Discovery and Development: Biomarkers and In Vitro/In Vivo Correlations
Edited by J. Andrew Williams, Richard Lalonde, Jeffrey Koup, and David D. Christ

Collaborative Innovation in Drug Discovery: Strategies for Public and Private Partnerships
Edited by Rathnam Chaguturu